Ion Mass Spectra

Ion Mass Spectra

ROBERT G. WILSON

Hughes Research Laboratories

Malibu, California

A WILEY-INTERSCIENCE PUBLICATION

John Wiley & Sons

New York • London • Sydney • Toronto

Library of Congress Cataloging in Publication Data:

Wilson, Robert G
Ion mass spectra.

"A Wiley-Interscicnce publication."
Includes bibliographical references.
1. Mass spectrometry. 2. Ions—Spectra.
I. Title.

QC454.M3W54 545′.33 74–17201
ISBN 0–471–94965–5

Printed in the United States of America

10 9 8 7 6 5 4 3 2 1

Preface

This book is a tabulation of ion mass spectra for approximately 100 elements and compounds ionized in a low-voltage, moderate pressure, hot filament electron bombardment ion source. The original purpose of this work was to determine how best to produce substantial currents of ions of many elements in the periodic table. In addition, a great deal of information concerning the dissociation of chemical compounds under electron bombardment and the formation of compound ions as well as elemental ions in a gas discharge resulted and is contained in the spectra reproduced here. These data are useful to chemists and to persons who wish to generate particular ions or ion beams for a variety of applications.

Workers in fields using ion beams will find this tabulation valuable for choosing materials to use in any ion source employing ionization by any form of electron bombardment to produce ions of selected elements or molecules that they may wish to generate. Reference to these data should lead them to materials that are known sources of the desired ion species without having to investigate a variety of materials themselves to determine which material may make any of the desired ions or which material produces the greatest current of the desired ion. Another valuable use of these data is to determine what ion species may exist near (in mass) any selected ion of interest and thereby help in determining the degree of mass separation required to produce a "pure" beam of the selected ion, or in determing the suitability of the material being considered. These data should be useful to persons working in the fields of ion implantation, mass spectrometry, and materials analysis.

The ion spectra given here are all actual experimental data, just as they were recorded. These data may not always agree exactly with experimental data obtained elsewhere because different ion sources, somewhat different experimental conditions, or different sources of ionizable materials may be used under different conditions. The data should be qualitatively the same, however. These data should also be qualitatively applicable to other electron bombardment or gas discharge ion sources such as cold cathode, rf, duoplasmatron, sputtering, and so on.

A cross-referenced index of selected major ions is included for ease in locating them in this book, although the index includes only a fraction of the thousands of ion species that are recorded in these spectra.

Some of the data included in this book were collected with the help of my associate, Douglas M. Jamba.

ROBERT G. WILSON

Contents

Ion Mass Spectra

INTRODUCTION

The data in this book are all actual recordings of the mass spectra of ion beams extracted from a low-voltage, moderate pressure, hot filament electron bombardment, gas discharge ion source. The ion beams were momentum analyzed by a 6-in. round magnet in a 150-kV ion implantation system with a resolving power from 100 to 300.*

Reasonably high resolution and accurate intensity (secondary-electron-suppressed) ion mass spectra are required for quality ion implantation work and other ion beam applications to (1) identify the ion species in the total beam, (2) locate the desired ion species accurately, and (3) determine adjacent beam impurity intensities and whether they will strike the target in the application of electrostatic scanned implantation.

Most of the data were obtained at 40 kV because this voltage for this system gave a convenient mass range that included most ions in the beams for most source materials. When higher masses exist in the beams, recordings are often included for lower voltages that allow the total ion spectra to be recorded. The voltage for which each ion mass spectrum was measured is indicated on each data page. The ion source anode voltage was 100 V unless otherwise noted on the data pages.

The experimental conditions under which these data were obtained and the ion source and feed systems used are described in following sections, and the experimental sensitivity and errors are also discussed. Methods of identifying and calibrating the masses of the ion components in these spectra are described. Selected data for elemental ions are tabulated in a later section, together with data on ionization potentials and vapor pressures of elements. Finally, a cross-referenced index of selected ion species and page numbers is included as an aid for the reader in locating ions of interest.

ION SOURCE AND MANIFOLD

The ion source used to obtain all of the data in this book is shown in the schematic of Fig. 1 and in the photographs of Figs. 2 and 3. The source is a low-voltage (~50 V) crossed-field (axial B and radial E) thermionic (tungsten filament) electron bombardment ion source that can operate with any pressure in the ionizing volume up to about 10^{-1} torr. Electrons emitted by the thermionic filament are accelerated radially toward a cylindrical anode and the axial magnetic field causes them to revolve, thus forming an ionizing electron cloud. Any pressure and flow rate of neutral

*Resolving power is defined here as $m \cdot \Delta w(\Delta m)/\mathrm{FWHM}(m) \cdot \Delta m$, where m is the mass of the ion; FWHM(m) is the width of the line in the spectrum at m; $\Delta w(\Delta m)$ is the separation between the m line and one removed by Δm; and Δm is the mass difference associated with $\Delta w(\Delta m)$.

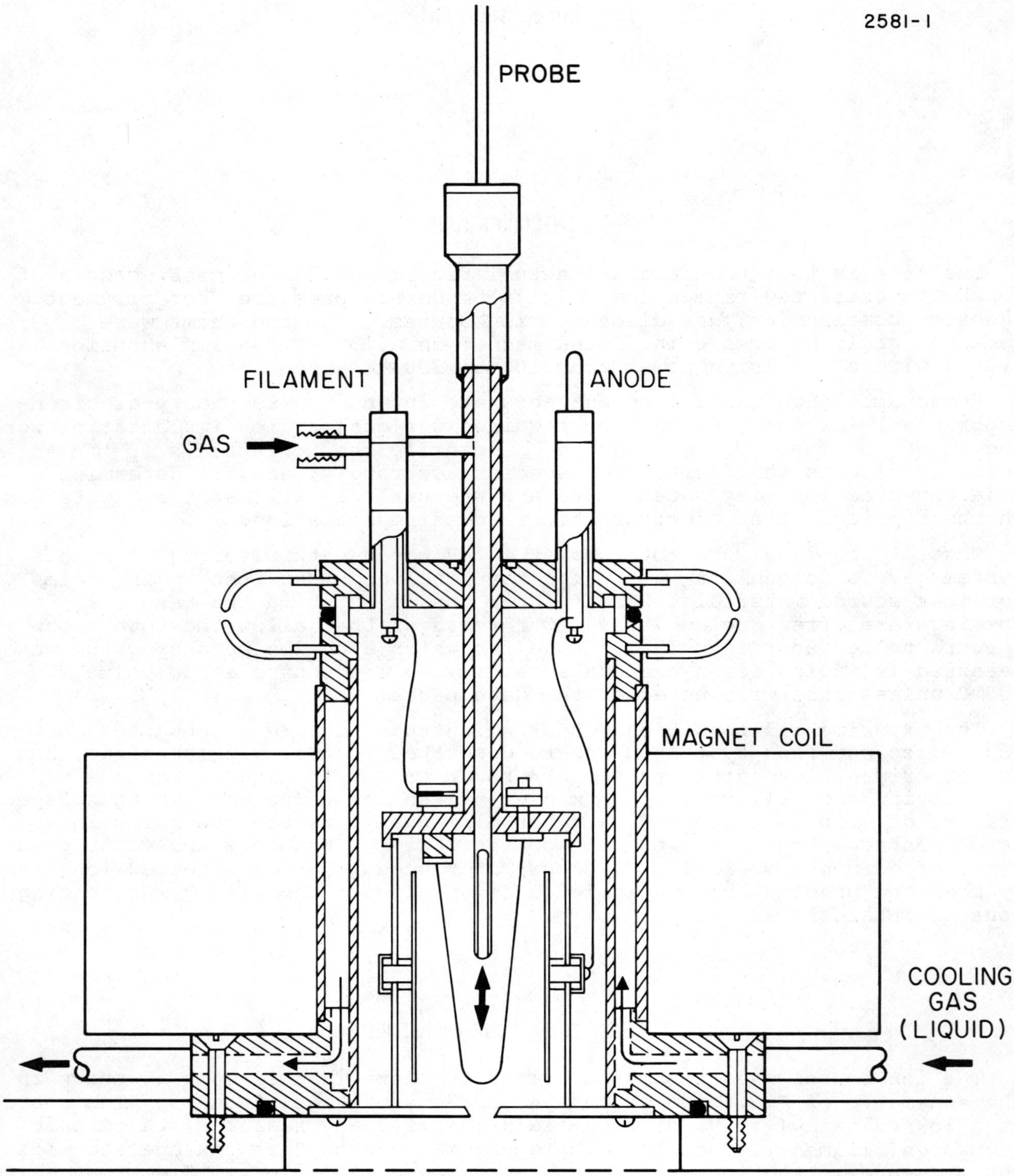

Fig. 1. Schematic of the ion source.

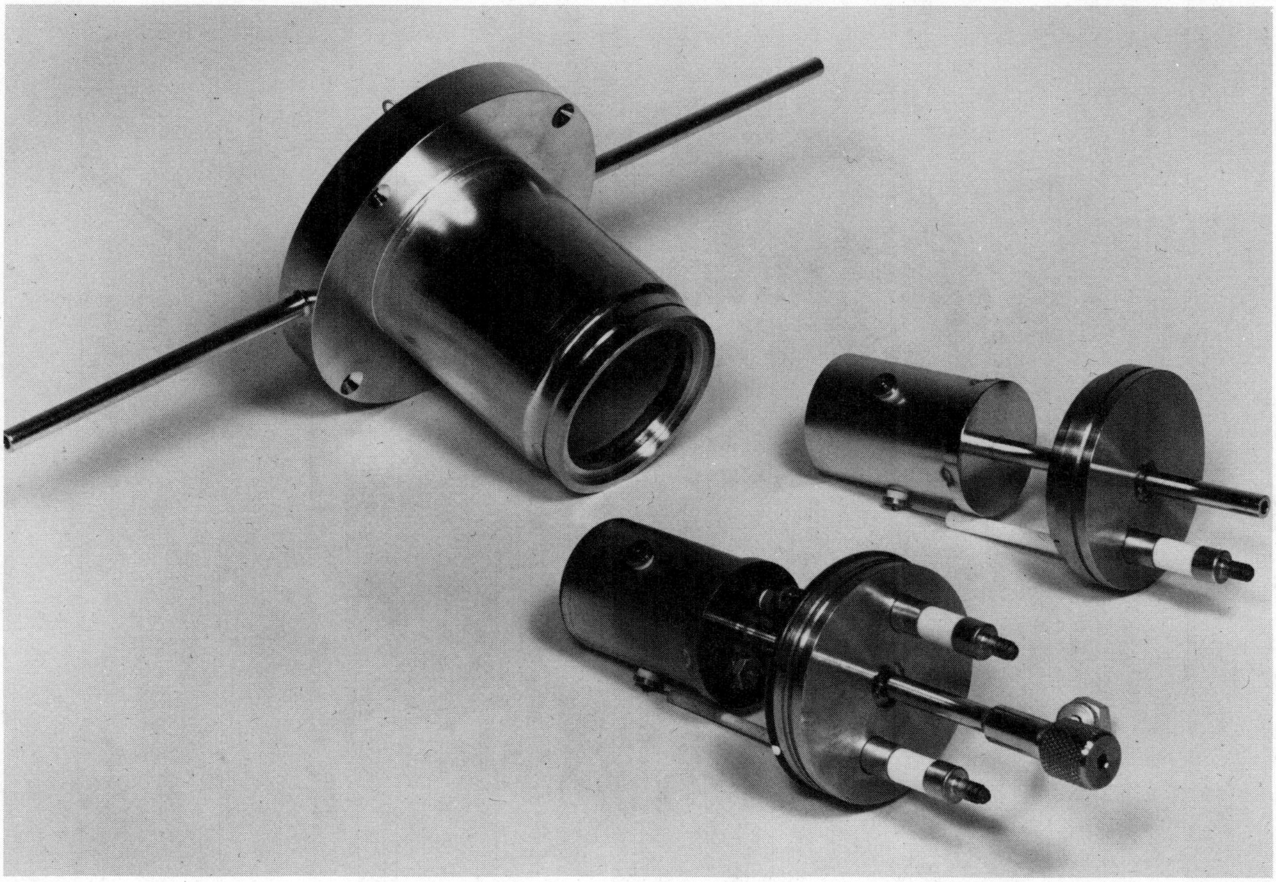

Fig. 2. The ion source with magnet removed and back flanges and insert withdrawn. The left-hand insert includes a filament and is the one used to obtain the data in this book.

Fig. 3. Ion source assembled, showing extraction aperture.

molecules may be introduced into this ionizing electron cloud. The ends of the cylindrical region are at cathode potential so that electrons are reflected from these surfaces, and ions created between cathode and anode potential are accelerated slightly toward the ends.

Ions are extracted from the ionizing volume on the axis at the filament tip end through optics designed from the results of analytical and experimental optimization studies. The ion beams used in this work were extracted at ~1 kV. The pressure in the ionizing volume can be ambient vacuum, say 10^{-7} or 10^{-6} torr (with the filament heated to emission), in which case a background ion spectrum is obtained with a beam current of the order of 100 nA, often composed largely of alkalis from the filament.

The vapor of a gas or liquid is introduced into the ionizing volume through a controlled leak valve, and a solid is heated in a probe moved along the axial thermal gradient in the source created by the filament. Thus, one filament in this source performs the two critical functions of supplying electrons to the bombarding revolving electron cloud and supplying the thermal gradient for vaporizing the lower vapor pressure materials to be ionized. Different diameters of filaments were used during the course of this work--from ~7 to 15 mils.

When the pressure in the ionizing volume becomes sufficiently high, a plasma or discharge is created and the ions are extracted from the resulting sheath formed at the extraction aperture (~4 mm). In this mode of operation the maximum total ion current is extracted, its value varying as $M^{-1/2}$ where M is some effective weighted-average mass for the material being ionized. Higher pressures cause the plasma to bulge out from the extraction aperture and create a discharge in the extraction electrode region. This situation ruins the optics which are designed for an optimum concave ion-emitting surface in the plasma.

One important feature of this source is the low value of energy spread of the ions in the extracted and accelerated ion beam. The value is difficult to meausre accurately because it is low, probably only a few volts. Low ion energy spread is one important factor in producing a small ion beam spot at the target after focusing, acceleration, and mass separation. A small target spot is important for good resolving power. Low transverse velocity spread is important for good spot size as well as low axial or potential energy spread. Therefore, low ripple in extraction, acceleration, and magnet power supplies is also important for good resolving power.

Two dc potentials are required to operate the ion source; namely, ~ 10 V and 12 A for the filament, and +100 V and 0.2 to 0.5 A for the anode. The current in the filament circuit decreases as the filament ages (becomes smaller from ion bombardment sputtering). The filament lifetime can be increased by operating the anode voltage at 30 to 50 V. A third dc voltage is required to operate the magnet; for example, 5 V and 1 or 2 A; about 100 G is required. Air cooling of the source is often helpful; a liquid coolant could be used (see Figs. 1 to 3).

All gaseous source materials were introduced through the gas feed line marked "gas" in Fig. 1. Liquid vapors were used in the same way as gases. In most cases probes were made of quartz; stainless steel was the alternative material. A probe can be present but it must be withdrawn while a gaseous material is being used. Materials that create the liquid acids HCl, HBr, and HI can attack stainless-steel probes and ion source components. HF does not cause such a problem because it is gaseous and pumps away quickly. However, HF attacks quartz readily. Thus, steel probes are better for fluorides because with a quartz probe, a large SiF_3^+ (masses 85, 86, 87) component is present in the ion spectra, as well as smaller lines from Si, SiF, and SiF_2. Although quartz probes could be cleaned and reused, they are so inexpensive to make (about 50¢) that it is best to discard them after use.

The versatile manifold for gaseous and liquid source materials used in this work is illustrated schematically in Fig. 4, and Fig. 5 is a photograph of such a system. A thermocouple gauge on or near the source can be used to monitor gas pressure in the ion source and it aids in reproducing previously determined good operating points. Exposing the leak valves to air after having exposed them to many source materials will cause their premature deterioration and failure. A section of dielectric tubing provides isolation across high voltage. The positive/negative pressure gauges in the feed system show the pressure in the feed lines at all times to allow monitoring of pressures to determine when the lines have been roughed out through the valves indicated. A hood enclosing the entire feed system provides safety and protection; the roughing pump exhaust must be vented and the same exhaust hood is convenient for this. Another valve and line can be used to couple in gas cylinders or wall gas lines.

EXPERIMENTAL CONDITIONS

The data recorded here were obtained for these many materials under the identical conditions (except where noted) of 100 V of cathode-anode potentail, a pressure in the discharge that produced 200 to 500 μA of total ion beam current with a few exceptions, and a sufficiently low energy spread that the resolution shown in these spectra could be achieved. Lower pressure in the discharge results in a larger ion energy spread that, in turn, results in lower resolution (or resolving power); this was determined by reducing the flow of a few selected gases to the ion source as mass spectra were recorded. The temperature of the probe (with the probe containing solid source materials inserted directly into the center of the discharge) was varied from 200 to 500°C depending on the temperature needed to produce the required vapor pressure. Materials requiring temperatures below approximately 200°C were kept withdrawn from the actual discharge volume on the axial probe traverse.

Gases, liquid vapors, and solids were all ionized in the same ion source; any material could be used that produced sufficient vapor pressure in the temperature range from ~60°C to 500°C. The ion source was cleaned before most materials were studied; scores of ion-source cleanings were required for this total work. For the cases where the source was not cleaned (often chlorides), some ions from a material that was previously studied exist in the spectra and some interesting combinations of these ions with the one intentionally under study are observed. Residual ions from materials used previously are usually identified as such in the spectra. Compound ions can be intentionally created by mixing two elements or compounds in this manner or by introducing them simultaneously into the ion source. Quite a number of combinations of materials were simultaneously introduced into the ion source for study in the original work, some of which are included in the spectra of this book.

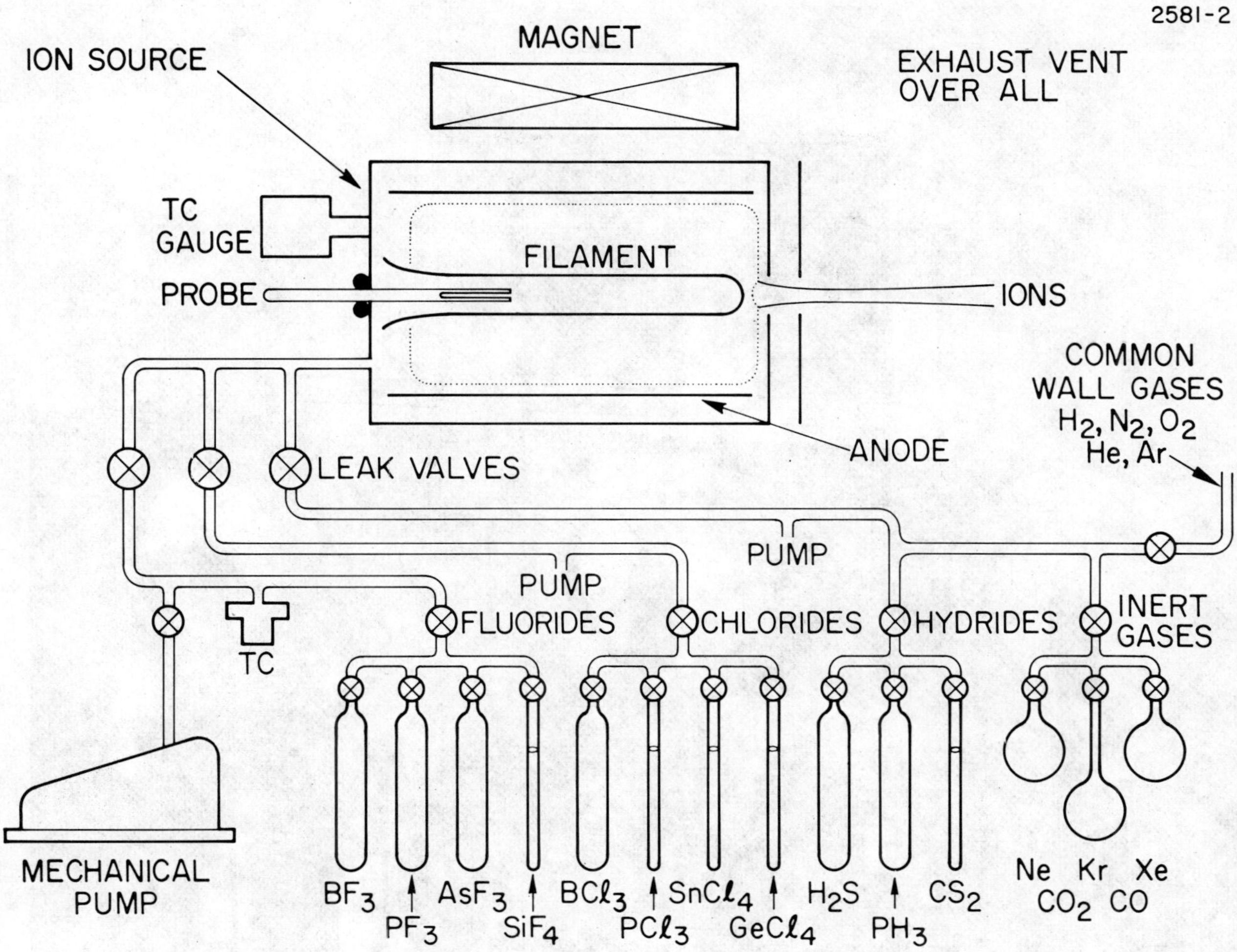

Fig. 4. Schematic of the versatile feed system manifold.

Fig. 5. Versatile ion source manifold.

COLLECTION AND PRESENTATION OF DATA

A small hole (0.1 in. in diameter) with a Faraday geometry for accurate secondary electron suppression (see Wilson and Brewer, 1973, for suitable dimensions), was used as the target collector for these data. The collected signal was fed to an electrometer by a shielded cable. The signal was plotted on the vertical axis of a recorder driven on the horizontal axis by a signal proportional to the magnetic field and obtained from a Hall probe. Mass spectra obtained in this way can be analyzed in detail almost as the data are recorded, after some familiarity with the technique is achieved.

The data for each material studied include records of several orders of magnitude in intensity. The relative sensitivities for the various data plots are indicated at the right-hand end of each curve so that all peak intensities within any one set of data curves can be directly compared by multiplying the individual peak intensities (heights) on any curve by the number at the right of the curve that is preceded by a large X. That is, these multiplying factors are the inverses of the sensitivities.

Some ion components have relative intensities as low as 0.1 ppm (10^{-7}). In a few cases, the data presented were selected from two to six different sets of data obtained at different times for certain materials of greater interest in the original work. The spectra selected for presentation here were usually those obtained after long periods of operation; thus, the spectra had stabilized. The major ions are identified on most of the spectra. Some ions are not identified and a user may search out ions of interest to him by calibration of the abcissa as described in the next section.

ION MASS ANALYSIS AND IDENTIFICATION

The general equation of motion for a charged particle of mass m, charge q, and velocity v moving on the radius ρ in the magnetic field B is

$$\frac{mv^2}{\rho} = qvB \quad \text{or} \quad \frac{mv}{q} = B\rho \qquad (1a)$$

where Bρ is a constant of the magnet (at fixed B). If the particle was accelerated through potential V_o with the same charge q, then $v = (2q\,V_o/m)^{1/2}$ and substituting in Eq. (1a) gives

$$\left(\frac{2mV_o}{q}\right)^{1/2} = B\rho \qquad (1b)$$

or

$$\left(\frac{2n_m M m_n V_o}{n_q e}\right)^{1/2} = B\rho \tag{1c}$$

where M is the molecular (atomic) weight of the fundamental chemical species, m_n is the average mass of a nucleon (1.67×10^{-27}gm), e is the basic unit of charge, n_q is the charge state of the ion (the number of elementary charges on the ion), and n_m is the multimeric state of the fundamental chemical species (monatomic, diatomic, etc.). Solving for B and simplifying gives

$$B = C_m \left(\frac{n_m M V_o}{n_q}\right)^{1/2} \tag{2}$$

where $C_m = (2m_n/\rho e)$. Thus

$$M = K_1 B^2 = \frac{K_2}{V_o} \tag{3}$$

Ion mass spectra are then interpreted based on Eq. (3). The lines in a mass spectrum can be identified by either of the following two techniques: K_1 can be calculated and B can be measured to allow direct calculation of M, or the mass spectrum can be calibrated directly by the location of known masses in the spectrum.* One great help in positive identification is knowledge of the ratios of naturally abundant isotopes of the elements and the scheme for multiple combinations of elements in compound molecules. These ratios are listed in Table 5 in the section entitled Data Tables.

The available range of B is associated with the fixed value of V_o. The value of V_o can be reduced to extend the range of mass associated with the maximum value of B for the magnet. This is because of the inherent mass-energy product of a simple magnet (bending, sector, or switching) separator. This voltage-reducing technique was employed to examine the higher mass regions.

The dependence of the location in B on system and ion parameters is seen in Eq. (2). Because $B \propto V_o^{1/2}$, the location of a given ion species will vary in B as the one-half power of the varied accelerating potential. If the number of charges (times ionized) on a given atom or molecule increases, the location on the B axis will vary as $n_q^{-1/2}$; that is, if the $n_q = 1$ ion peak (X^+) is located at B_x, then the $n_q = 2, 3, 4,$ --- ($X_2^+, X_3^+, X_4^+ \ldots$) ions will be located at 0.707 B_x, 0.577 B_x, 0.500 B_x,... Likewise, if the n_m peak (X_1^+) is located at B_x, then the $n_m = 2, 3, 4,$ --- ($X_2^+, X_3^+, X_4^+ \ldots$)

*To determine the masses of unidentified lines in the spectra of this book or to determine the location of a desired ion of known mass in any of these spectra, fix the location of a known mass on the A scale of a slide rule and set the position of the abscissa of the spectra (arbitrary) at that location by sliding the C scale. Then either an unknown mass can be determined by reading on the A scale at the location of the abscissa, or the location of a sought-for mass can be determined by reading the C scale corresponding to placing the slide at the desired mass on the A scale.

ions are located at 1.414 B_x, 1.732 B_x, 2.000 B_x, ... Complex combinations of n_q and n_m both greater than unity sometimes exist in implantation system spectra and can be identified from their location in terms of B_x, the location of the fundamental chemical species. For example, X_3^{+2} falls at $(3/2)^{1/2}B_x = 1.225\ B_x$. The probability of creating higher ionization sstates of multimers (multiatomic species) is low because the probability of dissociation becomes greater than the probability of higher ionization.

The problem of distinguishing between doubly ionized diatomic species (dimers) and singly ionized monatomic species (monomers) can be clearly defined. When every element in the molecule under consideration has only one isotope, then the two are indistinguishable. But when any one atom in the molecule has more than one isotope the diatomic molecule will have a different mass distribution than the monatomic molecule (which can be one atom); the doubly ionized dimer will have the dimer mass distribution and therefore will be distinguishable from the monomer. Rules for isotope and chemical combination are given in the following discussion.

The chemical elements have from one to ten isotopes of measured natural abundances that define a distribution of masses and relative intensities. When molecules comprising two or more atoms of the same or different chemical elements form, the resulting distributions of masses and relative intensities of the molecules are normally calculated in a straightforward manner as illustrated in the following simple example for two dissimilar elements.

If element M_1 with isotopes m_{11} and m_{12} of natural abundances I_{11} and I_{12} combines with element M_2 with isotopes m_{21} and m_{22} of natural abundances I_{21} and I_{22}, the resulting compound molecules have masses and relative intensities as follows:

Mass:	$m_{11} + m_{21}$	$m_{11} + m_{22}$	$m_{12} + m_{21}$	$m_{12} + m_{22}$
Intensity:	$I_{11}I_{21}$	$I_{11}I_{22}$	$I_{12}I_{21}$	$I_{12}I_{22}$

subject to the rules $\sum_i I_{1i} \equiv \sum_j I_{2j} \equiv$ unity and $\sum_{i,j} I_{1i} I_{2j} \equiv$ unity.

If any of the m_{1i}'s differ by the same integer as any of the m_{2j}'s, then some of the resulting combined masses coincide and the associated intensities are added. For example, in the simple illustration used above, if the differences between m_{11} and m_{12}, and m_{21} and m_{22} are both one or two, then $m_{11} + m_{22}$ and $m_{12} + m_{21}$ are identical. Simple specific illustrations of these cases of unequal and equal differences and the results described above follow. Case 1: Boron of masses 10 and 11 of approximate natural abundances 1/5 and 4/5 combines with Cl of masses 35 and 37 of approximate natural abundances 3/4 and 1/4. The compound molecule BCl has masses 45, 46, 47, and 48 with approximate relative intensities 3/20, 12/20, 1/20, and 4/20. Case 2: Cu of masses 63 and 65 of approximate natural abundances 7/10 and 3/10 combines with Cl. The compound molecule CuCl has masses 98, 100, and 102 of relative intensity 21/40, 16/40, and 3/40.

For multiply ionized species, the distributions of relative intensities are identical to the singly ionized intensities. Essentially every singly ionized species (except H) has been observed as a doubly ionized species. When diatomic elemental molecules are formed, the above rules are obeyed. Certain terms are identical and the intensities contain squared terms. As an example similar to the one used above, if m_{11}, m_{12}, and m_{13} and I_{11},

I_{12}, and I_{13} are the masses and relative abundances of an element, the diatomic molecular masses and intensities would be

Mass:	$2m_{11}$	$m_{11} + m_{12}$	$2m_{12}$	$m_{12} + m_{13}$	$2m_{13}$	$m_{11} + m_{13}$
Intensity:	I_{11}^2	$2I_{11}I_{12}$	I_{12}^2	$2I_{12}I_{13}$	I_{13}^2	$2I_{11}I_{13}$

If any of the monatomic mass differences are identical, then some of the diatomic masses will coincide. Conservation of particles requires that

$$I_{11}^2 + I_{11}I_{12} + I_{11}I_{13}: \ I_{12}^2 + I_{12}I_{13} + I_{11}I_{12}: \ I_{13}^2 + I_{11}I_{13} + I_{12}I_{13} \equiv I_{11}: I_{12}: I_{13}$$

Chlorine serves as a simple example with masses 35 and 37 and approximate relative intensities 3/4 and 1/4. Calculated diatomic masses and intensities are

Mass:	70	72	74
Intensity:	9/16	6/16	1/16

This mass distribution is observed experimentally for elemental chlorine as illustrated in Fig. 6 and for compound chlorides.

Lines unique to the doubly ionized diatomic distribution are seen among the singly ionized monatomic ions. The intensities of these lines allow the complete distribution of the doubly ionized diatomic ions to be calculated and subtracted from the total to yield the purely singly ionized monatomic intensities. Cl serves as a convenient illustration for this principle. Cl has two isotopes, 35 and 37. Diatomic Cl then has three masses at 70, 72, and 74. The singly ionized diatomic masses occur at 70, 72, and 74 and the doubly ionized lines exist at 35, 36, and 37 mixed with the singly ionized monatomic lines at 35 and 37. Mass 36 is unique to Cl_2^{2+}. Mass 36 is clearly seen in Fig. 6. A weaker line at mass 38 is $H^{37}Cl$ implying a portion of mass 36 is $H^{35}Cl$. The intensity of $H^{35}Cl$ can be calculated from the intensity of $H^{37}Cl$ and the relative abundances of ^{35}Cl and ^{37}Cl. The intensity of $H^{35}Cl$ can be subtracted from the intensity of mass 36 and the intensities of $^{35}Cl_2^{2+}$ and $^{37}Cl_2^{2+}$ can be calculated and subtracted from the intensities of masses 35 and 37. When all this is done, the relative intensities of $^{35}Cl^+$ and $^{37}Cl^+$ can be compared with the natural abundances for Cl and the distribution of masses 70, 72, and 74 can be checked against the calculated ones. In this case, and for Se (see Fig. 7) and others, the experimental and predicted distributions agree within experimental error. It is also noted that multiatomic molecules X_n of certain elements are observed up to n = 4, 6, 8, 10, and 12.

Multiatomic species of other elements with multiple isotopes can be identified by their calculable intensity distributions. Many of these intensity distributions are listed in one of the tabulations in the section entitled Data Tables. These data can be very helpful in identifying other molecular combinations of these species.

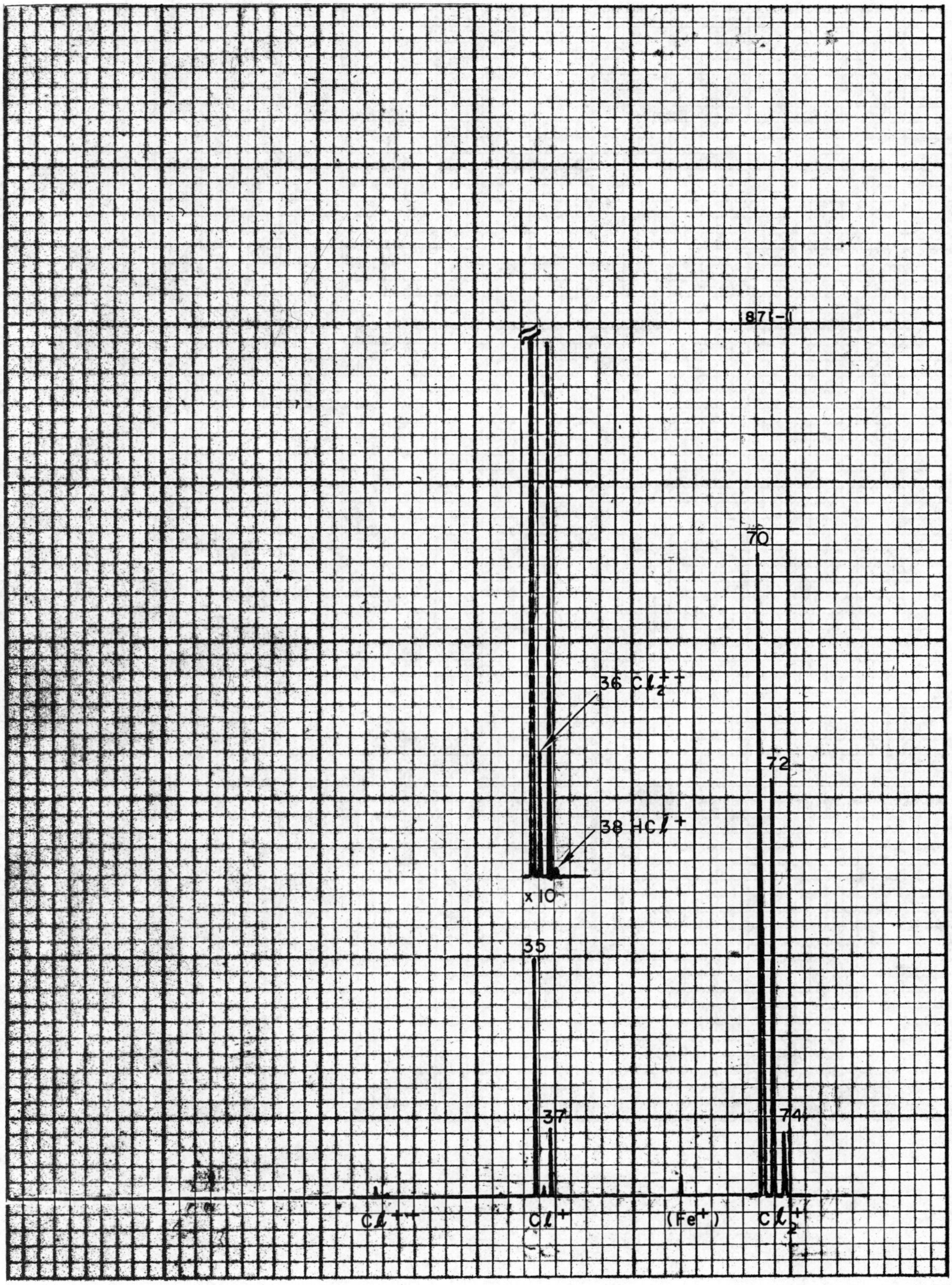

Fig. 6. Mass spectrum of chlorine, illustrating the natural abundances of Cl^+, the predicted diatomic intensity distribution for Cl_2^+ and the resolution of Cl_2^{2+} from Cl^+.

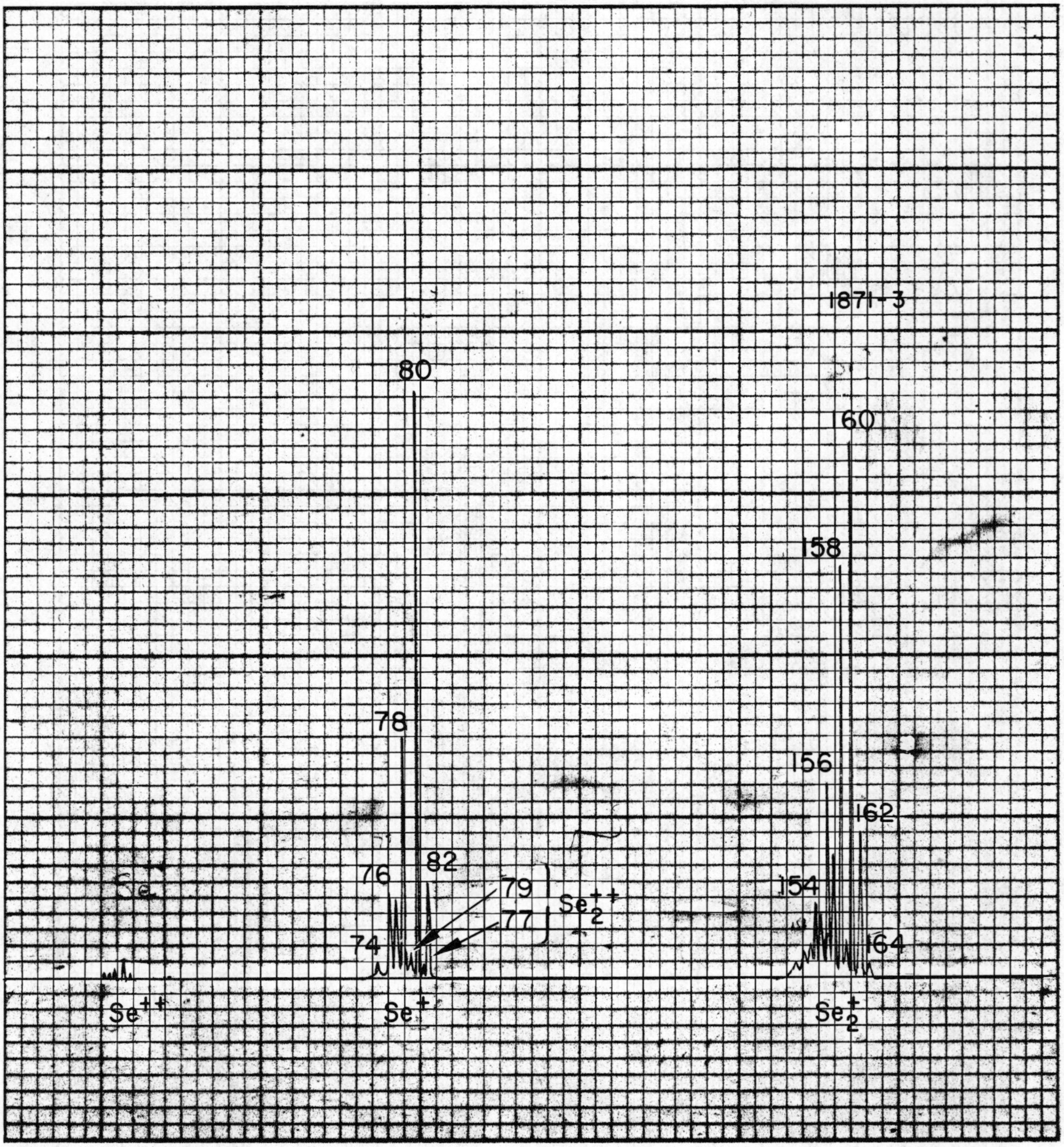

Fig. 7. Mass spectrum of Se, illustrating the diatomic mass distribution for a complex monatomic mass distribution and showing the presence of Se_2^{2+} in the Se^{+} spectrum.

EXPERIMENTAL SENSITIVITY AND ERRORS

The absolute sensitivity of these measurements is about 10^{-7} or 0.1 ppm when no intense line is so close that the tail of its distribution interferes. That is, a mass with no high intensity mass within a few mass units of it can be detected if its intensity is about one part in 10^{7} of the total beam and the total beam is $\geq \sim 100$ μA.

In this work, the skirt of a high intensity line reduces the detection sensitivity at adjacent and nearby masses approximately as follows. If a high intensity line occurs at mass 75 (viz., As^{+}) for example, then the detection sensitivity is ~1/3000 of the intensity of mass 75 at ±1 amu (74 and 76), ~1/30,000 of mass 75 at ±3 amu (72 and 78), and ~1/300,000 of mass 75 at ±5 amu (70 and 80), and so on. Mass determinations and assignments in this work are accurate to about 1% or to ~0.1 amu at mass 10 and ~1 amu at mass 100. The relative intensities are less accurate. All intensities were measured with an electrometer, accurate to probably 2% and scaled over many orders of sensitivity, and recorded on a high-quality recorder, probably responsive enough to reproduce the electrometer signal to 3% in a total measurement. However, recording a mass spectrum is a dynamic process and some error in intensity is introduced in the recorded spectrum by the scan rate because, for different intensity peaks, the recording rate varies and the recorder pen does not always track the peak heights with equal accuracy. Noise introduced through electrometer and recorder leads and some electrometer scaling inaccuracies are other sources of error in intensity. Relative intensities similar enough to be recorded on the same electrometer scale are more accurate relatively than those measured between lines differing in intensity by orders of magnitude for which greatly different electrometer scales are used.

Potentially the largest error could be introduced by beam spot-Faraday cup misalignment. Fluctuations in ion source or beam handling potentials can cause small beam spot position changes as the mass spectrum is scanned from zero to maximum. This effect was minimized in this work by making the analyzed beam shape a narrow vertical line. Actual beam intensity variations may occur during the time required to record the total spectrum slowly enough to minimize errors that result from recording rate. About 20 minutes was required to measure a spectrum over 10^{6} or 10^{7} in total intensity range. Errors resulting from this effect were minimized by checking the total beam intensity or the intensity of a major line before each scale recording and, if necessary, by adjusting ion source variables to maintain the reference intensity constant.

The total error introduced from the summation of all these effects is estimated to make the accuracy of the relative intensities reported here 20 to 30% over an intensity range of 10^{4}, with a value more nearly 2 or 3% being applicable to intensities within about 10, that is, on the same scale setting.

REFERENCES

Honig, R. E., and D. A. Kramer, RCA Review 30, 285 (1969).

Wilson, R. G., and G. R. Brewer, Ion Beams Wiley-Interscience, New York, 1973.

DATA TABLES

Certain data for the chemical elements are useful for analyzing mass spectra. Some of these data are tabulated in this section, including the first and second ionization energies of selected chemical elements, vapor pressure curves for many of the elements (Honig and Kramer, 1969), isotopic natural abundances for selected elements,and calculated multiatomic elemental mass intensity distributions.

TABLE 1. First Ionization Potentials of the Elements

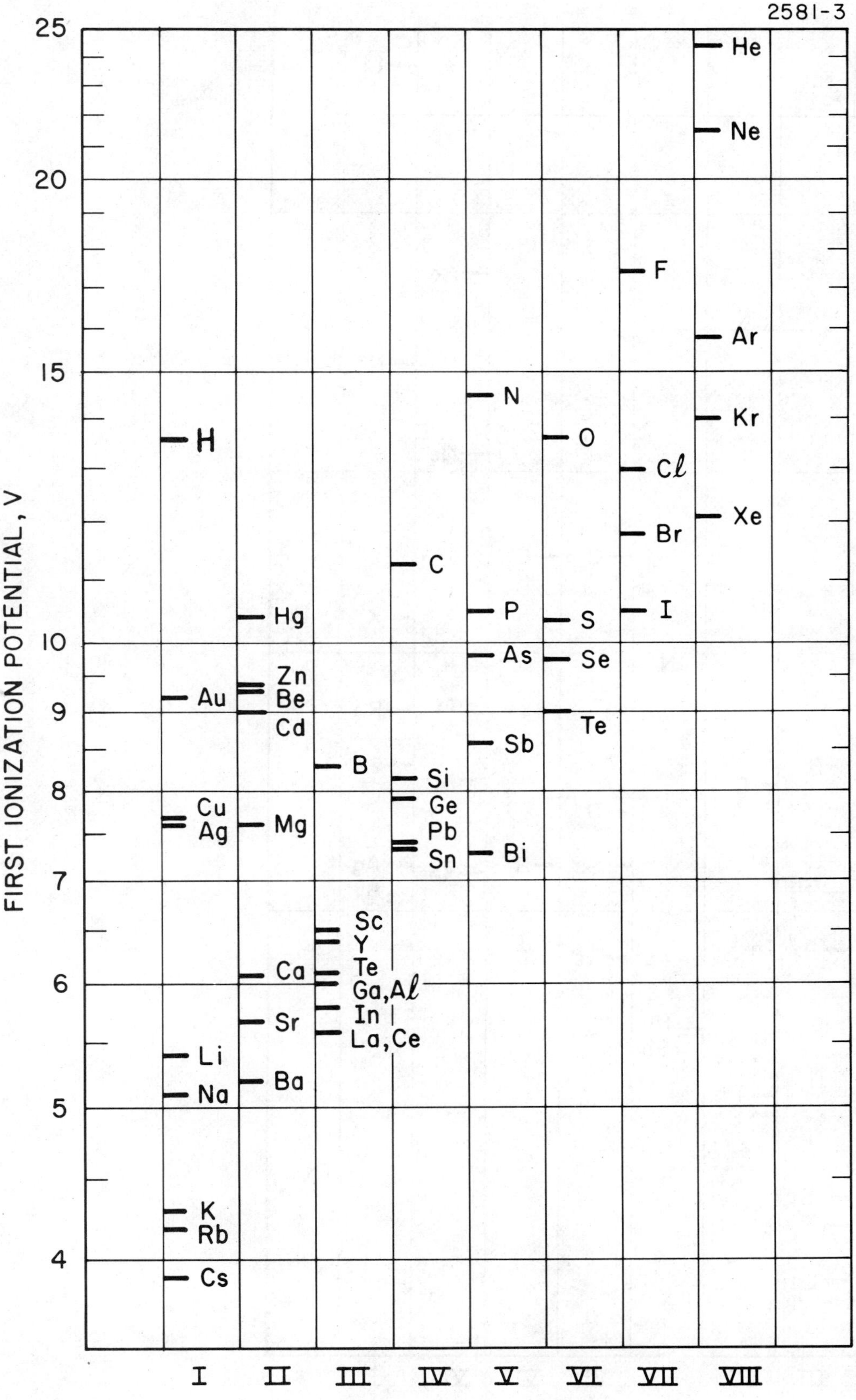

TABLE 2. Second Ionization Potentials of the Elements

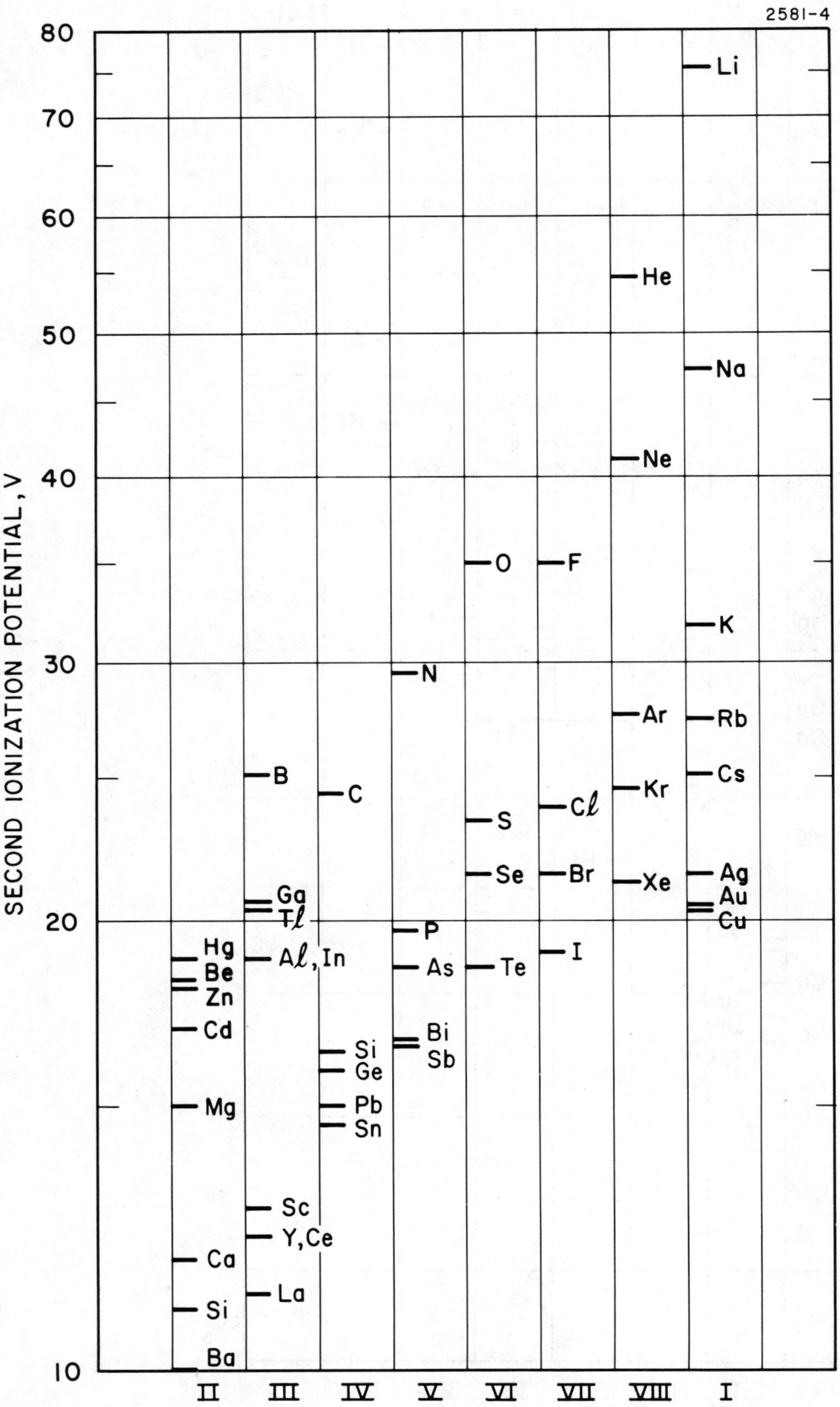

TABLE 3. Vapor Pressure Curves for Elements

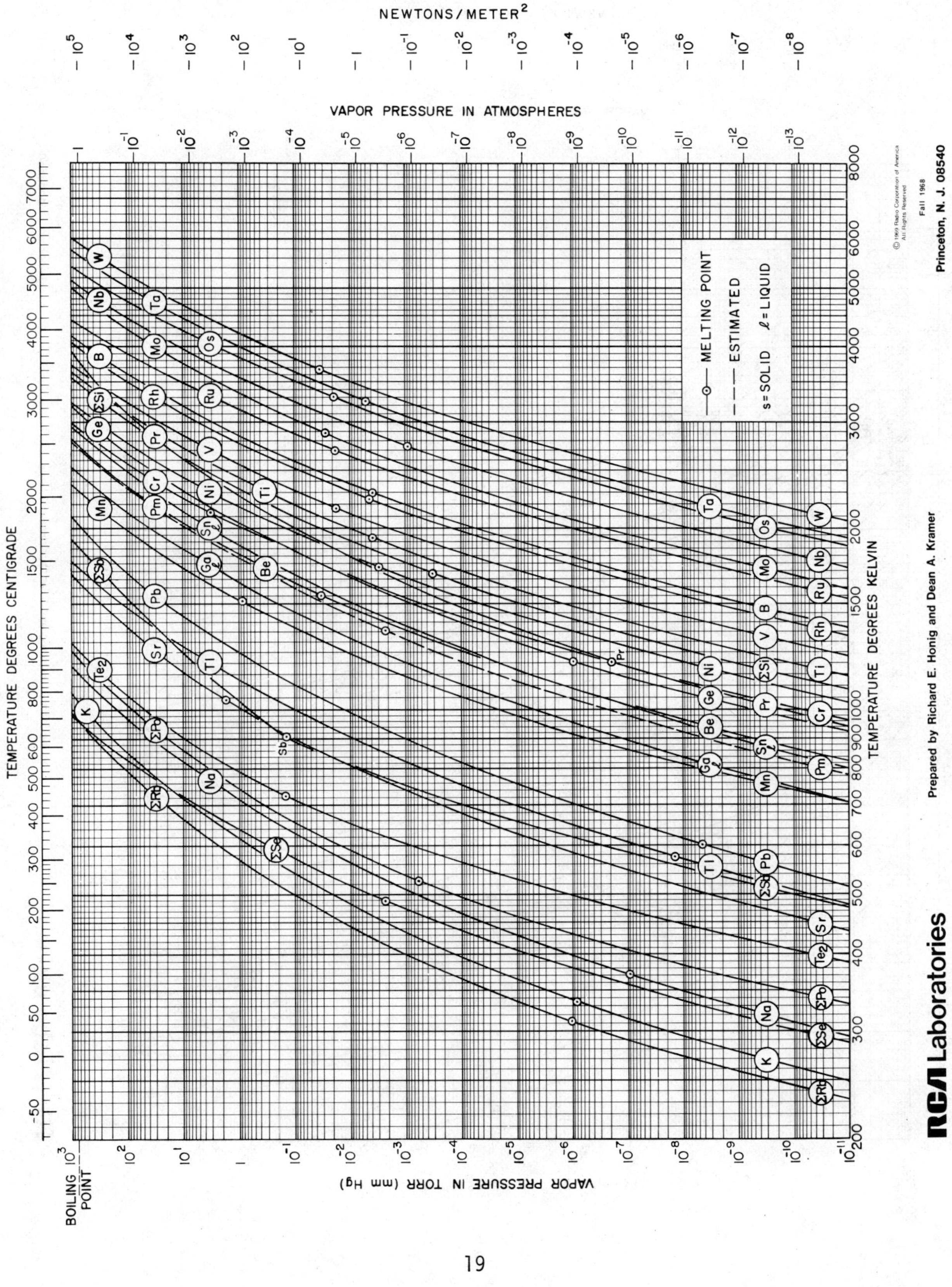

TABLE 4. Vapor Pressure Curves for Elements

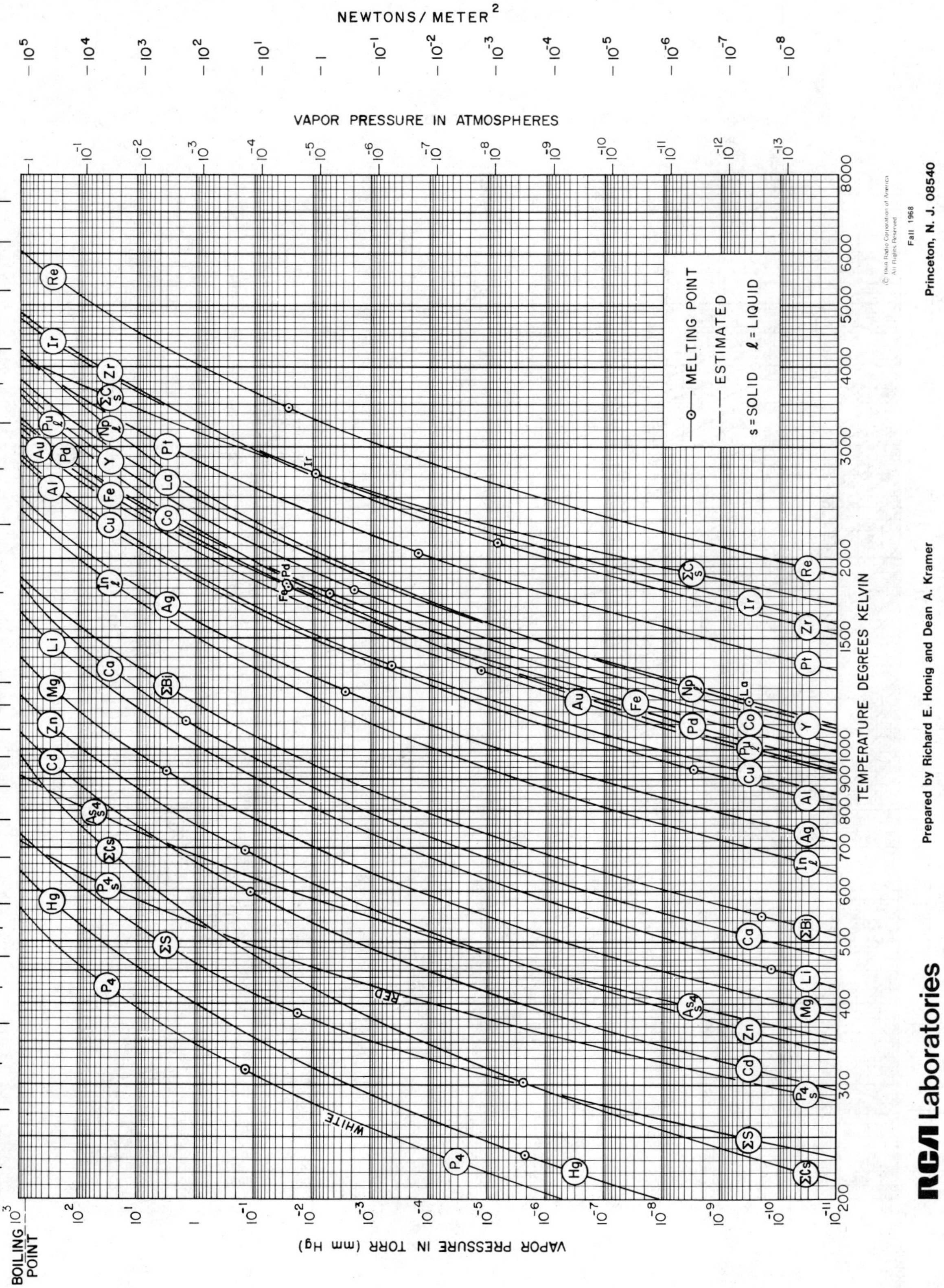

TABLE 5. Natural Abundances and Calculated Multiatomic Elemental Mass Intensity Distributions for Selected Elements

H	1	99.985
	2	0.015
H_2	2	99.97
	3	0.030
	4	0.0002
He	3	0.00013
	4	100^{-}
Li	6	7.42
	7	92.58
Be	9	100
B	10	19.78
	11	80.22
B_2	20	3.912
	21	31.735
	22	64.348
C	12	98.89
	13	1.11
C_2	24	97.792
	25	2.1954
	26	0.0123
N	14	99.63
	15	0.37
N_2	28	99.261
	29	0.7373
	30	0.0014
O	16	99.759
	17	0.037
	18	0.204
O_2	32	99.5176
	33	0.07482
	34	0.40704
	35	0.00015
	36	0.00042
F	19	100
F_n	19n	100
Ne	20	90.92
	21	0.257
	22	8.82
Na	23	100
Mg	24	78.70
	25	10.13
	26	11.17
Mg_2	48	61.937
	49	15.945
	50	18.608
	51	2.263
	52	1.248
Al	27	100
Si	28	92.21
Si	29	4.70
	30	3.09
Si_2	56	85.027
	57	8.668
	58	5.920
	59	0.290
	60	0.095
P	31	100
P_n	31n	100
S	32	95.00
	33	0.76
	34	4.22
	36	0.014
S_2	64	90.25
	65	1.44
	66	8.01
	67	0.06^{+}
	68	0.20^{+}
	70	0.001
S_3	96	85.74
	97	2.05
	98	11.43
	99	0.122
	100	0.528
	101	0.002
	102	0.008
S_4	128	82.45
	129	2.55
	130	14.49
	131	0.287
	132	0.98
	133	0.011
	134	0.030
Cl	35	75.53
	37	24.47
Cl_2	70	57.048
	72	36.962
	74	5.988
Cl_3	105	(27)
	107	(27)
	109	(9)
	111	(1)
Cl_4	140	(81)
	142	(108)
	144	(54)
	146	(12)
	148	(1)
Cl_5	175	(243)
	177	(405)
Cl_5	179	(270)
	181	(90)
	183	(15)
	185	(1)
Cl_6	210	(729)
	212	(1458)
	214	(1215)
	216	(477)
	218	(135)
	220	(18)
	222	(1)
Ar	36	0.337
	38	0.063
	40	99.60
K	39	93.10
	40	0.012
	41	6.88
K_2	78	86.676
	79	0.022
	80	12.825
	81	0.002
	82	0.473
Ca	40	96.97
	42	0.64
	43	0.145
	44	2.06
	46	0.0033
	48	0.18
Sc	45	100
Ti	46	7.93
	47	7.28
	48	73.94
	49	5.51
	50	5.34
V	50	0.24
	51	99.76
Cr	50	4.31
	52	83.76
	53	9.55
	54	2.38
Mn	55	100
Fe	54	5.82
	56	91.66
	57	2.19
	58	0.33
Co	59	100
Ni	58	67.88
	60	26.23
	61	1.19

TABLE 5. (continued)

Ni	62	3.66
	64	1.08
Cu	63	69.09
	65	30.91
Cu_2	126	47.83
	128	42.62
	130	9.55
Cu_3	189	33.05
	181	44.23
	193	19.77
	195	2.95
Zn	64	48.89
	66	27.81
	67	4.11
	68	18.57
	70	0.62
Zn_2	128	23.90
	130	27.20
	131	4.02
	132	25.87
	133	2.28
	134	11.10
	135	1.53
	136	3.79
	137	0.05
	138	0.23
	140	0.004
Ga	69	60.4
	71	39.6
Ga_2	138	36.48
	140	47.84
	142	15.68
Ge	70	20.53
	72	27.43
	73	7.76
	74	36.54
	76	7.76
As	75	100
As_n	75n	100
Se	74	0.87
	76	9.02
	77	7.58
	78	23.52
	80	49.82
	82	9.19
Se_2	148	0.007
	150	0.16
	151	0.13
	152	1.22
	153	1.37
Se_2	154	5.69
	155	3.57
	156	14.68
	157	7.55
	158	25.09
	159	1.39
	160	24.82
	162	13.48
	164	0.85
Br	79	50.54
	81	49.46
Br_2	158	(1)
	160	(2)
	162	(1)
Br_3	237	(1)
	239	(3)
	241	(3)
	243	(1)
Kr	78	0.35
	80	2.27
	82	11.56
	83	11.55
	84	56.90
	86	17.37
Rb	85	72.15
	87	27.85
Sr	84	0.56
	86	9.86
	87	7.02
	88	82.56
Y	89	100
Zr	90	51.46
	91	11.23
	92	17.11
	94	17.40
	96	2.80
Nb	93	100
Mo	92	15.84
	94	9.04
	95	15.72
	96	16.53
	97	9.46
	98	23.78
	100	9.63
Ru	96	5.51
	98	1.87
	99	12.72
	100	12.62
	101	17.07
	102	31.61
Ru	104	18.58
Rh	103	100
Pd	102	0.96
	104	10.97
	105	22.23
	106	27.33
	108	26.71
	110	11.81
Ag	107	51.82
	109	48.18
Cd	106	1.22
	108	0.88
	110	12.39
	111	12.75
	112	24.07
	113	12.26
	114	28.86
	116	7.58
In	113	4.28
	115	95.72
In_2	126	0.18
	128	8.19
	130	91.62
Sn	112	0.96
	114	0.66
	115	0.35
	116	14.30
	117	7.61
	118	24.03
	119	8.58
	120	32.85
	122	4.72
	124	5.94
Sb	121	57.25
	123	42.75
Sb_2	242	(16)
	244	(24)
	246	(9)
Sb_3	363	(64)
	365	(144)
	367	(108)
	369	(27)
Sb_4	484	(256)
	486	(768)
	488	(864)
	490	(432)
	491	(81)
Te	120	0.089
	122	2.46
	123	0.87

TABLE 5. (continued)

Te	124	4.61	Xe	136	8.87	Ir	191	37.3
	125	6.99	Cs	133	100		193	62.7
	126	18.71	Ba	130	0.101	Pt	192	0.78
	128	31.79		132	0.097		194	32.9
	130	34.48		134	2.42		195	33.8
Te_2	242	0.004		135	6.59		196	25.3
	243	0.002		136	7.81		198	7.21
	244	0.07		137	11.32	Au	197	100
	245	0.055		138	71.66	Hg	196	0.146
	246	0.26	La	138	0.089		198	10.02
	247	0.42		140	99.911		199	16.84
	248	1.31	Hf	174	0.18		200	23.13
	249	0.97		176	5.20		201	13.22
	250	3.84		177	18.50		202	29.80
	251	3.17		178	27.14		204	6.85
	252	8.13		179	13.75	Tl	203	29.50
	253	0.60		180	35.24		205	70.50
	254	19.62	Ta	180	0.0123	Pb	204	1.48
	255	4.82		181	99.988		206	23.6
	256	23.00	W	180	0.14		207	22.6
	258	21.91		182	26.41		208	52.3
	260	11.88		183	14.40	Pb_2	408	0.02
I	127	100		184	30.64		410	0.70
I_n	127n	100		186	28.41		411	0.67
Xe	124	0.096	Re	185	37.07		412	7.12
	126	0.090		187	62.93		413	10.67
	128	1.92	Os	184	0.018		414	29.80
	129	26.44		186	1.59		415	23.64
	130	4.08		187	1.64		416	27.35
	131	21.18		188	13.3	Bi	209	100
	132	26.89		189	16.1	Bi_n	209n	100
	134	10.44		190	26.4			

INDEX OF IONS

As an aid to the reader, the page numbers of spectra that include the listed elemental and compound ion species are tabulated. The system used for the ion species does not distinguish the number of times one element occurs in a compound ion, only the different elemental species. That is, AB is used to indicate ions where elements A and B are combined and AB, AB_2, AB_3, A_2B, and so on, may all occur in the spectra on the pages noted. Not all combinations that occur in the spectra are included in the index. If the reader is seeking one not listed, he can use some logic in searching it out by using combinations that are listed.

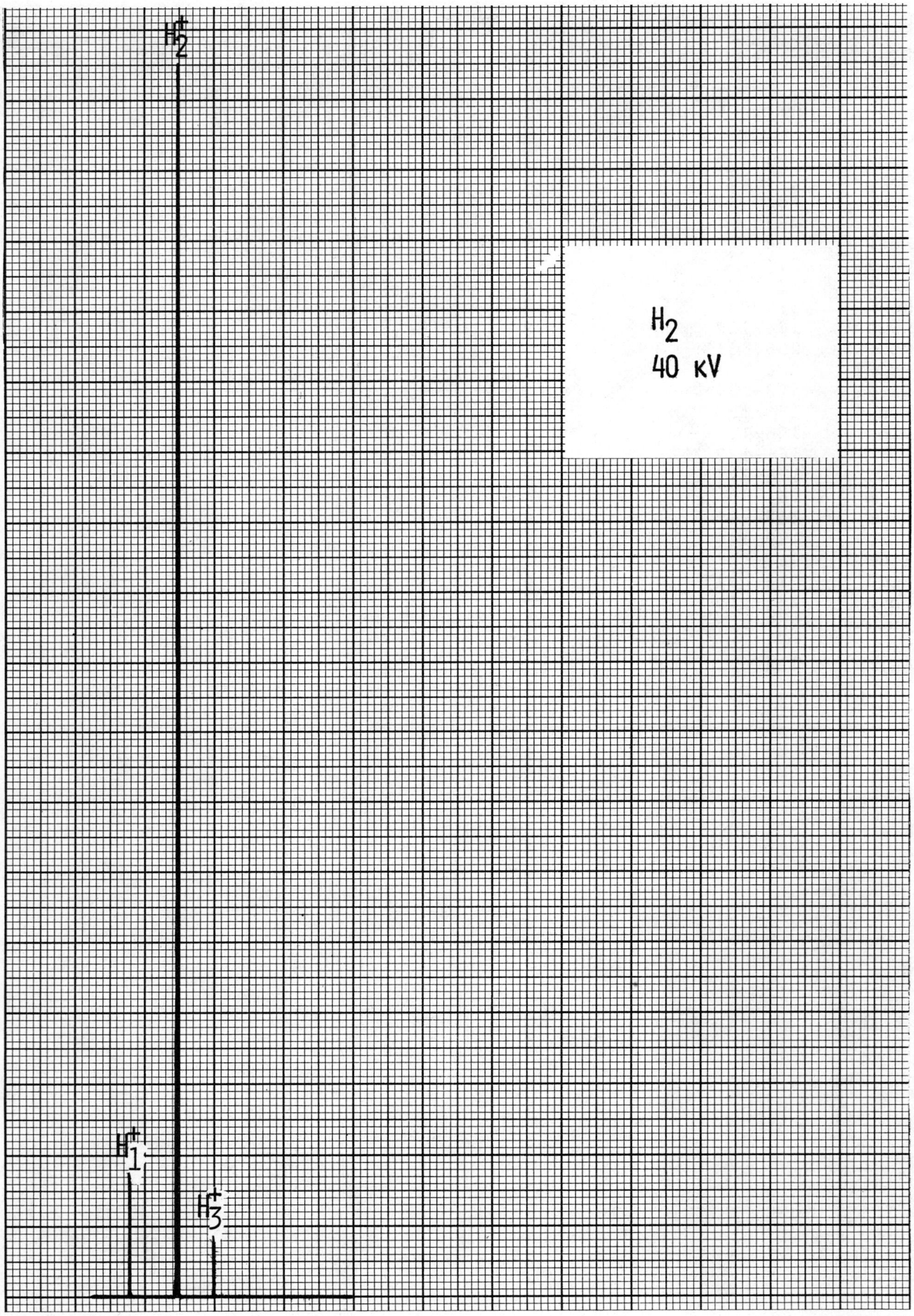
H_2^+
H_1^+
H_3^+
H_2
40 kV

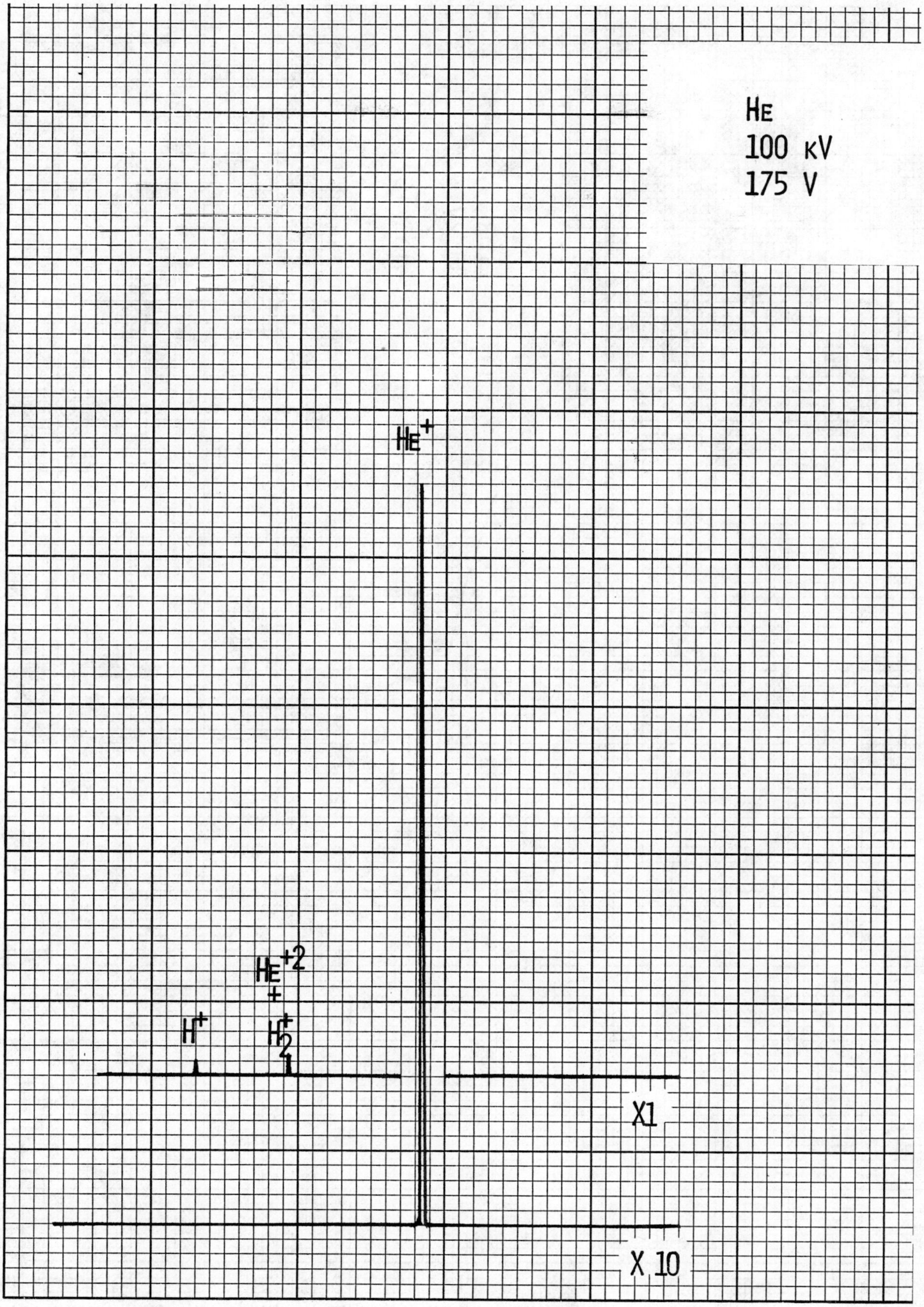
He
100 kV
175 V
He+
He+2
+
H+
H2+
X1
X 10

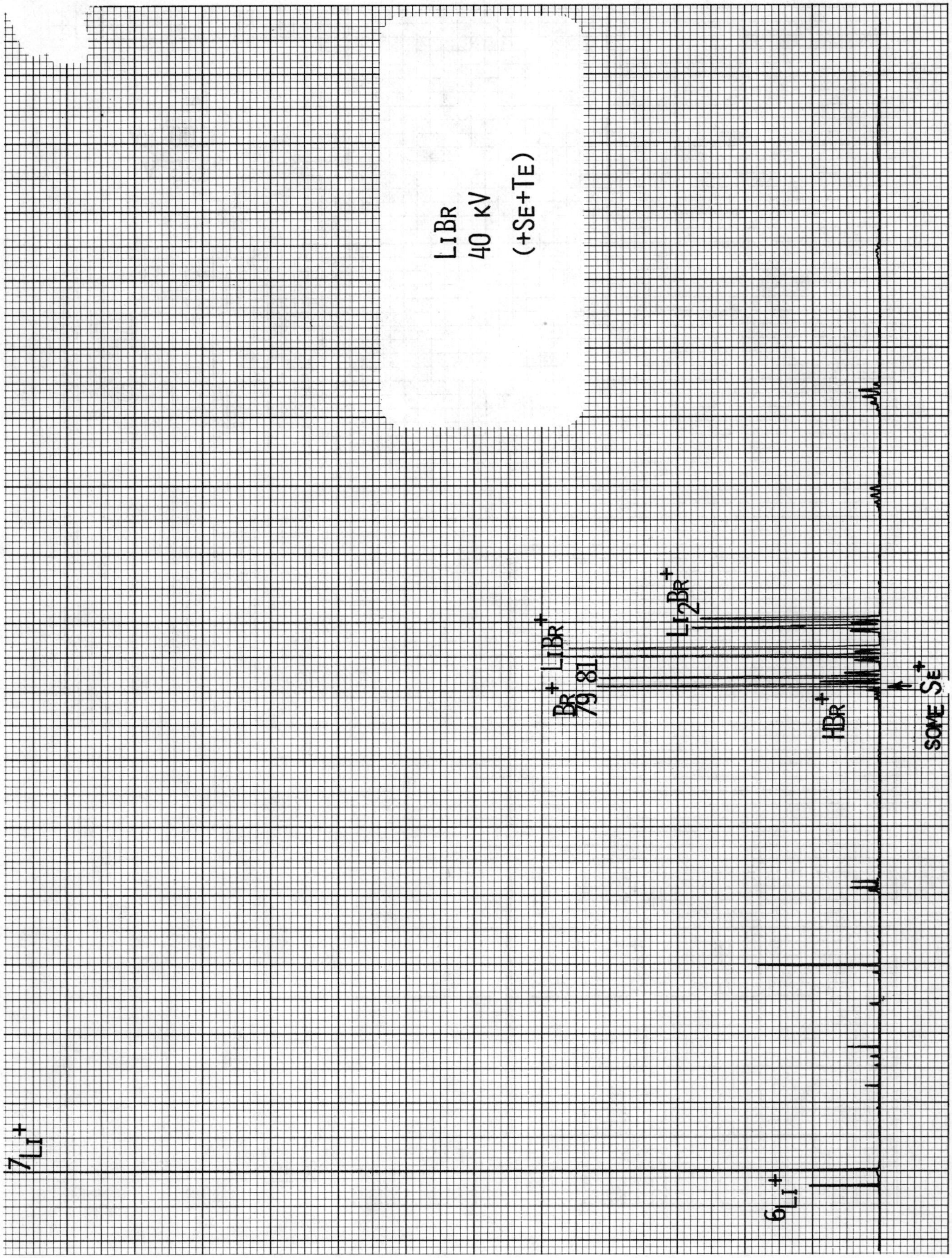
LiBr
40 kV
(+Se+Te)
$^7Li^+$
$^6Li^+$
Br^+
79 81
$LiBr^+$
Li_2Br^+
HBr^+
some Se^+

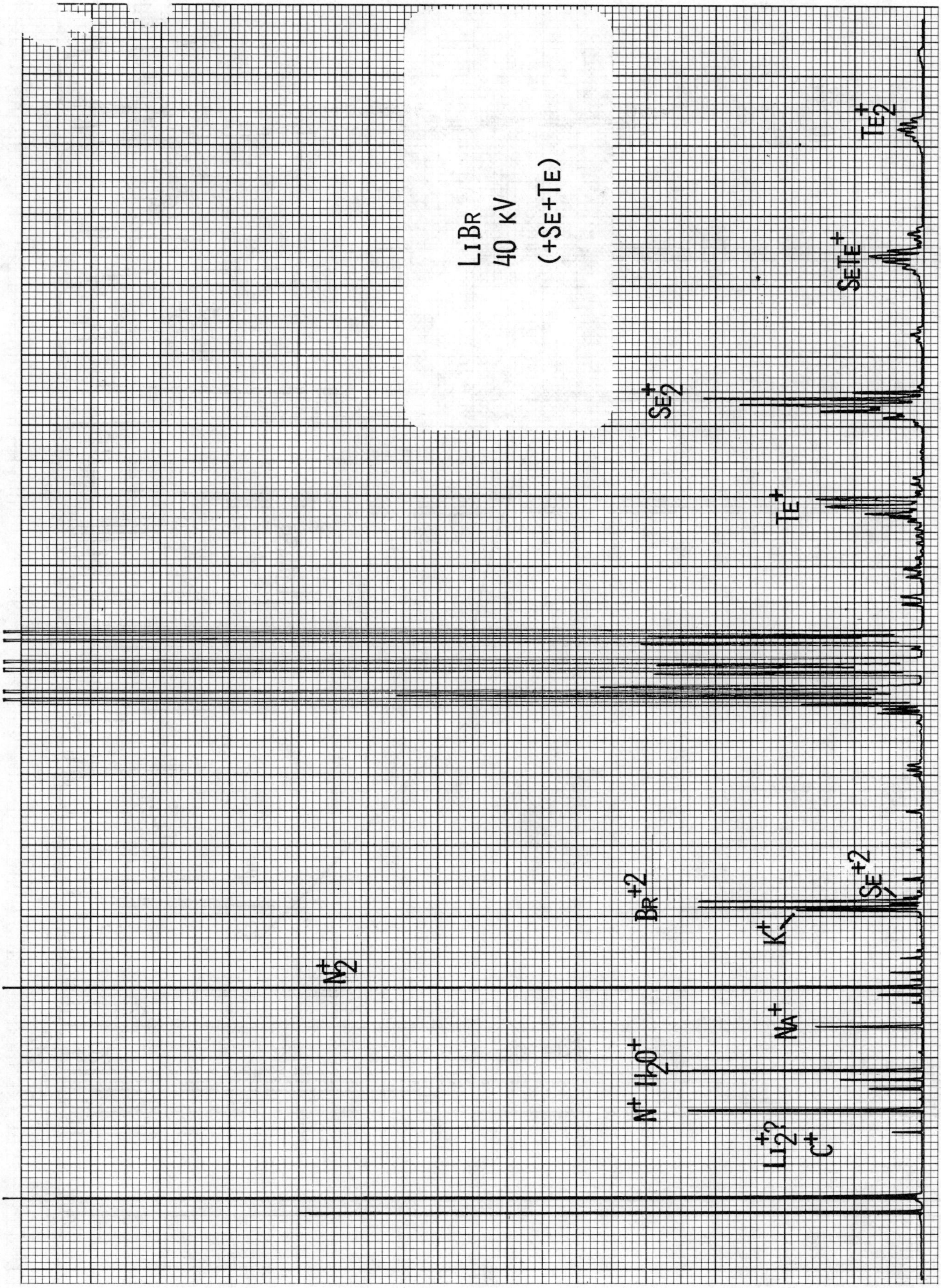
LiBr
40 kV
(+Se+Te)
Te2+
SeTe+
Se2+
Te+
Se+2
Br+2
K+
N2+
Na+
N+ H2O+
Li2+2
C+

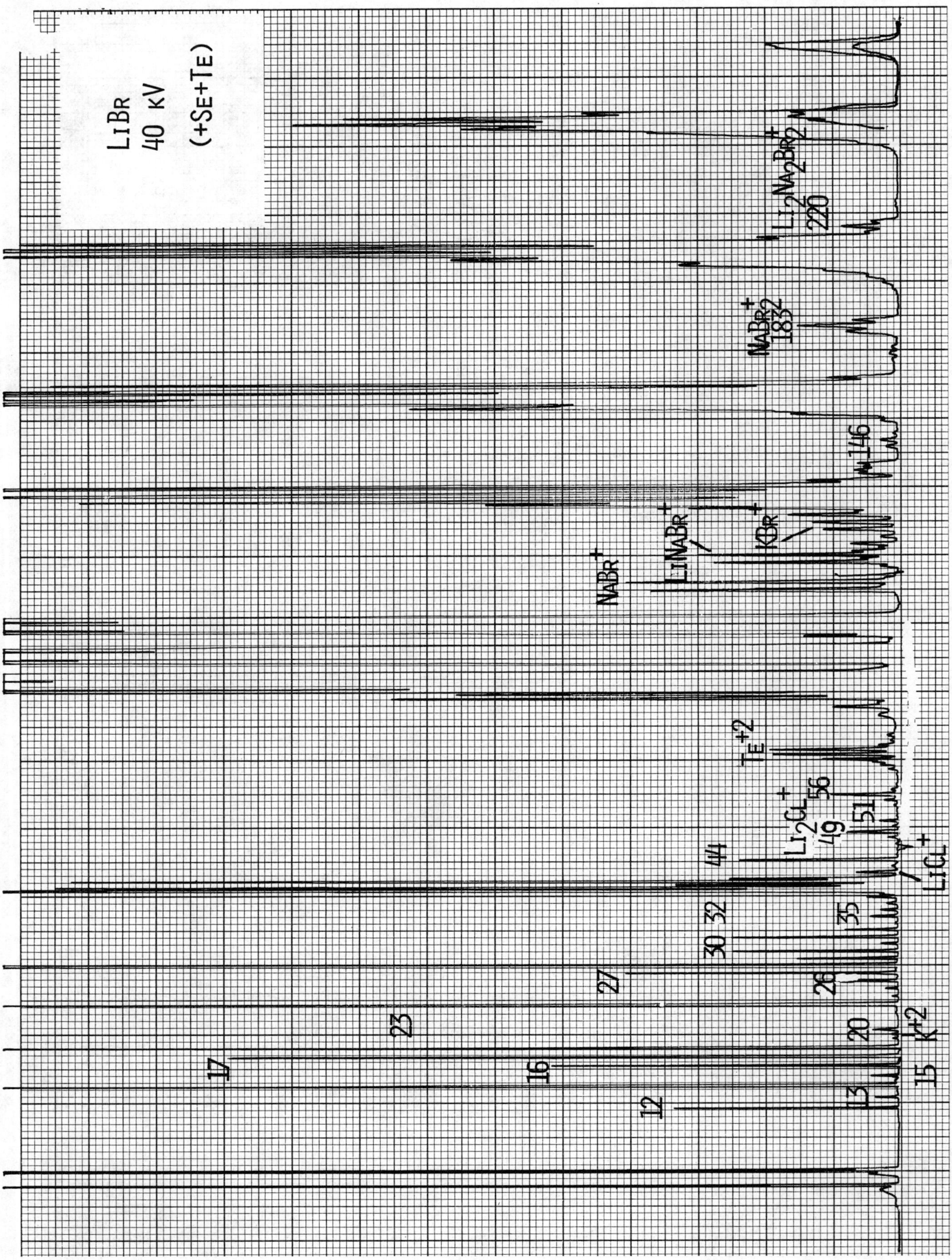

LiBr
40 kV
(+Se+Te)
Li2Na2Br2+
220
NaBr2+
183
146
LiNaBr+
KBr+
NaBr+
Te+2
Li2Cl+
49
56
51
44
LiCl+
30
32
35
27
26
23
20
K+2
17
16
15
12
13

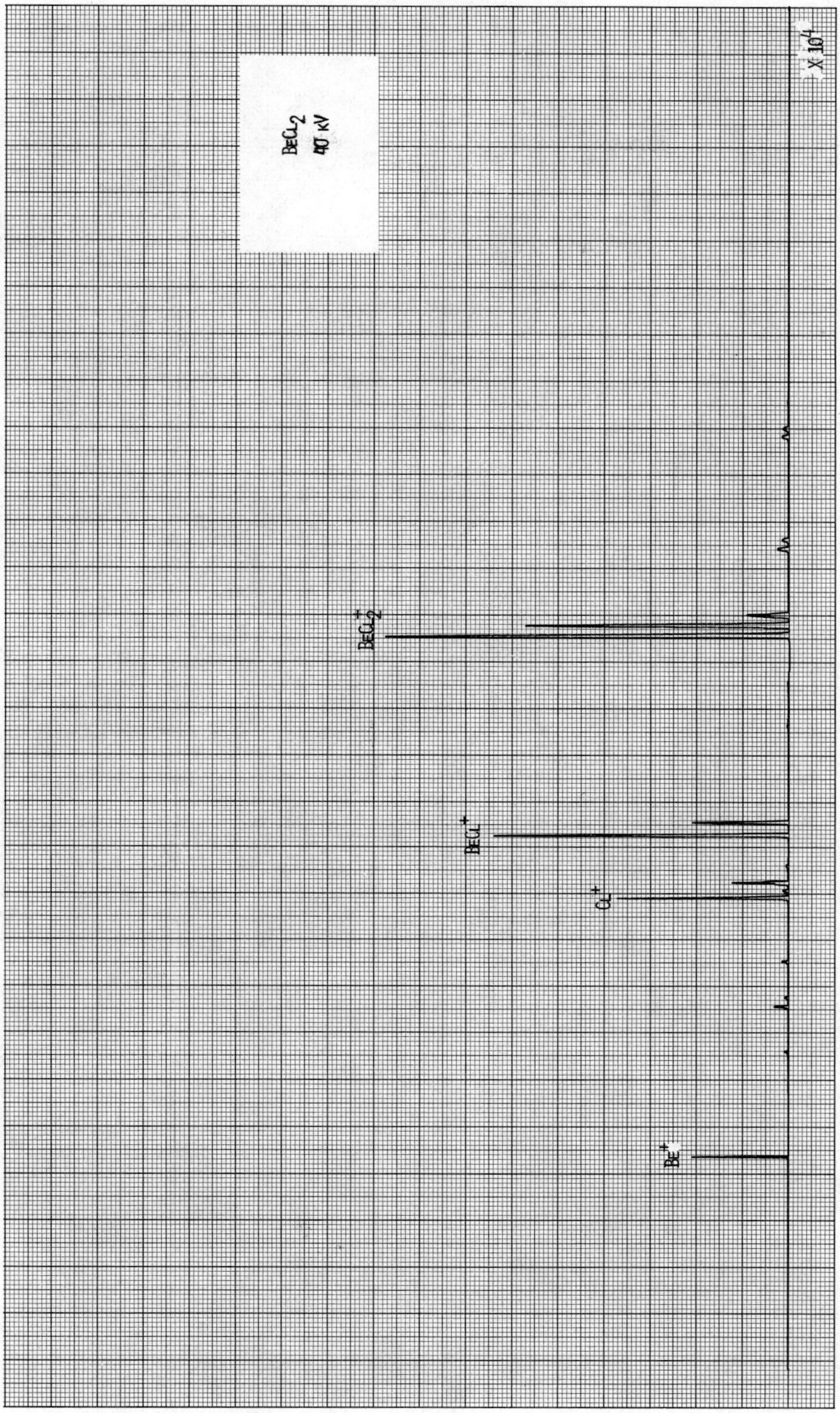
$BeCl_2$
40 kV
$BeCl_2^+$
$BeCl^+$
Cl^+
Be^+
X 10^4

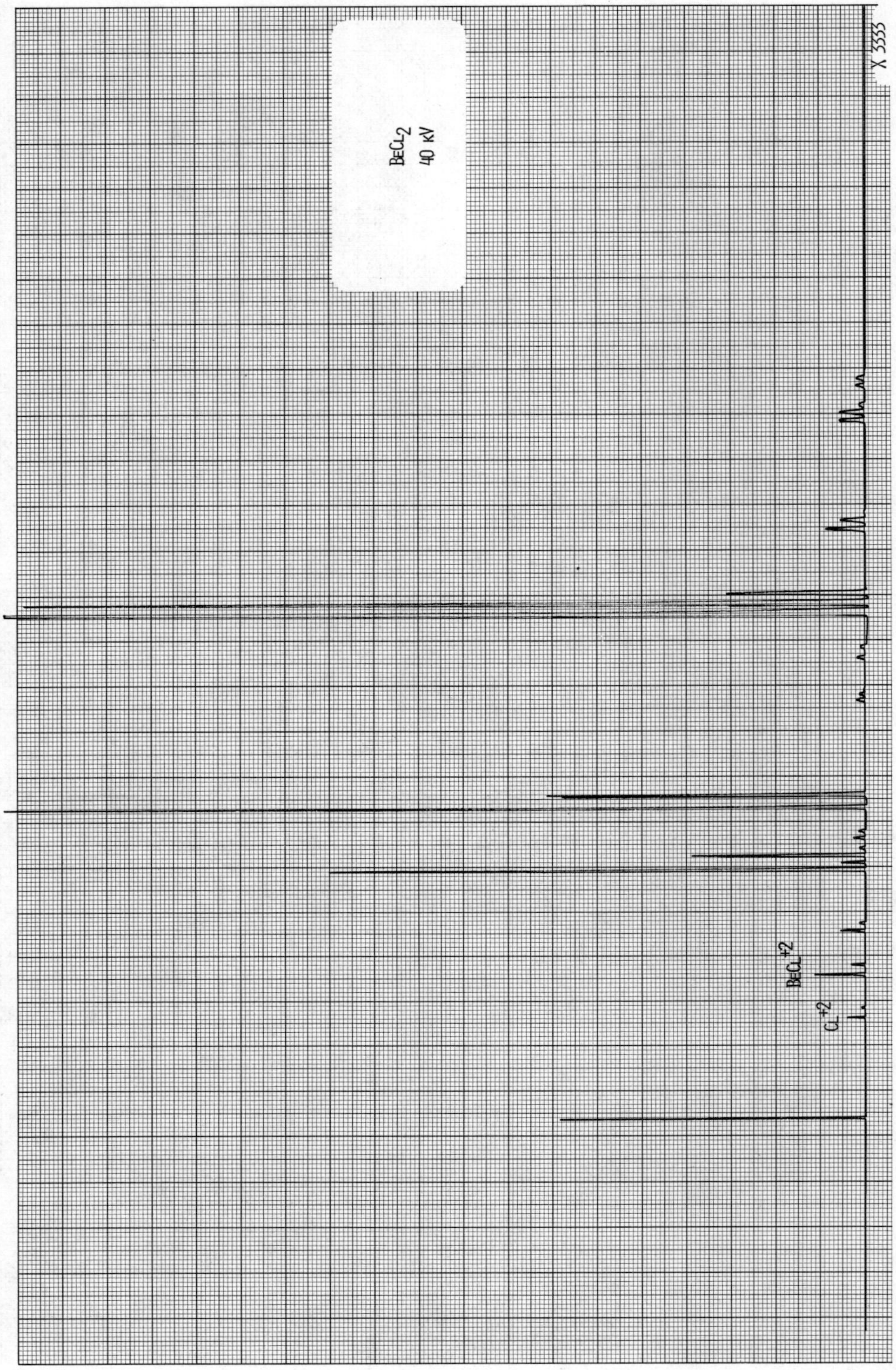
X 3333
BeCl$_2$
40 kV
BeCl^{+2}
C^{+2}

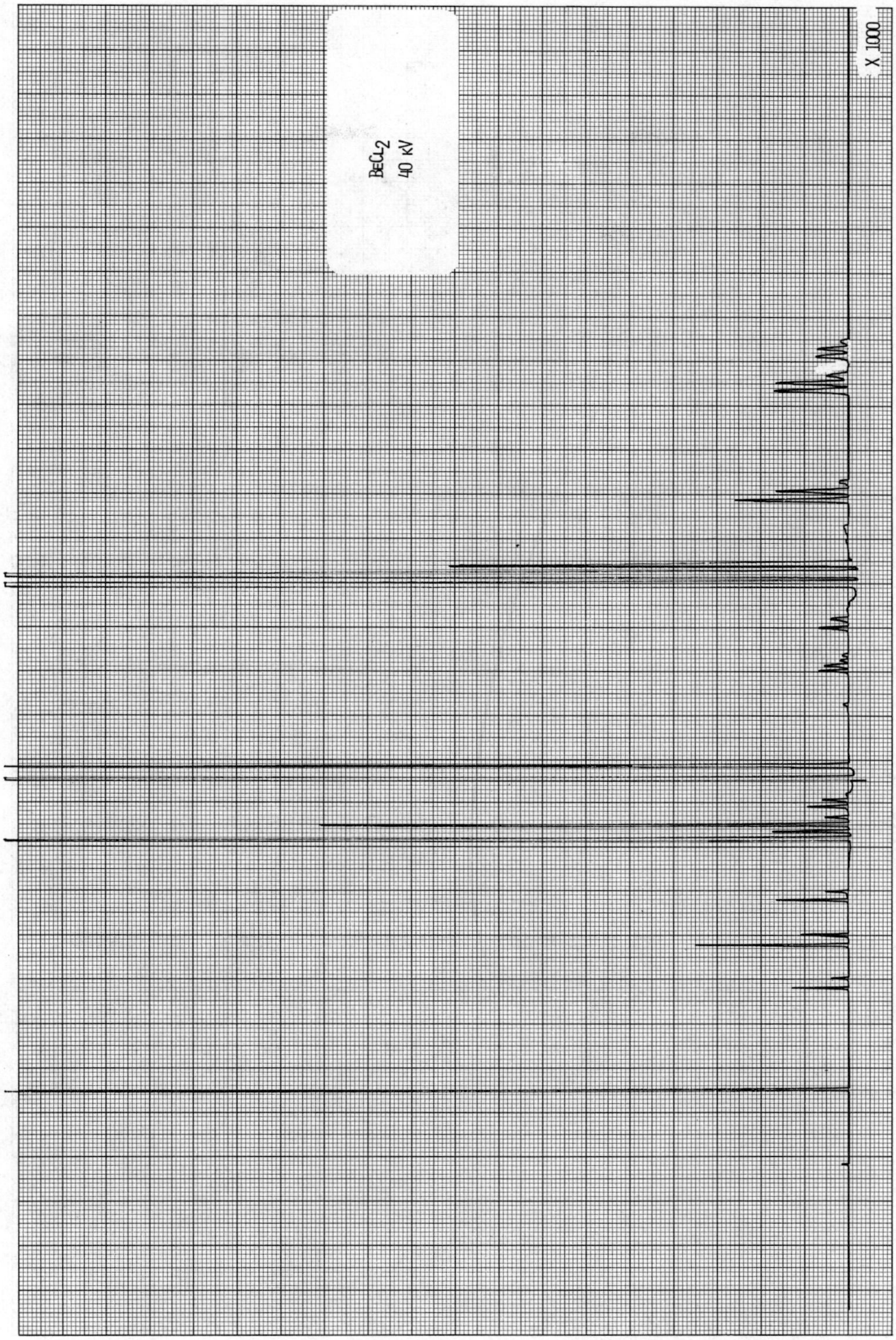
BeCl2
40 kV
X 1000

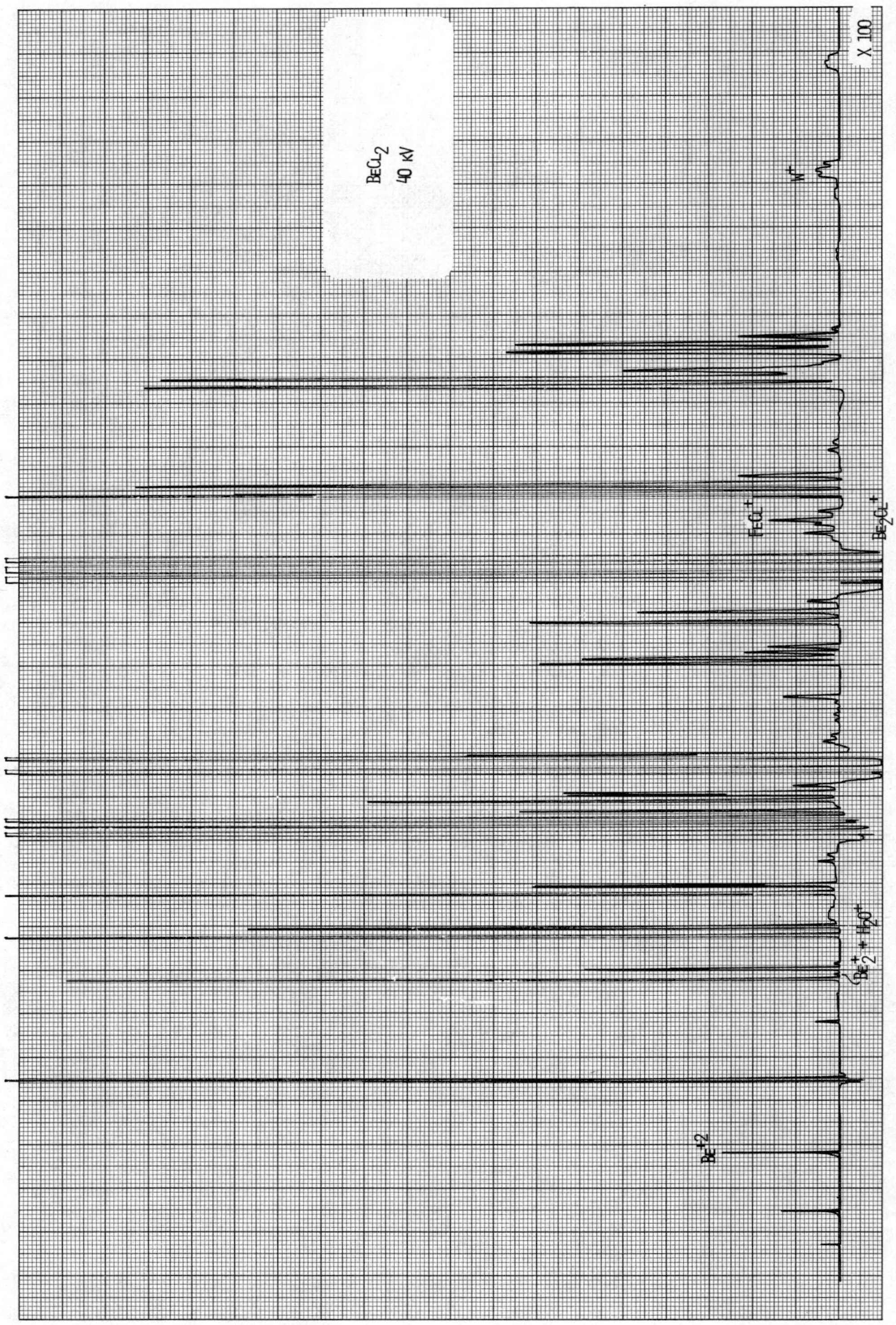
BeCl2
40 kV
X 100
W+
FeCl+
Be2Cl+
Be2+ + H2O+
Be+2

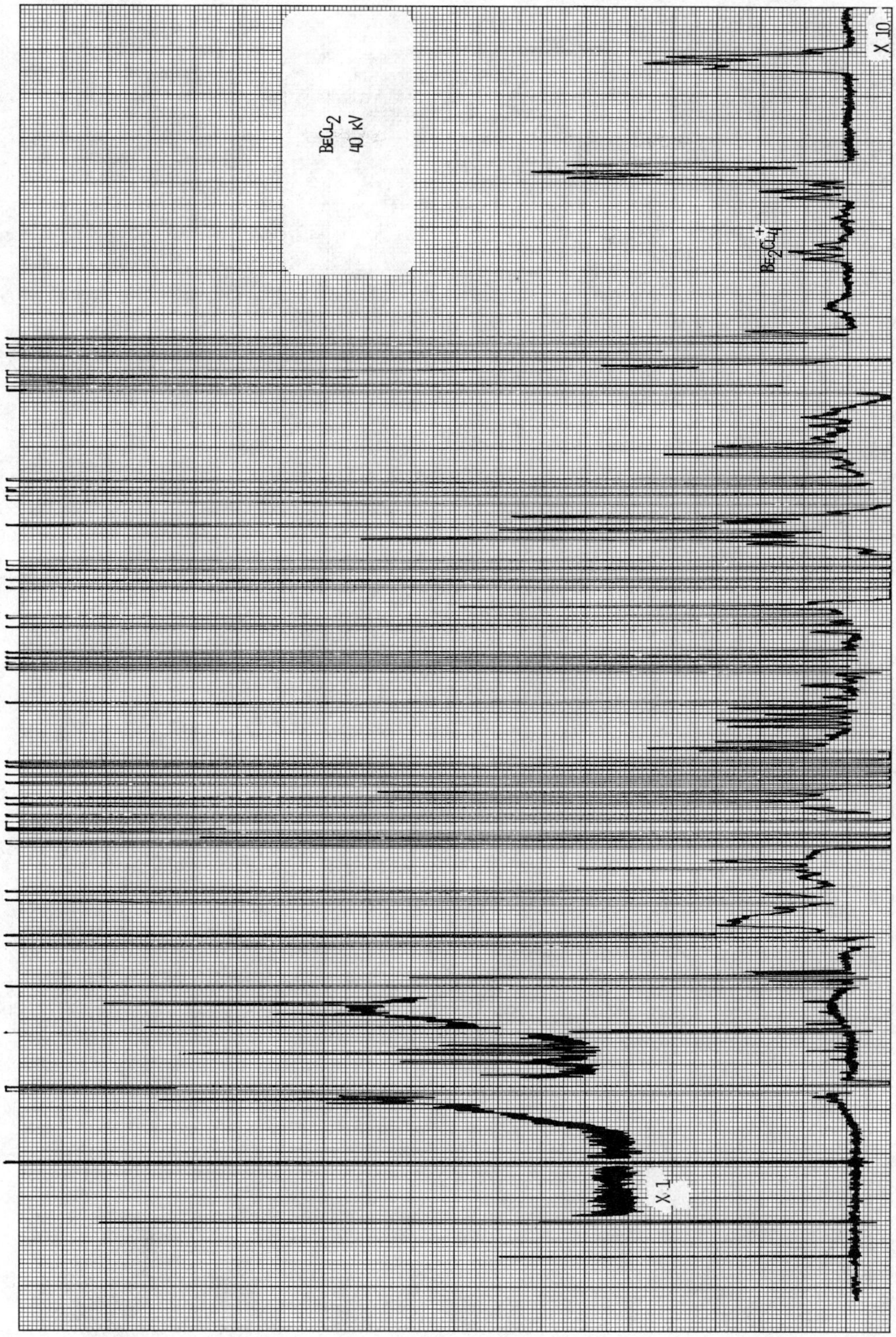
BeCl2
40 kV
X 10
Be2Cl4+
X 1

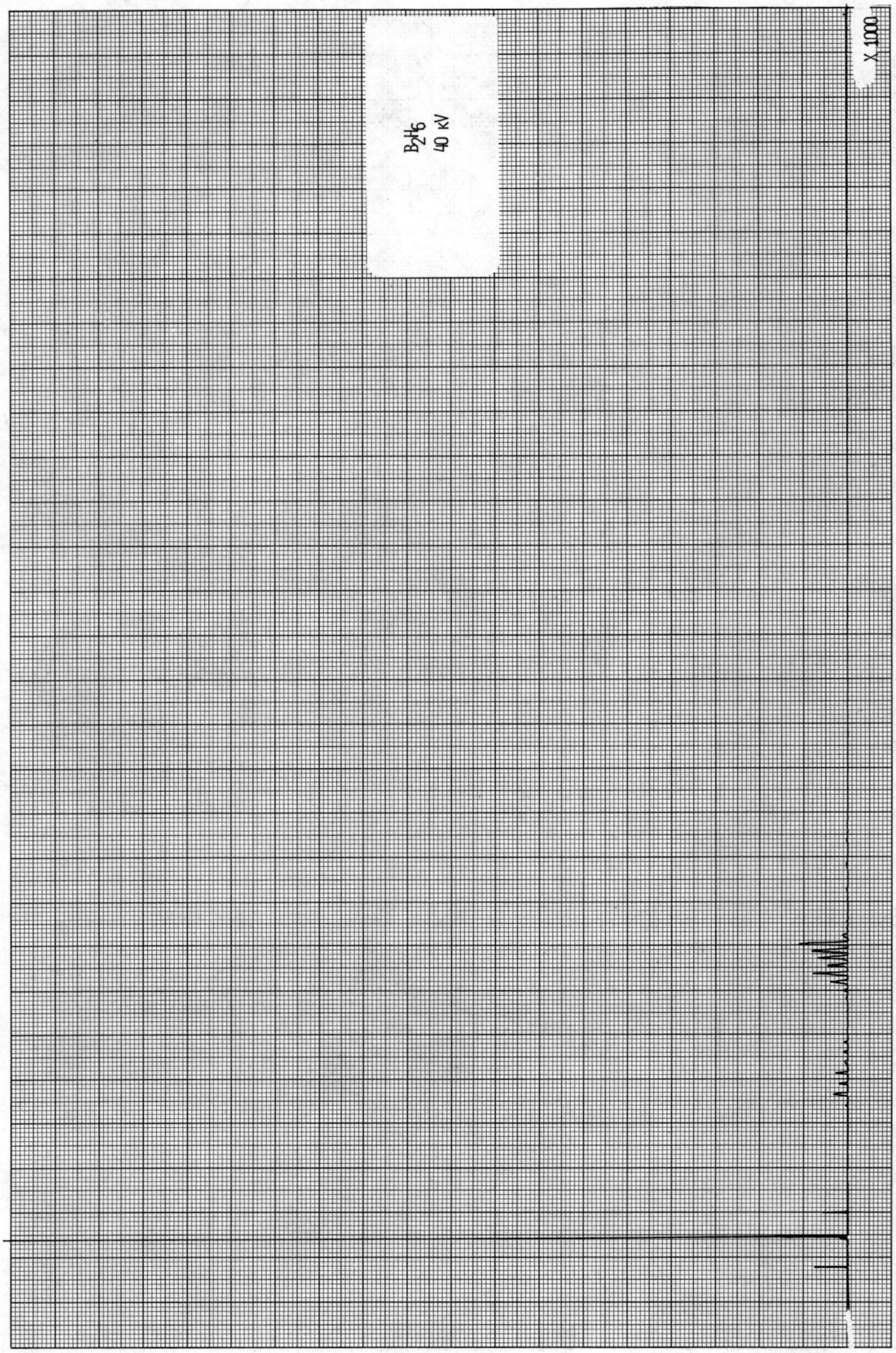
B_2H_6
40 kV
X 1000

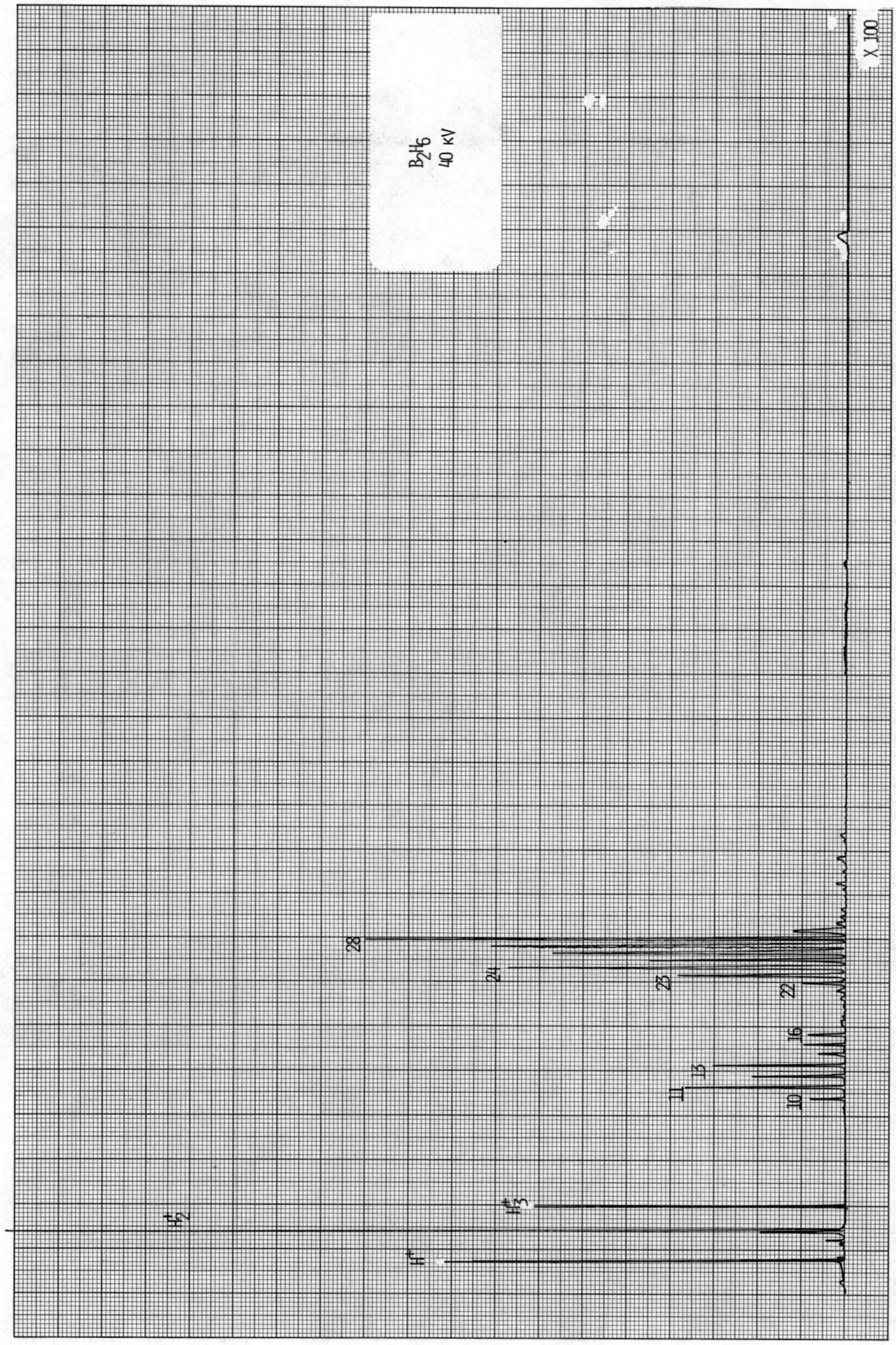

B_2H_6
40 kV
X 100
H^+
H_2^+
H_3^+
10
11
13
16
22
23
24
28

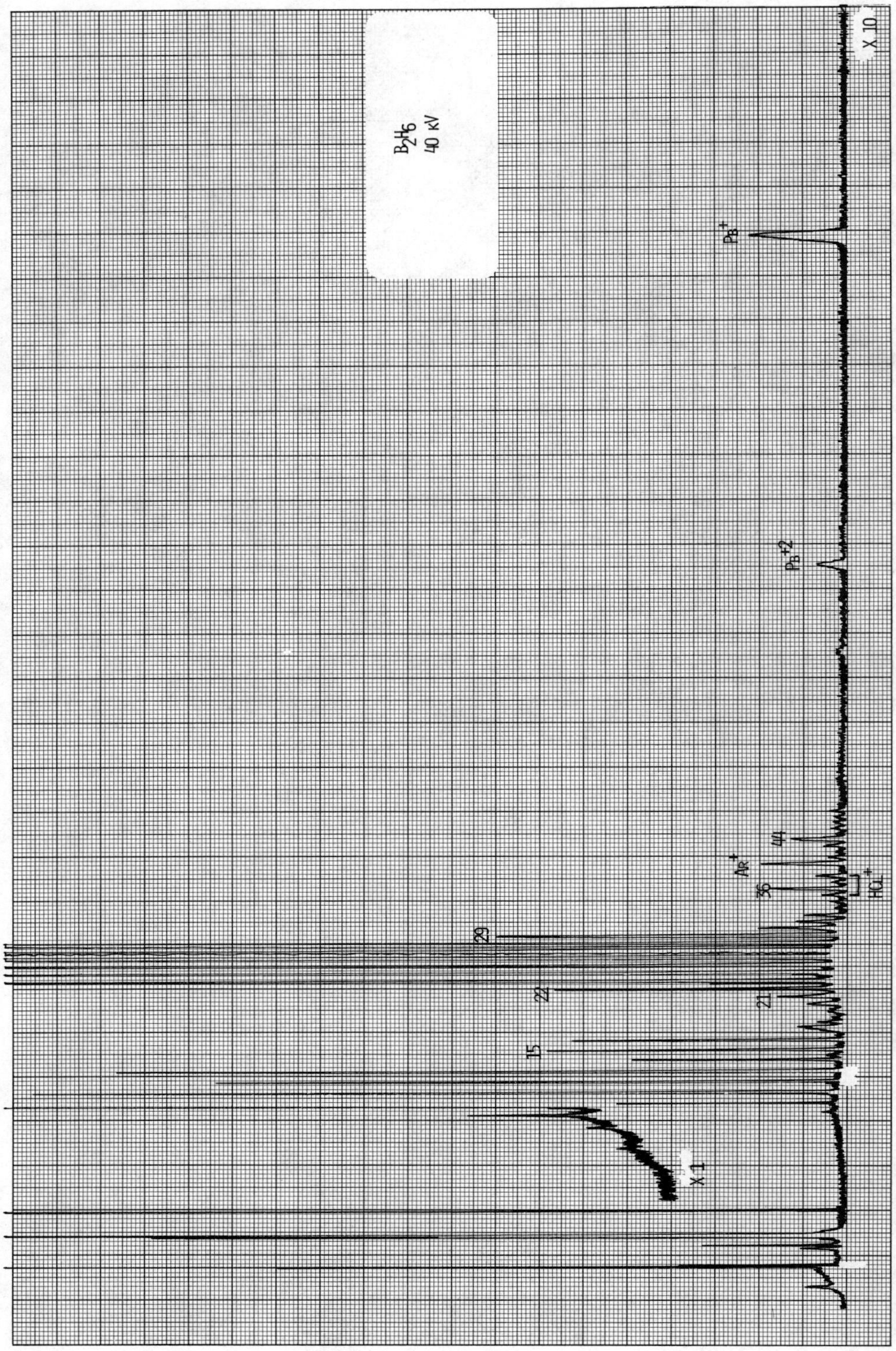
B_2H_6
40 kV
X 10
Pb^+
Pb^{+2}
44
Ar^+
36
HCl^+
29
22
21
15
X 1

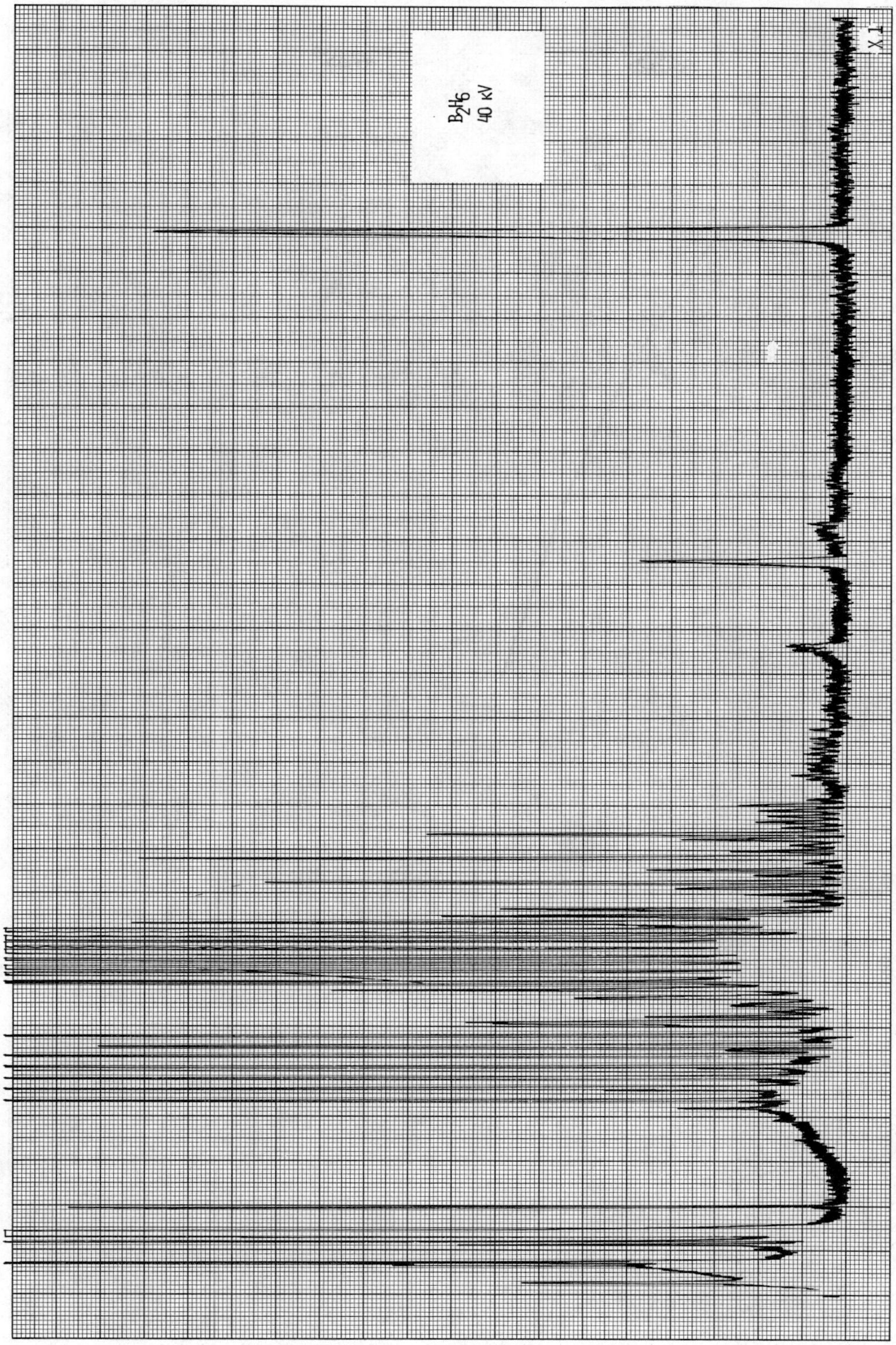
B2H6
40 kV
X 1

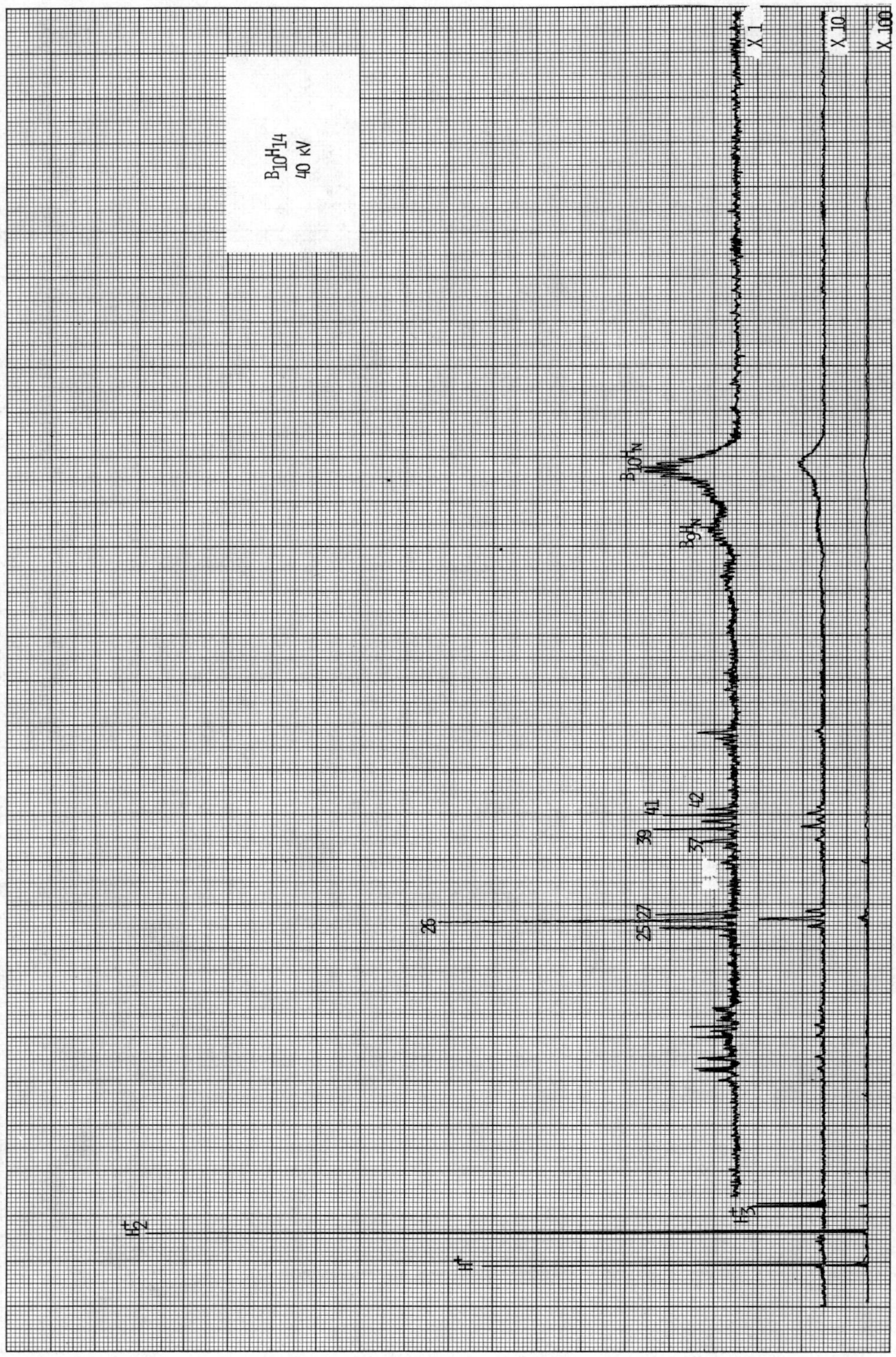

$B_{10}H_{14}$
40 kV
H^+
H_2^+
H_3^+
25
26
27
37
39
41
42
B_9H_x
$B_{10}H_x$
X 1
X 10
X 100

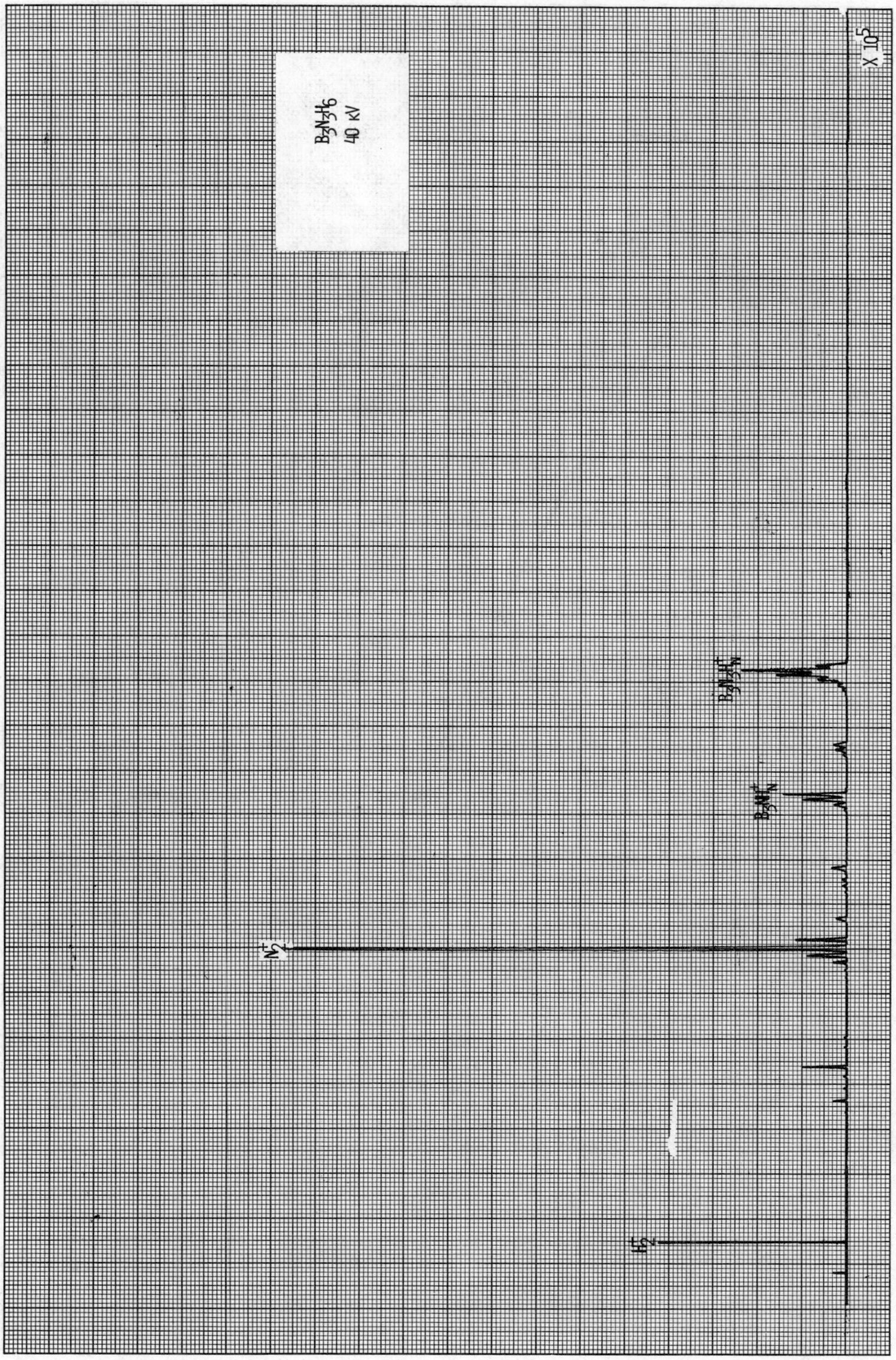
$B_3N_3H_6$
40 kV
X 10^5
$B_3N_3H_N^+$
$B_3N_N^+$
N_2^+
H_2^+

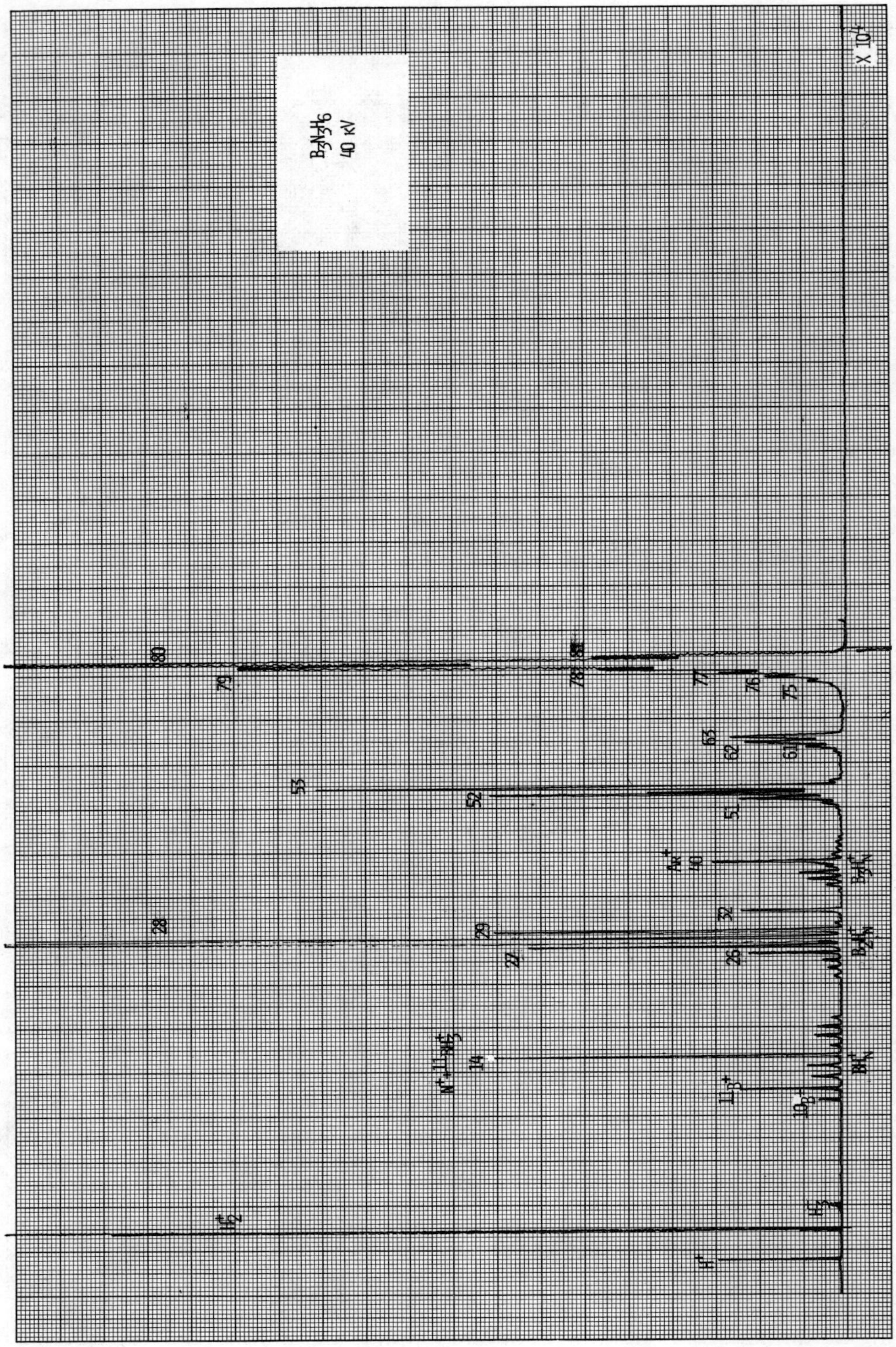
B3N3H6
40 kV
80
79
78
53
52
28
29
27
14

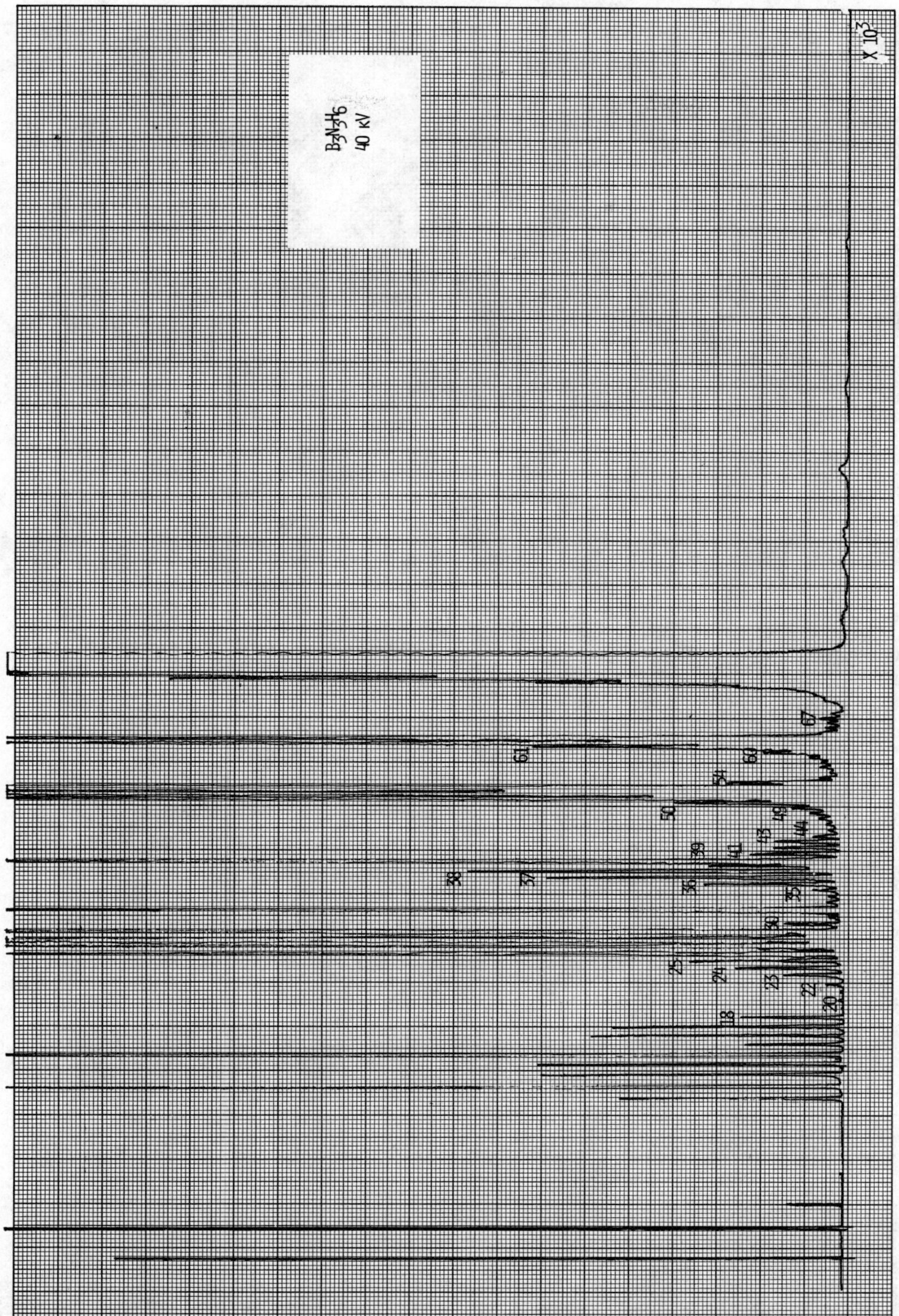

B3N3H6
40 kV
X 10 3

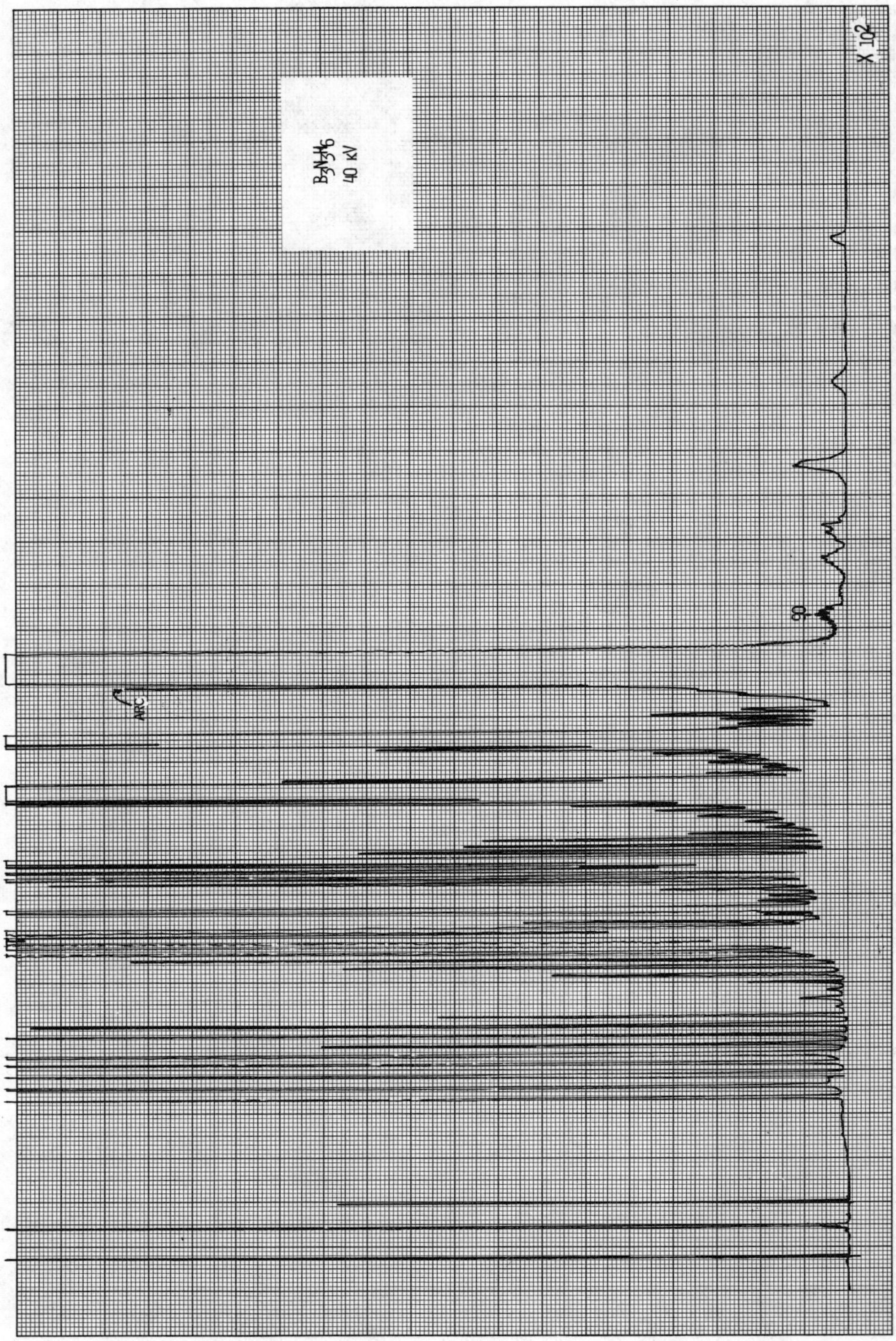

B3N3H6
40 kV
X 10

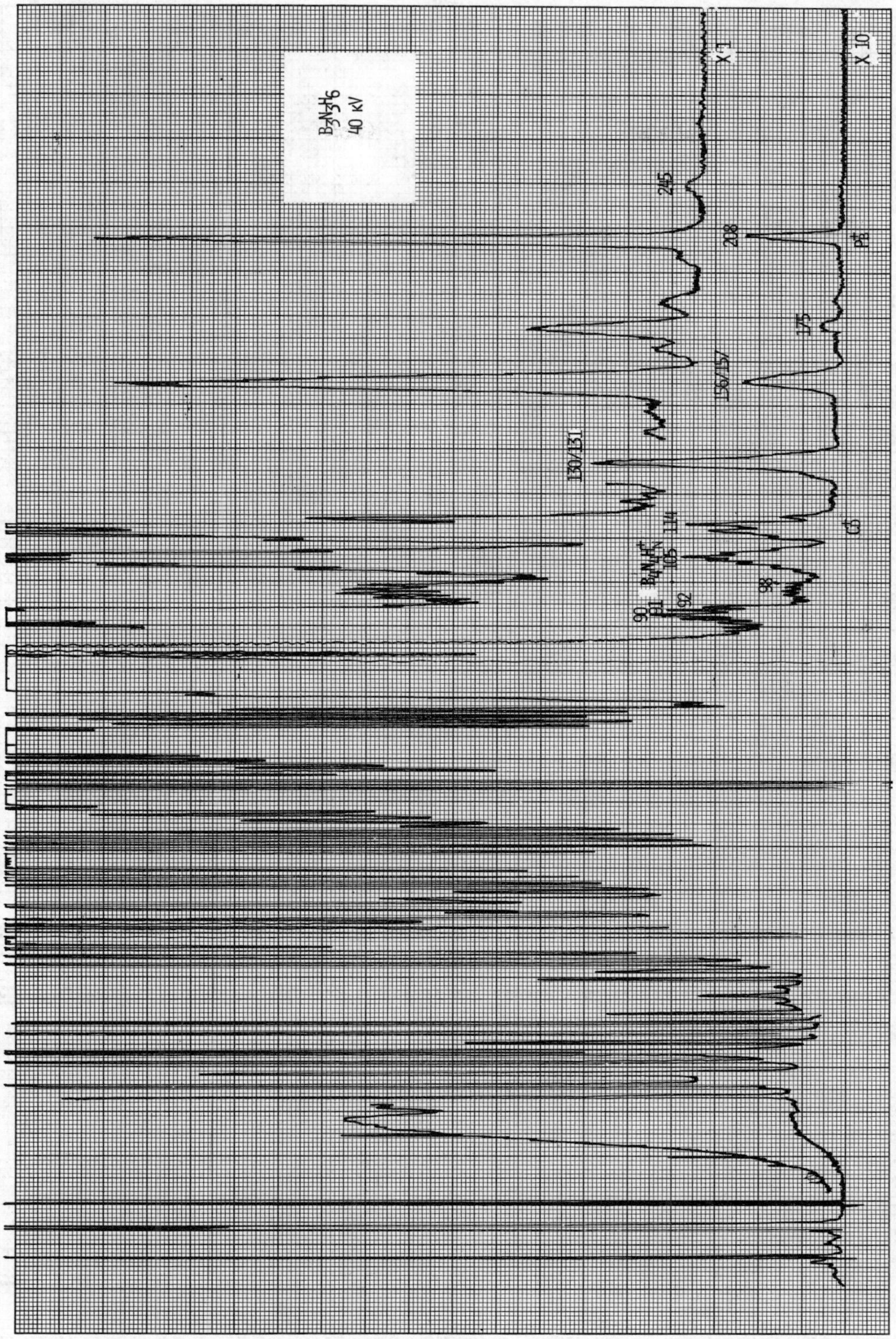

$B_3N_3H_6$
40 kV
X 1
X 10
245
208
175
156/157
130/131
114
105
98
92
90

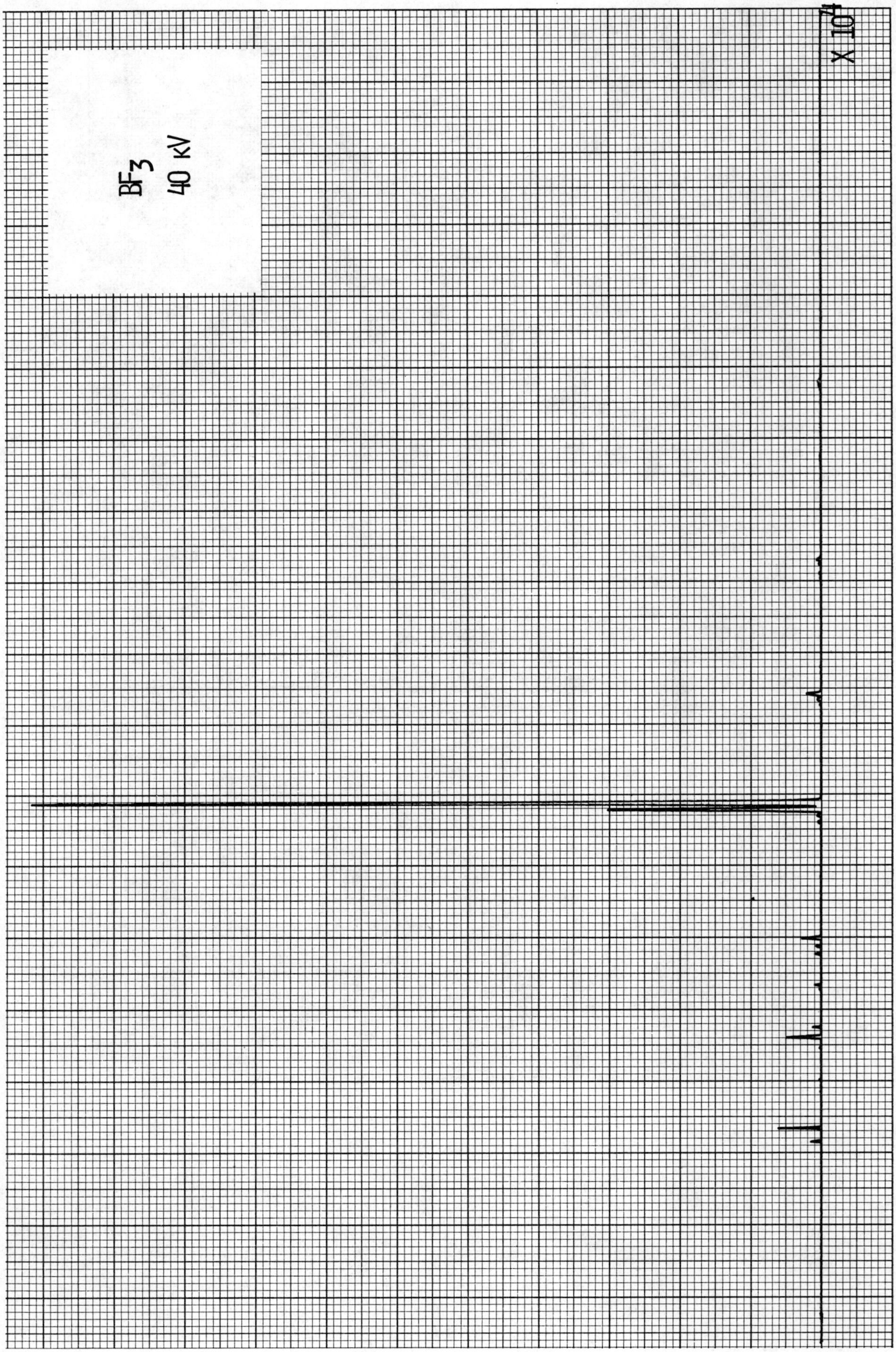

BF3
40 kV
X 10^4

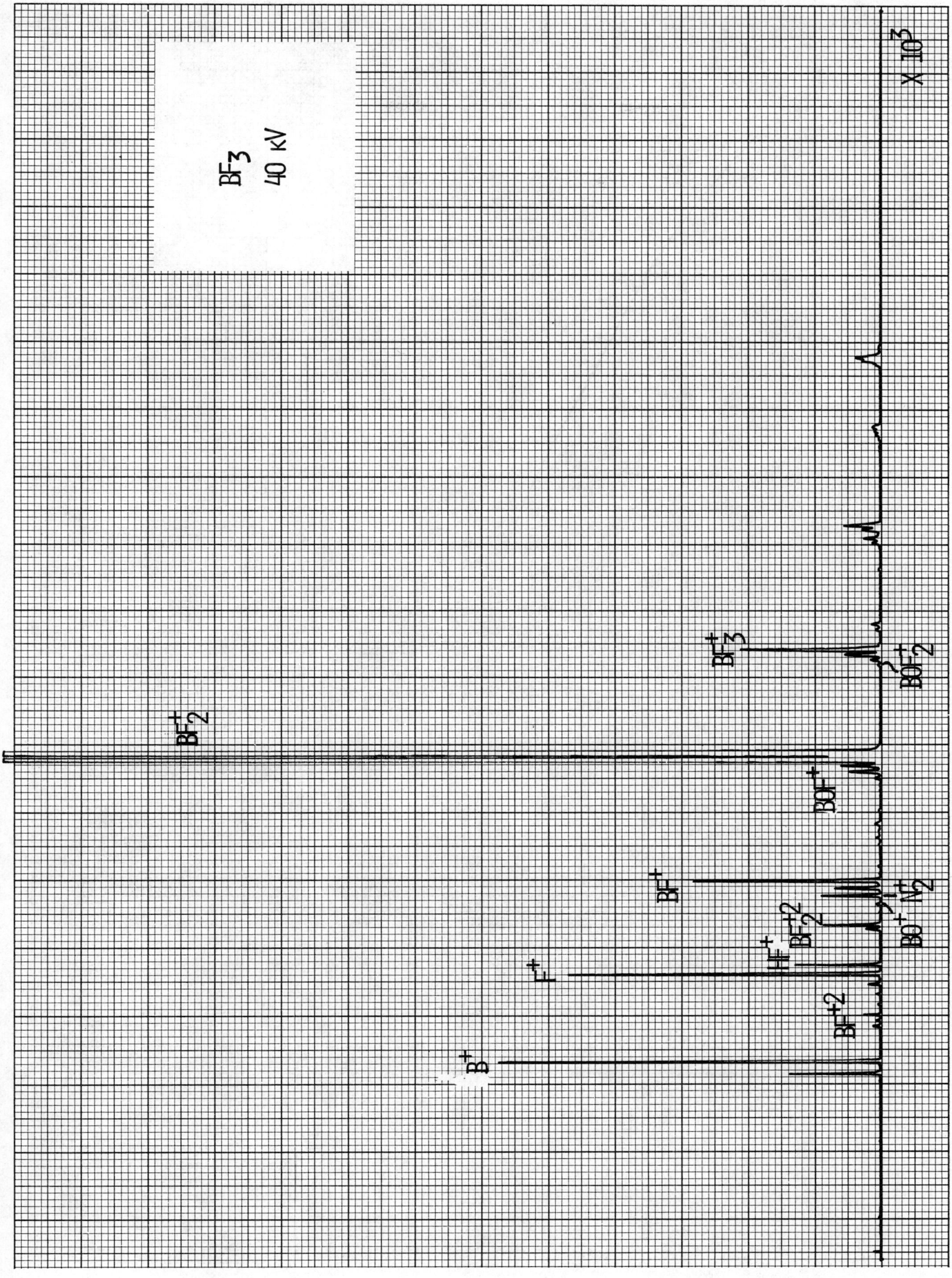
BF3
40 kV
X 10^3
BF2+
BF3+
BOF2+
BOF+
BF+
N2+
BO+
BF2+2
HF+
F+
BF+2
B+

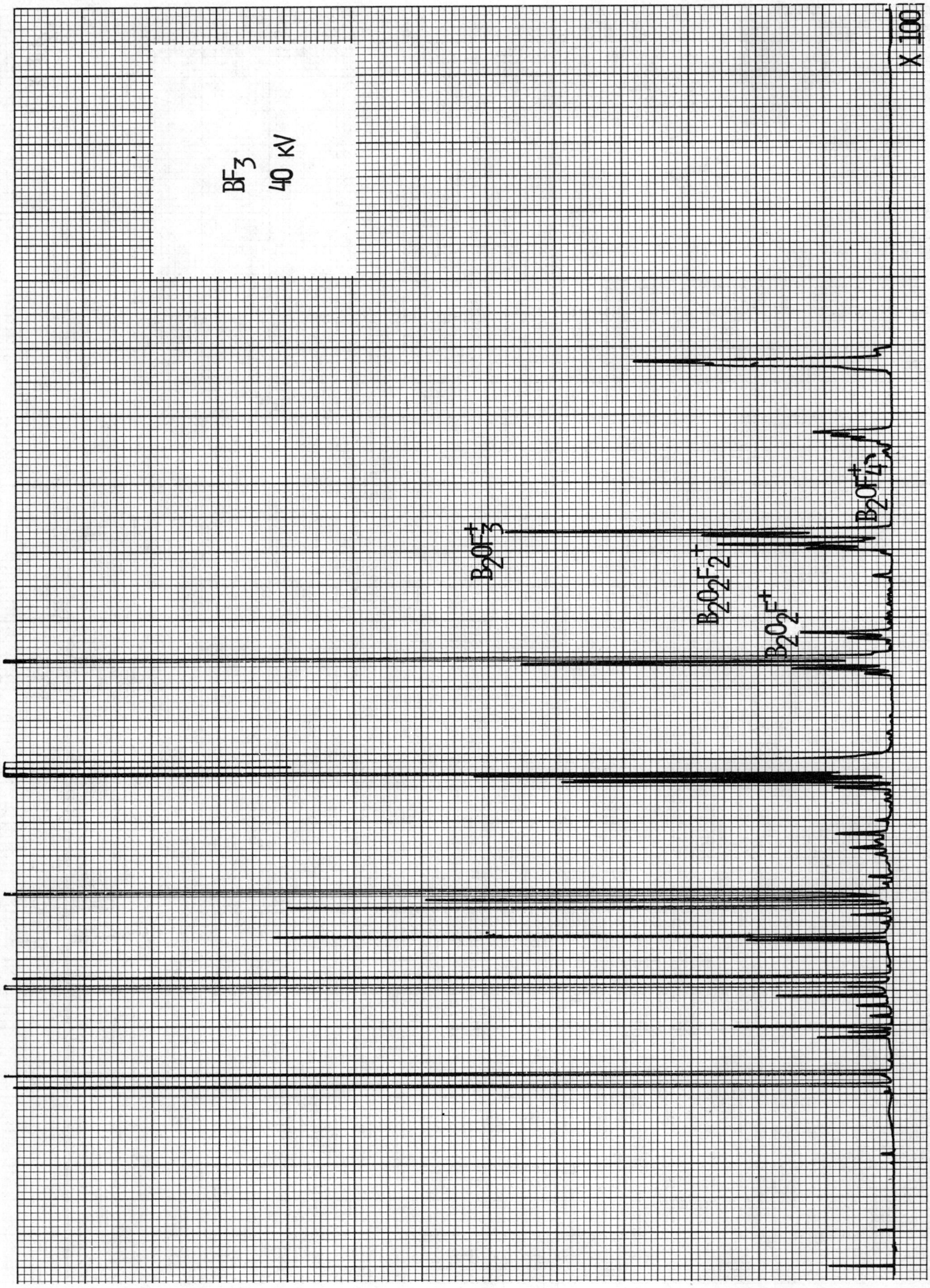

BF_3
40 kV
$B_2OF_3^+$
$B_2O_2F_2^+$
$B_2O_2F^+$
$B_2OF_4^+$
X 100

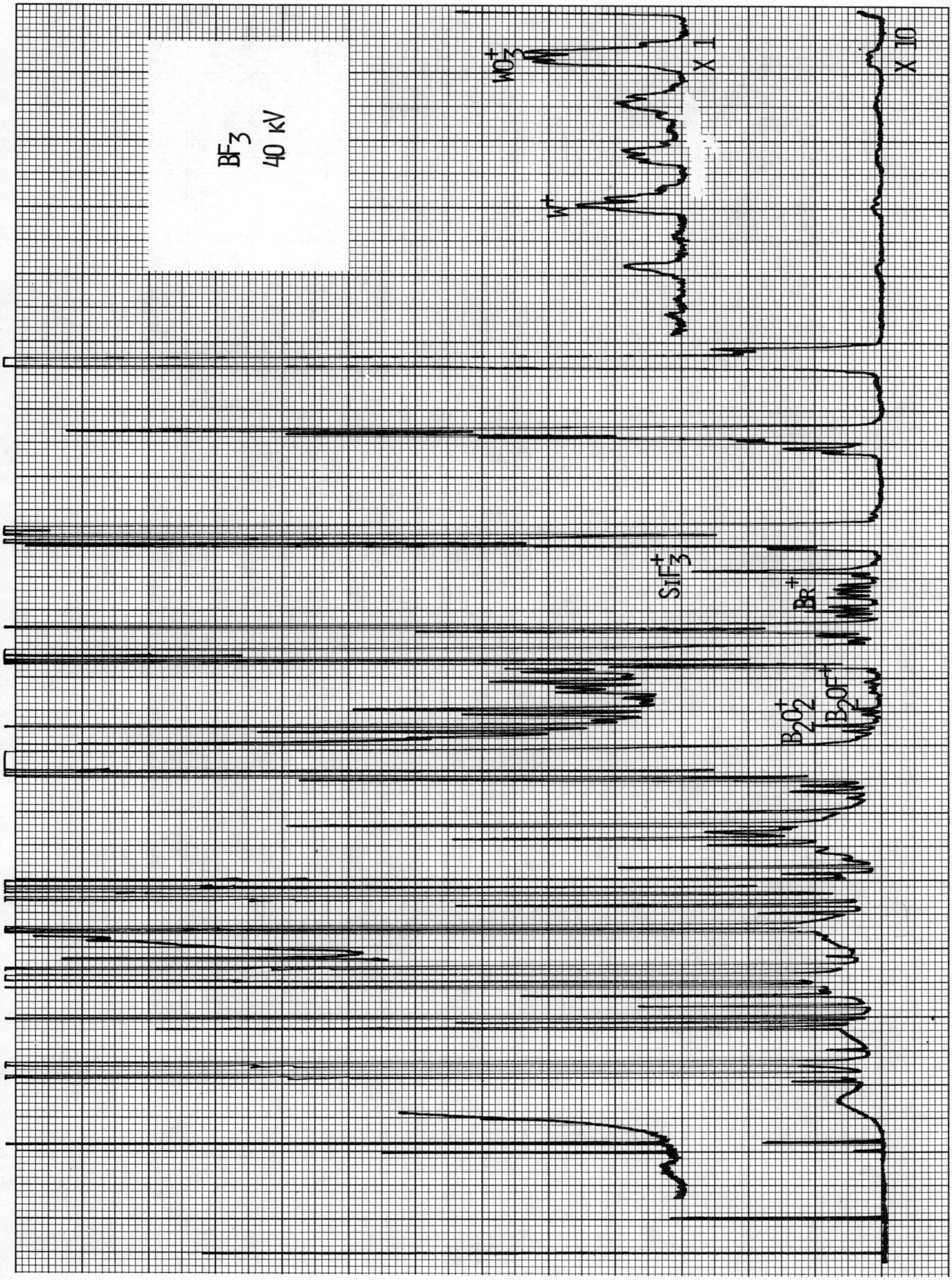
BF_3
40 kV
WO_3^+
W^+
X 1
X 10
SiF_3^+
Br^+
$B_2O_2^+$
$B_2OF_2^+$

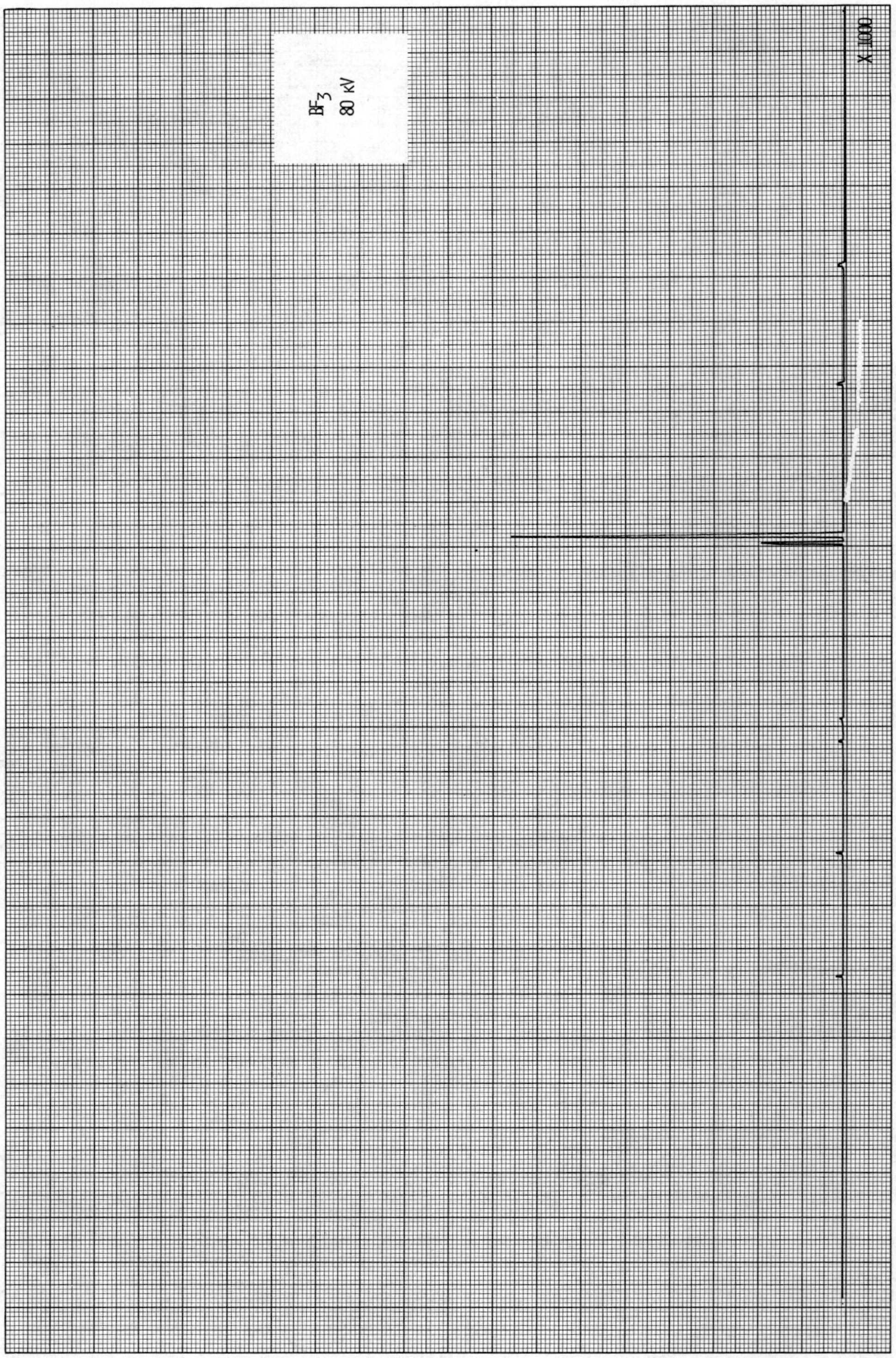
BF3
80 kV
X 1000

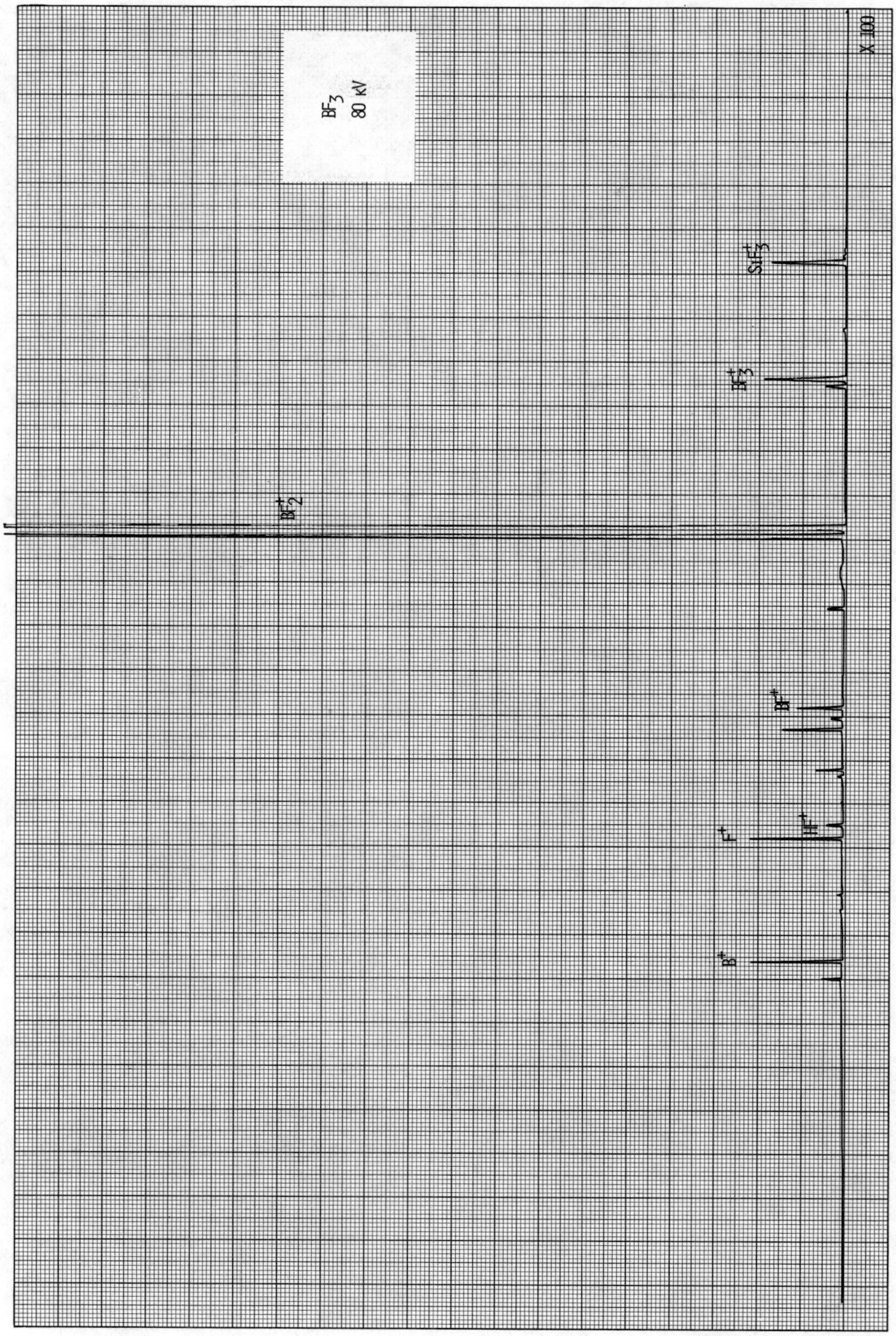

BF_3
80 kV
X 100
SiF_3^+
BF_3^+
BF_2^+
BF^+
F^+
HF^+
B^+

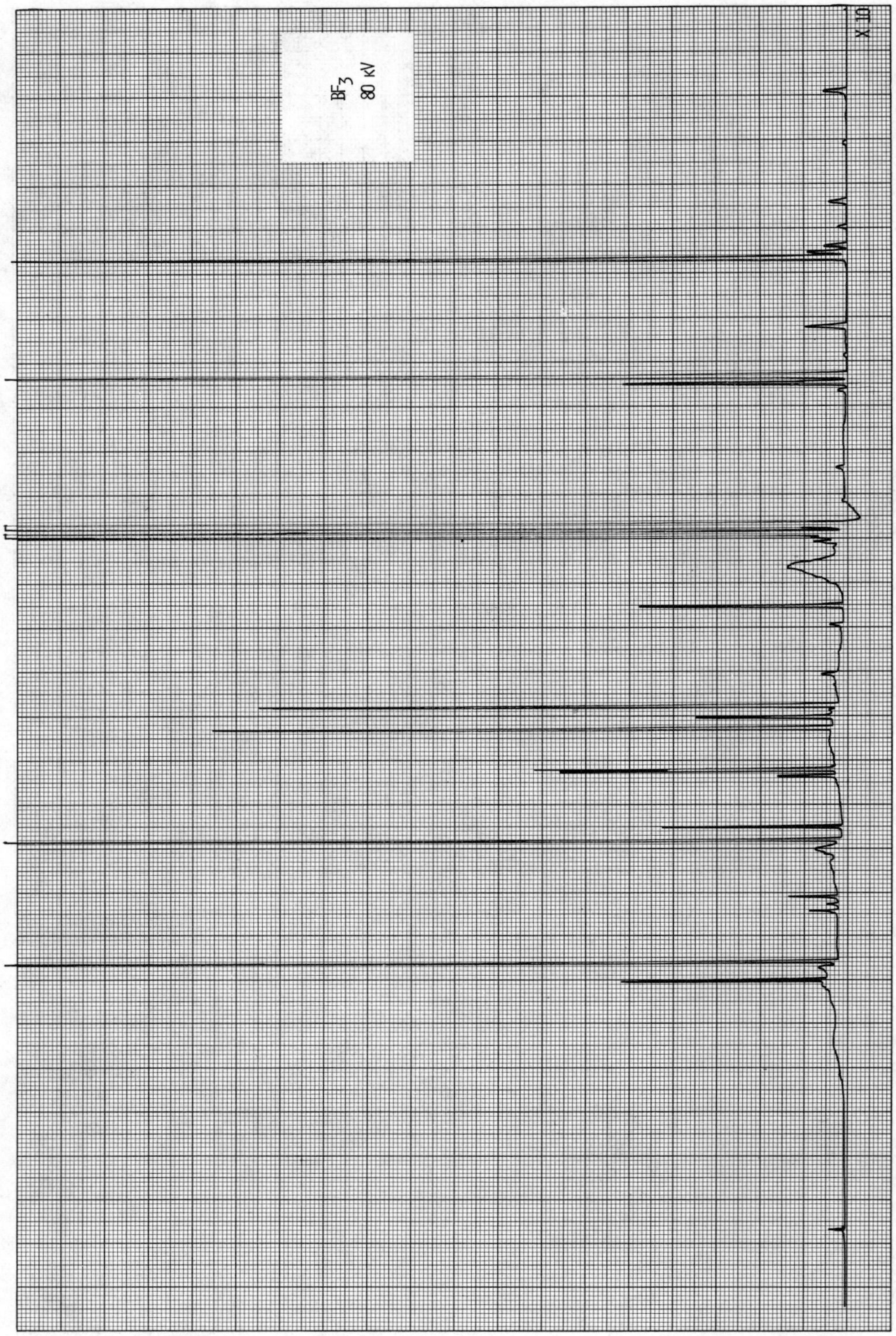
BF_3
80 kV
X 10

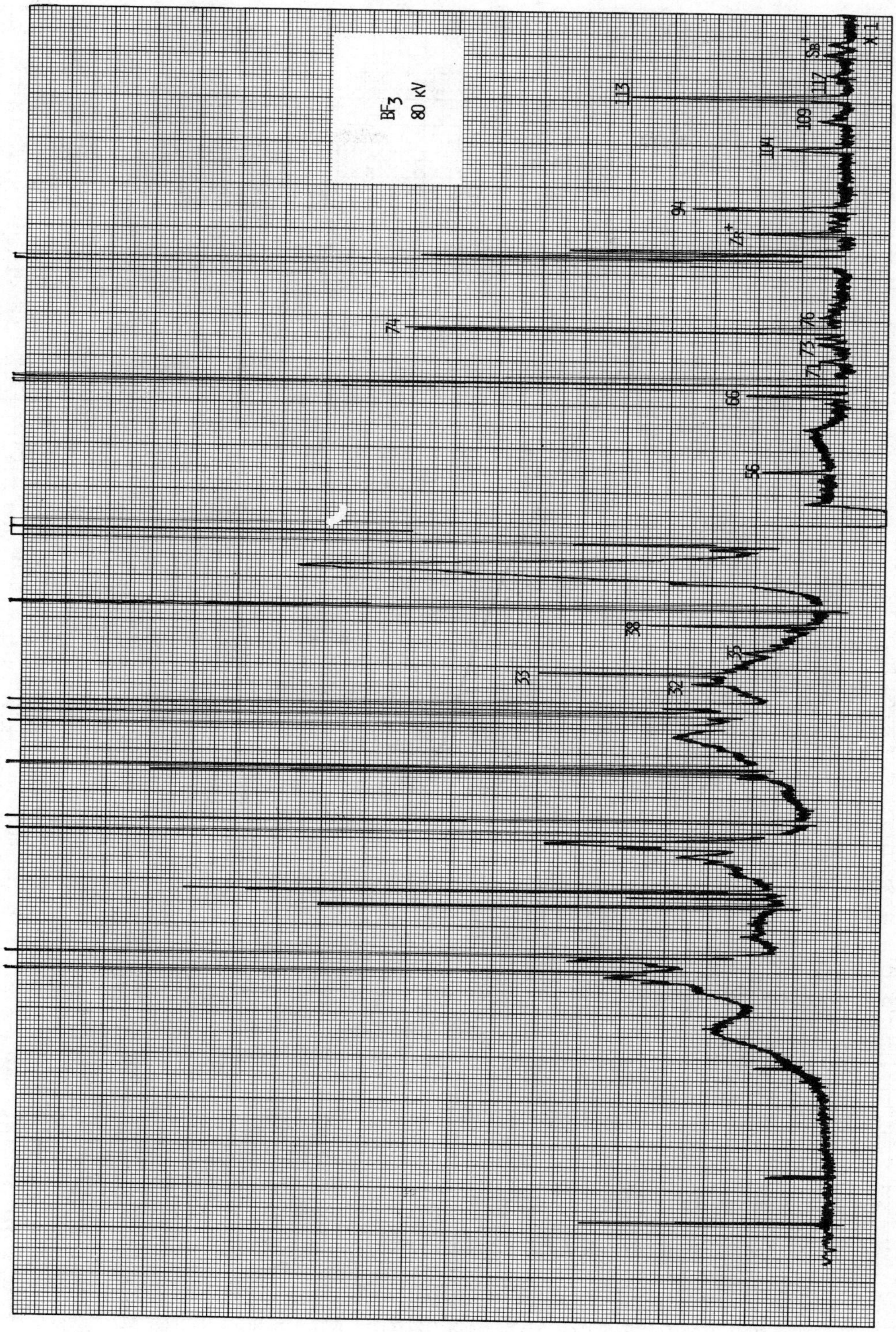
BF3
80 kV
x 1
Sb8+
117
113
109
104
94
Zr+
74
76
73
71
69
56
38
35
33
32

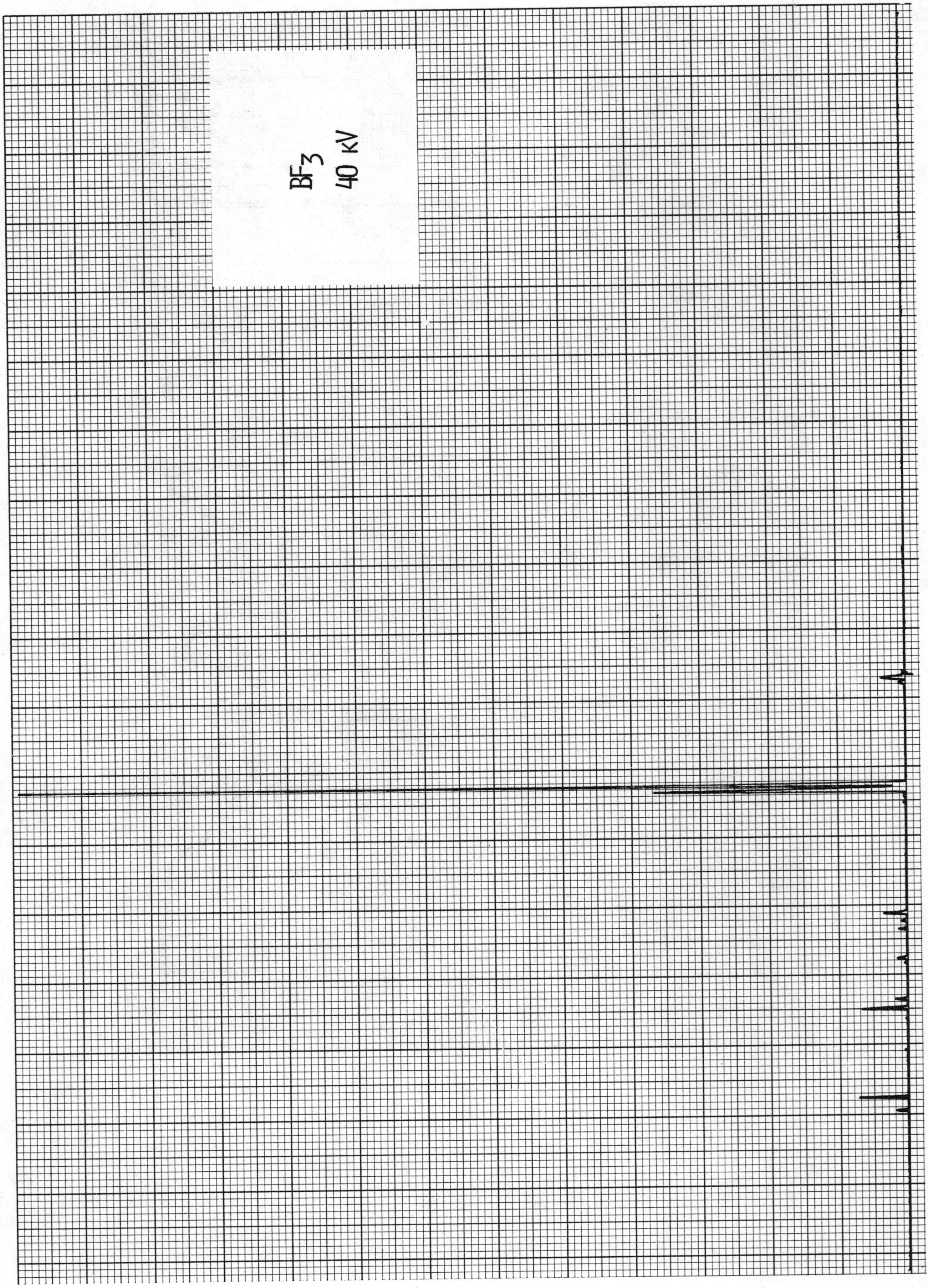
BF3
40 kV

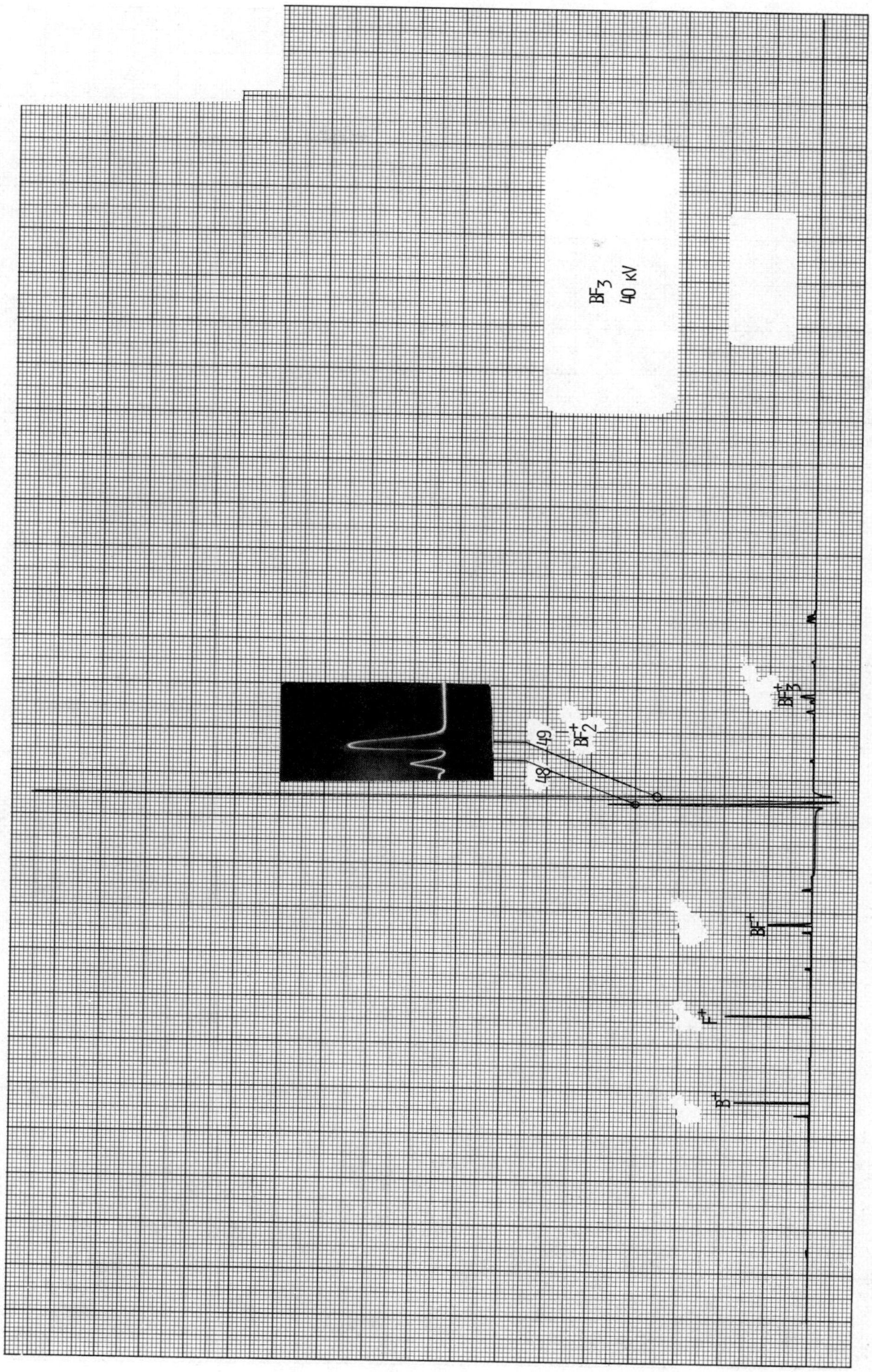

BF_3
40 kV
48
49
BF_2^+
BF_3^+
BF^+
F^+
B^+

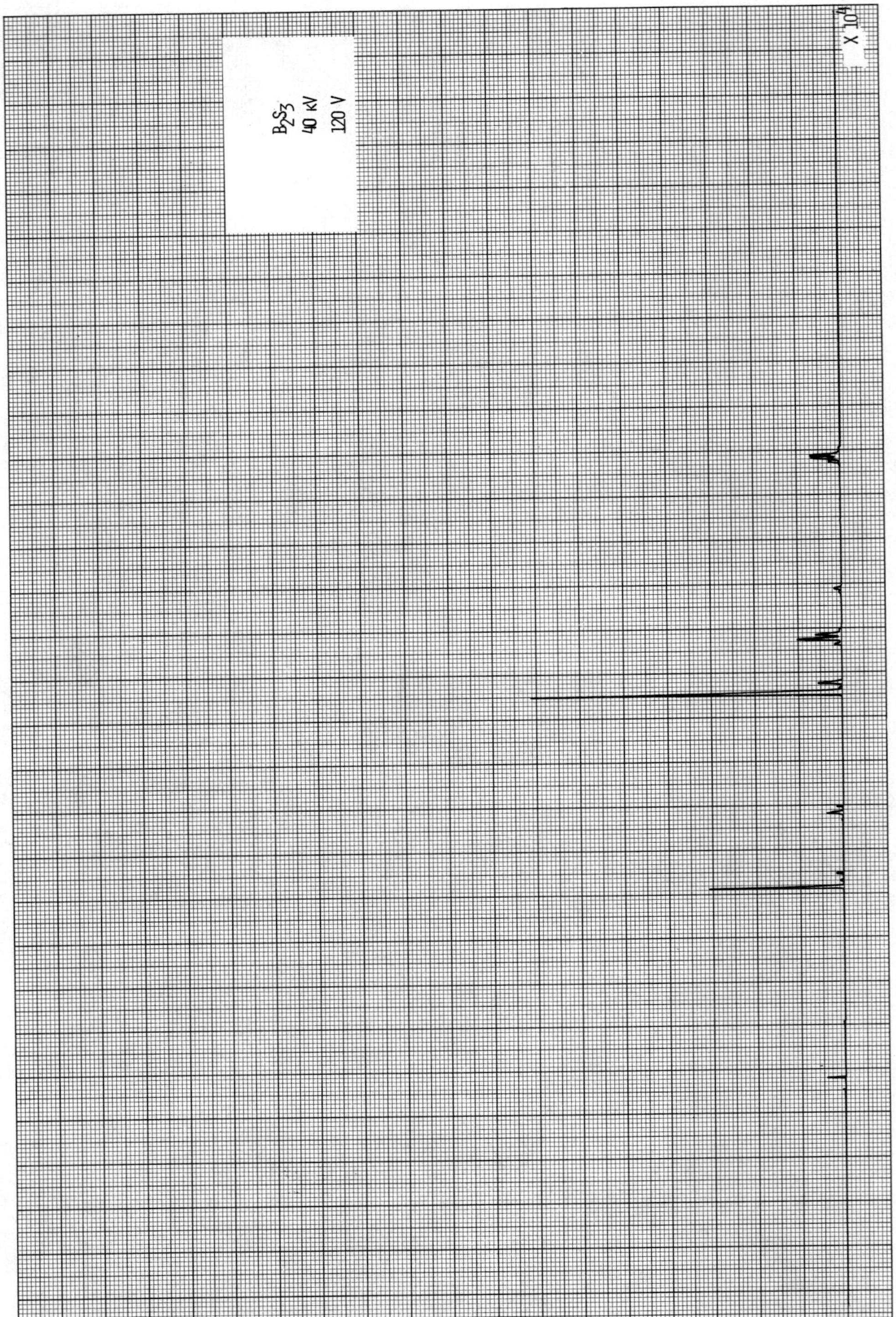

B_2S_3
40 kV
120 V
X 10^4

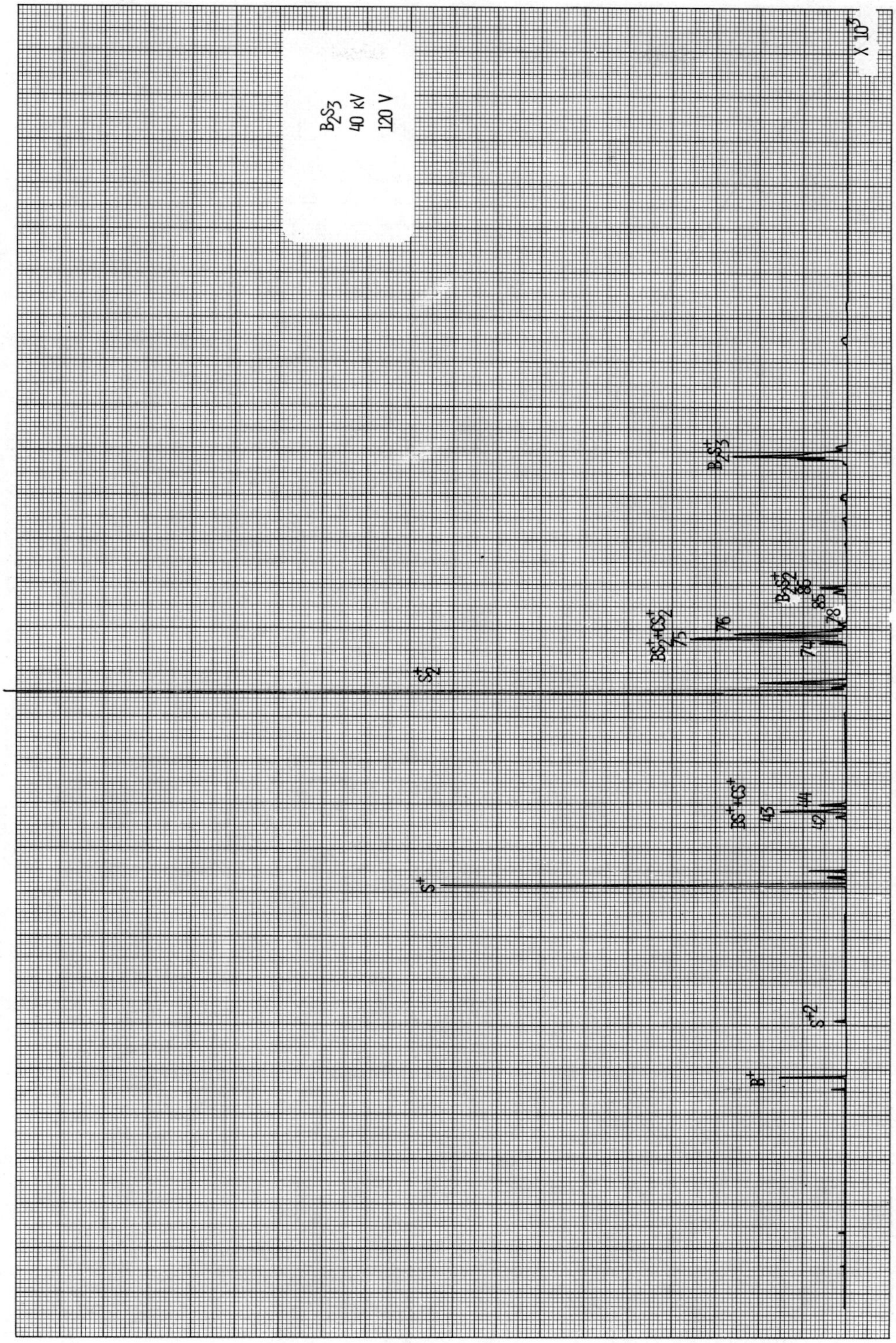
B2S3
40 kV
120 V
X 10³
B⁺
S⁺²
S⁺
BS⁺+CS⁺
42
43
44
S₂⁺
BS₂⁺+CS₂⁺
74
75
76
78
85
86
B₂S₂⁺
B₂S₃⁺

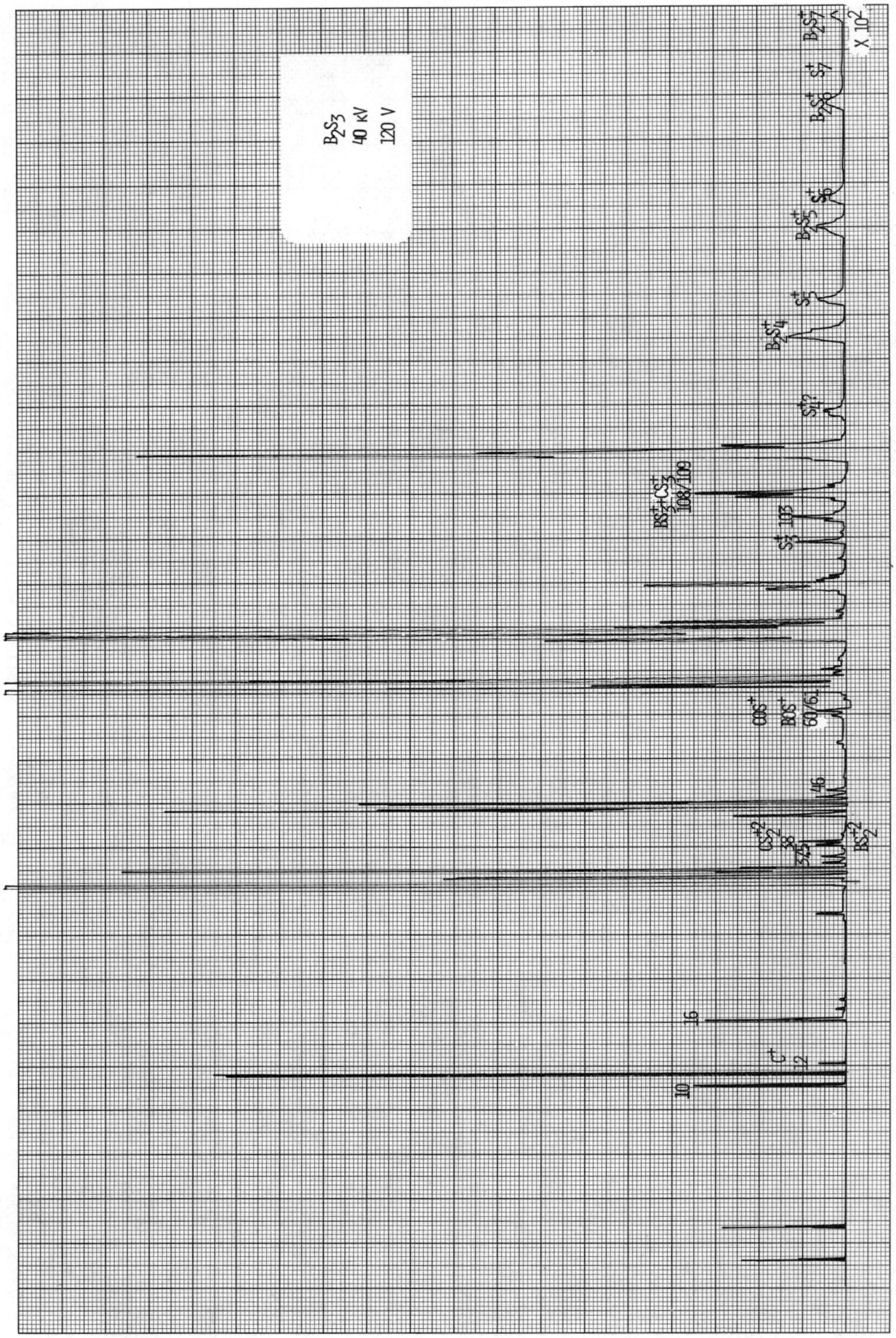

B2S3
40 kV
120 V
X 10^2
10
16
46

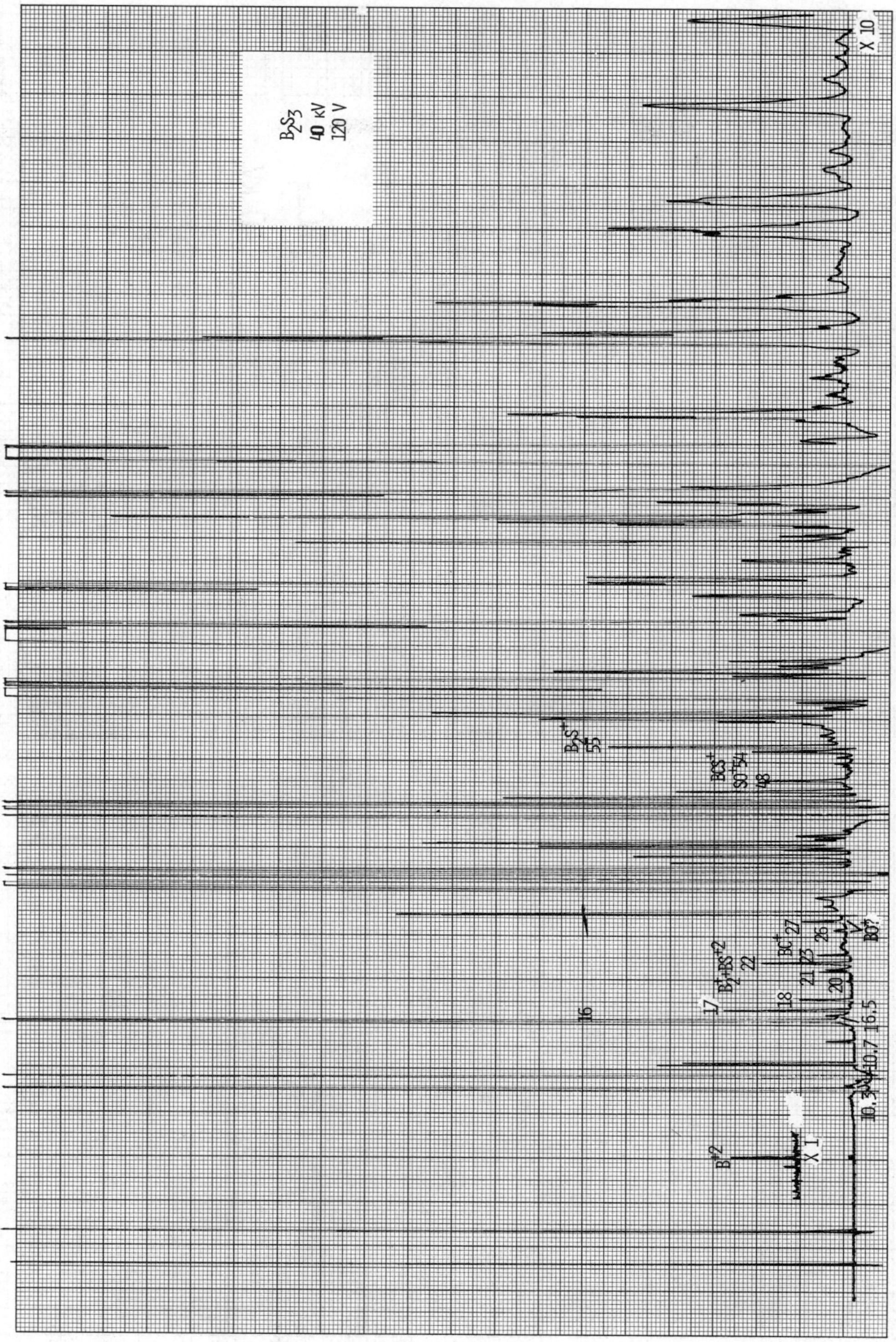

B_2S_3
40 kV
120 V
X 10
X 1

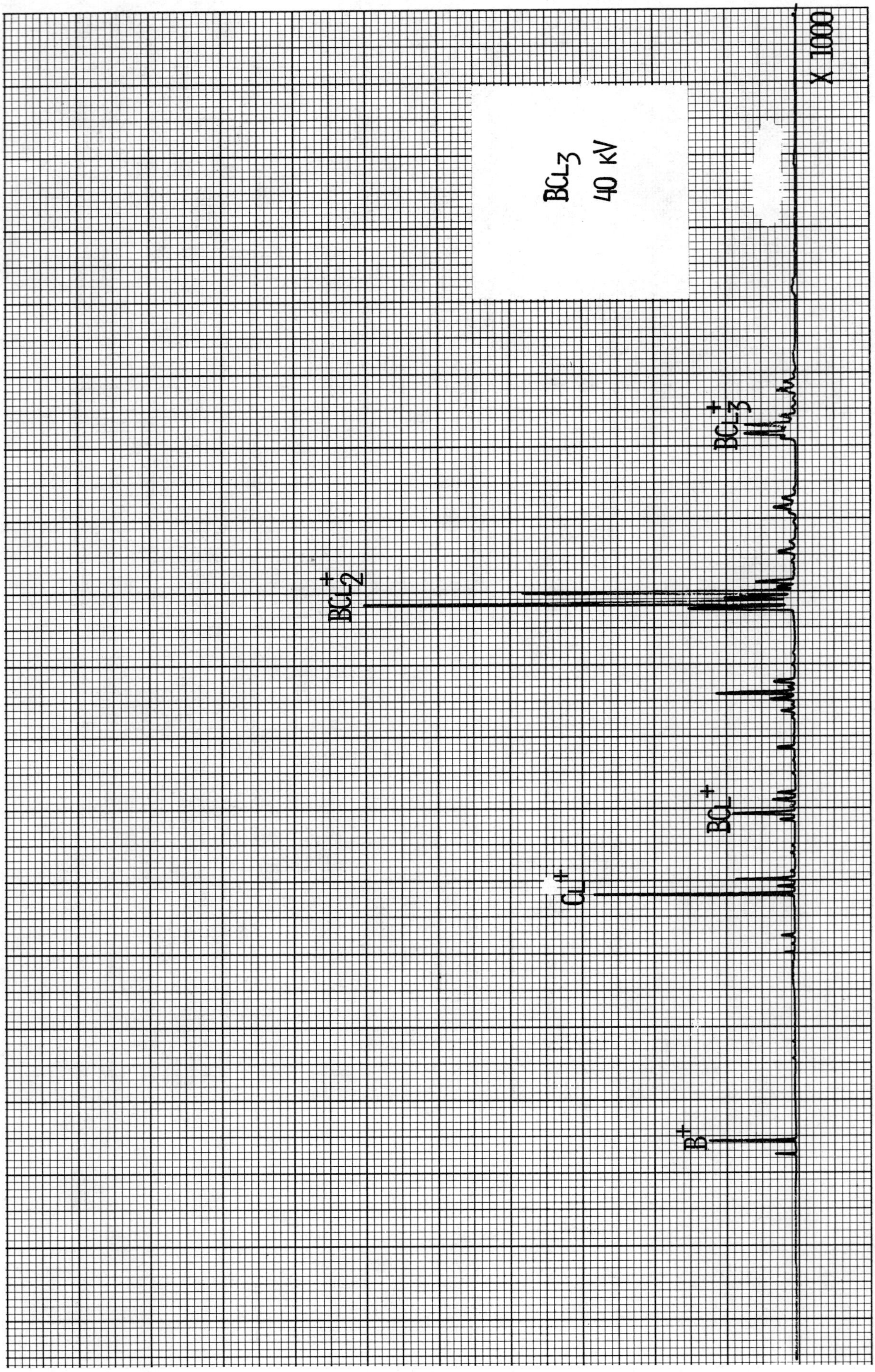
X 1000
BCl3
40 kV
BCl2+
BCl3+
BCl+
Cl+
B+

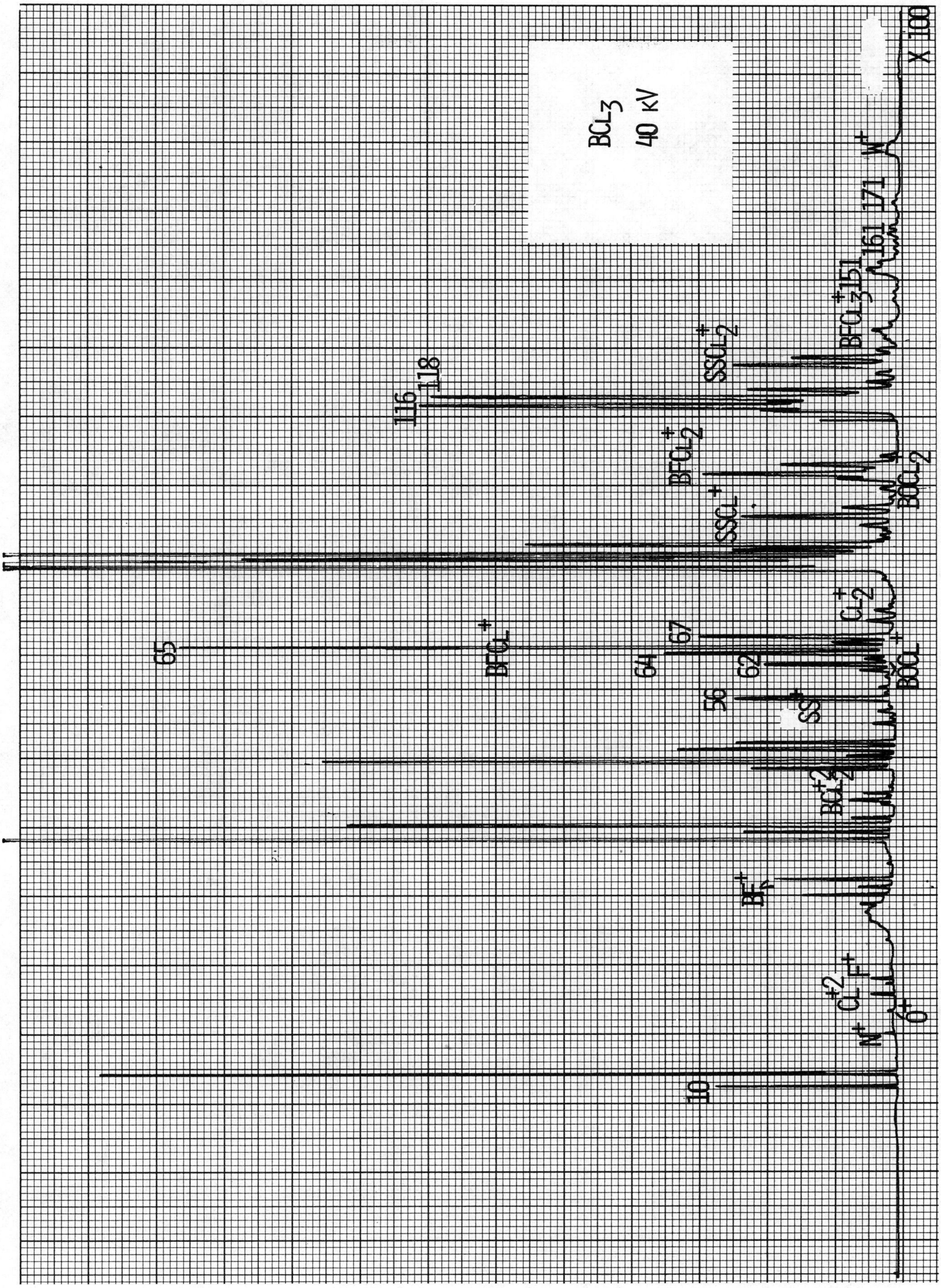
BCL3
40 kV
X 100
10
65
64
67
56
62
116
118
151
161
171

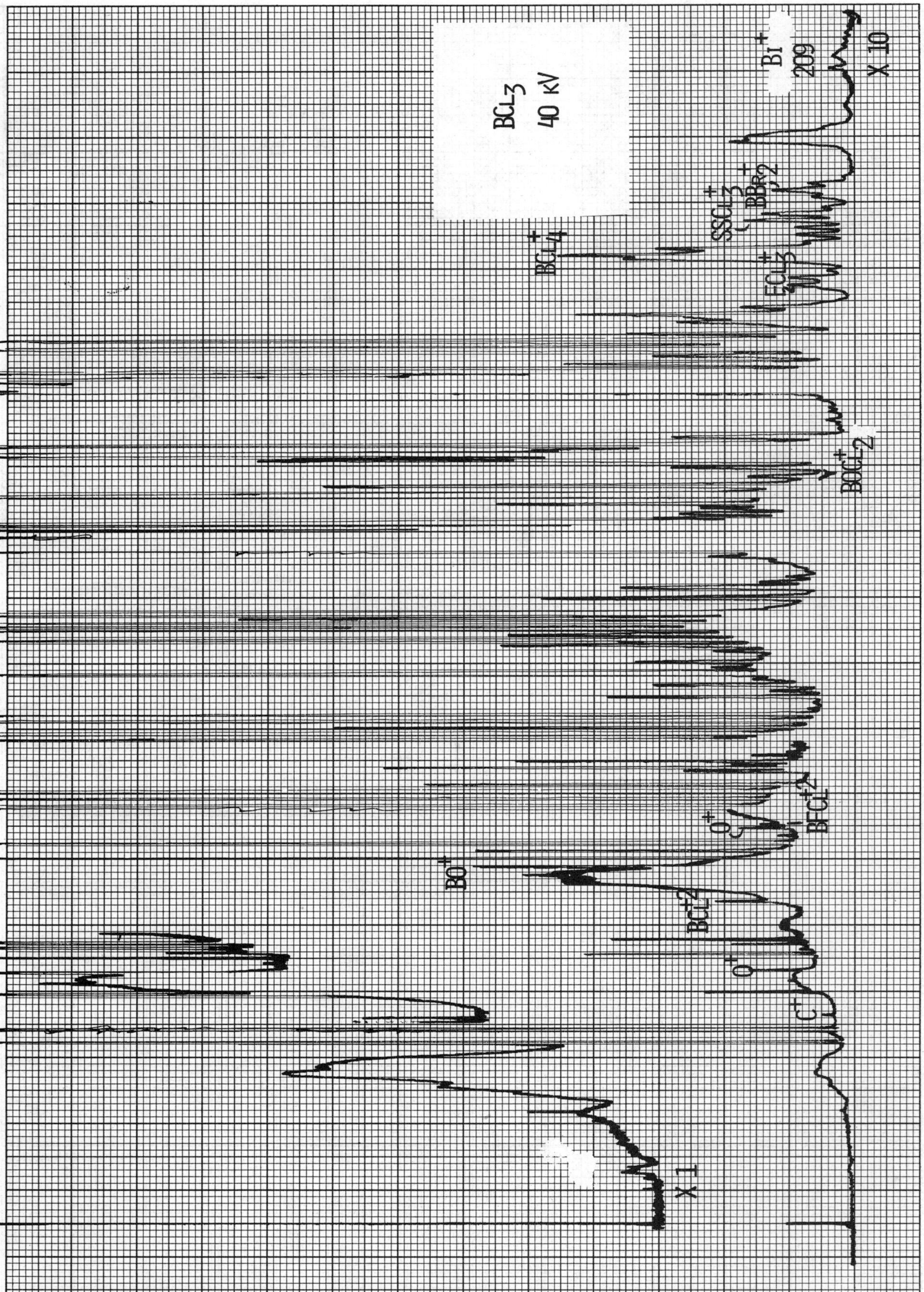
BCl3
40 kV
X 1
X 10
C+
O+
BO+
BCl4+
209

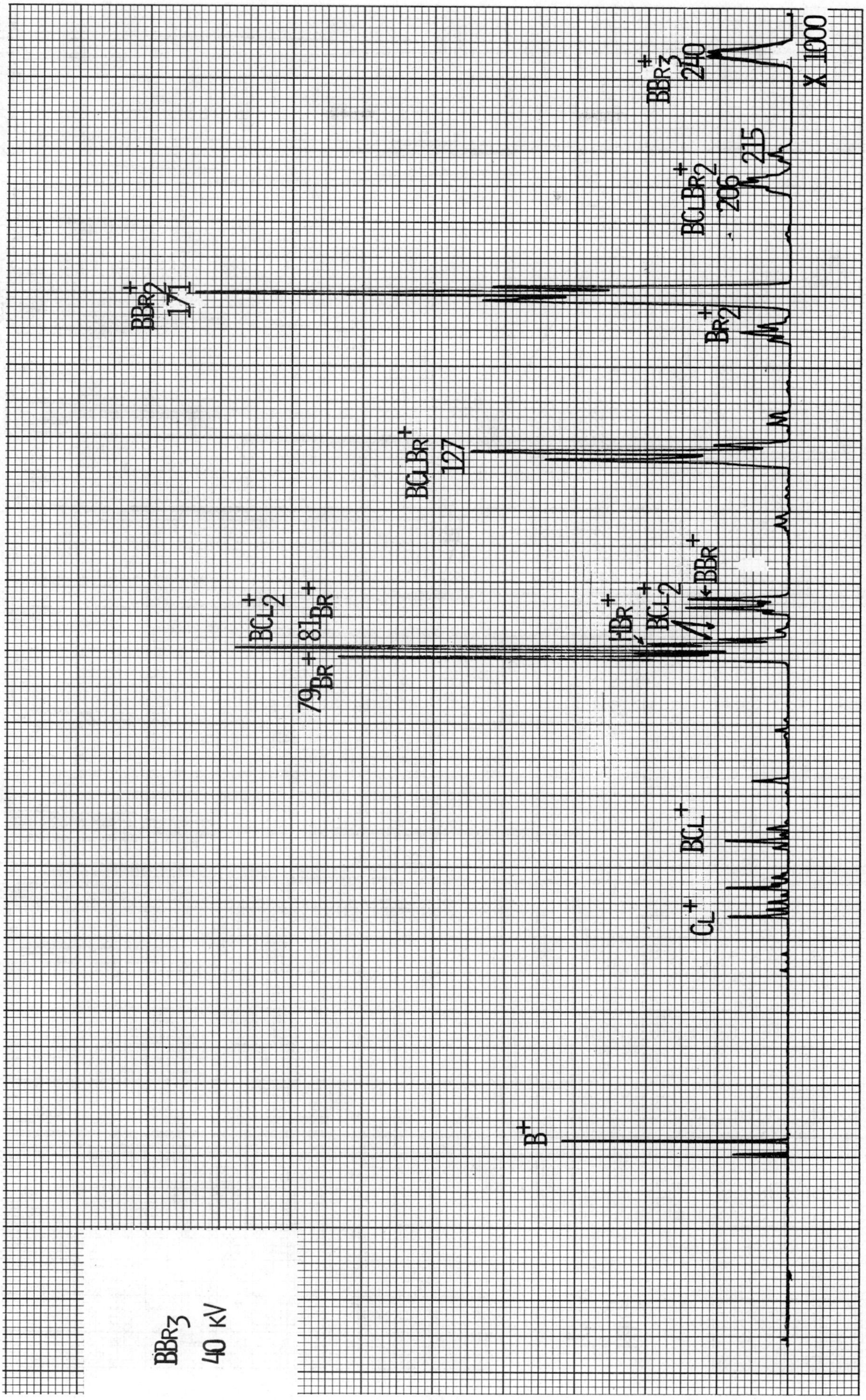

BBr3
40 kV
B+
Cl+
BCl+
79Br+
BCl2+
81Br+
HBr+
BCl2+
BBr+
BClBr+
127
BBr2+
171
Br2+
BClBr2+
206
215
BBr3+
240
X 1000

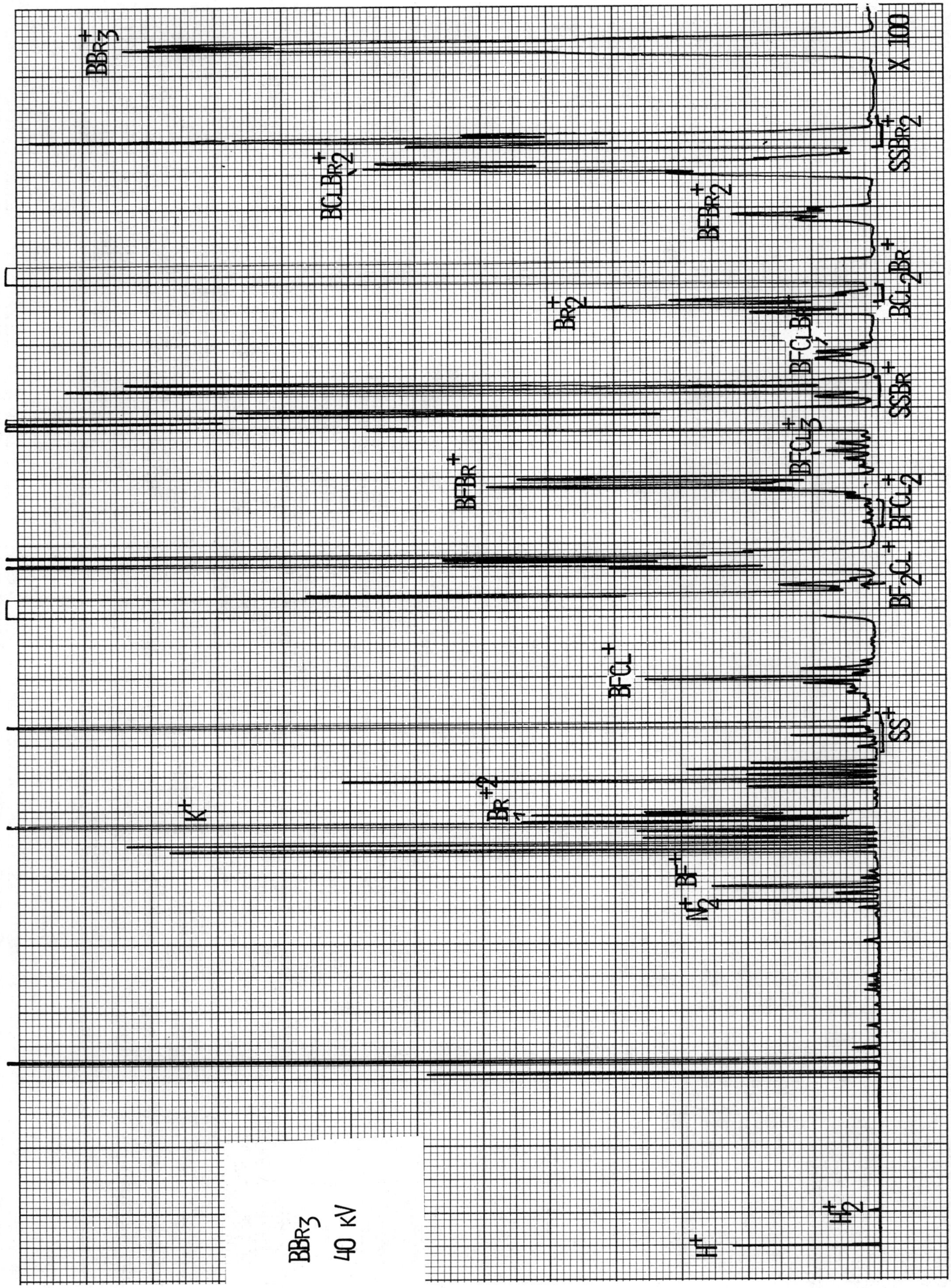
BBr3+
X 100
SSBr2+
BClBr2+
BFBr2+
BCl2Br+
Br2+
BFClBr+
SSBr+
BFCl3+
BFBr+
BFCl2+
BF2Cl+
BFCl+
SS+
Br+2
Br+
K+
N2+
BF+
H2+
H+
BBr3
40 kV

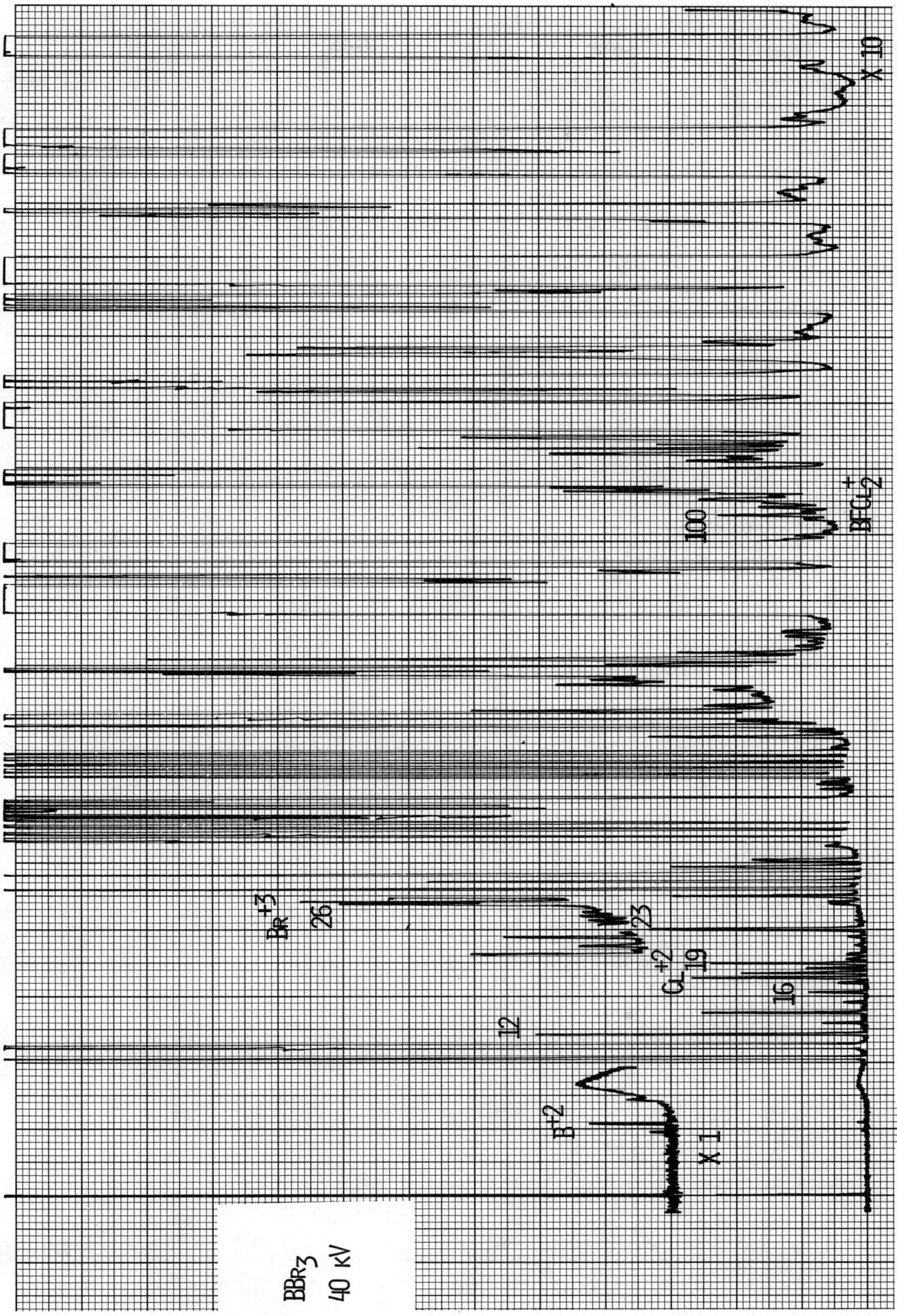
X 10
100
BCl2+
Br+3
26
23
Cl+2
19
16
12
B+2
X 1
BBr3
40 kV

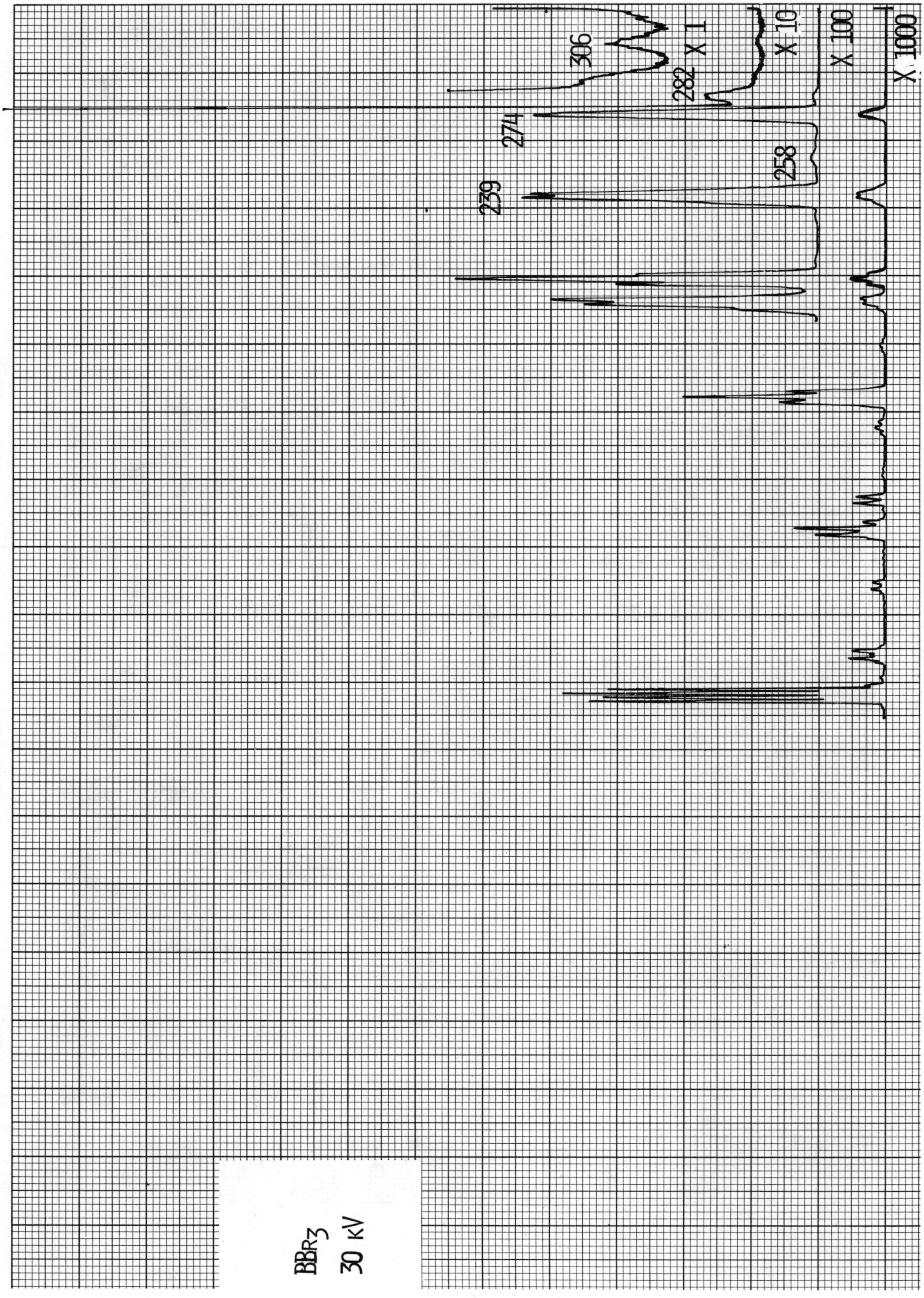
306
282
274
239
258
X 1
X 10
X 100
X 1000
BBr3
30 kV

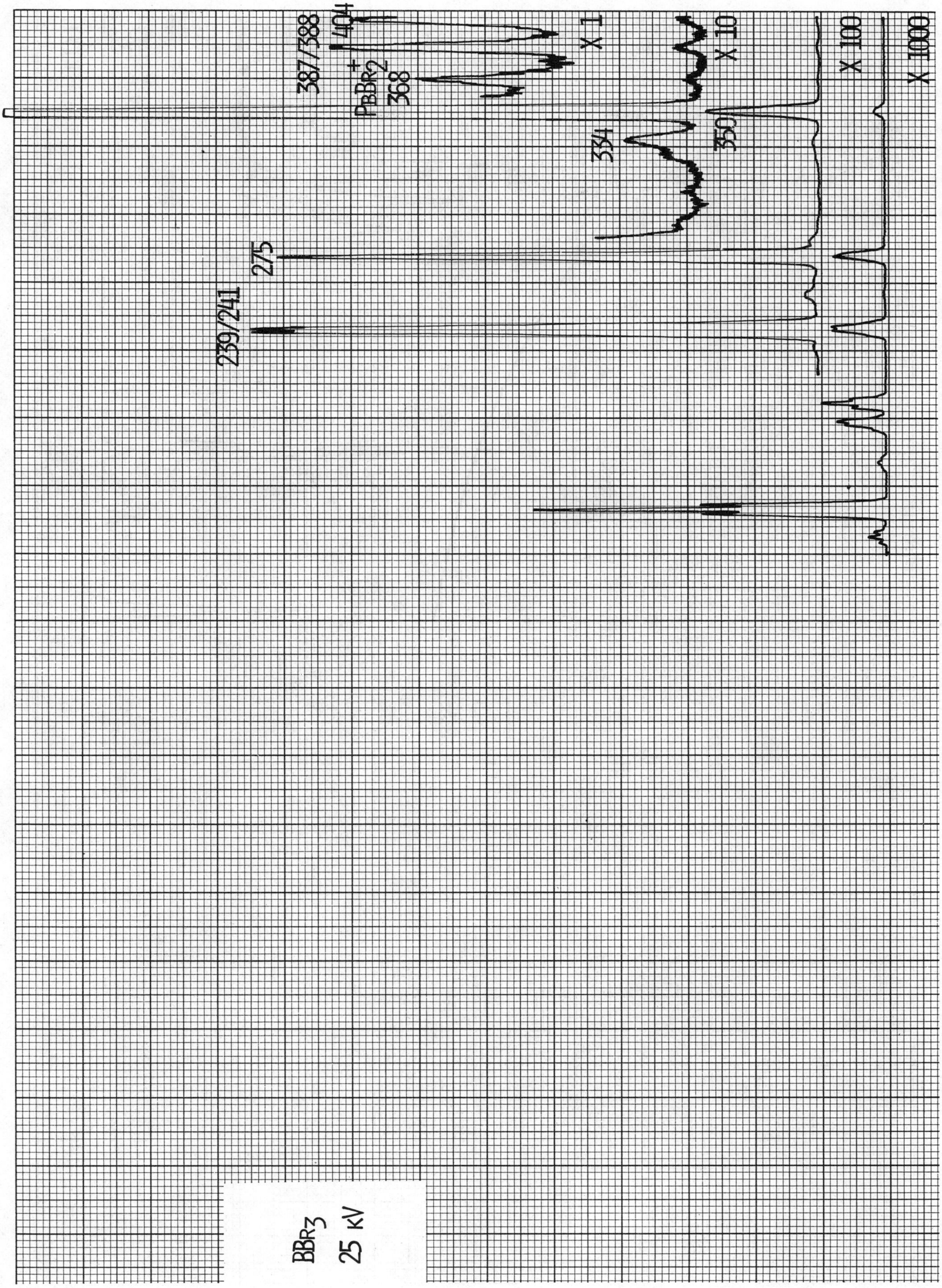
387/388
404
$PBBr_2^+$
368
X 1
X 10
334
350
X 100
X 1000
275
239/241
BBr_3
25 kV

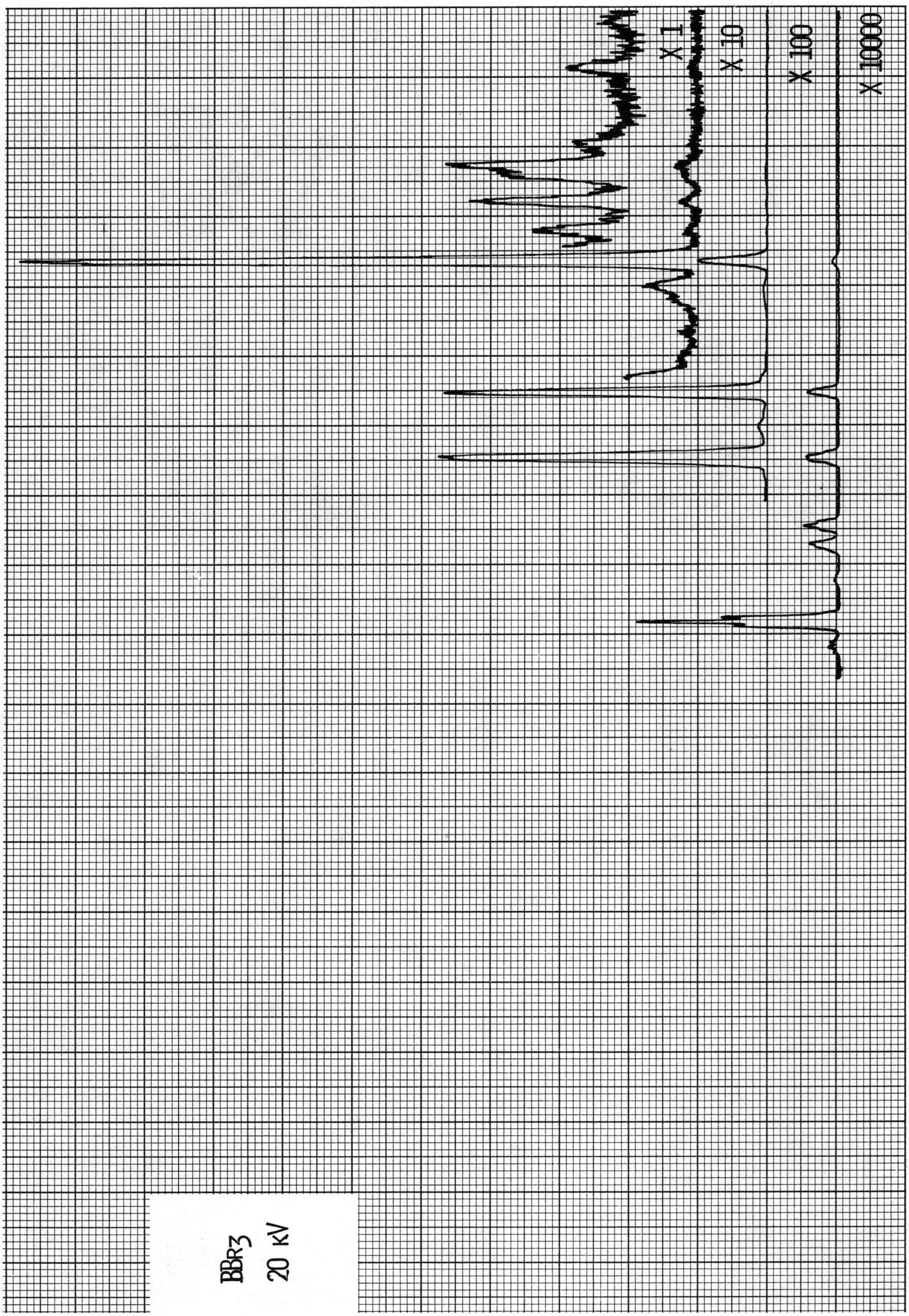
X 1
X 10
X 100
X 1000
BBr3
20 kV

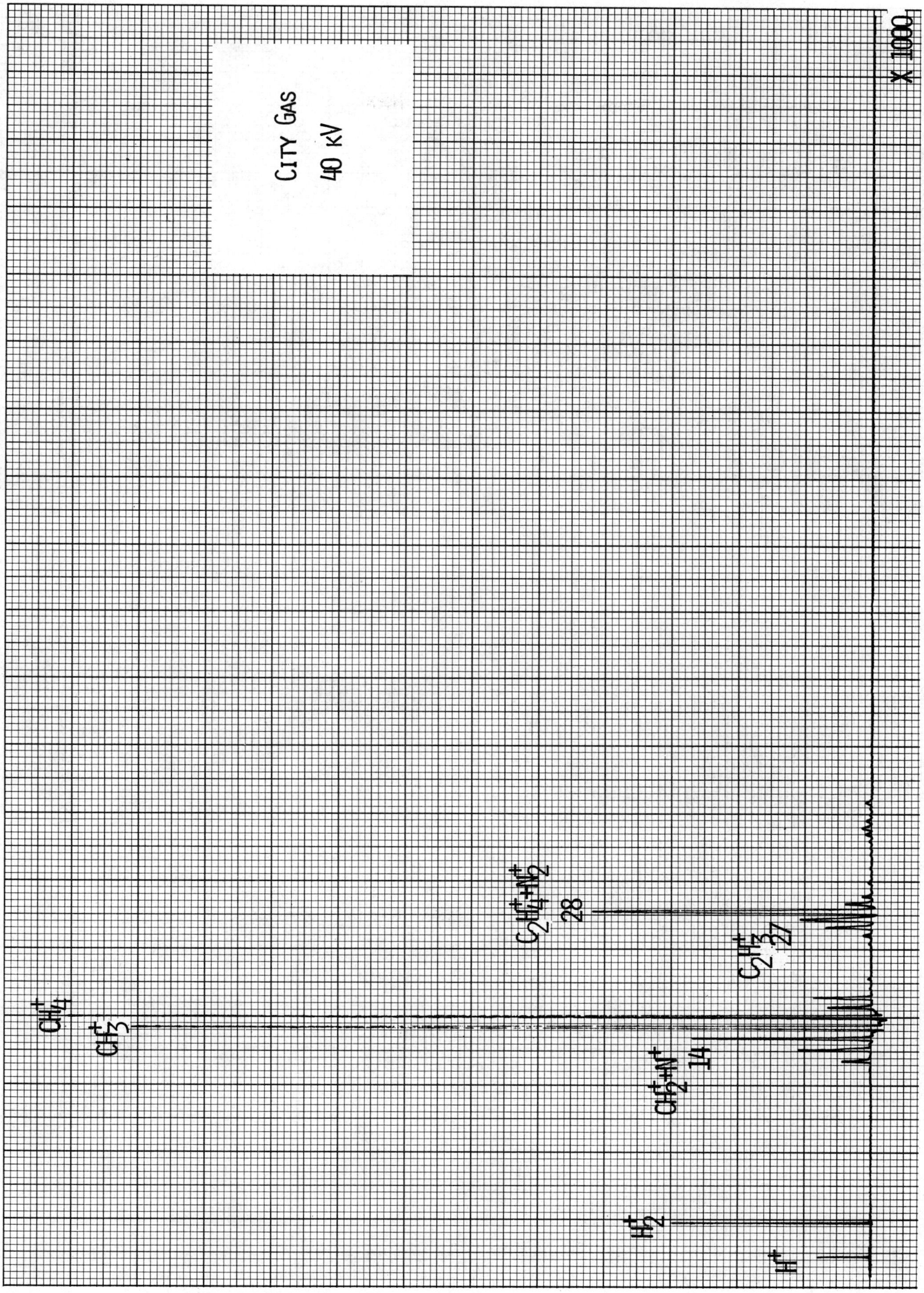
City Gas
40 kV
X 1000
CH_4^+
CH_3^+
$C_2H_4^+ + N_2^+$
28
$C_2H_3^+$
27
$CH_2^+ + N^+$
14
H_2^+
H^+

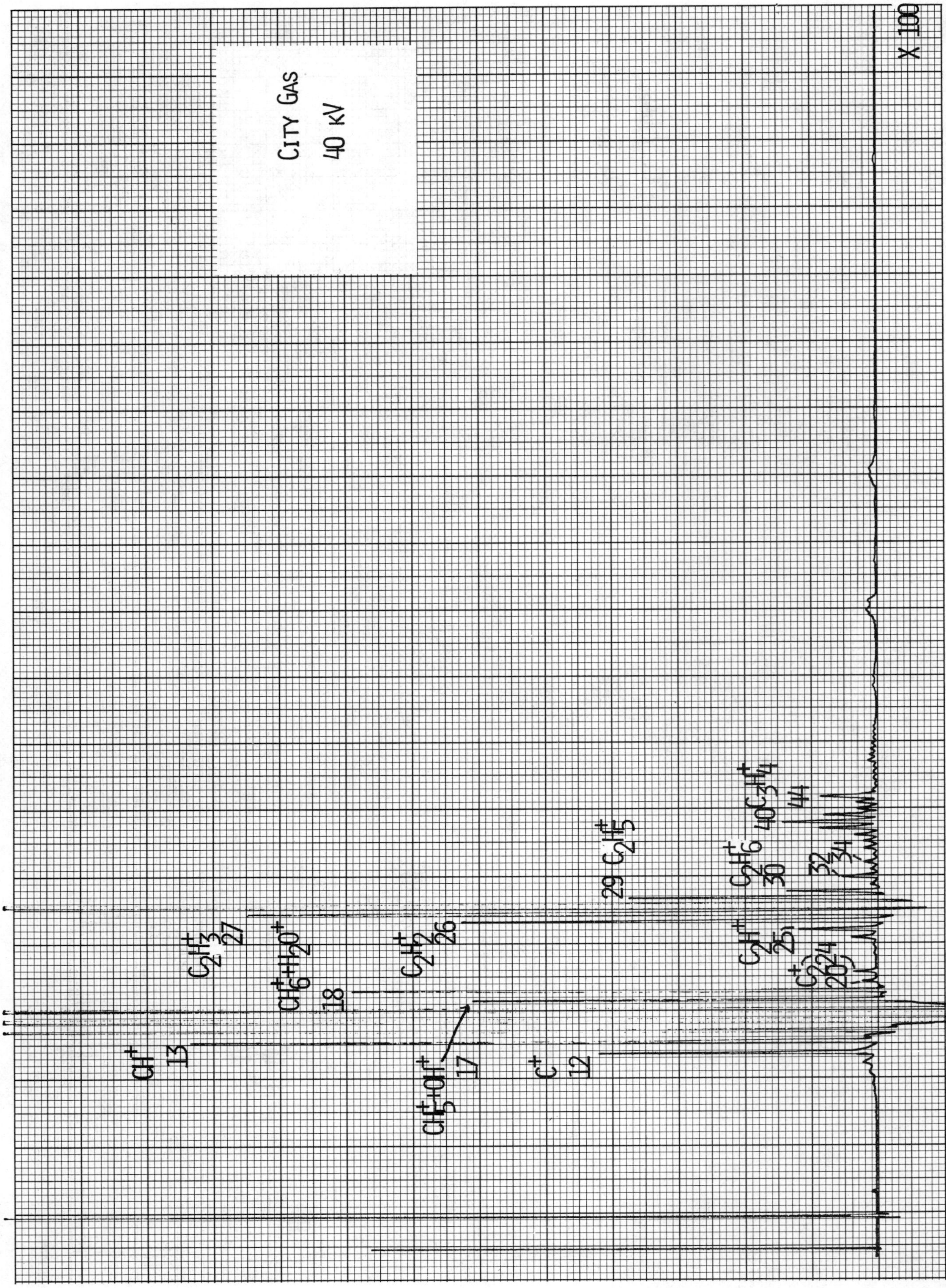
City Gas
40 kV
X 100
CH^+ 13
$C_2H_3^+$ 27
$CH_6^+ + H_2O^+$ 18
$CH_5^+ + OH^+$ 17
$C_2H_2^+$ 26
C^+ 12
29 $C_2H_5^+$
$C_2H_6^+$ 30
C_2H^+ 25
C_2^+ 24
20
32
34
40 $C_3H_4^+$
44

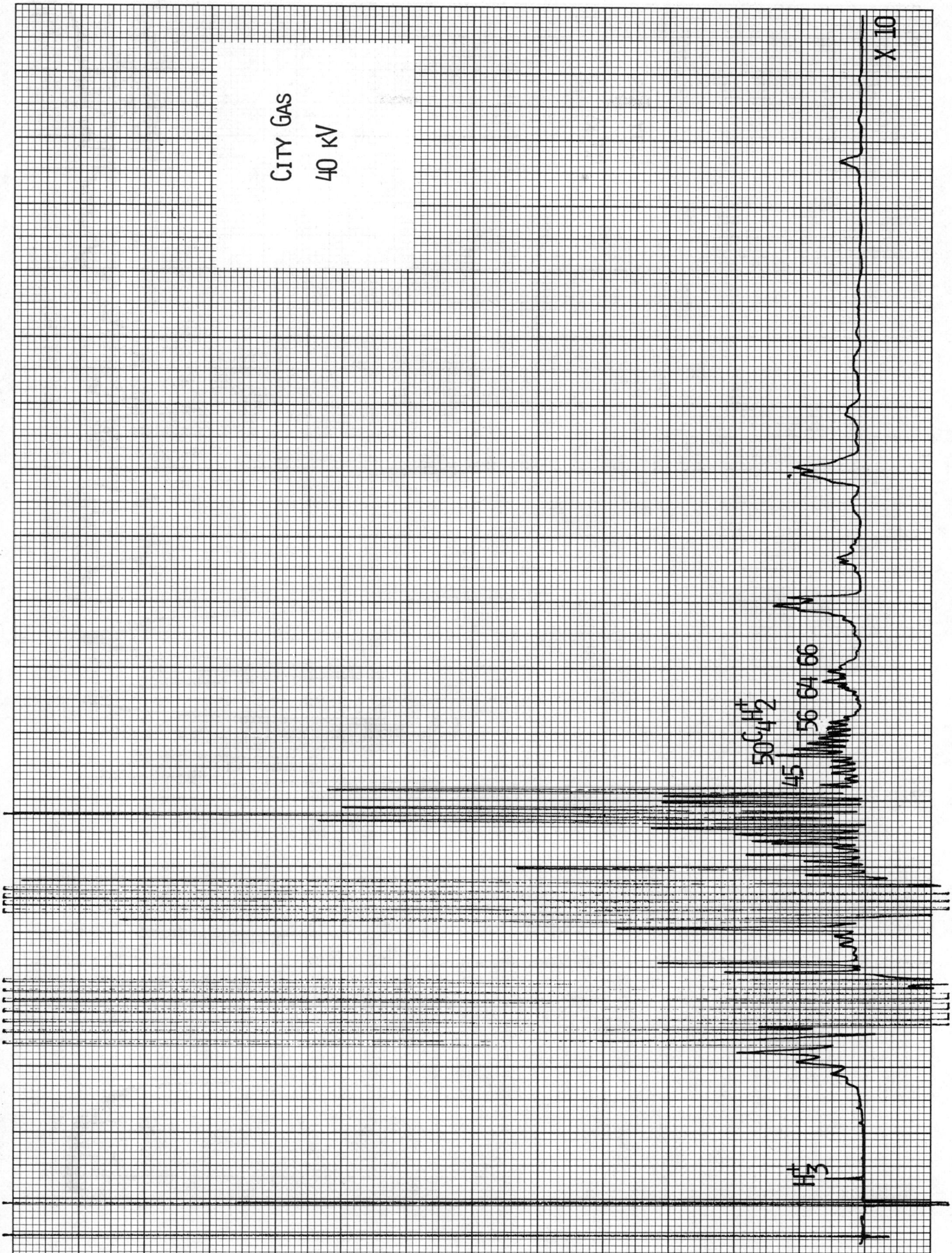
City Gas
40 kV
X 10
50 $C_4H_2^+$
45
56 64 66
H_3^+

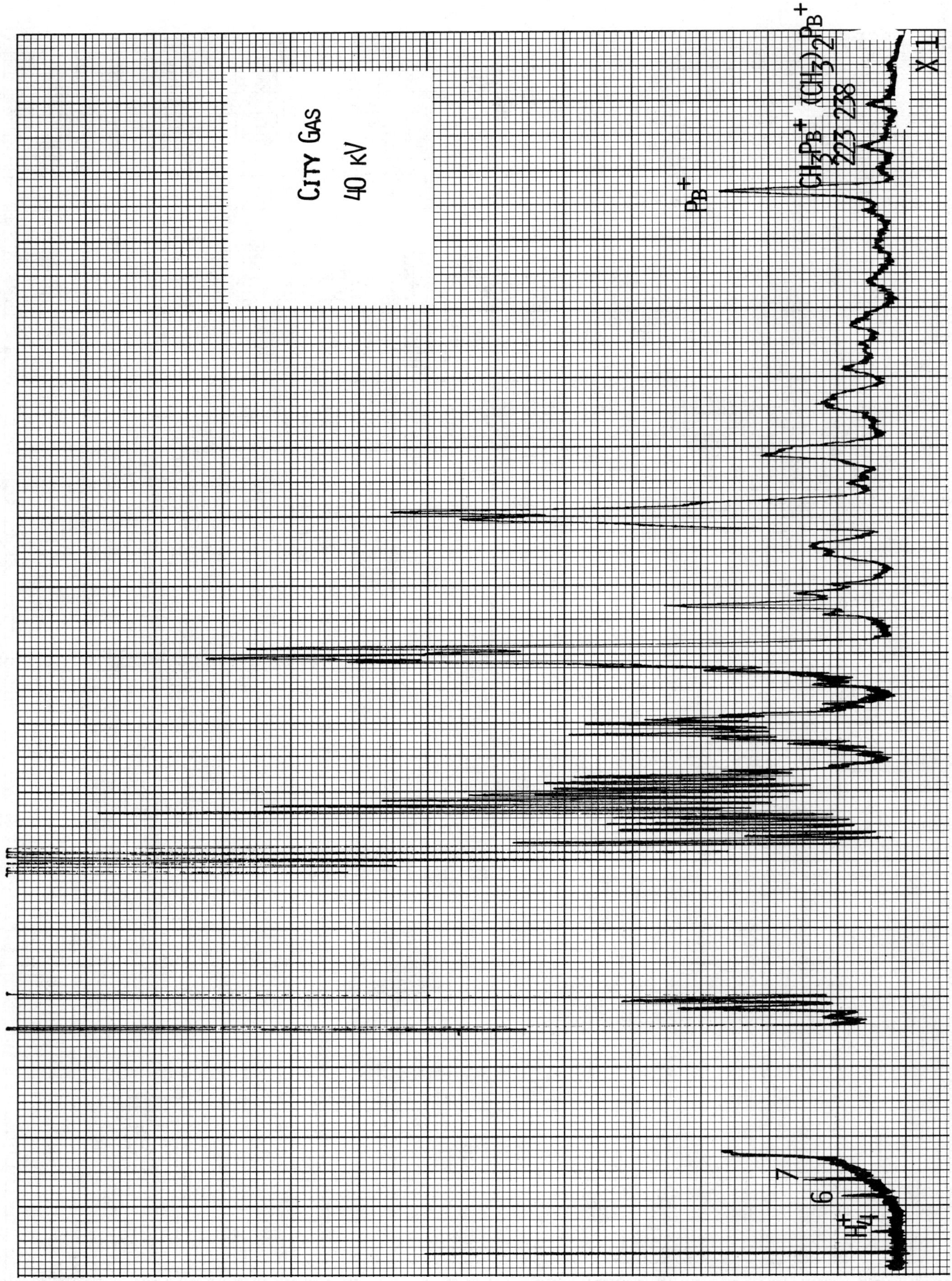
City Gas
40 kV
P_B^+
$CH_3P_B^+$
$(CH_3)_2P_B^+$
223
238
X 1
H_4^+
6
7

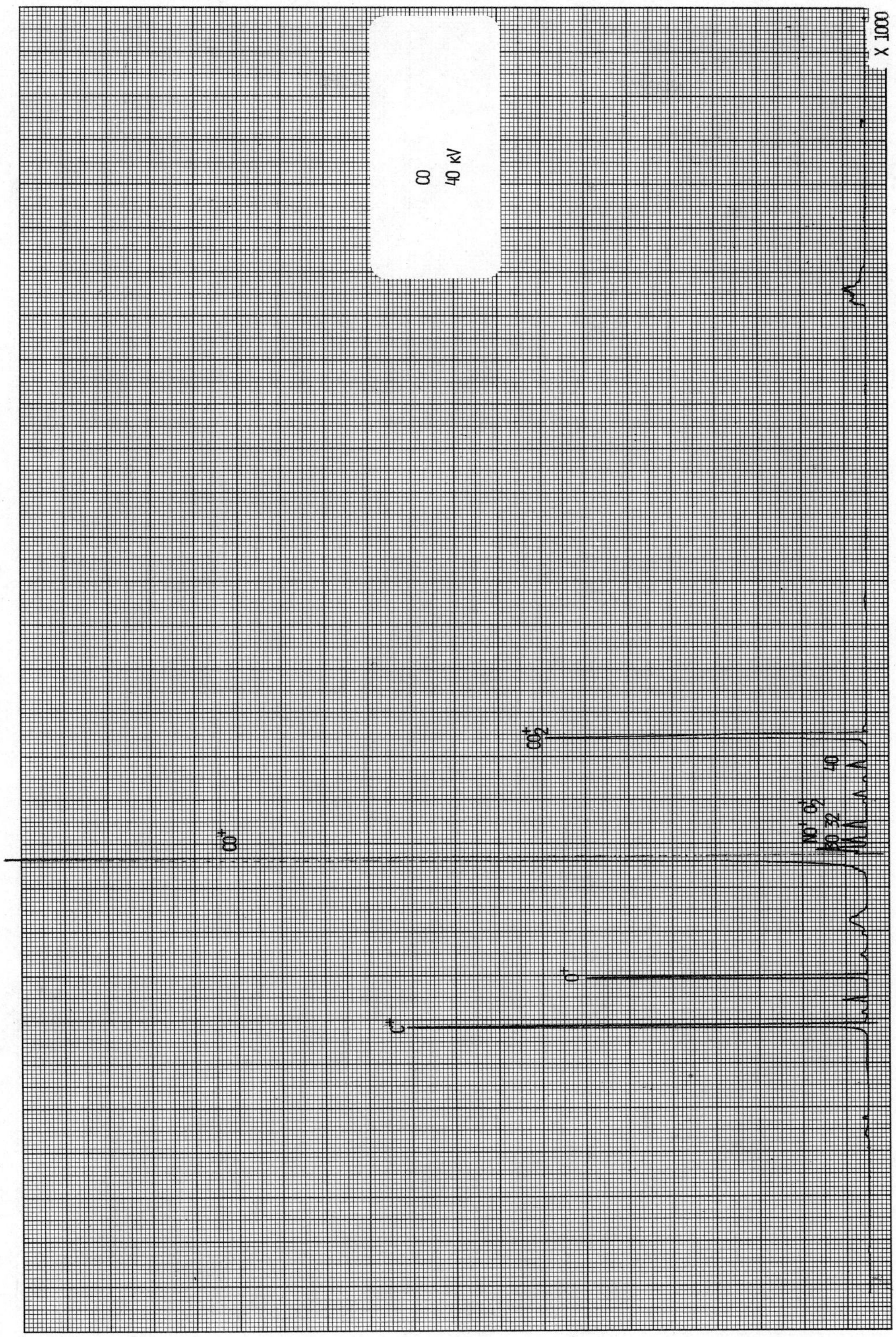
CO
40 kV
X 1000
CO^+
C^+
O^+
CO_2^+
40

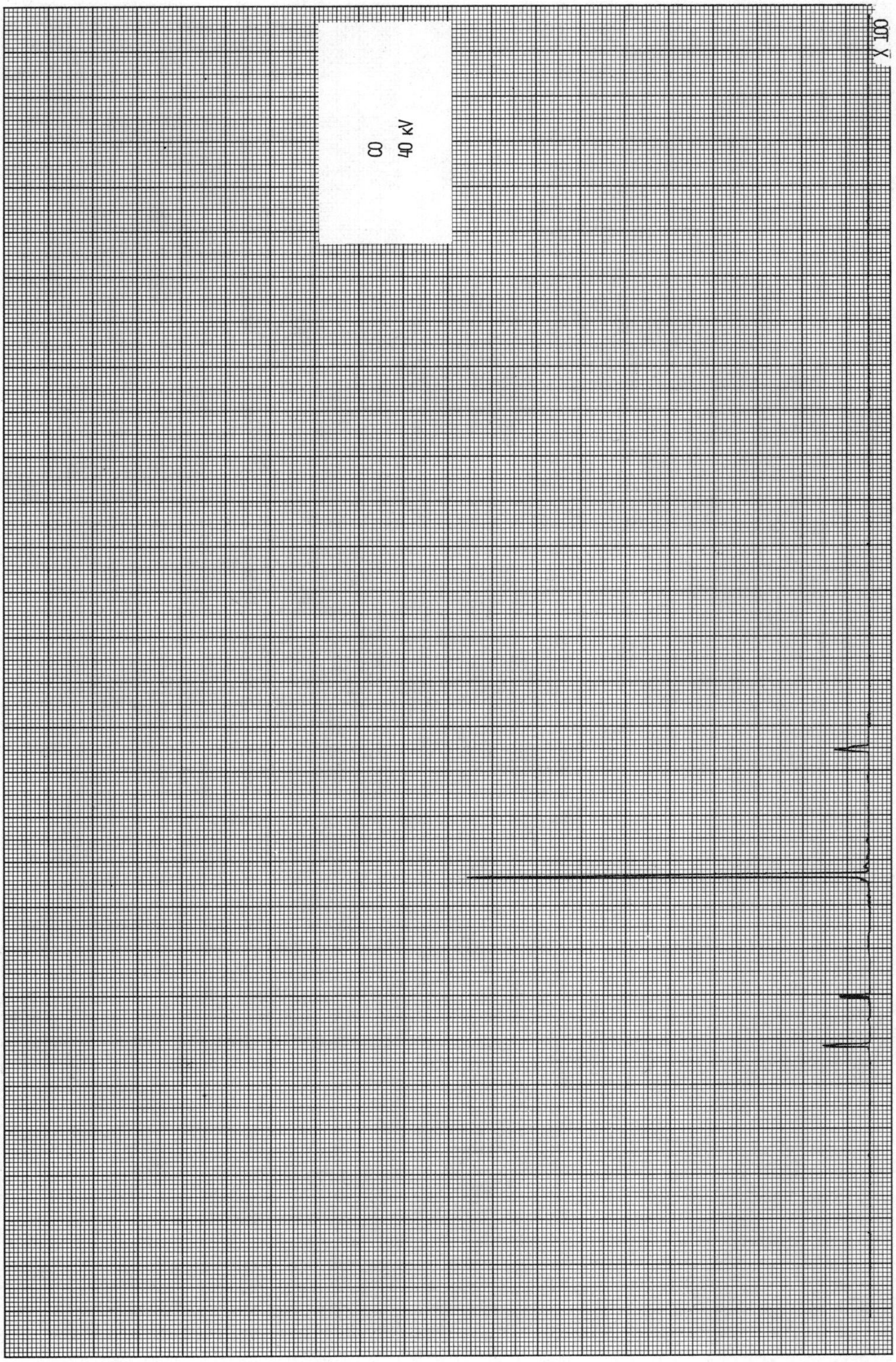
CO
40 kV
X 100

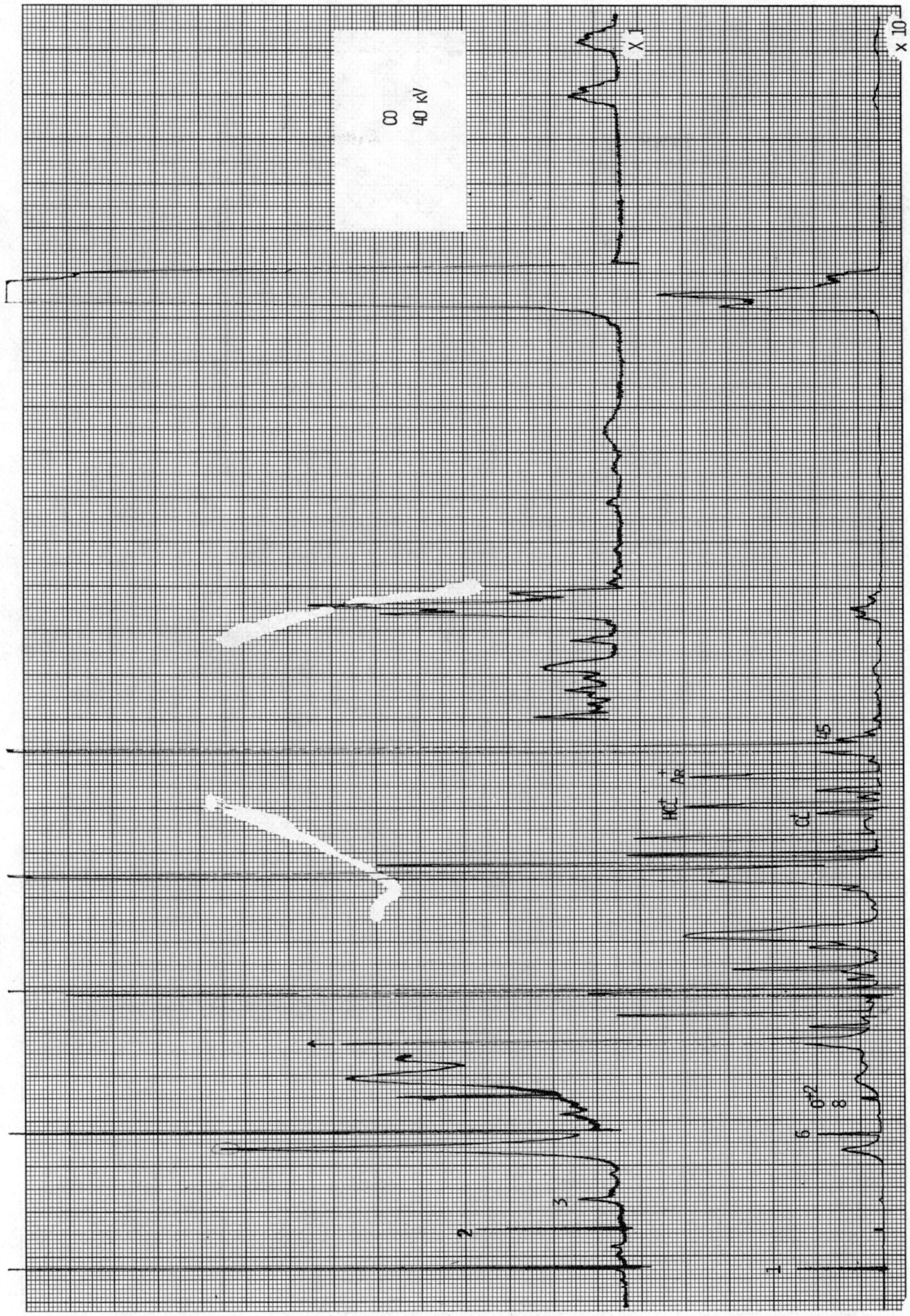
CO
40 kV
X 1
x 10

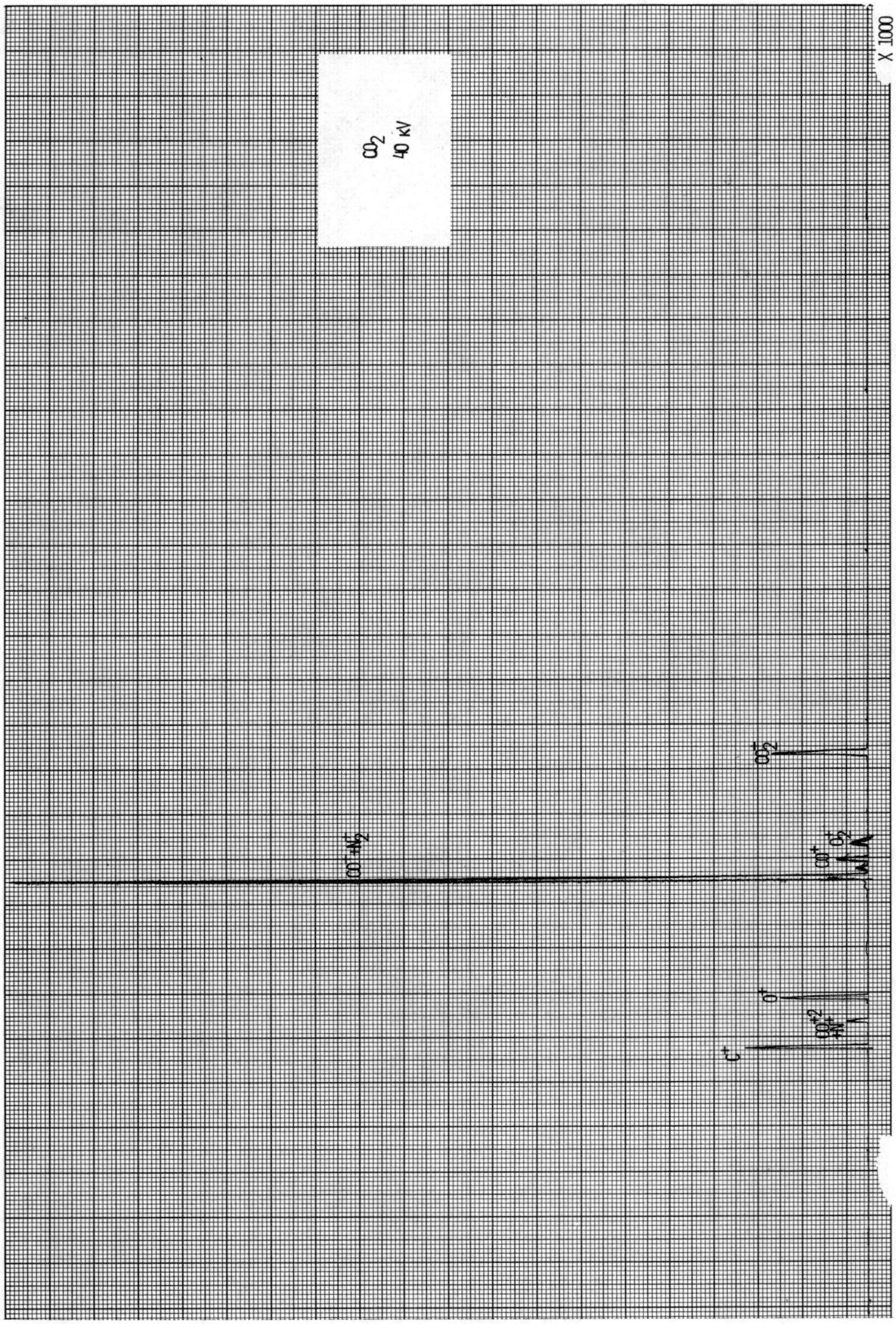
CO2
40 kV
X 1000

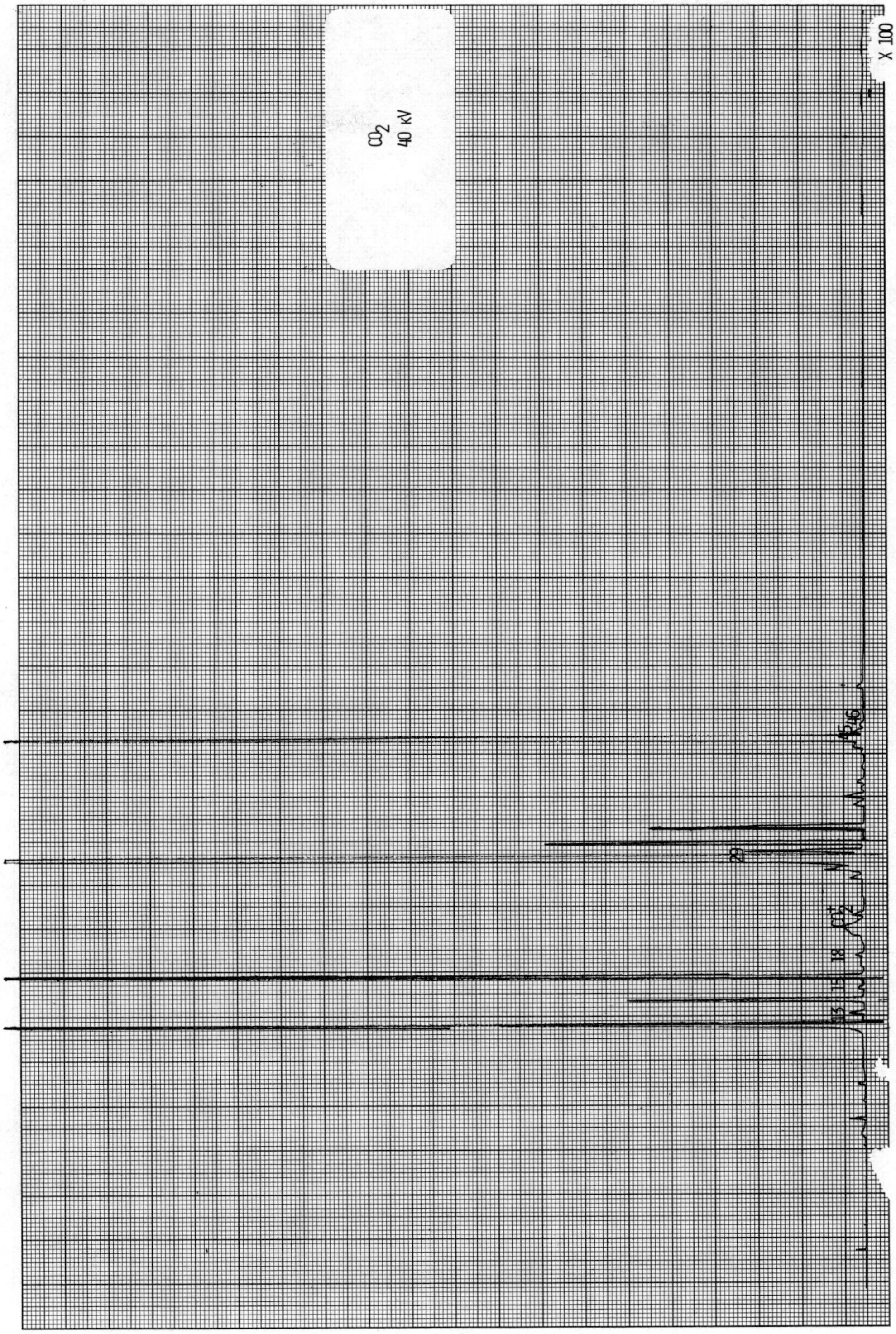
CO_2
40 kV
X 100

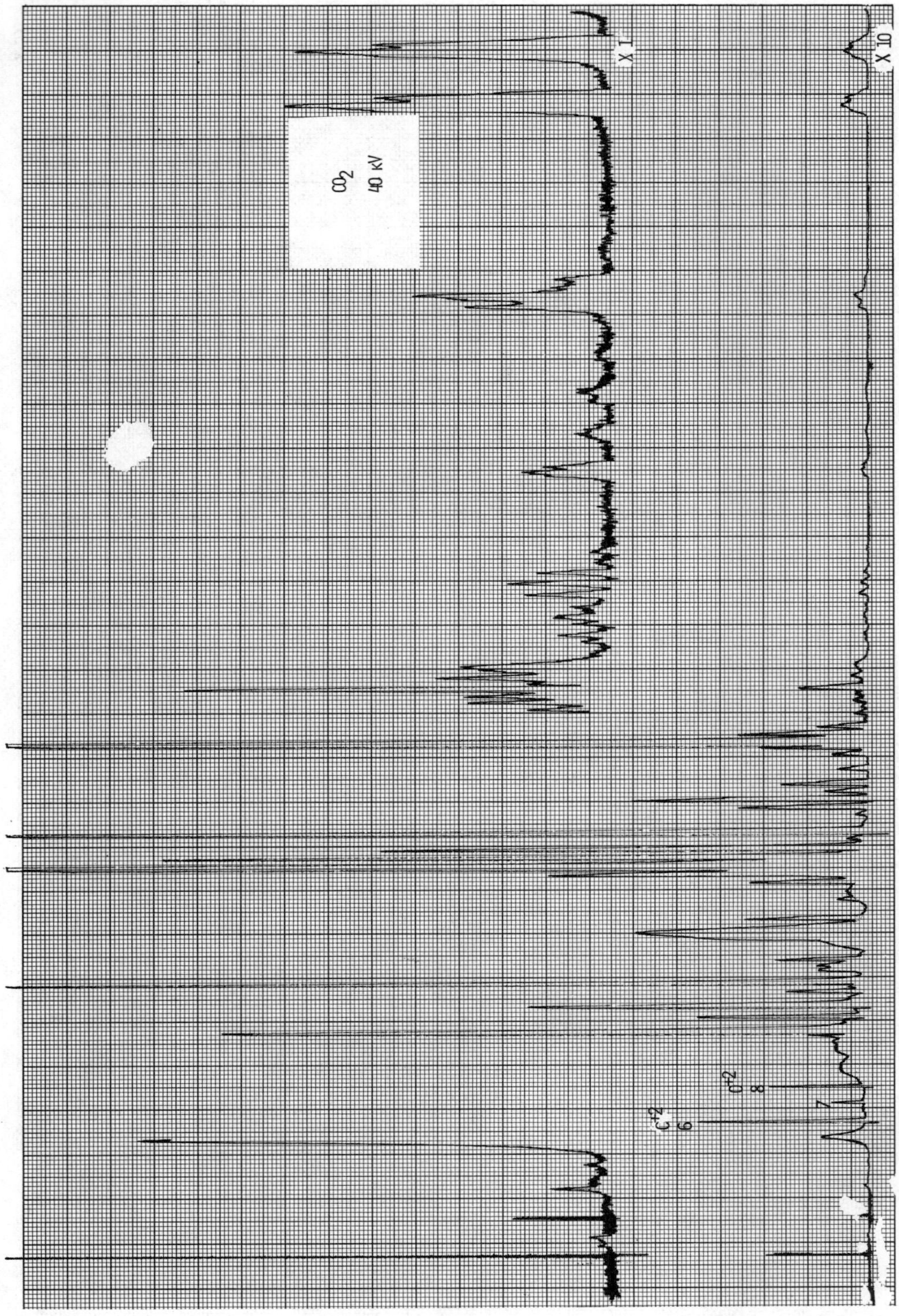
CO_2
40 kV
X 1
X 10

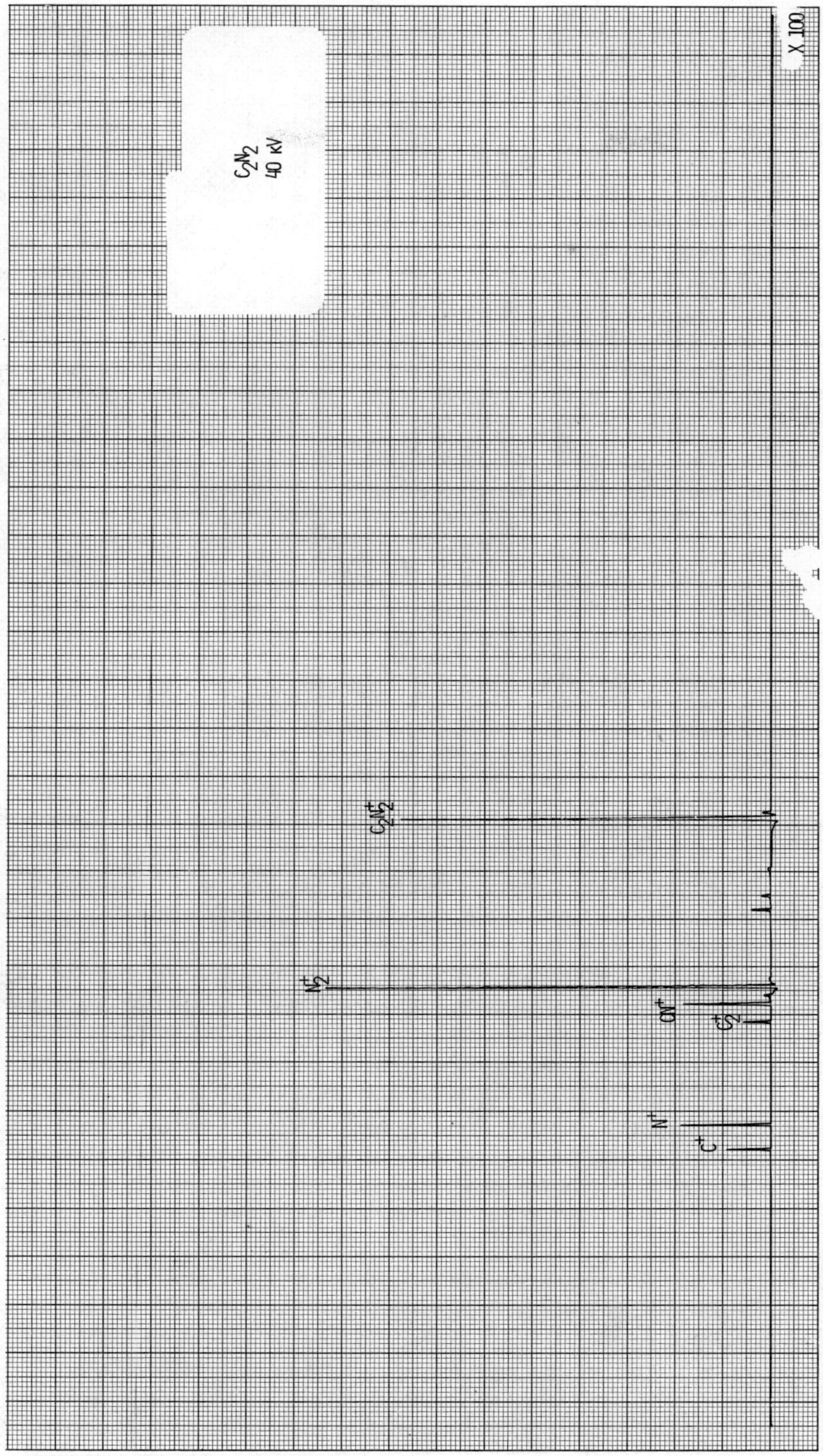
C_2N_2
40 kV
X 100
$C_2N_2^+$
N_2^+
CN^+
C_2^+
N^+
C^+

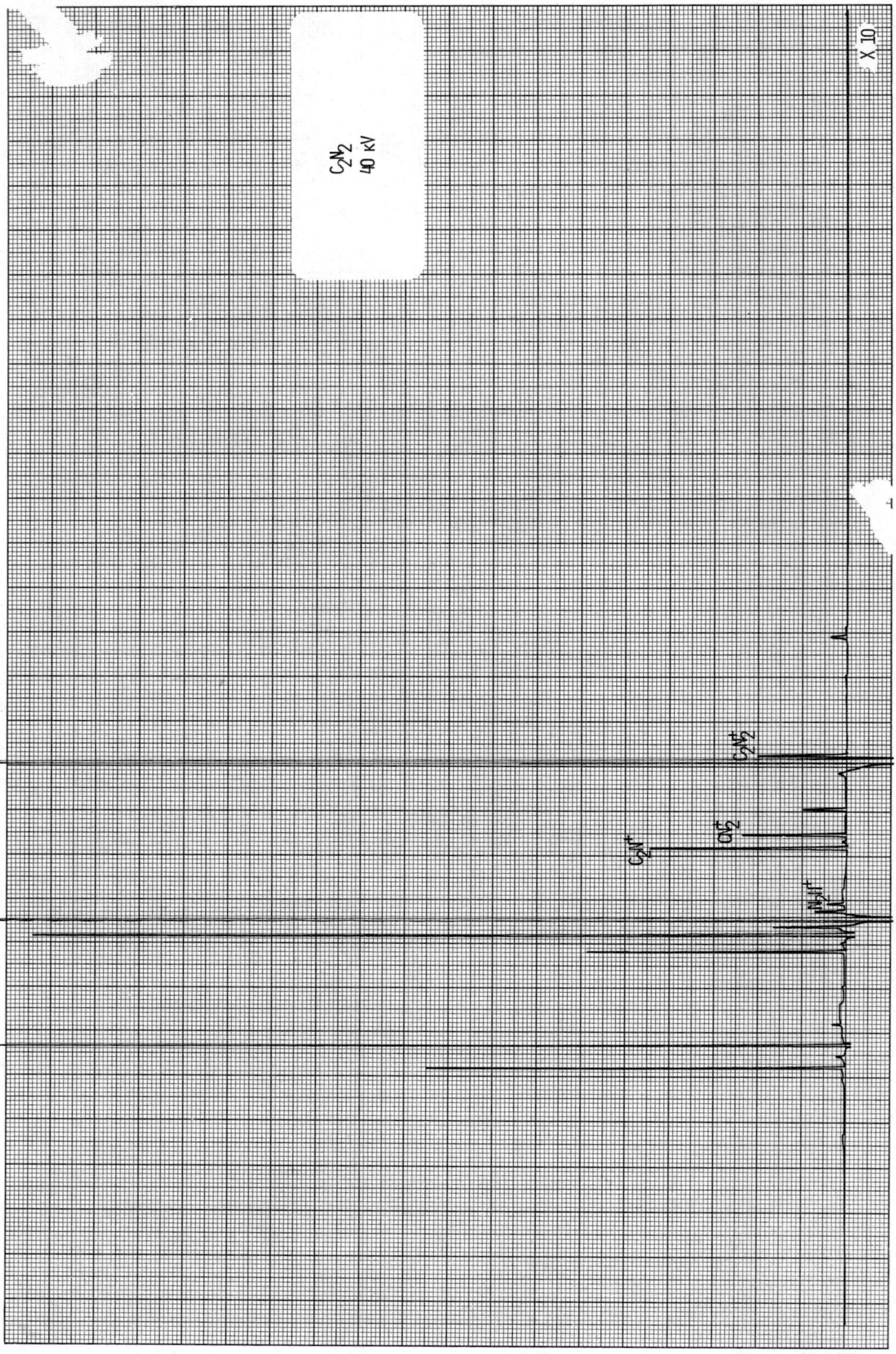
C_2N_2
40 kV
X 10
$C_2N_2^+$
C_2N^+

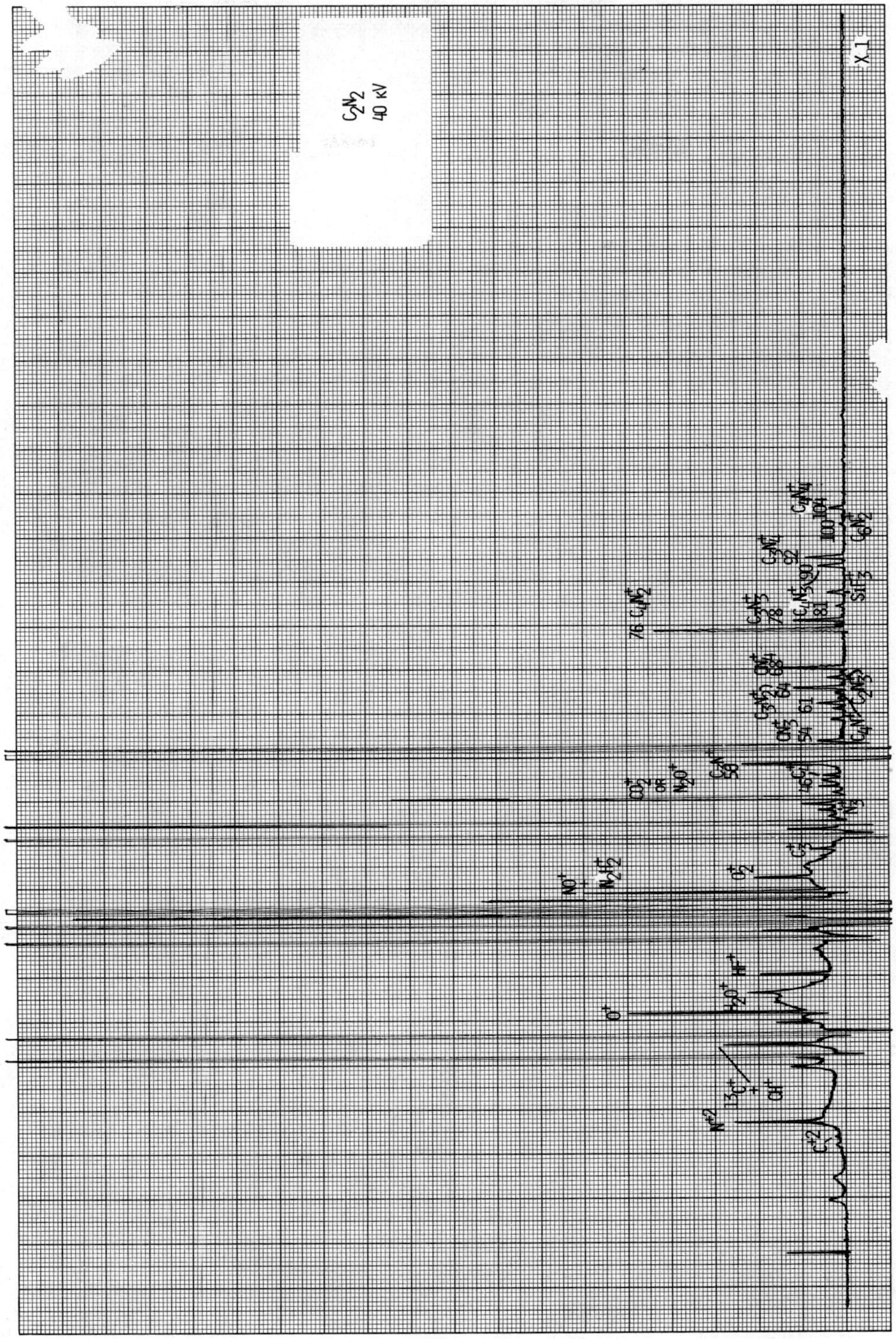
C2N2
40 kV
X 1

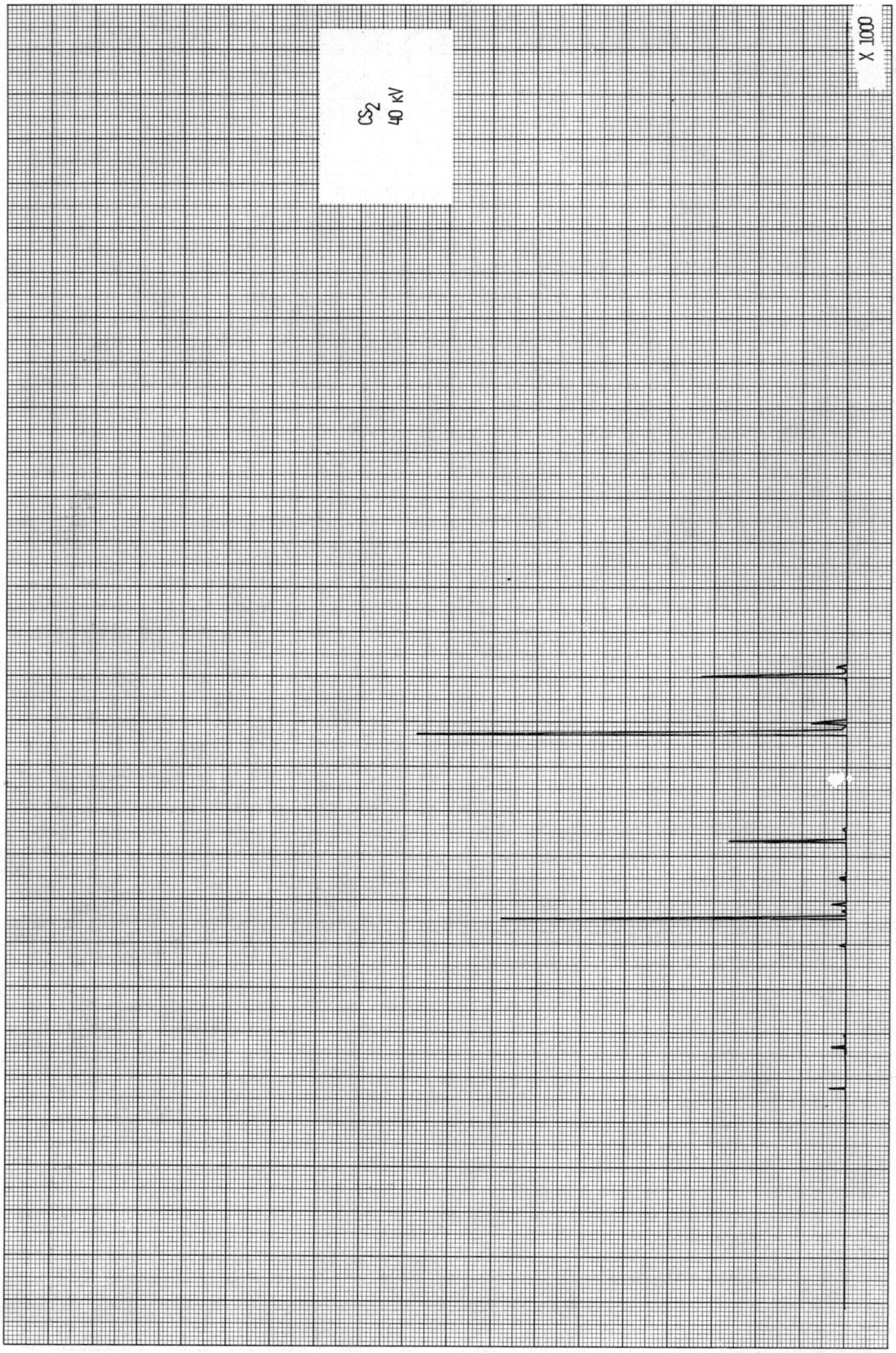
X 1000
CS2
40 kV

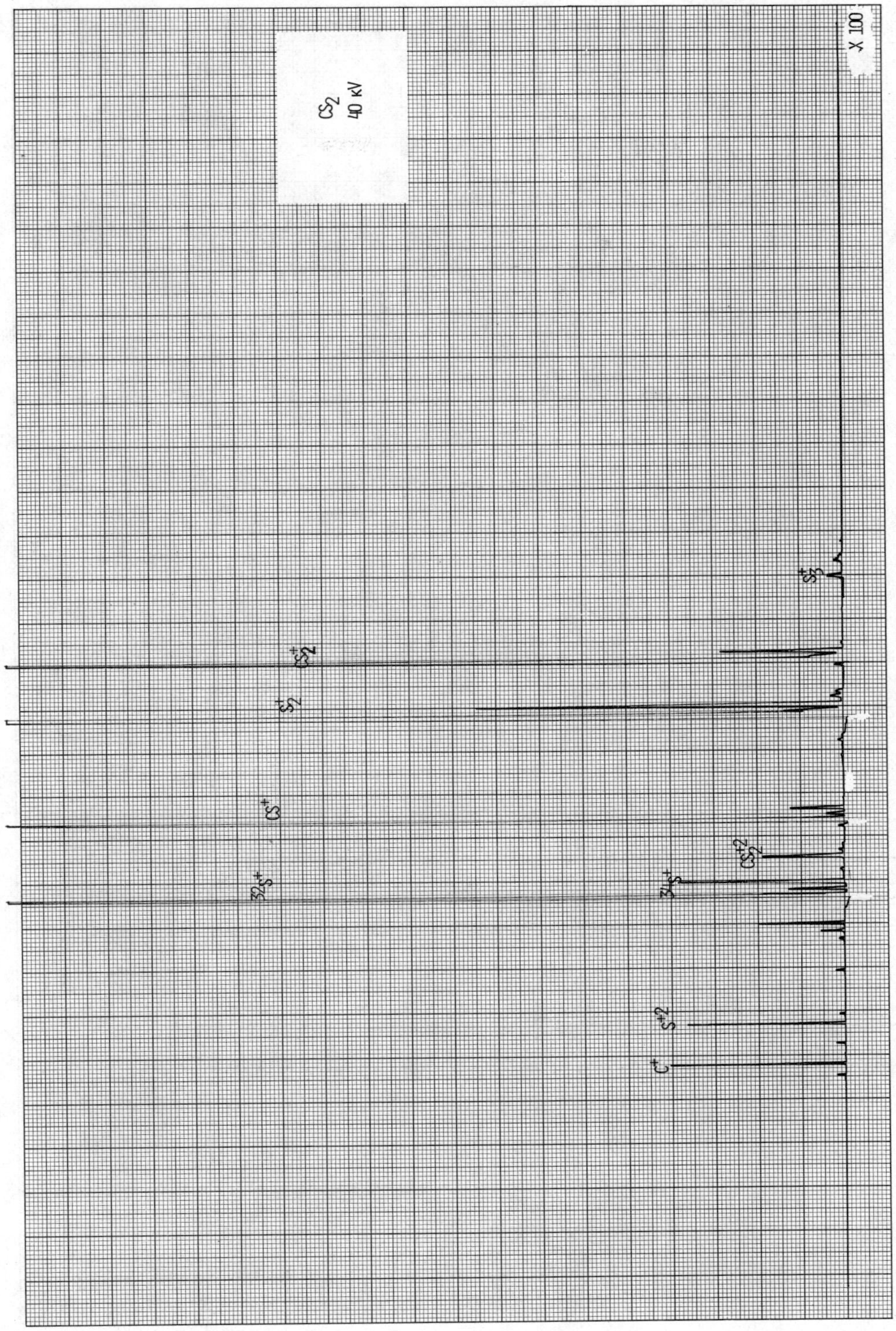
CS_2
40 kV
X 100
S_3^+
CS_2^+
S_2^+
CS^+
$^{32}S^+$
$^{34}S^+$
CS_2^{+2}
S^{+2}
C^+

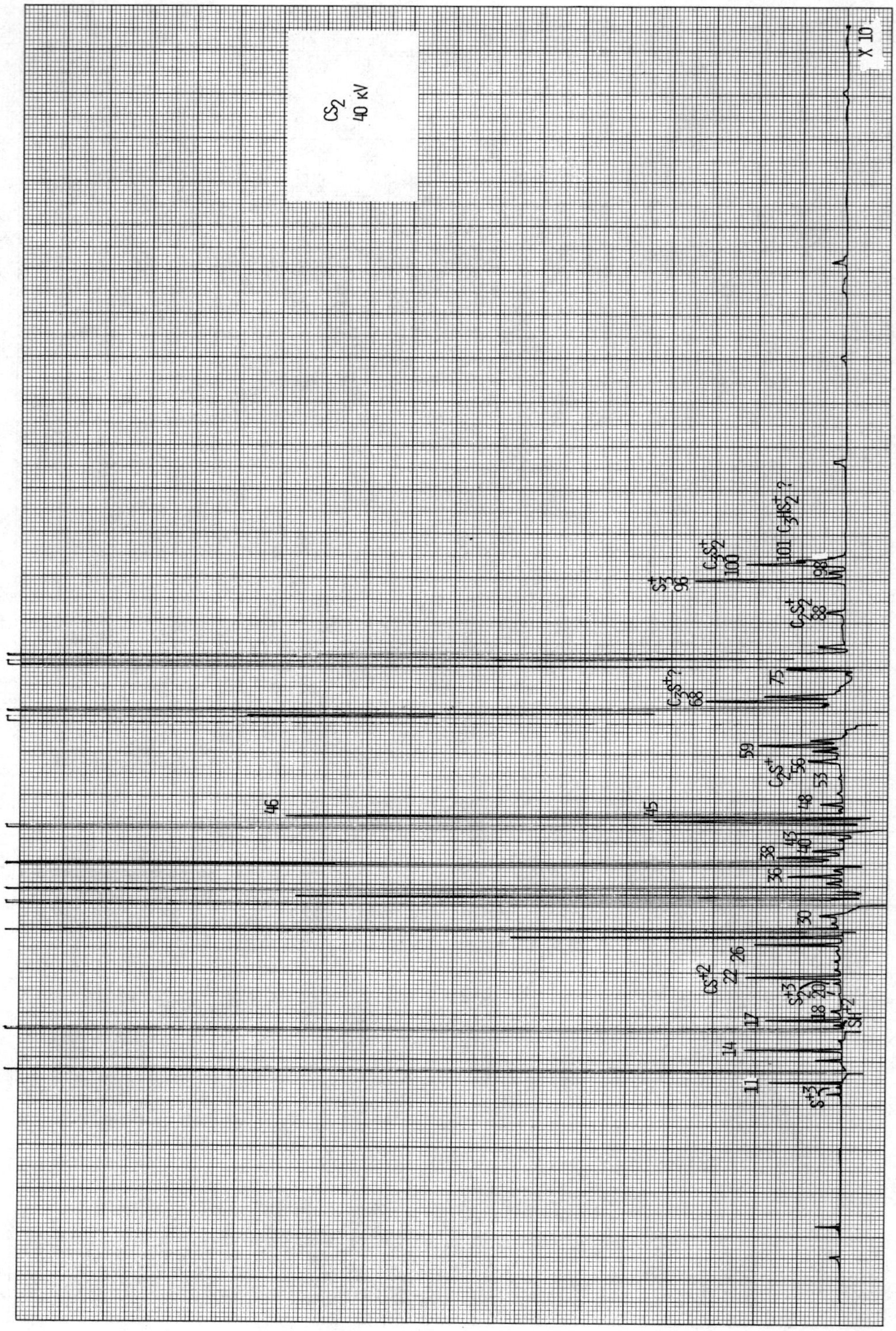

CS2
40 kV
X 10
11
14
17
22
26
30
36
38
43
45
46
48
53
56
59
68
75
88
96
98
100
101

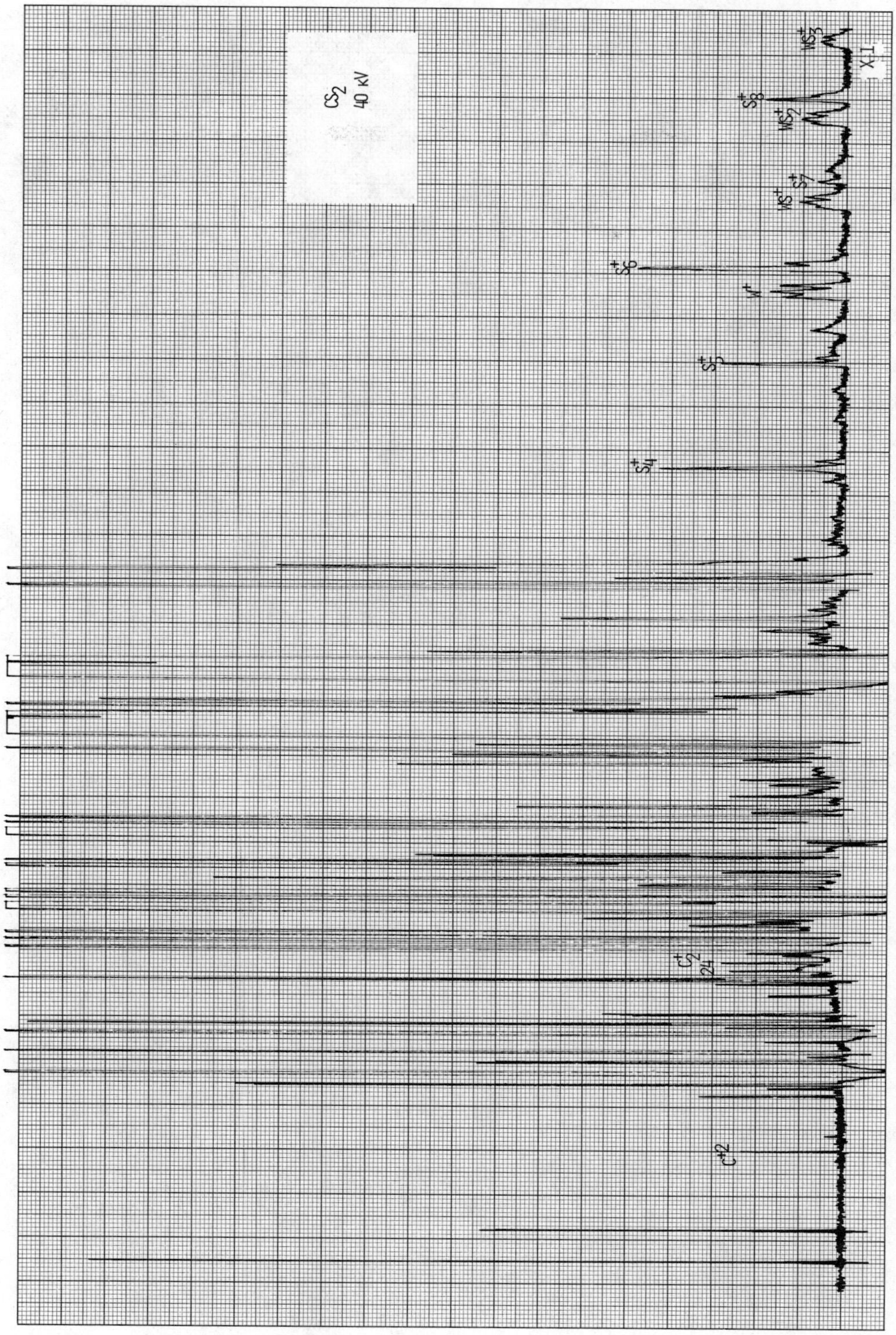

CS2
40 kV
S4+
S5+
S6+
S8+
K+

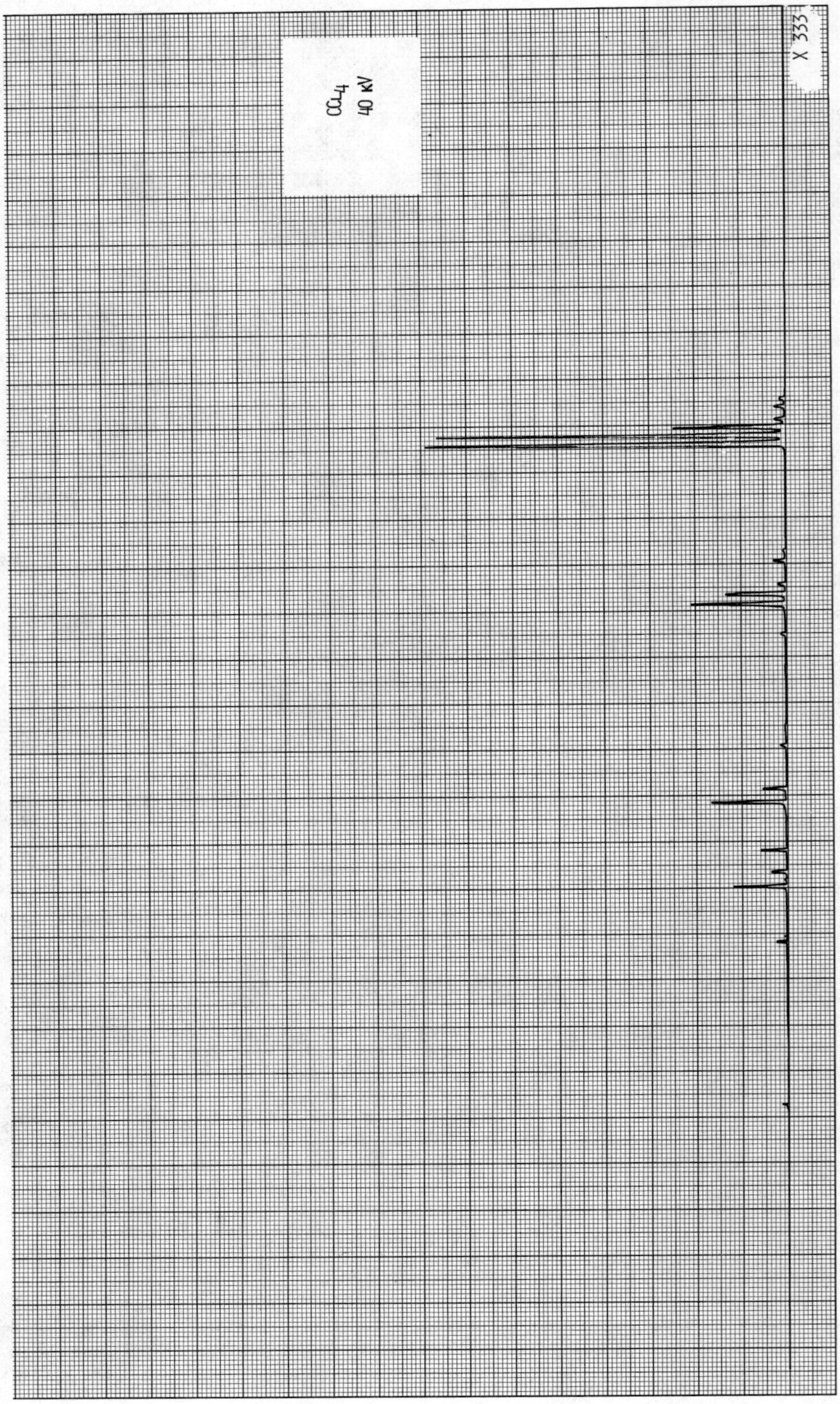

CCl4
40 kV
X 333

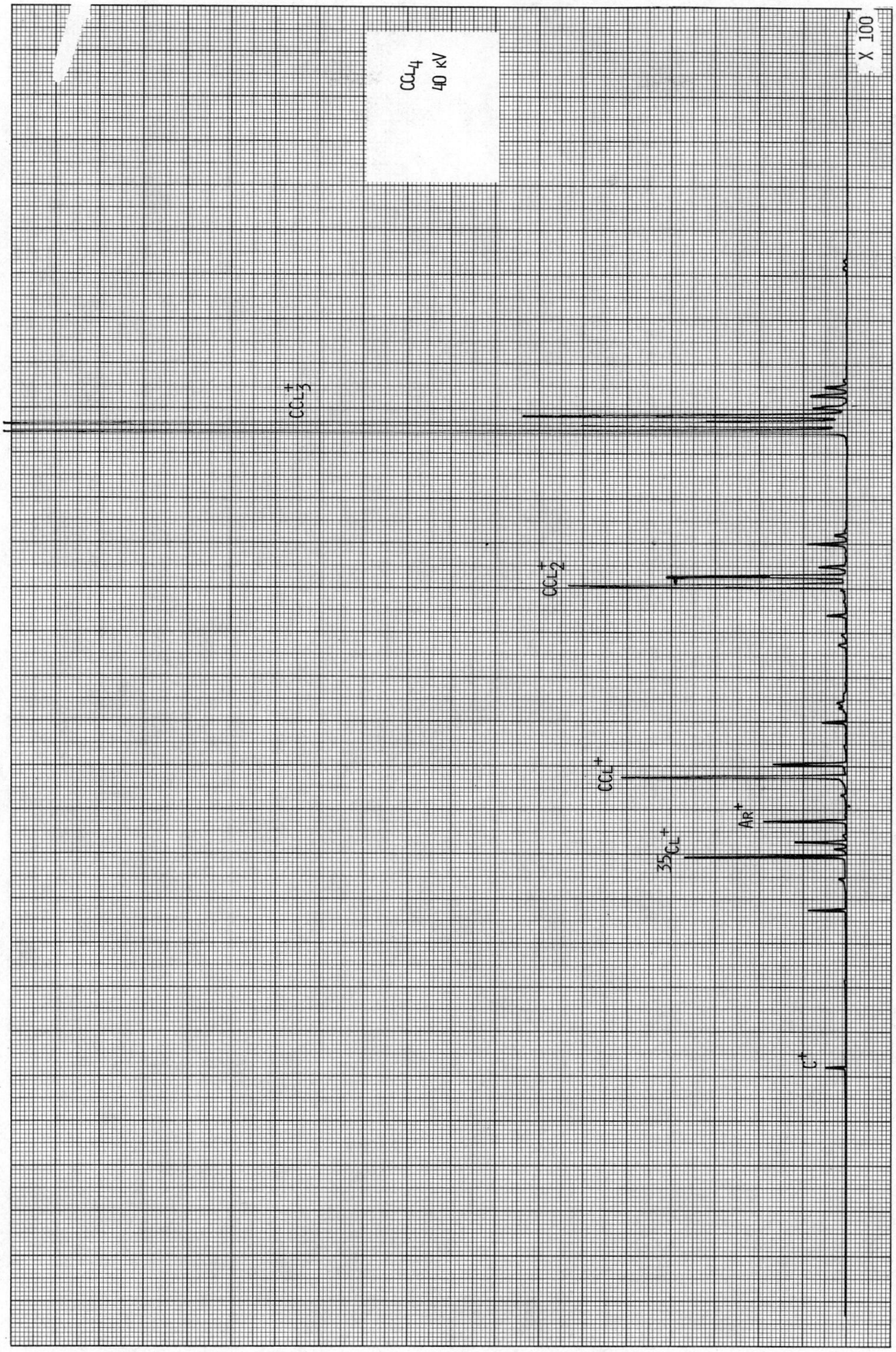

CCl_4
40 kV
X 100
CCl_3^+
CCl_2^+
CCl^+
Ar^+
$^{35}Cl^+$
C^+

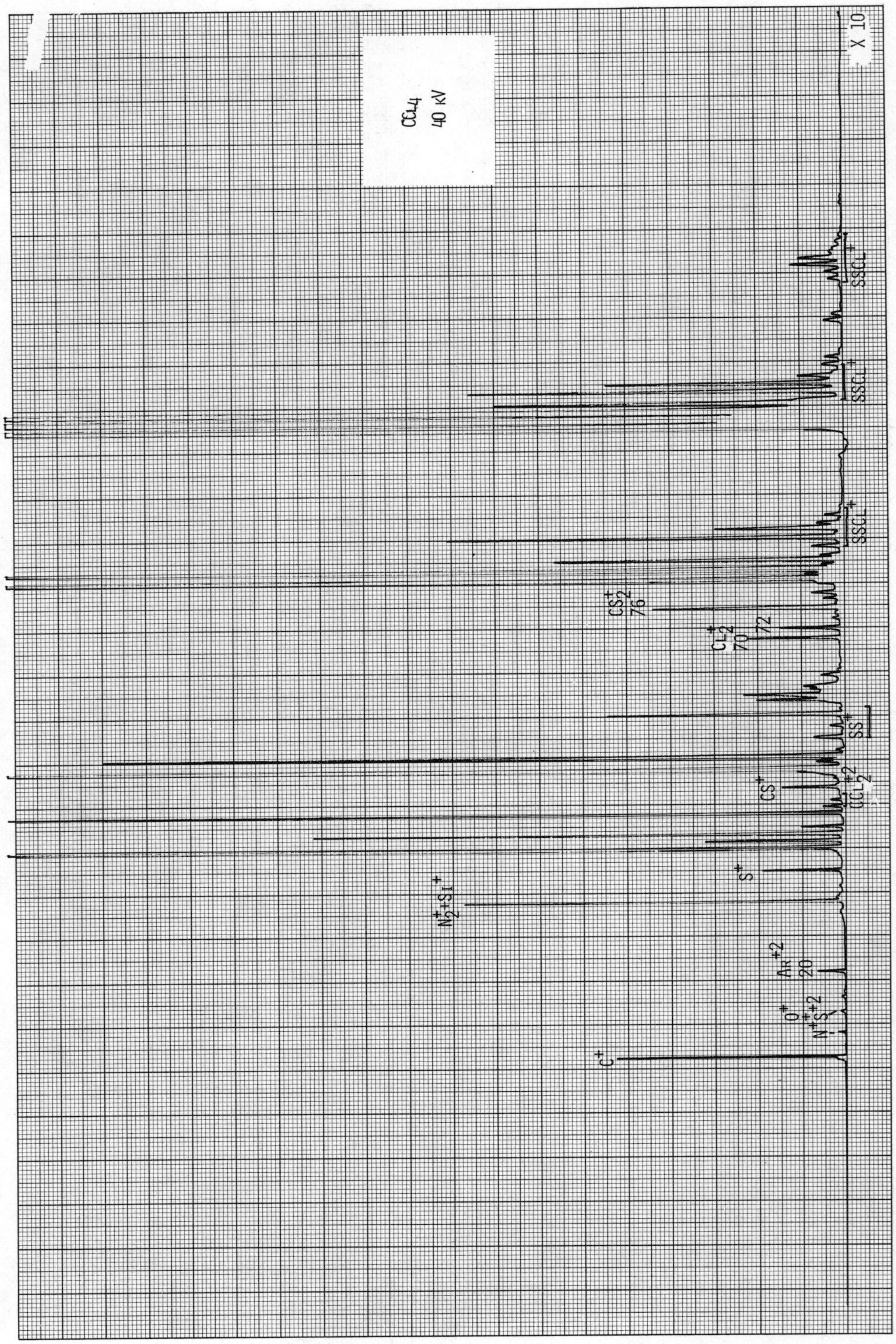
CCl4
40 kV
X 10
C+
O+
N+ S+2
Ar+2
20
N2+ + Si+
S+
CS+
CCl2+2
SS+
CL2+
70
72
CS2+
76
SSCL+
SSCL+
SSCL+

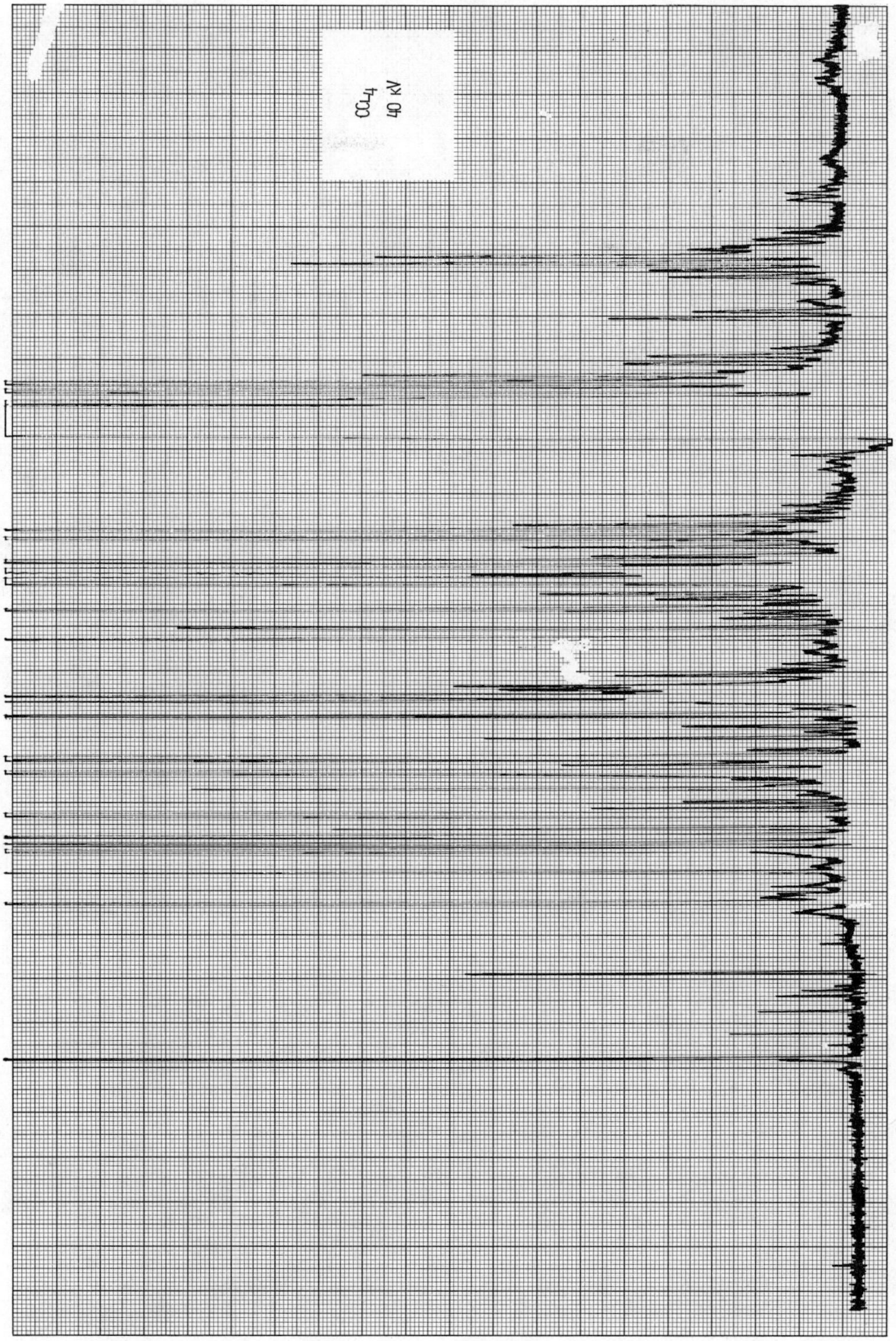
CCl_4
40 kV

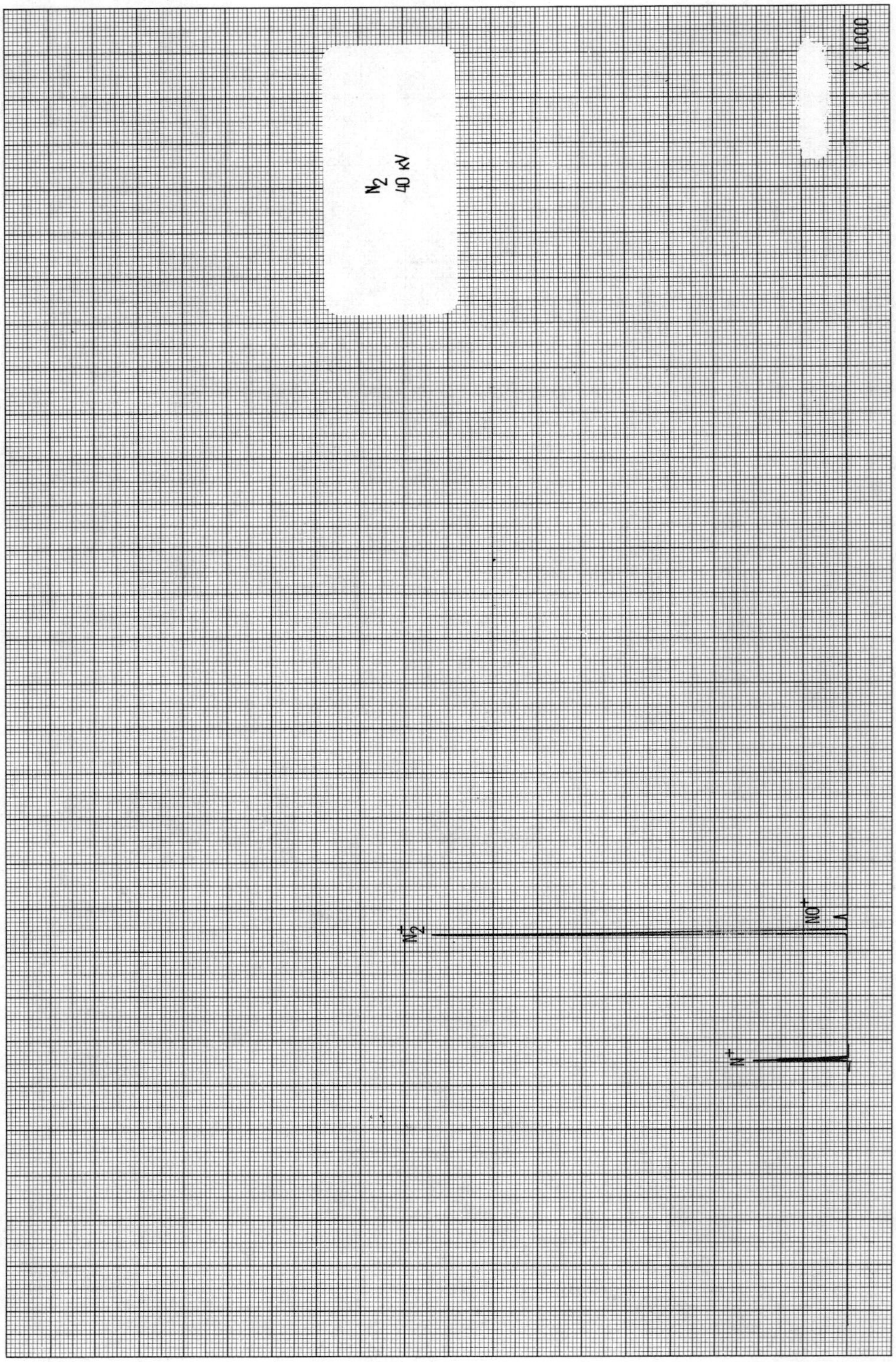
N_2
40 kV
X 1000
N_2^+
NO^+
N^+

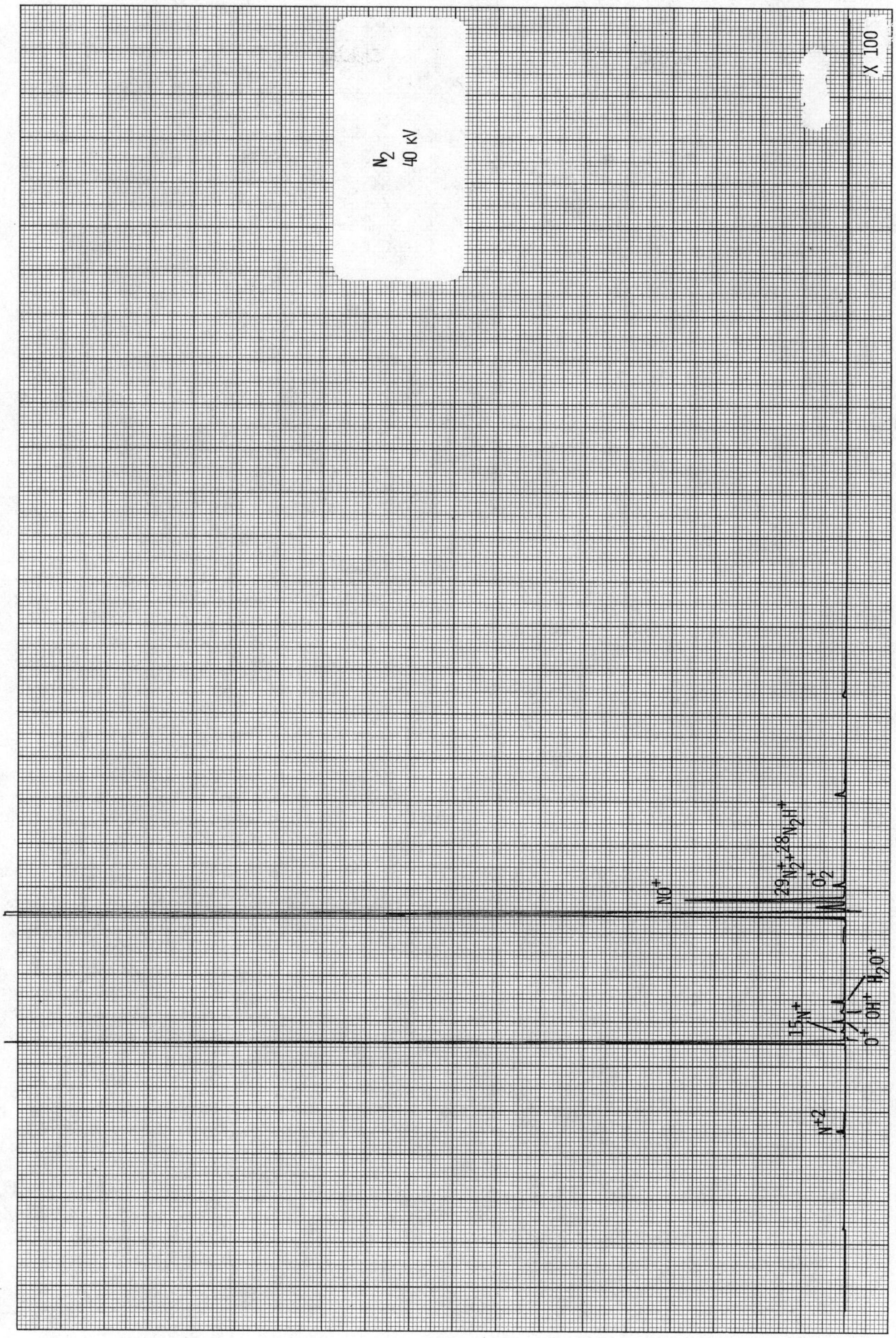
N2
40 kV
X 100
NO+
29N2+ + 28N2H+
O2+
15N+
O+
OH+
H2O+
N+2

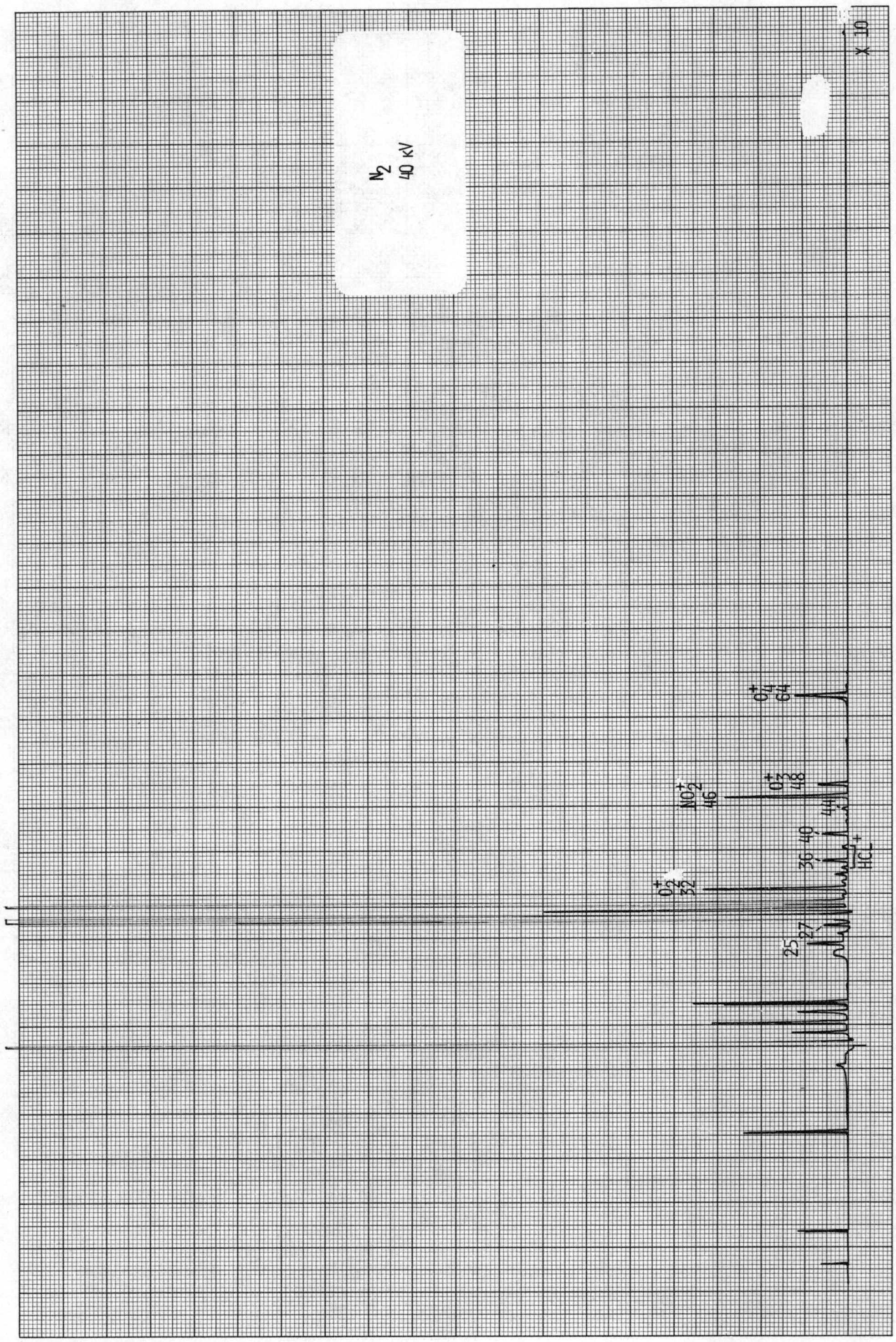
N_2
40 kV
X 10
O_2^+
32
25
27
36
40
NO_2^+
46
O_3^+
48
44
HC^+
O_4^+
64

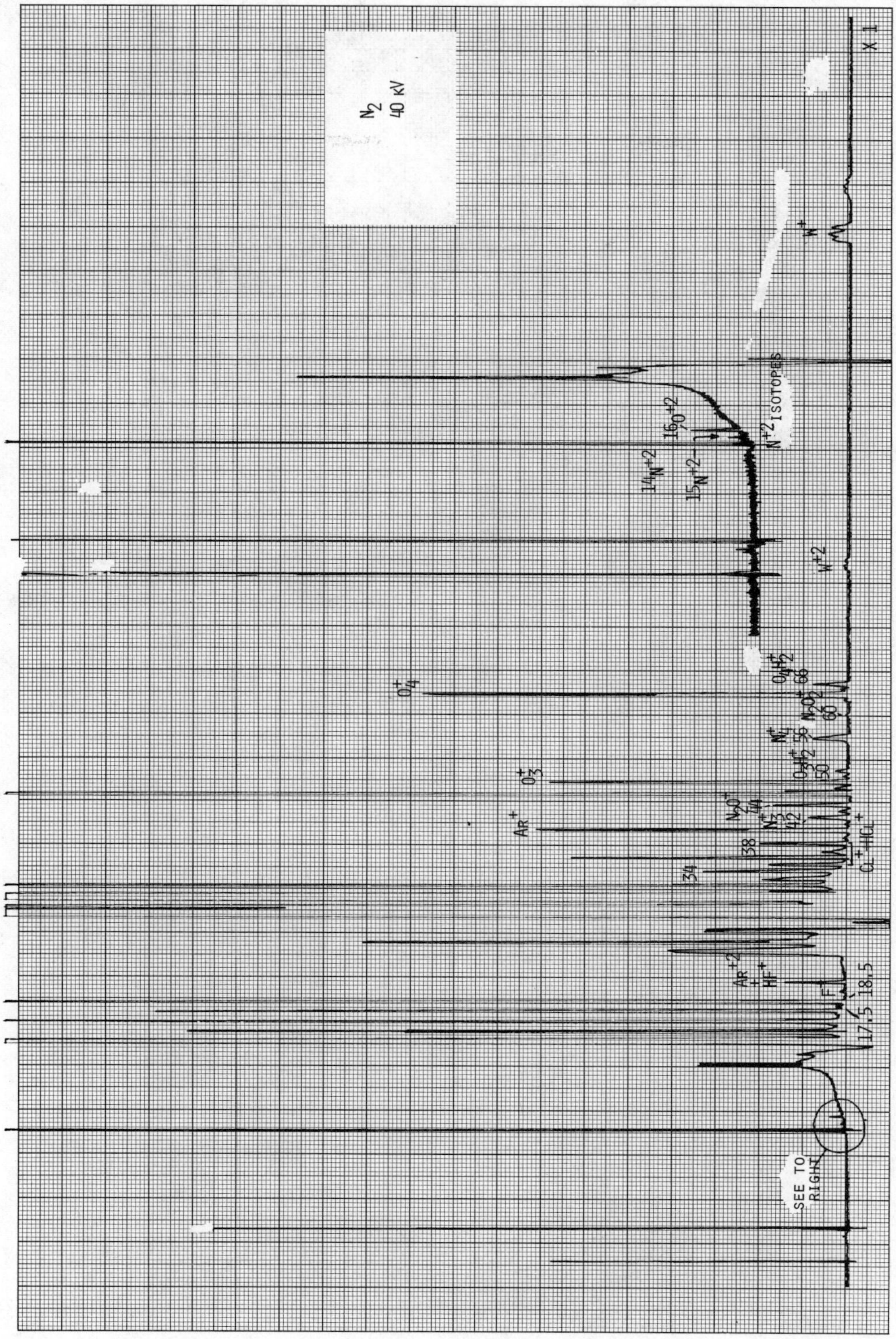

N2
40 kV
X 1
N+
N+2 ISOTOPES
14N+2
16O+2
15N+2
N+2
O4+
O3+
Ar+
34
38
Ar+2
HF+
17.5
18.5
SEE TO RIGHT

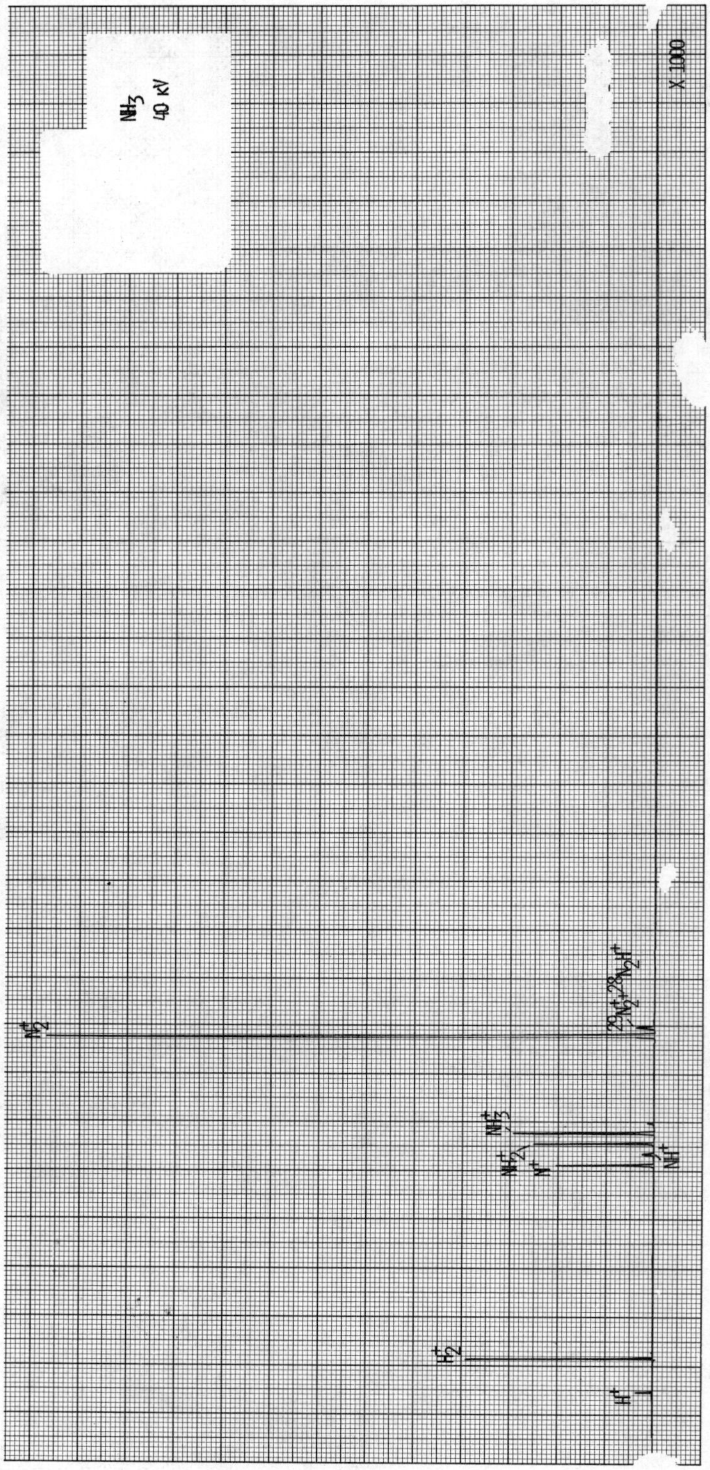

NH_3
40 kV
X 1000
N_2^+
NH_3^+
NH_2^+
N^+
NH^+
H_2^+
H^+

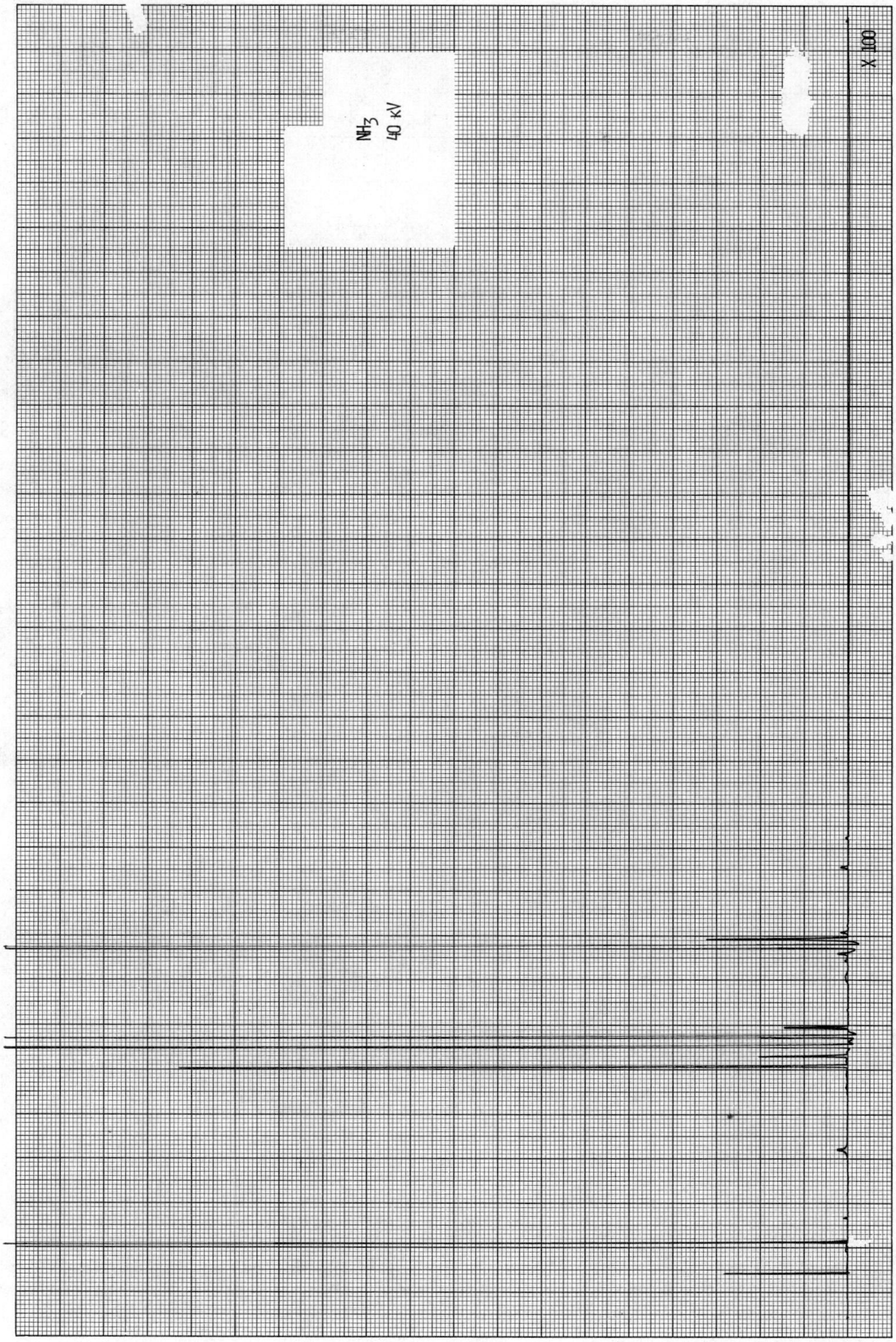
NH3
40 kV
X 100

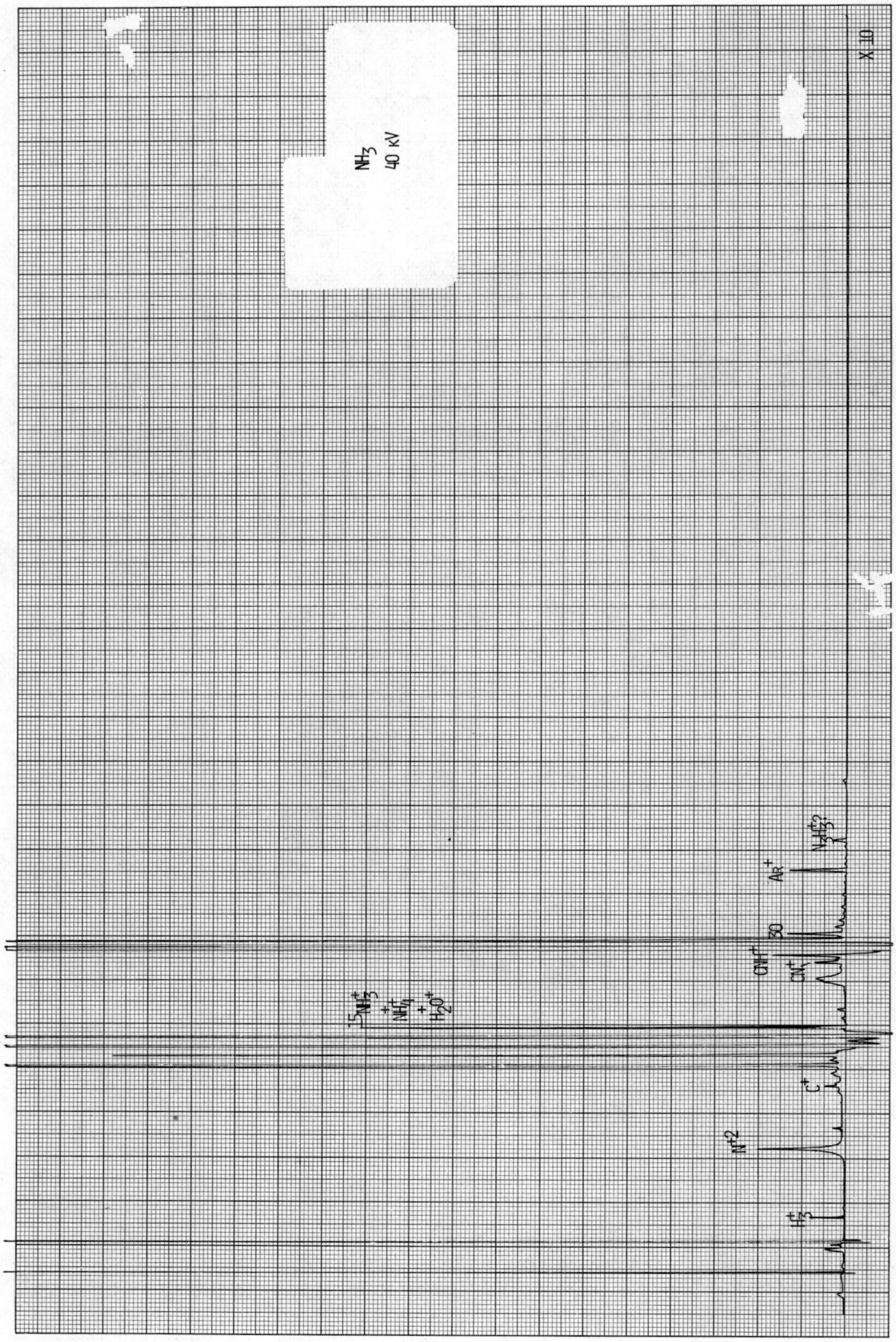

NH_3
40 kV
X 10

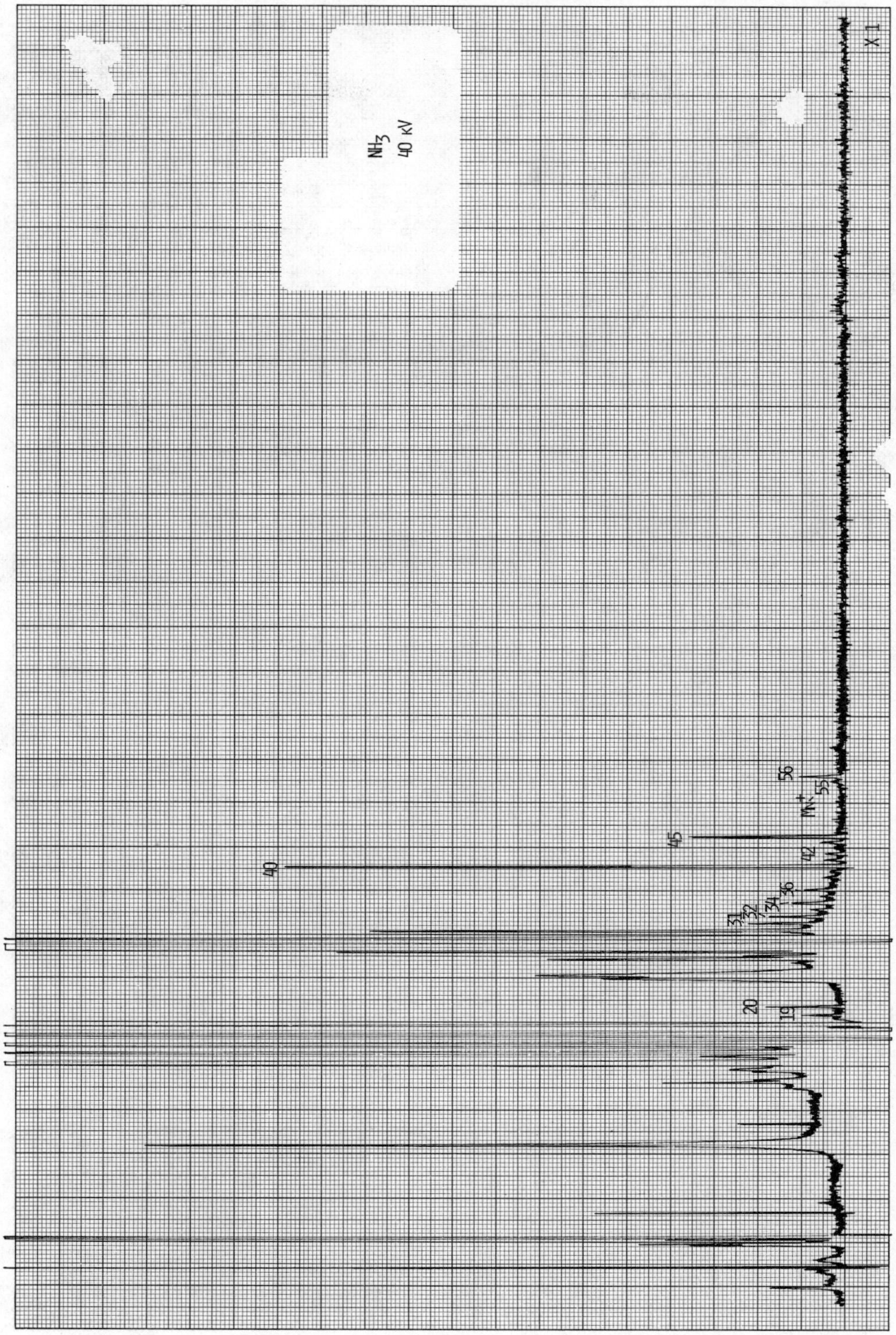
NH3
40 kV
X 1
40
45
42
36
20
19
56
55

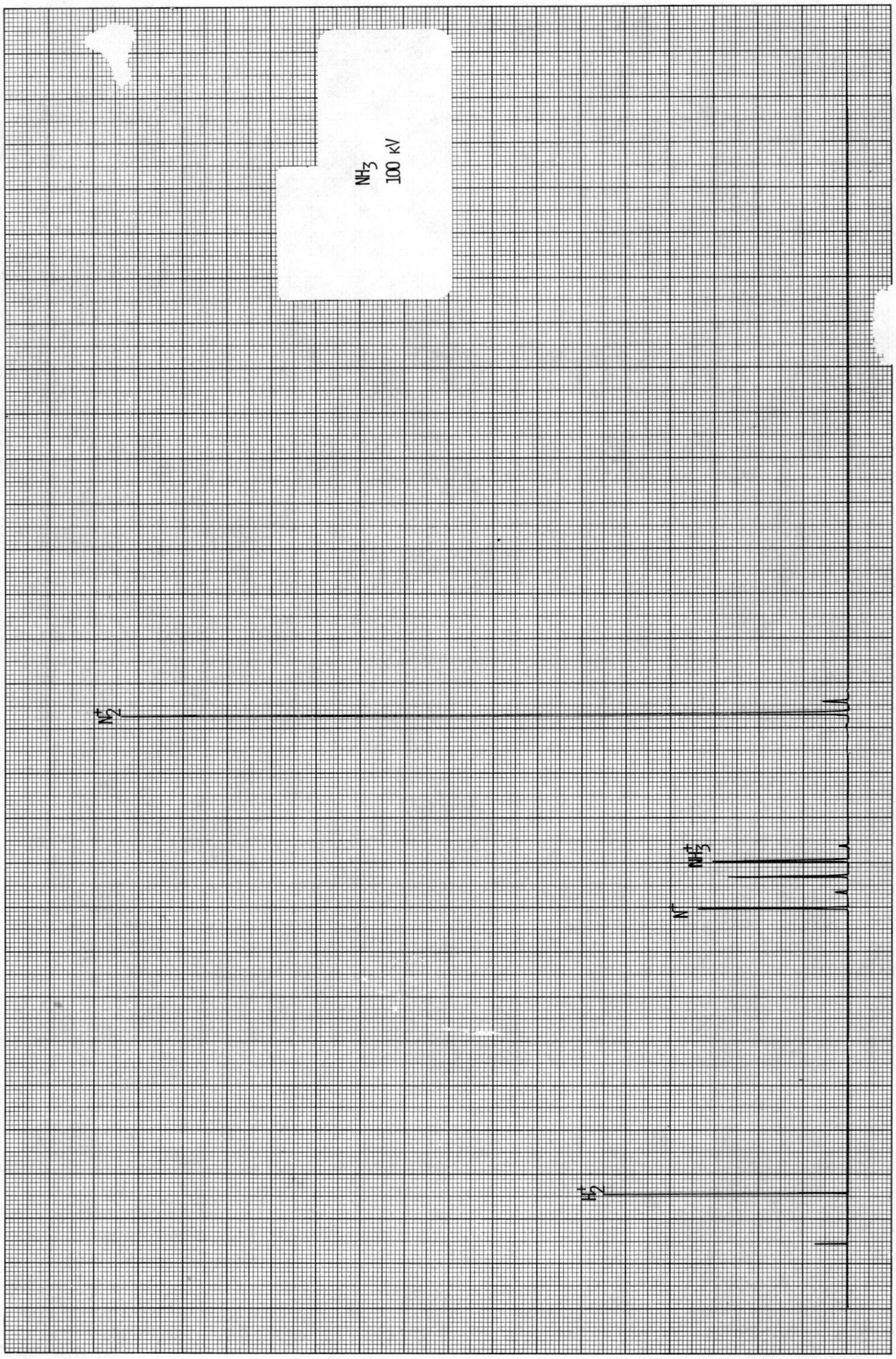

NH_3
100 kV
N_2^+
NH_3^+
N^+
H_2^+

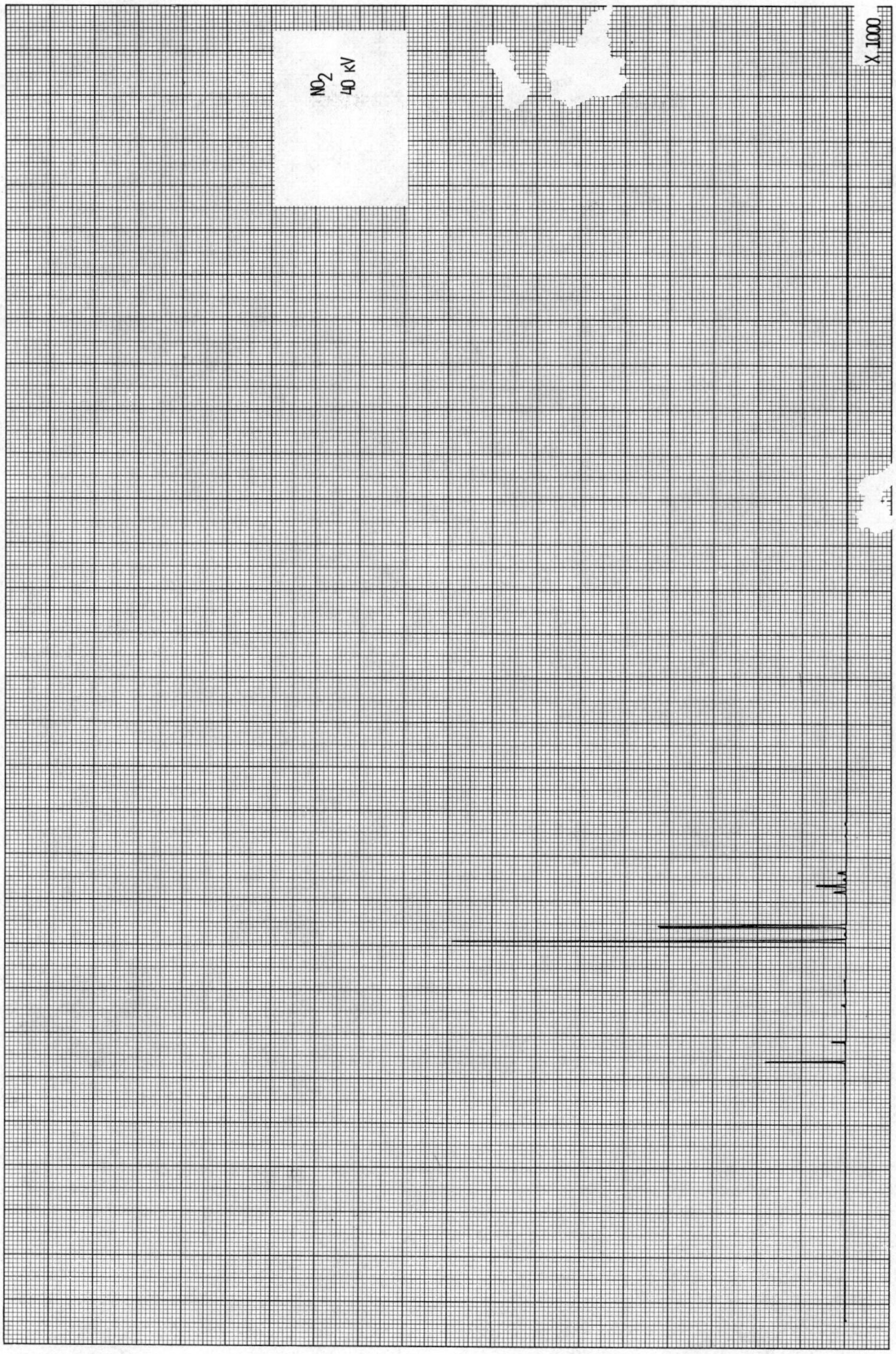
NO_2
40 kV
X 1000

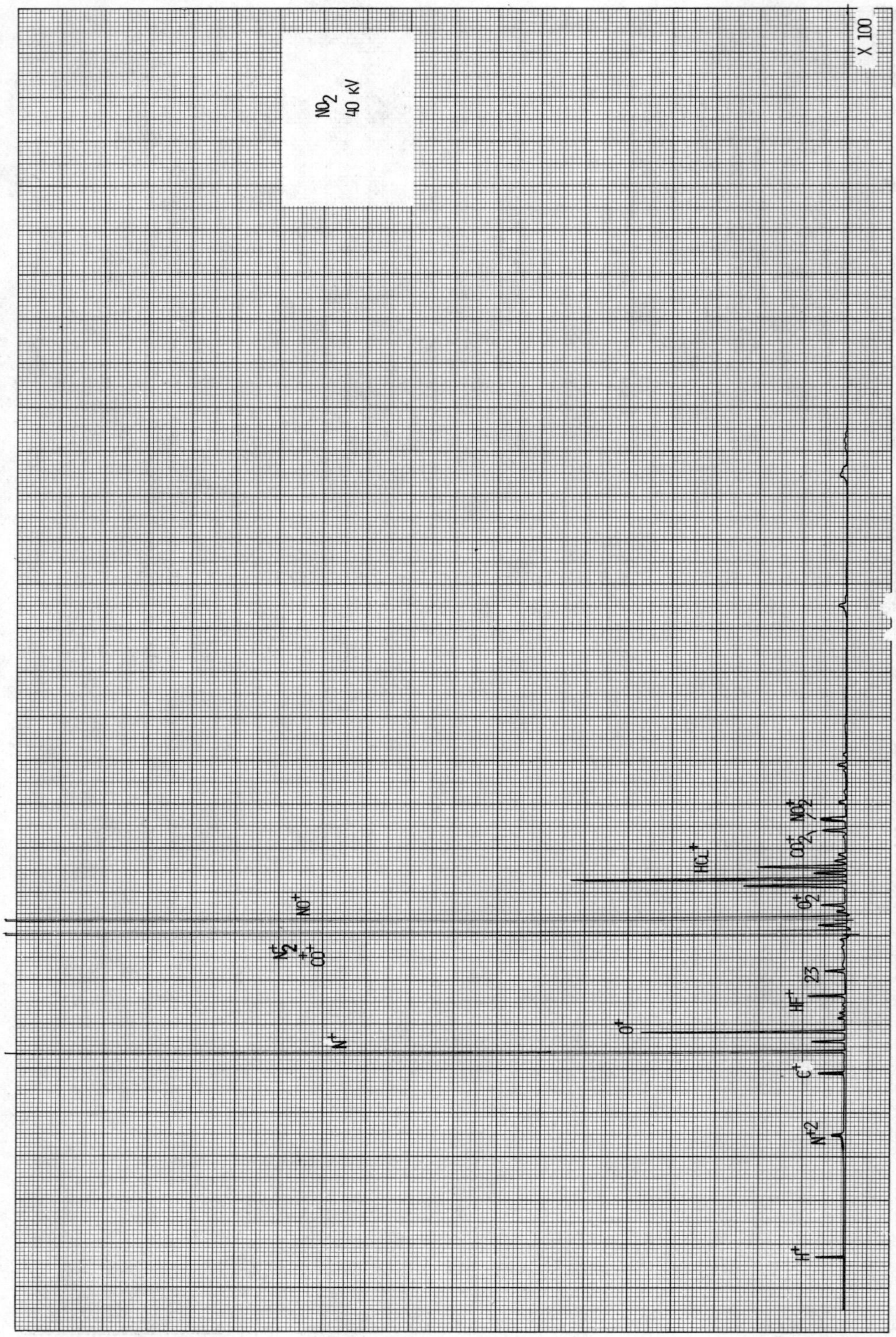

NO2
40 kV
X 100
H+
N+2
C+
N+
O+
HF+
23
N2+
CO+
NO+
O2+
HCl+
CO2+
NO2+

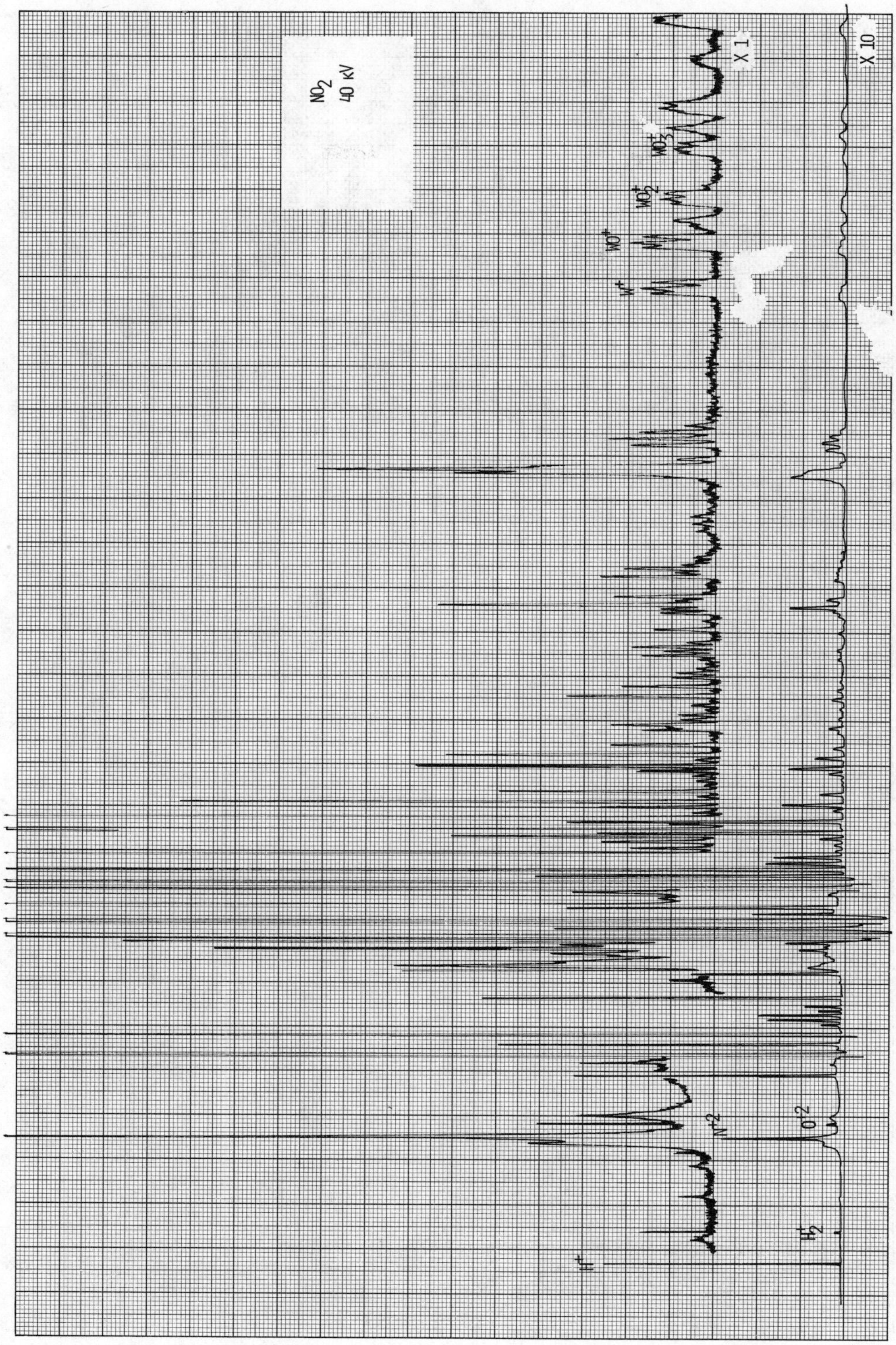

NO_2
40 kV
X 1
X 10
NO^+
NO_2^+
NO_3^+

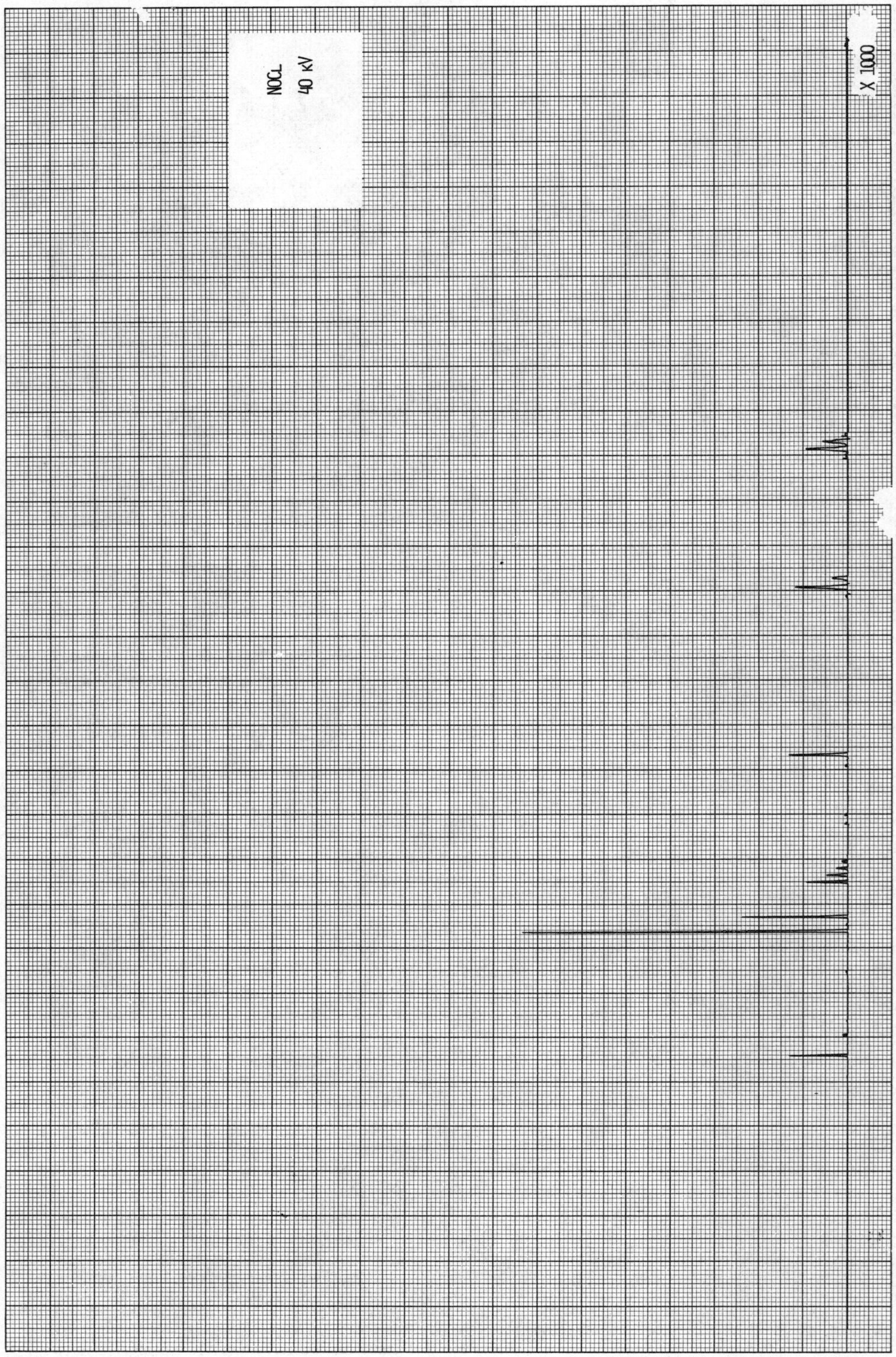
NOCL
40 kV
X 1000

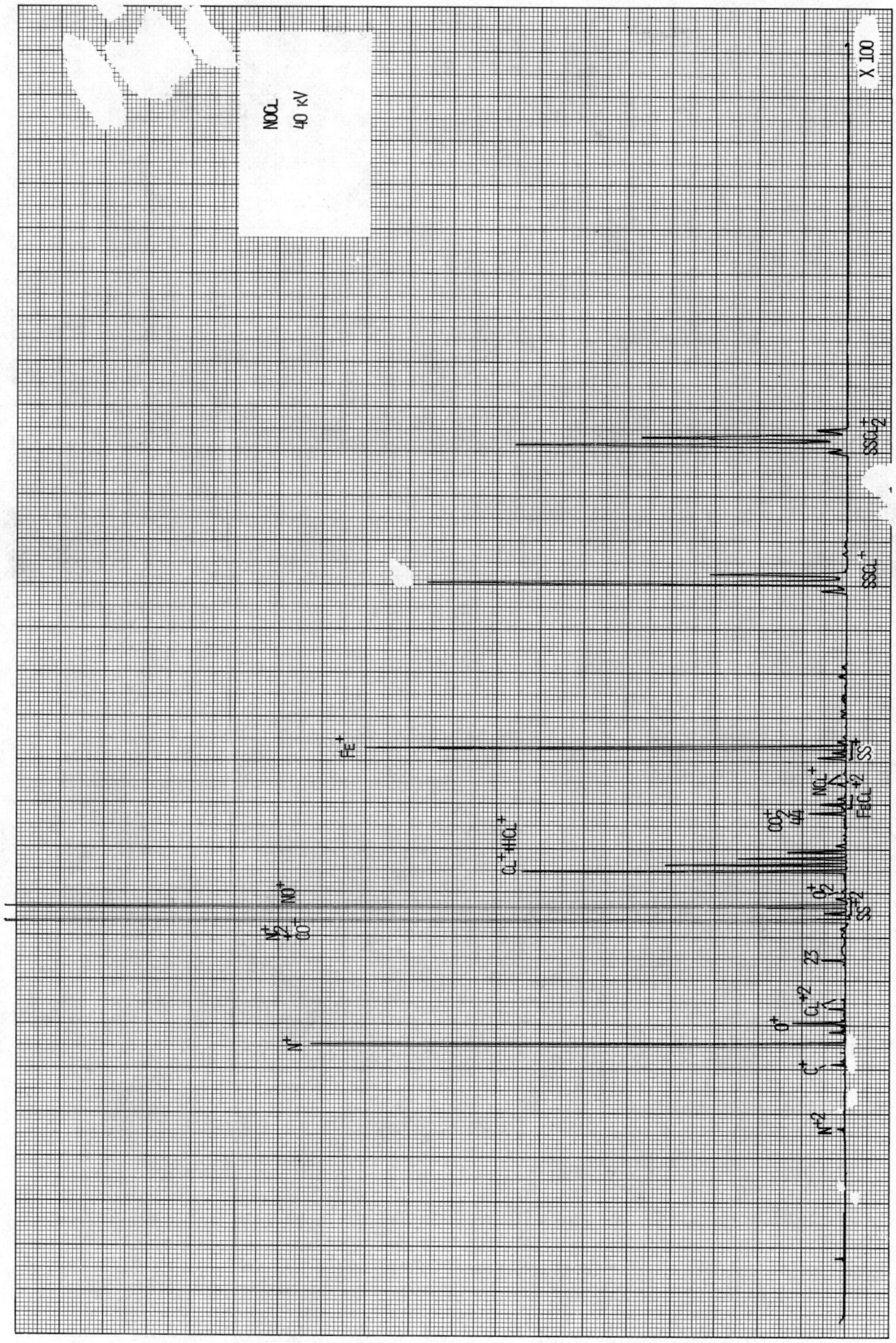
NOCL
40 kV
X 100
N+
N+2
C+
O+
CL+2
23
N2+
CO+
NO+
SS+2
O2+
CL+ HCL+
CO2+
44
FeCL+2
NCL+
SS+
Fe+
SSCL+
SSCL2+

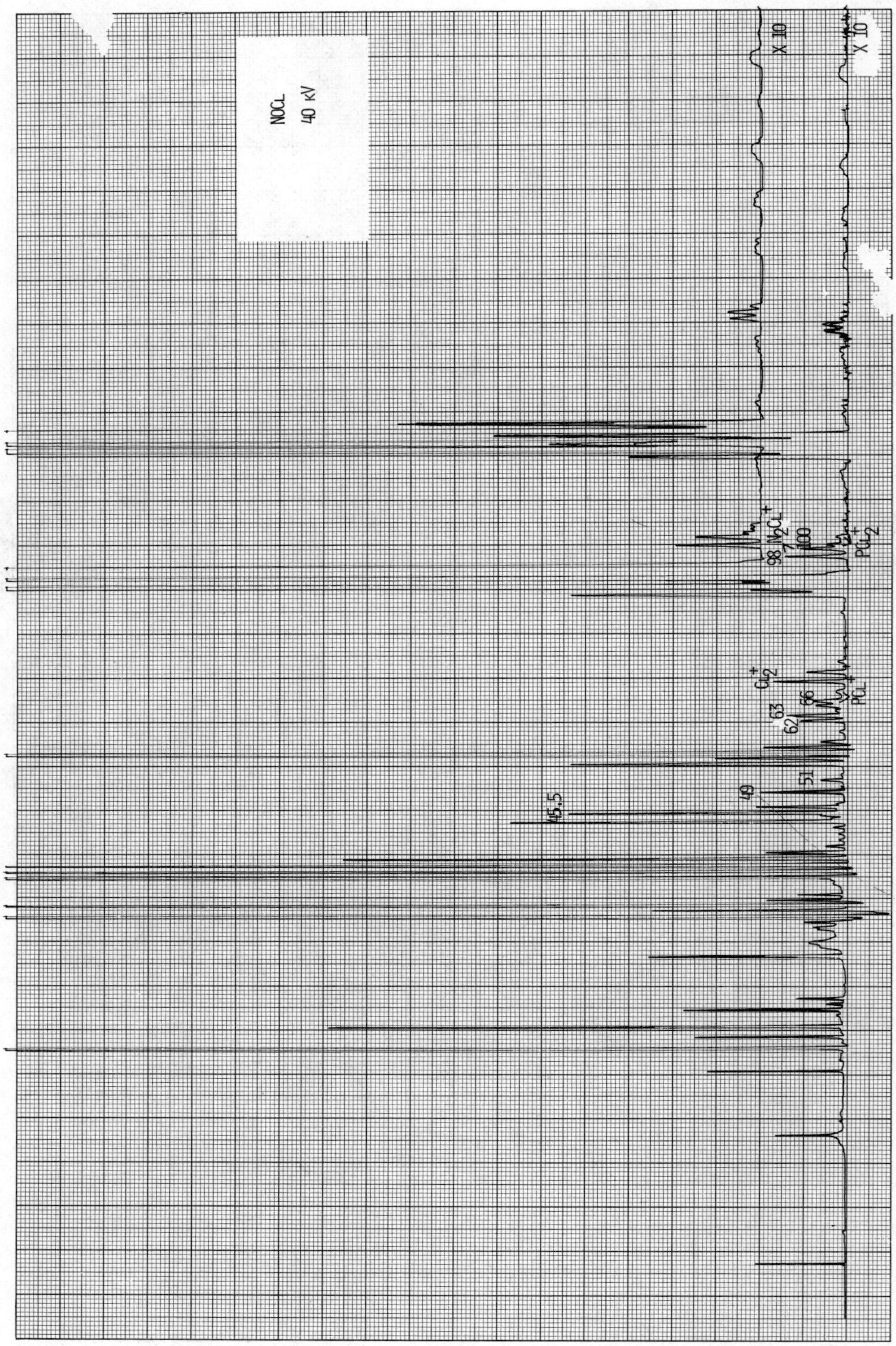
NOCL
40 kV
X 10
X 10
45.5
49
51
62
63
66
Cl_2^+
PCl^+
98
100
N_2Cl^+
PCl_2^+

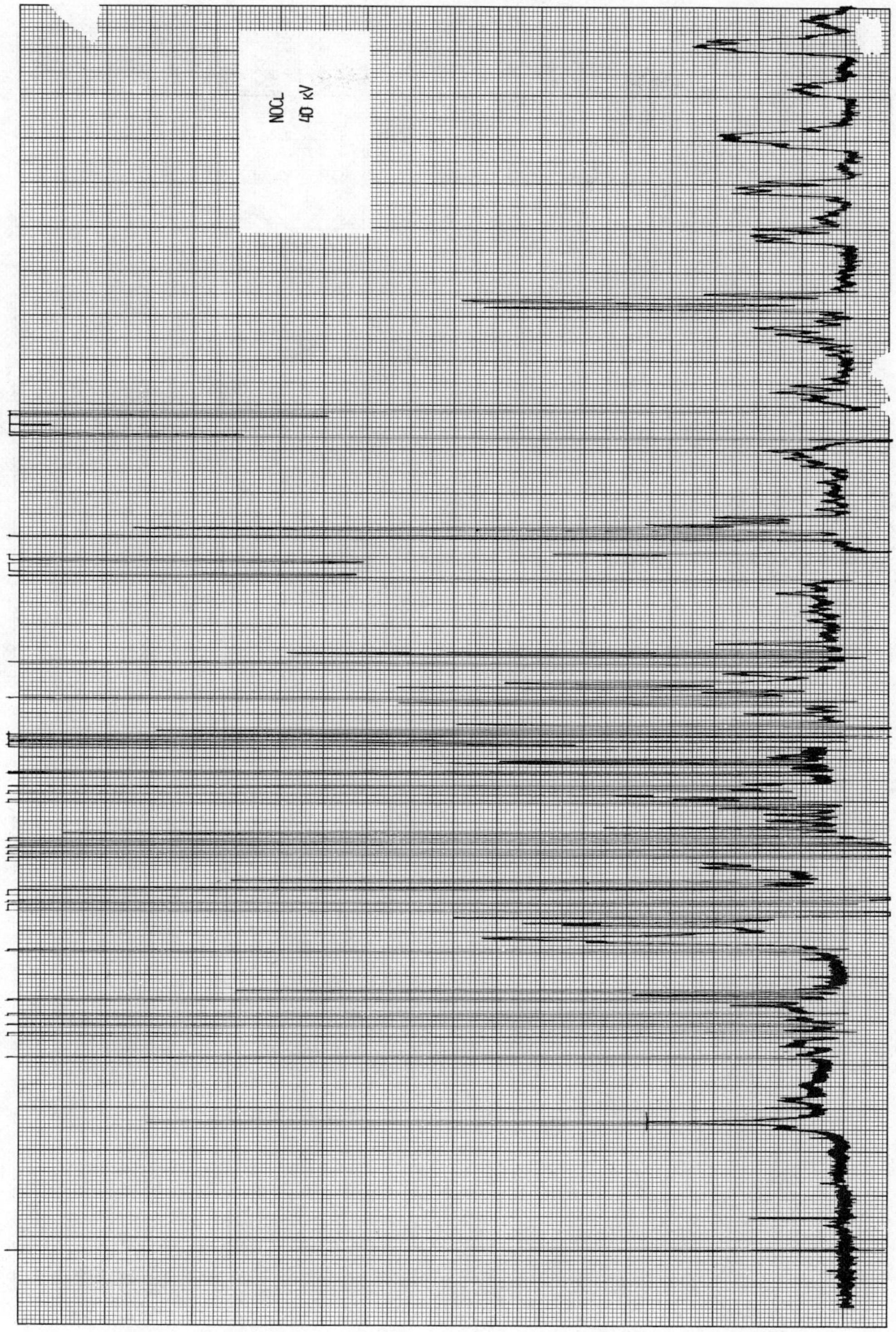
NOCL
40 kV

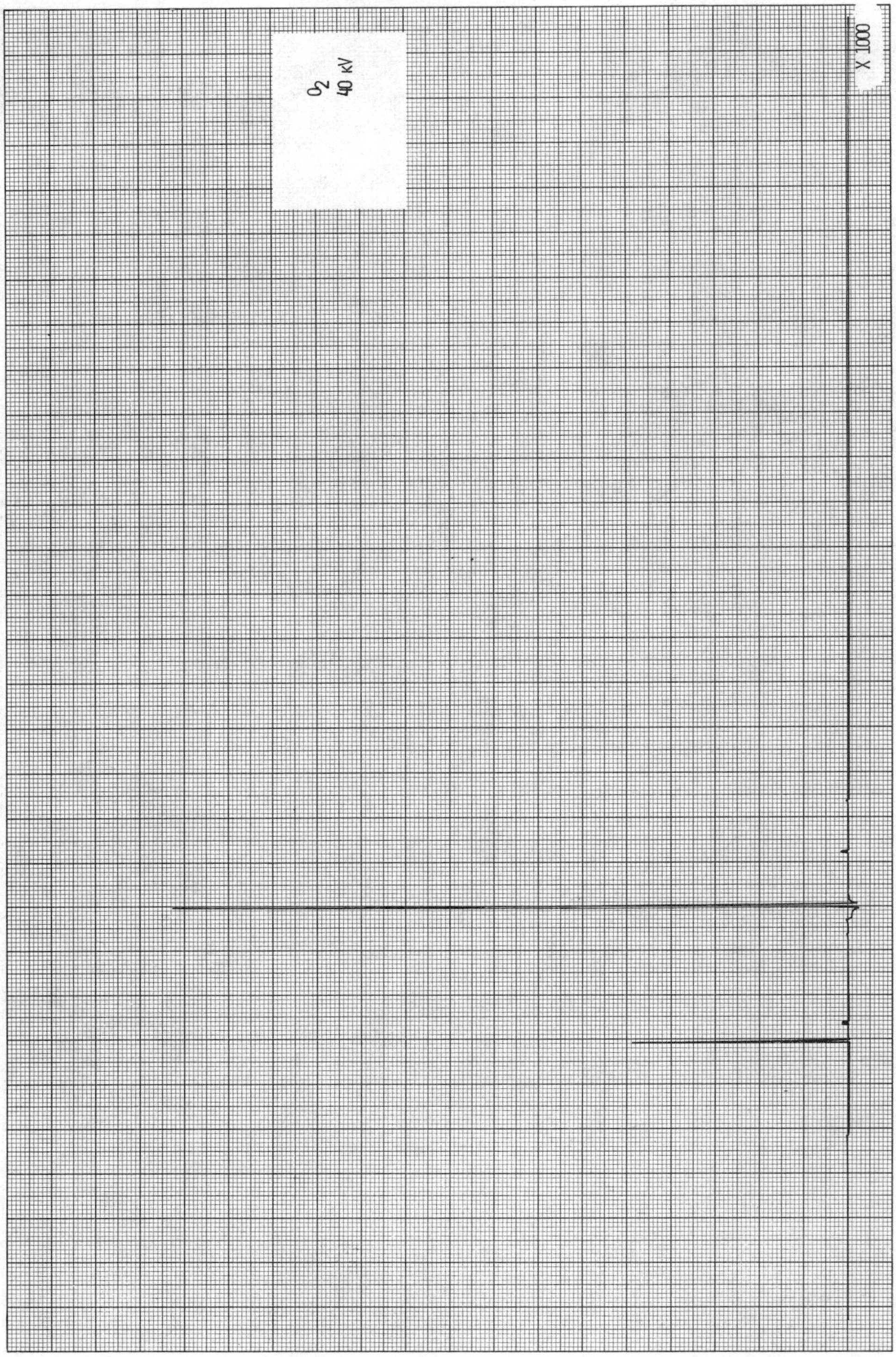

O_2
40 kV
X 1000

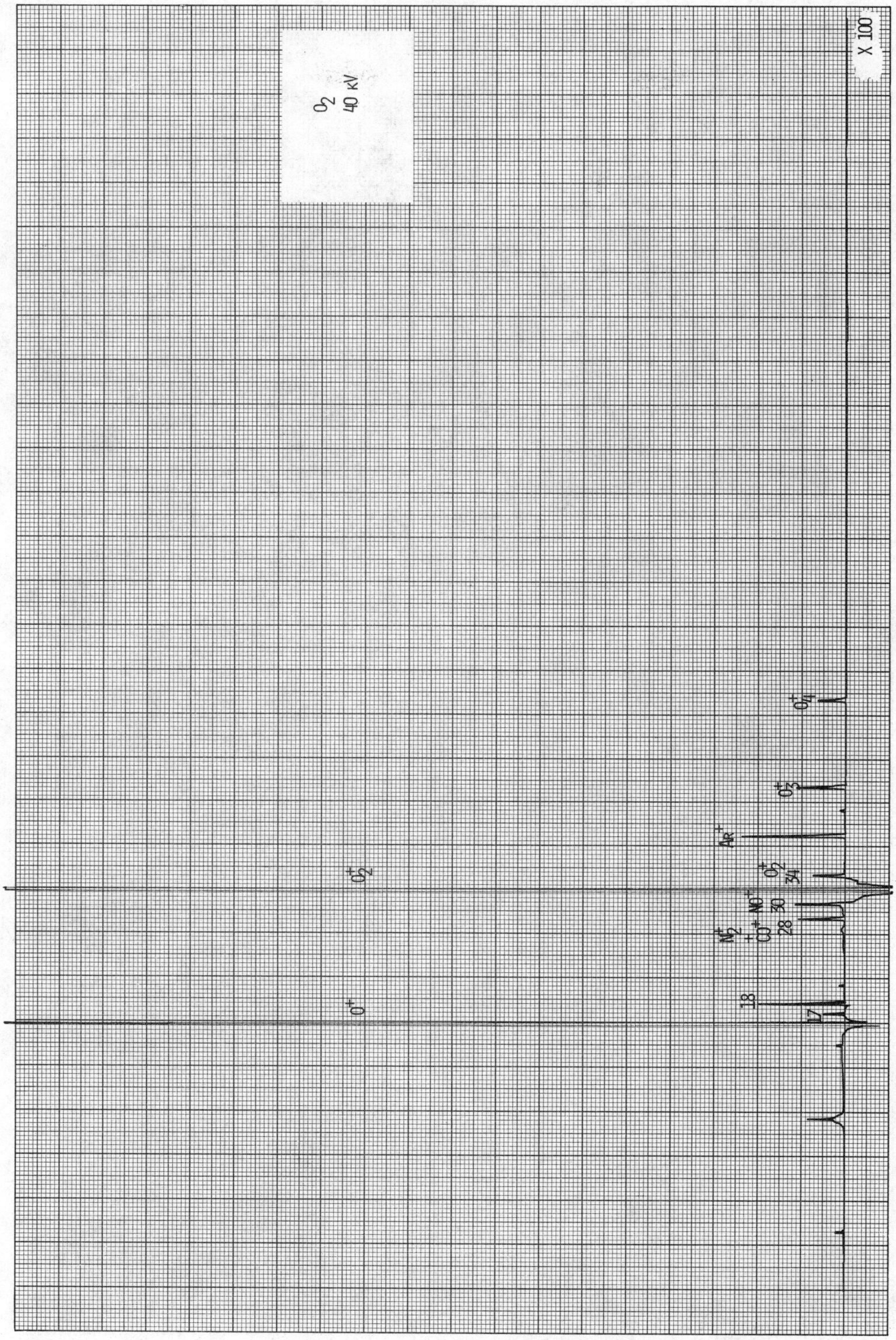
O_2
40 kV
X 100
O^+
O_2^+
17
18
N_2^+
CO^+
NO^+
28
30
O_2^+
34
Ar^+
O_3^+
O_4^+

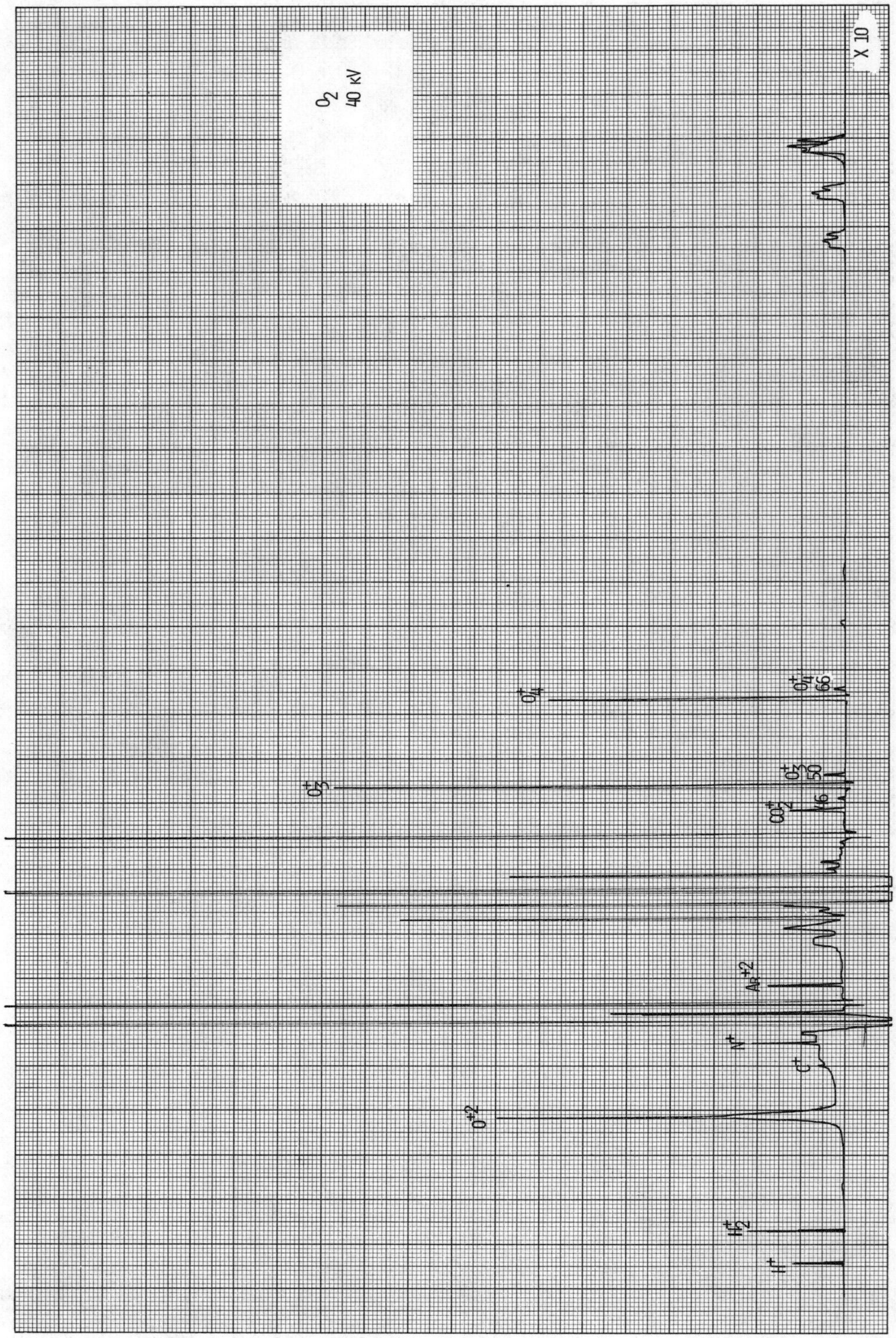

O_2
40 kV
X 10
H^+
H_2^+
O^{+2}
C^+
N^+
Ar^{+2}
CO_2^+
46
O_3^+
O_3^+
50
O_4^+
O_4^+
66

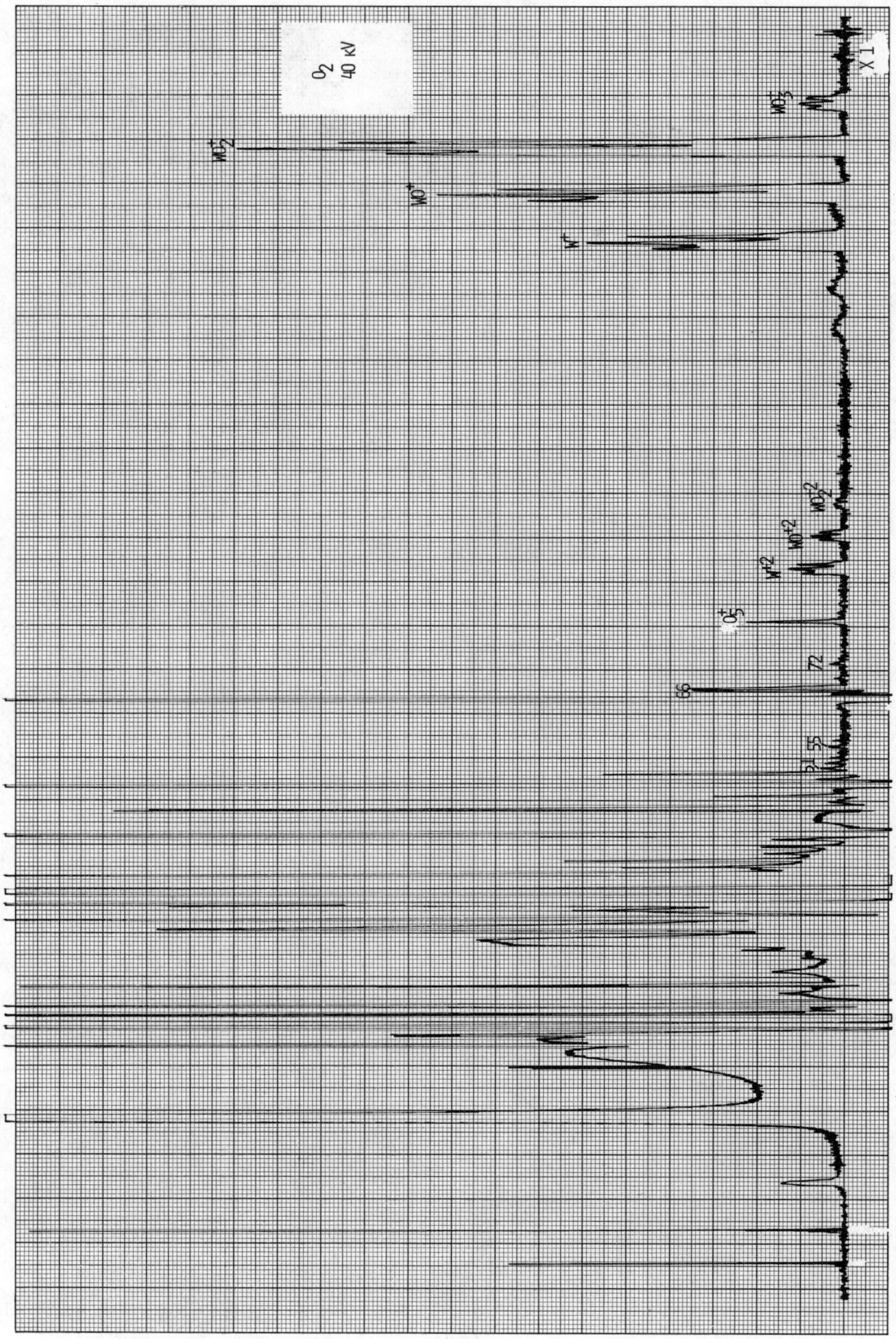
O2
40 kV
X 1
66
72

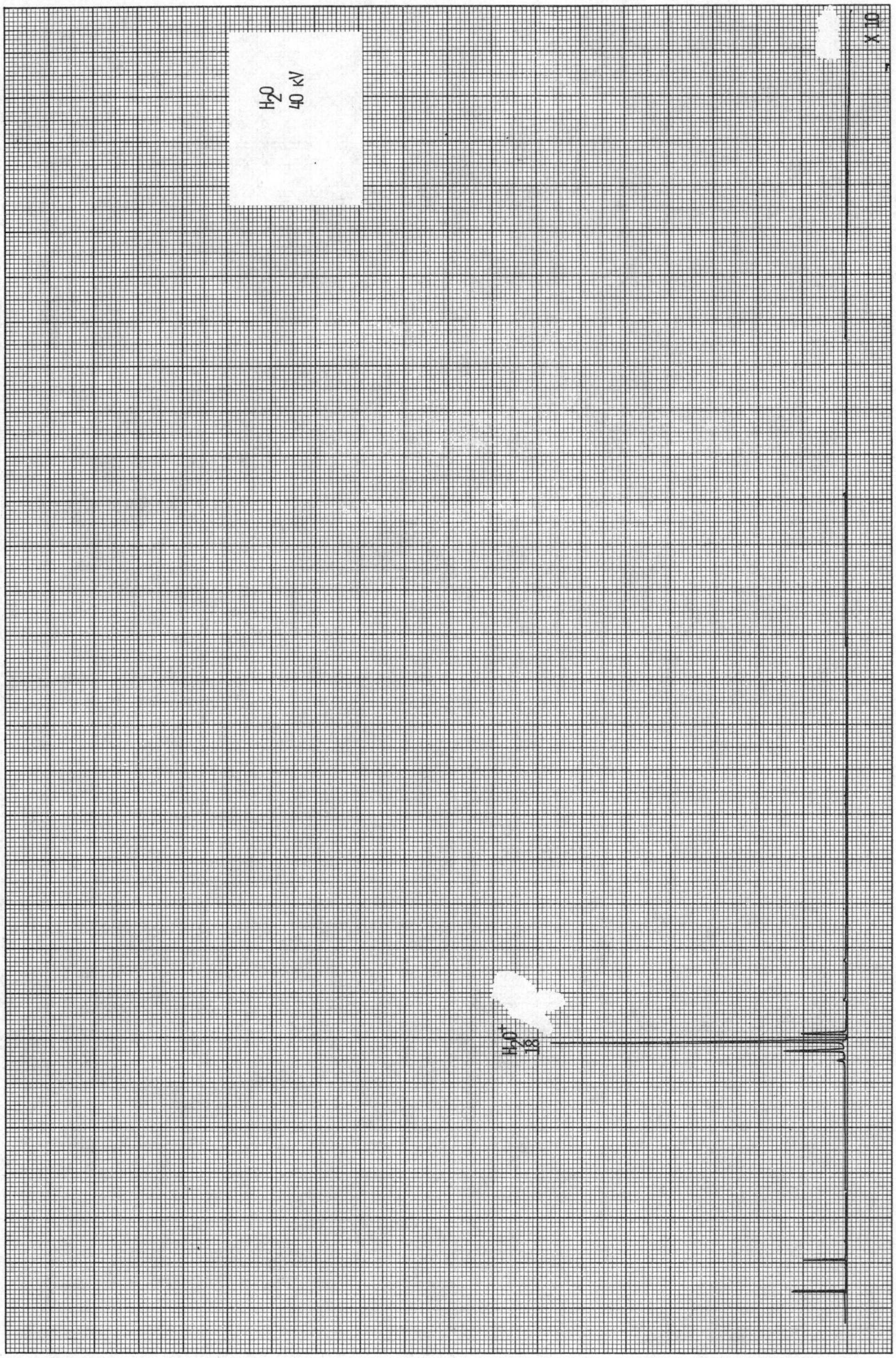
H_2O
40 kV
X 10
H_2O^+
18

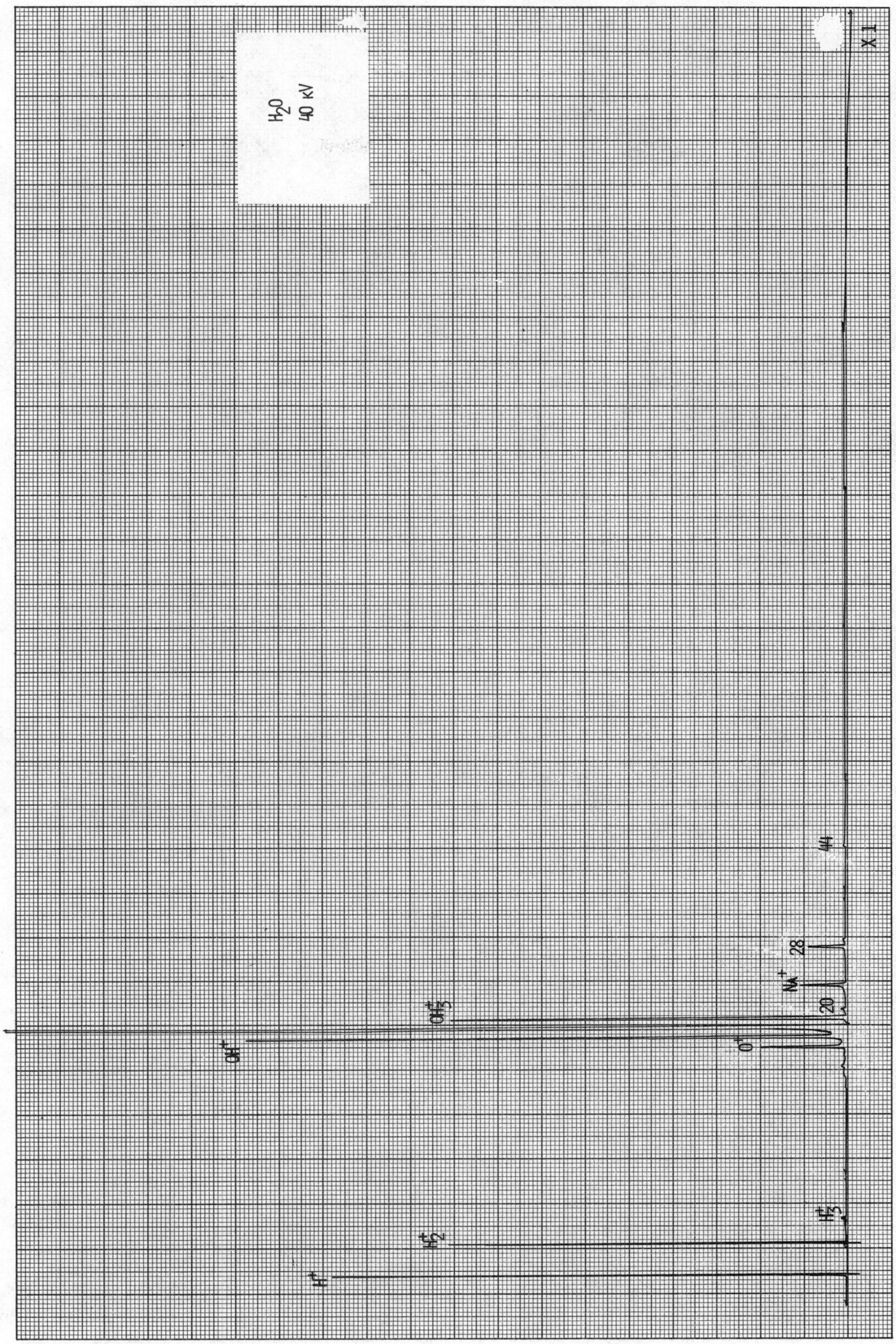
H2O
40 kV
X 1
H+
H2+
H3+
O+
OH+
OH2+
Na+
20
28
44

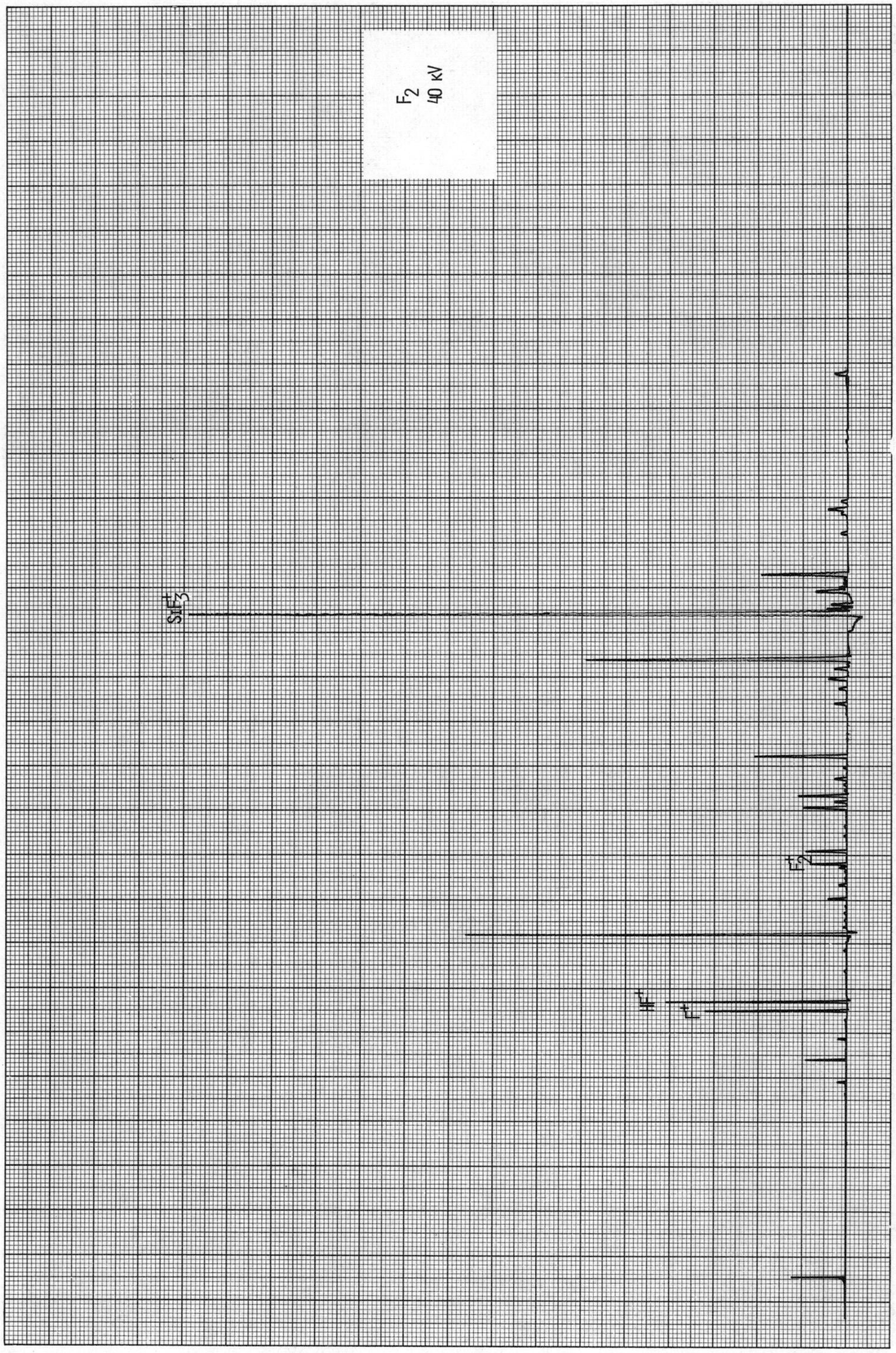
F_2
40 kV
SiF_3^+
F_2^+
HF^+
F^+

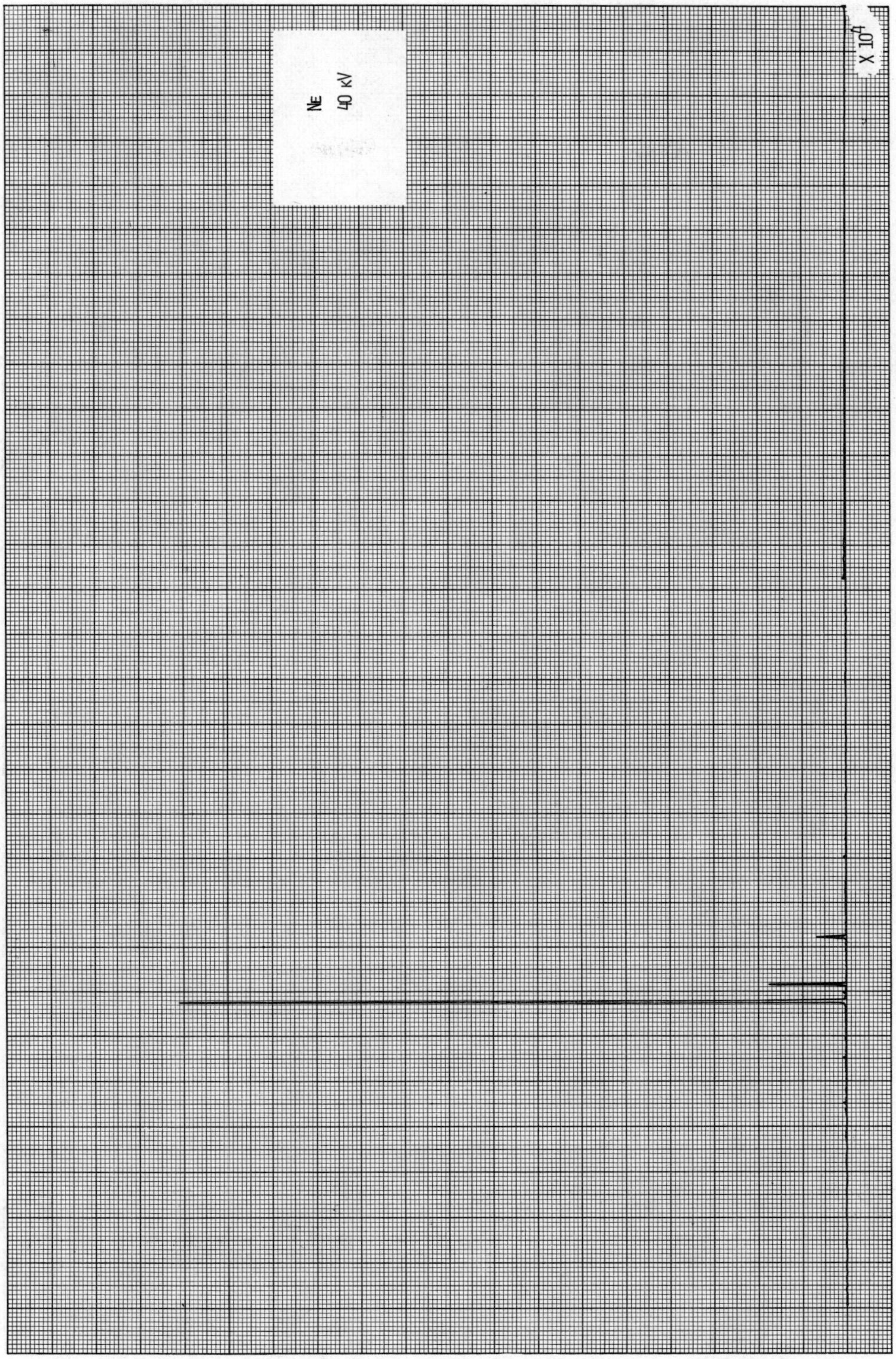
NE
40 kV
X 10^4

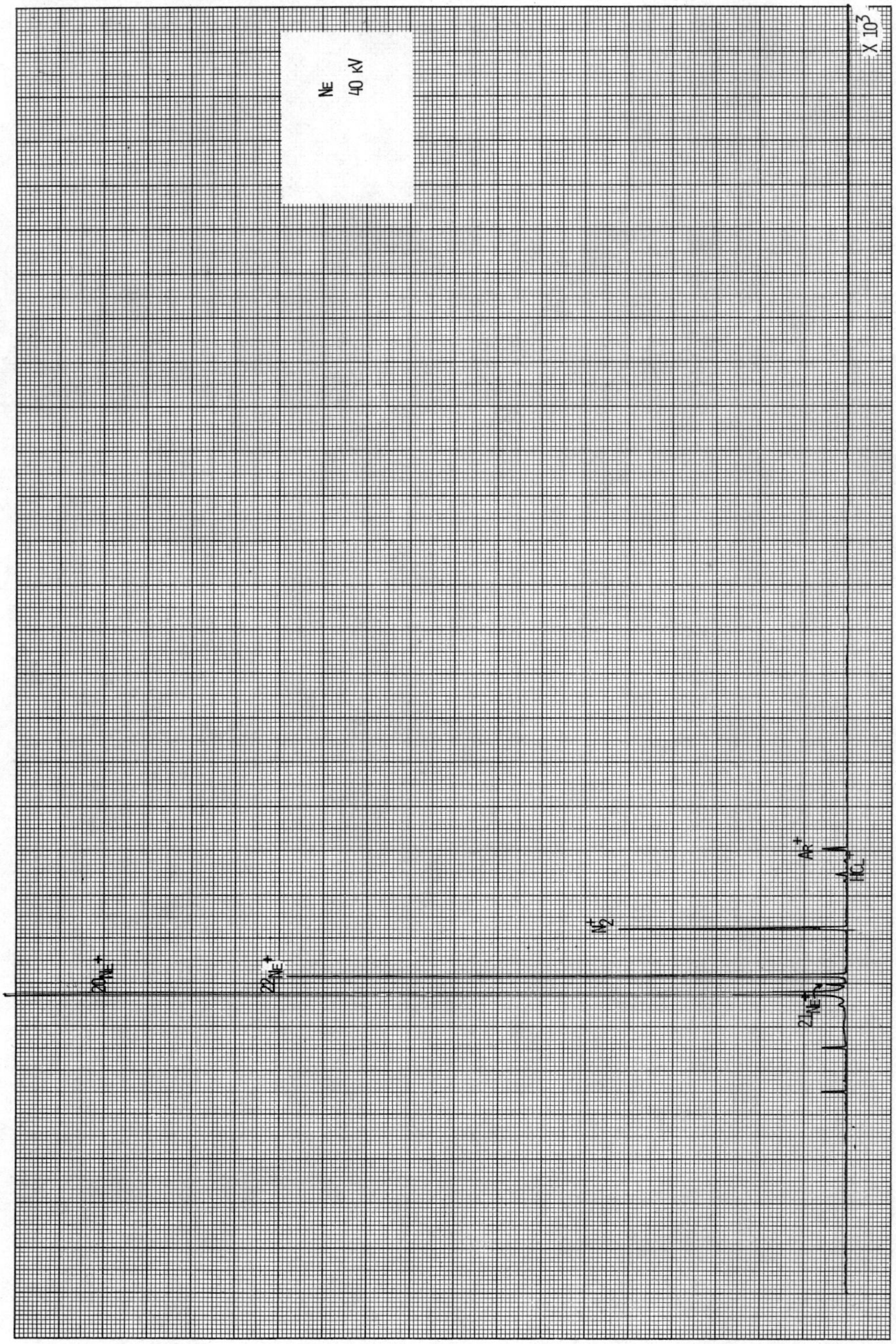

Ne
40 kV
X 10^3
20Ne+
22Ne+
21Ne+
N2+
Ar+
HCl+

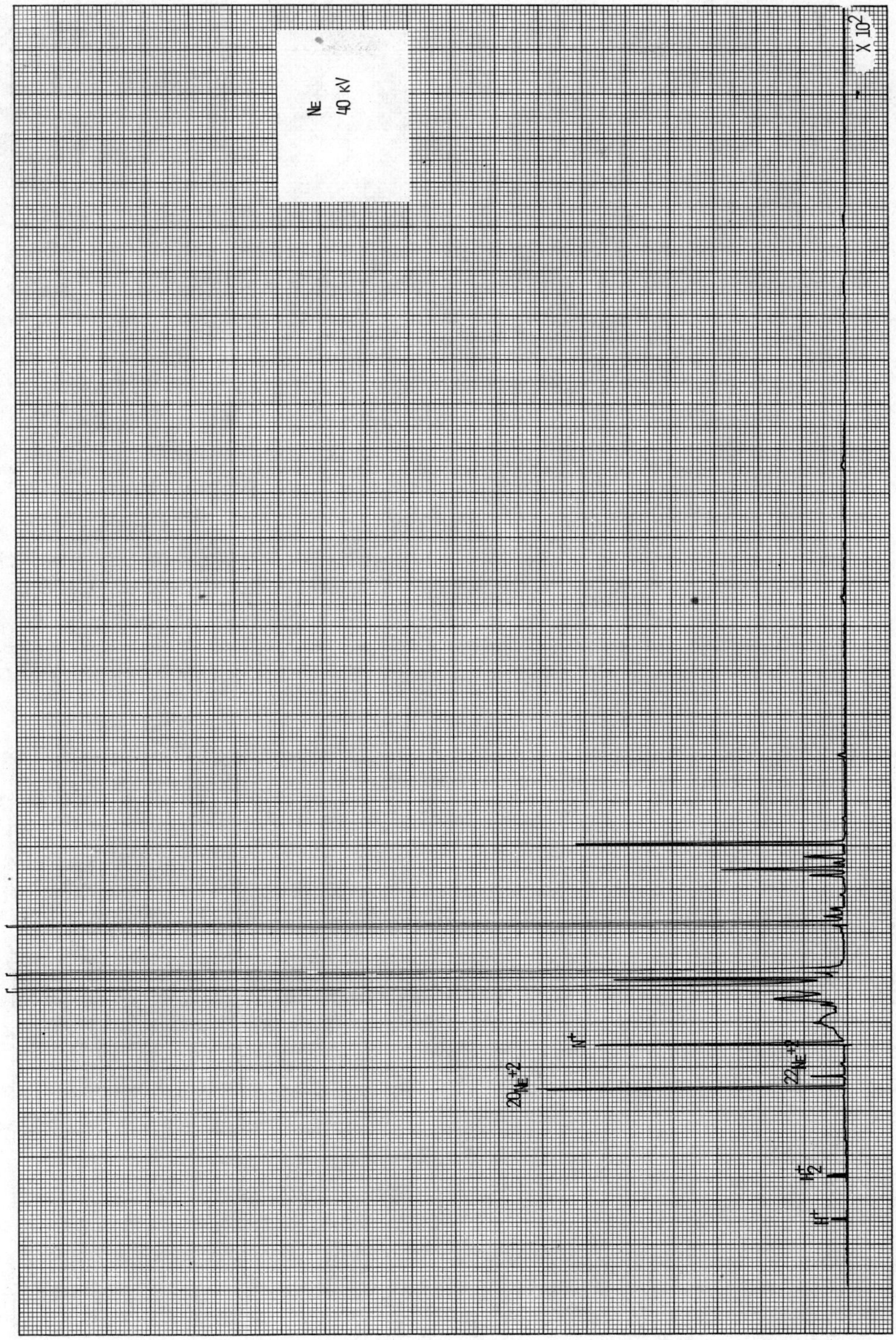
Ne
40 kV
X 10-2
N+
20Ne+2
22Ne+2
H+
H2+

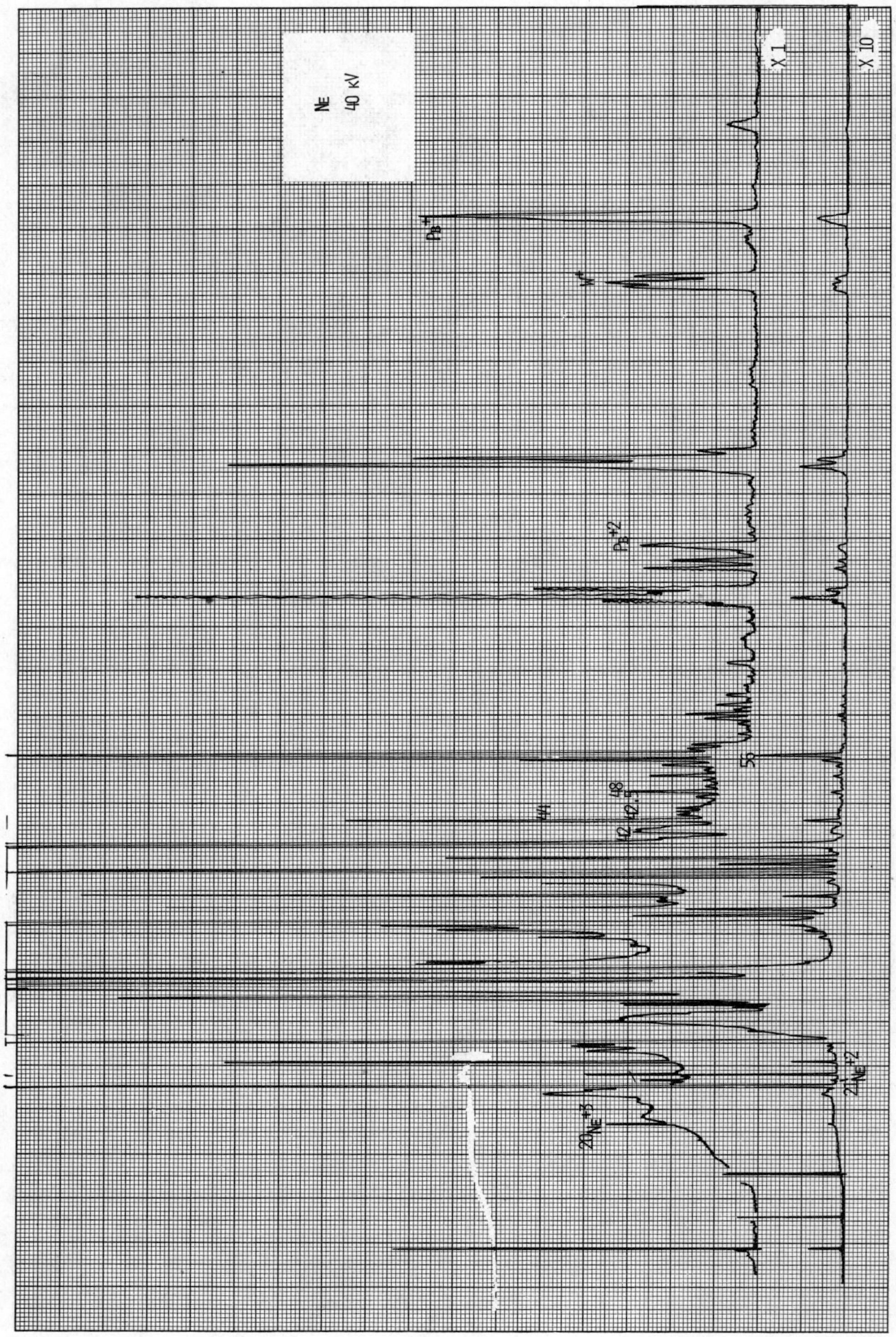
Ne
40 kV
X 1
X 10
56
44
48
42.5
42

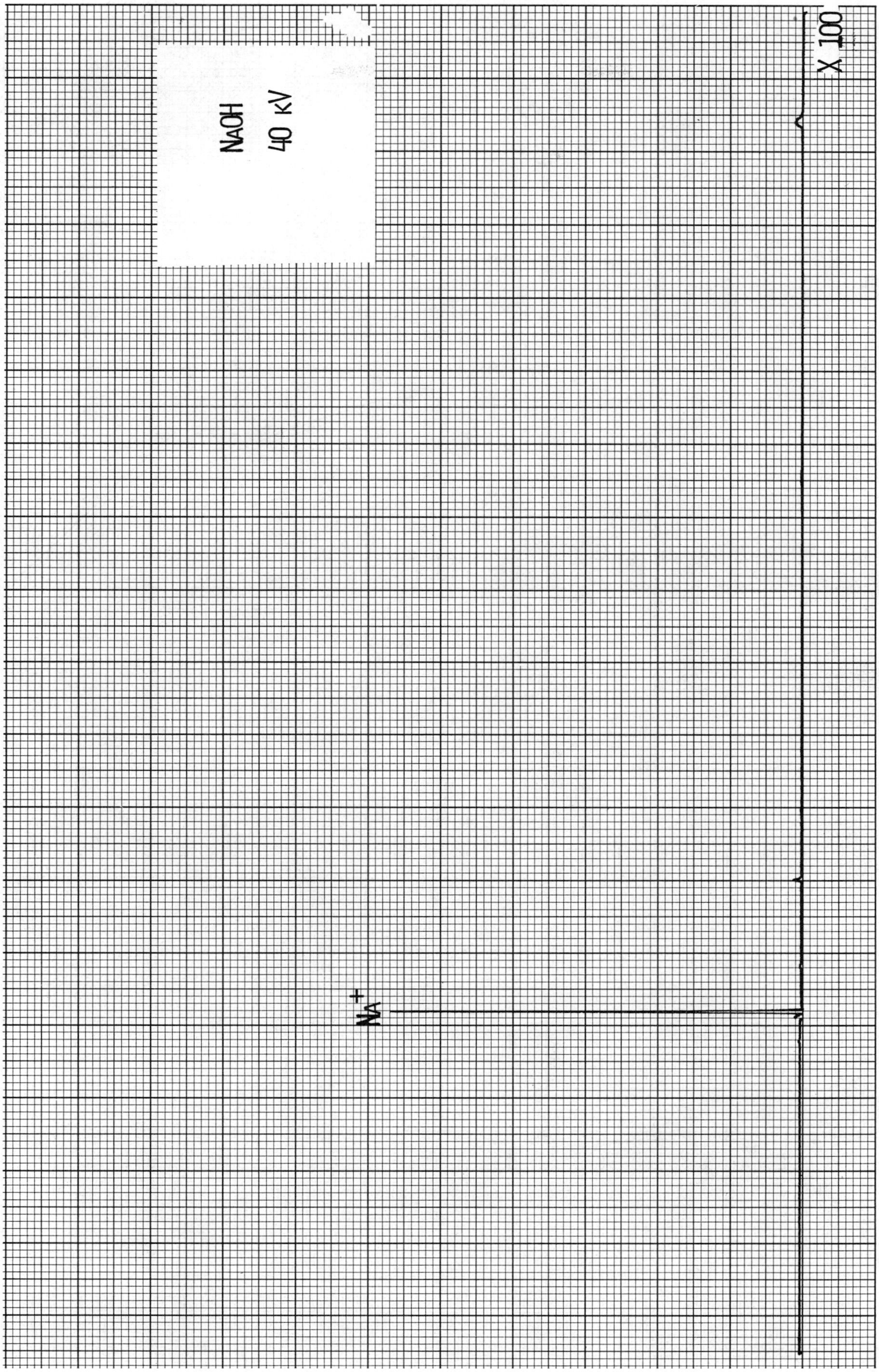
NaOH
40 kV
X 100
Na+

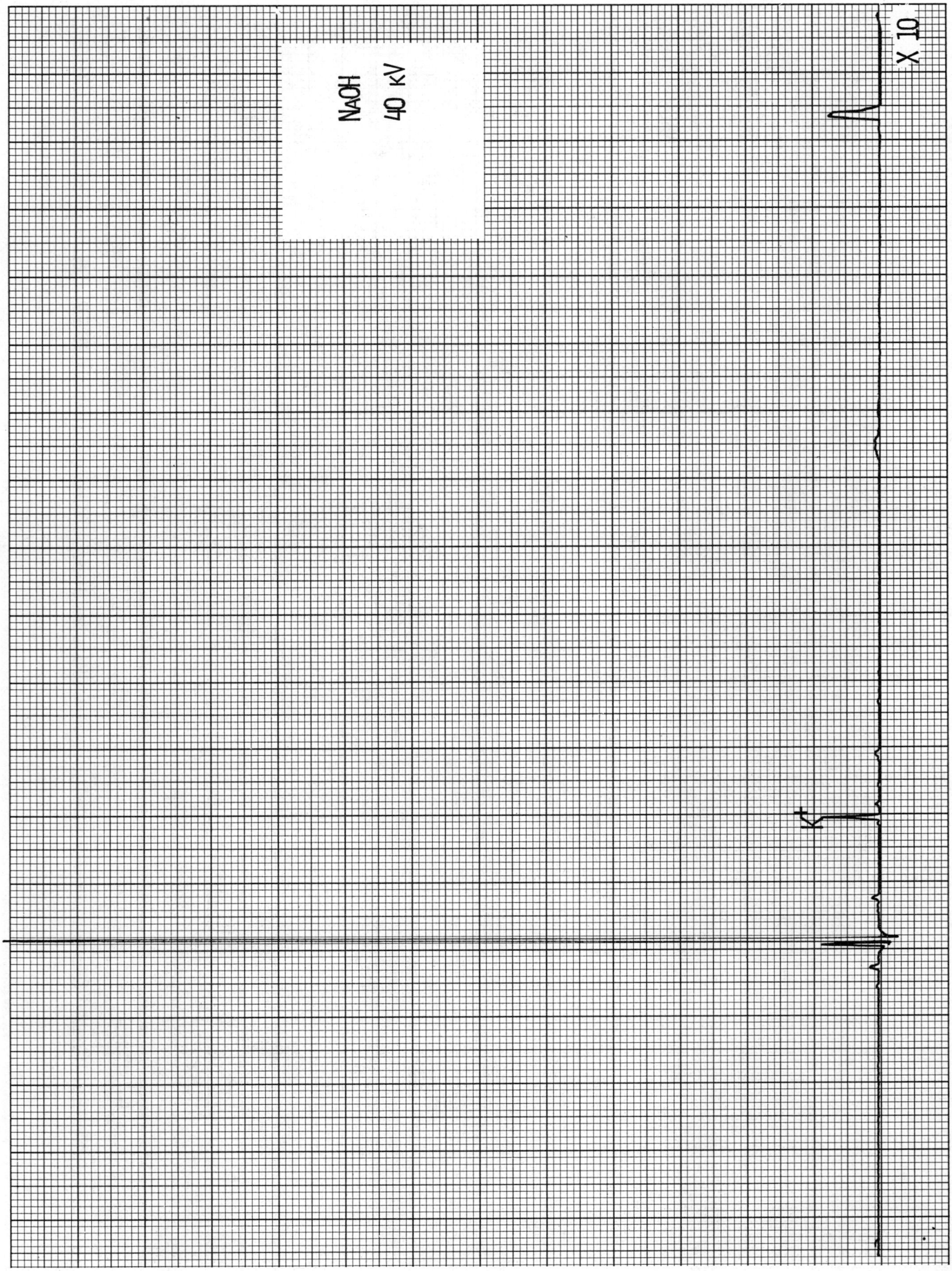
NaOH
40 kV
X 10
K

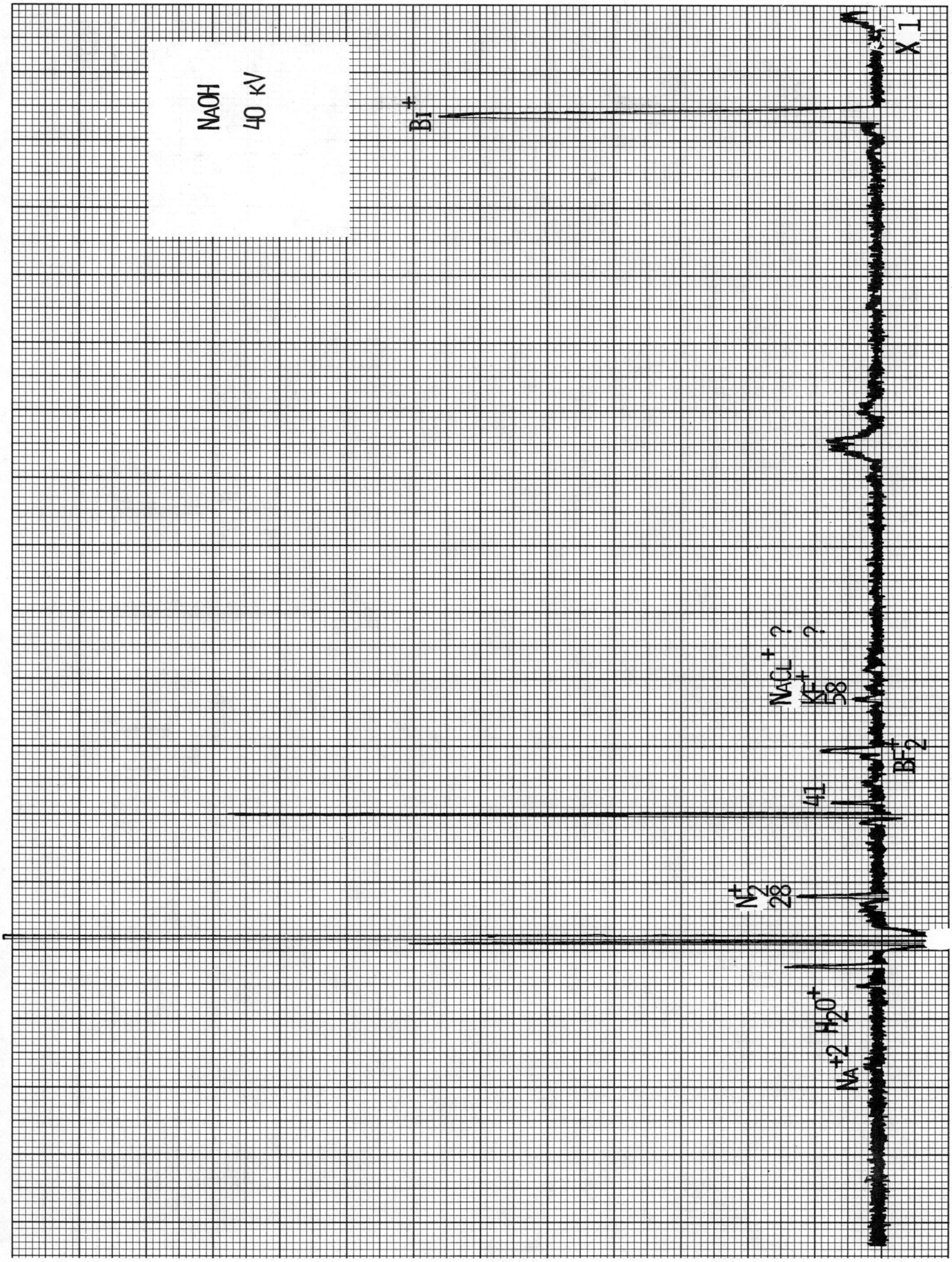
NaOH
40 kV
Bi+
X 1
NaCl+ ?
?
58
Bi2+
41
N2+
28
H2O+
Na+2

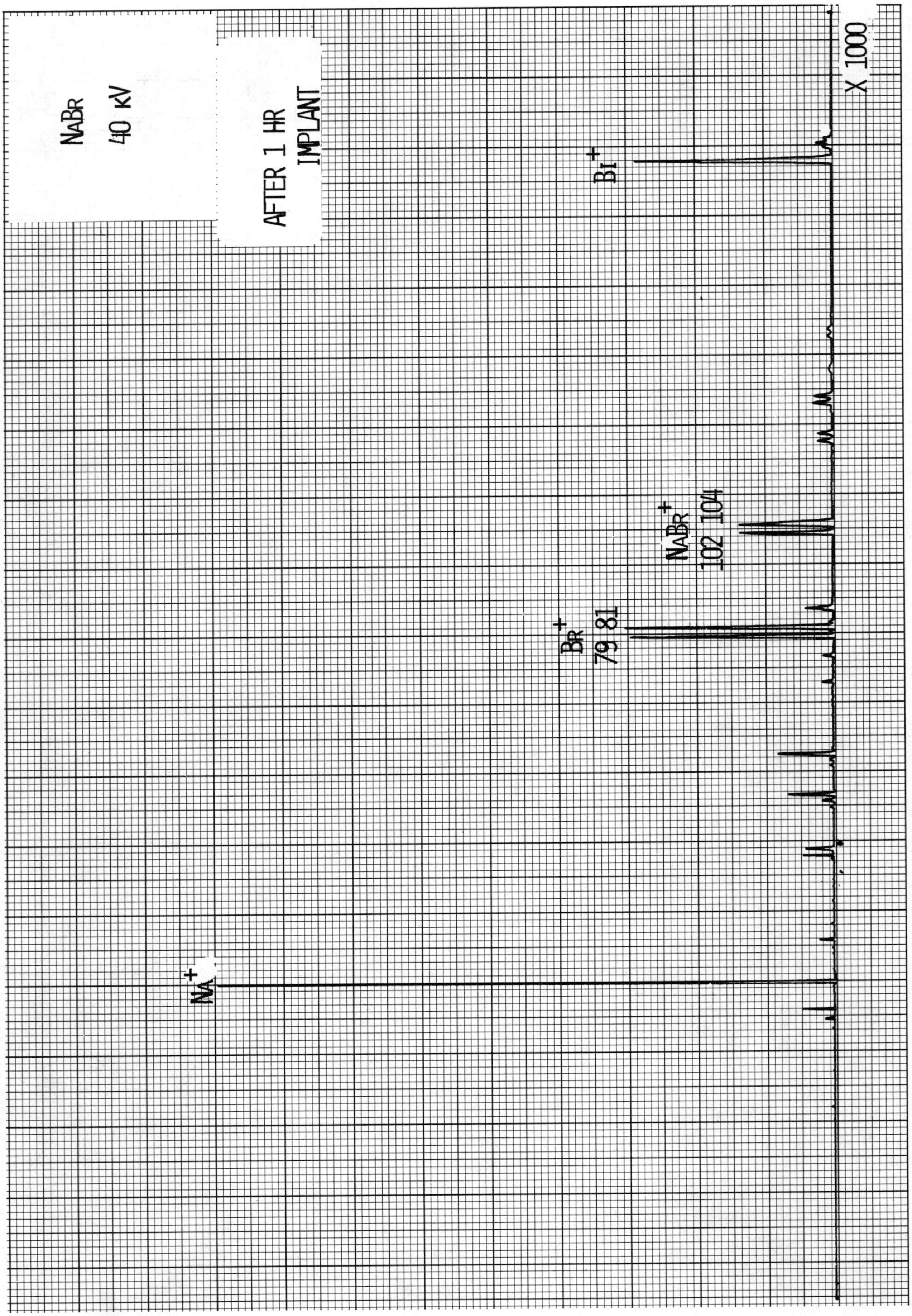
NaBr
40 kV
AFTER 1 HR
IMPLANT
X 1000
Bi+
NaBr+
102 104
Br+
79 81
Na+

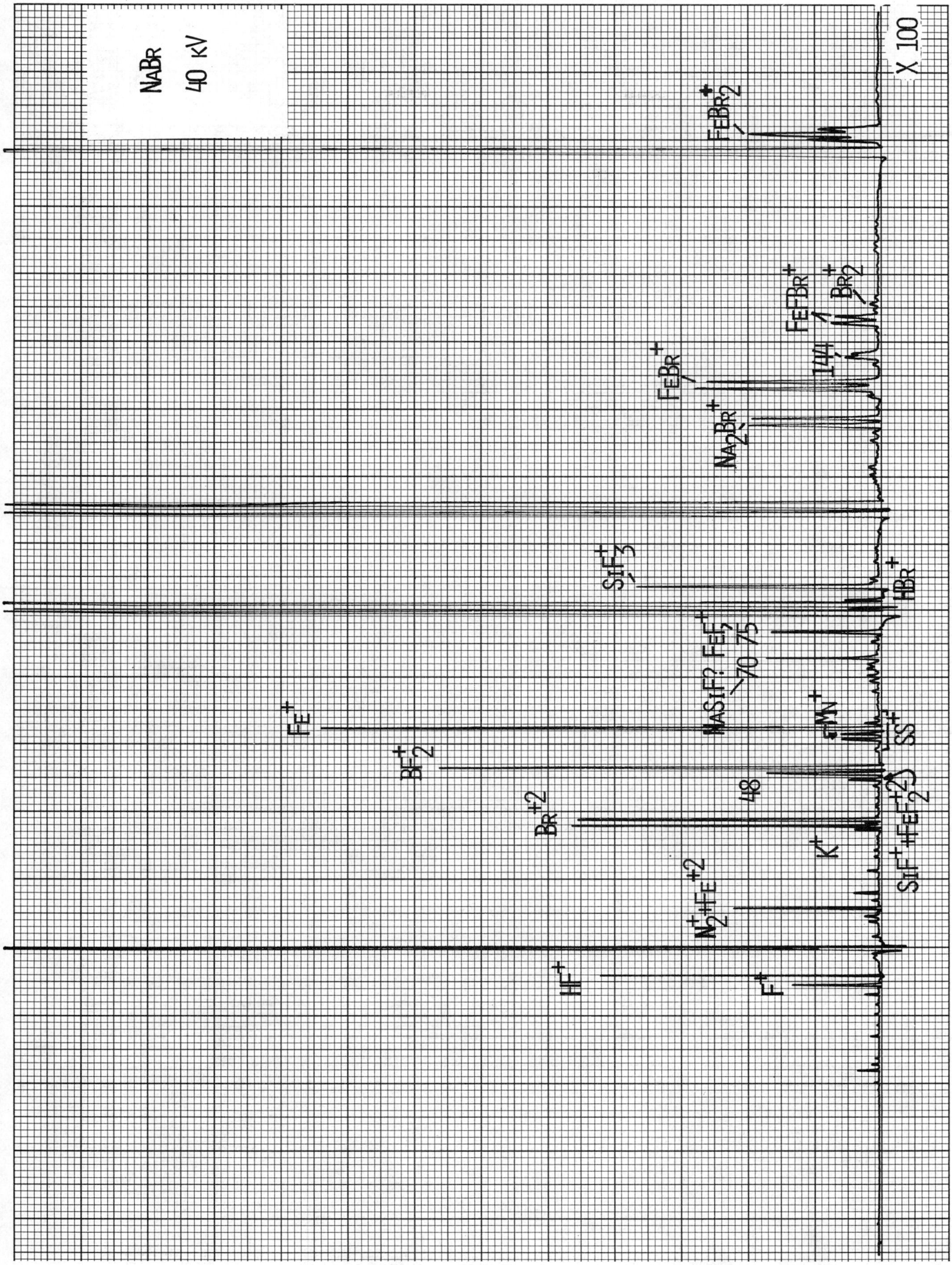
NaBr
40 kV
X 100
$FeBr_2^+$
$FeFBr^+$
Br_2^+
144
$FeBr^+$
Na_2Br^+
SiF_3^+
HBr^+
NaSiF? FeF^+
70 75
Fe^+
Mn^+
SS^+
BF_2^+
48
Br^{+2}
$SiF^+ + FeF_2^{+2}$
K^+
$N_2^+ + Fe^{+2}$
HF^+
F^+

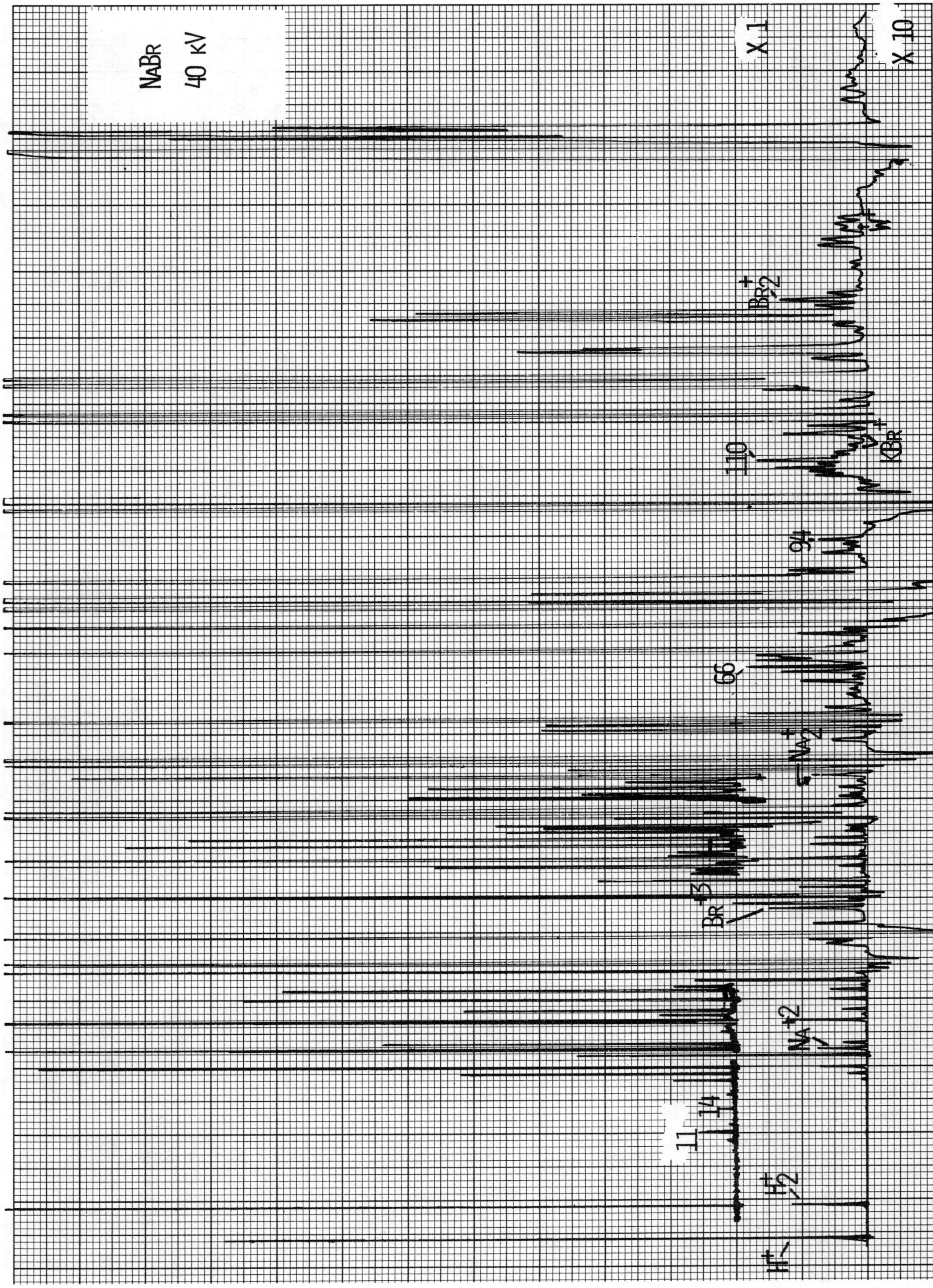
NaBr
40 kV
X 1
X 10
Br_2^+
110
KBr^+
94
66
Na_2^+
Br^{+3}
Na^{+2}
14
11
H_2^+
H^+

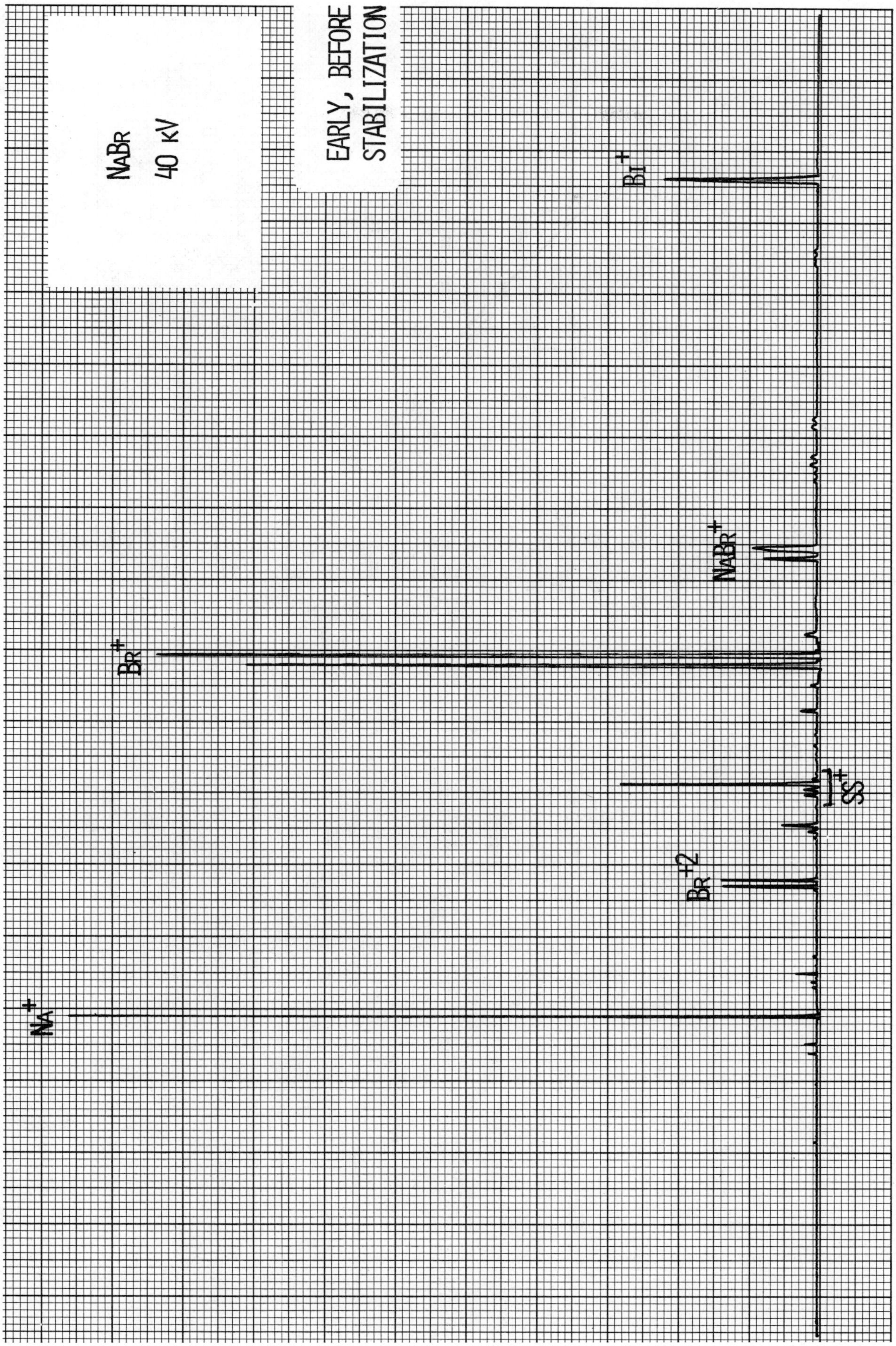

NaBr
40 kV
EARLY, BEFORE
STABILIZATION
Bi+
NaBr+
Br+
SS+
Br+2
Na+

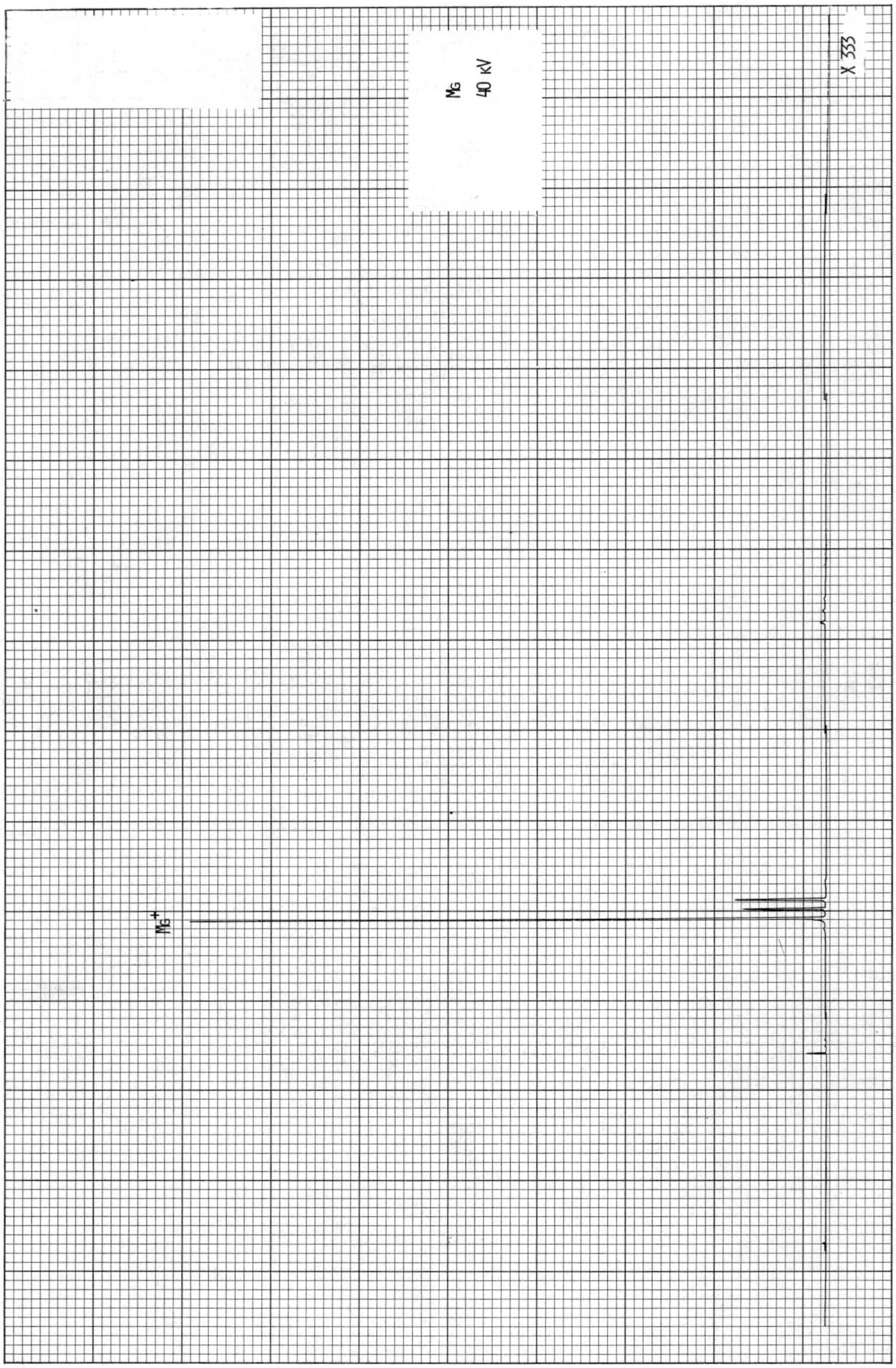
Mg
40 kV
X 333
Mg+

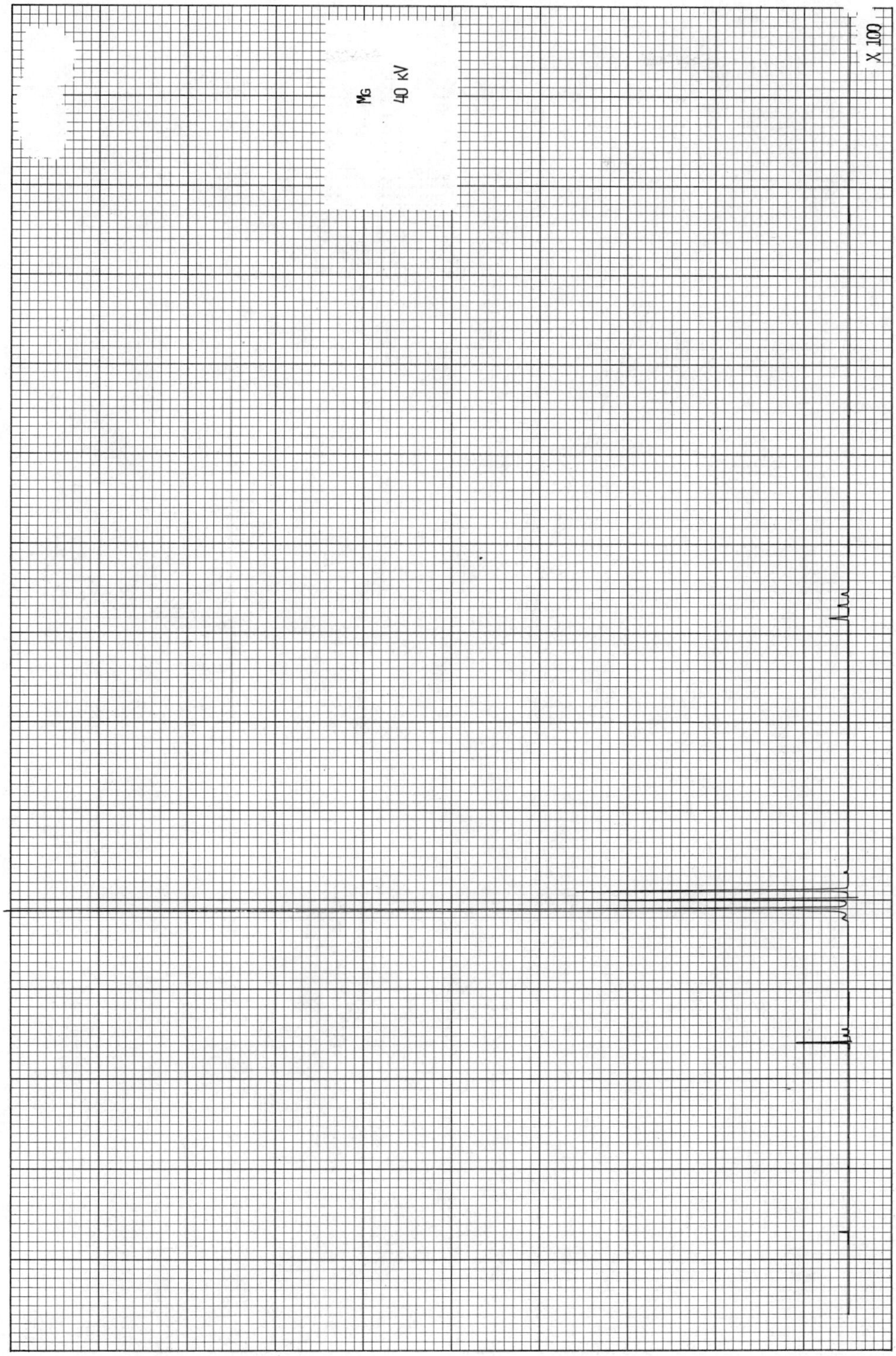
Mg
40 kV
X 100

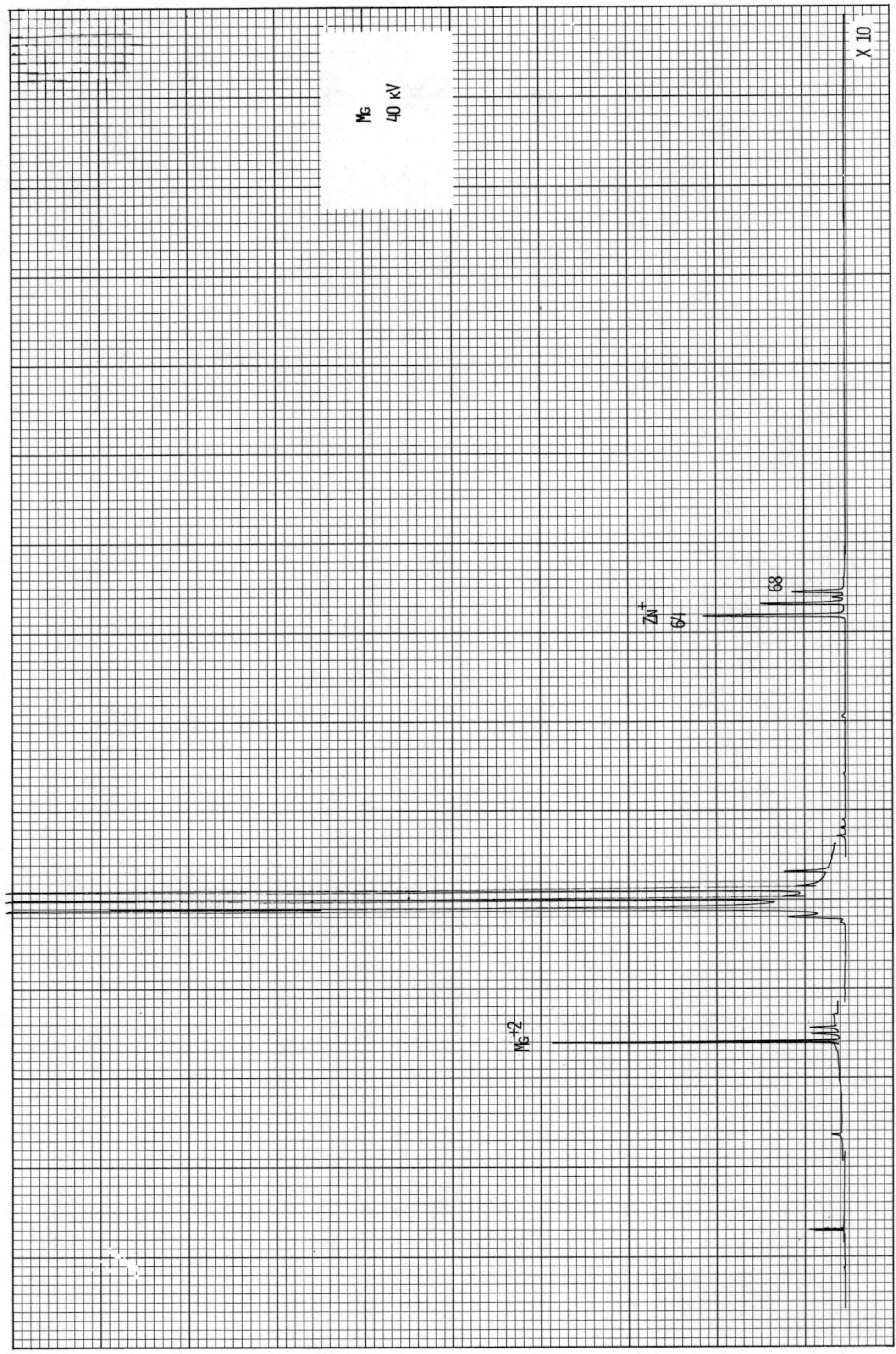
Mg
40 kV
X 10
Zn+
64
68
Mg+2

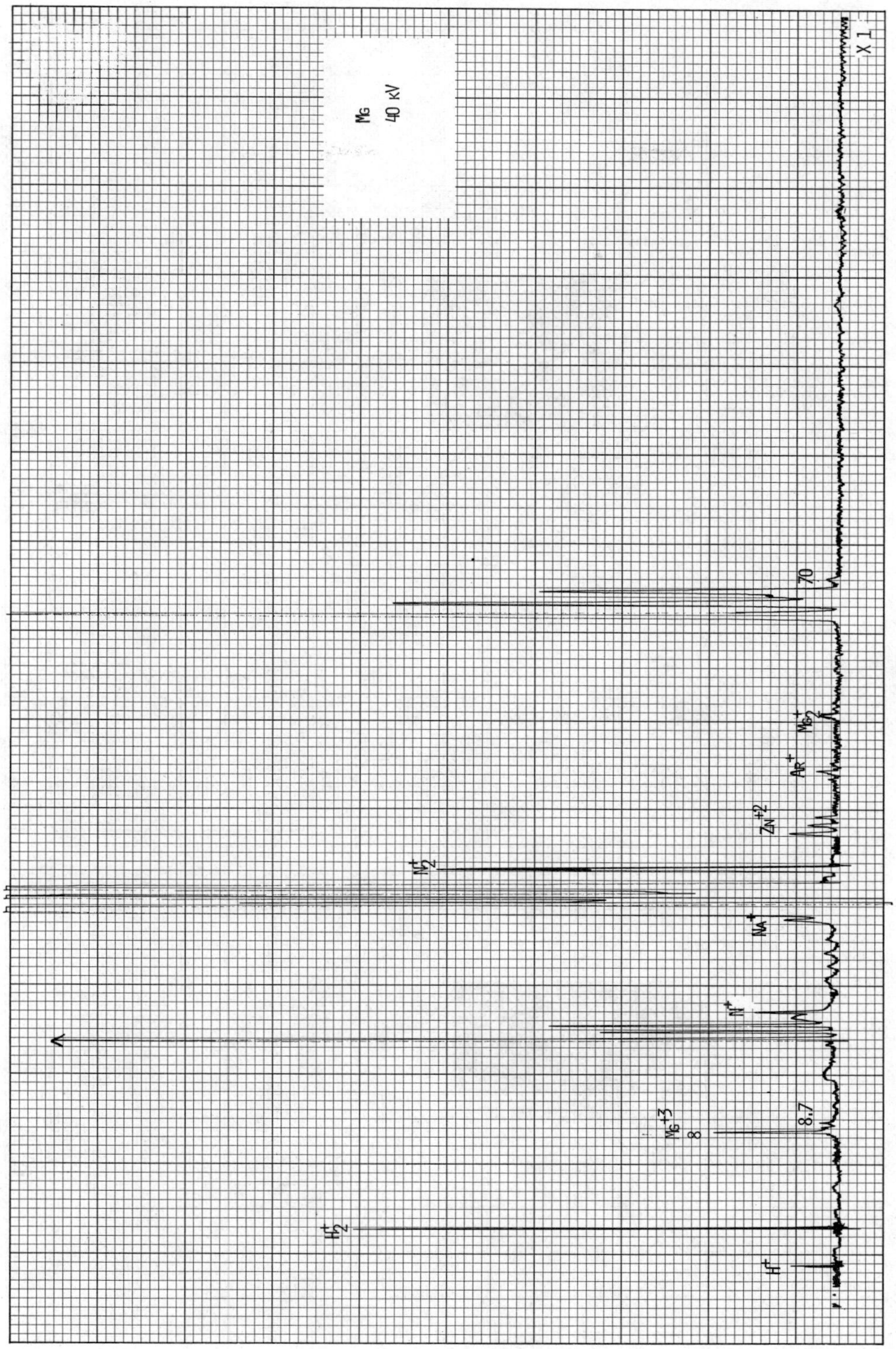
Mg
40 kV
X 1
70
Mg_2^+
Ar^+
Zn^{+2}
N_2^+
Na^+
N^+
Mg^{+3}
8
8.7
H_2^+
H^+

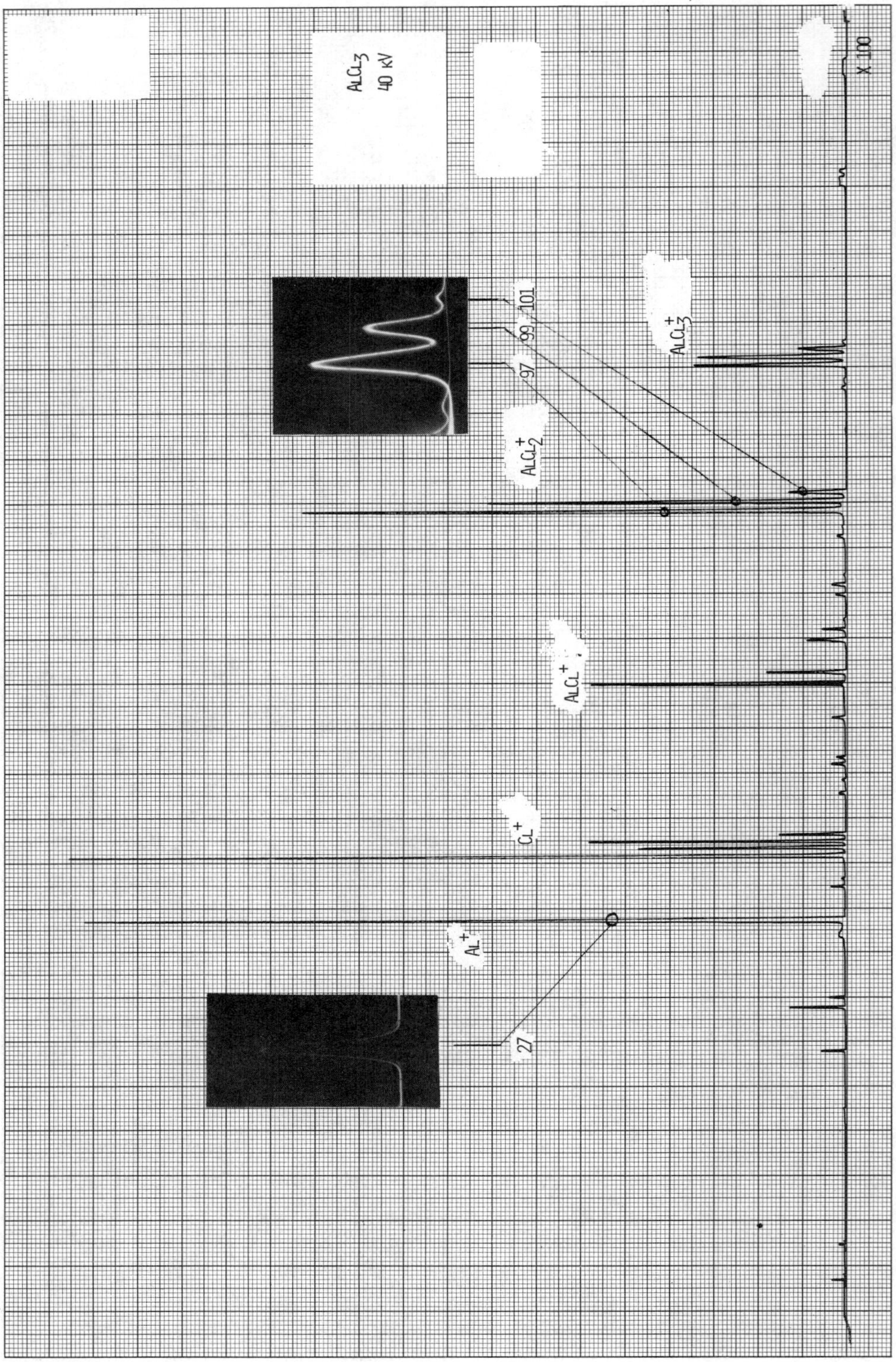
ALCL3
40 kV
X 100
ALCL3+
97
99
101
ALCL2+
ALCL+
CL+
AL+
27

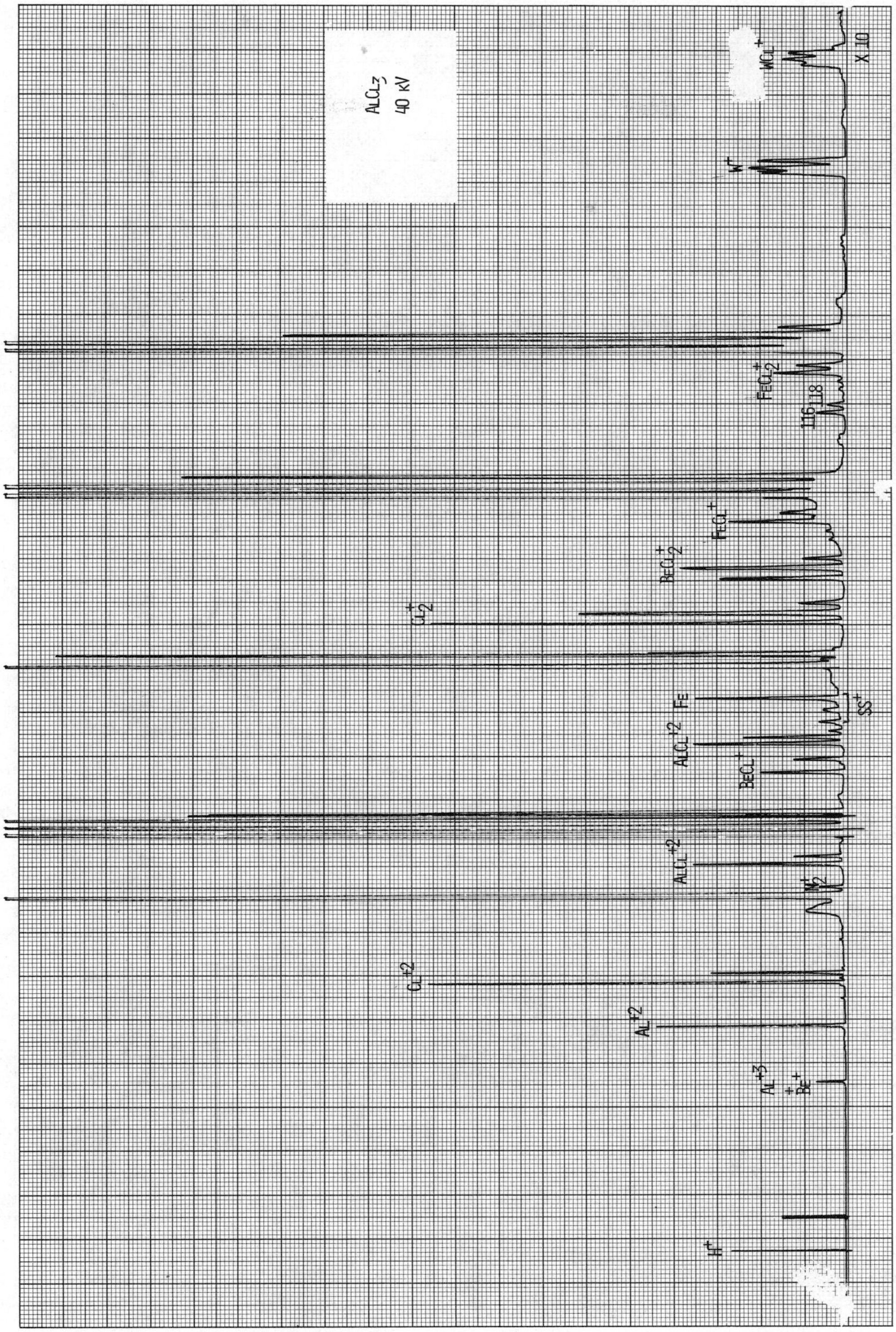

AlCl3
40 kV
H+
Al+3
Be+
Al+2
Cl+2
N2+
AlCl+2
BeCl+
AlCl+2
Fe
SS+
Cl2+
BeCl2+
FeCl+
116
118
FeCl2+
W+
WCl+
X 10

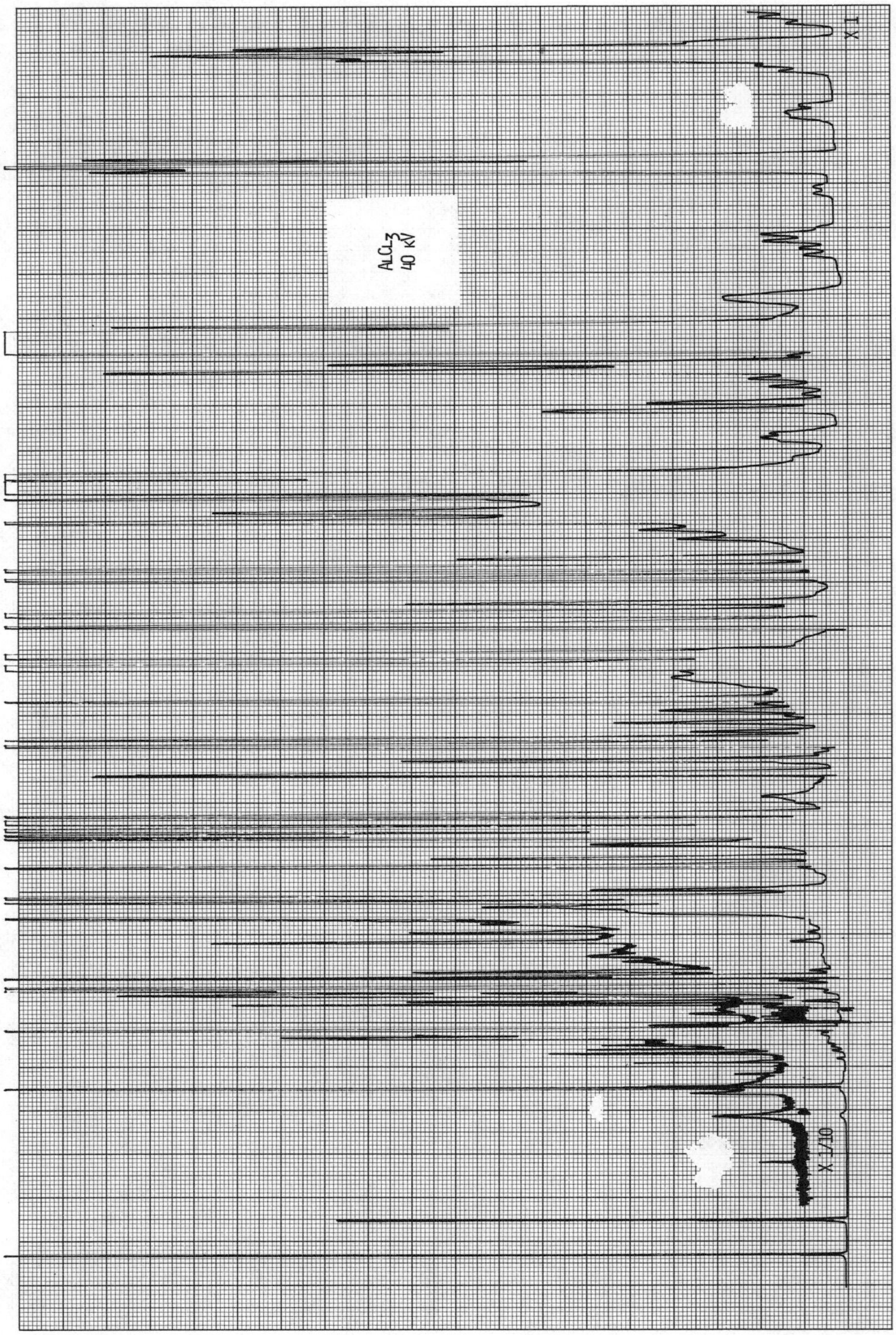
AlCl3
40 kV
X 1
X 1/10

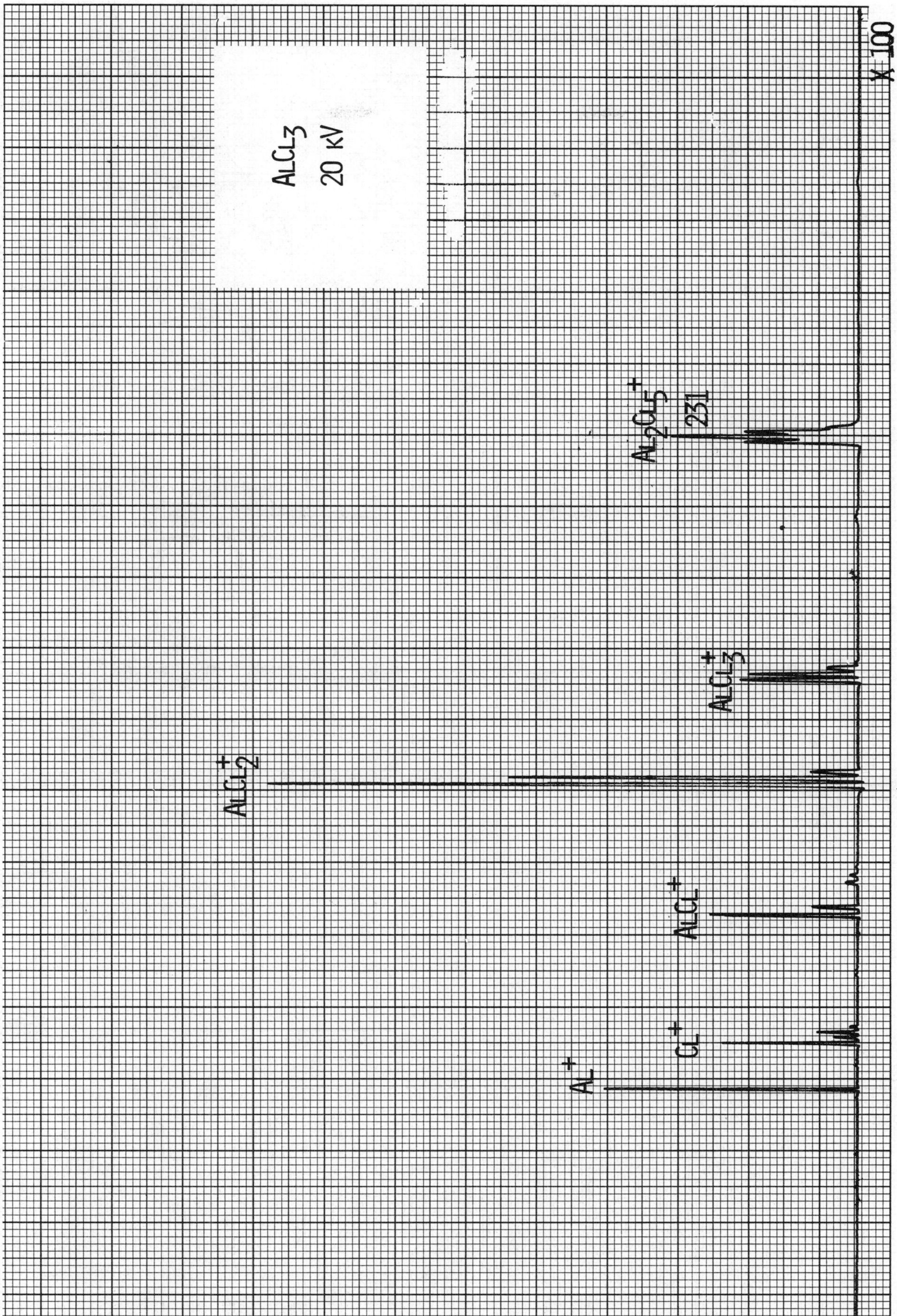
AlCl3
20 kV
X 100
Al2Cl5+
231
AlCl3+
AlCl2+
AlCl+
Cl+
Al+

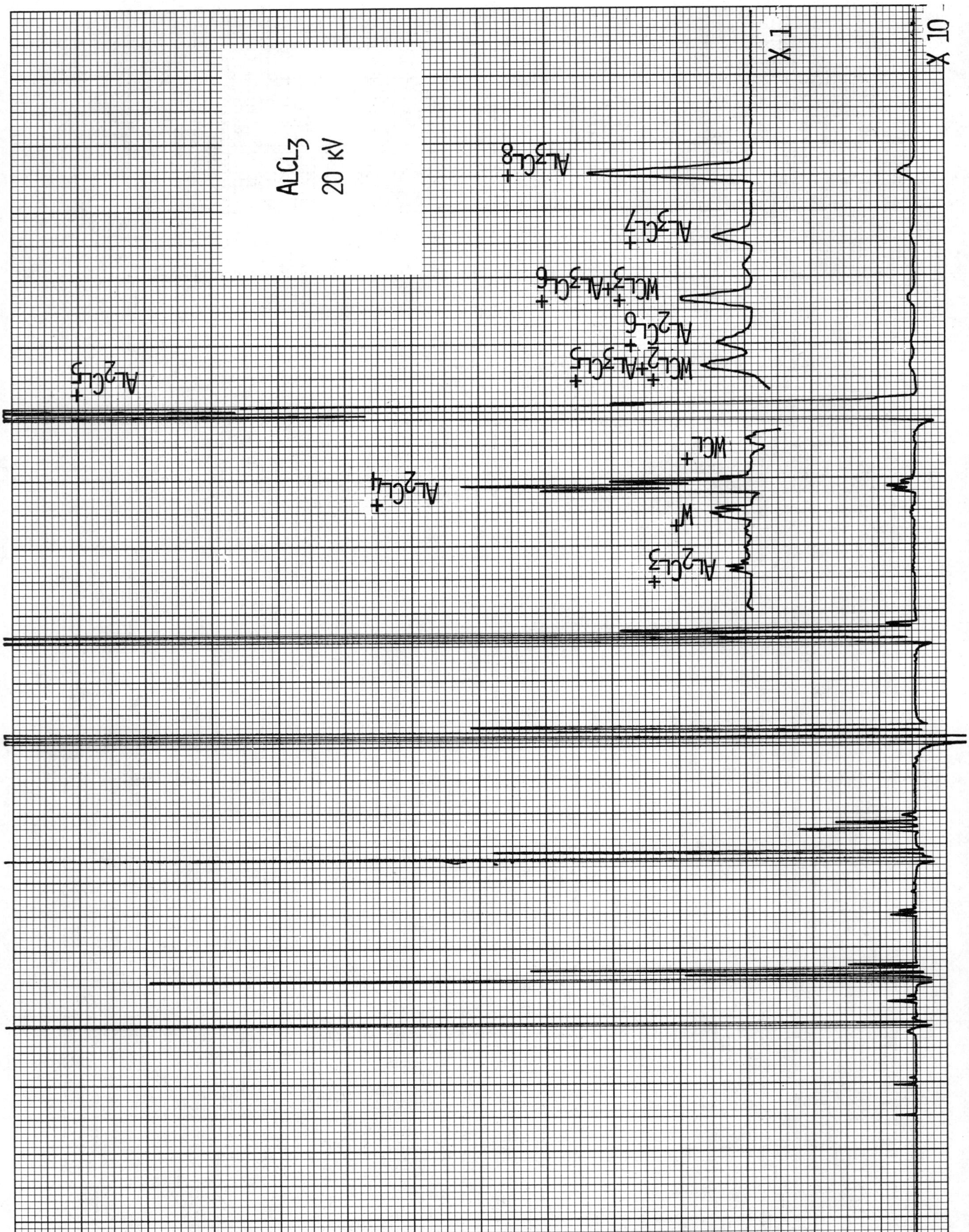

$AlCl_3$
20 kV
X 1
X 10
$Al_3Cl_8^+$
$Al_3Cl_7^+$
$WCl_3^+ + Al_3Cl_6^+$
$Al_2Cl_6^+$
$WCl_2^+ + Al_3Cl_5^+$
$Al_2Cl_5^+$
WCl^+
$Al_2Cl_4^+$
W^+
$Al_2Cl_3^+$

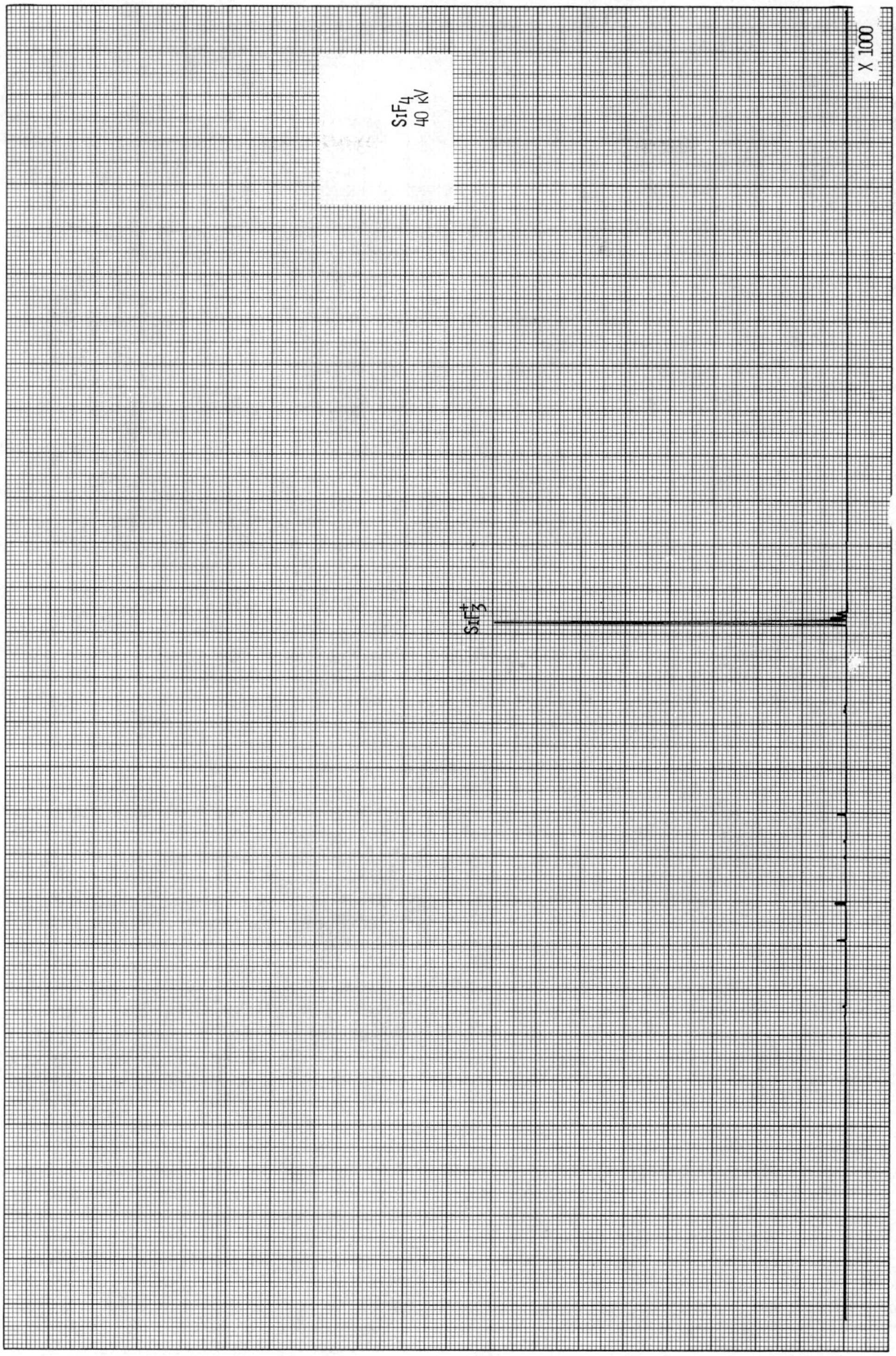

SiF$_4$
40 kV
SiF$_3^+$
X 1000

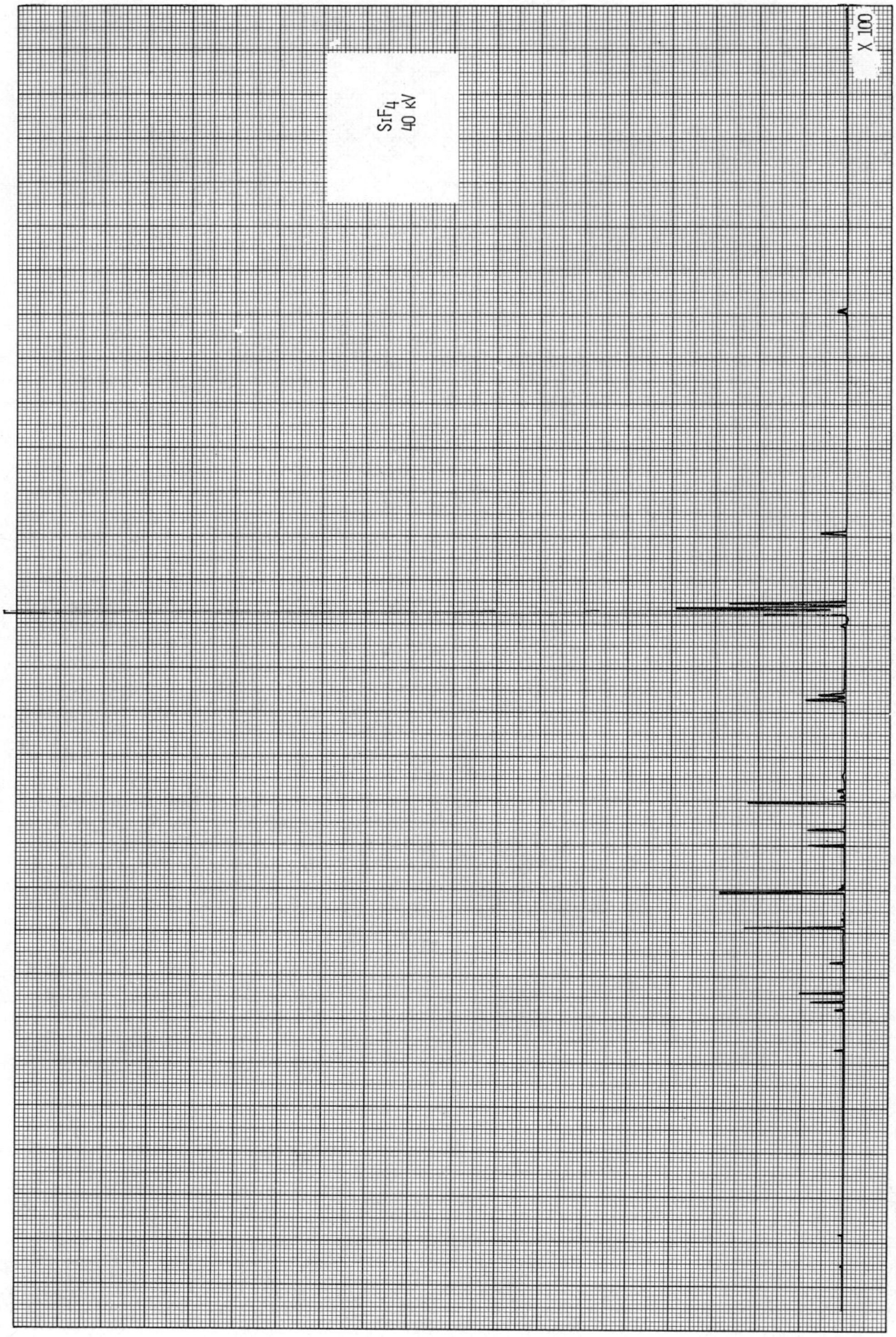
SiF$_4$
40 kV
X 100

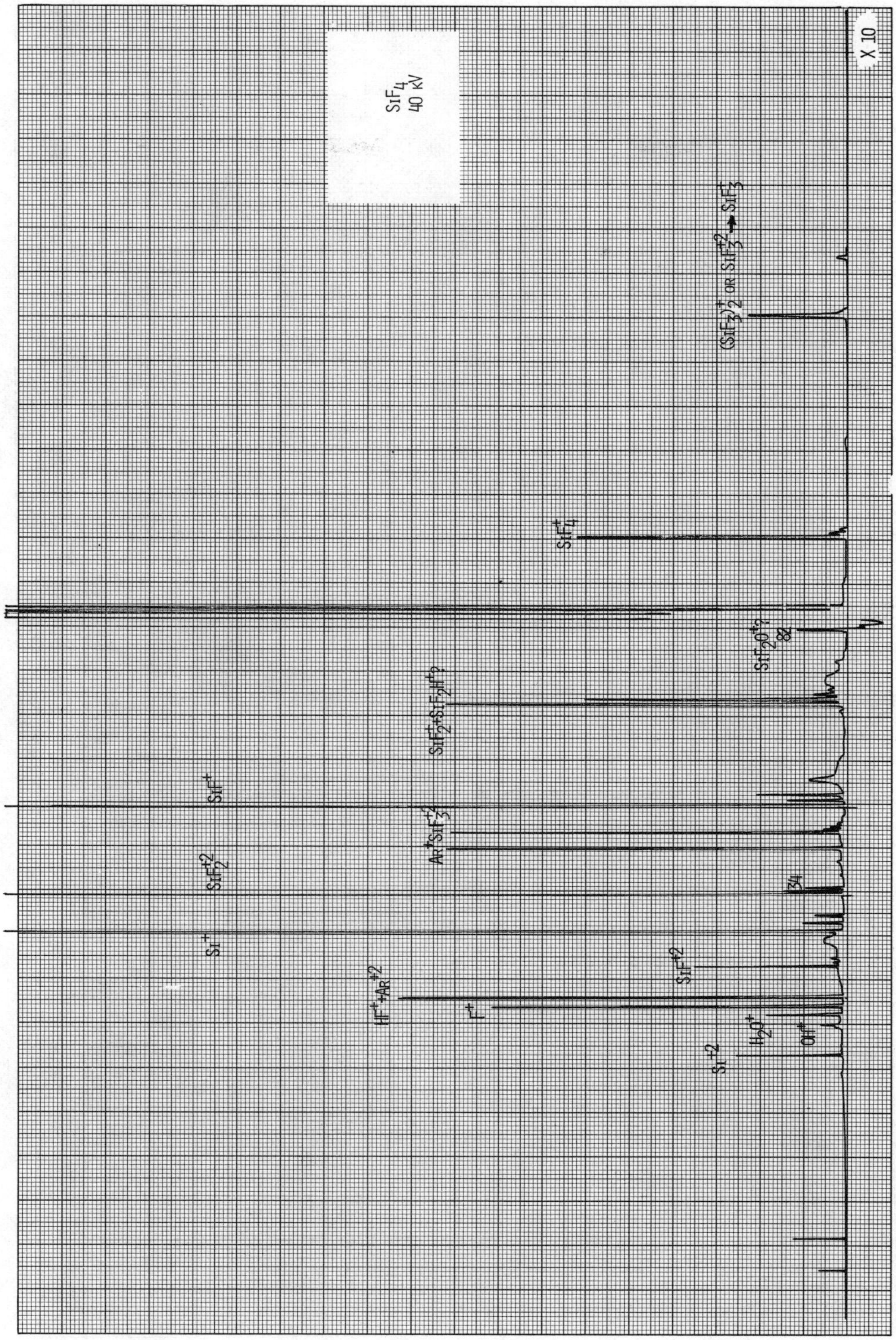
SiF_4
40 kV
X 10
$(SiF_3)_2^+$ OR $SiF_3^{+2} \rightarrow SiF_3^+$
SiF_4^+
SiF_2O^+?
82
$SiF_2^+ + SiF_2H^+$?
SiF^+
$Ar^+ SiF_3^{+2}$
SiF_2^{+2}
74
Si^+
SiF^{+2}
$HF^+ + Ar^{+2}$
F^+
Si^{+2}
H_2O^+
OH^+

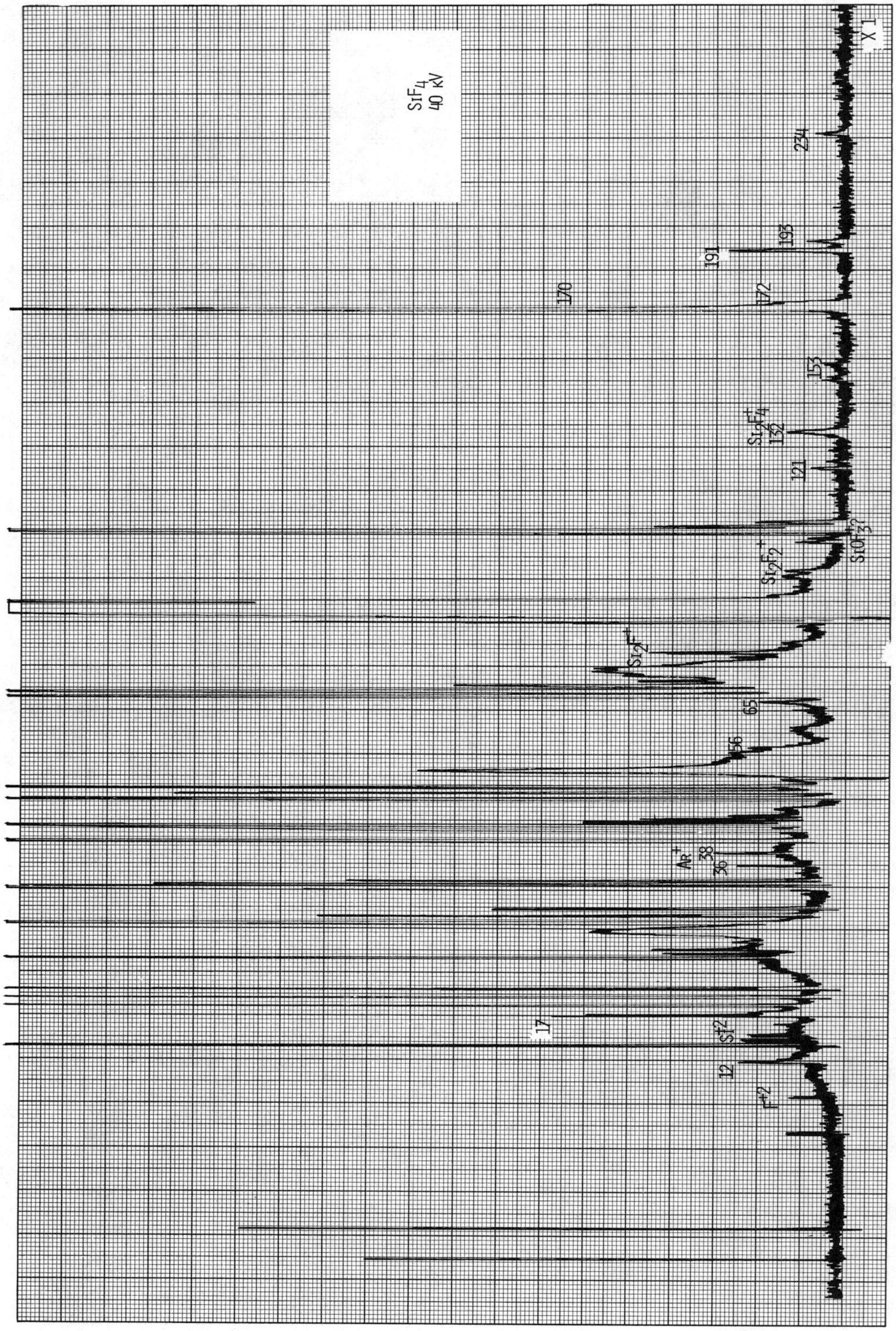

SiF4
40 kV
X 1
234
193
191
170
172
153
132
121
65
56
Ar+
38
36
17
12

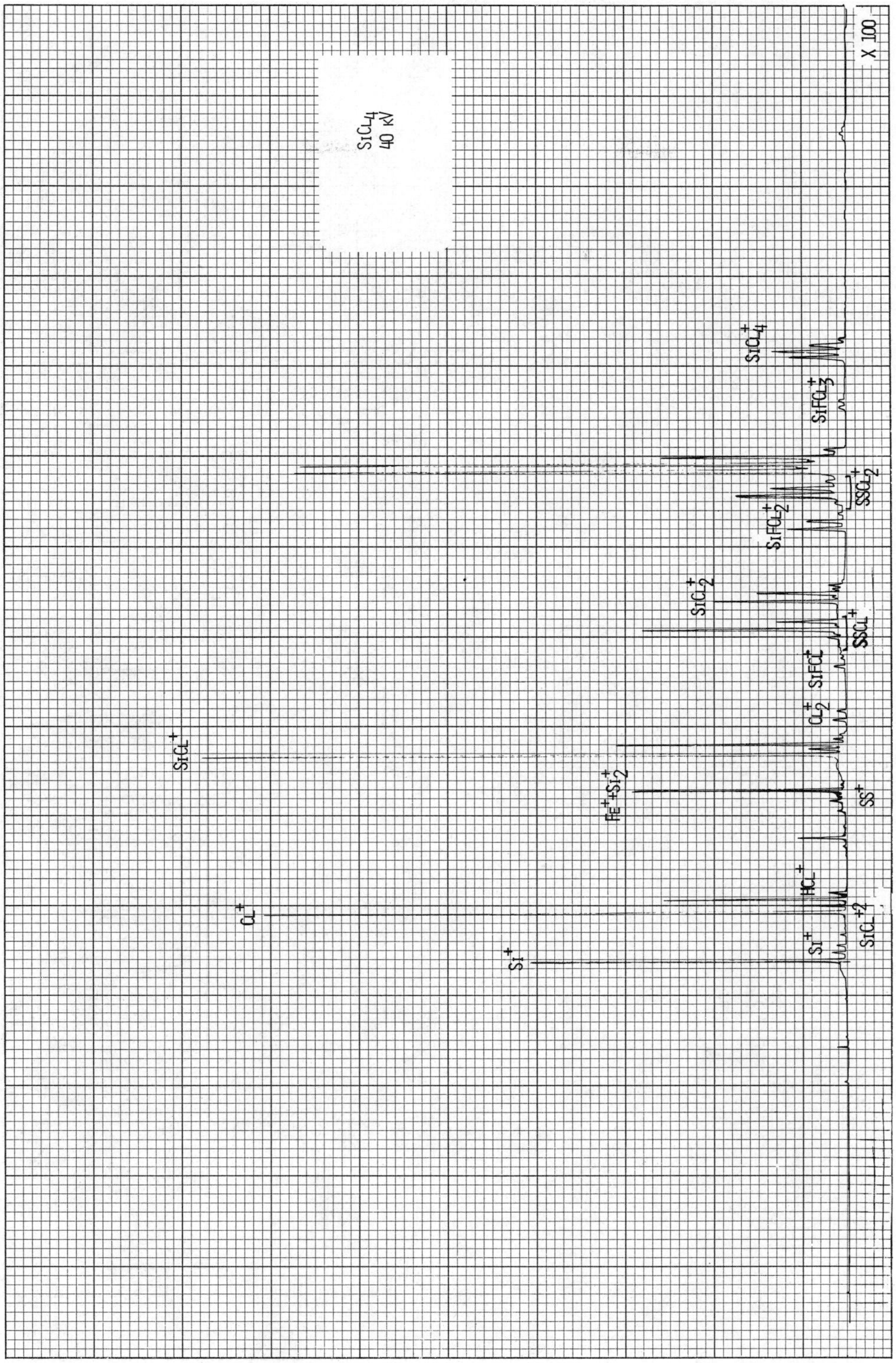
SiCl4
40 kV
X 100
Si+
Cl+
HCl+
SiCl+2
Fe+ + Si2+
SS+
SiCl+
Cl2+
SiFCl+
SSCl+
SiCl2+
SiFCl2+
SSCl2+
SiFCl3+
SiCl4+

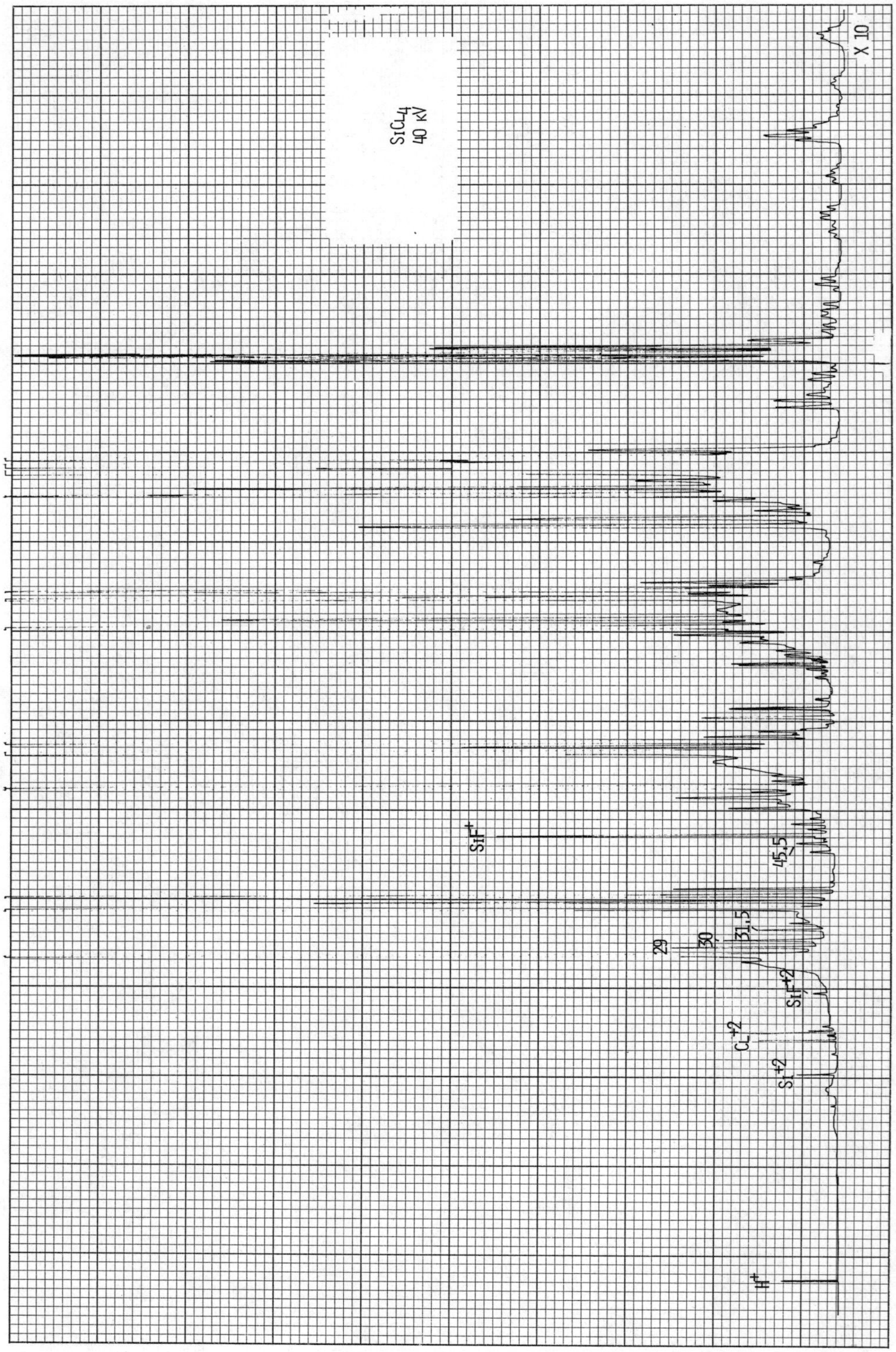
SiCl4
40 kV
X 10
SiF+
45.5
31.5
30
29
SiF+2
Cl+2
Si+2
H+

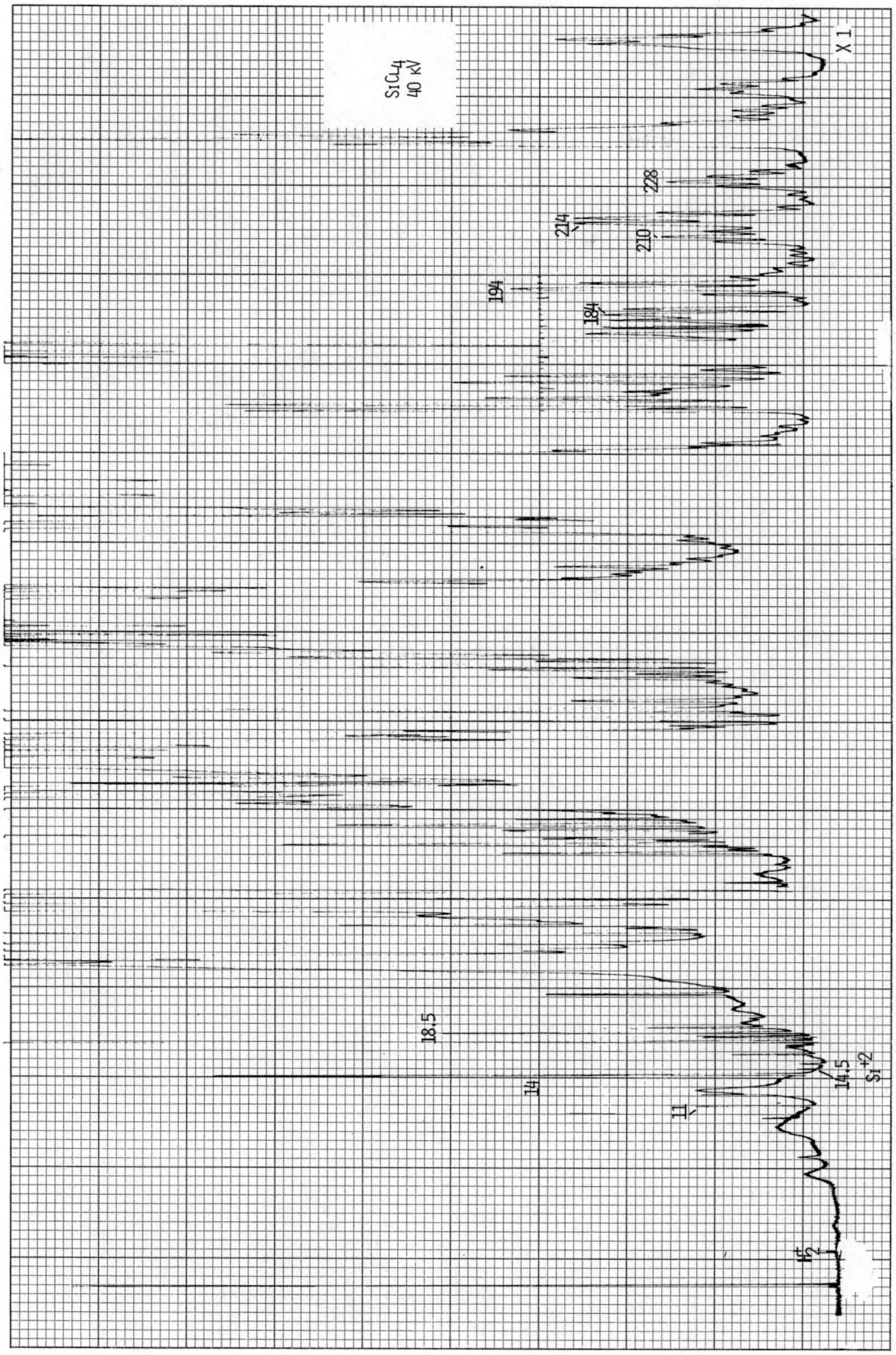
SiCl4
40 kV
X 1
228
214
210
194
184
18.5
14
11
14.5
Si+2

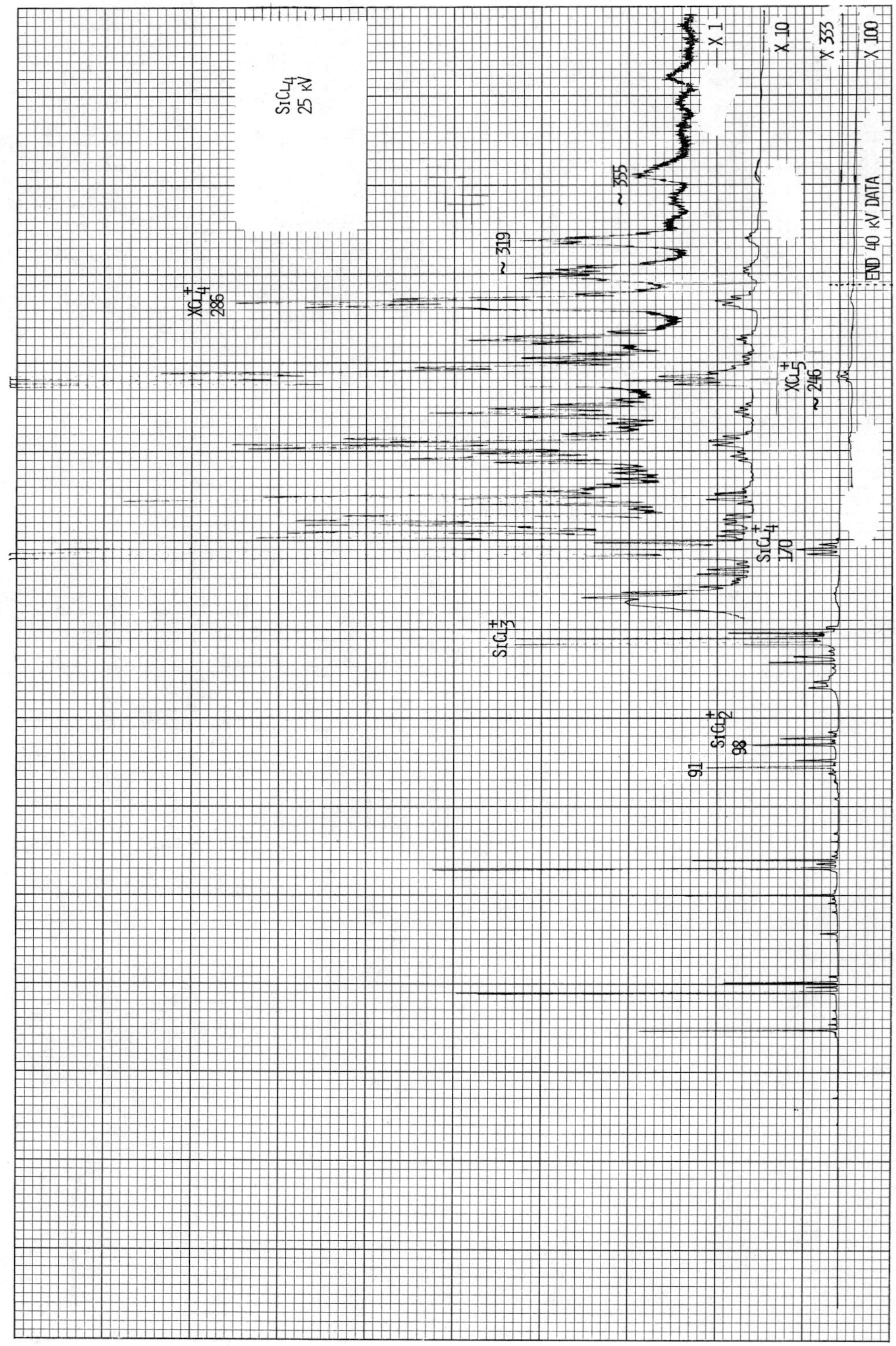
SiCl4
25 kV
X 1
X 10
X 333
X 100
END 40 kV DATA
~ 355
~ 319
XCl4+
286
XCl5+
~ 246
SiCl4+
170
SiCl3+
SiCl2+
98
91

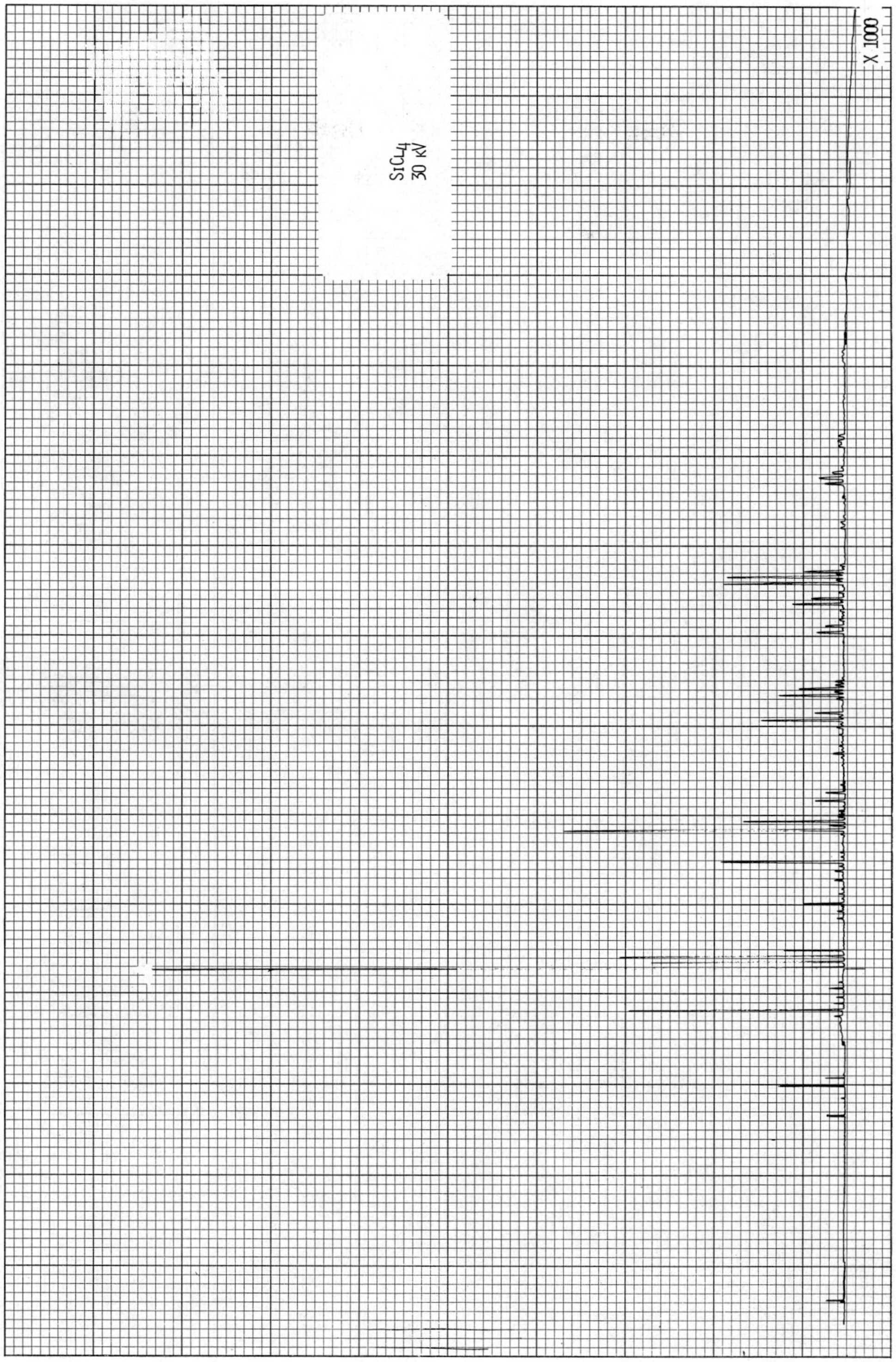
SiCl4
30 kV
X 1000

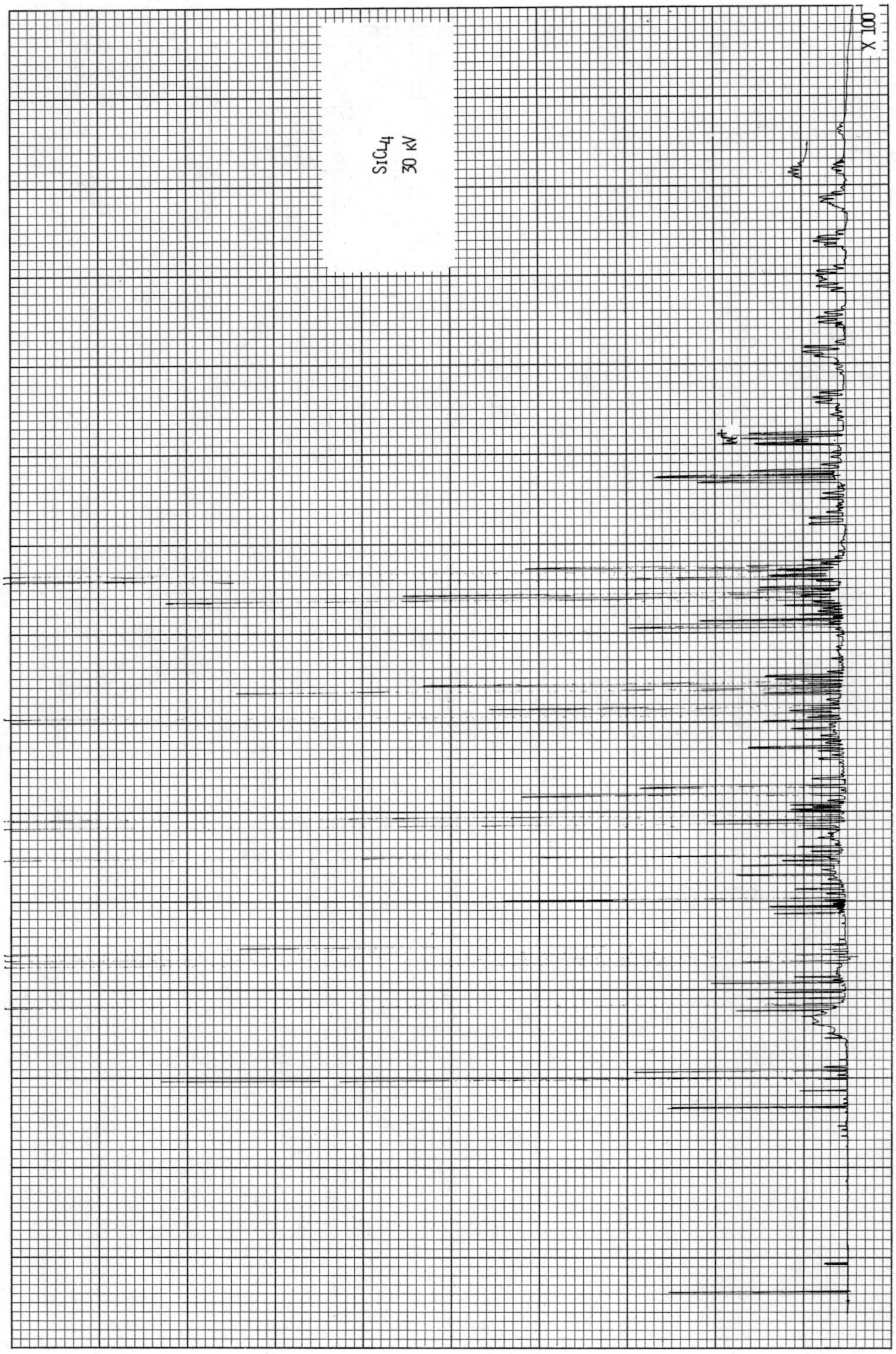
$SiCl_4$
30 kV
X 100

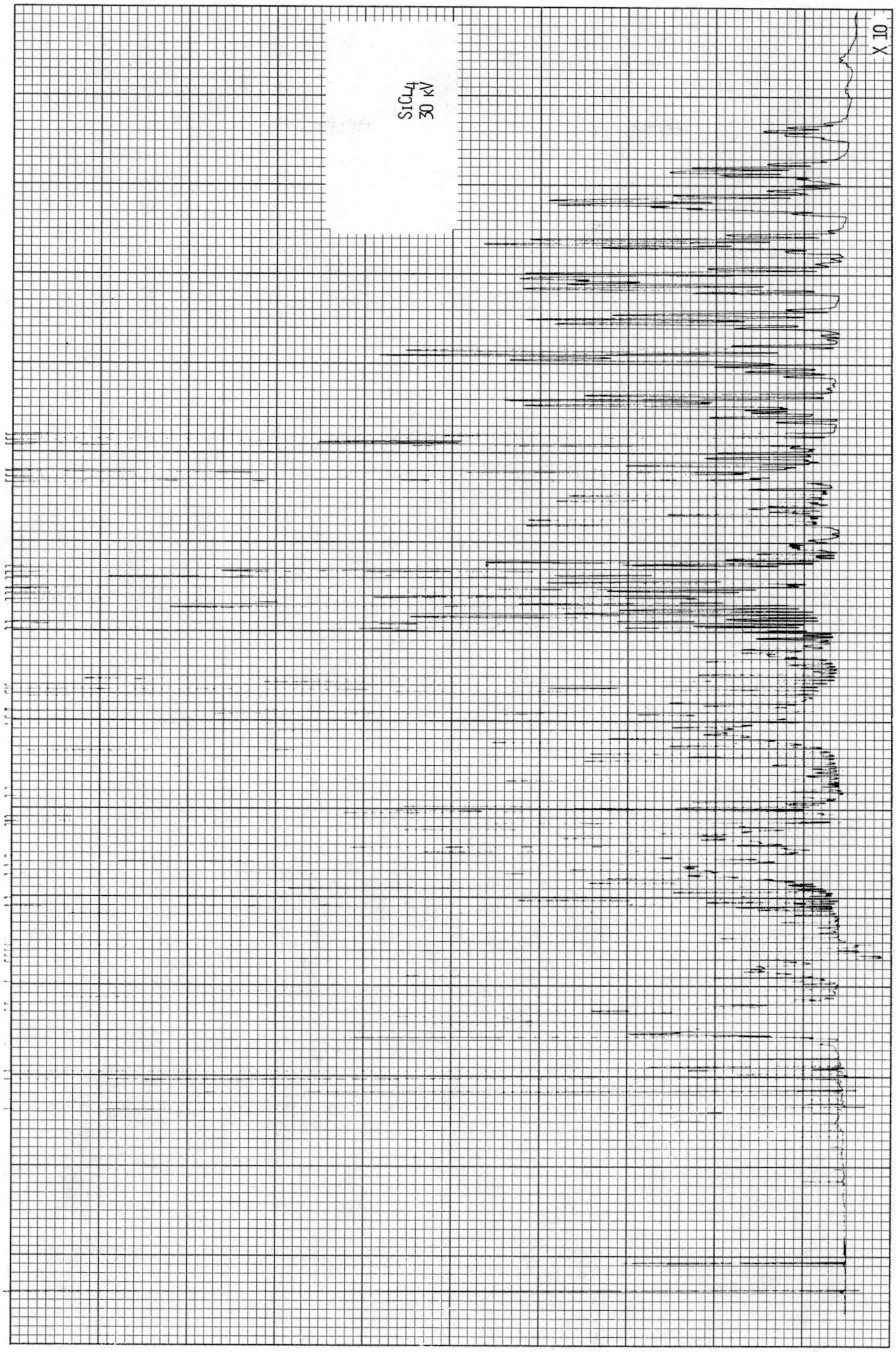

$SiCl_4$
30 kV
X 10

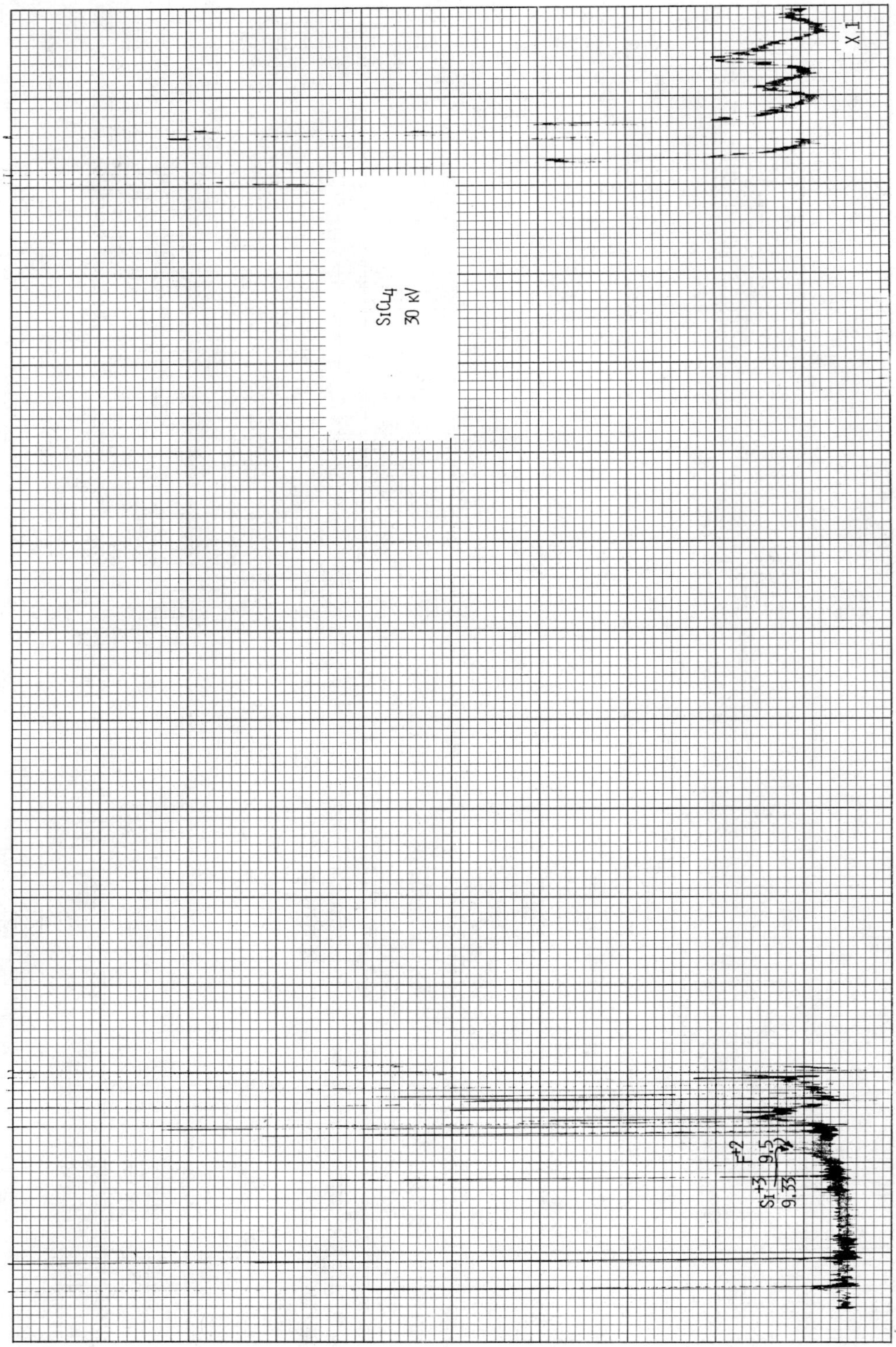
SiCl4
30 kV
X 1
Si+3
9.33
F+2
9.5

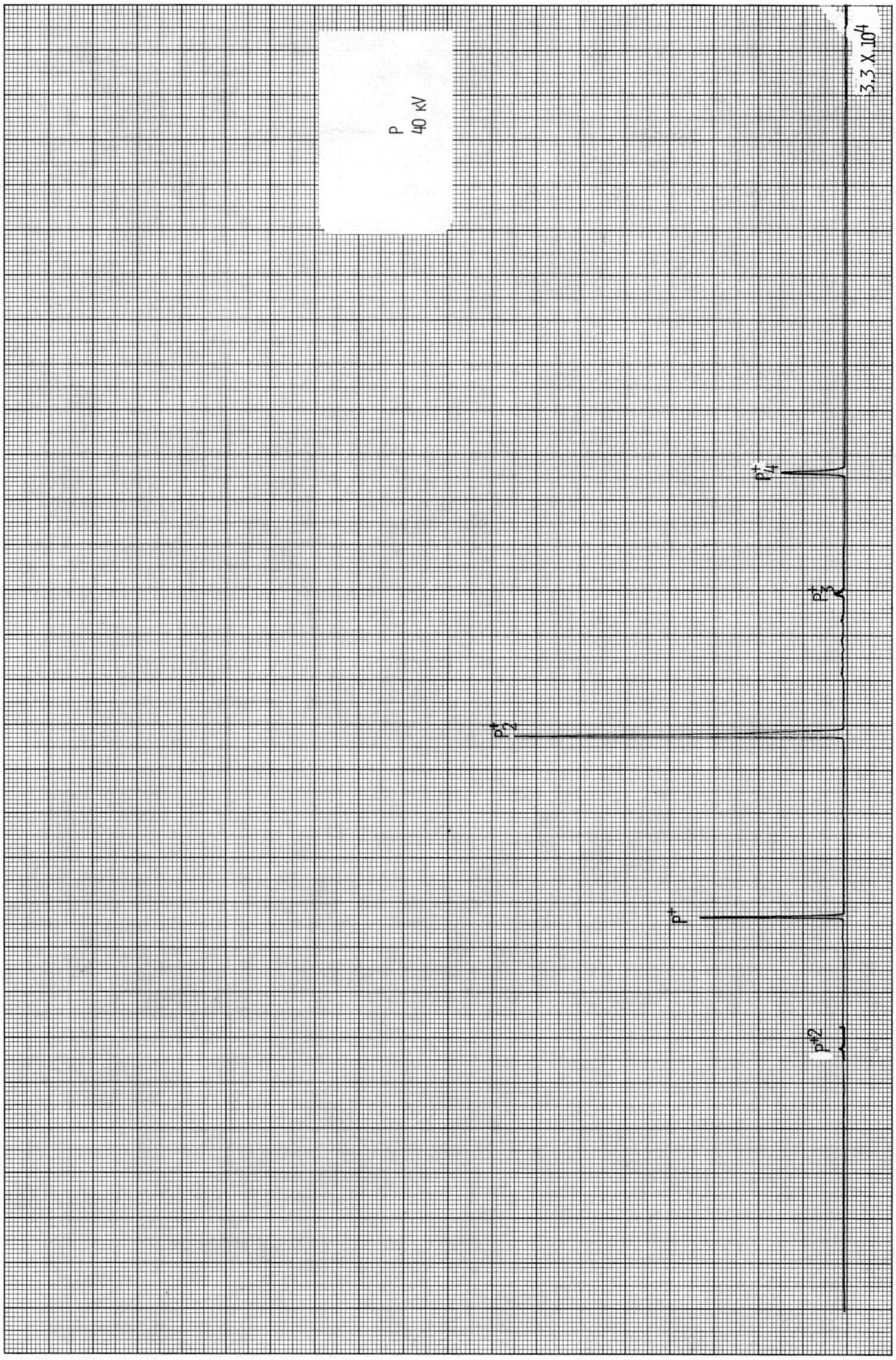
P
40 kV
3,3 X 10^4
P4+
P3+
P2+
P+
P+2

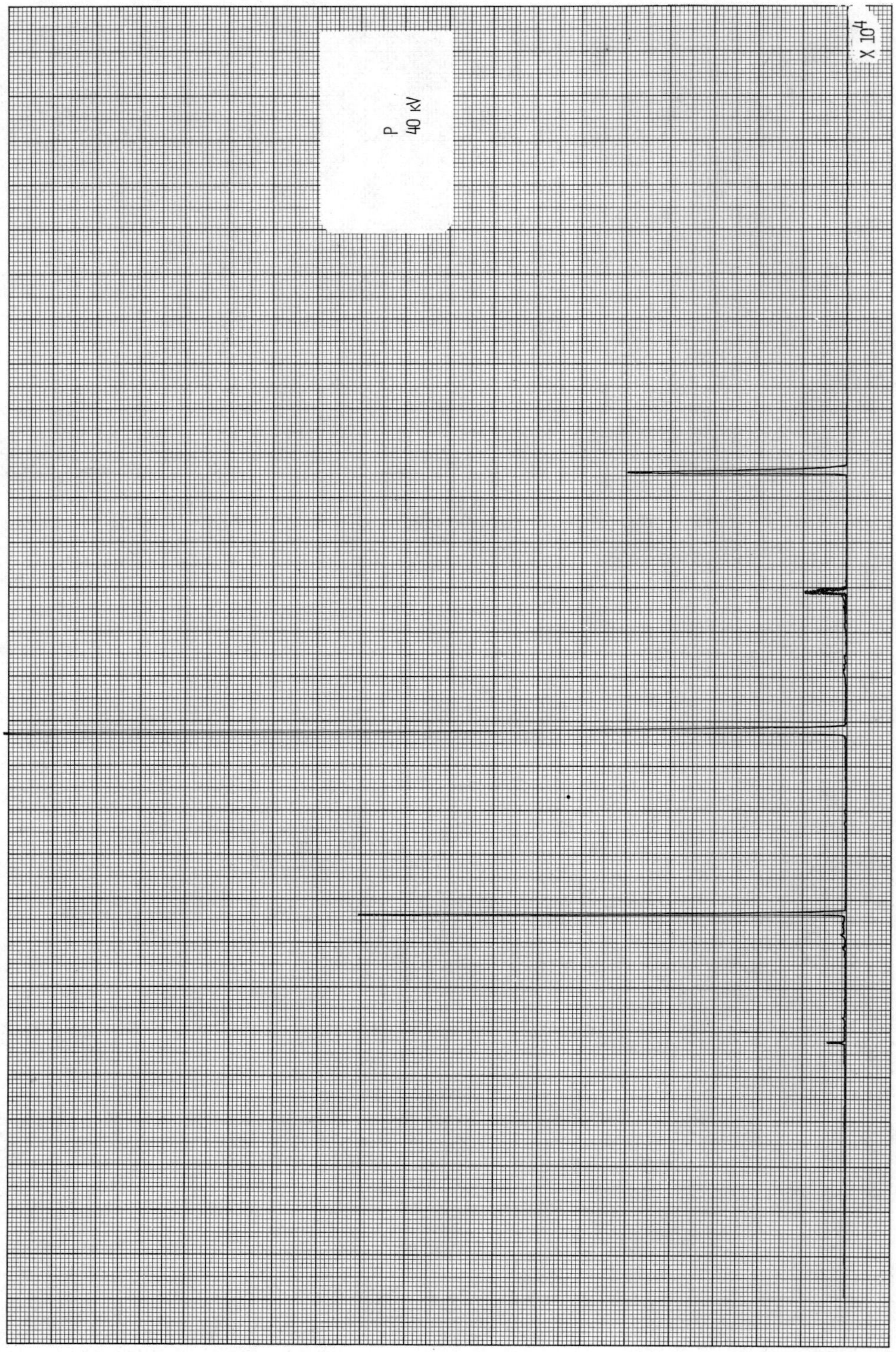
P
40 kV
X 10^4

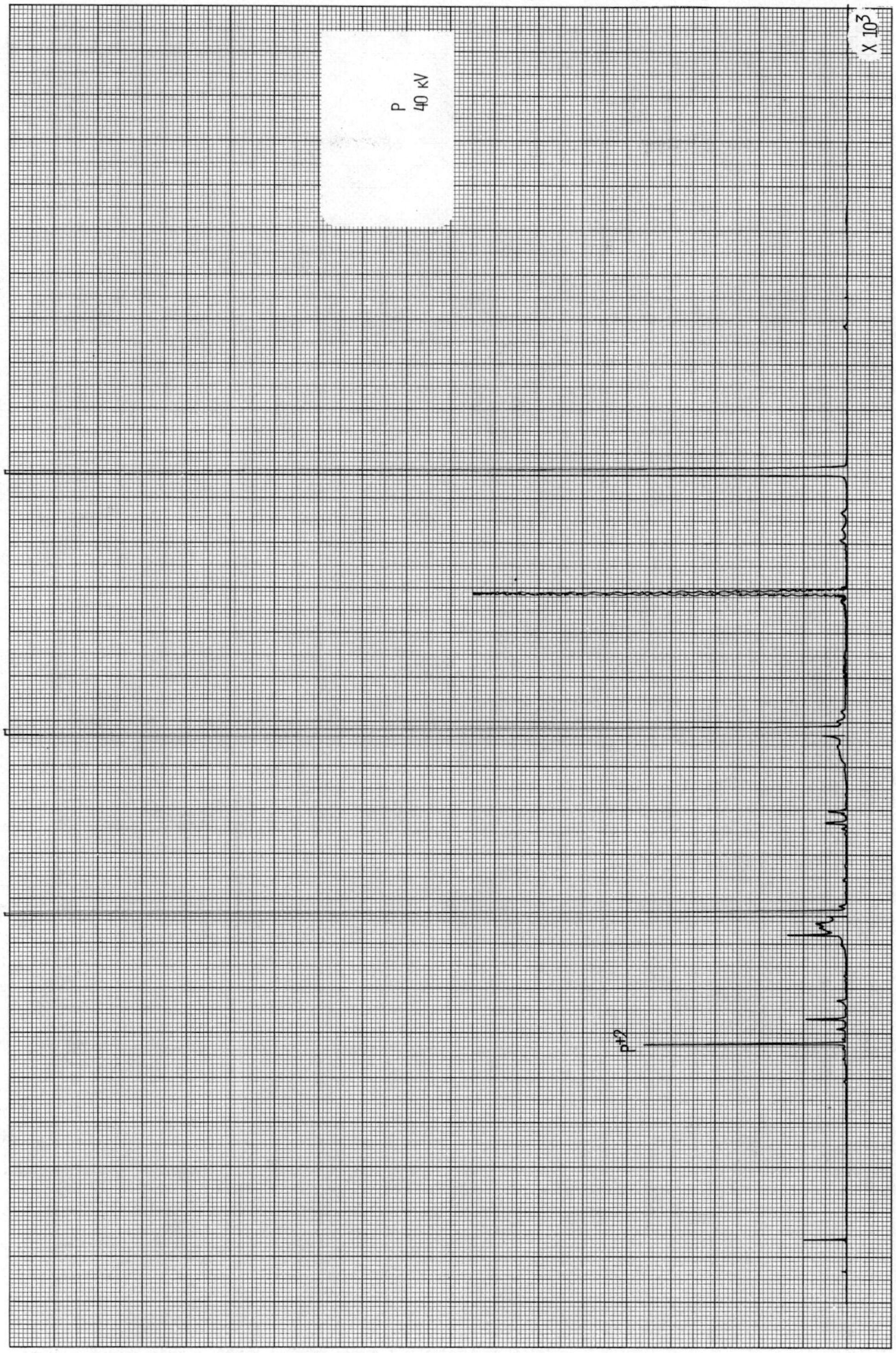
P
40 kV
X 10³
P+2

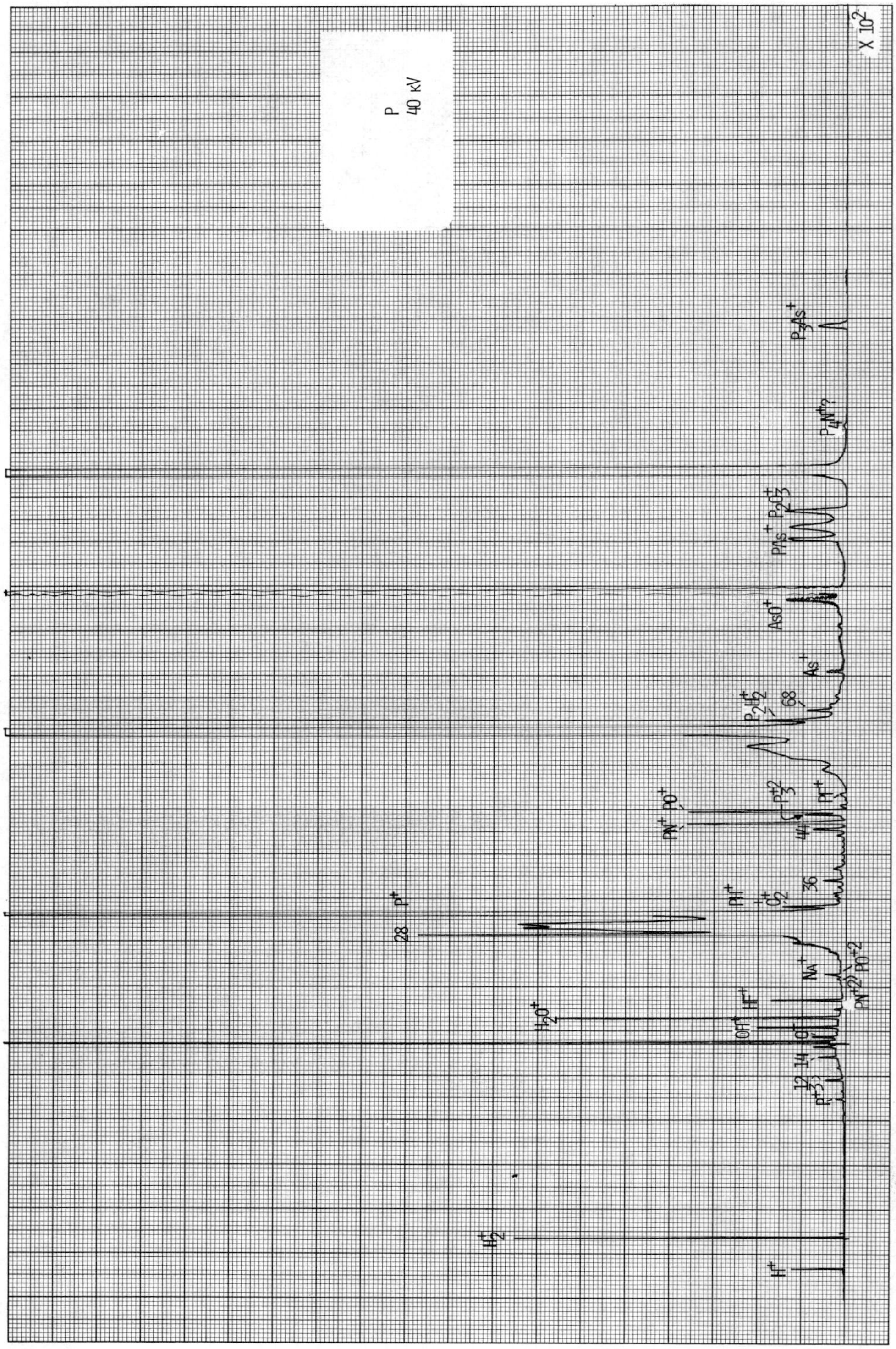
P
40 KV
X 10²
H+
28
P+
As+
AsO+
P3As+

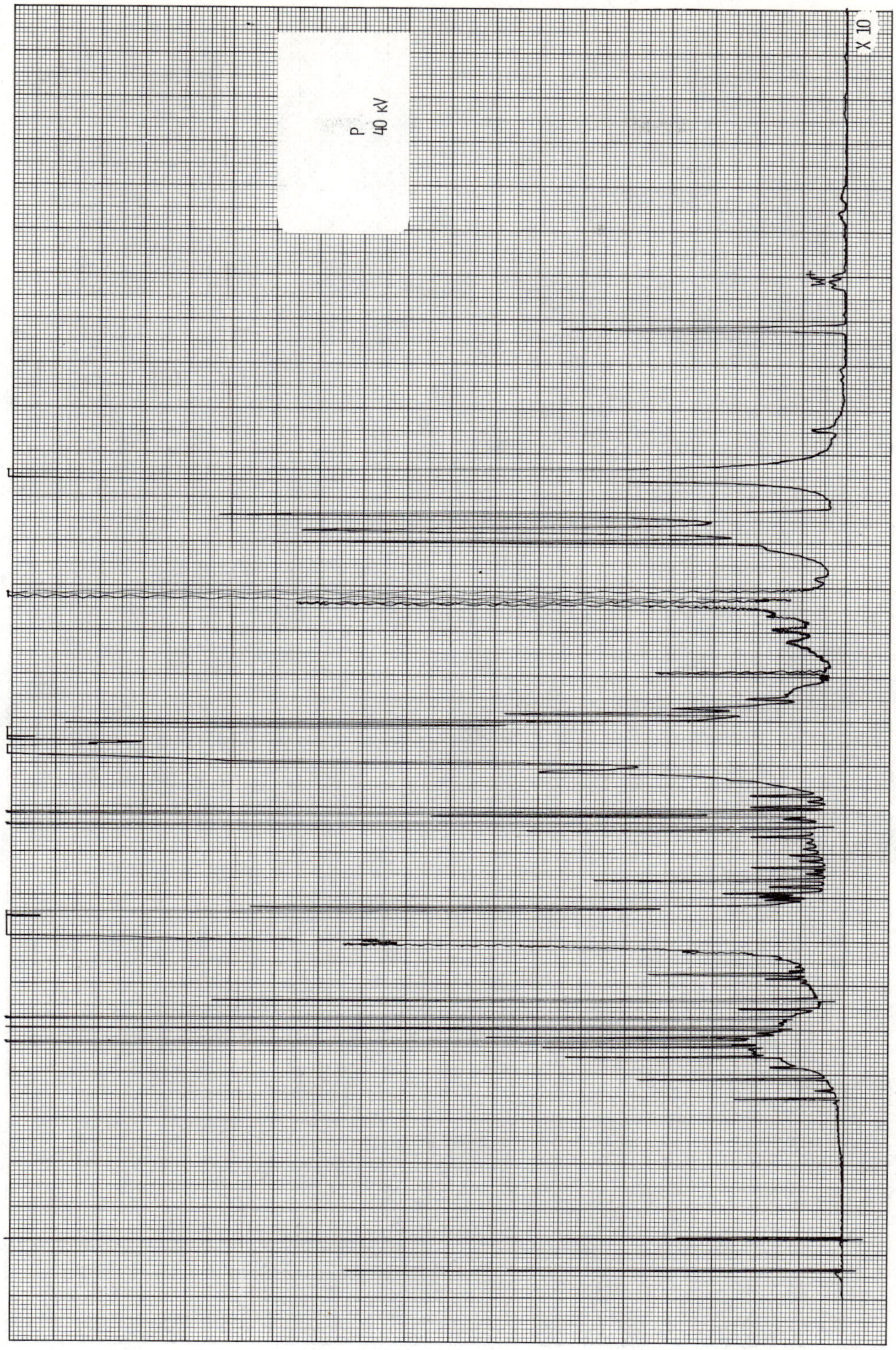
P
40 kV
X 10

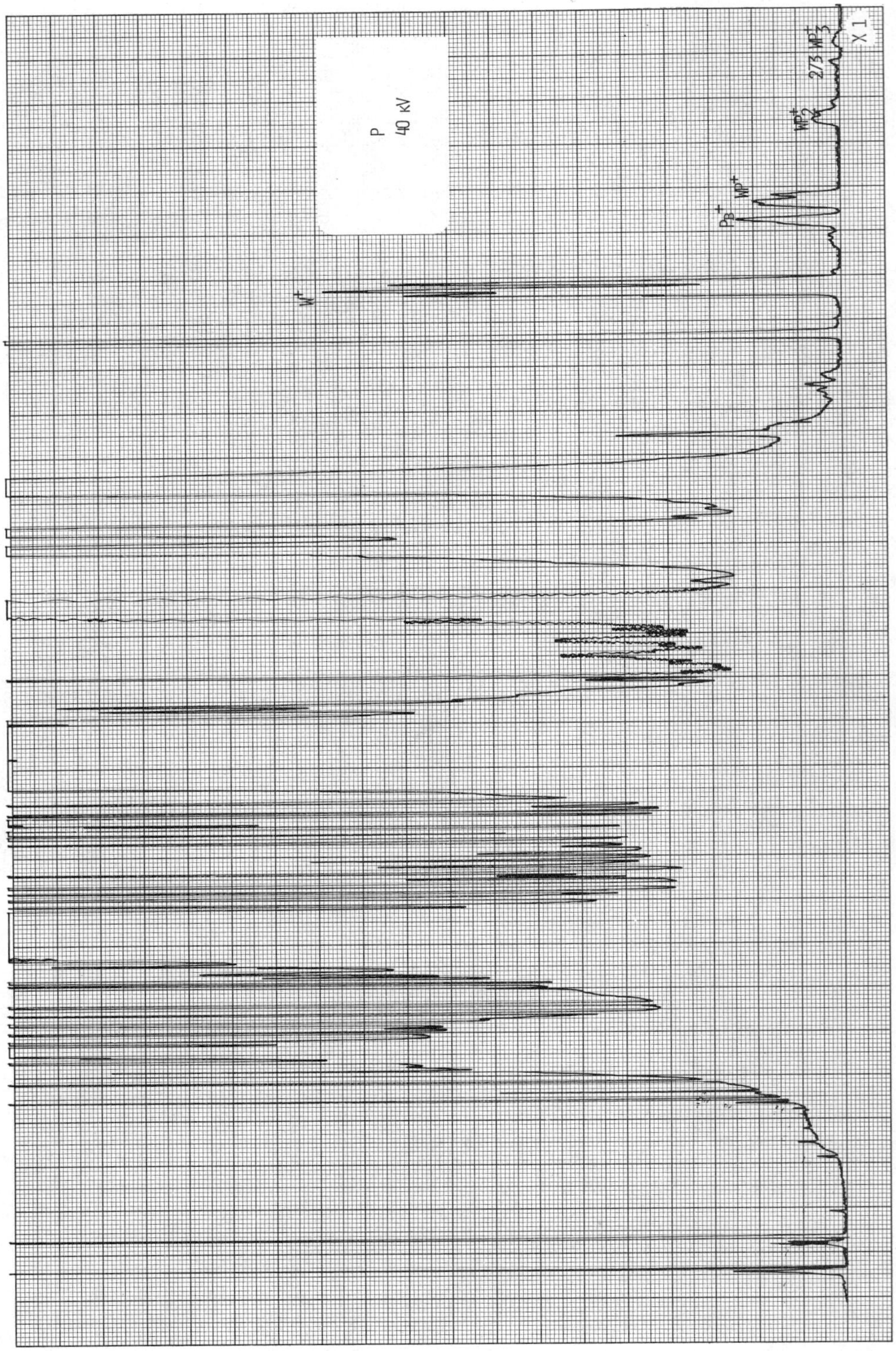
P
40 kV
X 1

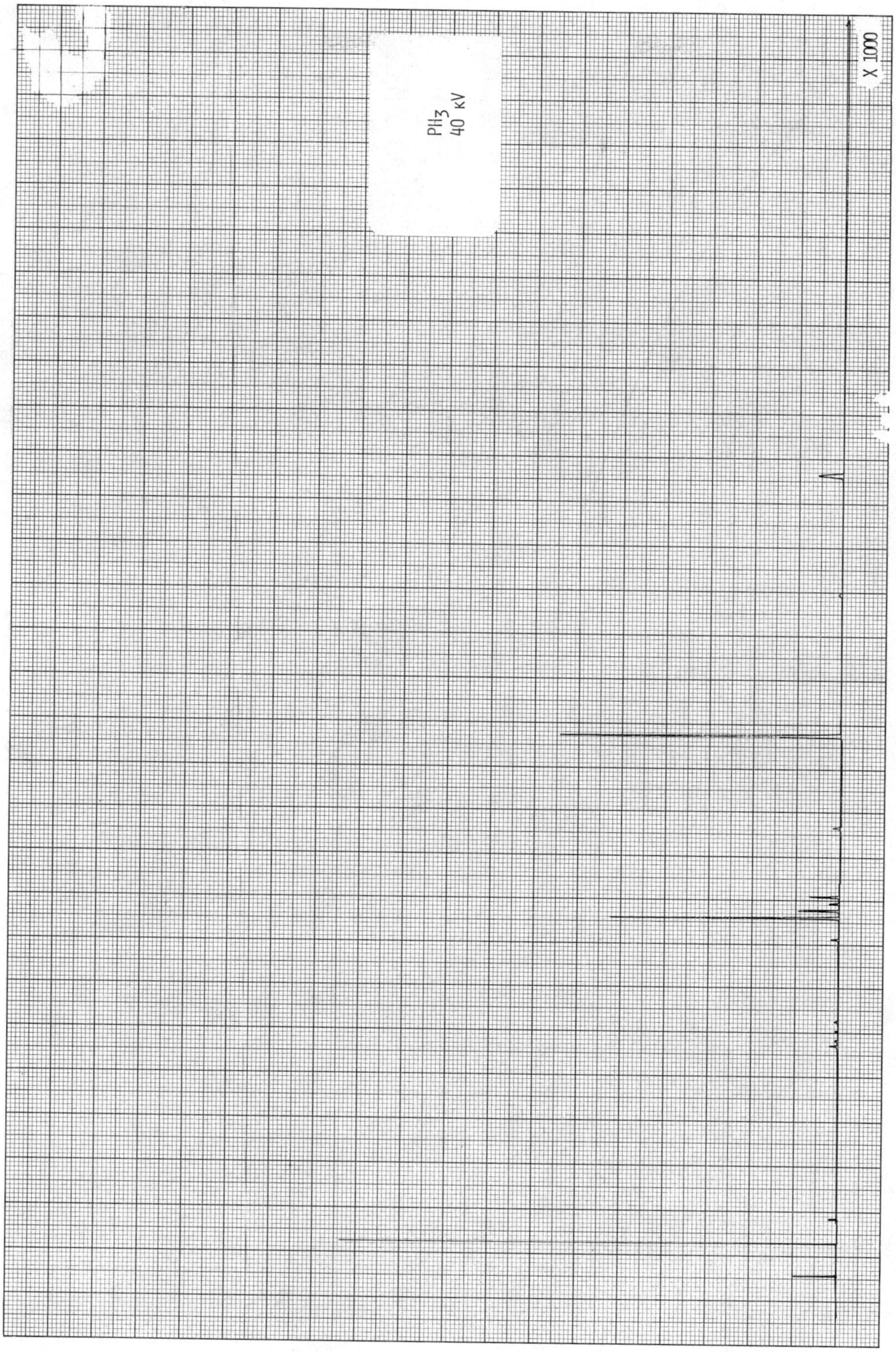
PH3
40 kV
X 1000

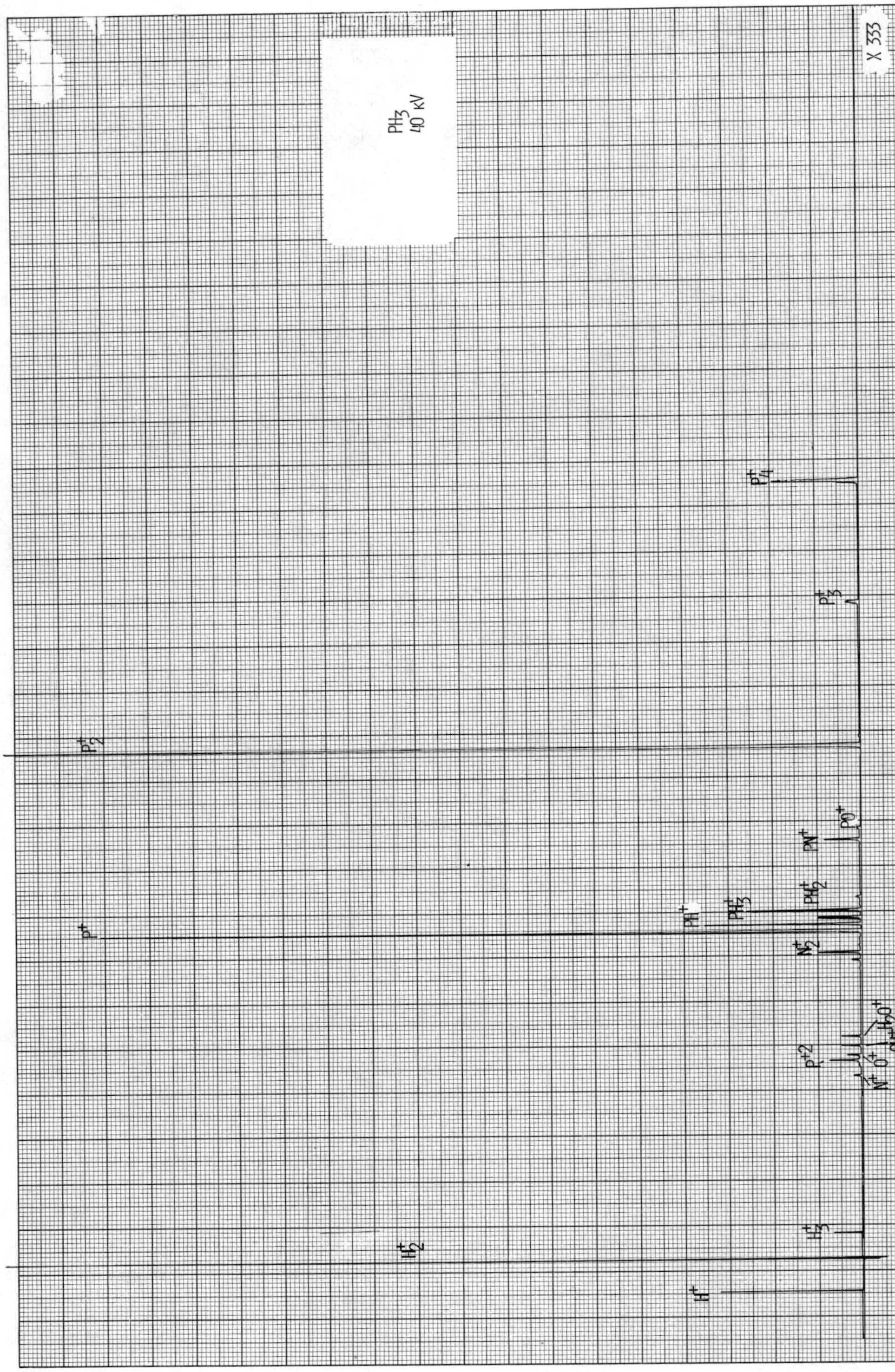
PH3
40 kV
X 333
P4+
P3+
P2+
PO+
PN+
PH2+
PH3+
PH+
P+
N2+
P+2
N+
O+
OH+
H2O+
H3+
H2+
H+

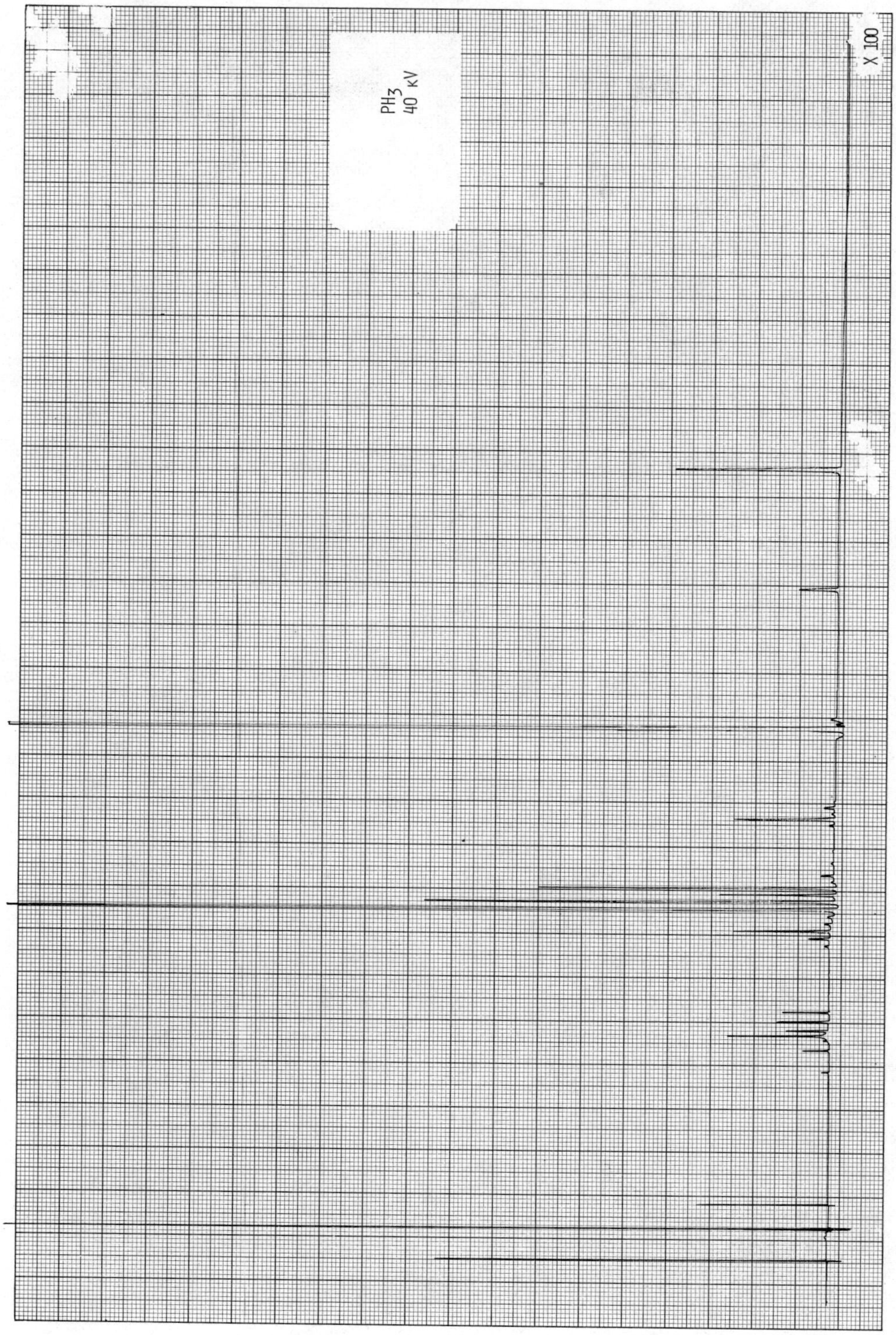
PH_3
40 kV
X 100

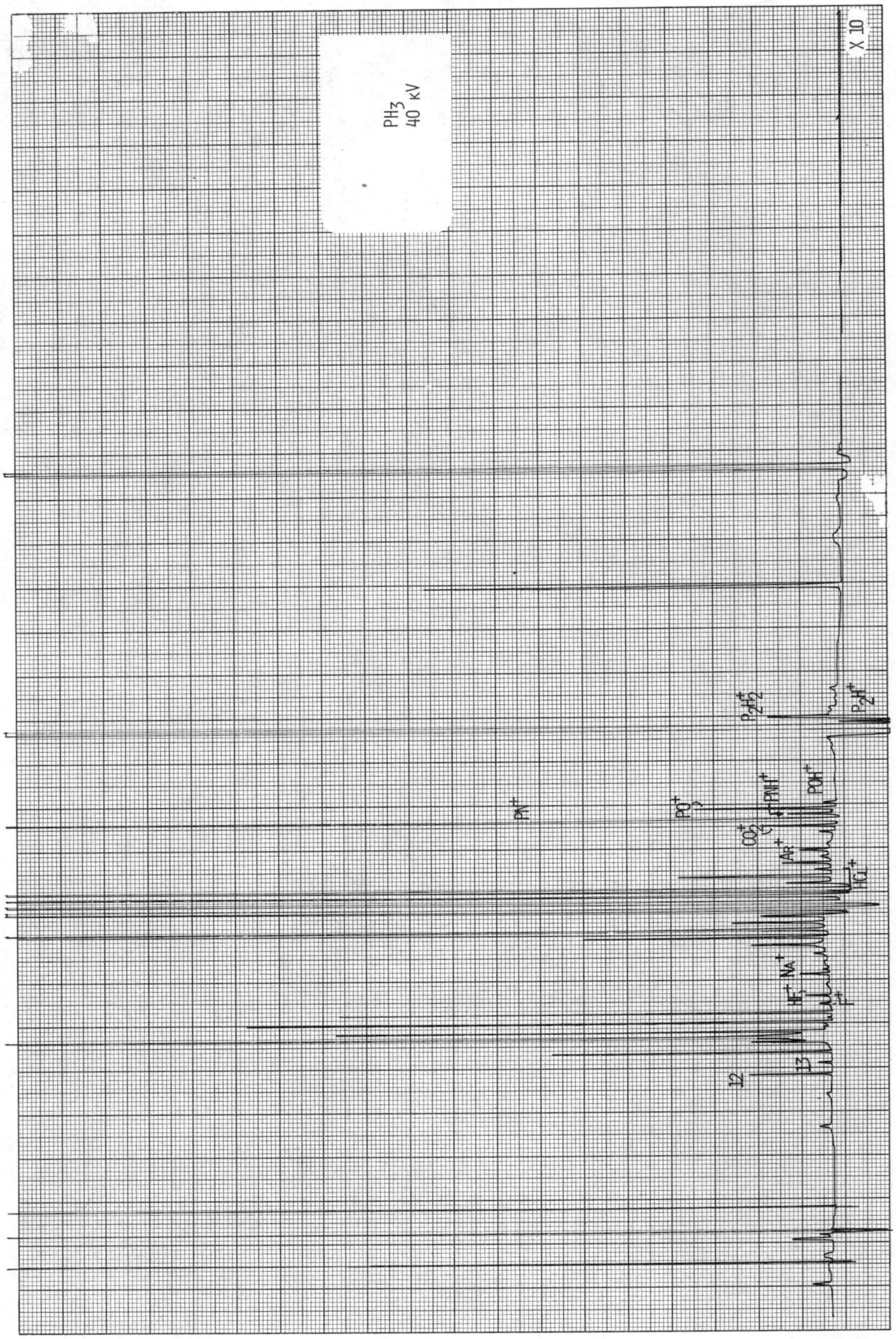
PH3
40 kV
X 10
PN+
PO+
Ar+
PNH+
POH+
HCl+
Na+
12
13

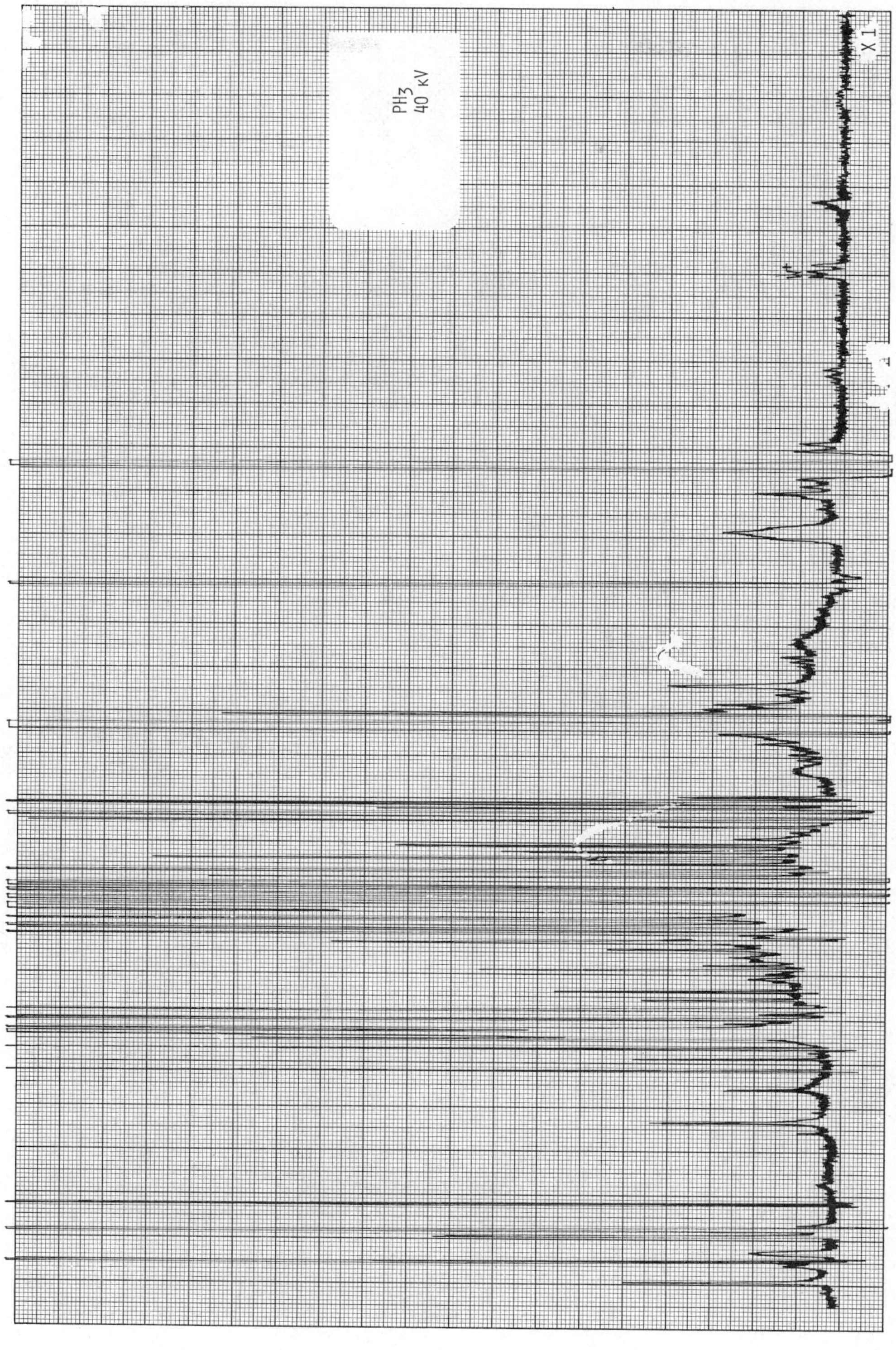
PH_3
40 kV
X 1

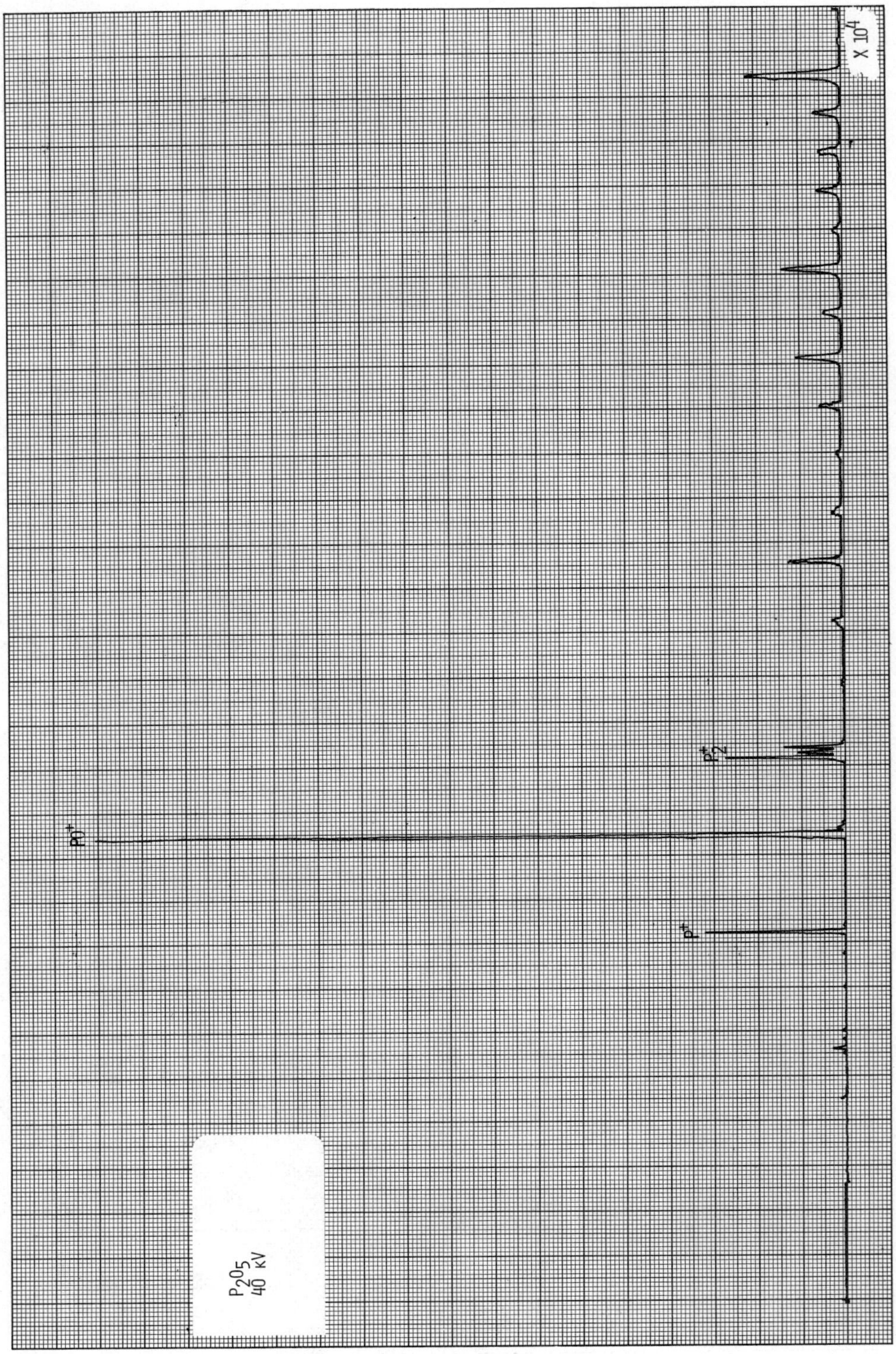
X 10^4
PO+
P+
P2+
P2O5
40 kV

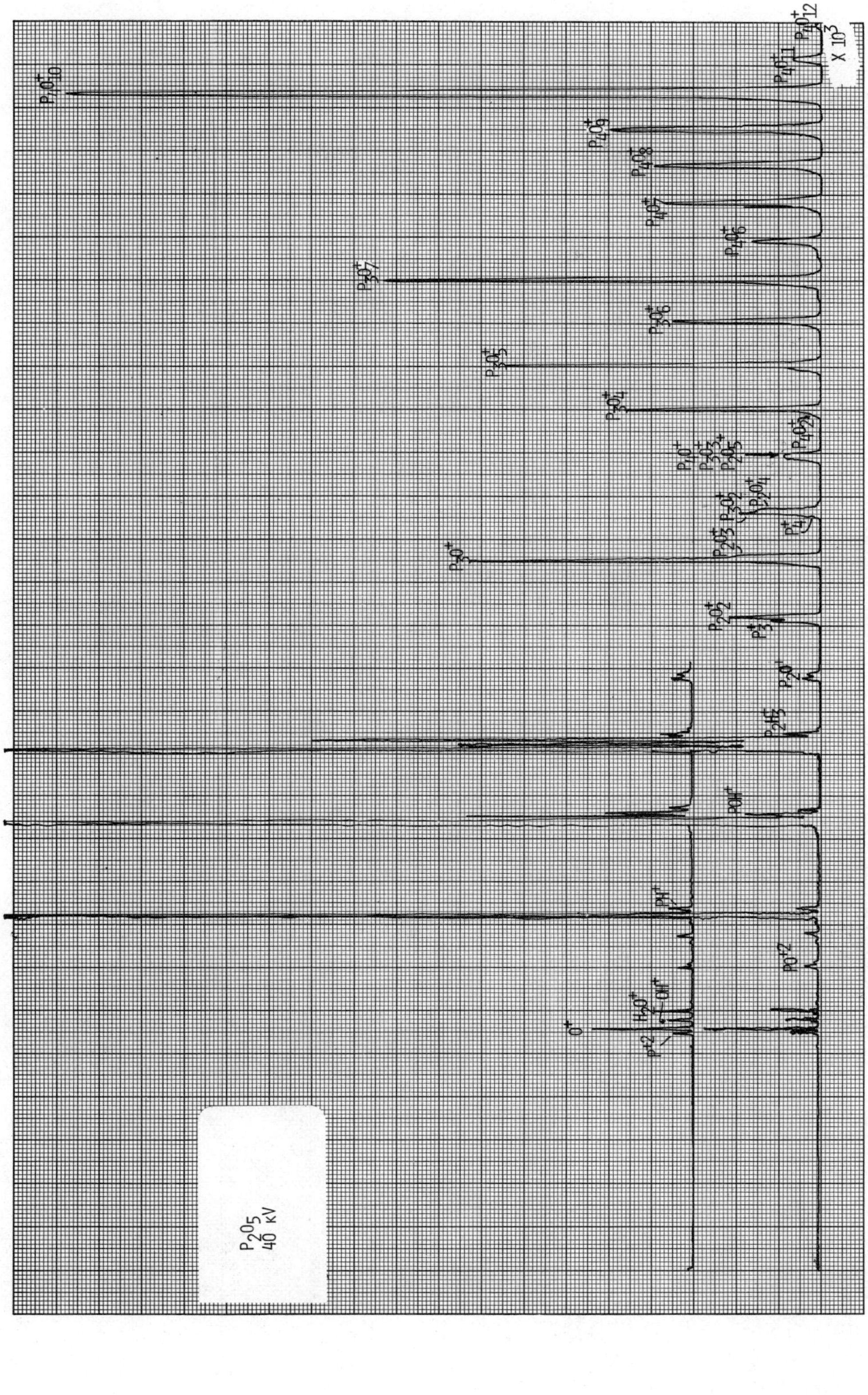

P2O5
40 kV
X 10^3

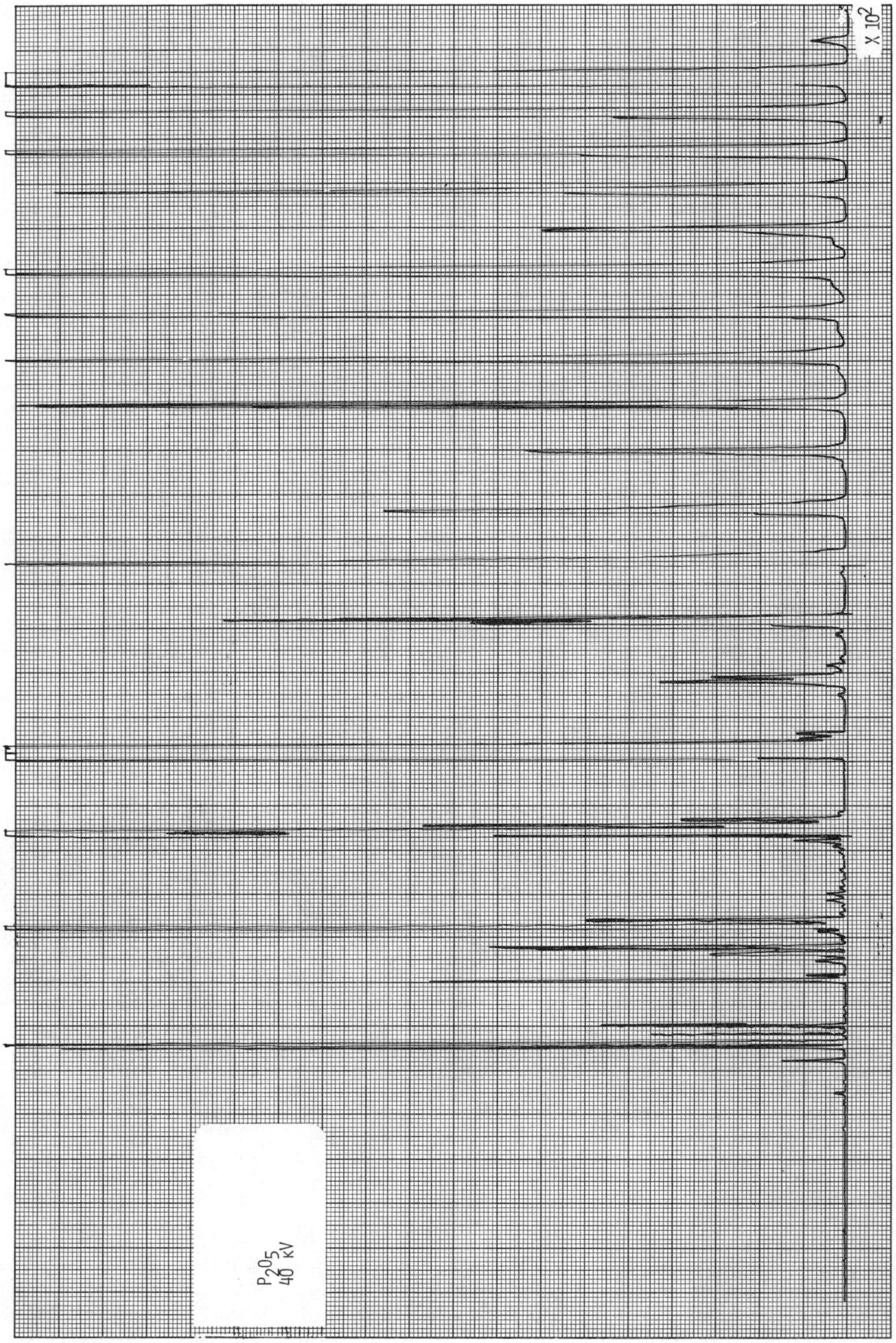
P_2O_5
40 kV
X 10^2

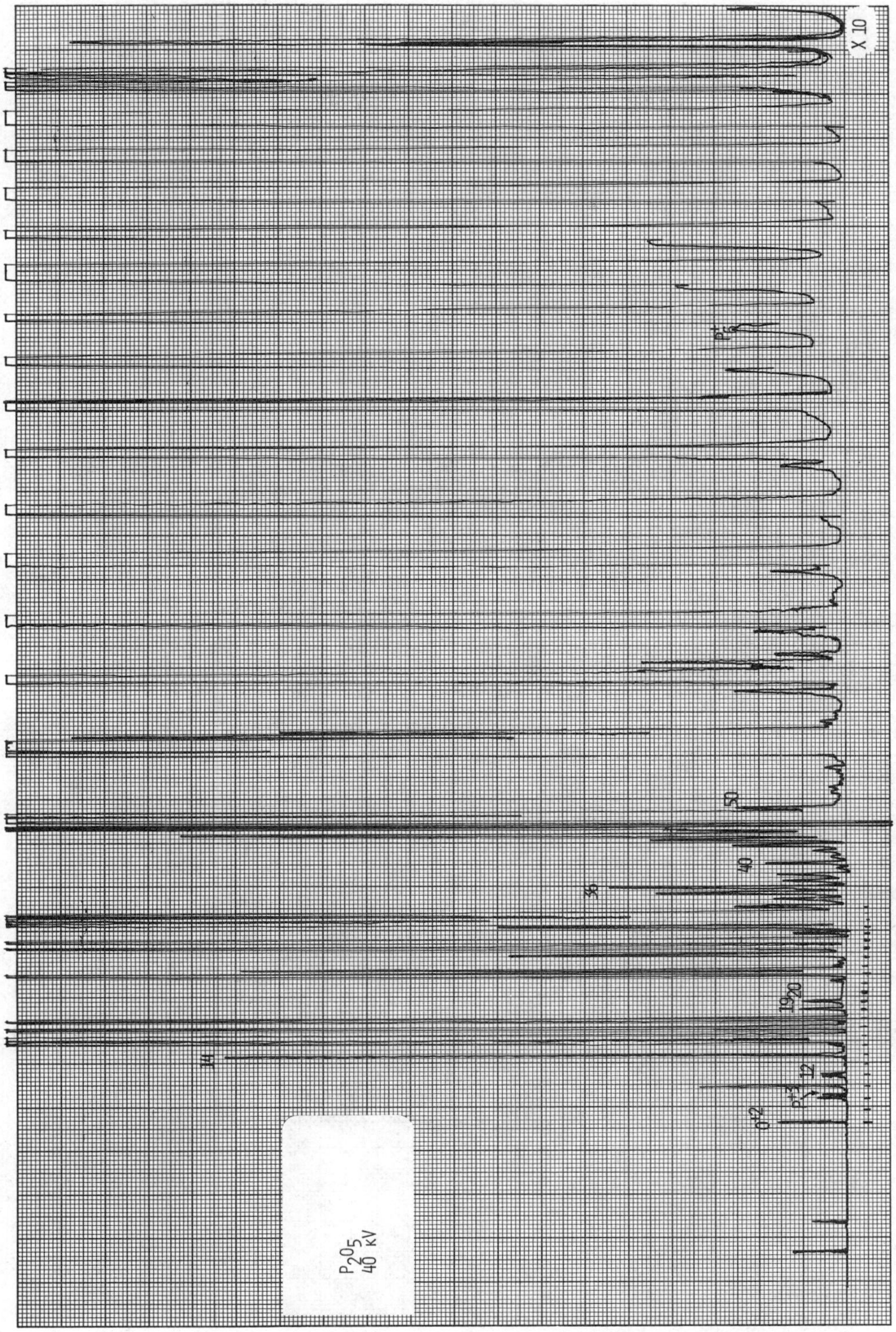
P2O5
40 kV
X 10
14
36
40
50

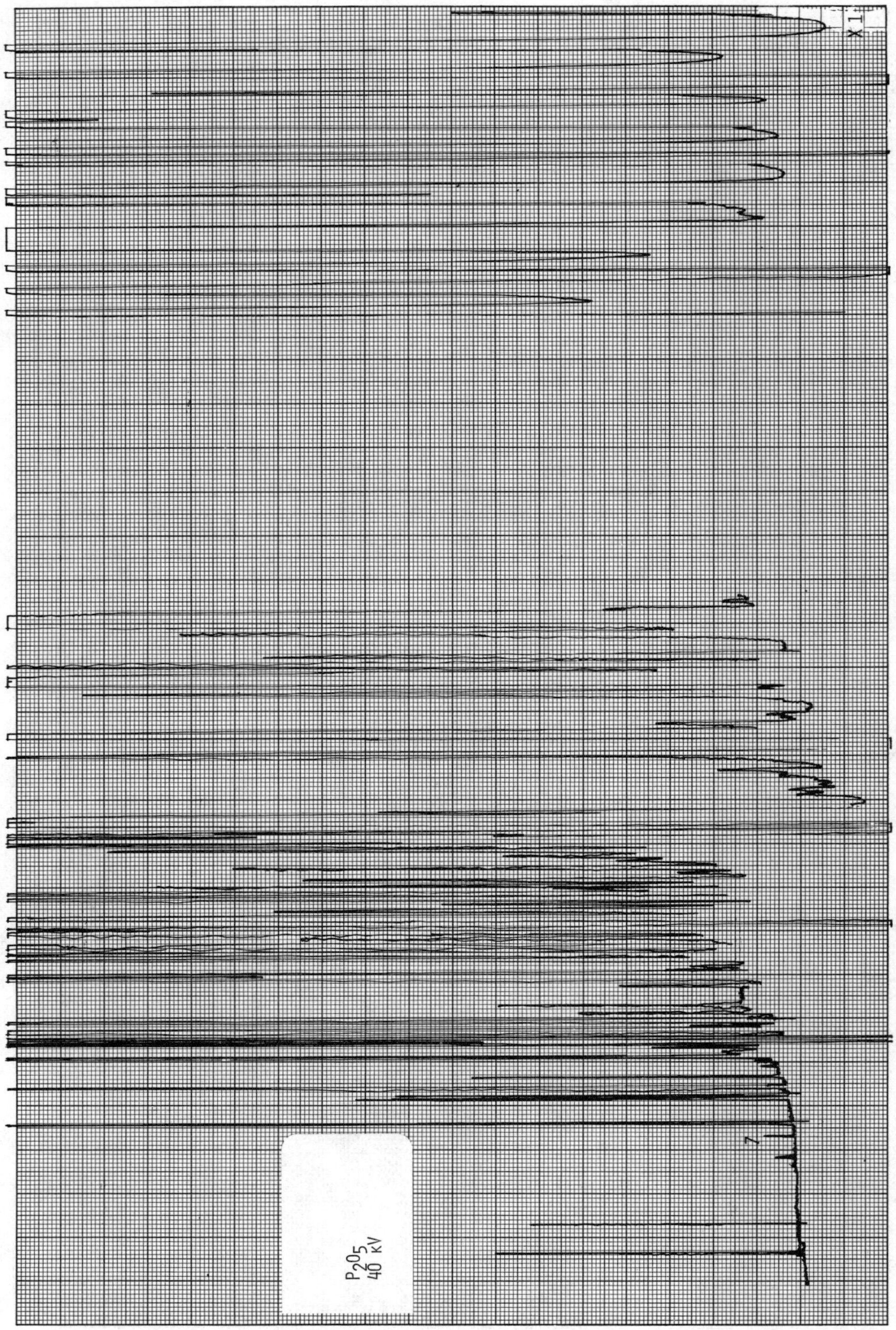

P_2O_5
40 kV
X 1

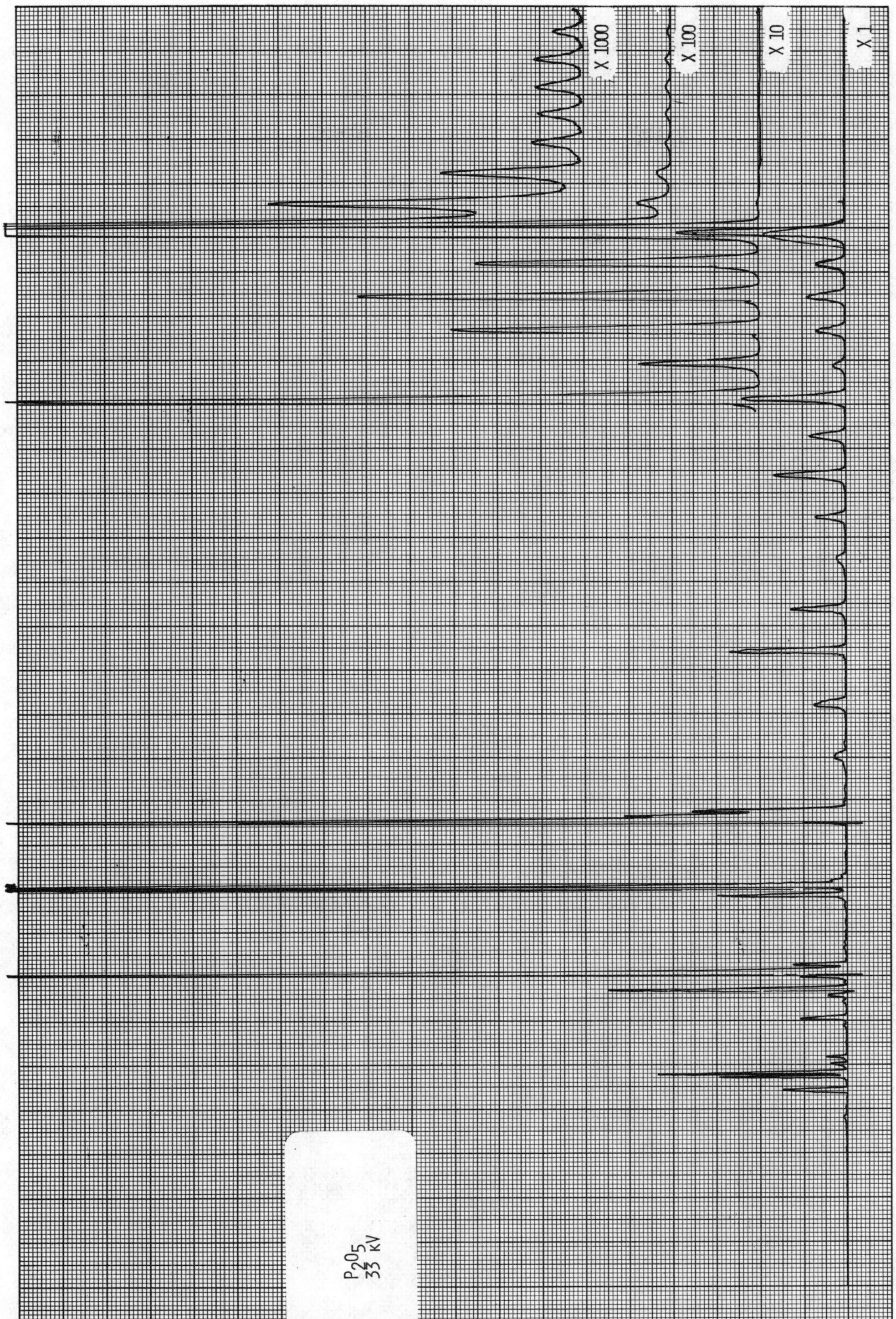
X 1000
X 100
X 10
X 1
P_2O_5
33 kV

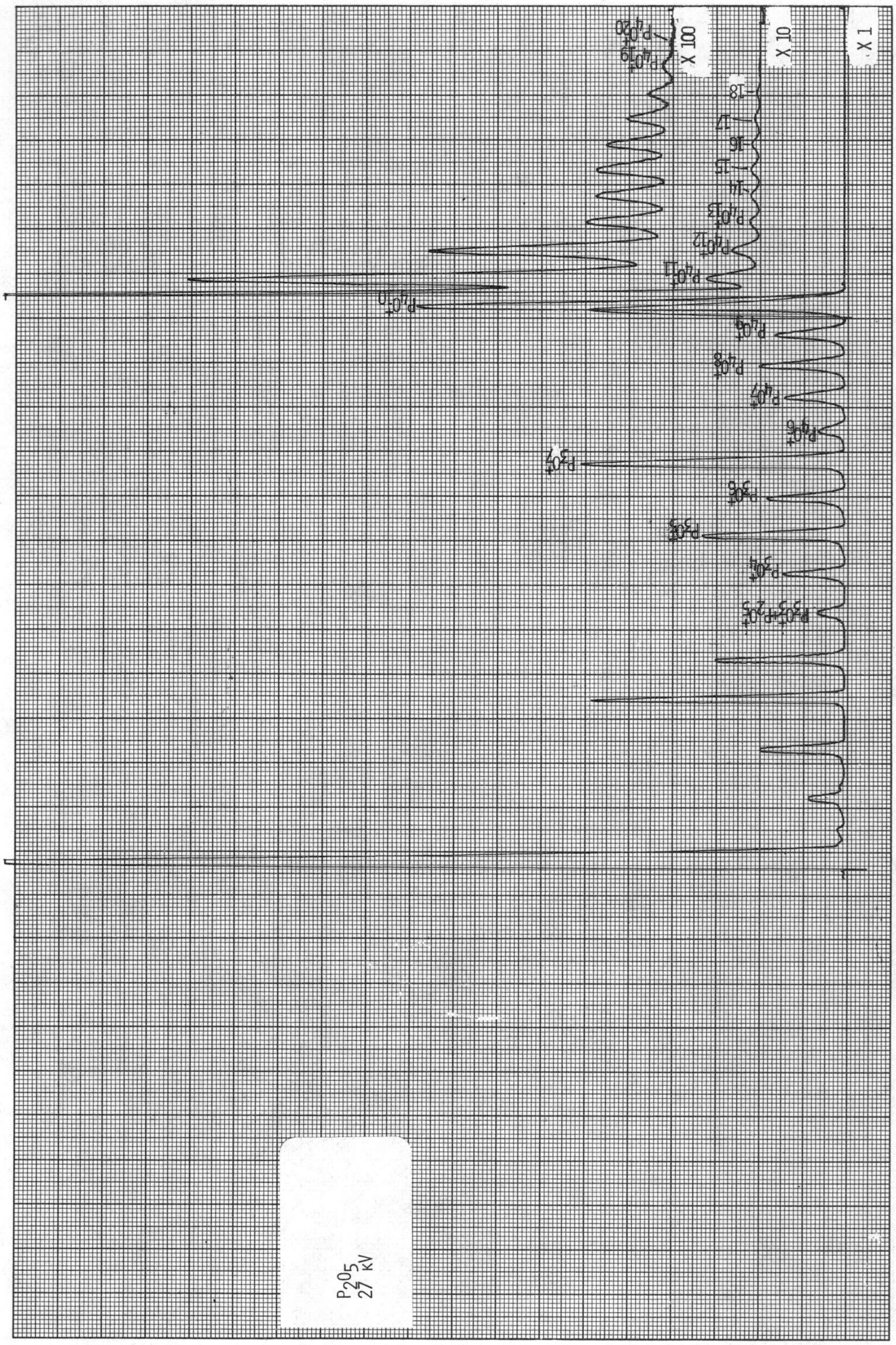
P2O5
27 kV
X 100
X 10
X 1

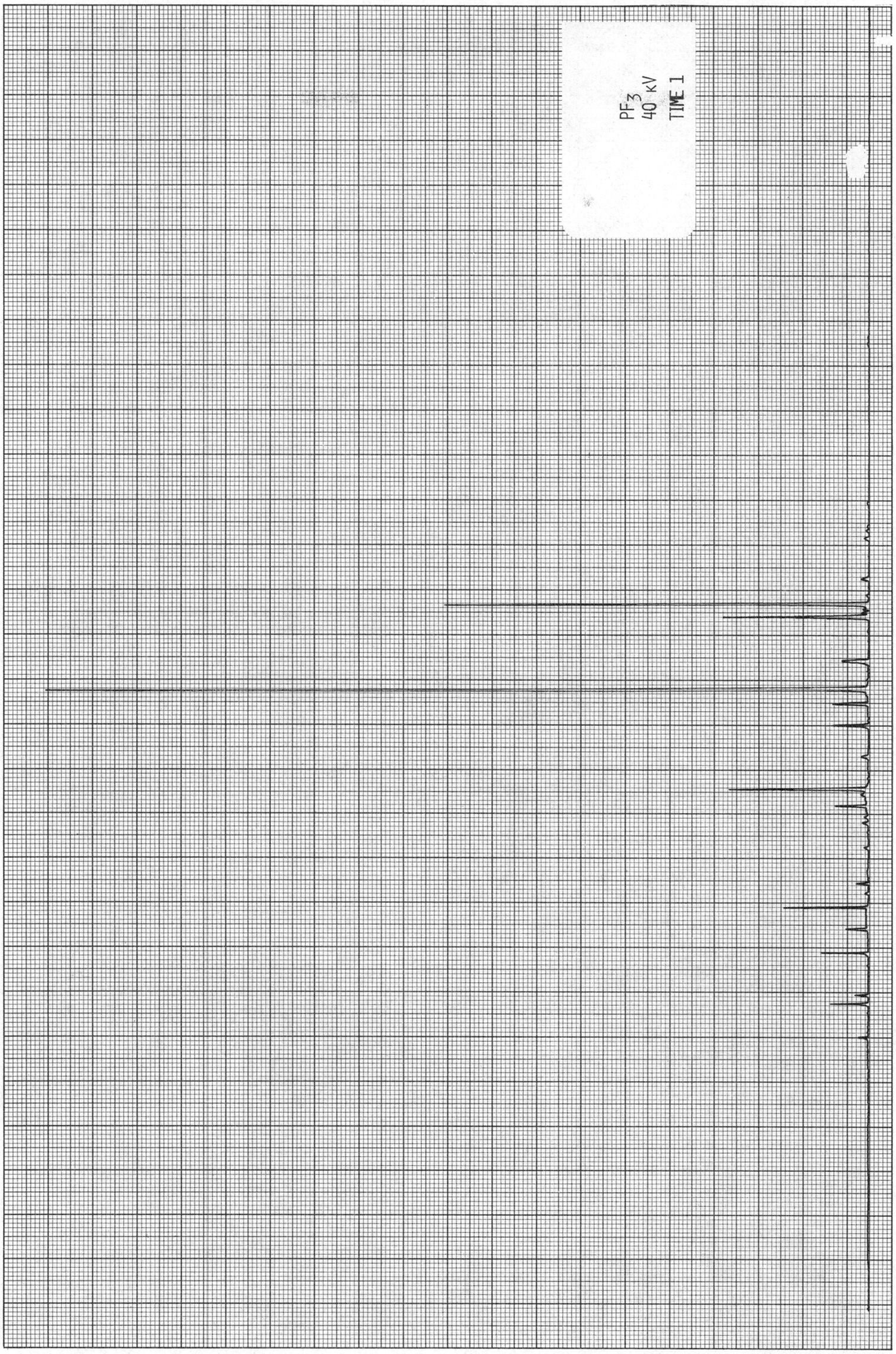
PF3
40 kV
TIME 1

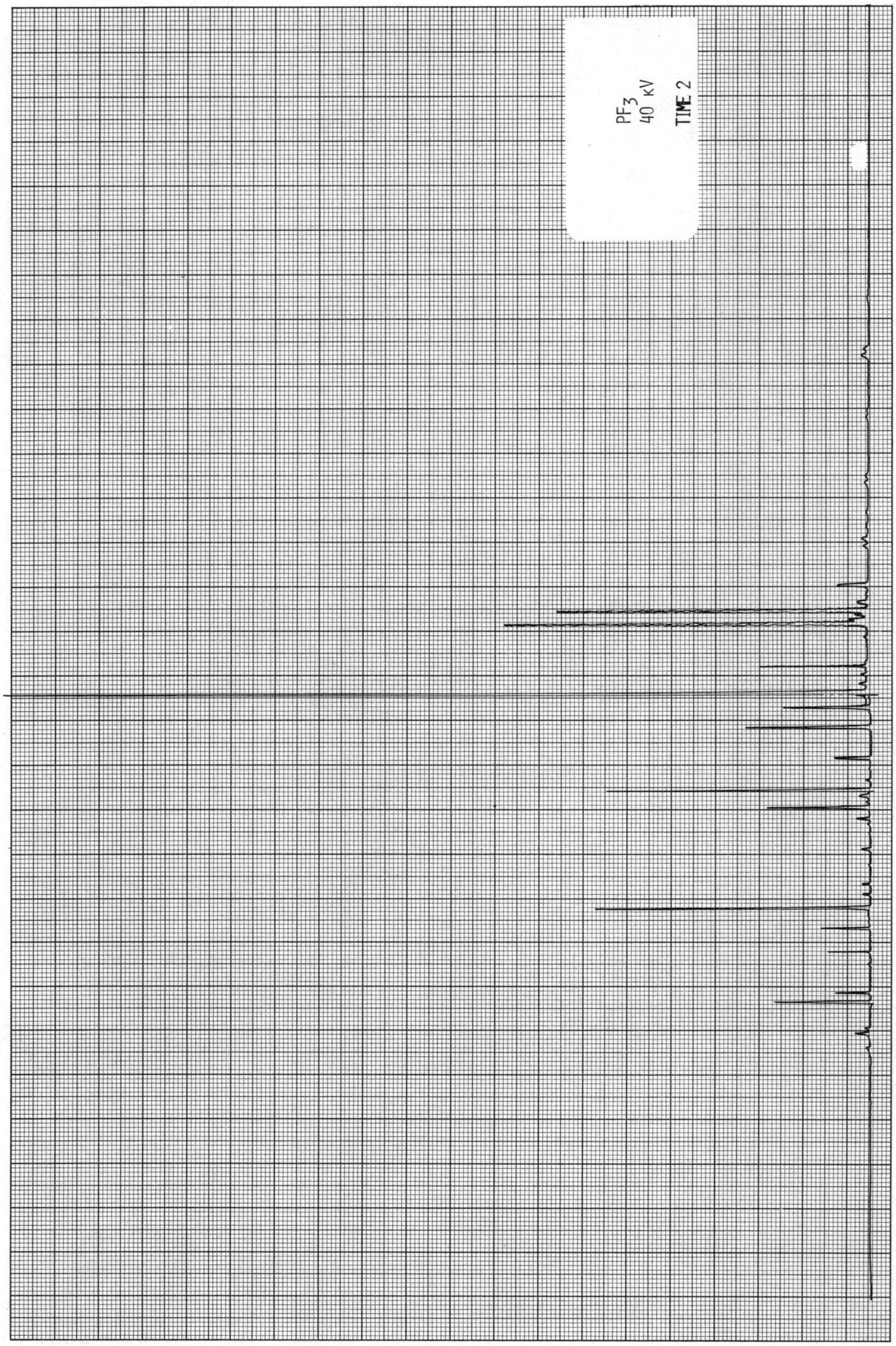
PF3
40 kV
TIME 2

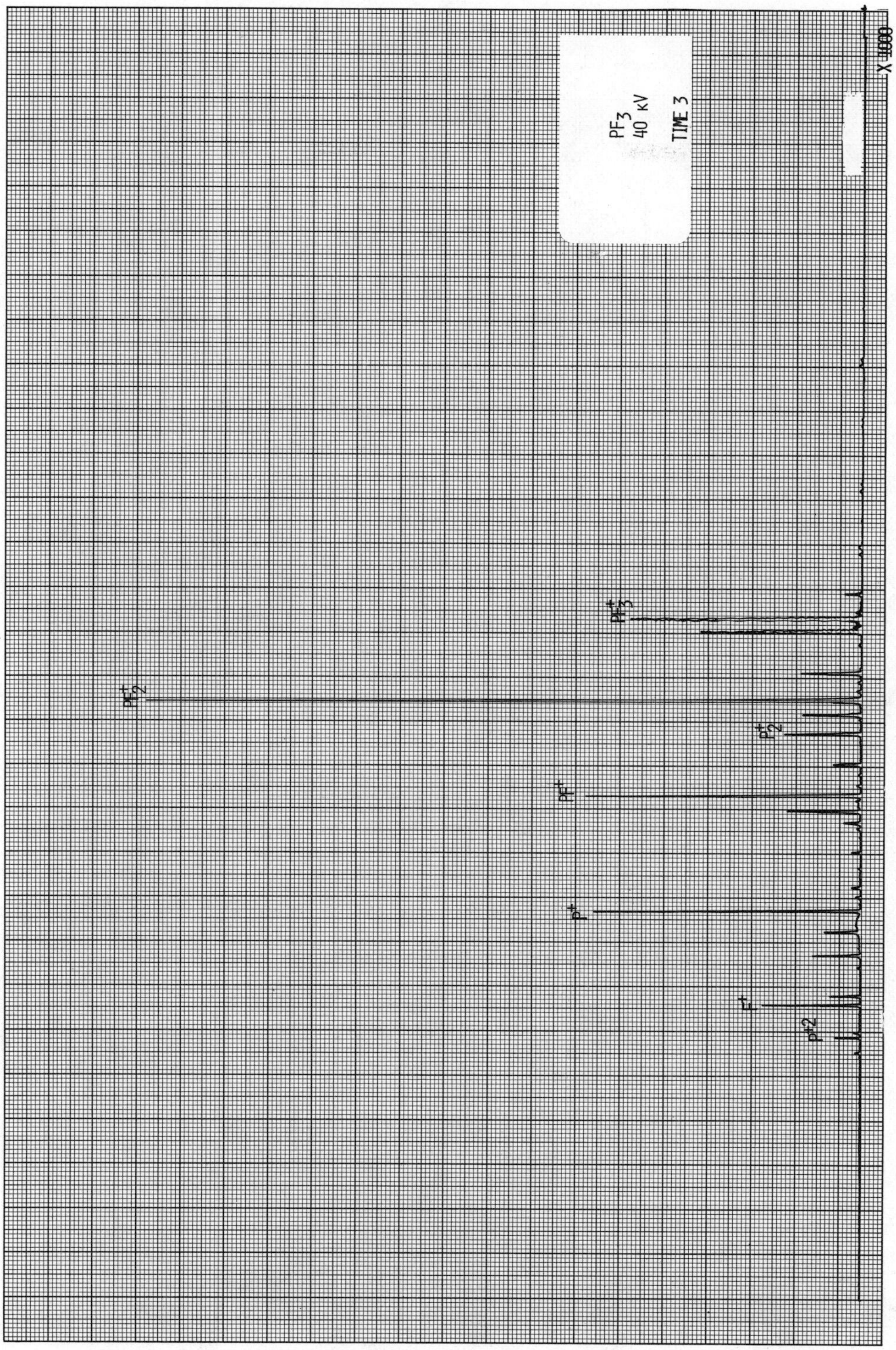
PF3
40 kV
TIME 3
X 1000
PF3+
PF2+
P2+
PF+
P+
F+
P+2

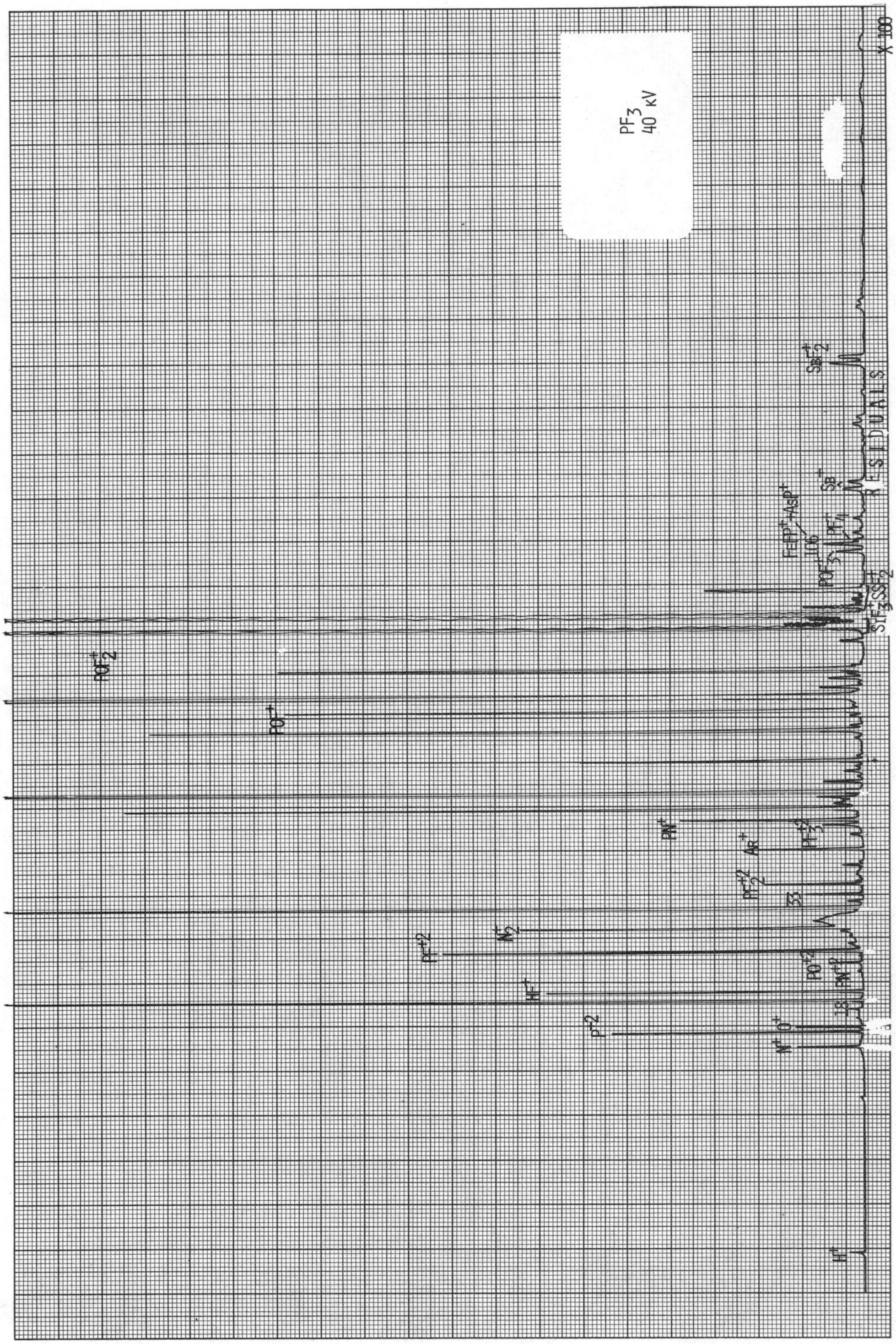

PF_3
40 kV
X 100
R E S I D U A L S

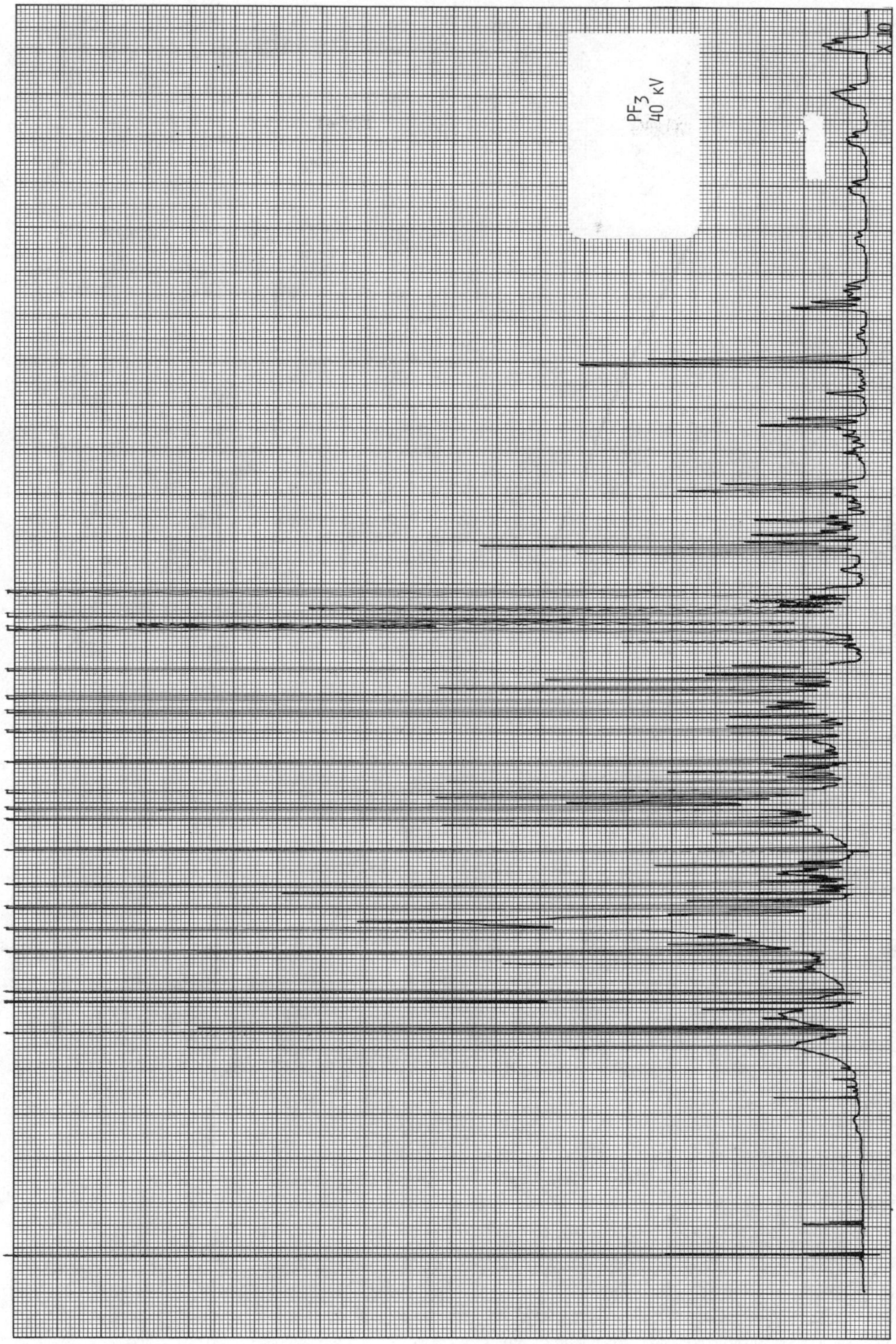
PF3
40 kV
X 10

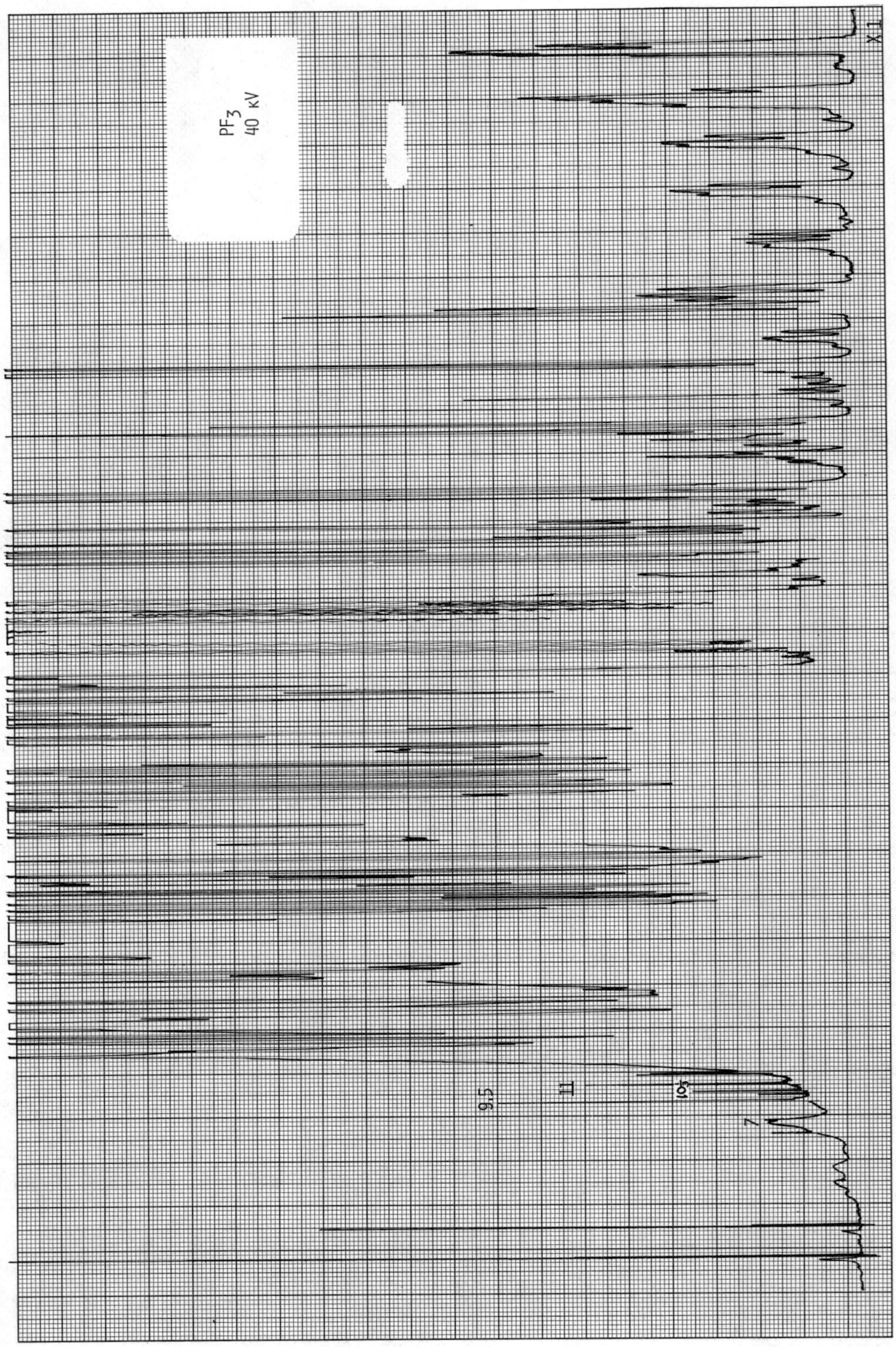

PF_3
40 kV
X 1
9.5
11

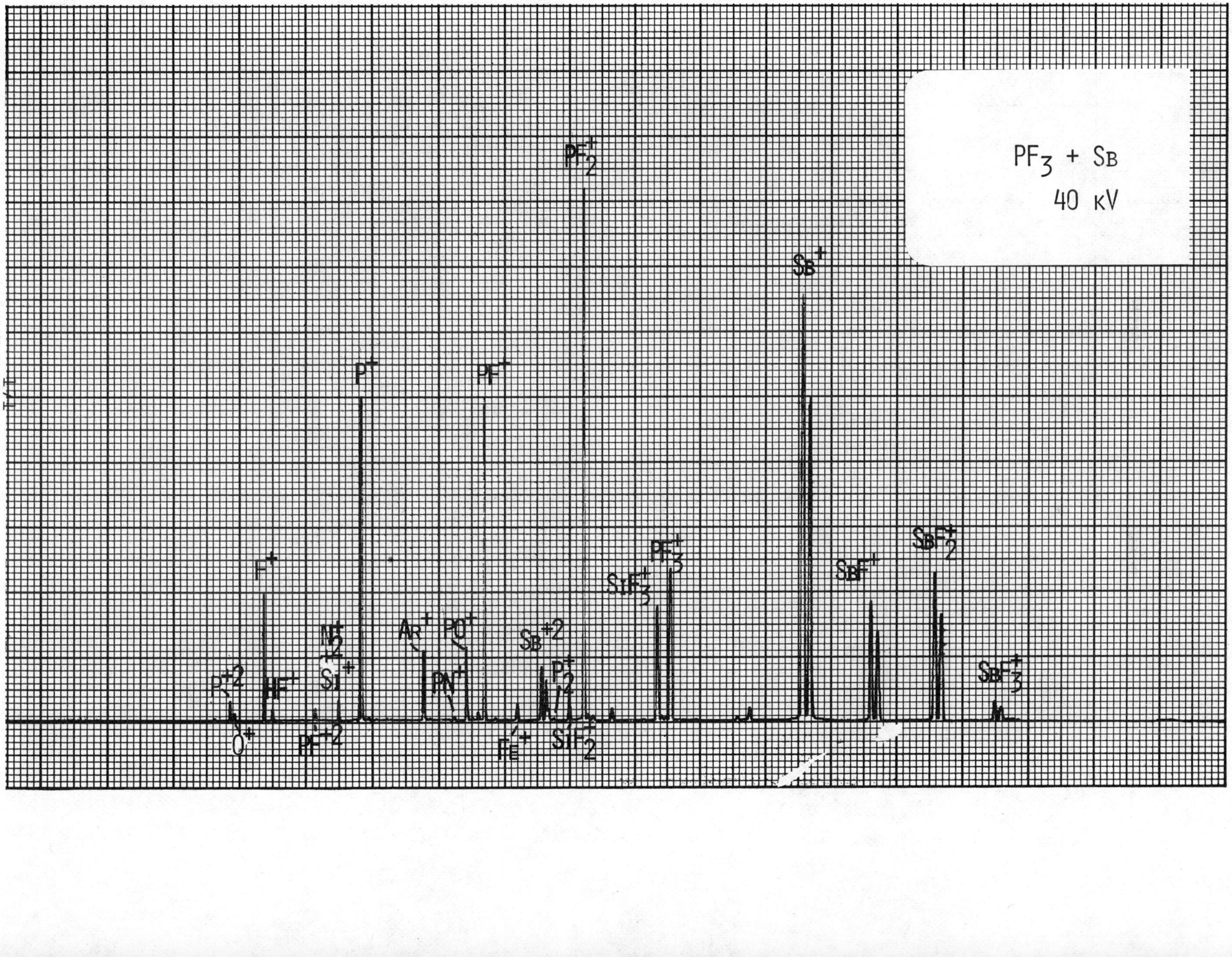
PF3 + SB
40 kV
P+2
O+
F+
HF+
PF+2
N2+
Si+
P+
Ar+
PO+
PN+
PF+
Fe+
SB+2
P2+
SiF2+
PF2+
SiF3+
PF3+
SB+
SBF+
SBF2+
SBF3+

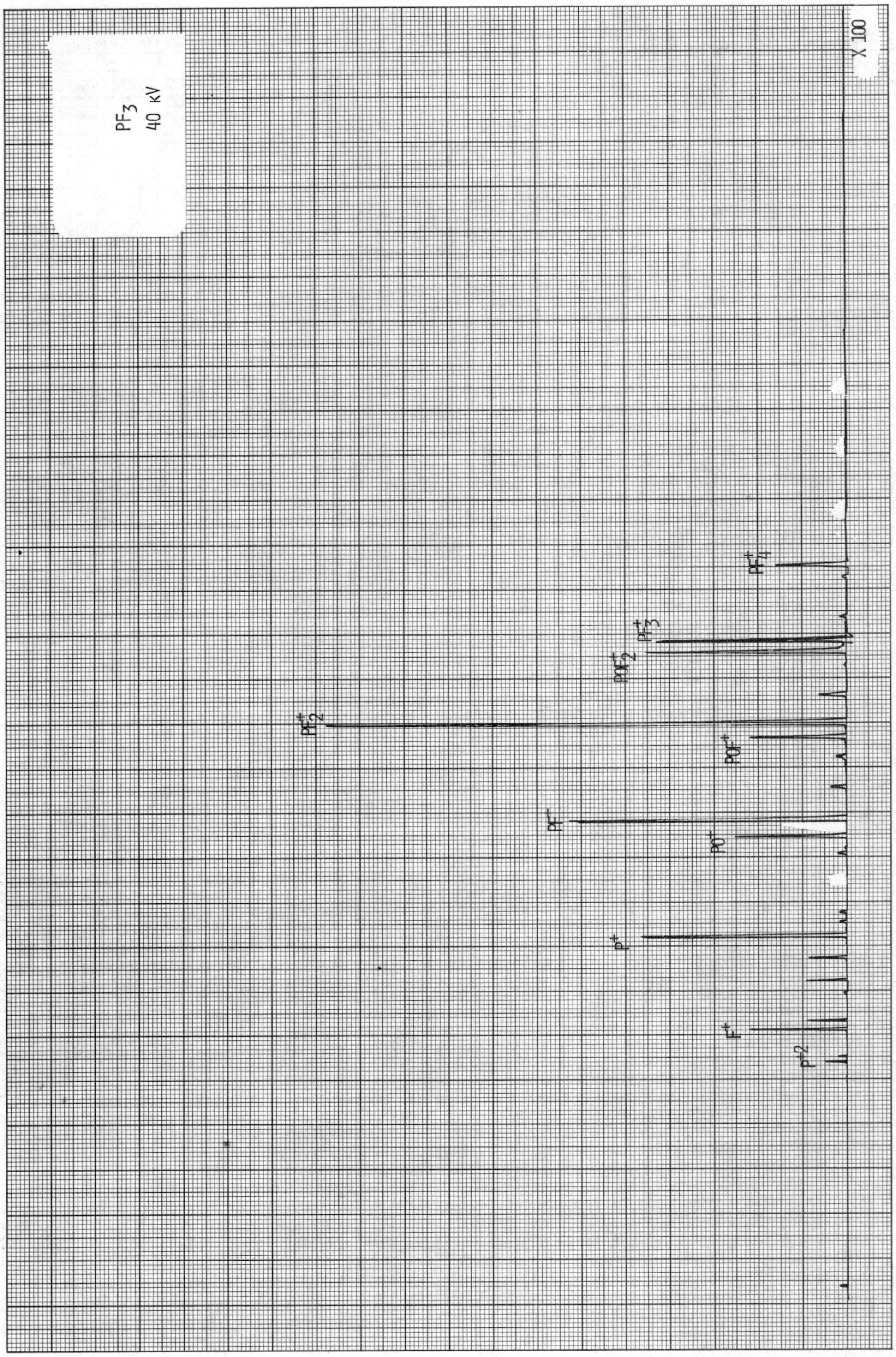
PF3
40 kV
X 100
PF2+
POF2+
PF3+
PF4+
POF+
PF+
PO+
P+
F+
P+2

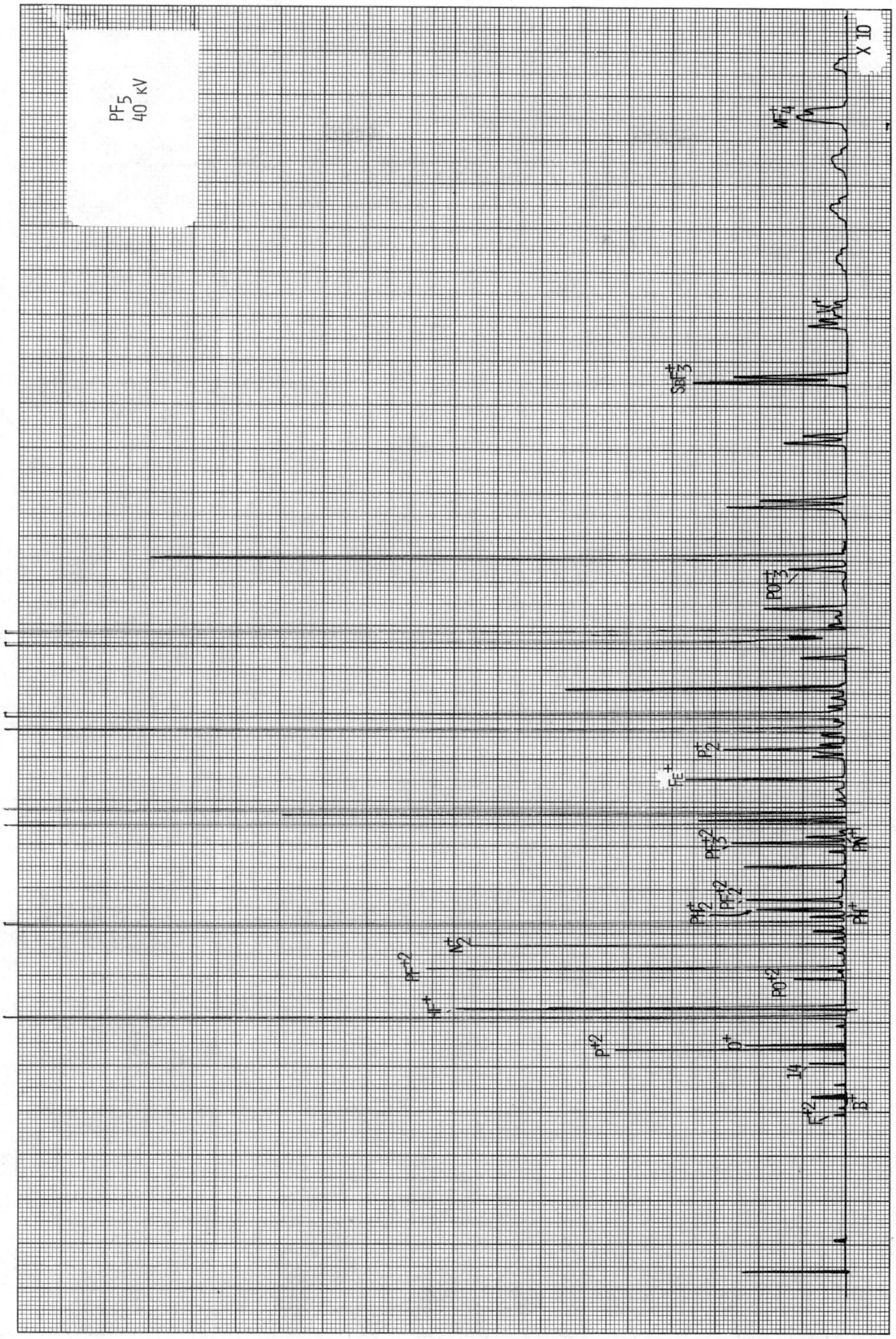
PF$_5$ 40 kV
X 10

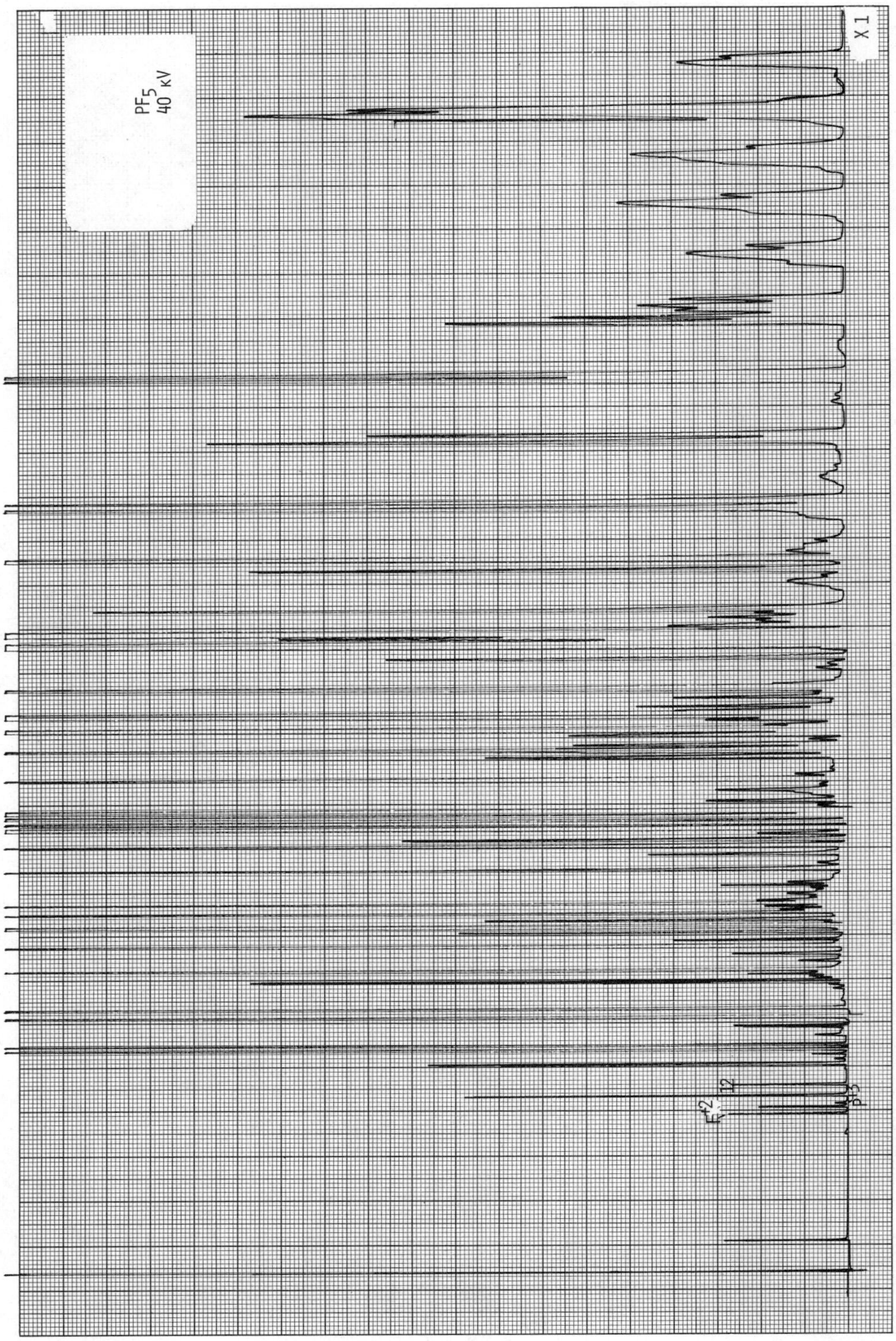
PF5
40 kV
X 1

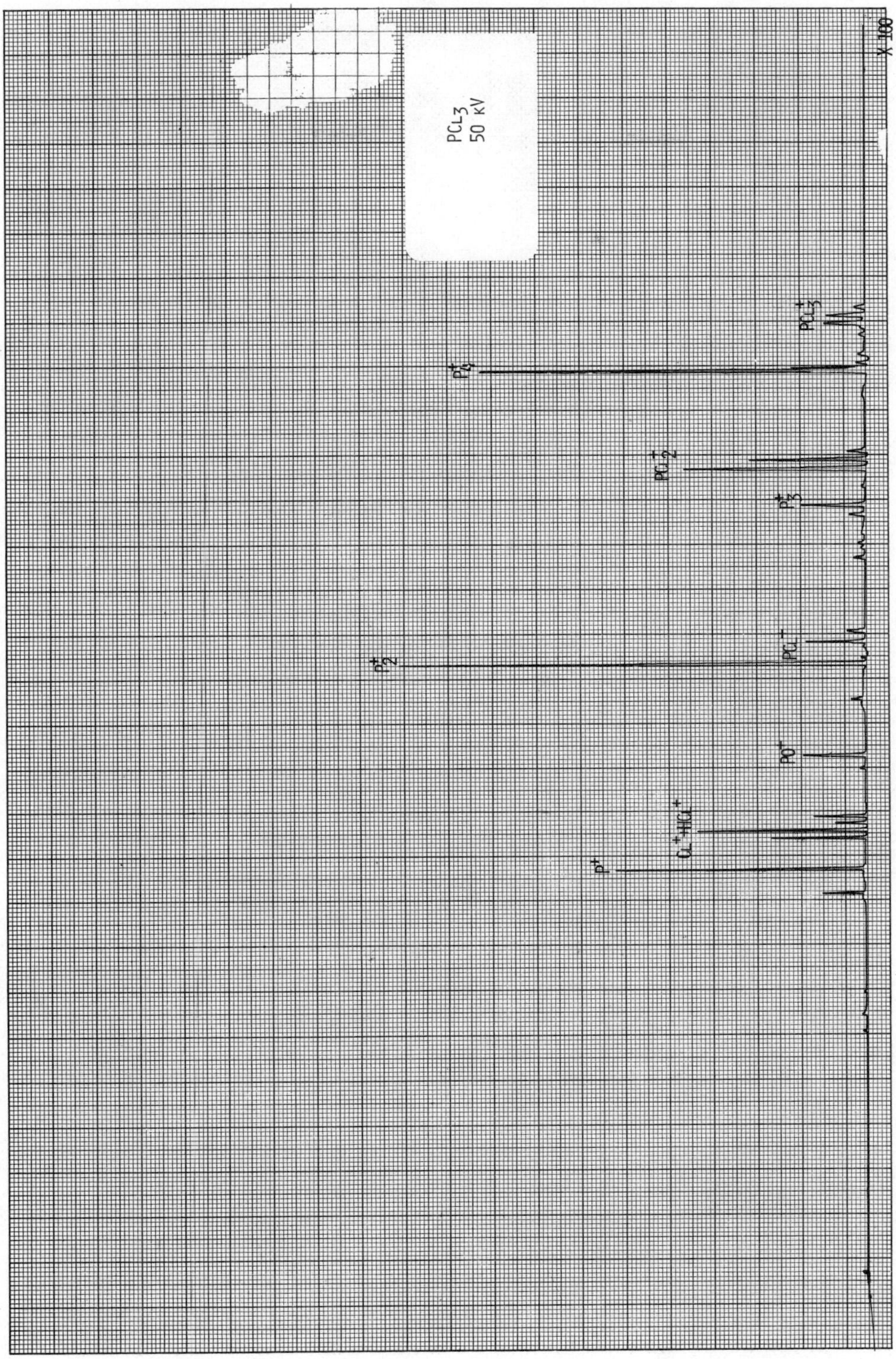

PCl_3
50 kV
X 100
PCl_3^+
P_4^+
PCl_2^+
P_3^+
PCl^+
P_2^+
PO^+
$Cl^+ + HCl^+$
P^+

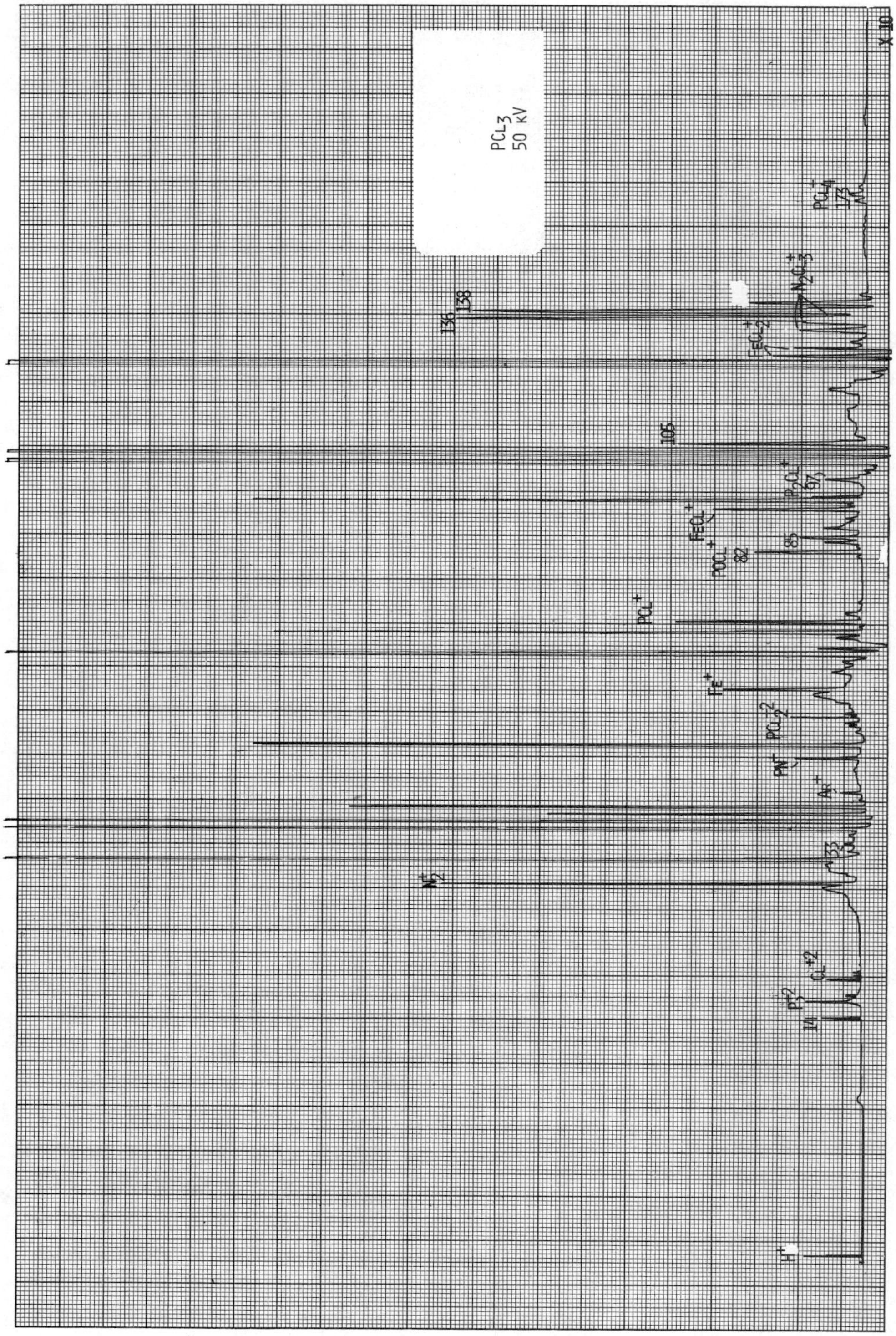
PCl3
50 kV
X 10
136
138
106
82
85
97
173
14
33

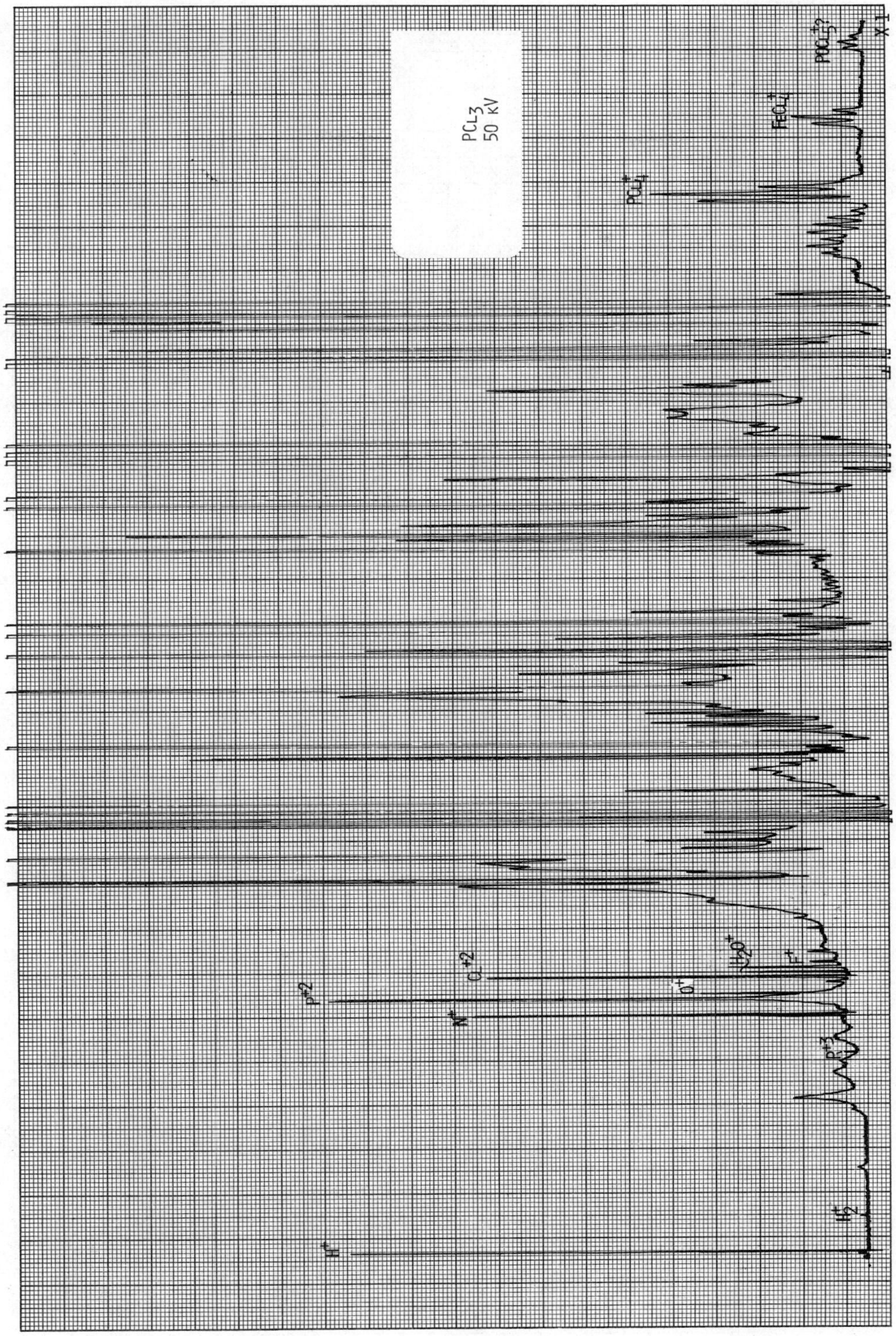
PCl3
50 kV
PCl4+
FeCl4+
H+
H2+
N+
P+2
O+2
O+
H2O+
F+

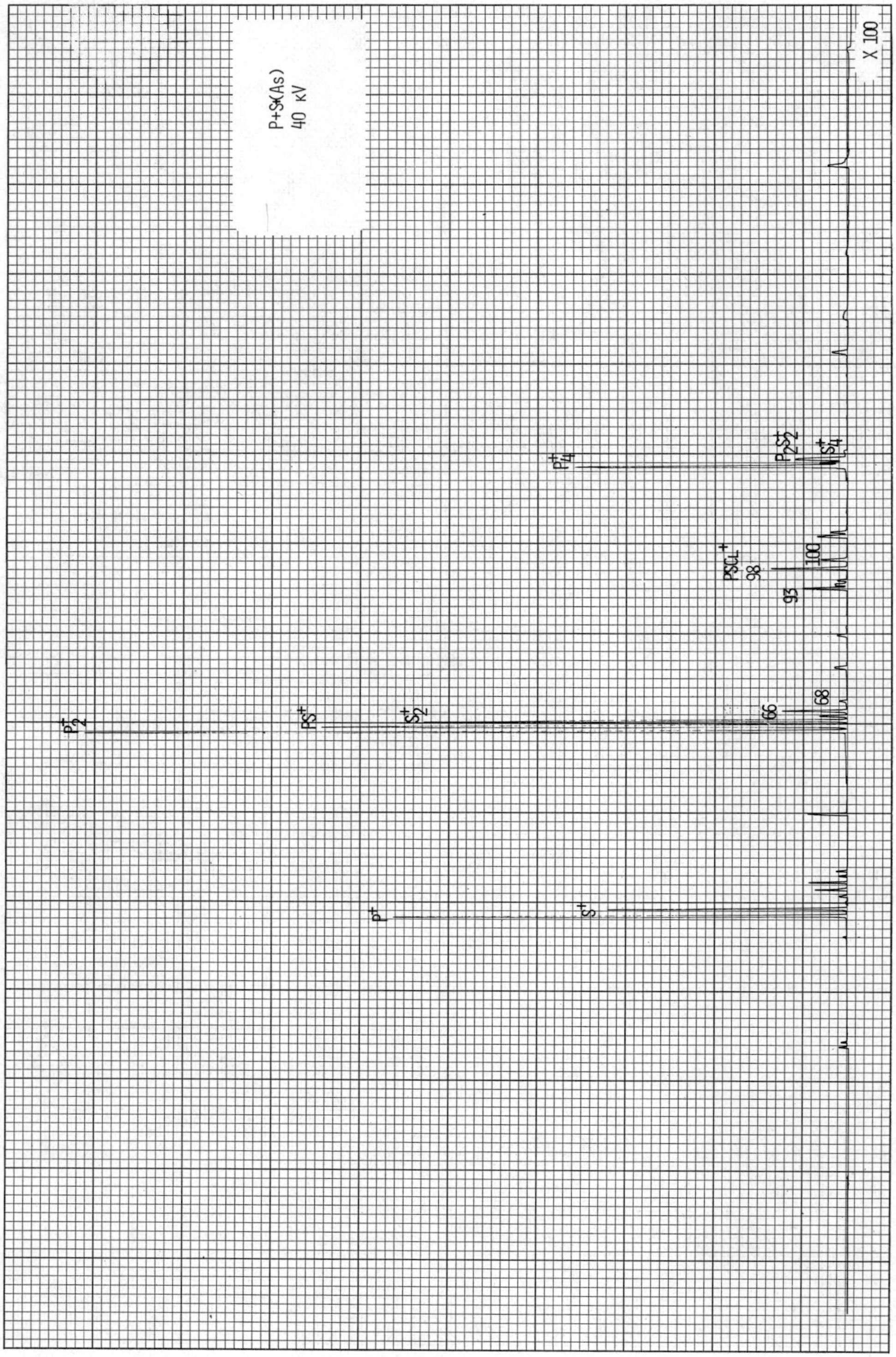
P+S(As)
40 kV
X 100
P2+
PS+
S2+
66
68
PSCl+
98
93
100
P4+
P2S2+
S4+
P+
S+

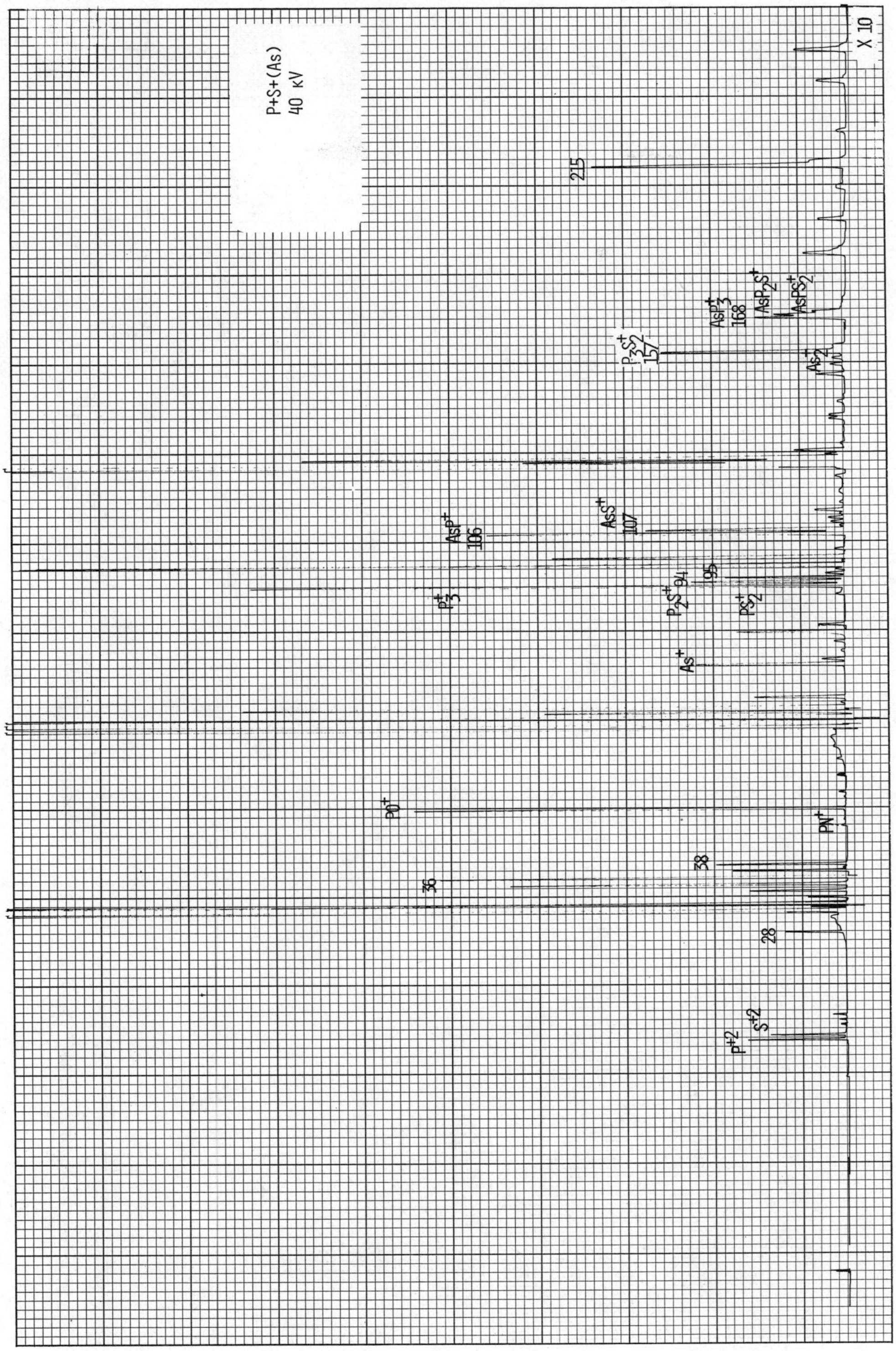
P+S+(As)
40 kV
X 10
215
$P_3S_2^+$
157
AsP_3^+
168
AsP_2S^+
$AsPS_2^+$
As_2^+
AsP^+
106
AsS^+
107
P_3^+
P_2S^+ 94
95
PS_2^+
As^+
PO^+
PN^+
36
38
28
P^{+2}
S^{+2}

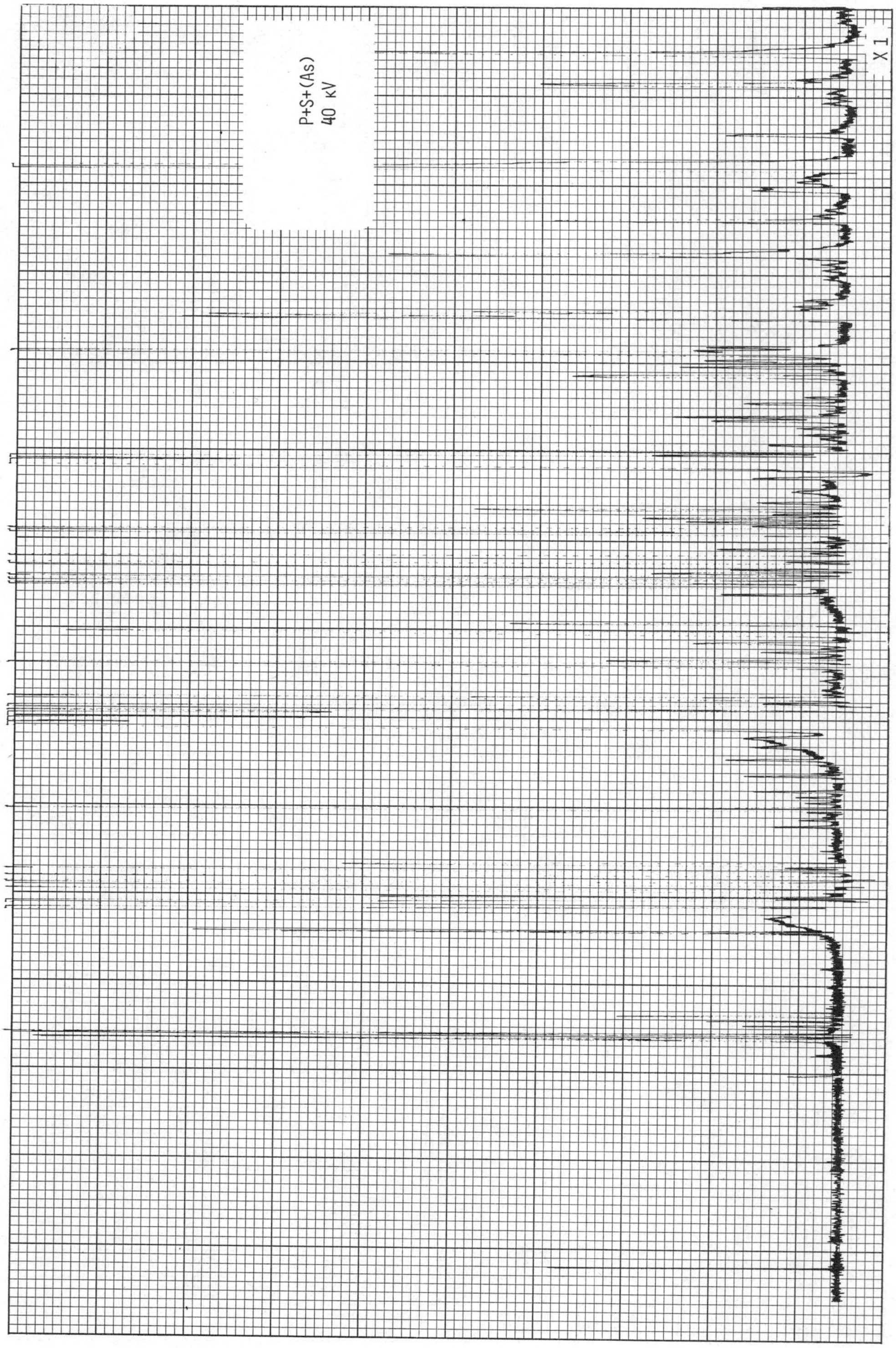
P+S+(As)
40 kV
X 1

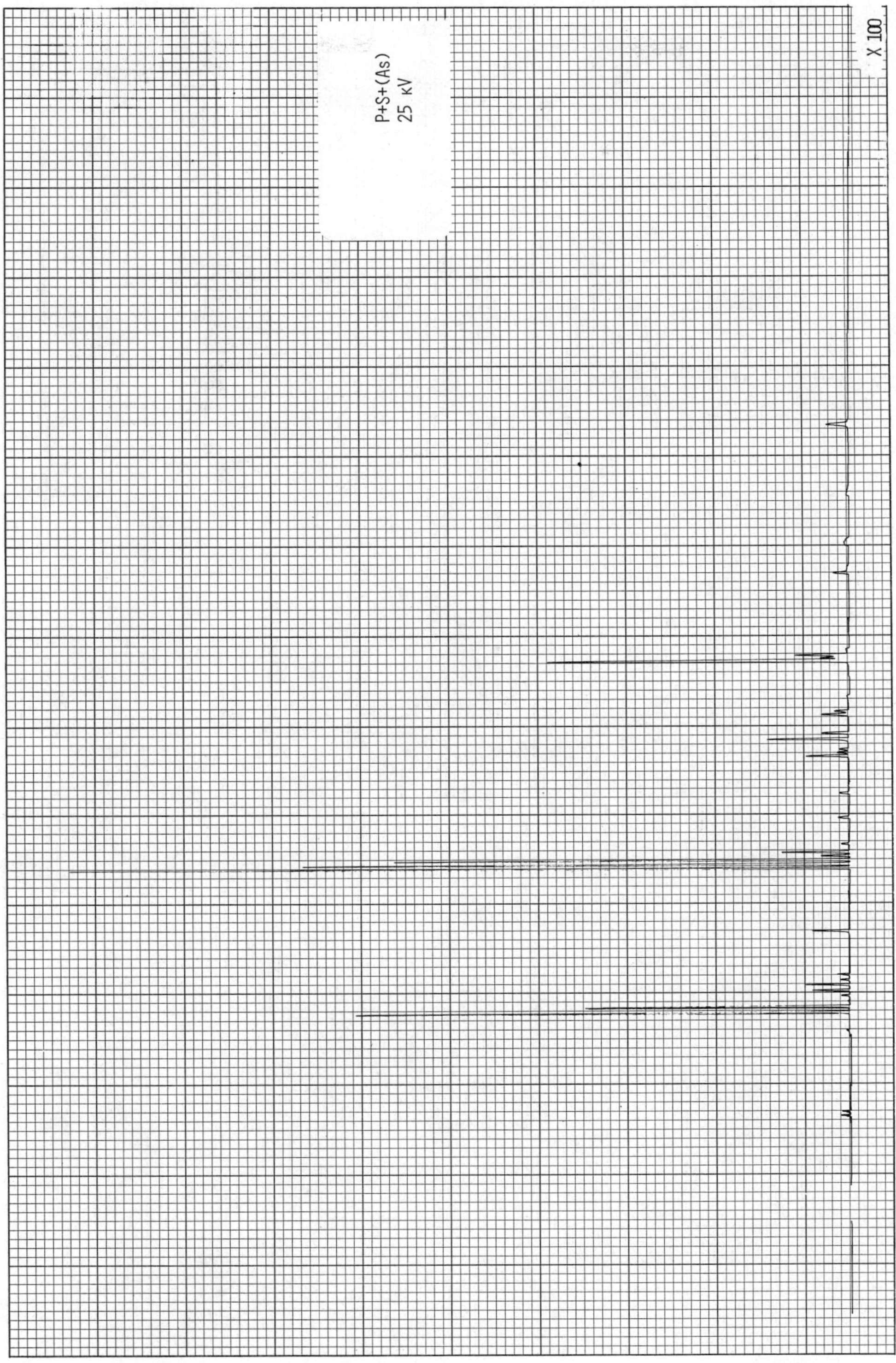

P+S+(As)
25 kV
X 100

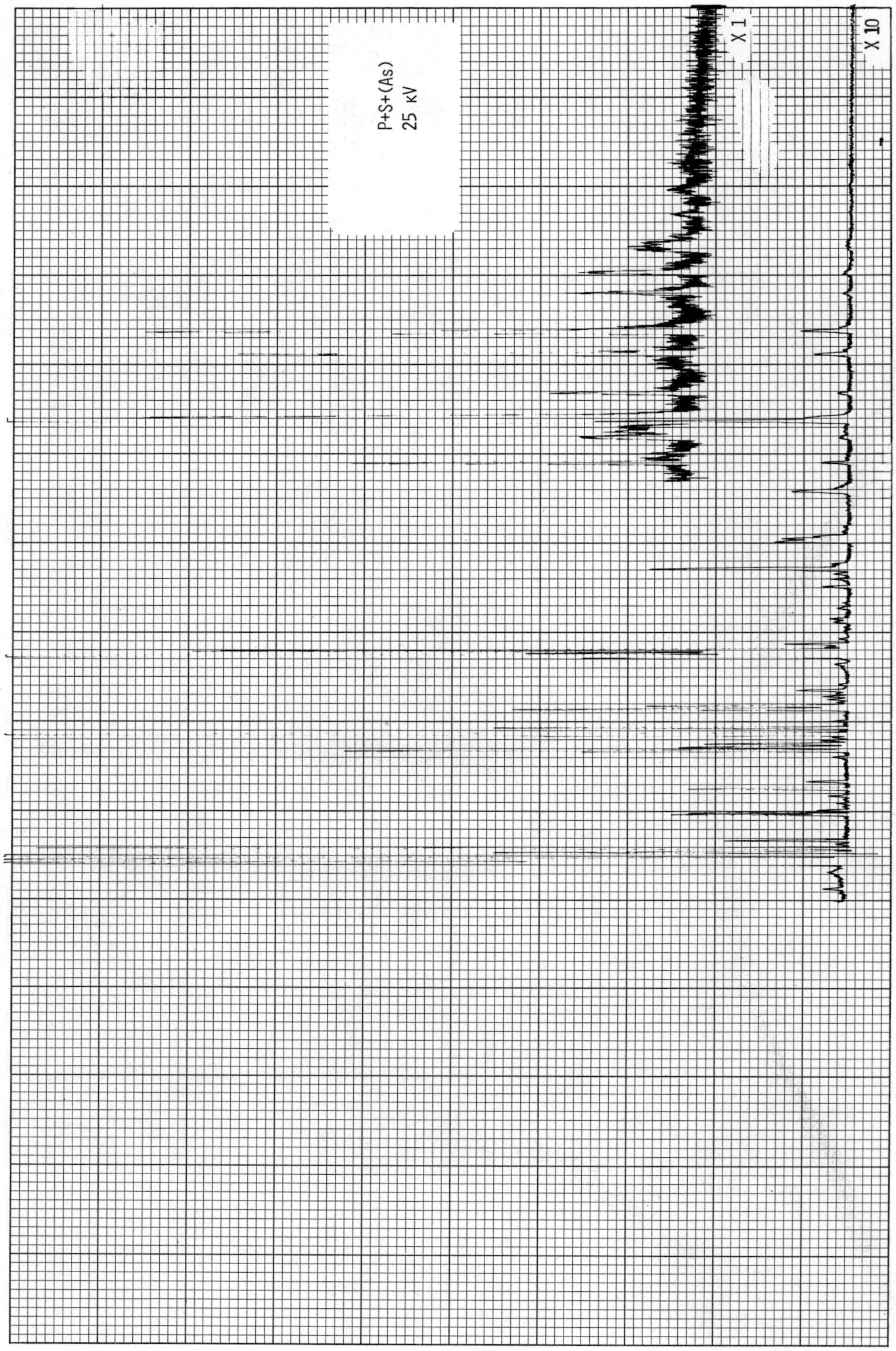
P+S+(As)
25 kV
X 1
X 10

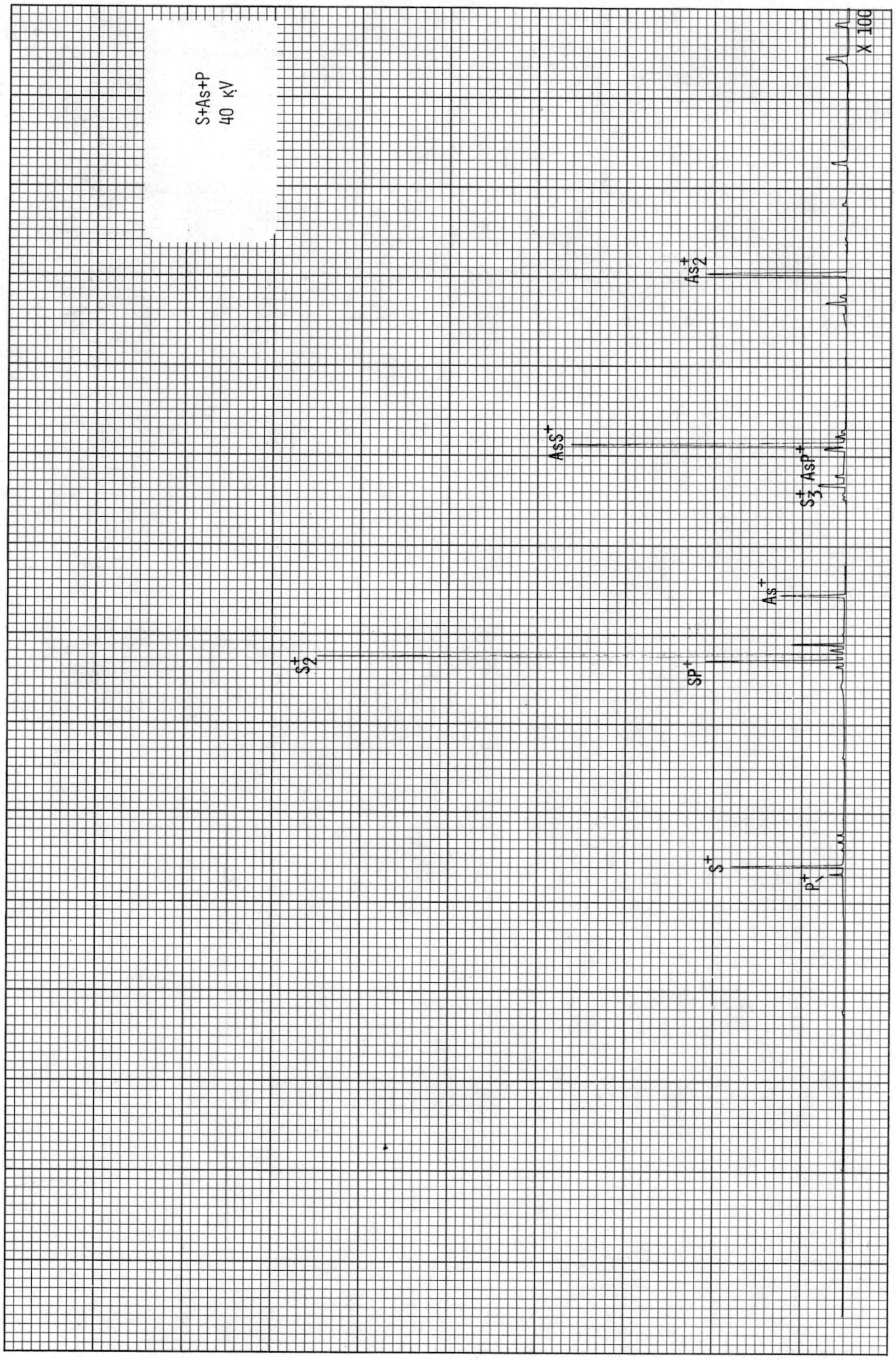

S+As+P
40 kV
X 100
As_2^+
AsS^+
S_3^+, AsP^+
As^+
S_2^+
SP^+
S^+
P^+

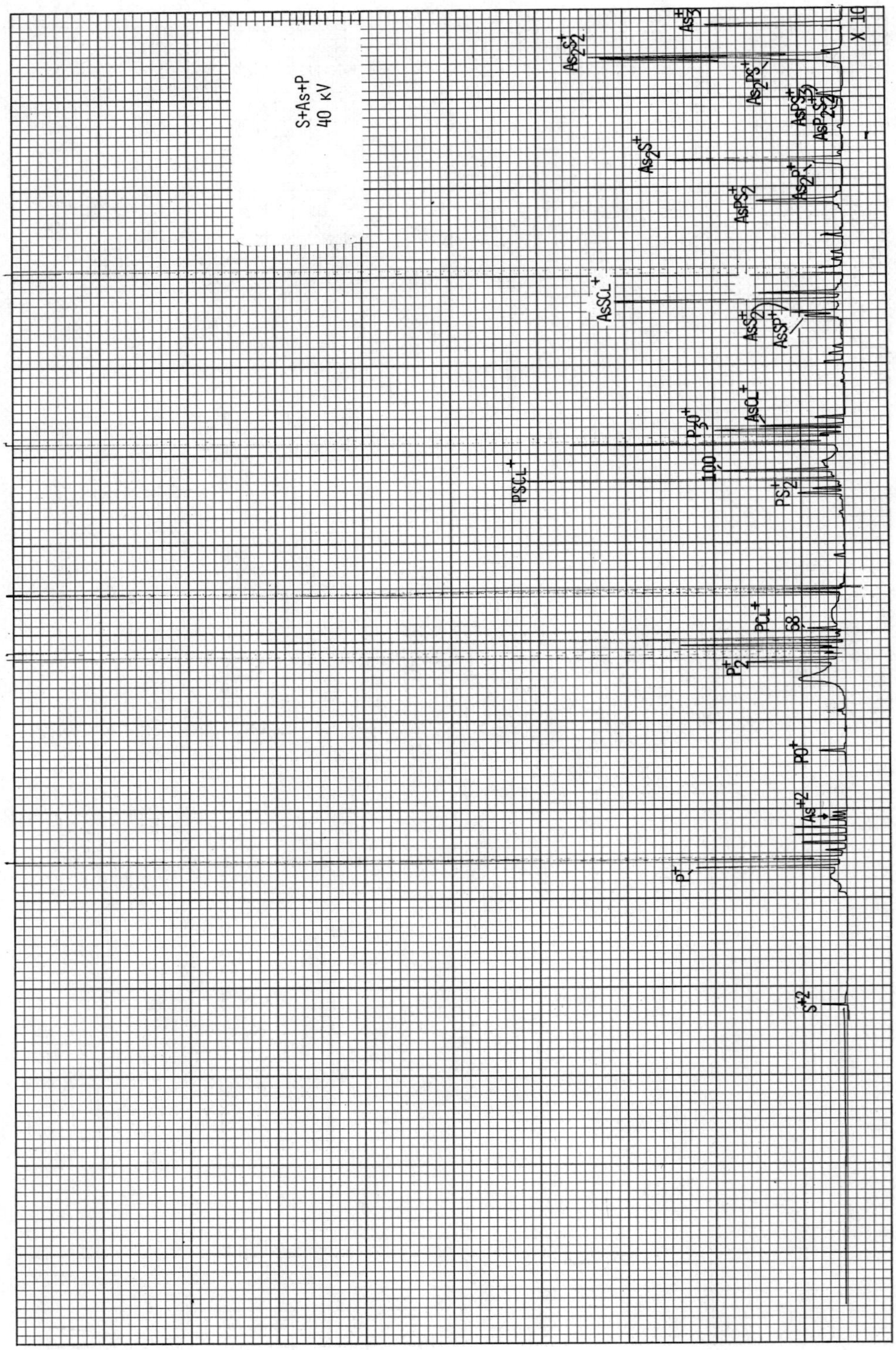
S+As+P
40 kV

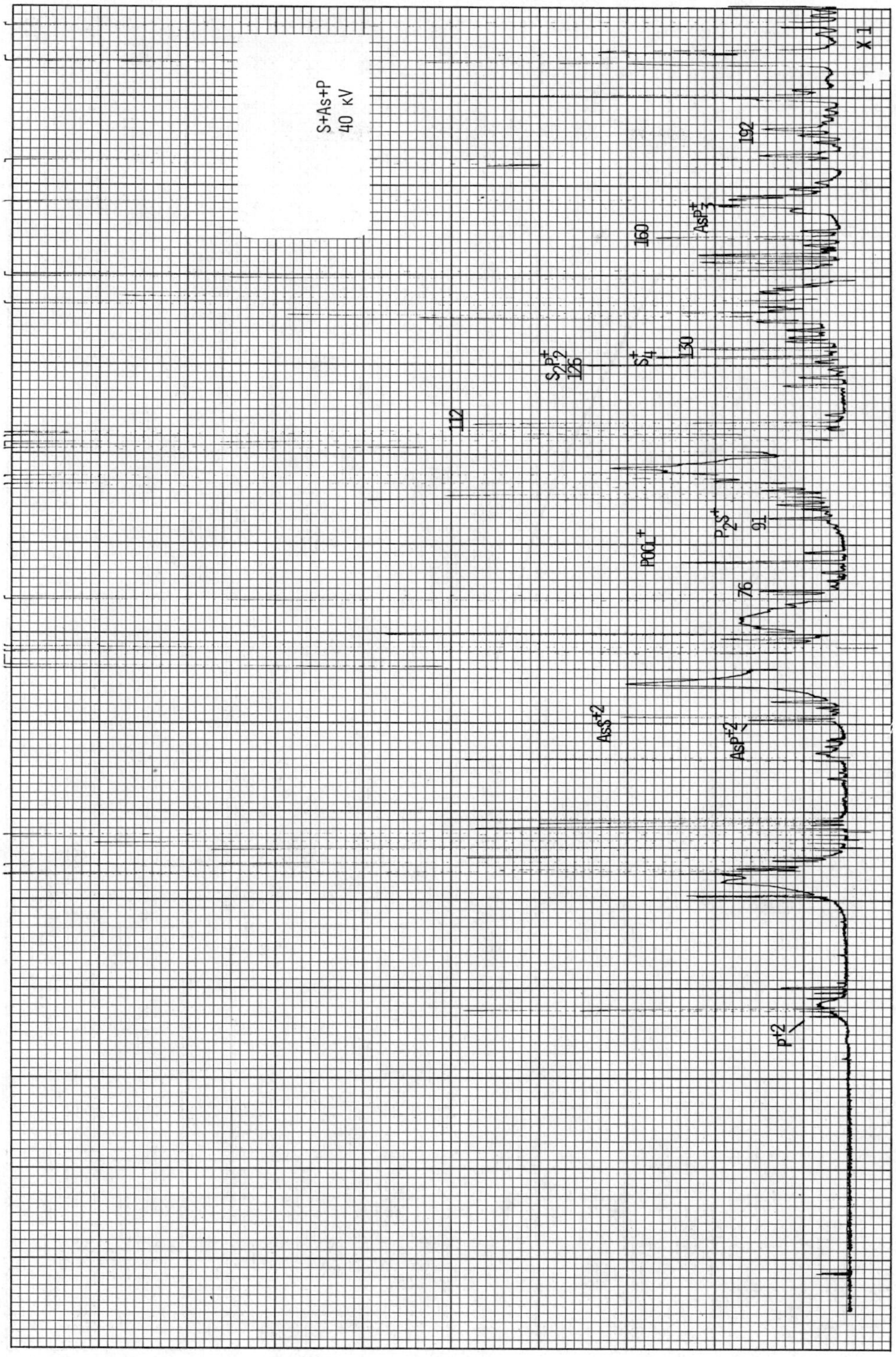
S+As+P
40 kV
X 1
192
AsP3+
160
S2P2+
126
S4+
130
112
POCl+
P2S+
91
76
AsS+2
AsP+2
P+2

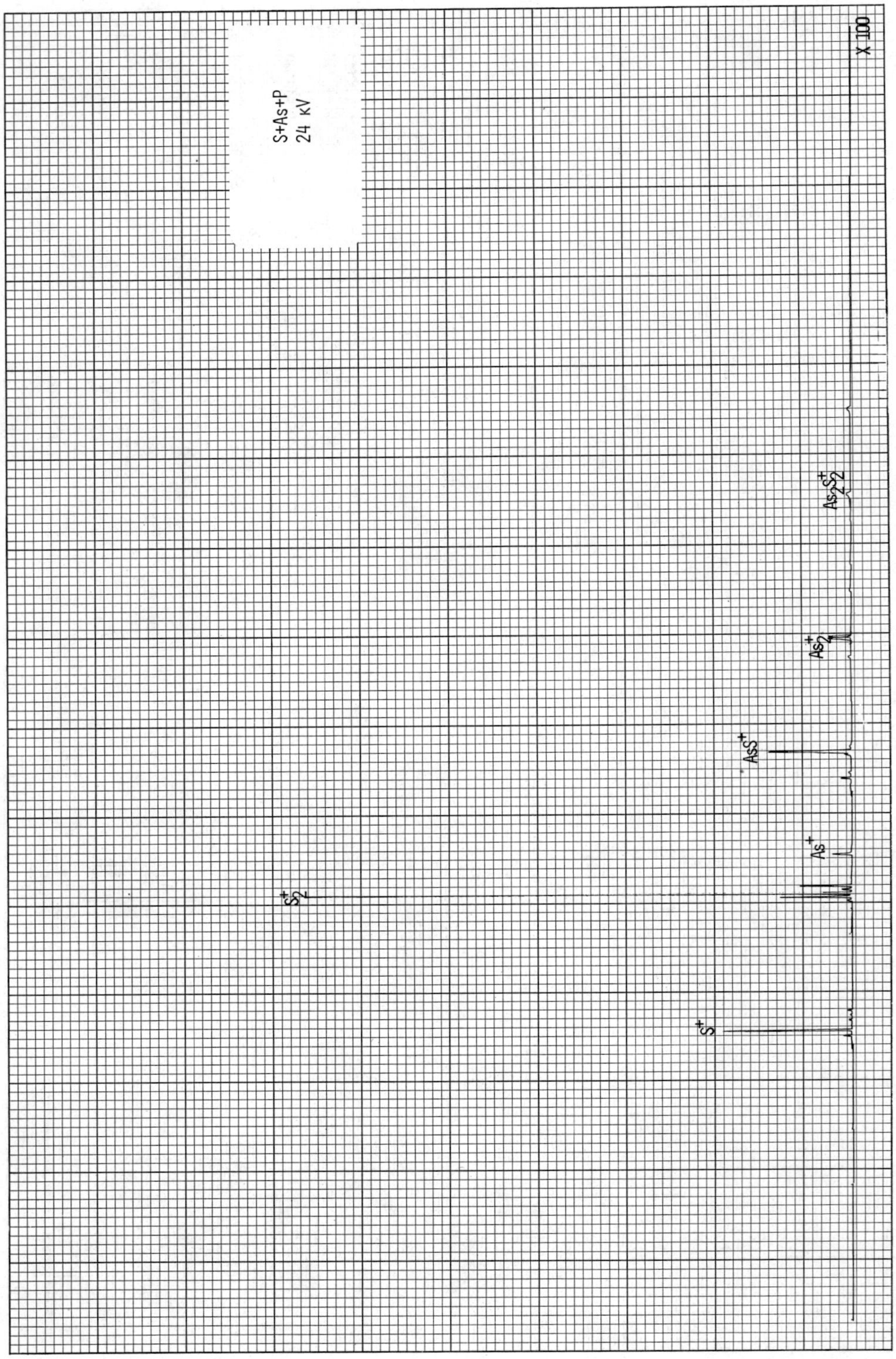
S+As+P
24 kV
X 100
S^+
S_2^+
As^+
AsS^+
As_2^+
$As_2S_2^+$

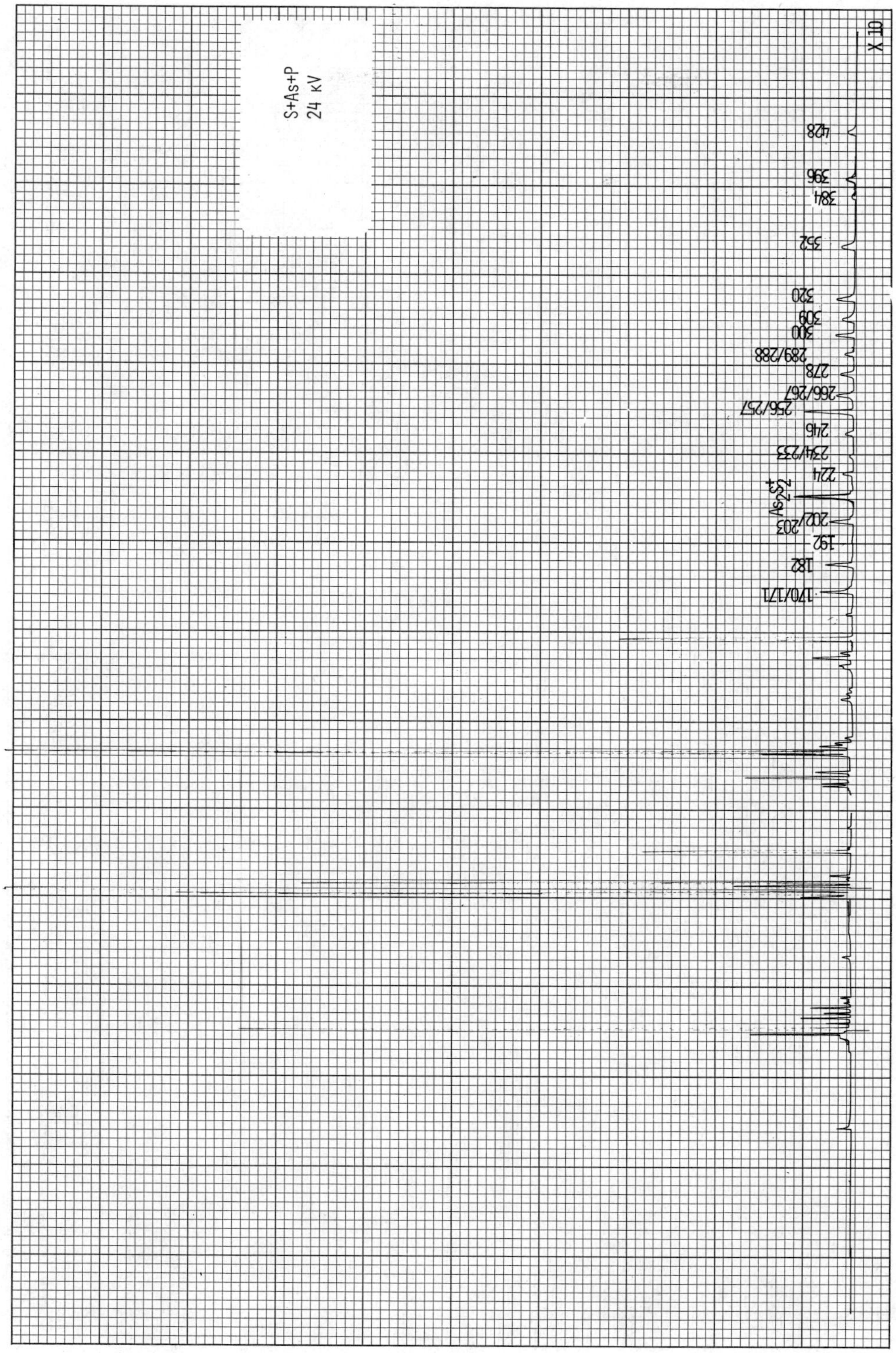
S+As+P
24 kV
X 10
428
396
384
352
320
309
300
289/288
278
266/267
256/257
246
234/233
224
$As_2S_3^+$
202/203
192
182
170/171

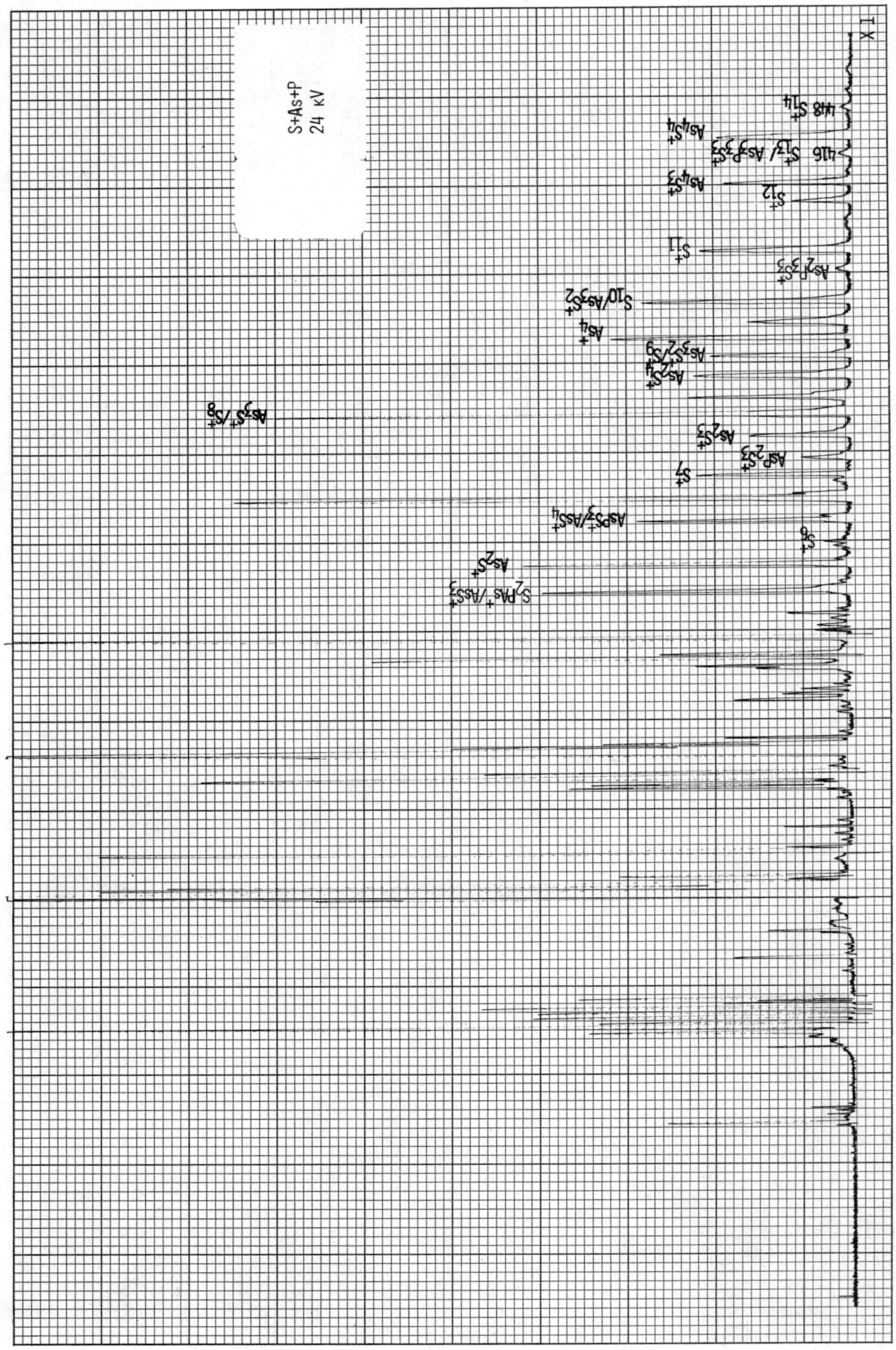
S+As+P
24 kV

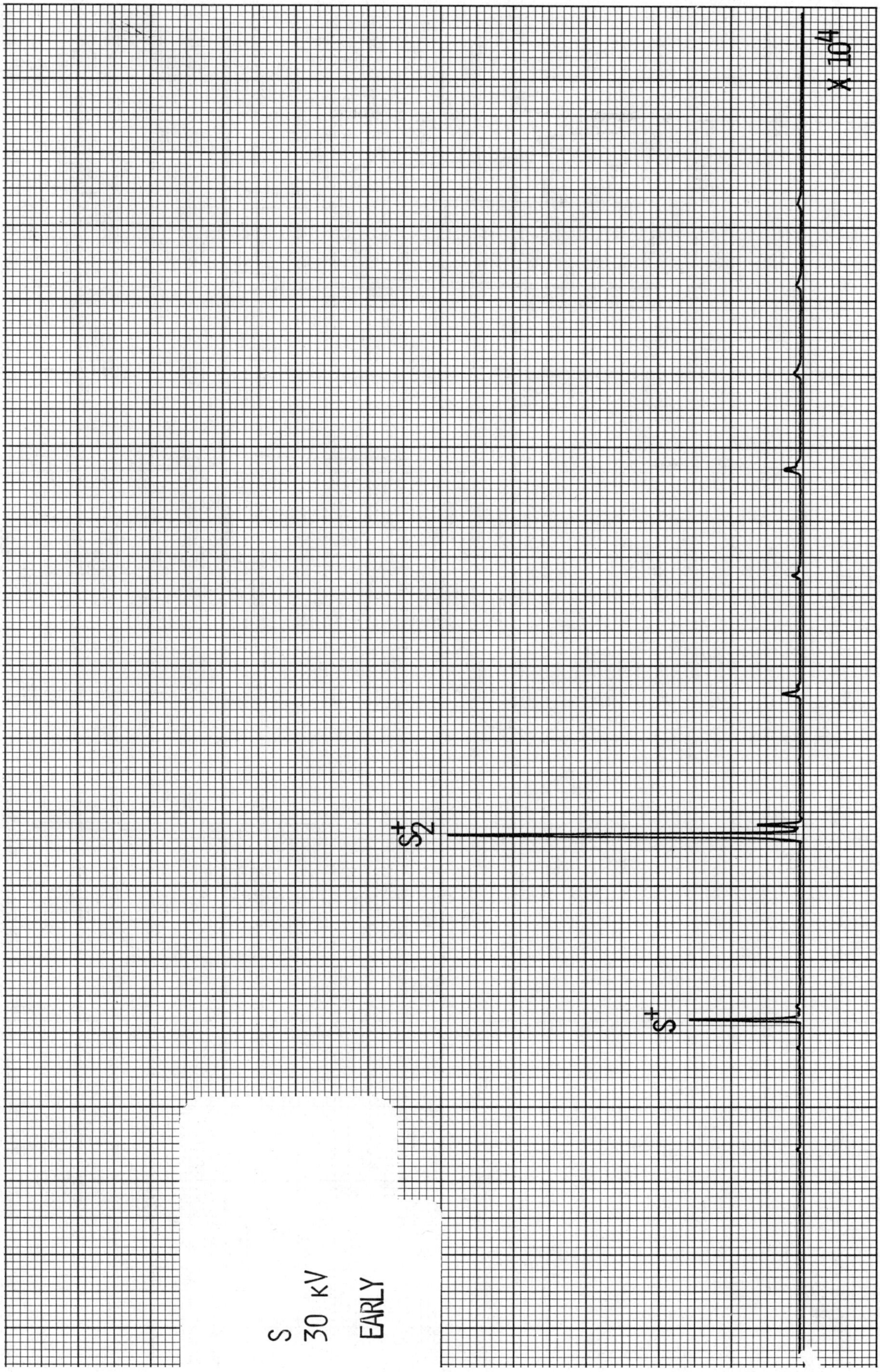
S
30 kV
EARLY
S_2^+
S^+
X 10^4

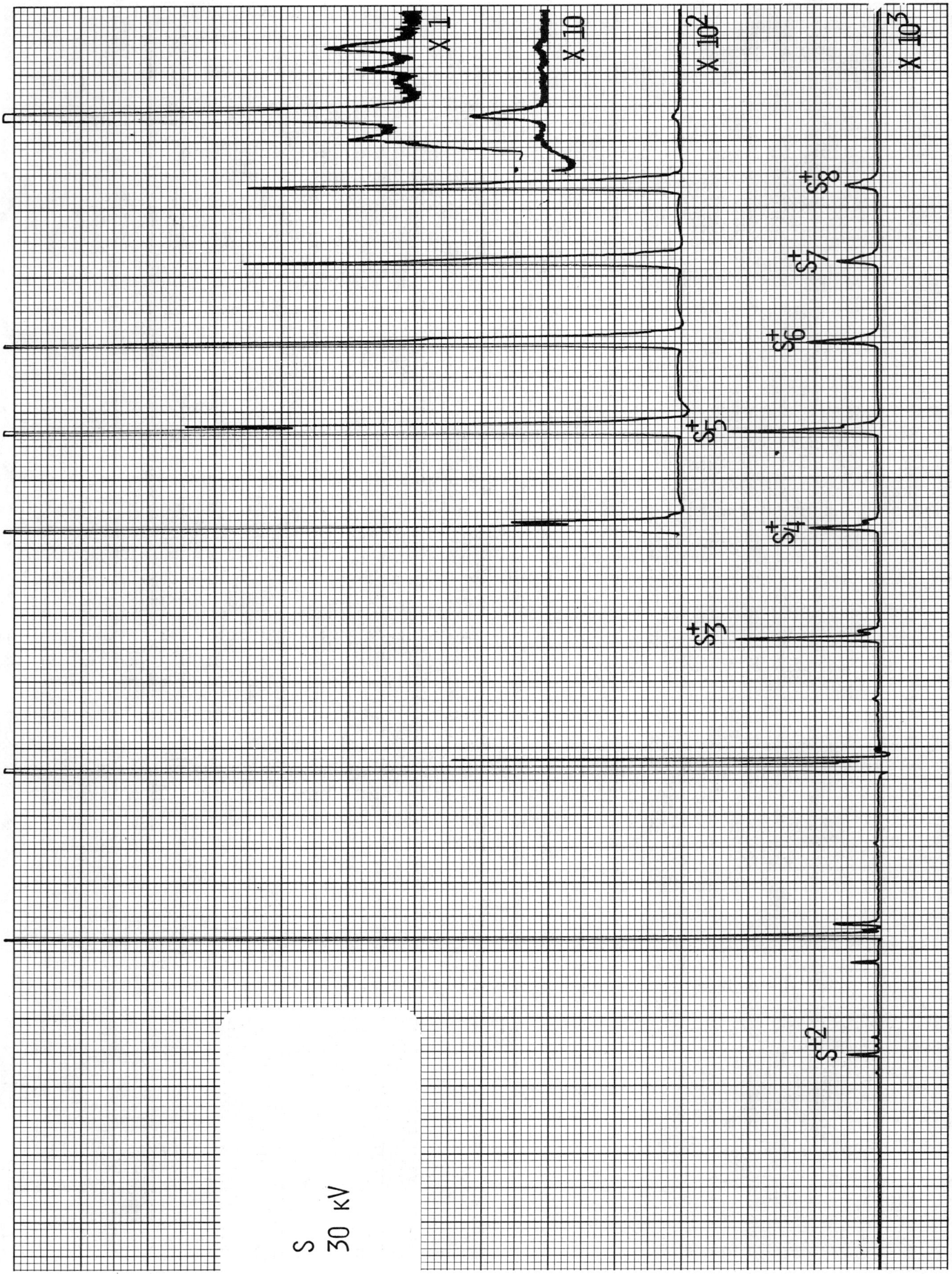
S
30 kV
X 1
X 10
X 10²
X 10³
S₂⁺
S₃⁺
S₄⁺
S₅⁺
S₆⁺
S₇⁺
S₈⁺
S⁺²

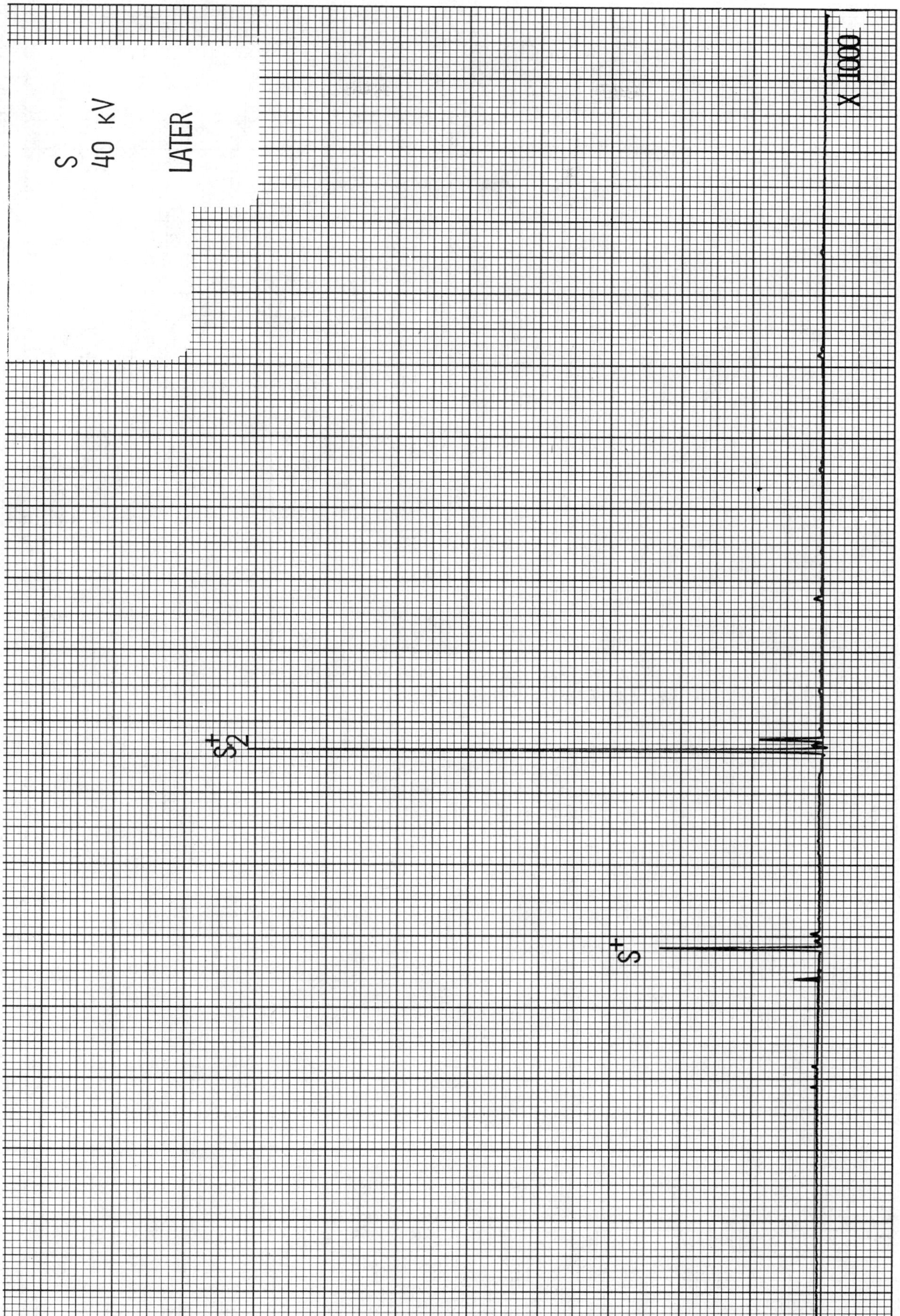
S
40 kV
LATER
X 1000
S_2^+
S^+

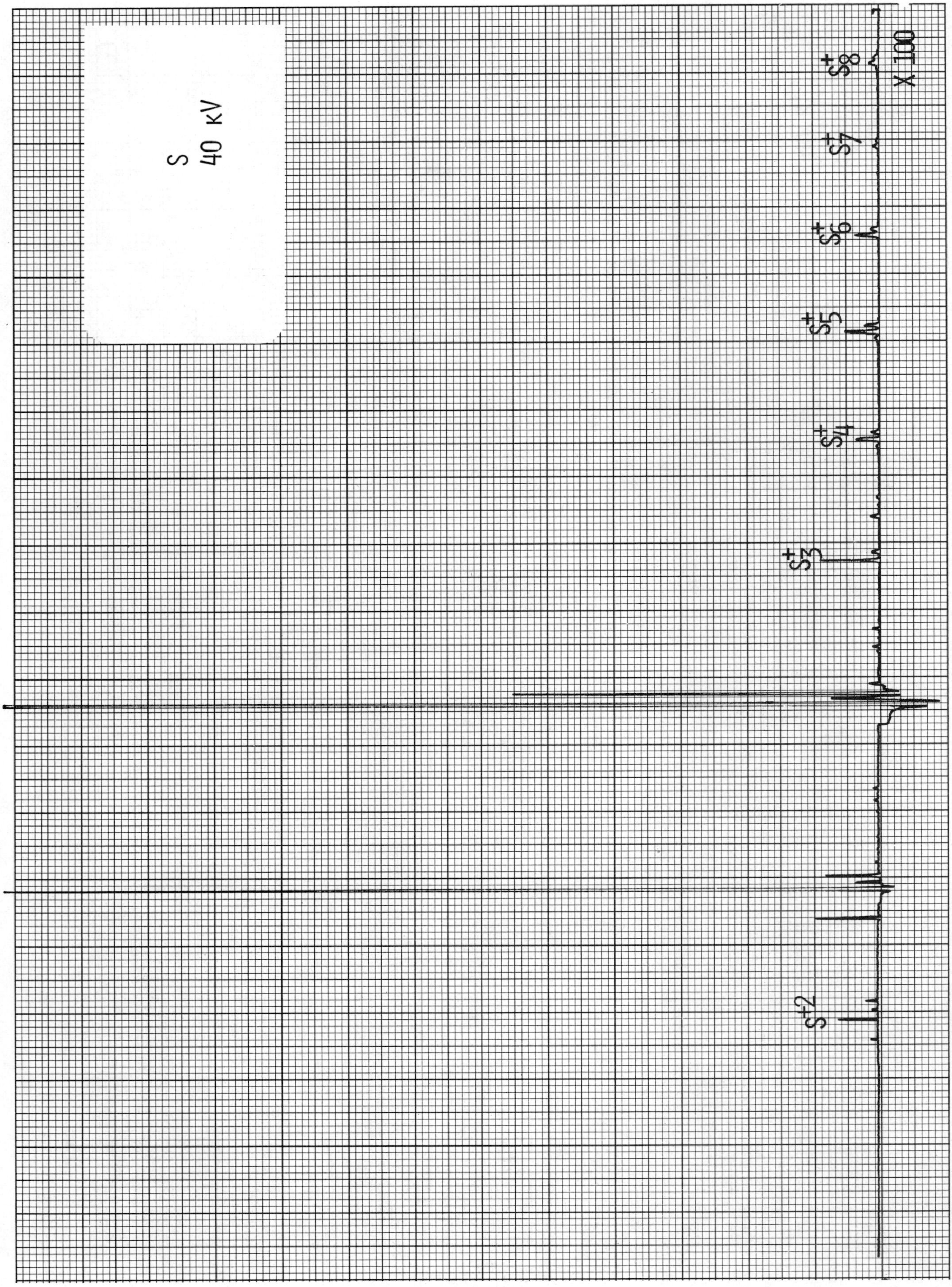
S
40 kV
S_2^+
S_3^+
S_4^+
S_5^+
S_6^+
S_7^+
S_8^+
X 100

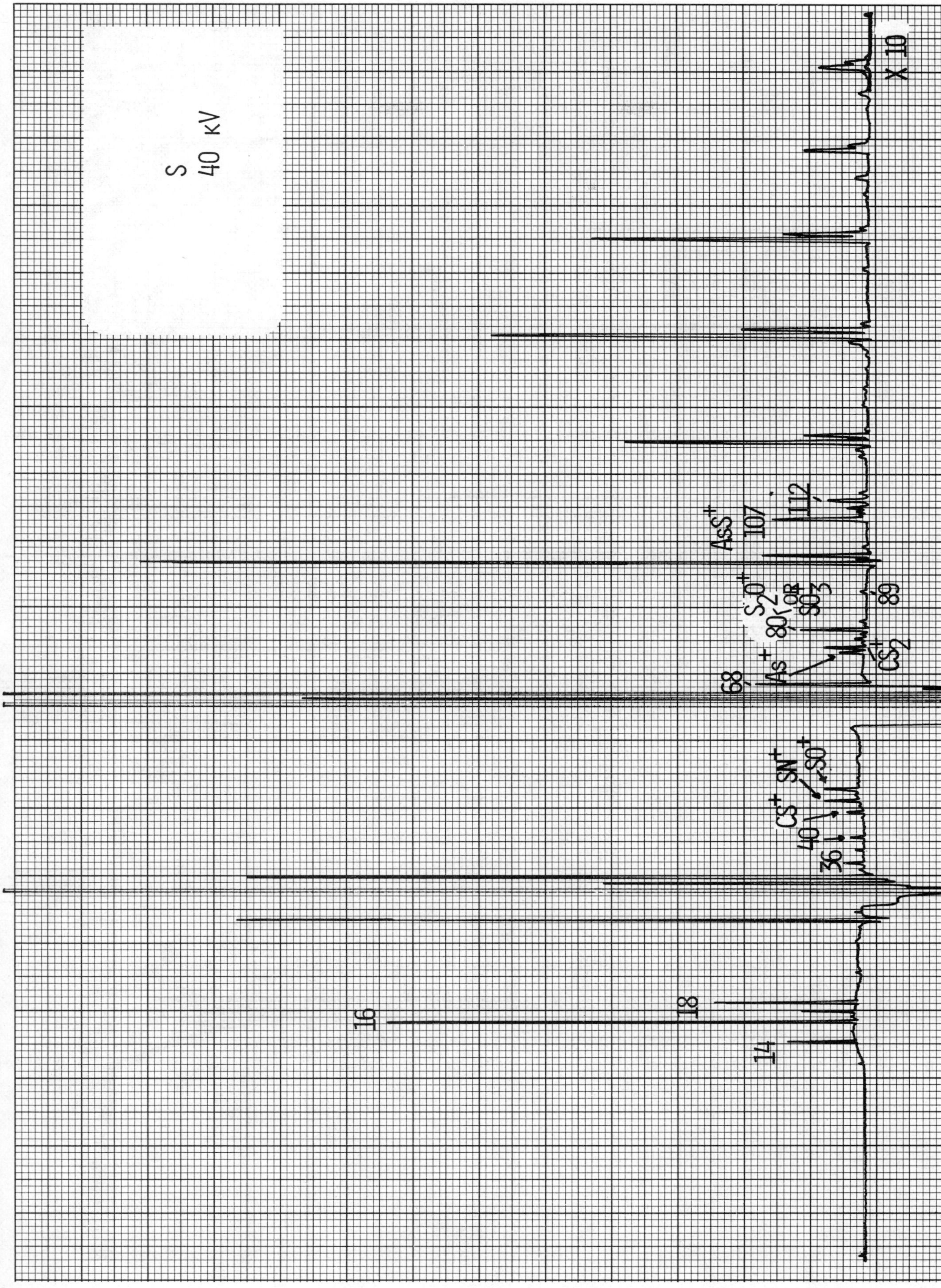

S
40 kV
X 10
14
16
18
36
40
CS^+
SN^+
SO^+
68
As^+
80
S_2O^+
OR
SO_3^+
CS_2^+
89
AsS^+
107
112

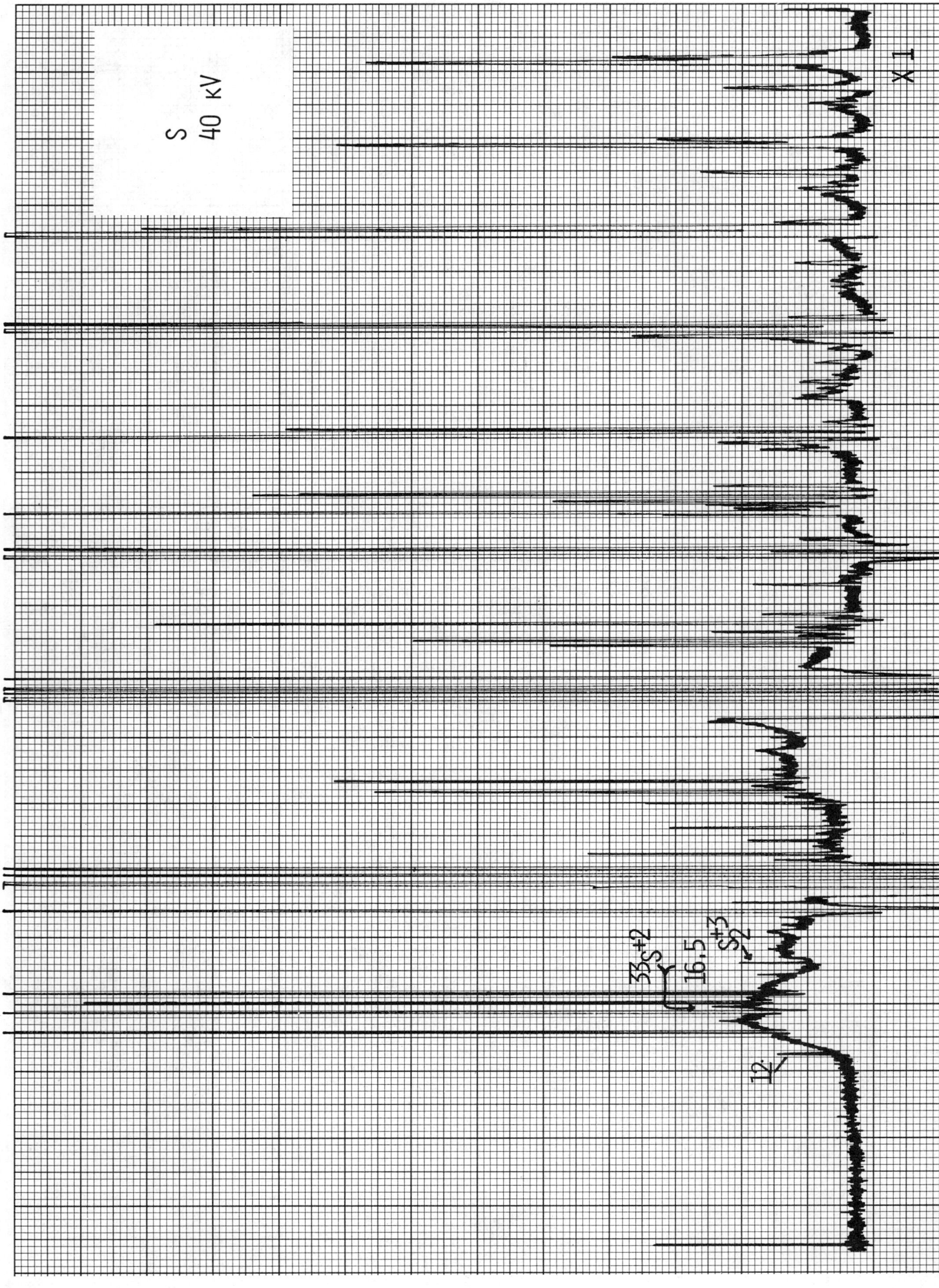
S
40 kV
X 1
16.5
12

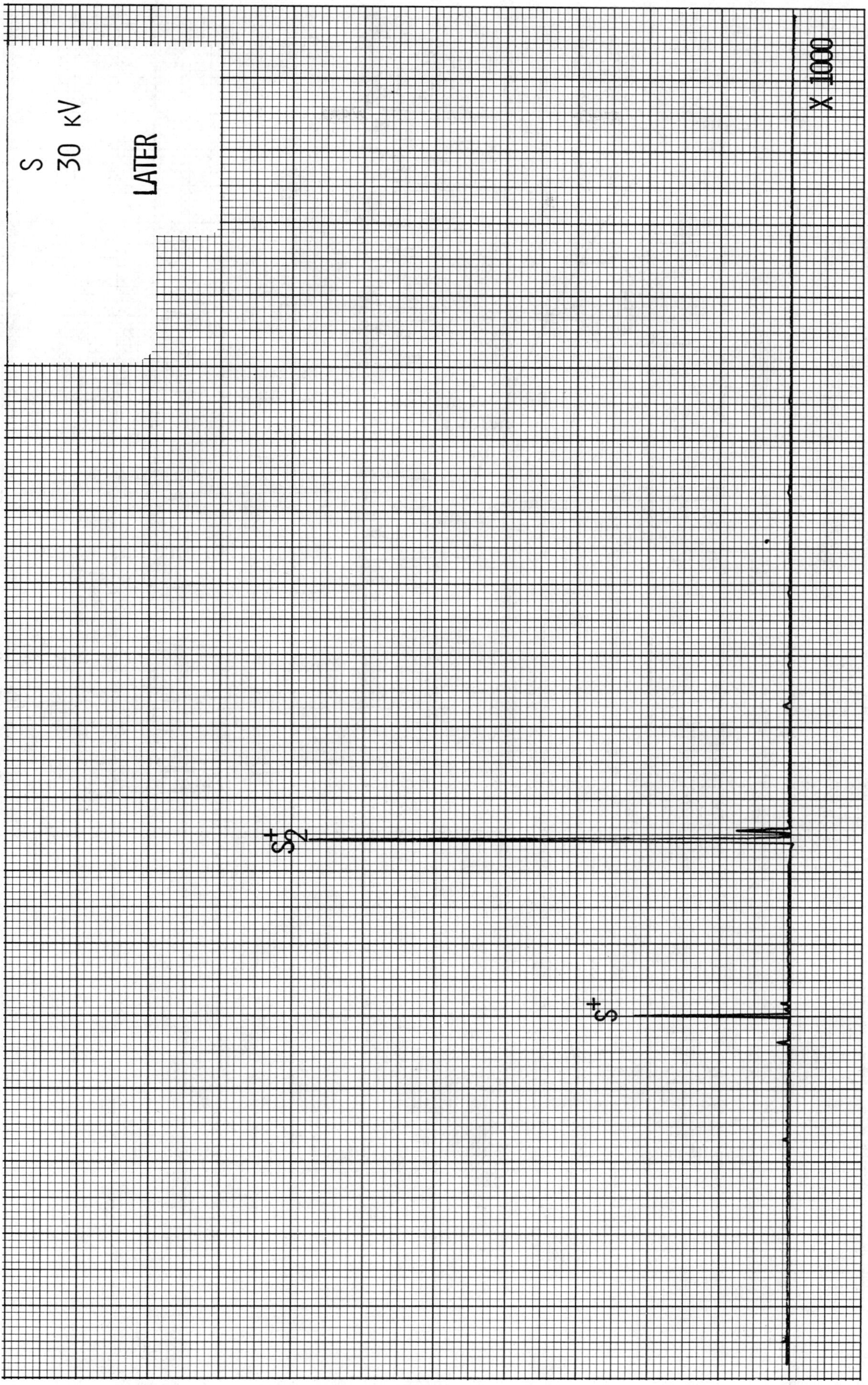

S
30 kV
LATER
X 1000
S_2^+
S^+

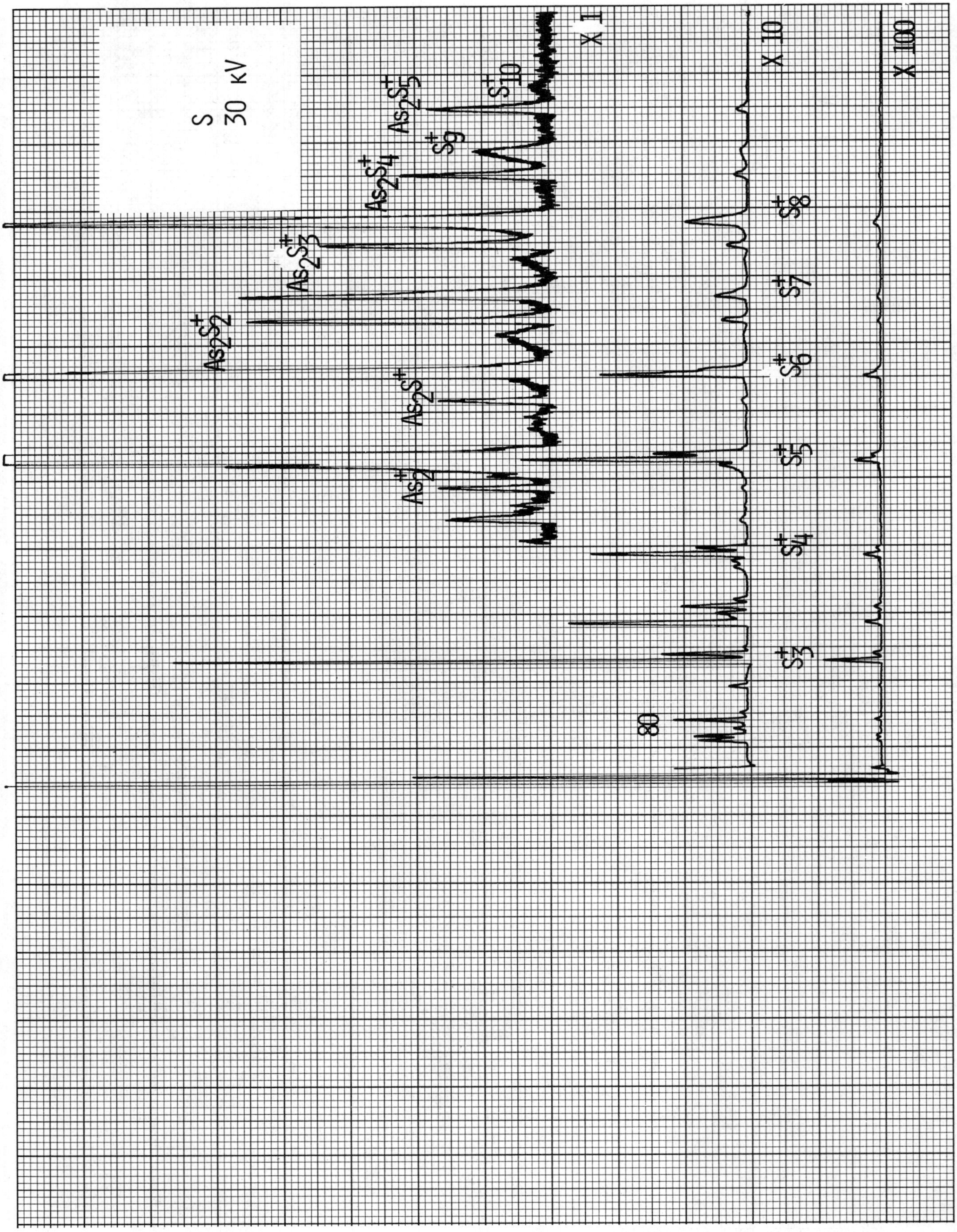
S
30 kV
X 1
X 10
X 100
As_2^+
As_2S^+
$As_2S_2^+$
$As_2S_3^+$
$As_2S_4^+$
$As_2S_5^+$
S_3^+
S_4^+
S_5^+
S_6^+
S_7^+
S_8^+
S_9^+
S_{10}^+
80

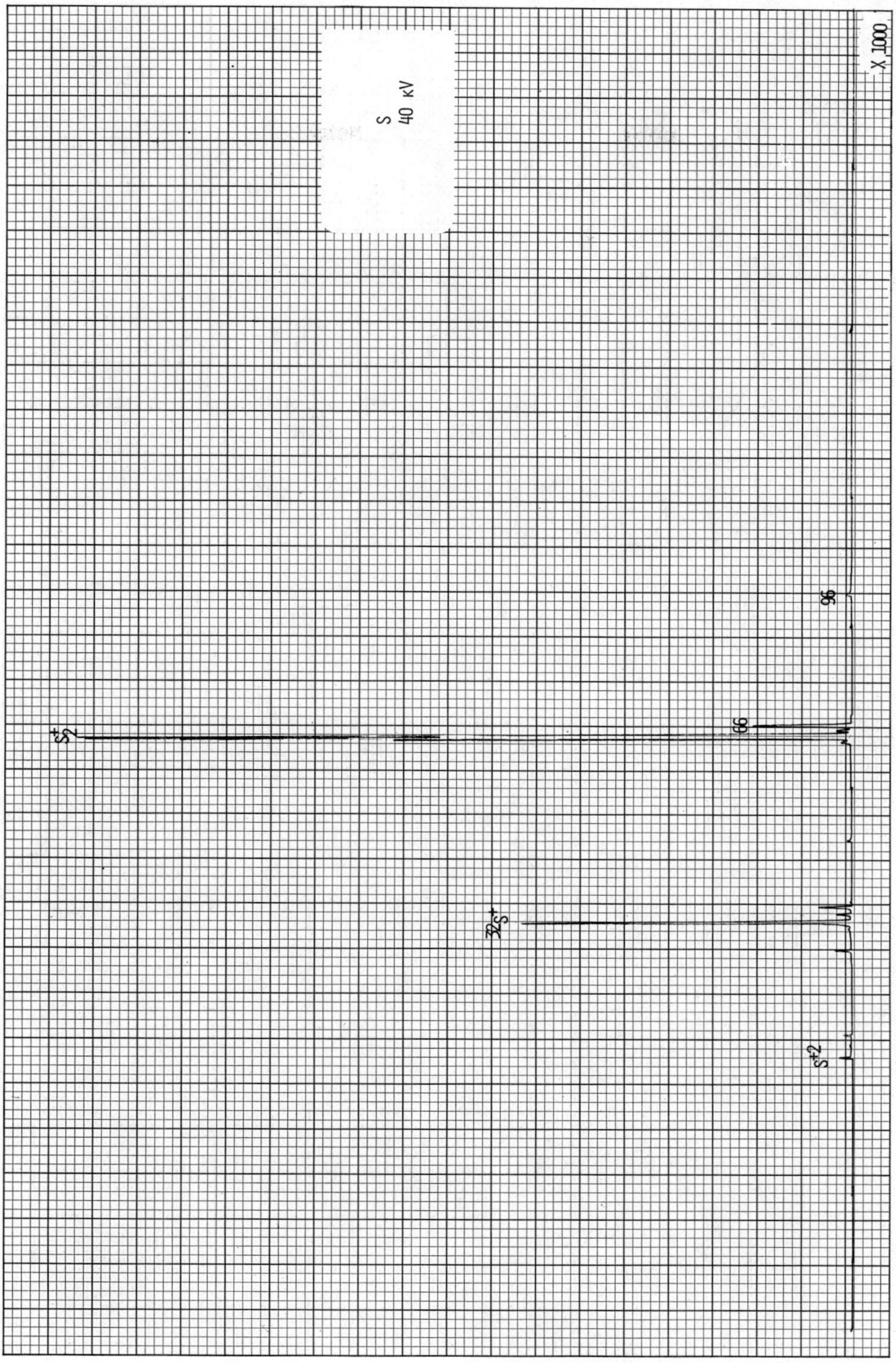

S
40 kV
X 1000
96
66
S_2^+
$^{32}S^+$
S^{+2}

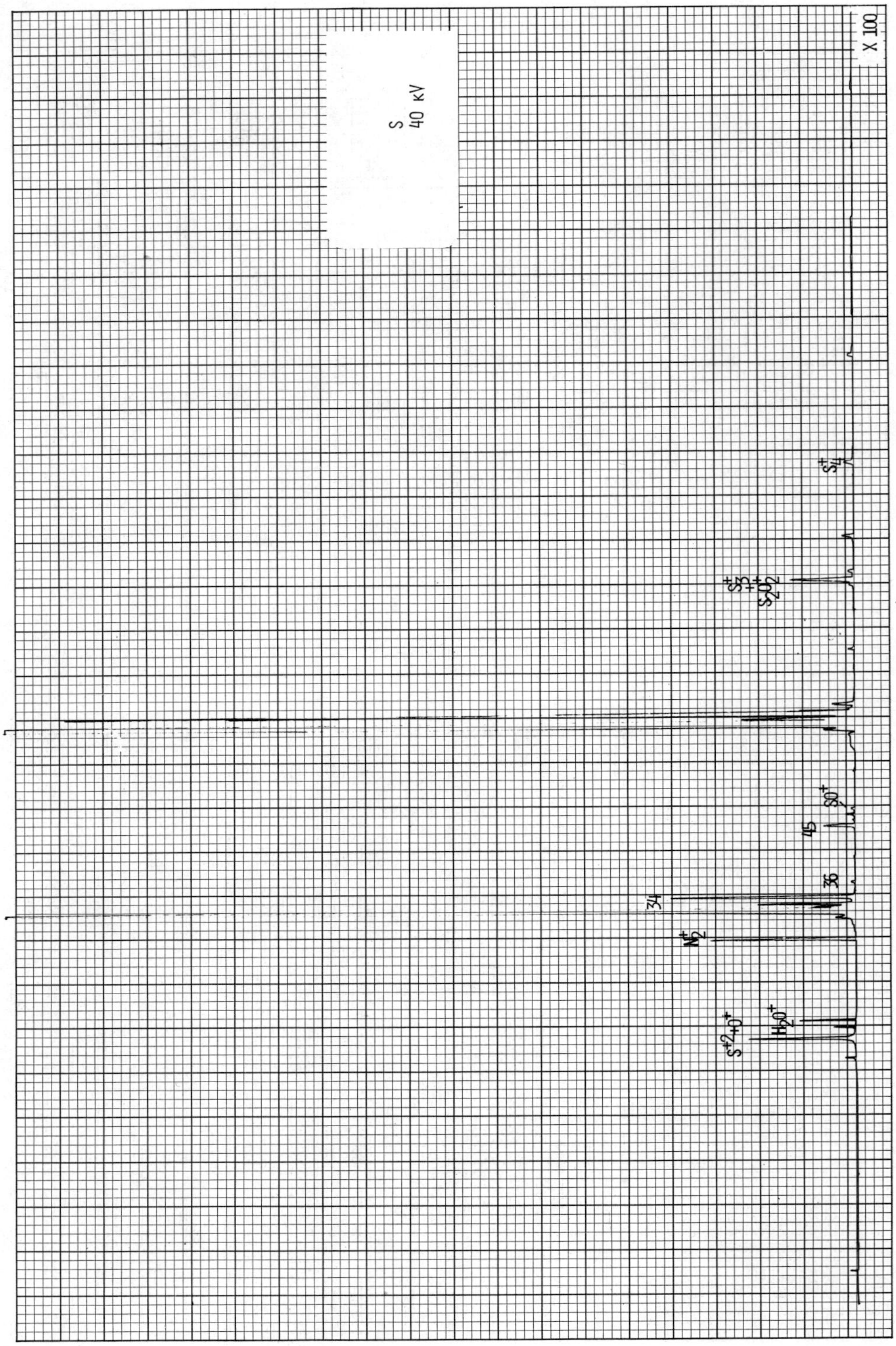
S
40 kV
X 100

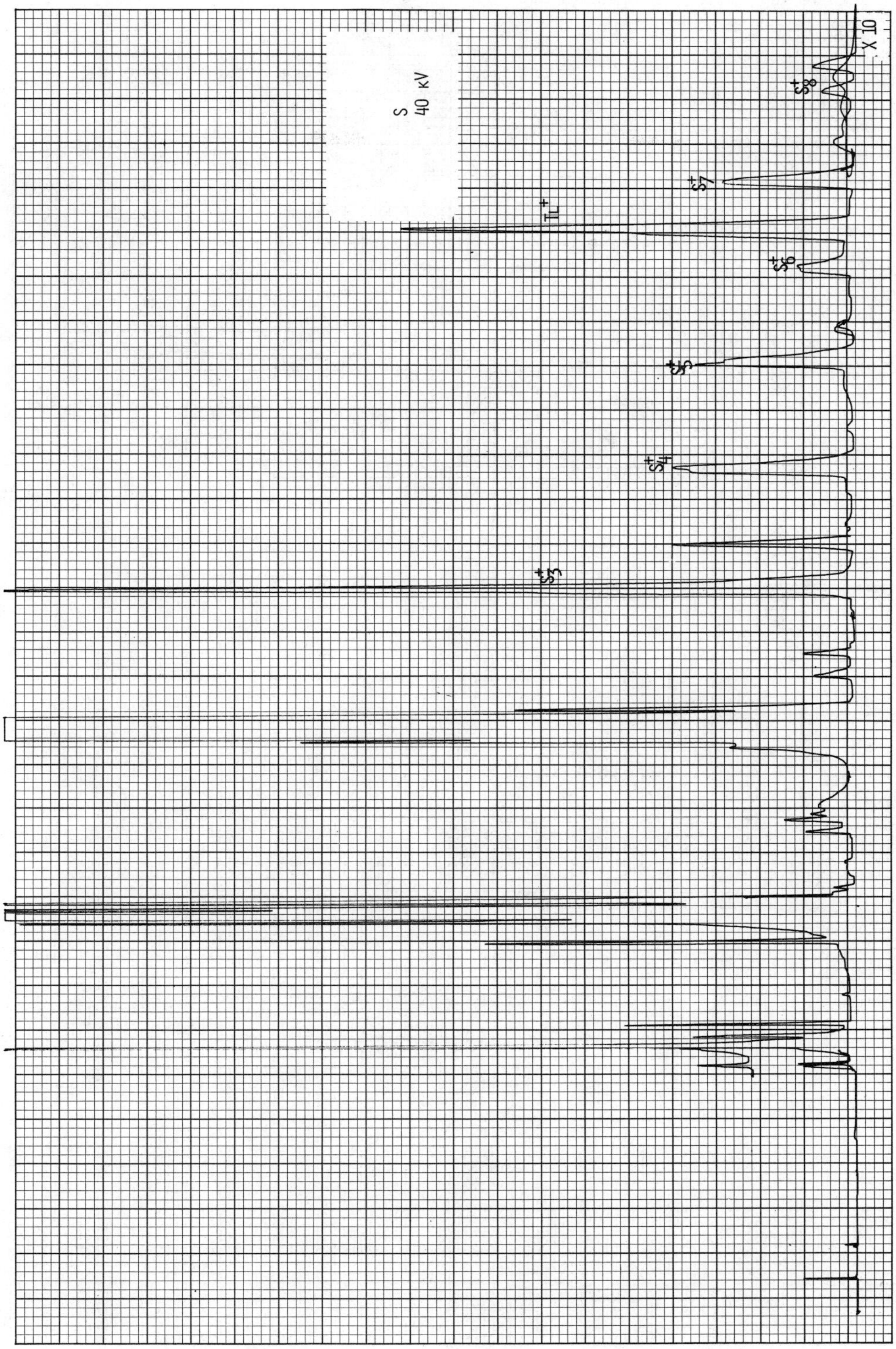

S
40 kV
X 10
S_8^+
S_7^+
Tl^+
S_6^+
S_5^+
S_4^+
S_3^+

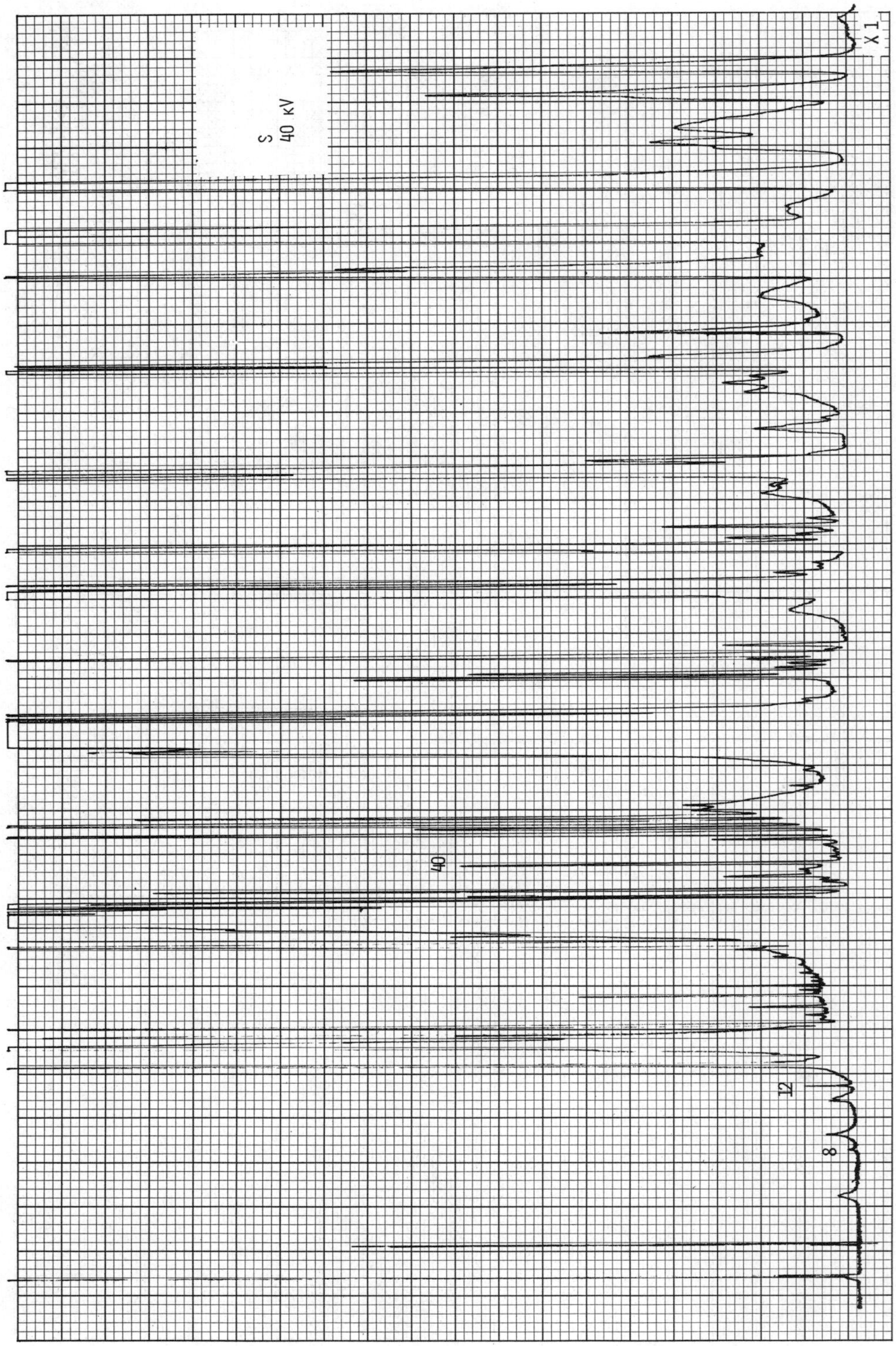
S
40 kV
X 1
40
12
8

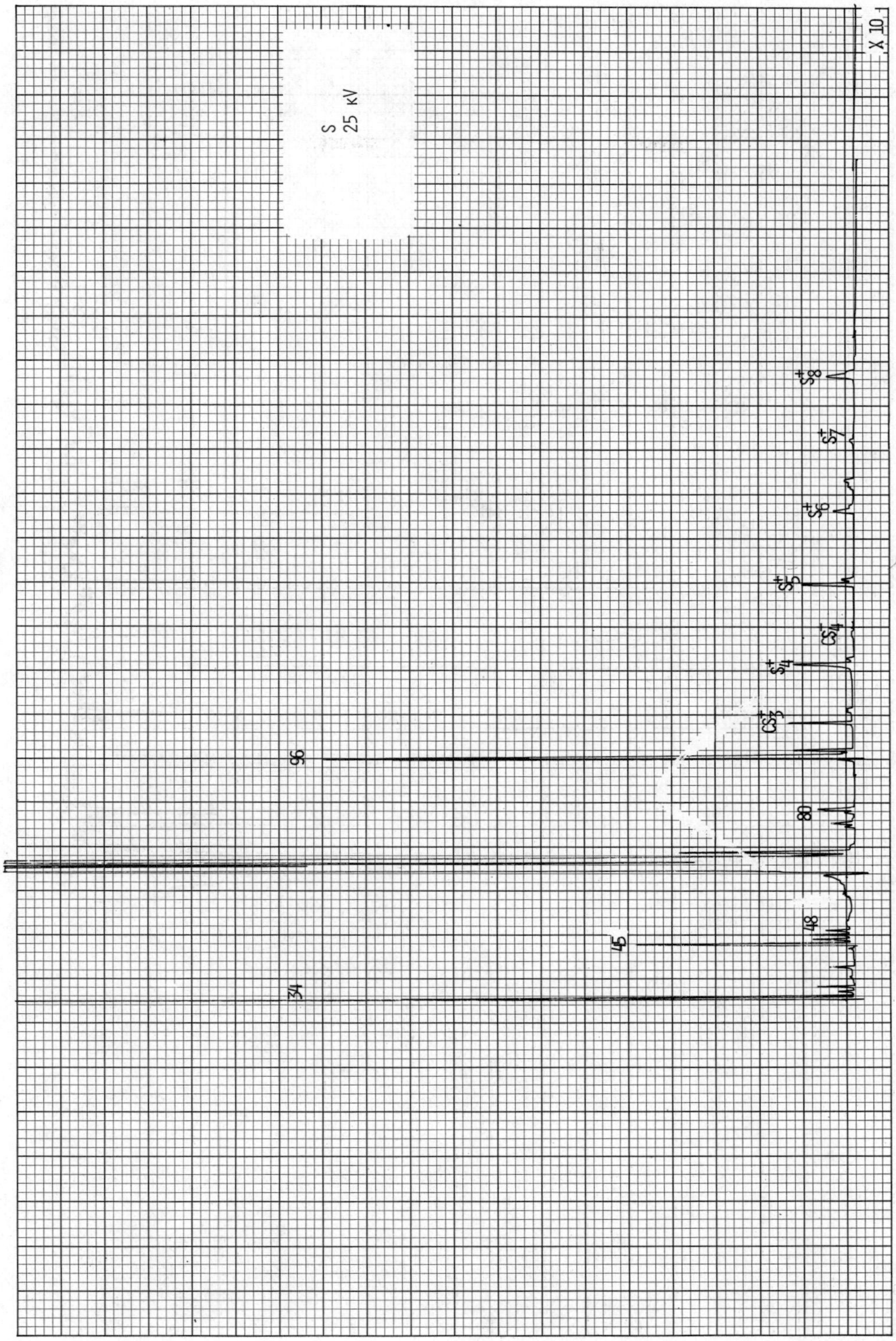
S
25 kV
X 10
S_8^+
S_7^+
S_6^+
S_5^+
CS_4^+
S_4^+
CS_3^+
96
80
48
45
34

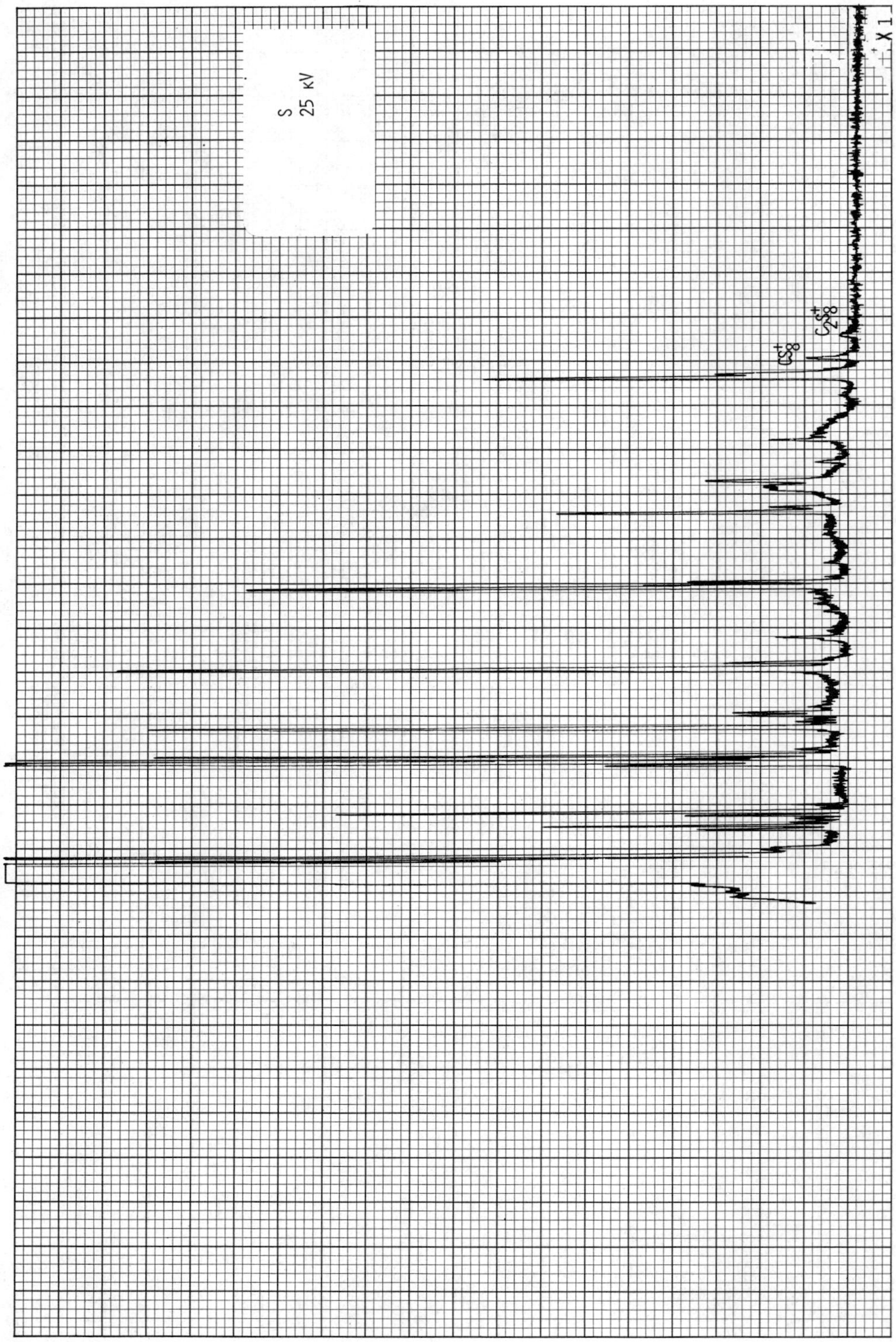

S
25 kV
CS_8^+
$C_2S_8^+$
X 1

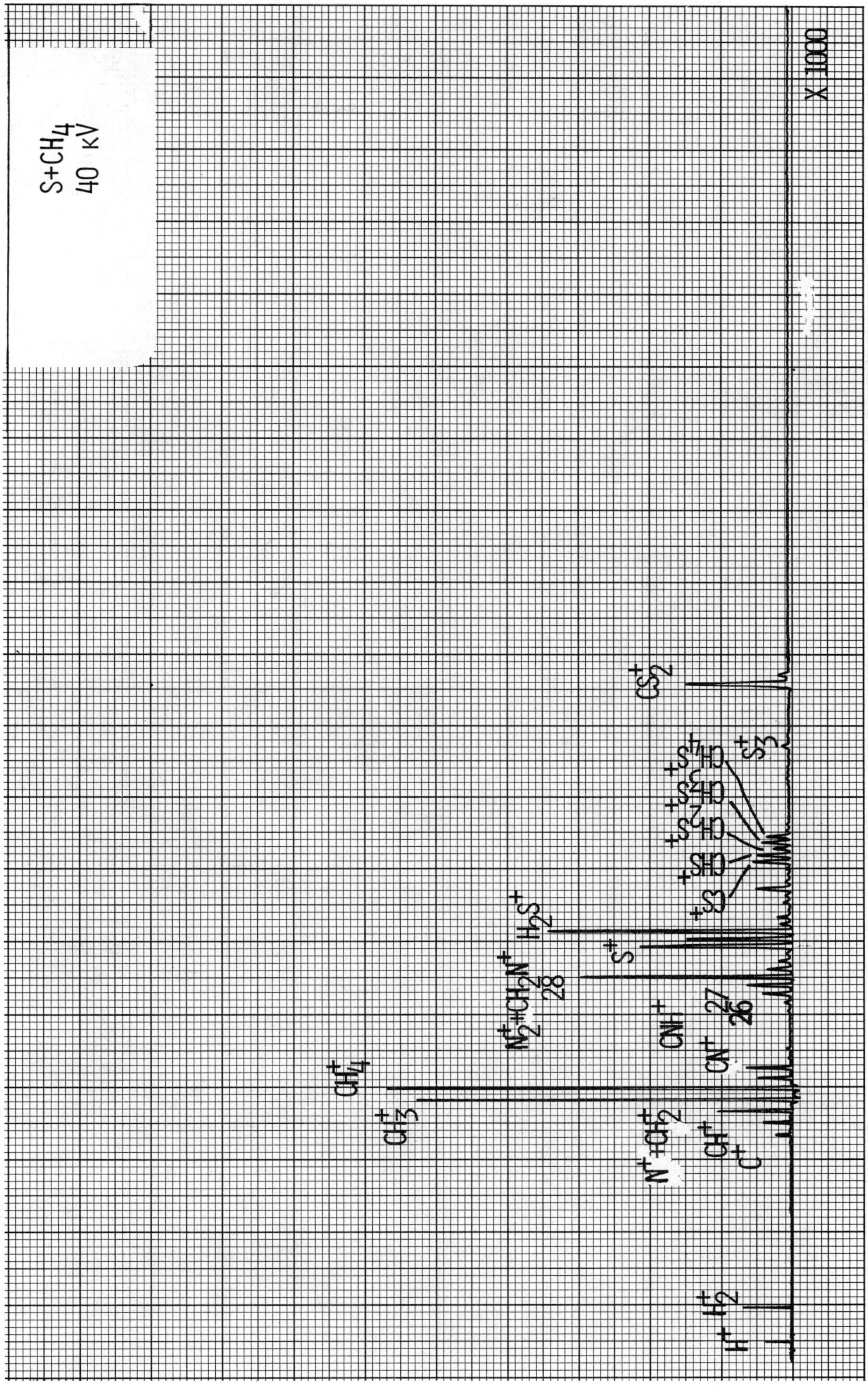
S+CH4
40 kV
X 1000
H+
H2+
C+
CH+
N+ +CH2+
CH3+
CH4+
CN+
CNH+
27
26
N2+ +CH2N+
28
S+
H2S+
CS+
CHS+
CH2S+
CH3S+
CH4S+
S3+
CS2+

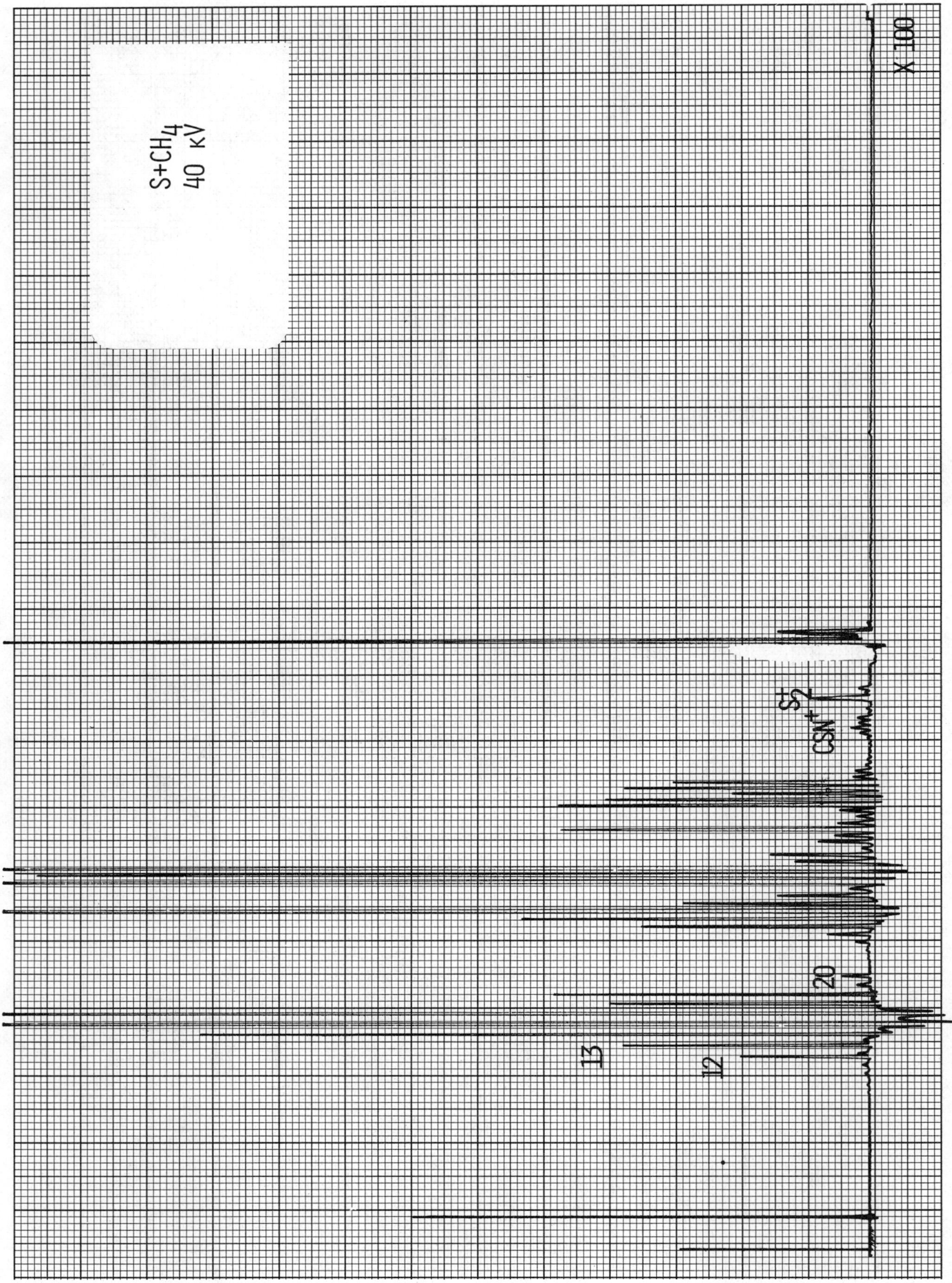
S+CH$_4$
40 kV
X 100
S$_2^+$
CSN$^+$
20
13
12

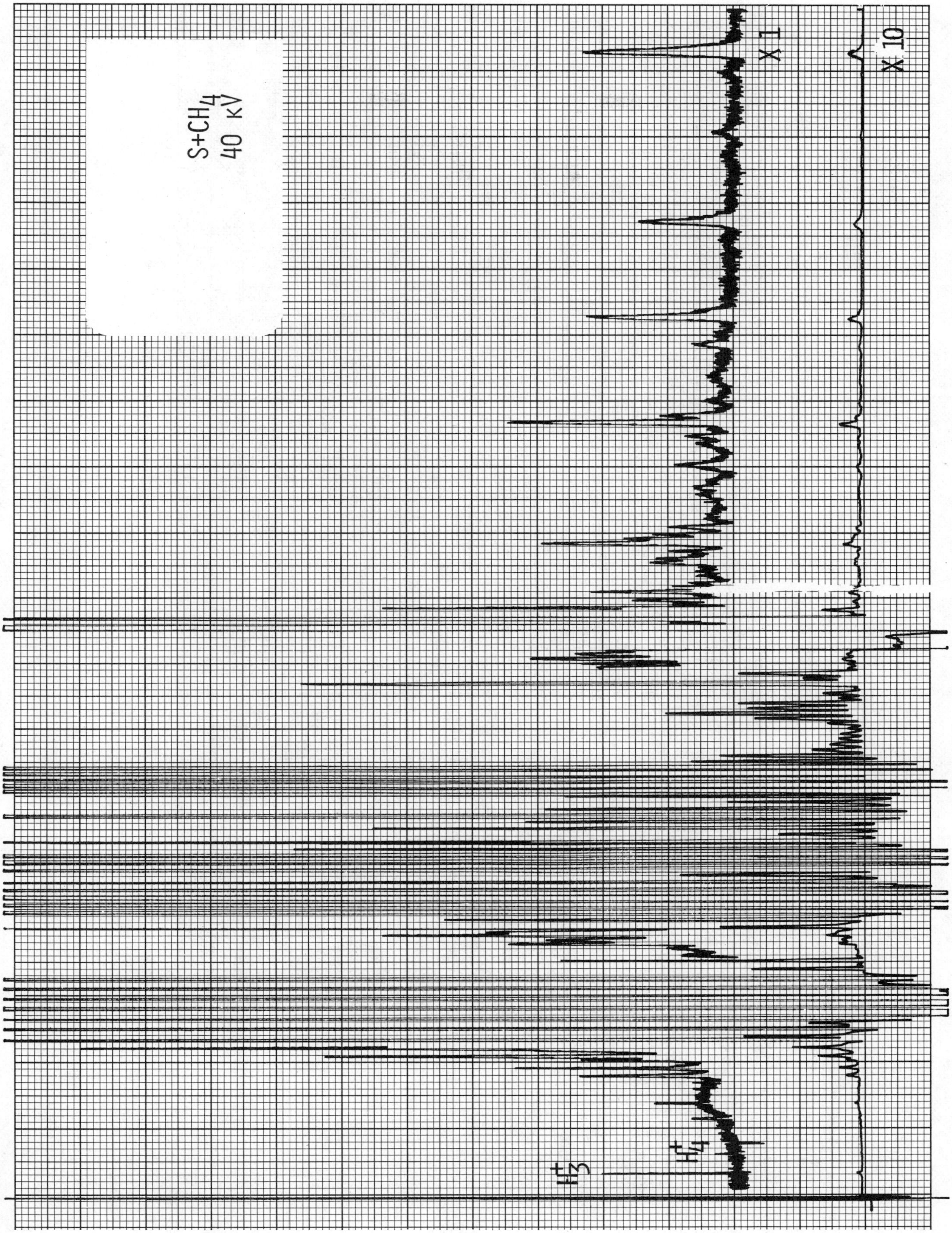
S+CH$_4$
40 kV
X 1
X 10
H_3^+
H_4^+

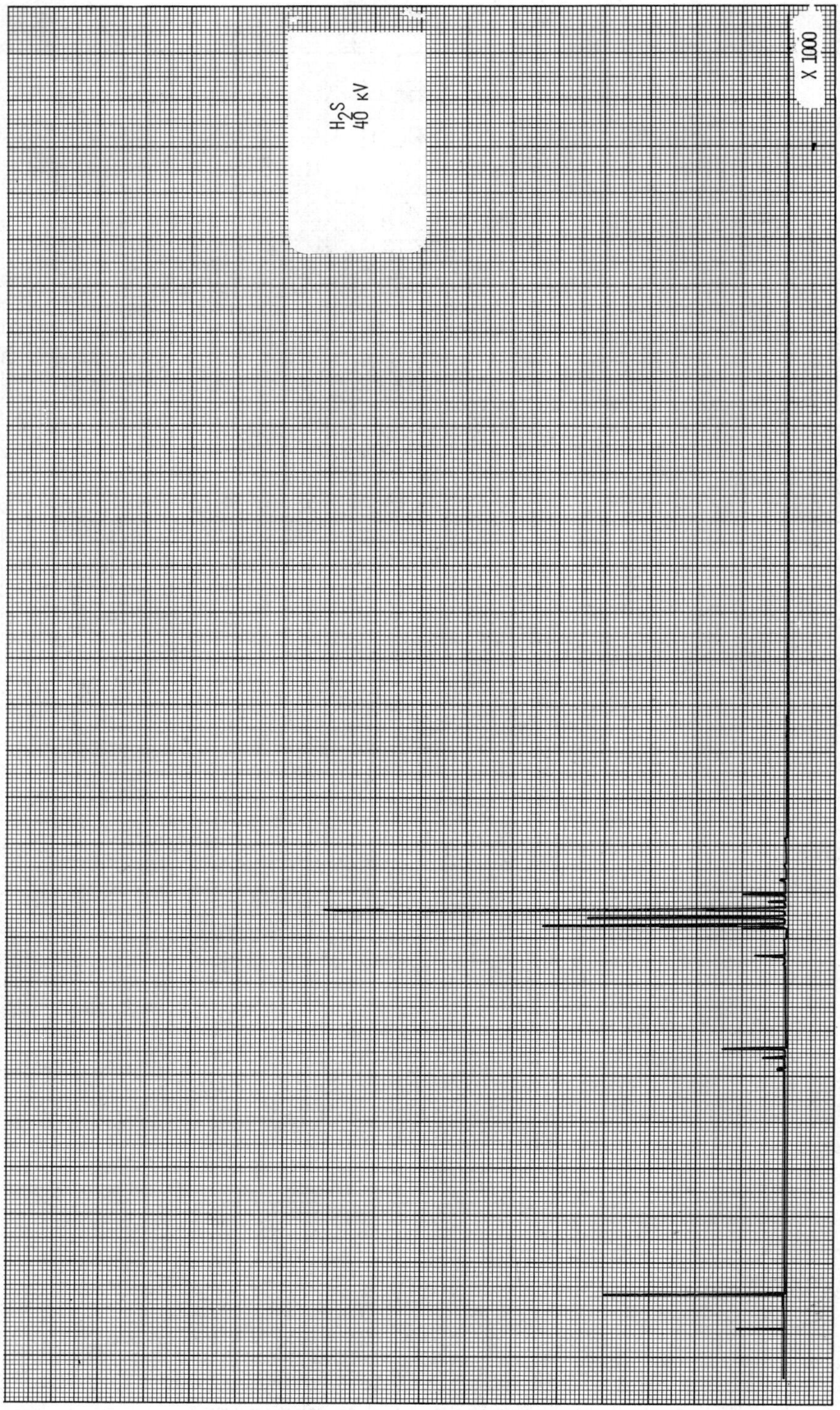
H_2S
40 kV
X 1000

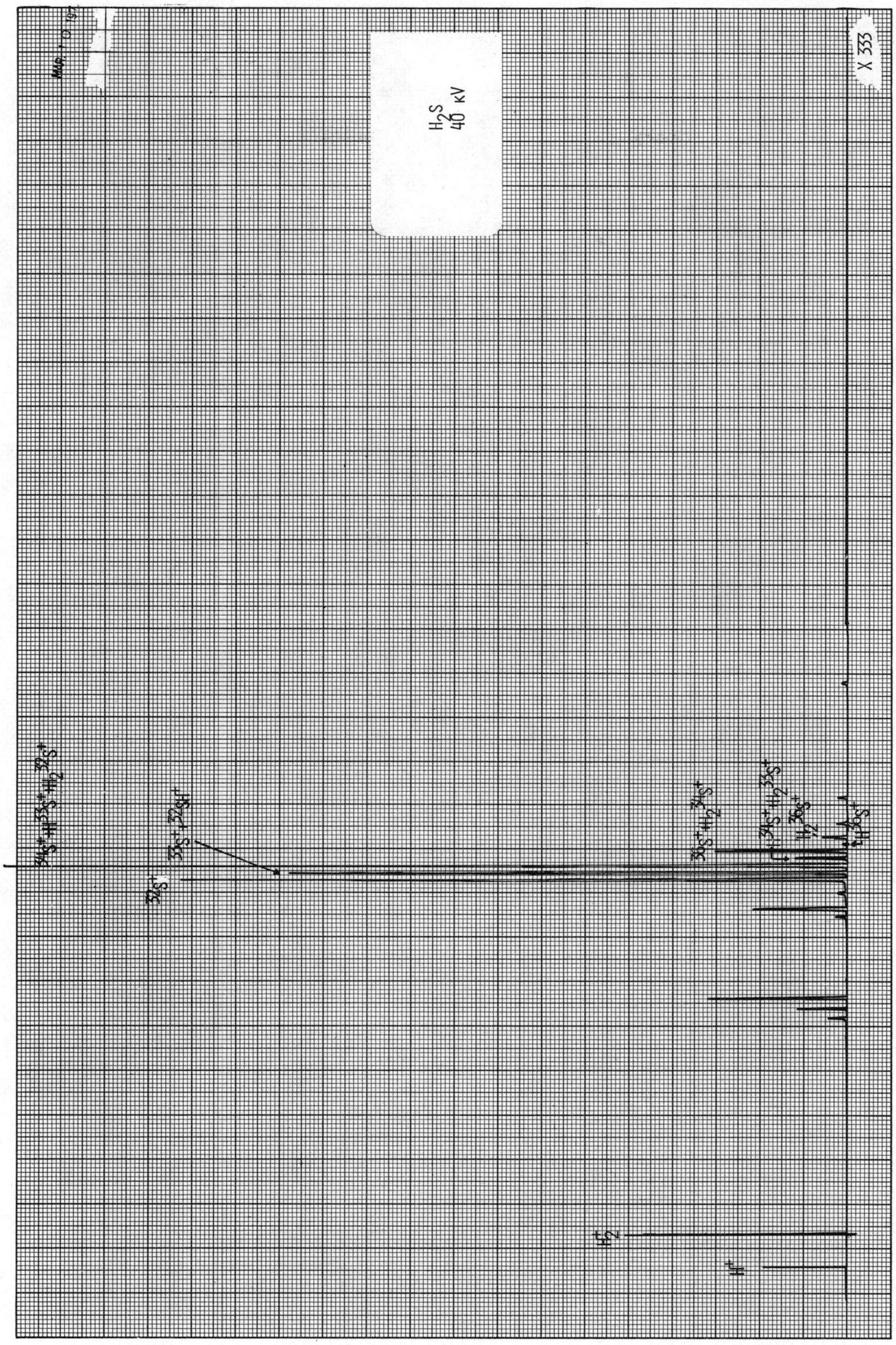
H2S
40 kV
X 333
H+
H2+
32S+

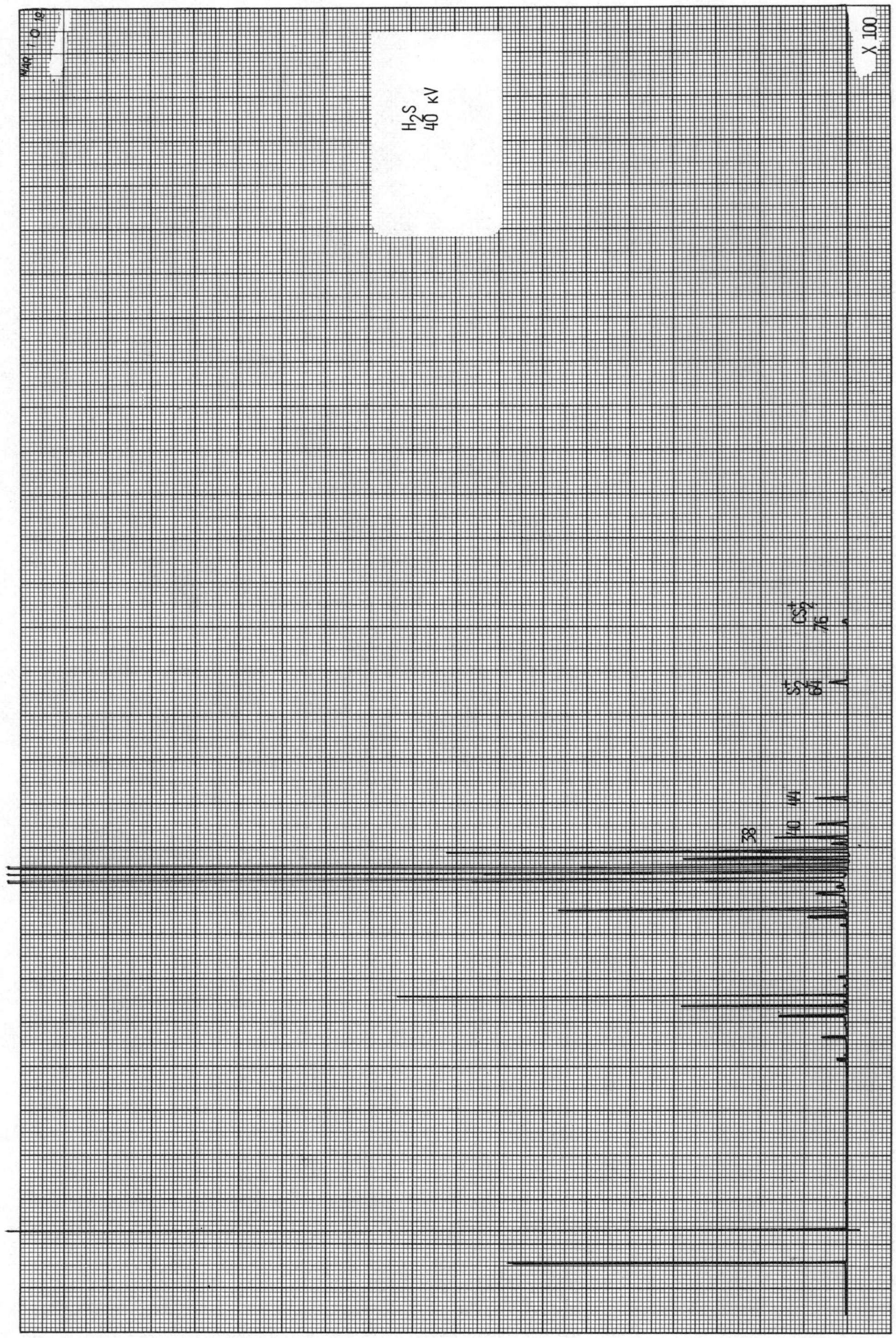
H_2S
40 kV
X 100
38
40
44
64 S_2^+
76 CS_2^+

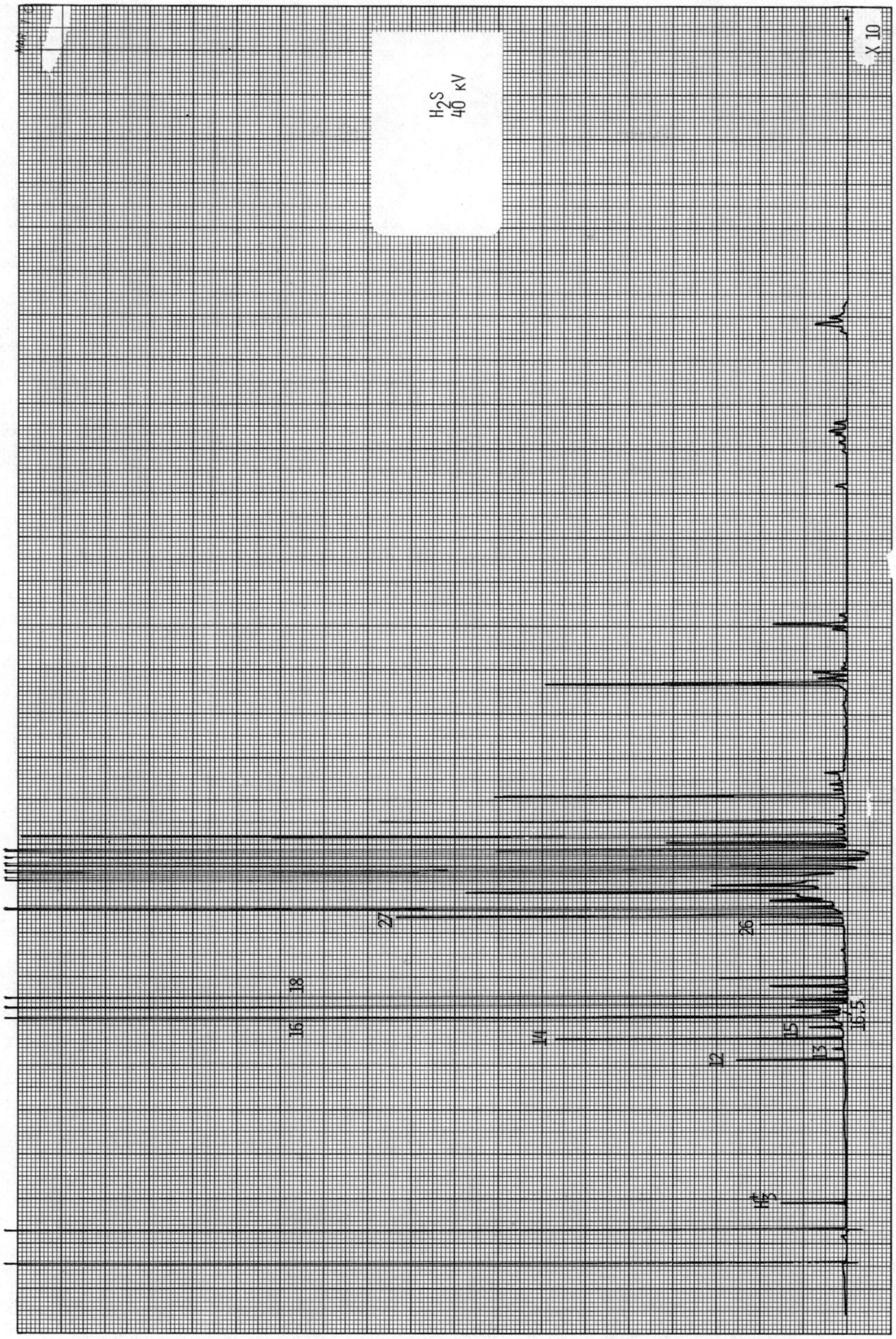
H$_2$S
40 kV
X 10
27
26
18
16
14
15
13
12
16.5

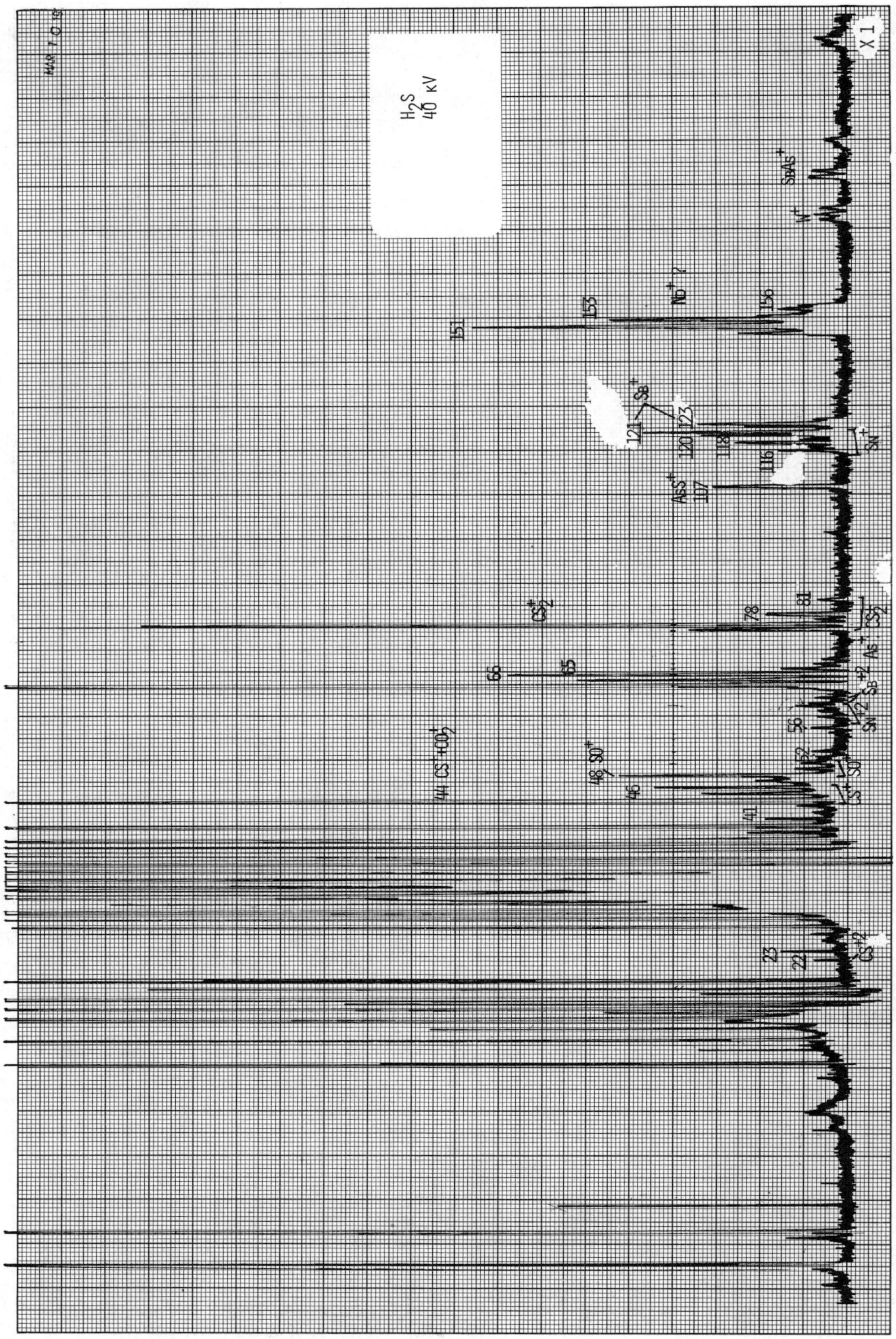
H2S
40 kV
X 1
151
153
66
65
78
81
44 CS+ + CO2+
48 SO+
46
41
23
22
107
116
118
120
121
123
SbAs+

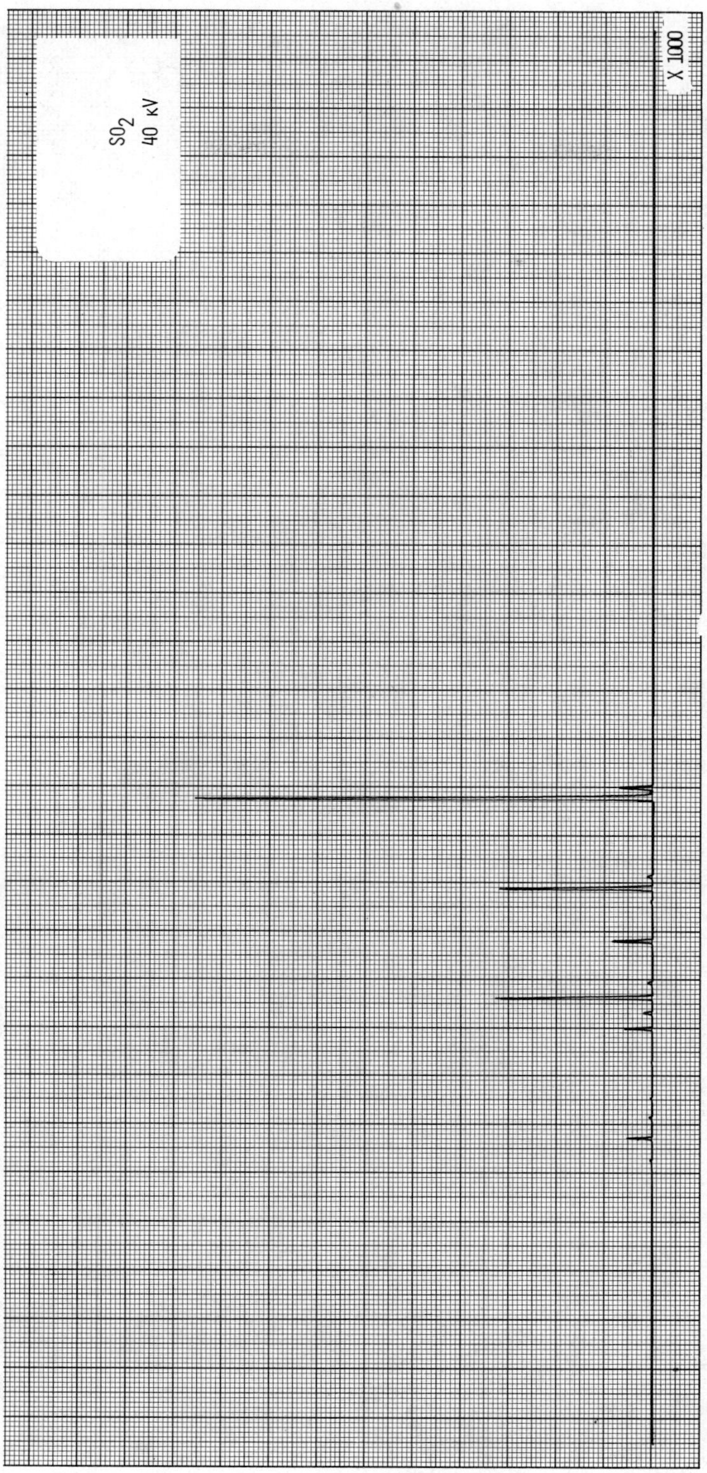
SO_2
40 kV
X 1000

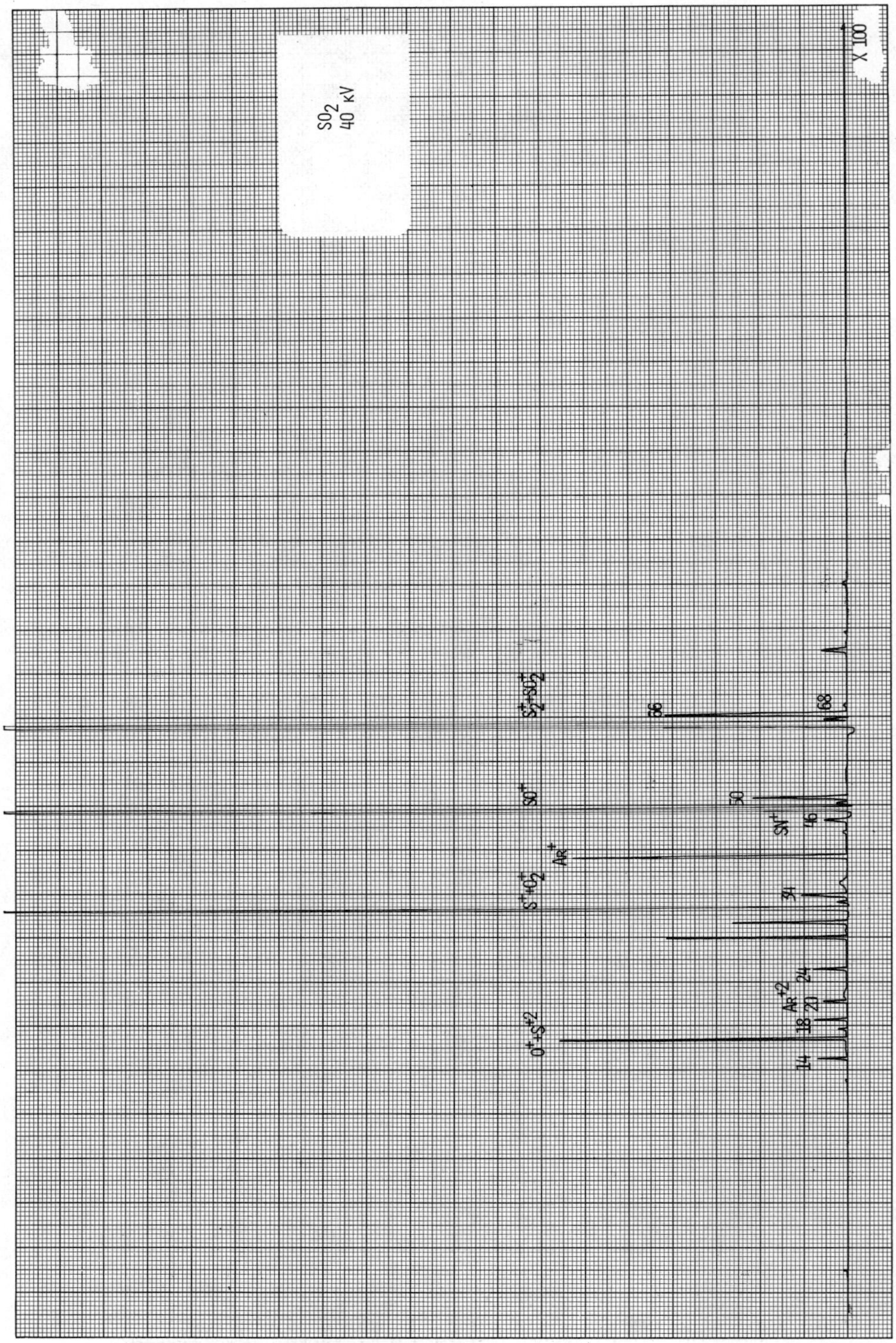

SO2
40 kV
X 100
O+ +S+2
S+ +O2+
Ar+
SO+
S2+ +SO2+
14
18
20
Ar+2
24
34
SN+
46
50
66
68

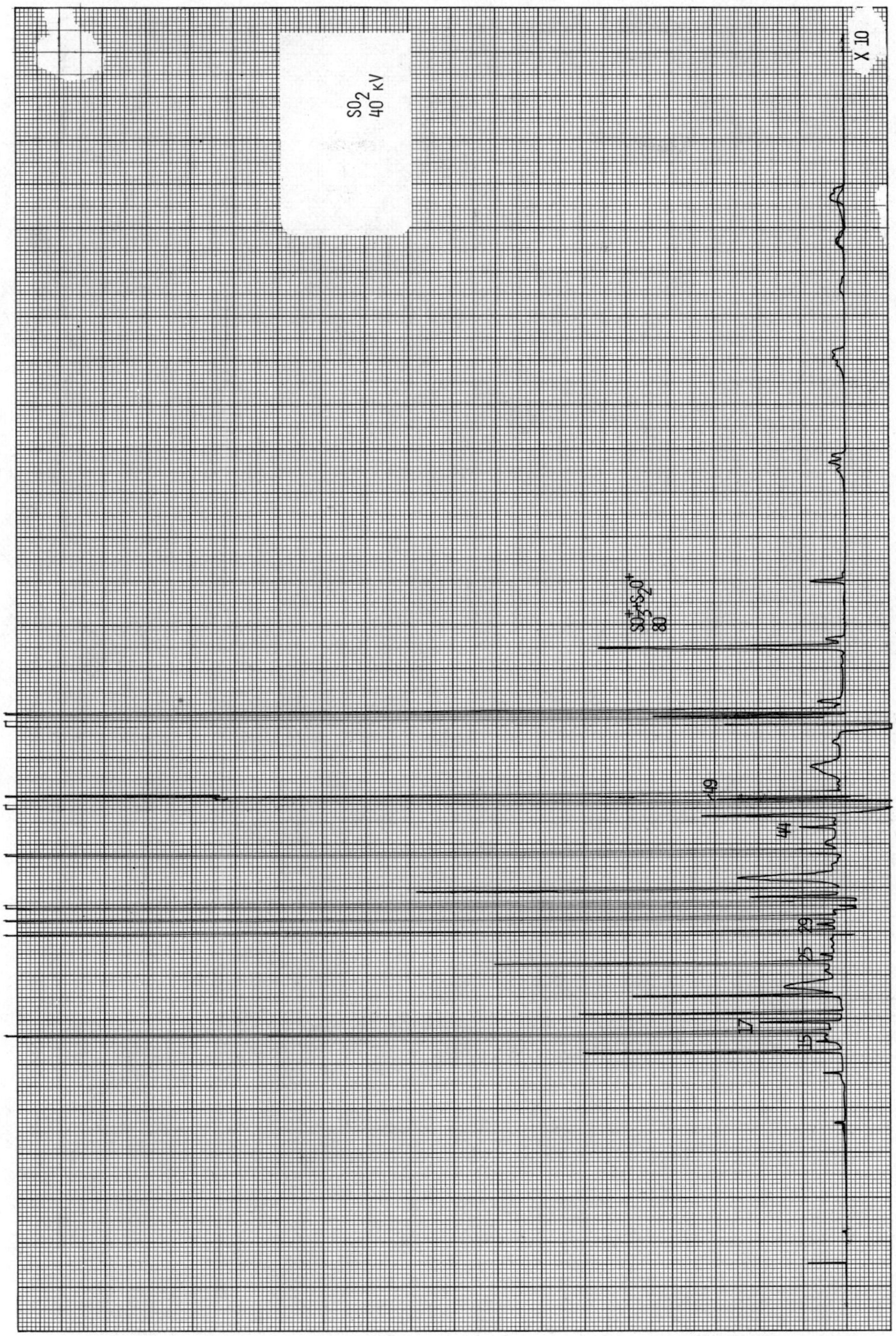
SO_2
40 kV
X 10
SO_3^+ S_2O^+
80
49
44
29
25
17
15

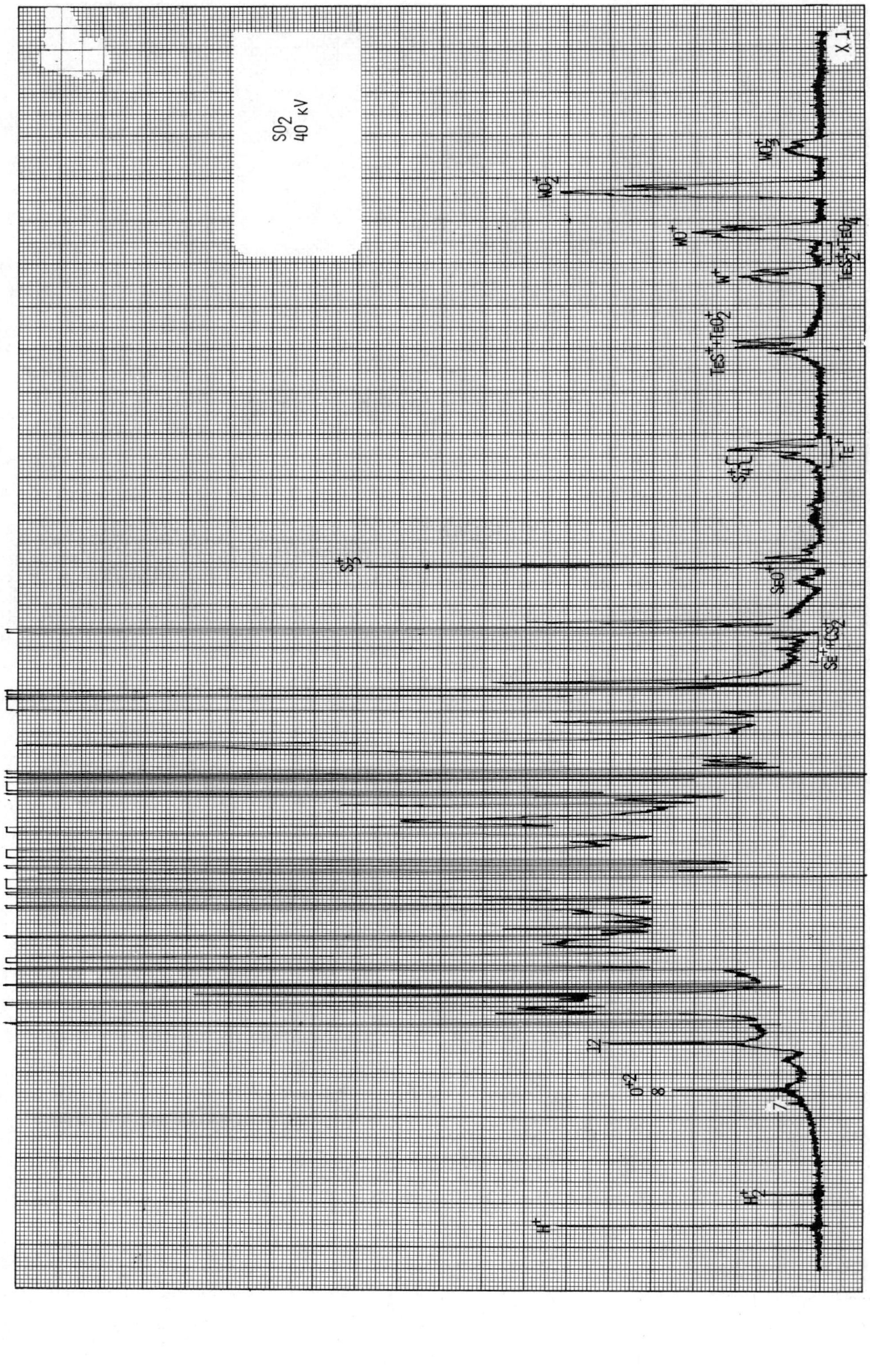
SO2
40 kV
X 1

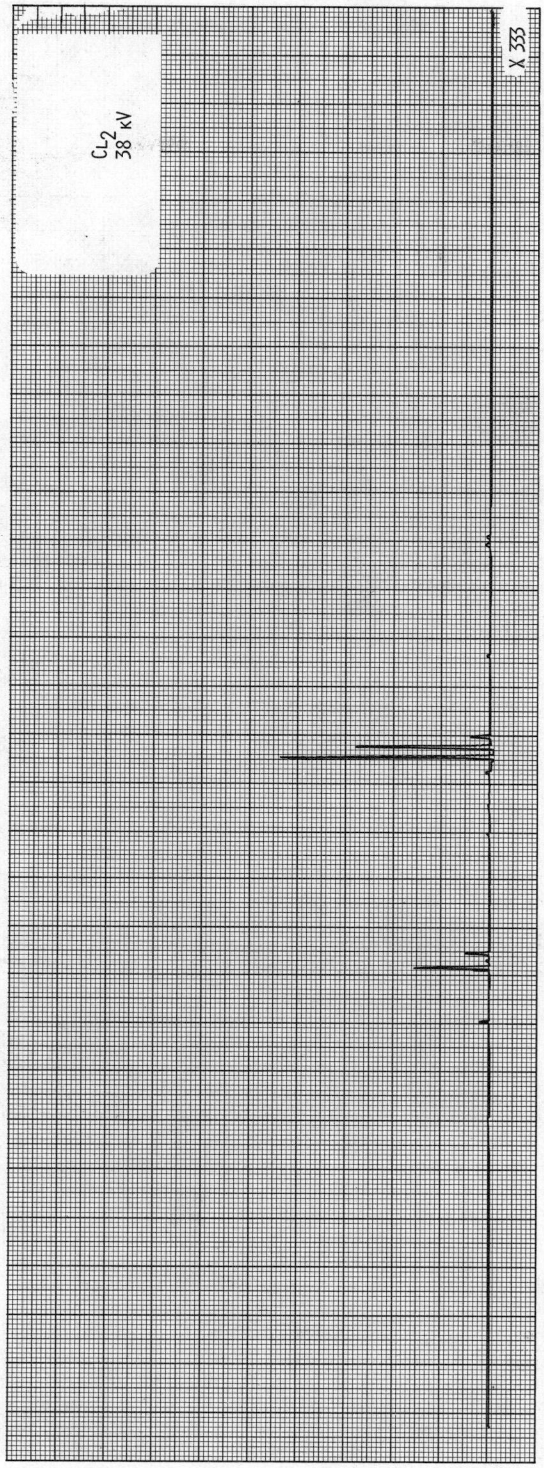

CL2
38 kV
X 333

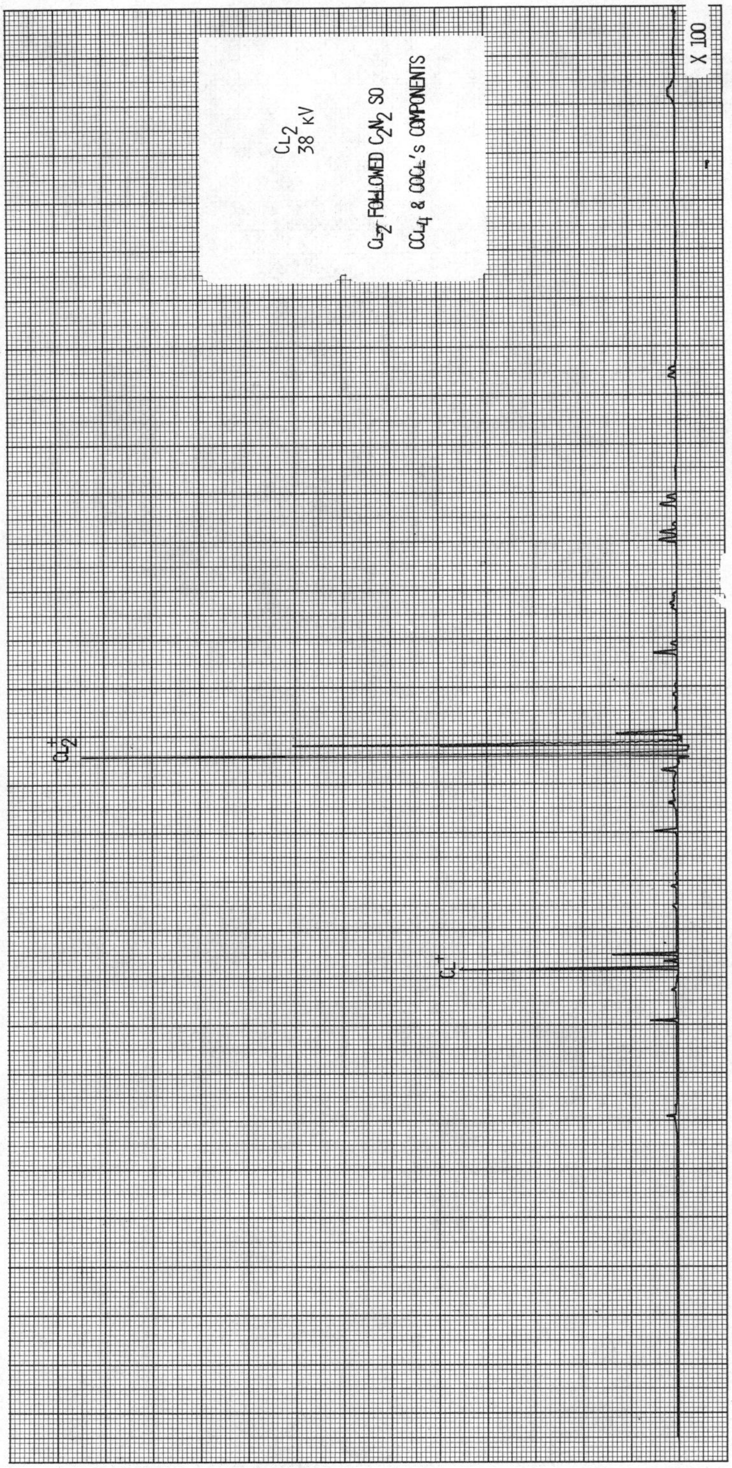

CL2
38 kV
CL2 FOLLOWED C2N2 SO
CCL4 & COCL2'S COMPONENTS
X 100
CL2+
CL+

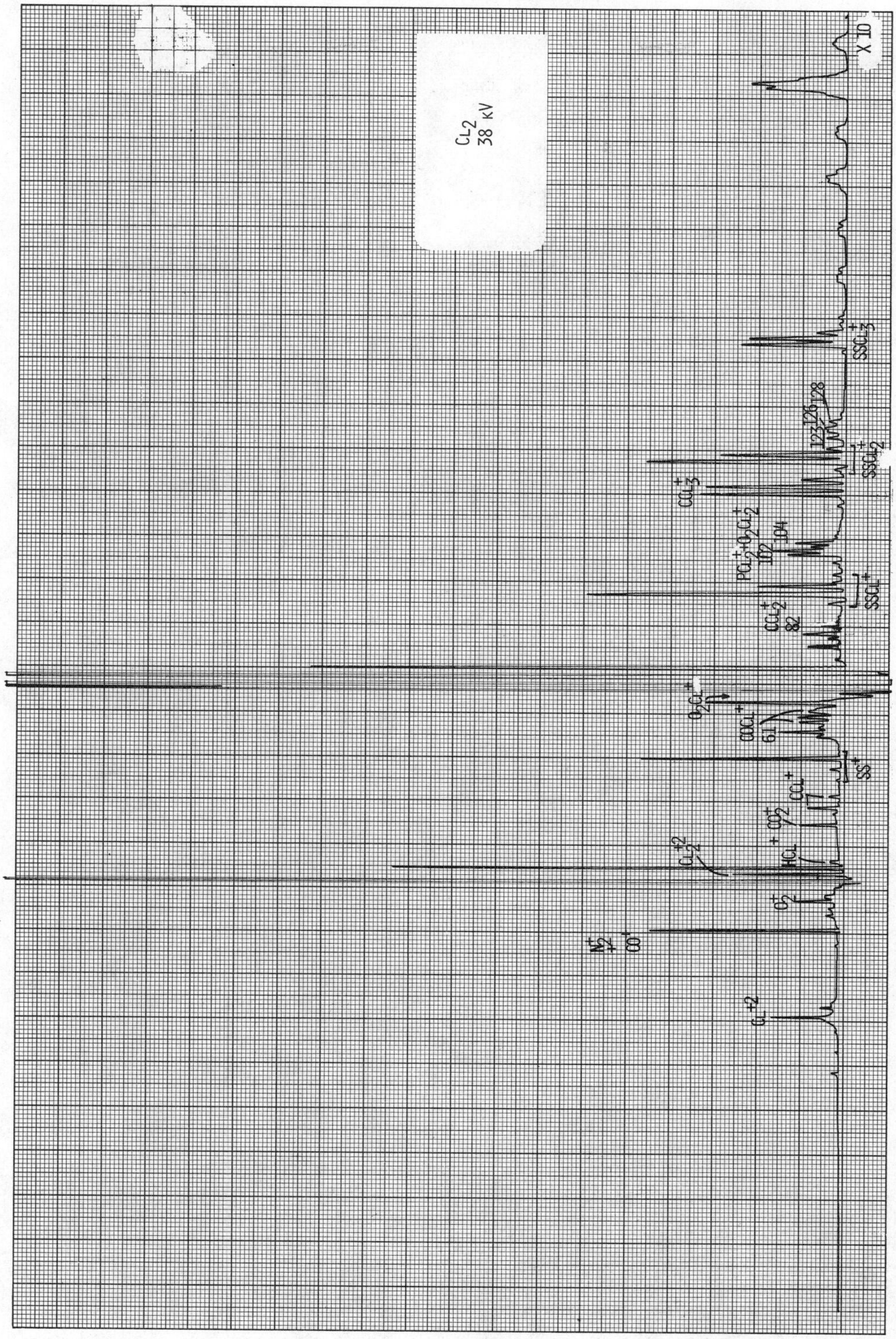

CL2
38 kV
X 10
Cl+2
N2+ + CO+
O2+
Cl2+2
HCl+
CO2+
CCl+
SS+
COCl+
61
O2Cl+
CCl2+
82
SSCl+
PCl2+ + O2Cl2+
102
104
CCl3+
123
126
128
SSCl2+
SSCl3+

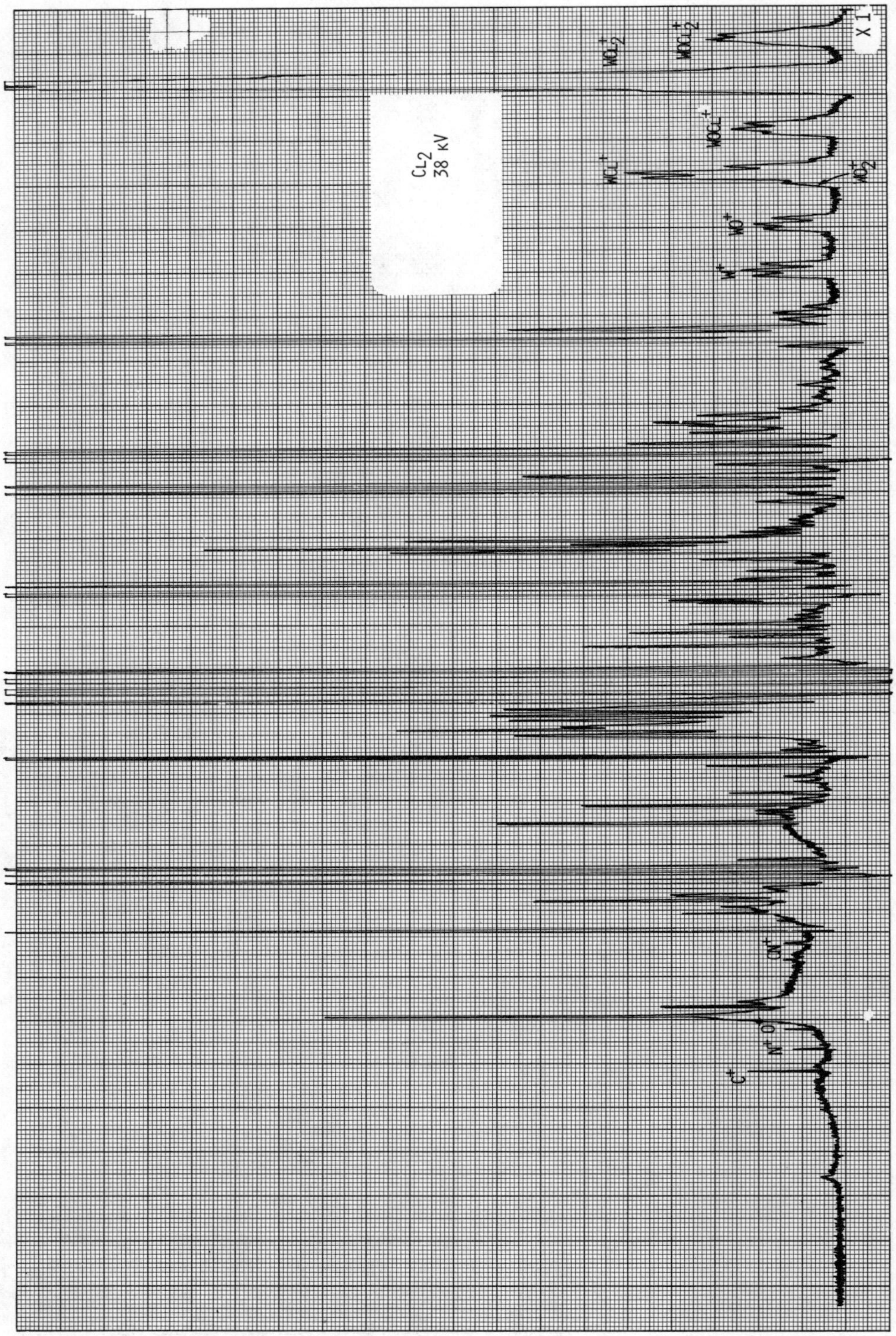
CL2
38 kV
X 1
WCL2+
WOCL2+
WCL+
WOCL+
W+
WO+
WO2+
C+
N+ O
OH+

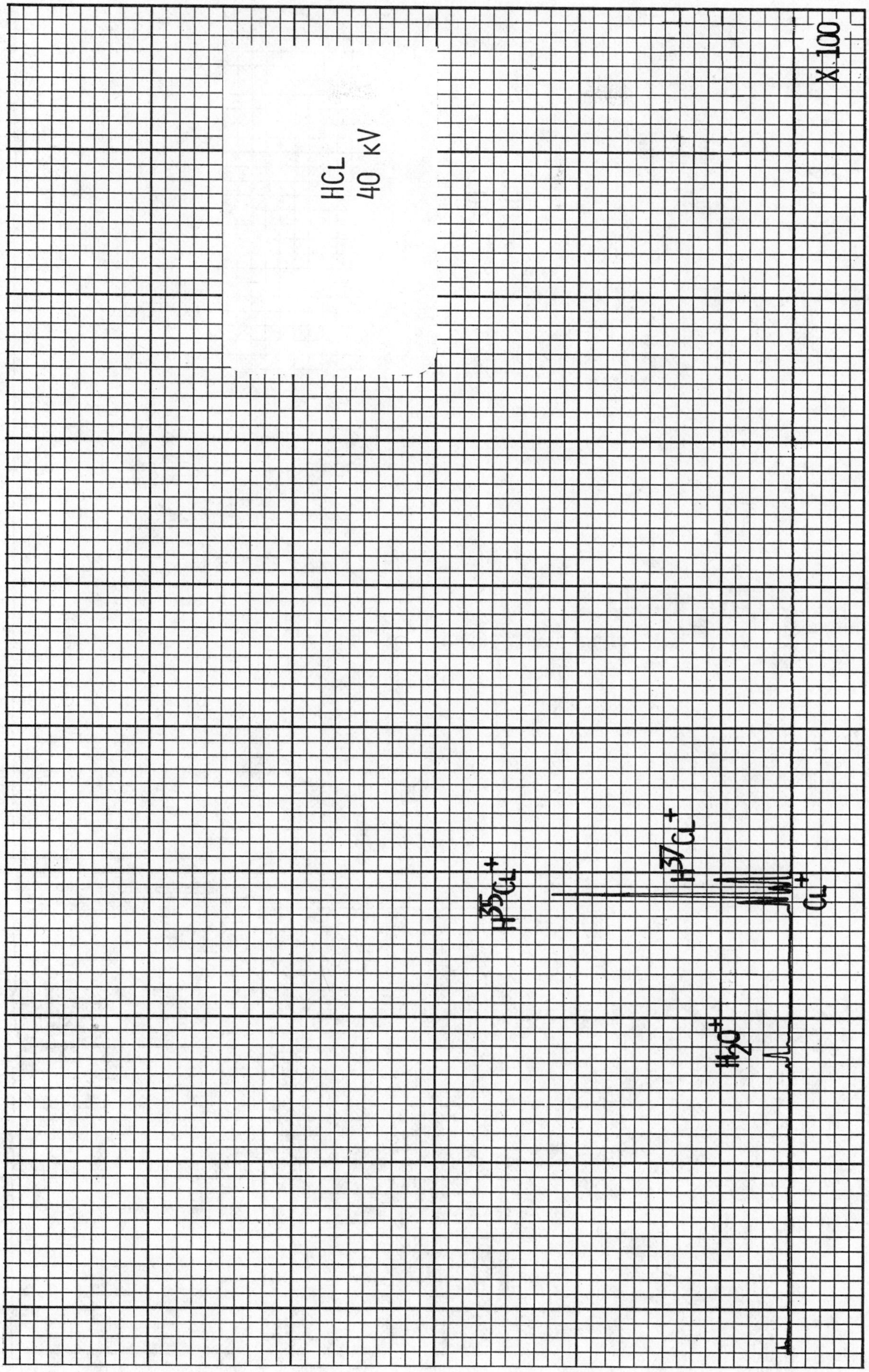

HCL
40 KV
X 100
H35CL+
H37CL+
CL+
H2O+

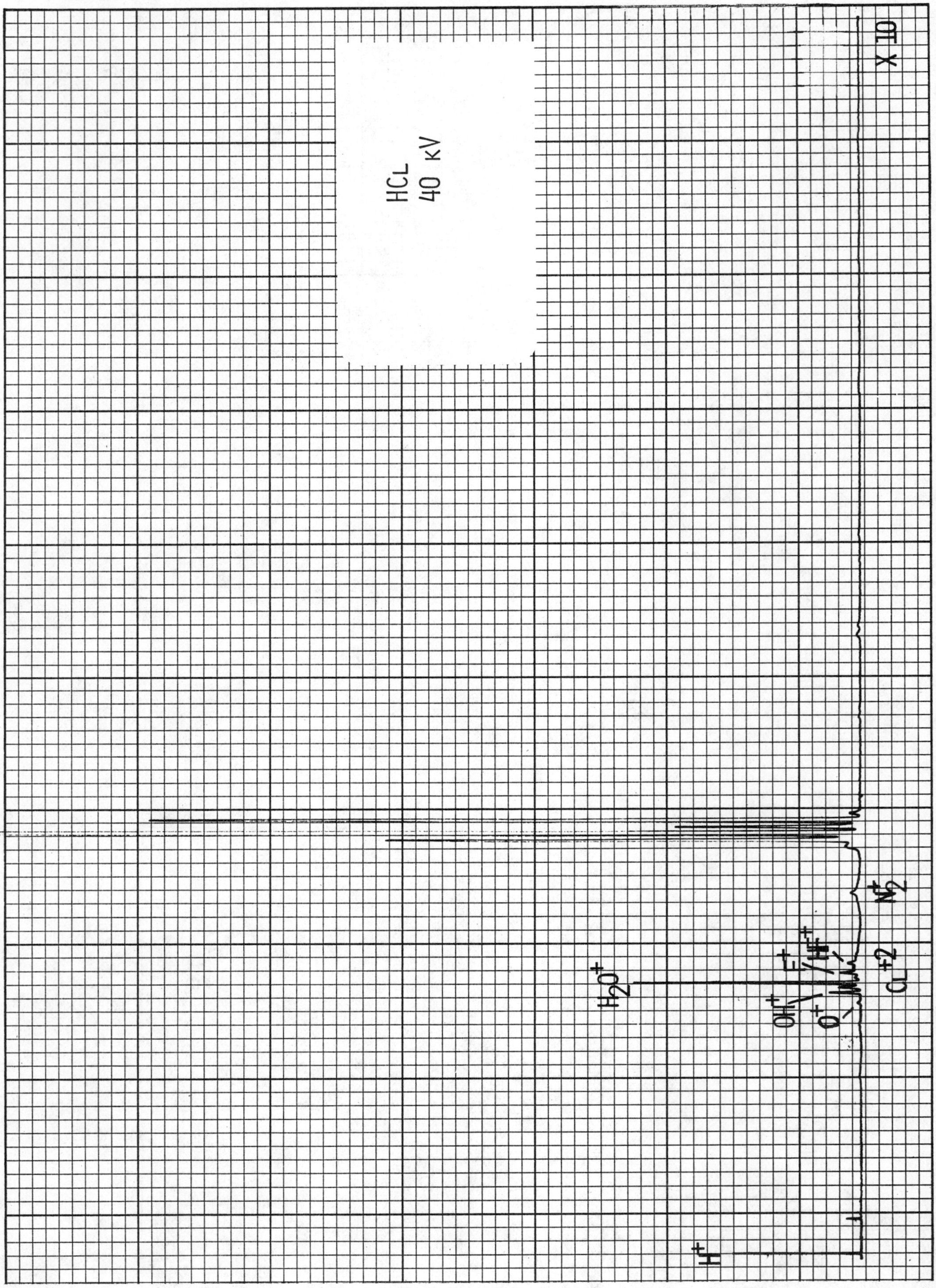
HCL
40 kV
X 10
H+
H2O+
OH+
O+
F+
HF+
CL+2
N2+

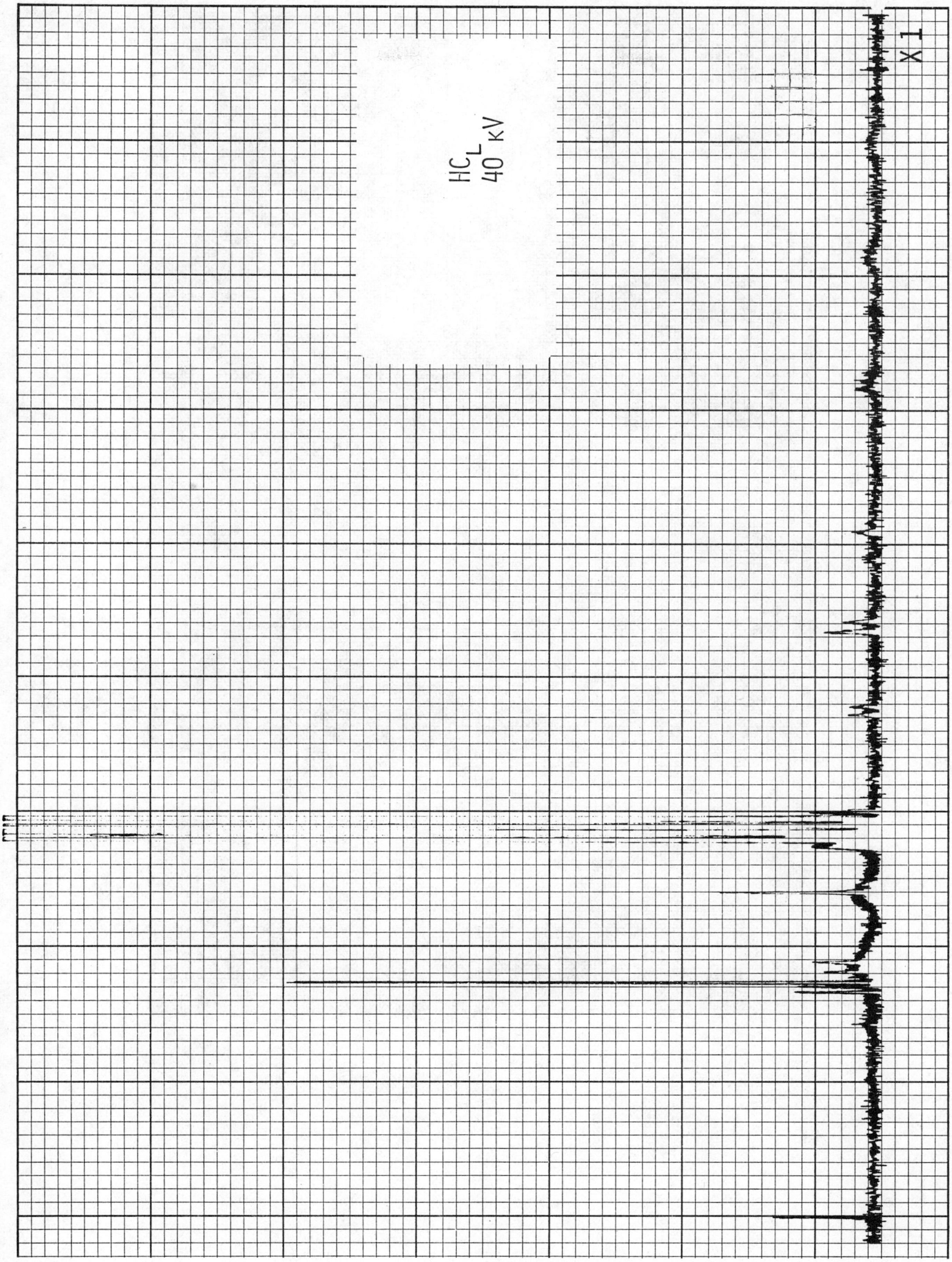
HC$_L$
40 kV
X 1

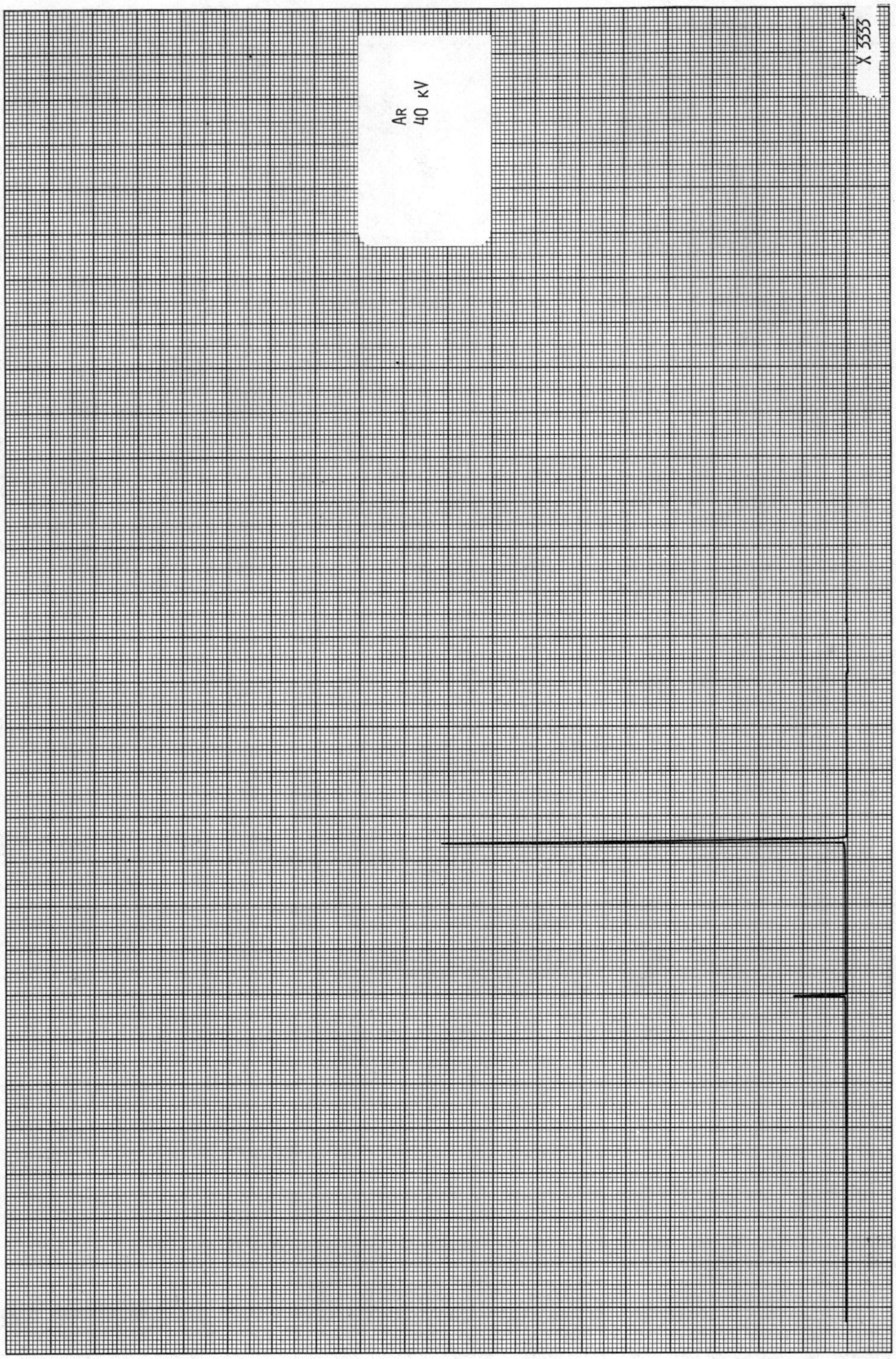
Ar
40 kV
X 3333

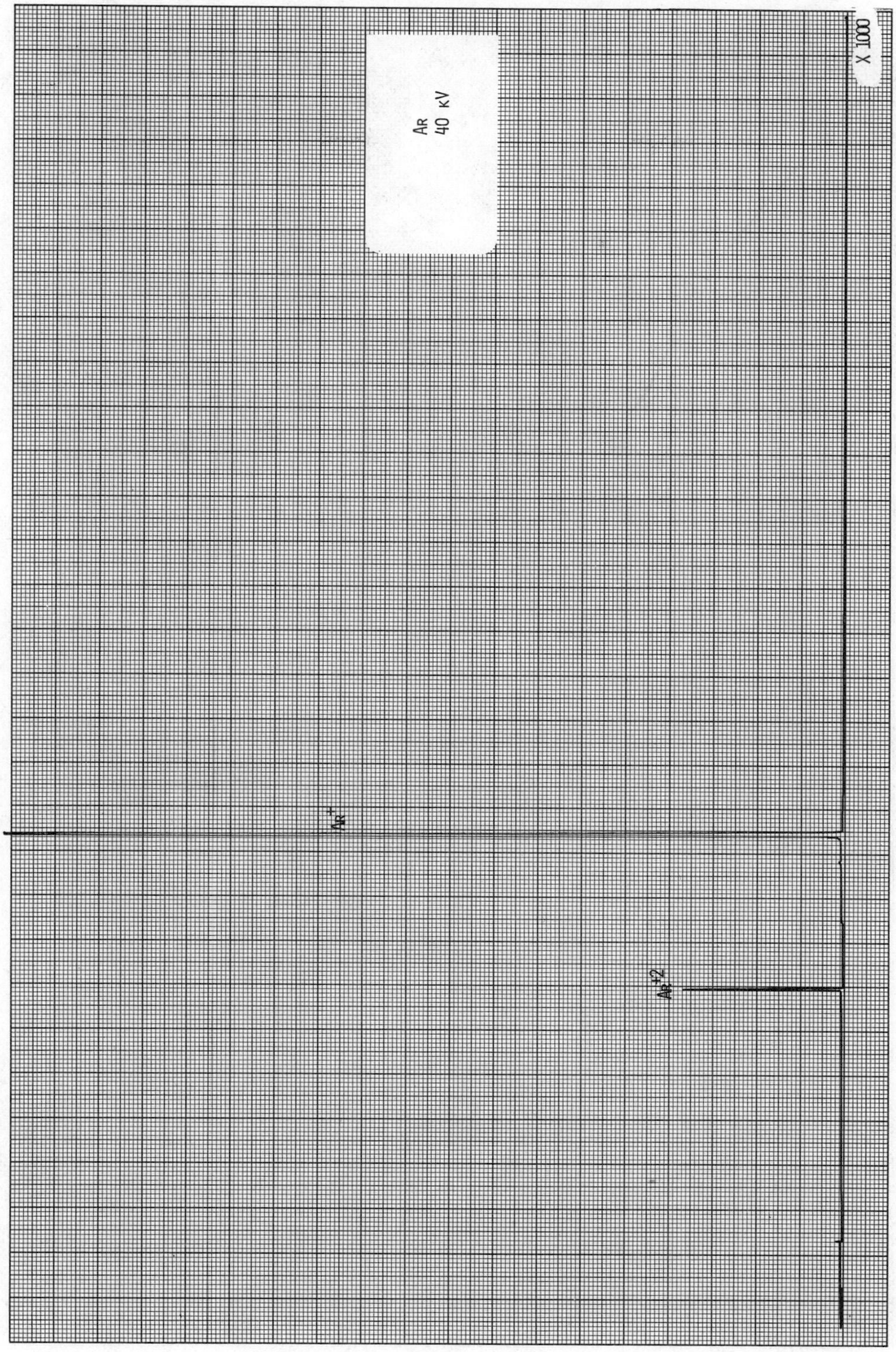
X 1000
Ar
40 kV
Ar^{+}
Ar^{+2}

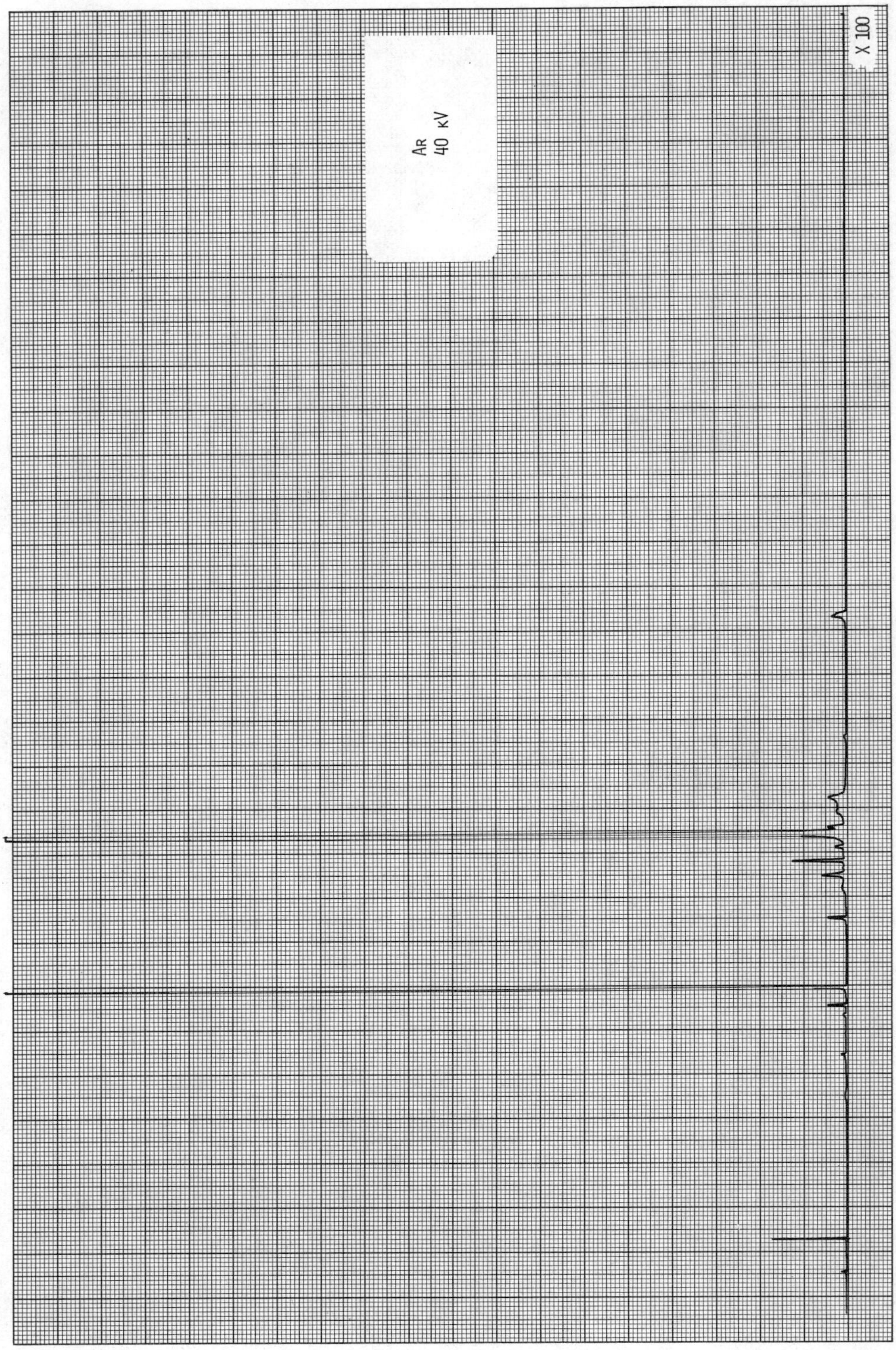
Ar
40 kV
X 100

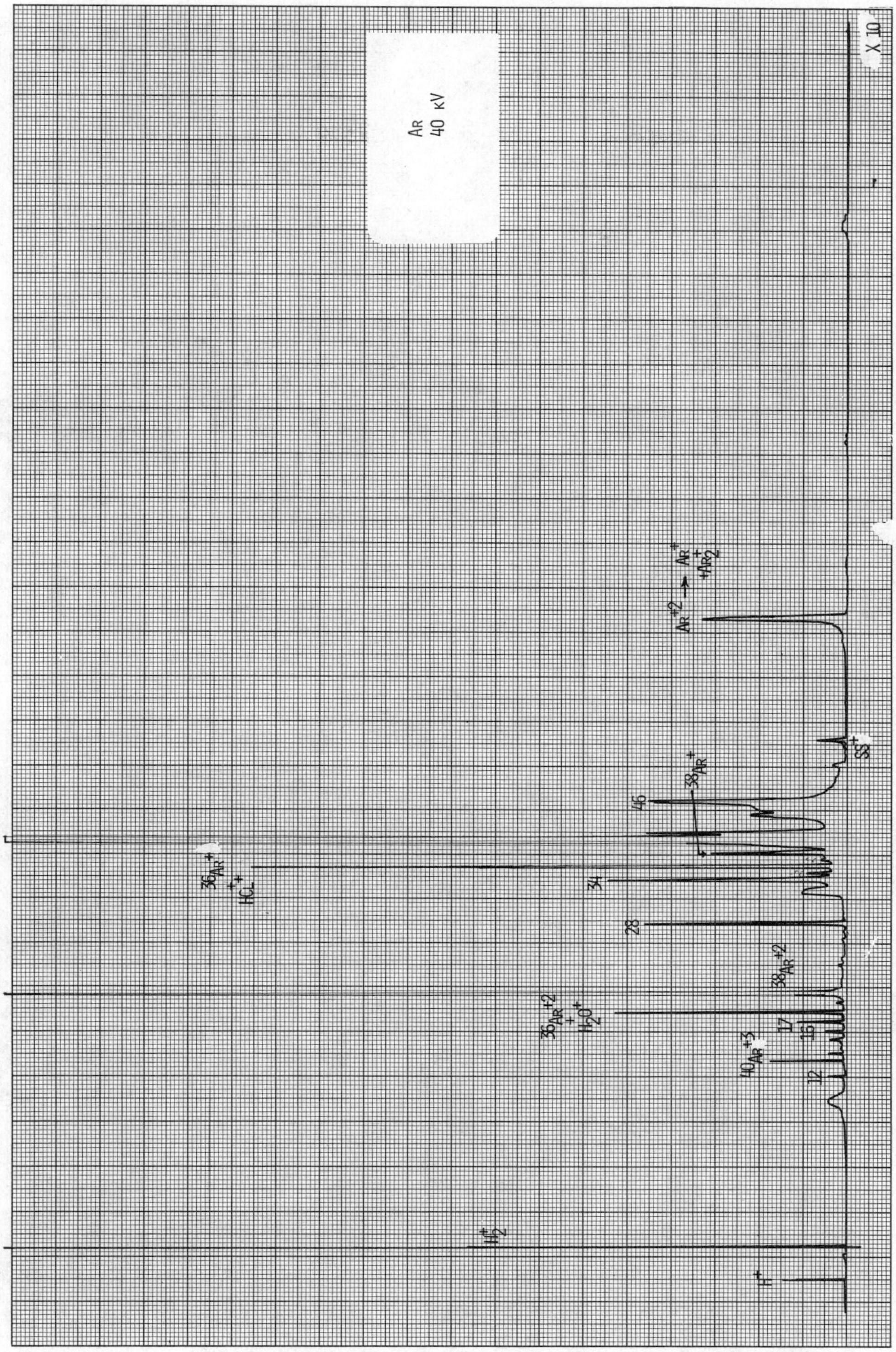
Ar
40 kV
X 10
H+
H2+
36Ar+ + HCl+
36Ar+2 + H2O+
40Ar+3
12
17
38Ar+2
28
34
46
38Ar+
SS+
Ar+2 → Ar+ + Ar2+

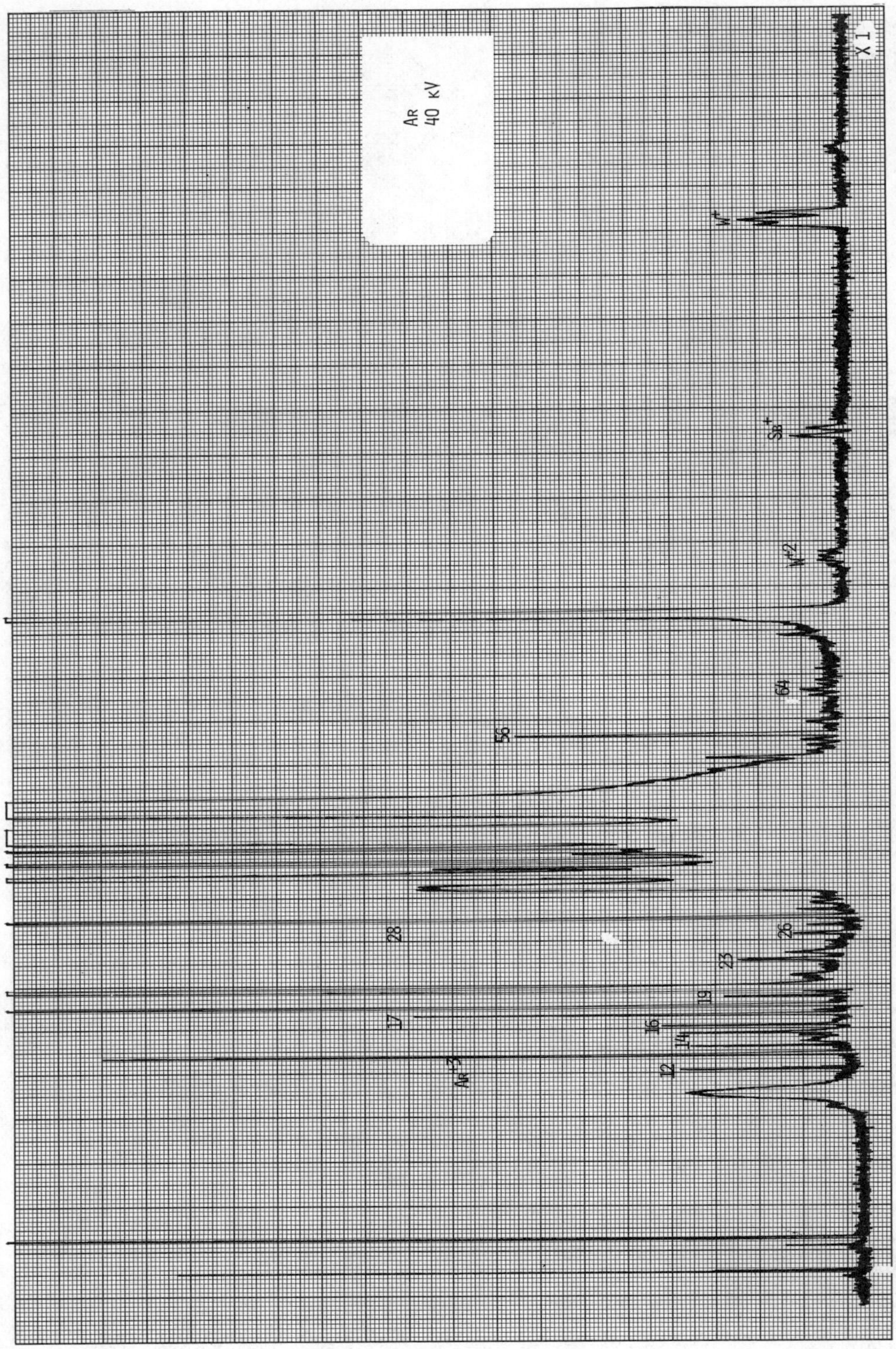
Ar
40 kV
X 1
W+
Sb+
W+2
64
56
28
26
23
19
17
16
14
12
Ar+3

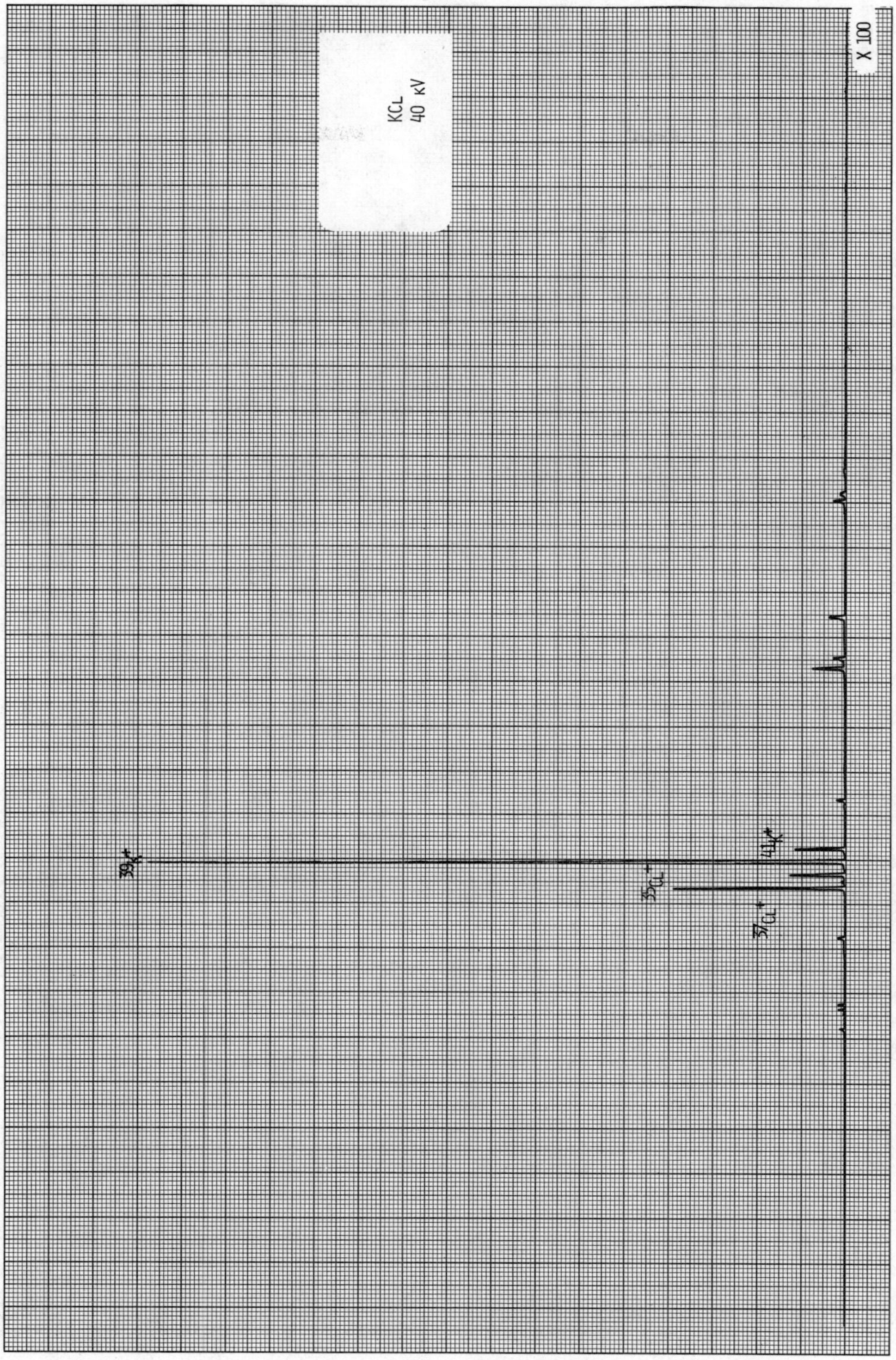
KCL
40 kV
X 100
39K+
35CL+
41K+
37CL+

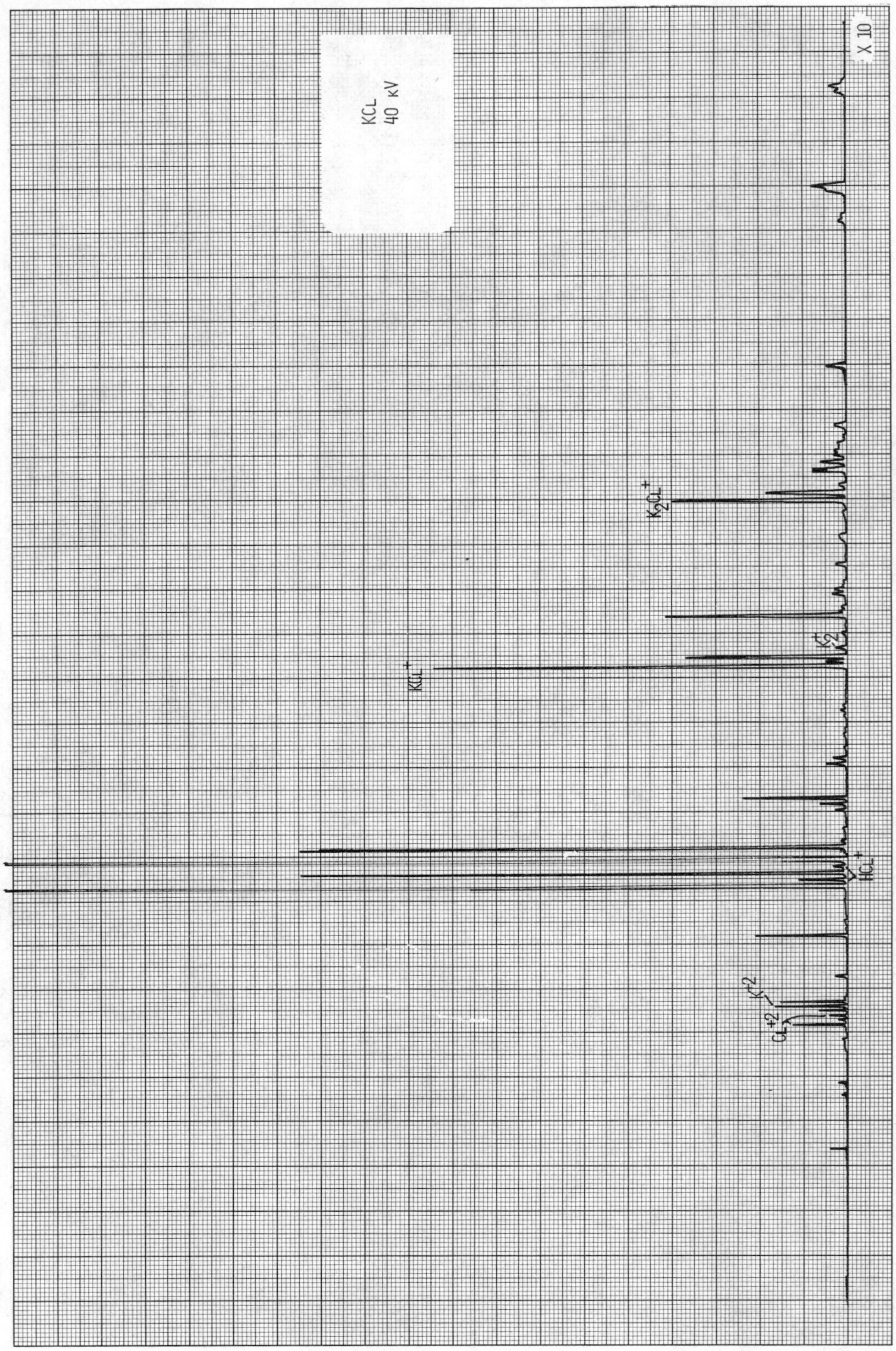
KCL
40 kV
X 10

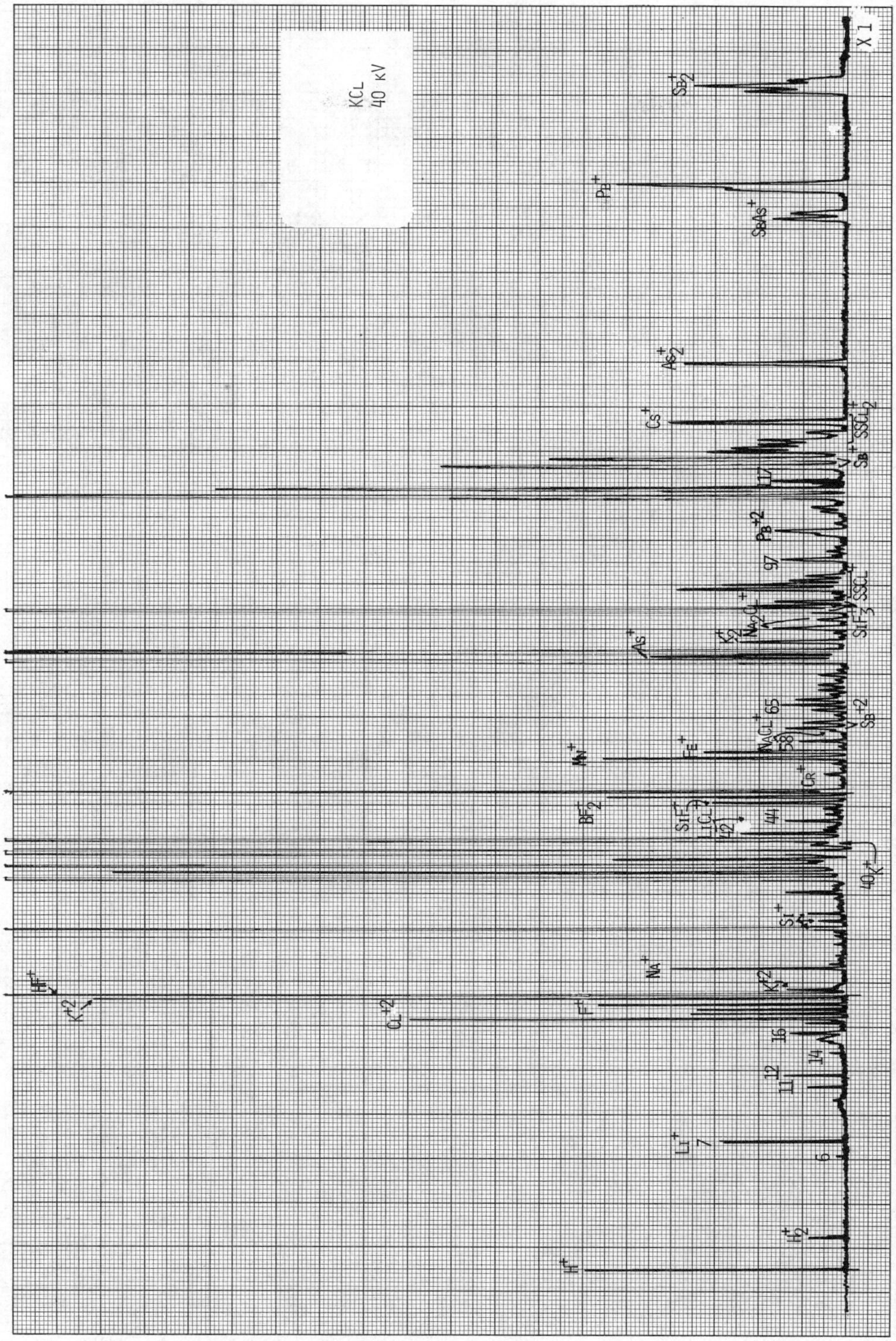
KCL
40 kV
X 1

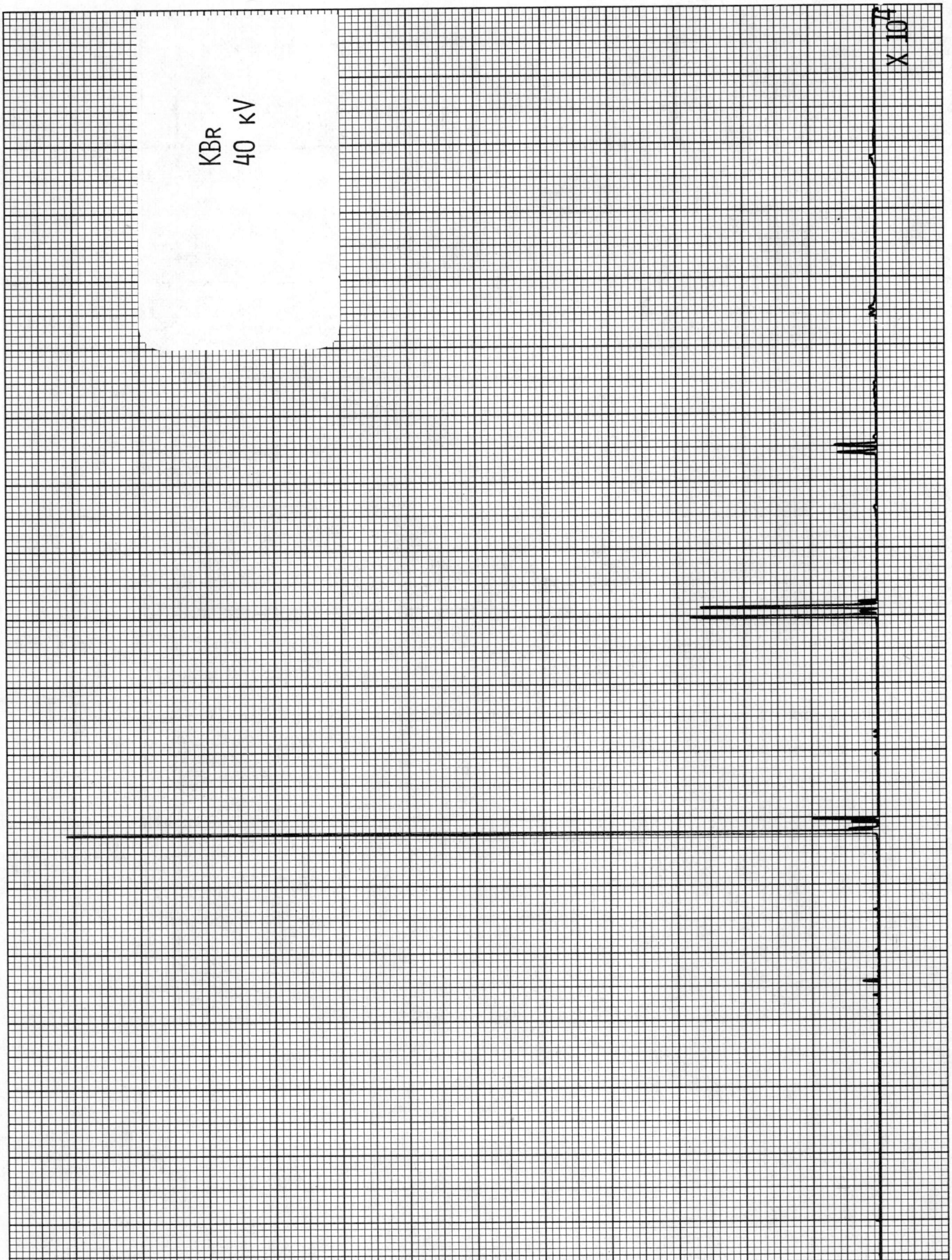
KBr
40 kV
X 10⁴

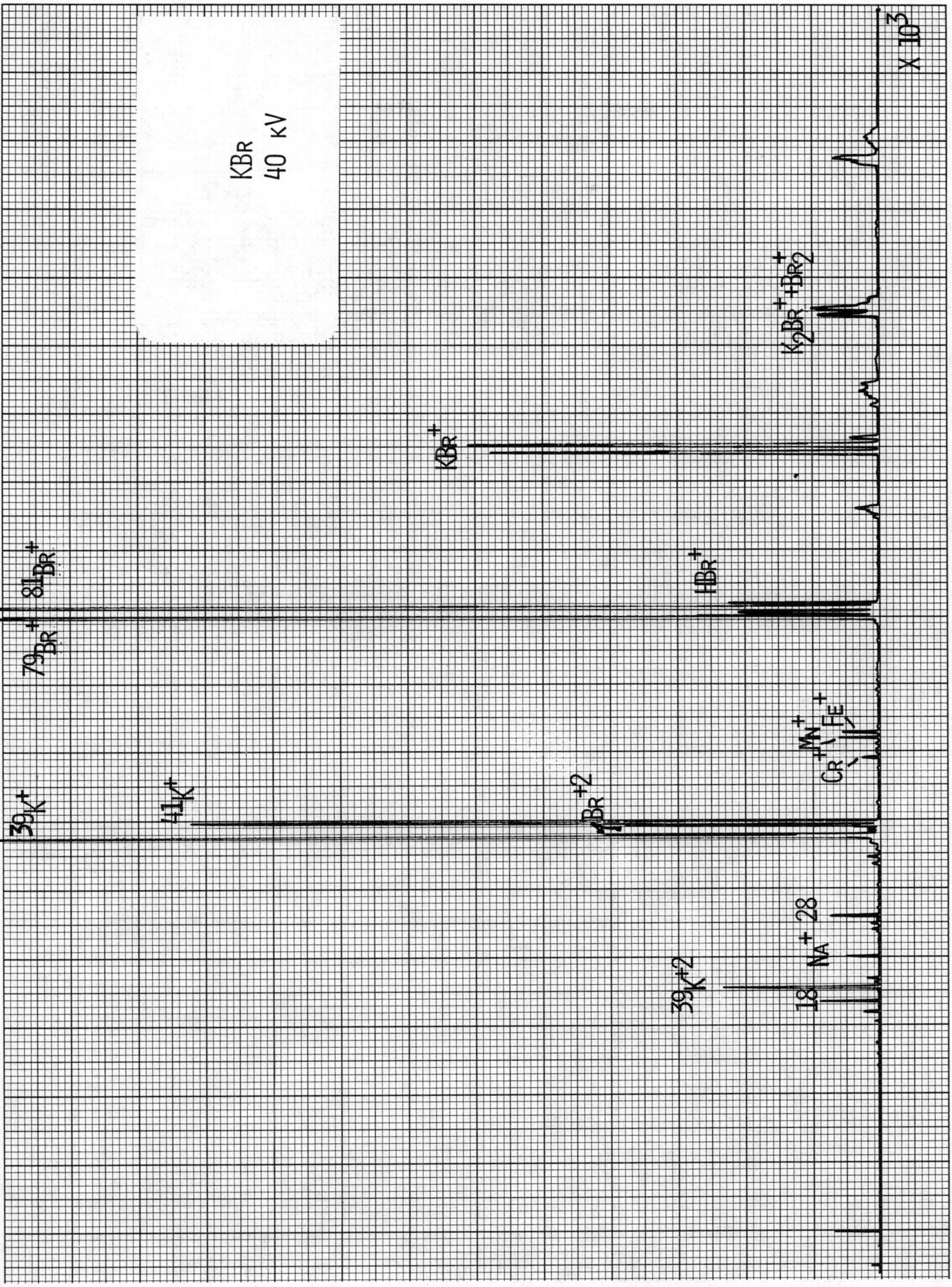

KBr
40 kV
$^{39}K^+$
$^{41}K^+$
$^{79}Br^+$
$^{81}Br^+$
Br^{+2}
$^{39}K^{+2}$
18
Na^+
28
Cr^+
Mn^+
Fe^+
HBr^+
KBr^+
K_2Br^+
Br_2^+
X 10^3

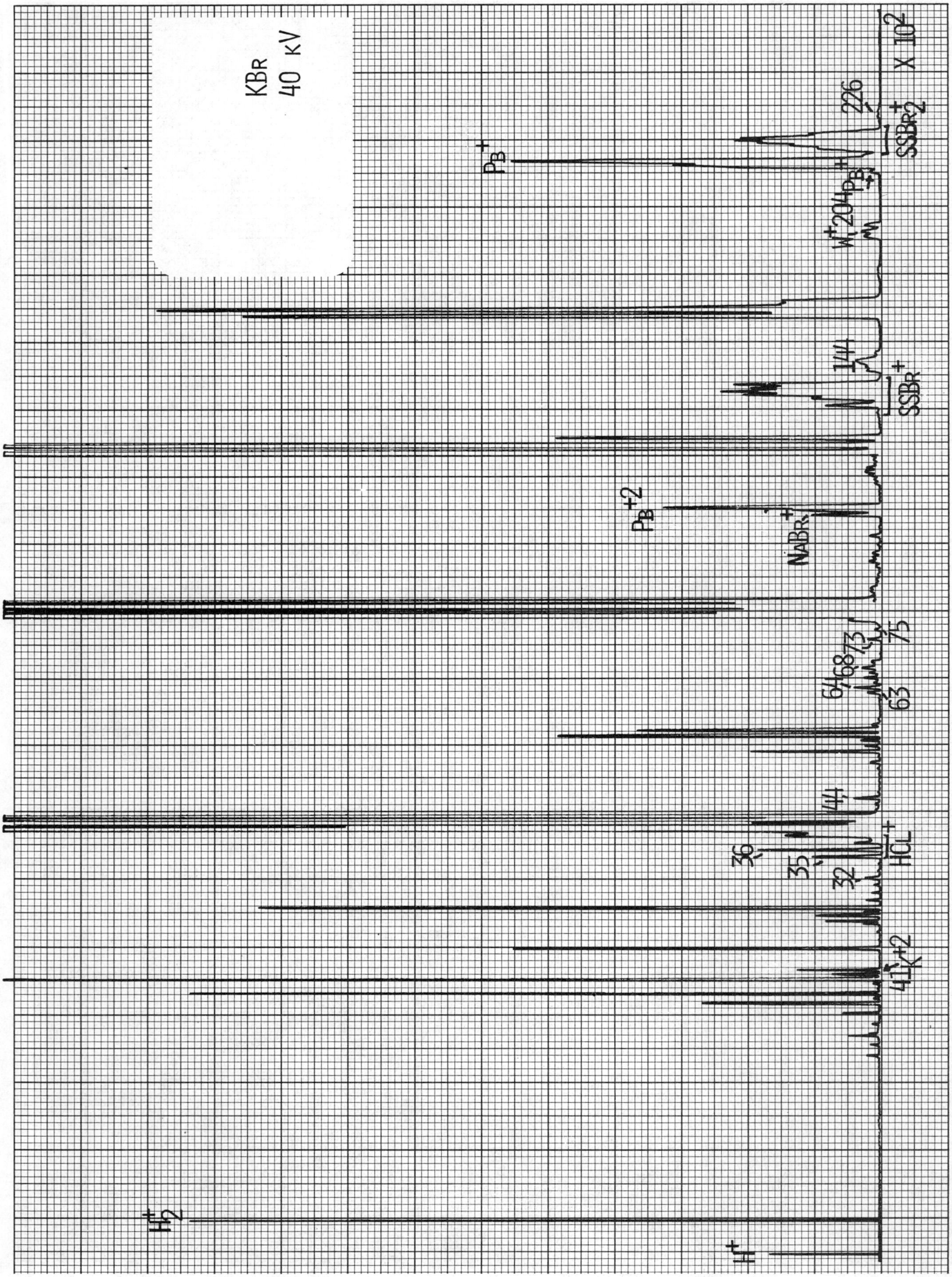
KBr
40 kV
X 10^2
226
SSBr$_2^+$
P$_B^+$
204
W$^+$
144
SSBr$^+$
P$_B^{+2}$
NaBr$^+$
64
68
73
63
75
44
36
35
32
HCl$^+$
41
K^{+2}
H$_2^+$
H$^+$

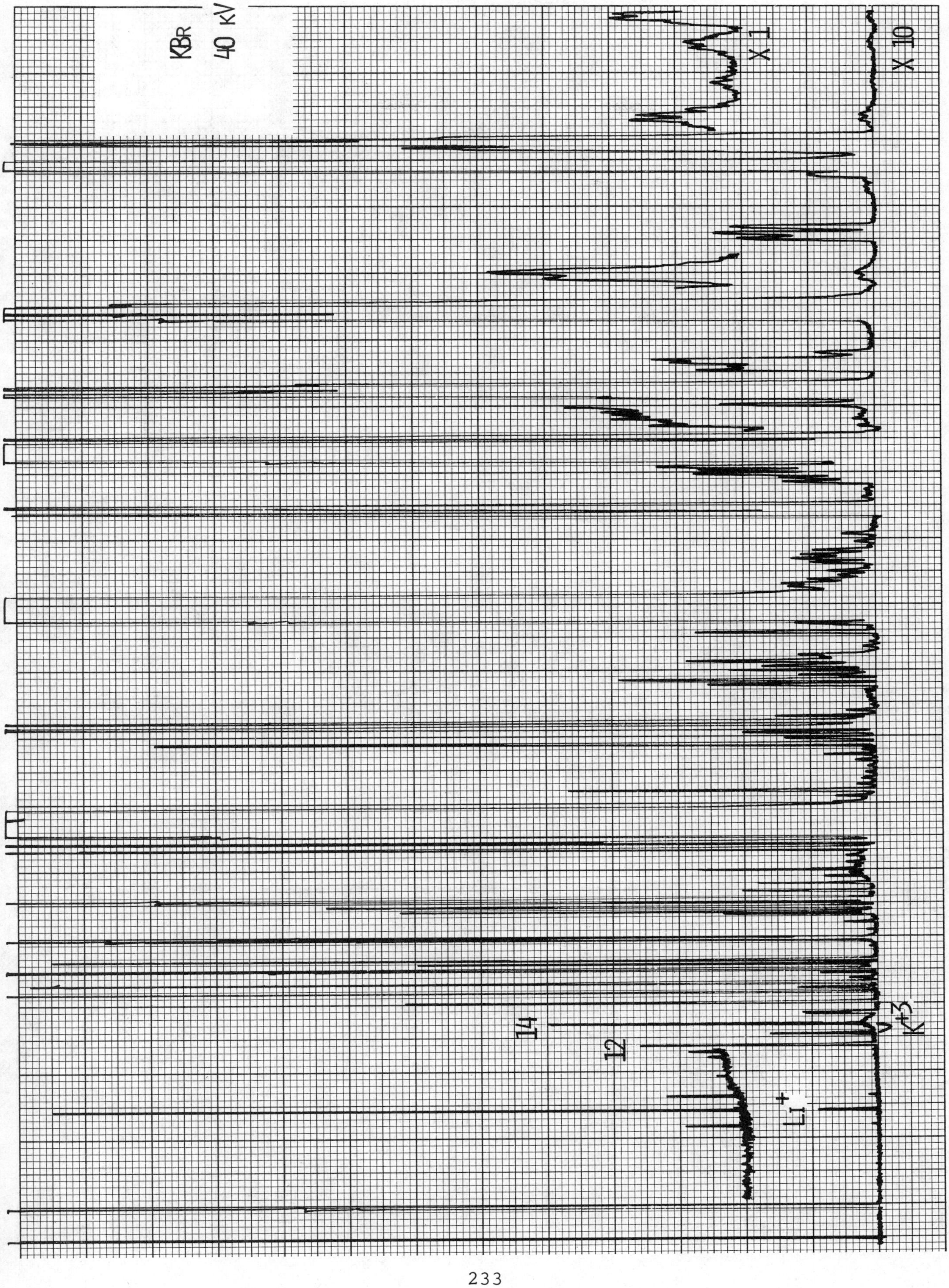
KBr
40 kV
X 1
X 10
14
12
Li+

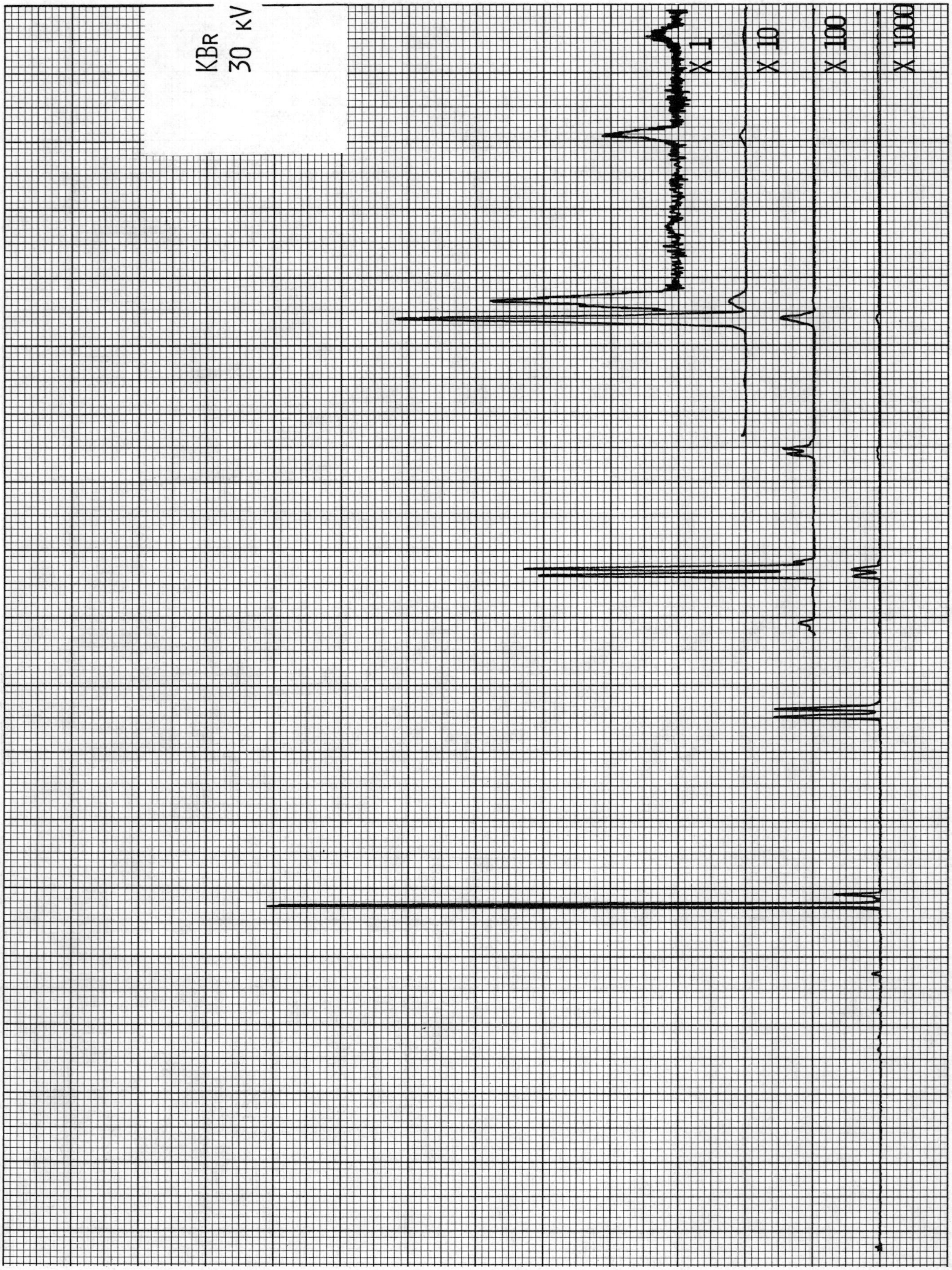
KBr
30 kV
X 1
X 10
X 100
X 1000

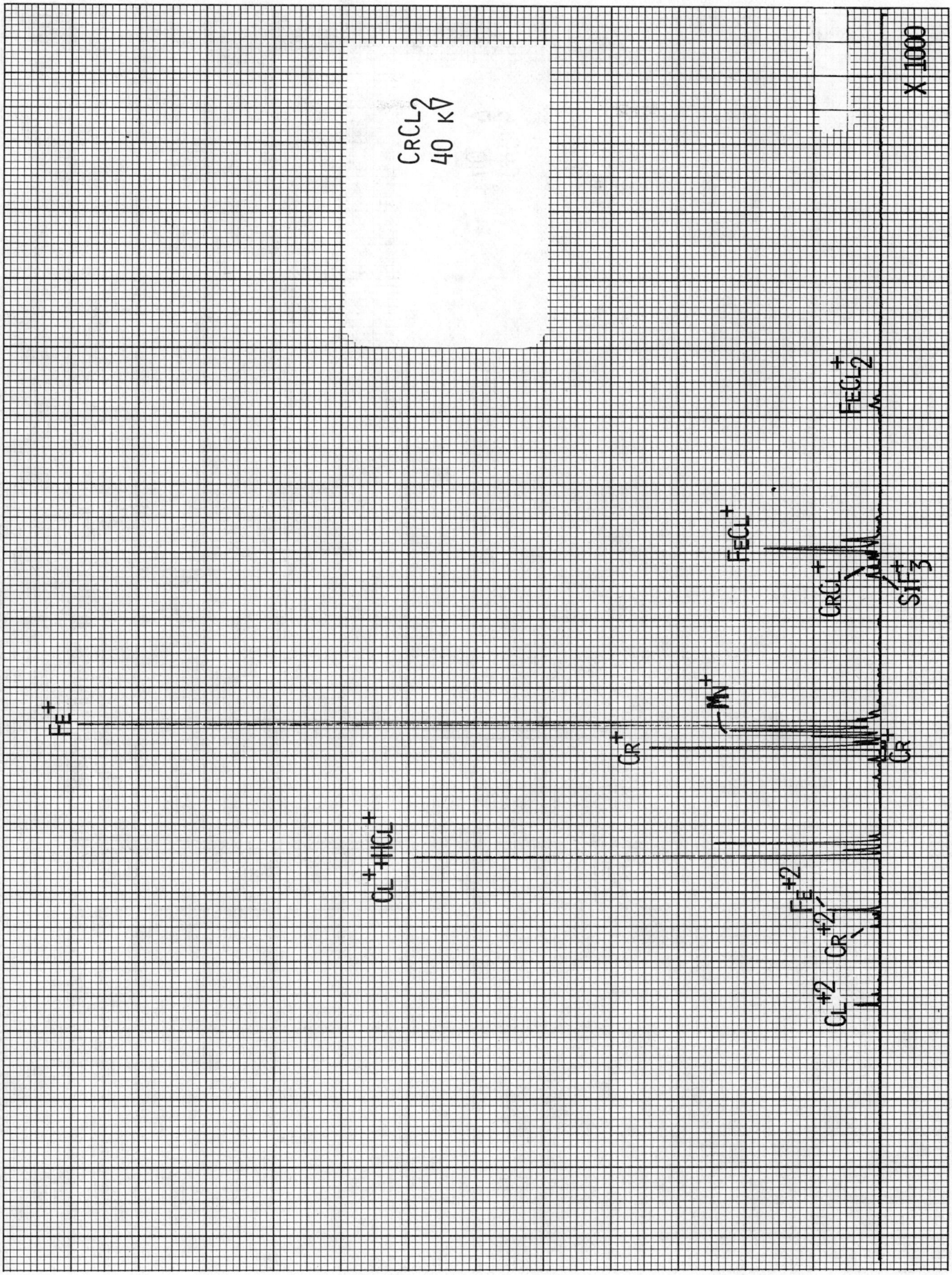

CrCl2
40 kV
X 1000
Fe+
Cr+
Mn+
FeCl+
CrCl+
SiF3+
FeCl2+
Cr+
Cl+ +HCl+
Fe+2
Cr+2
Cl+2

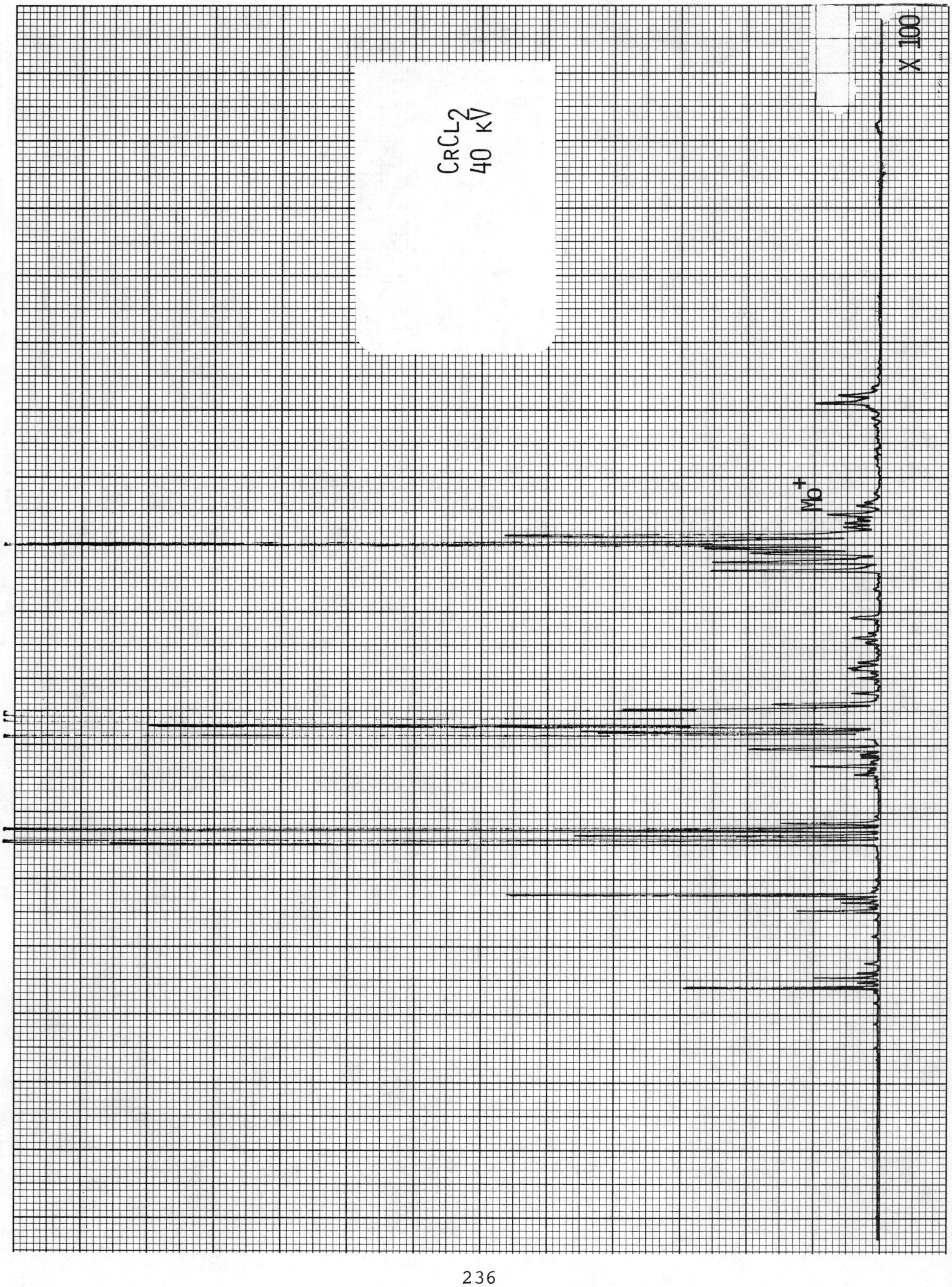

CrCl2
40 kV
X 100
Mo+

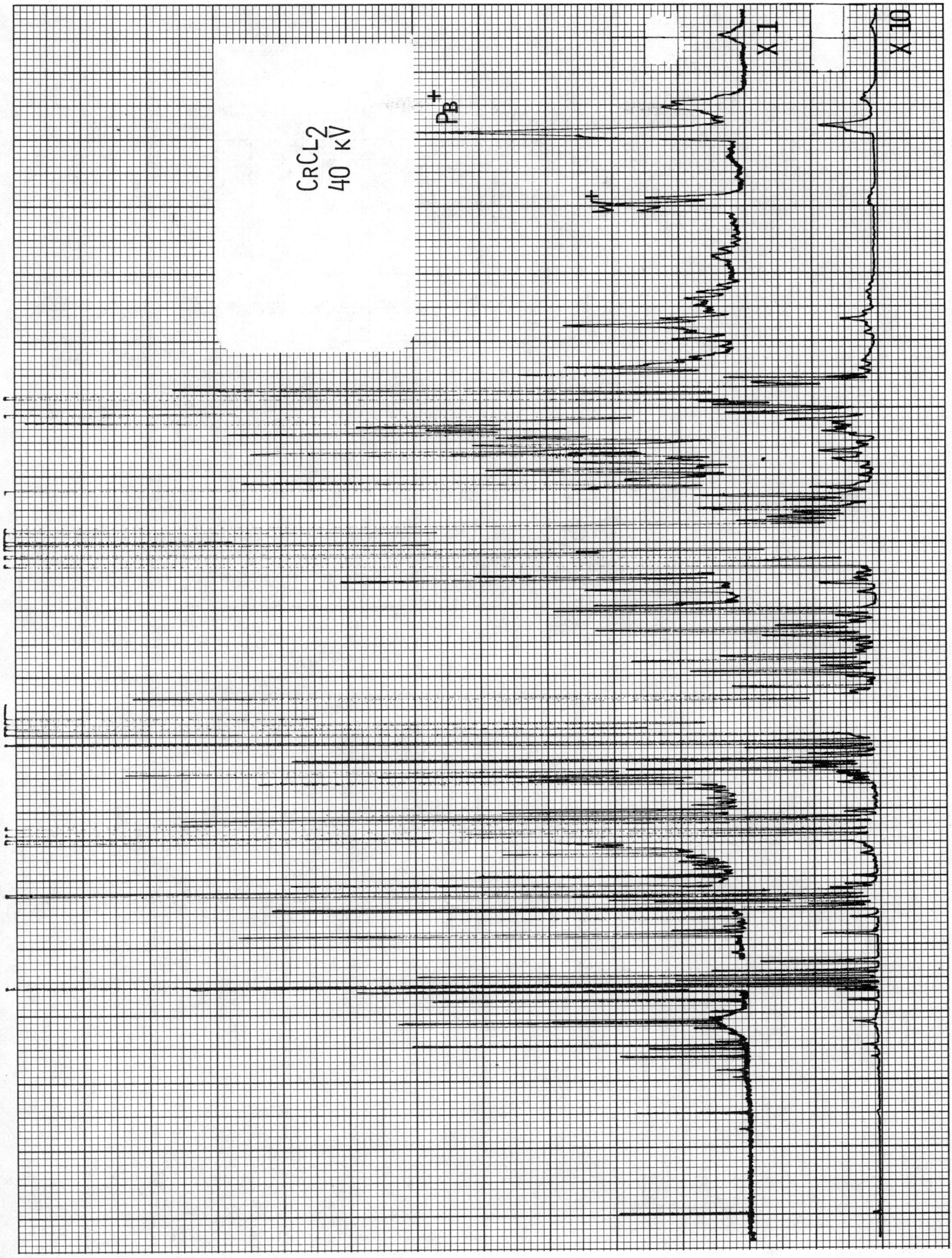
CrCl2
40 kV
Pb+
V+
X 1
X 10

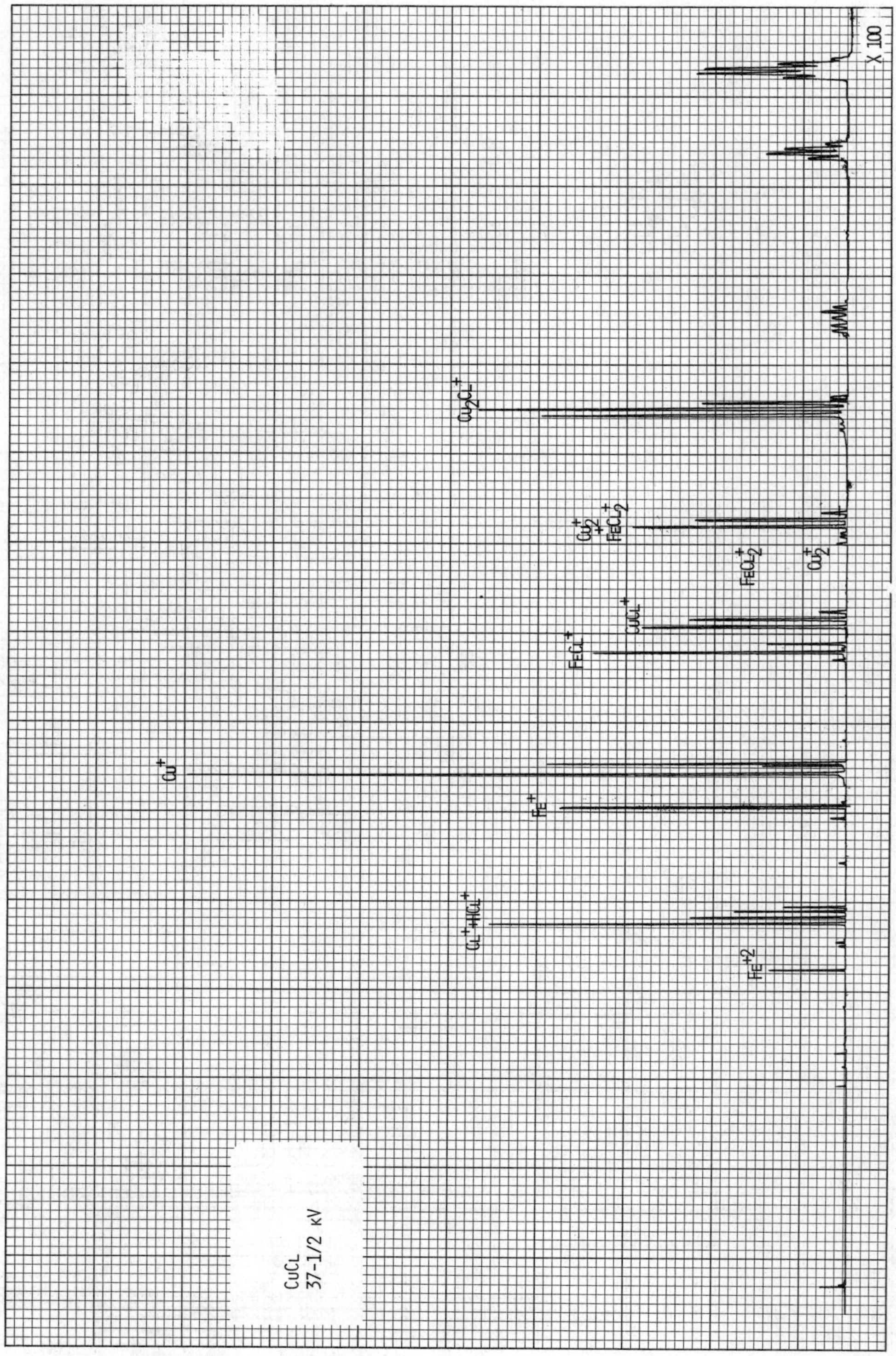

X 100
Cu_2Cl^+
Cu_2^+ + $FeCl_2^+$
$FeCl_2^+$
Cu_2^+
$CuCl^+$
$FeCl^+$
Cu^+
Fe^+
Cl^+ + HCl^+
Fe^{+2}
CuCl
37-1/2 kV

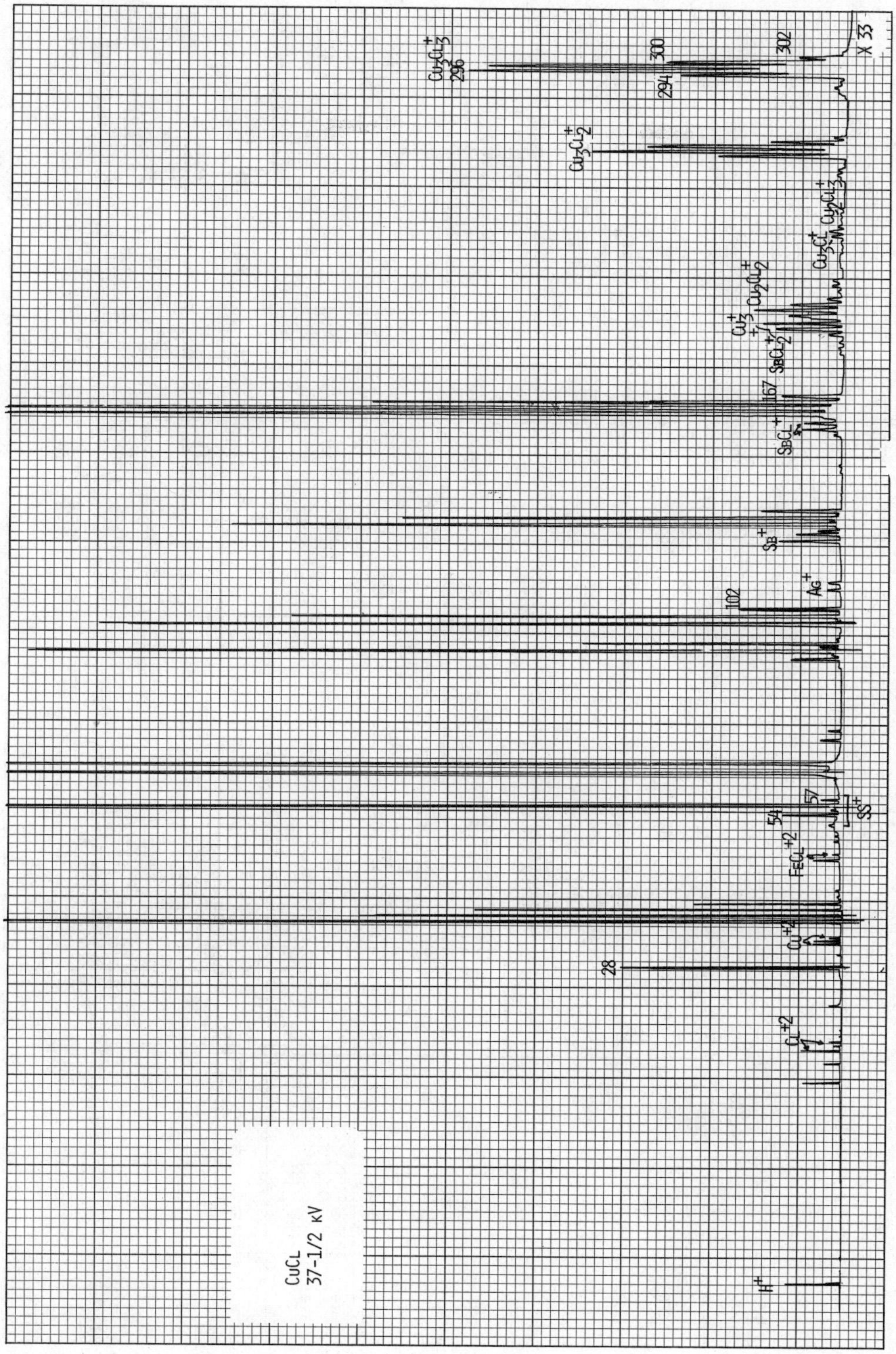
CuCl
37-1/2 kV
H+
28
54
57
SS+
102
Ag+
Sb+
SbCl+
167
294
300
296
302
X 33

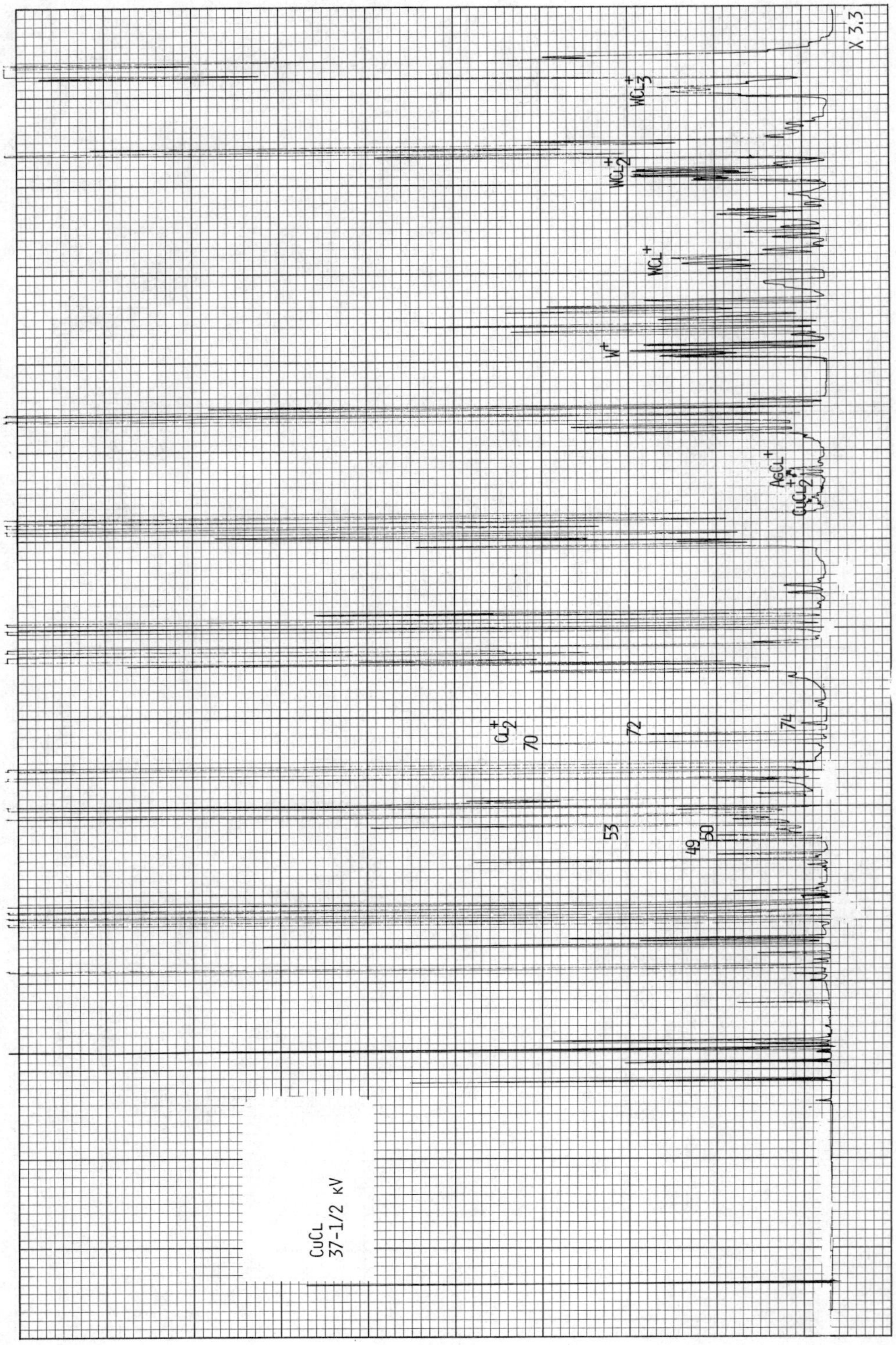

X 3.3
WCL3+
WCL2+
WCL+
W+
AsCL+
CuCL2+
Cu2+
70
72
74
53
49
50
CuCL
37-1/2 kV

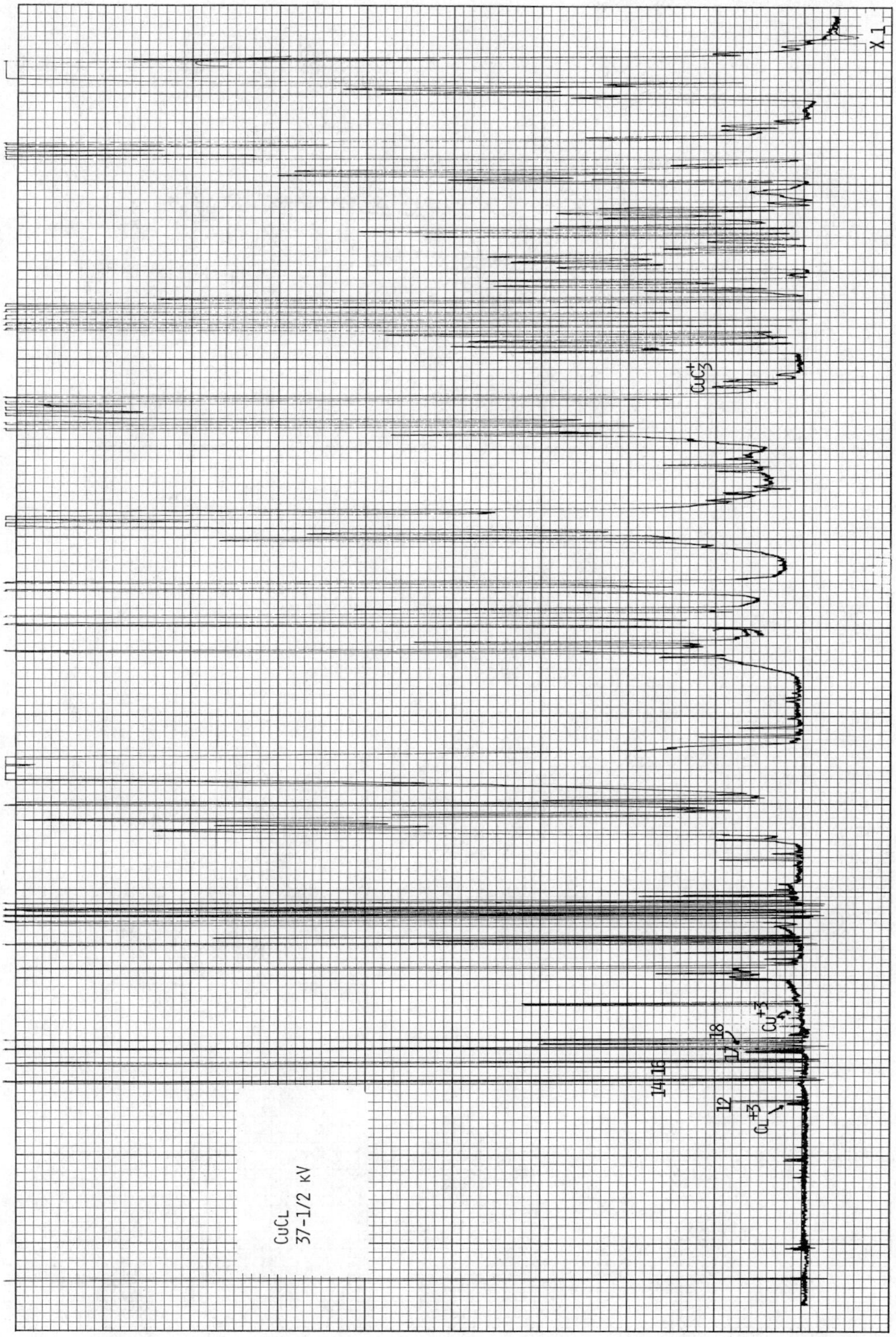
CuCL
37-1/2 kV
X 1

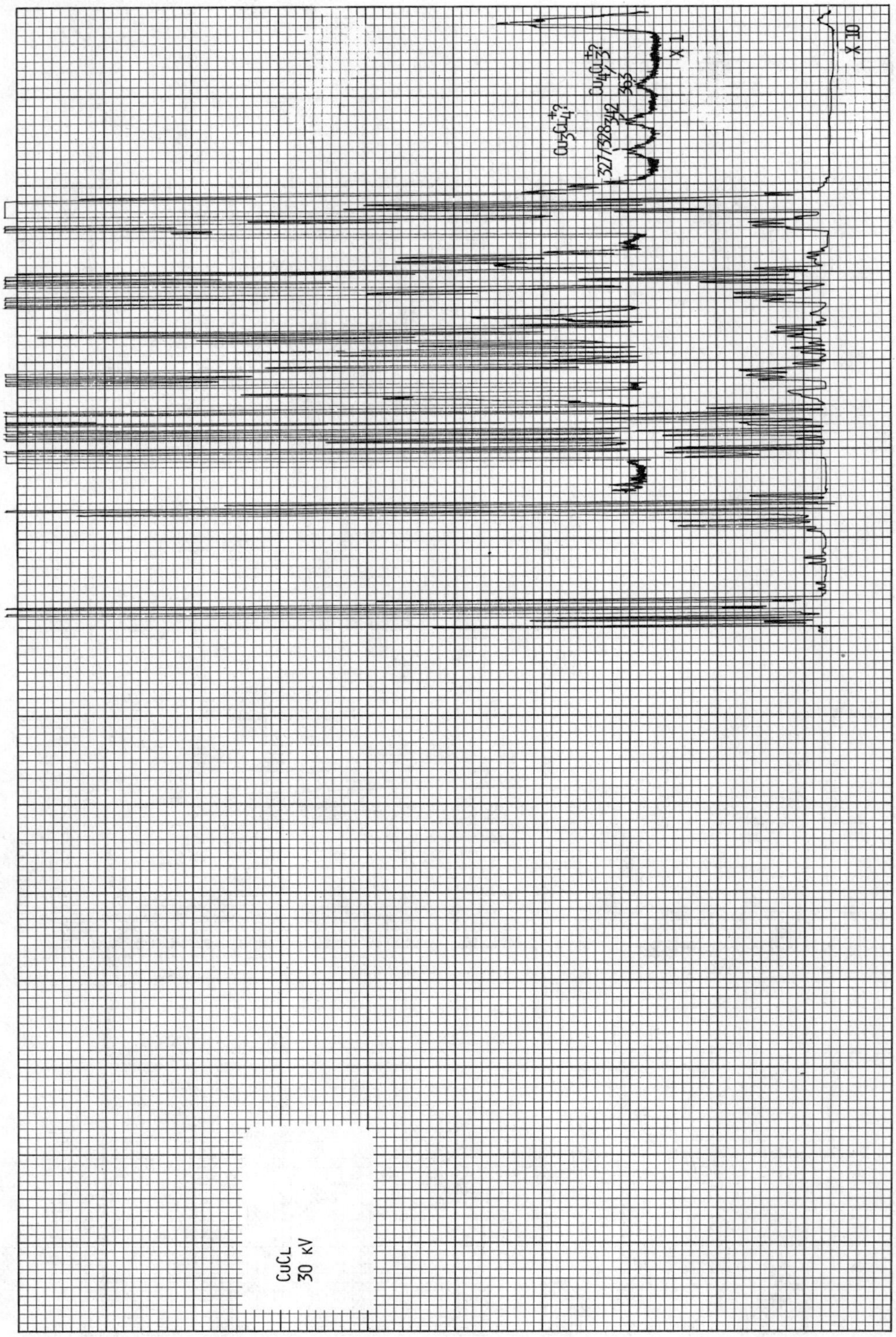
CuCl
30 kV
X 1
X 10

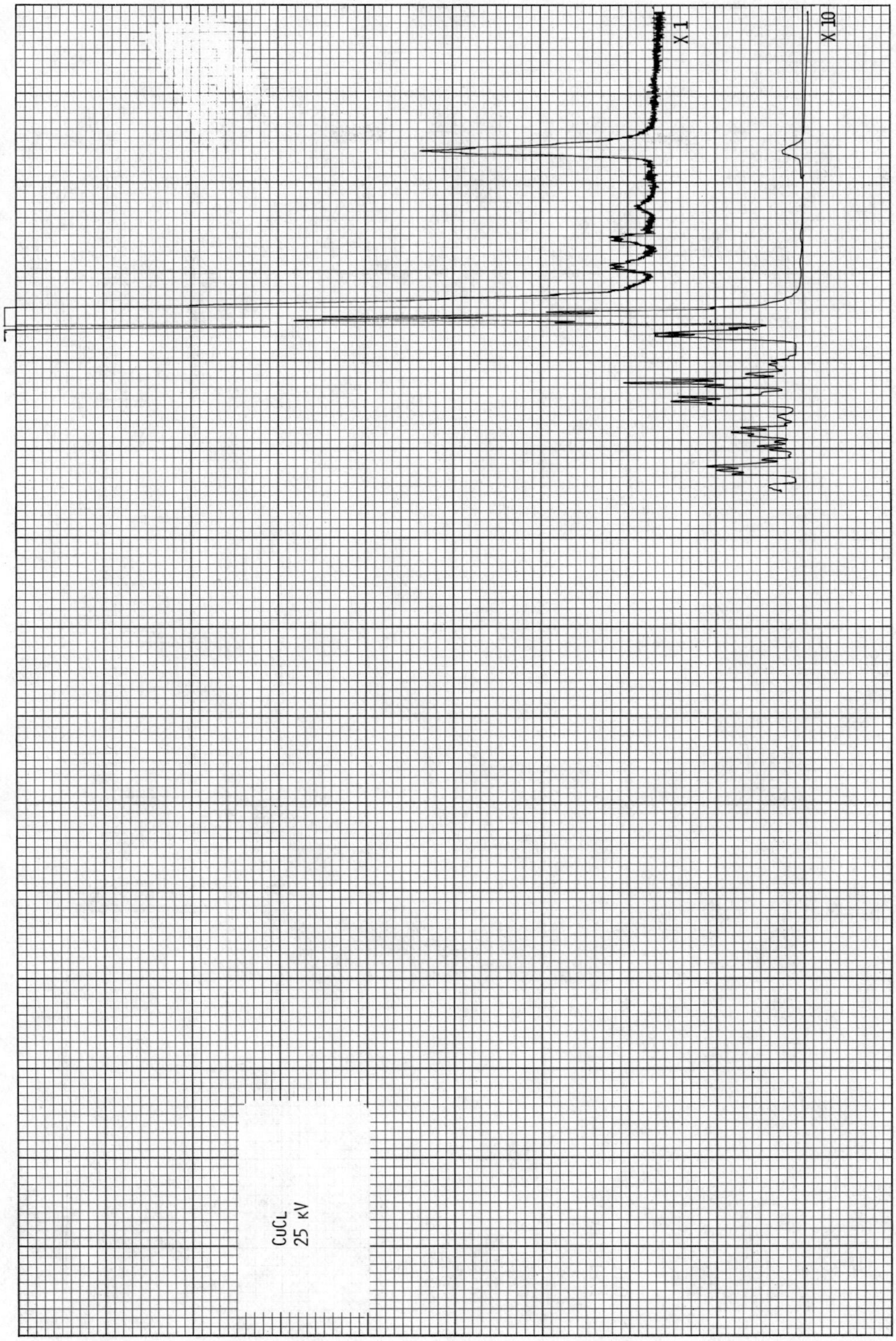
X 1
X 10
CuCL
25 kV

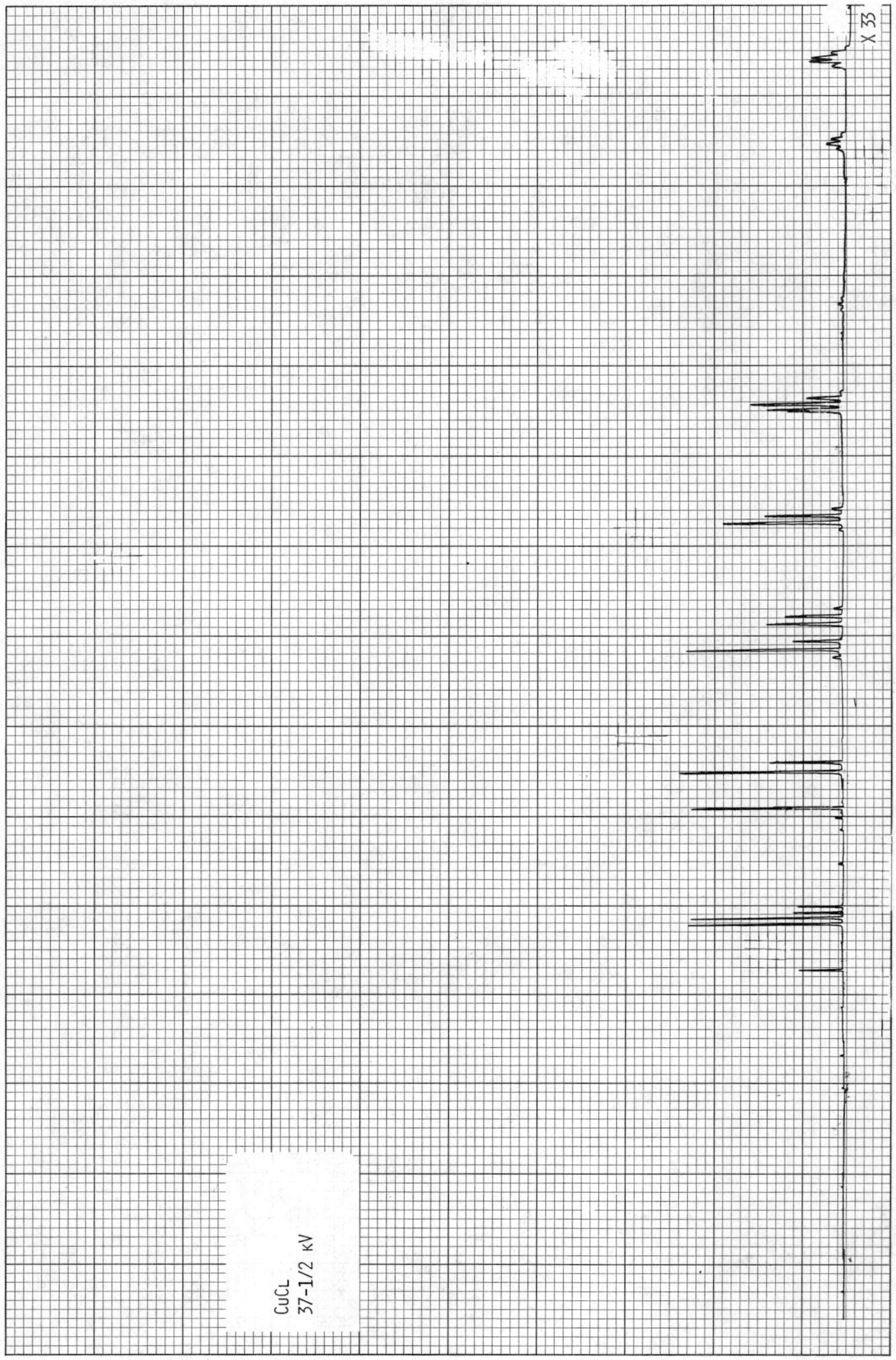
CuCL
37-1/2 kV
X 33

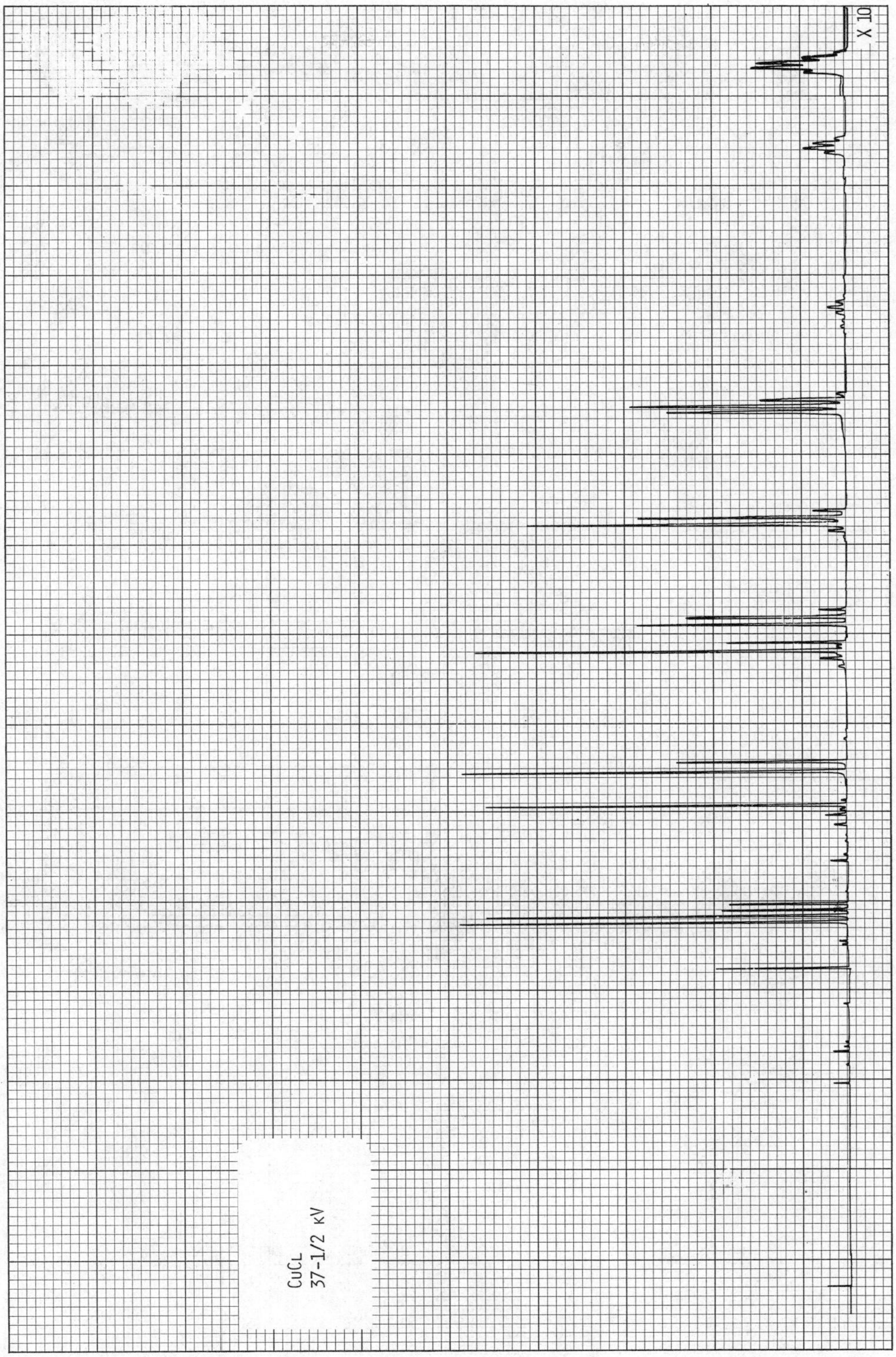
X 10
CuCl
37-1/2 kV

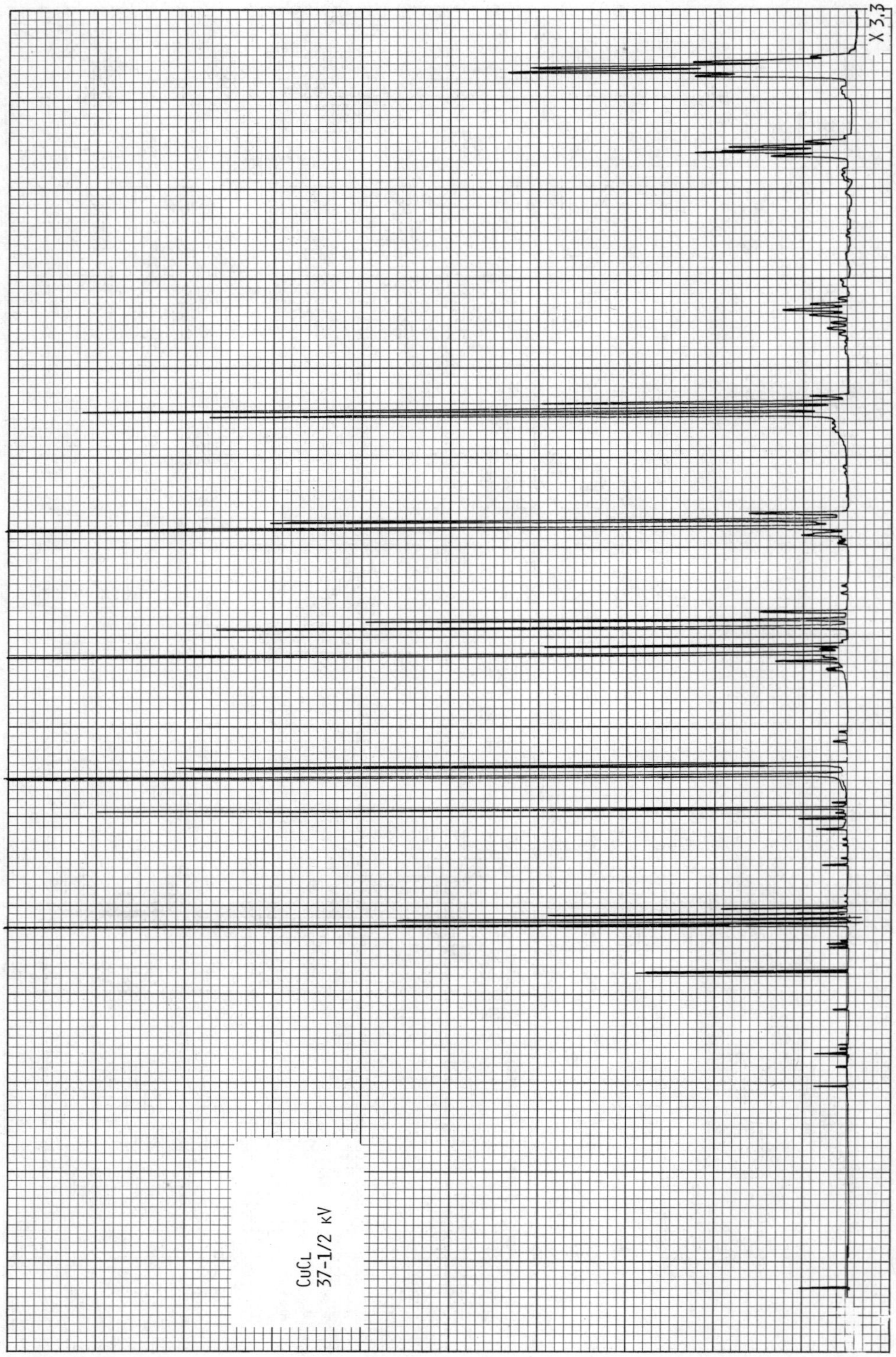
X 3.3
CuCl
37-1/2 kV

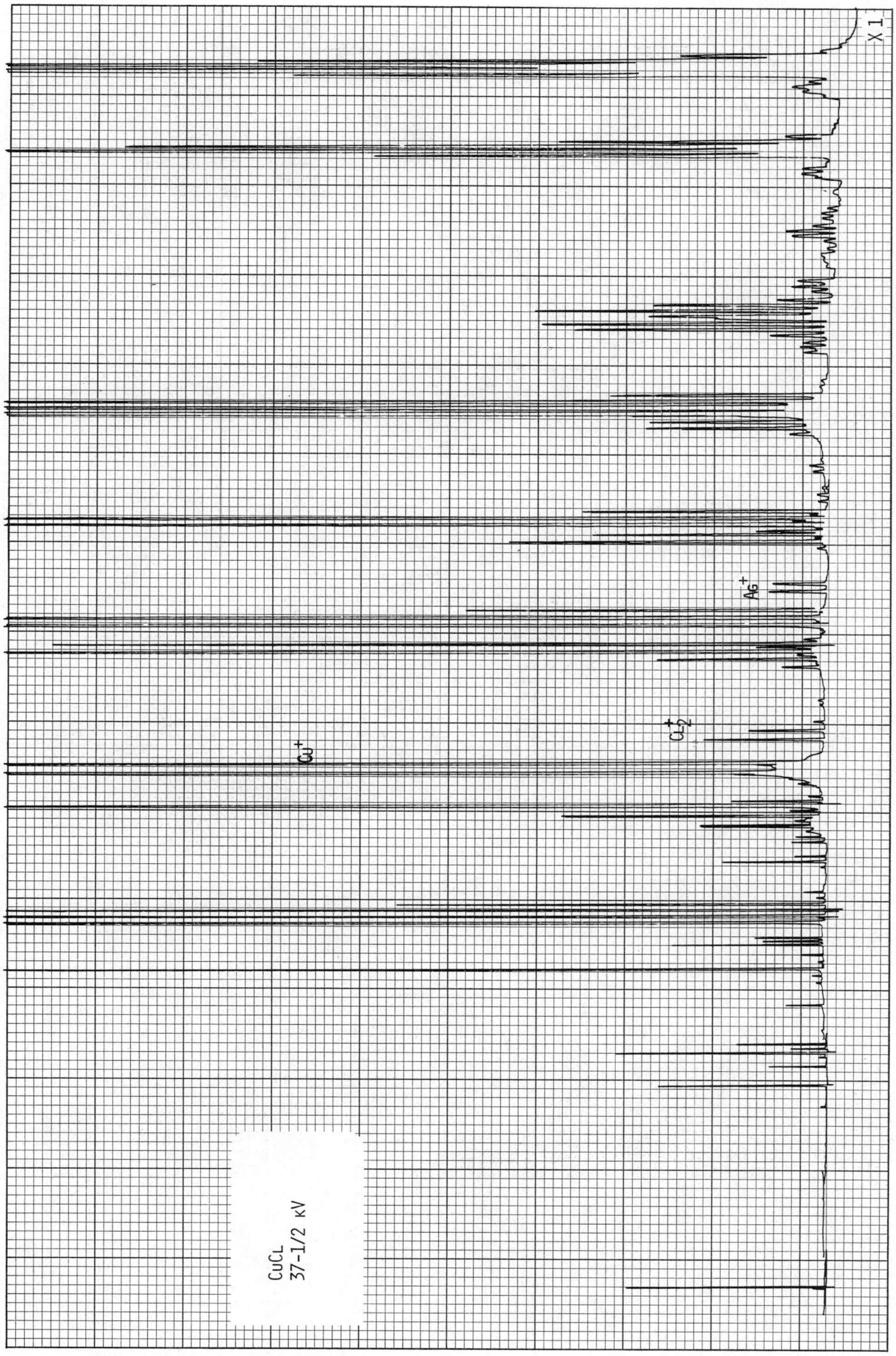
X 1
Ag+
Cl2+
Cu+
CuCl
37-1/2 kV

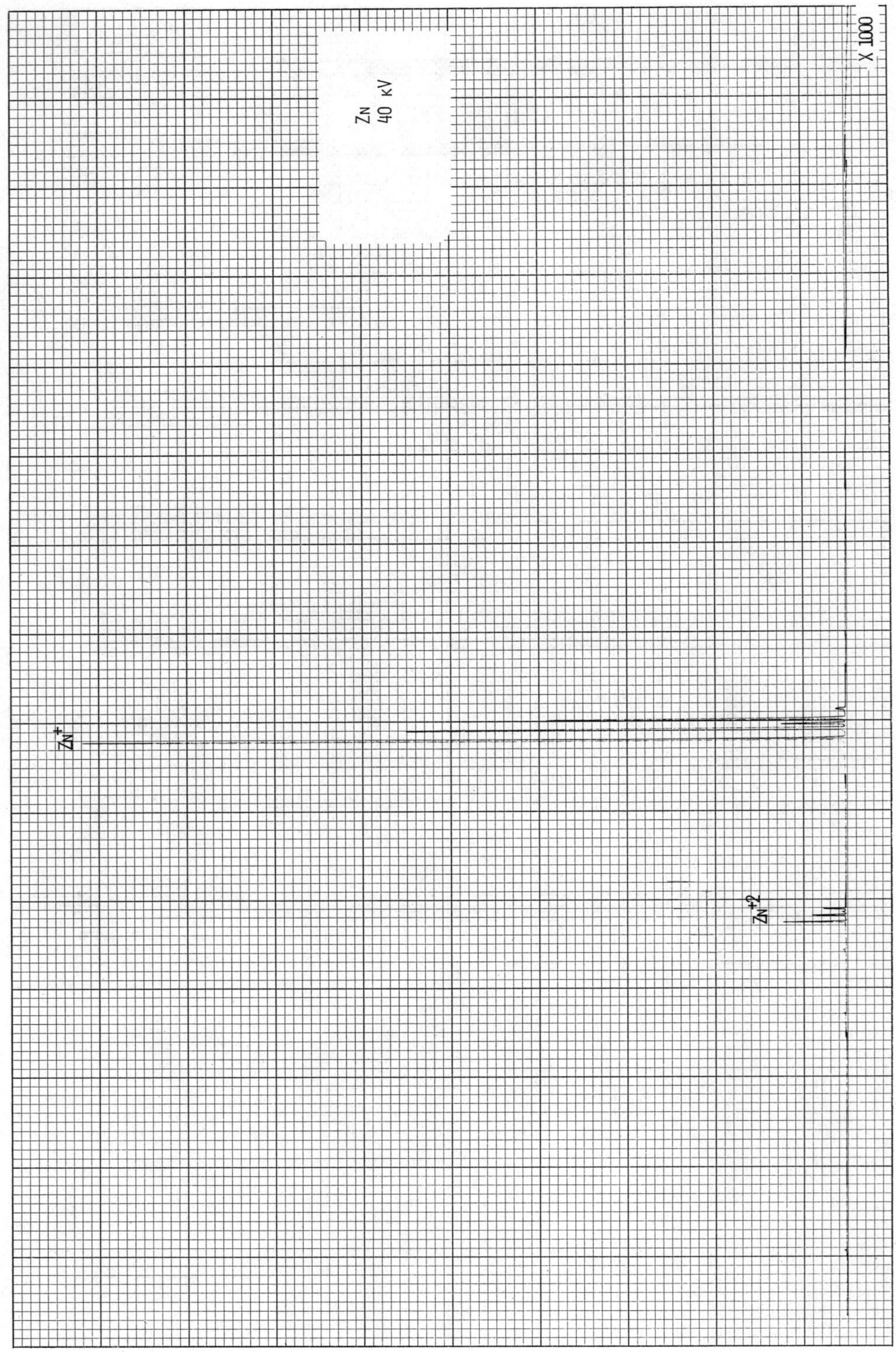
ZN
40 kV
X 1000
ZN+
ZN+2

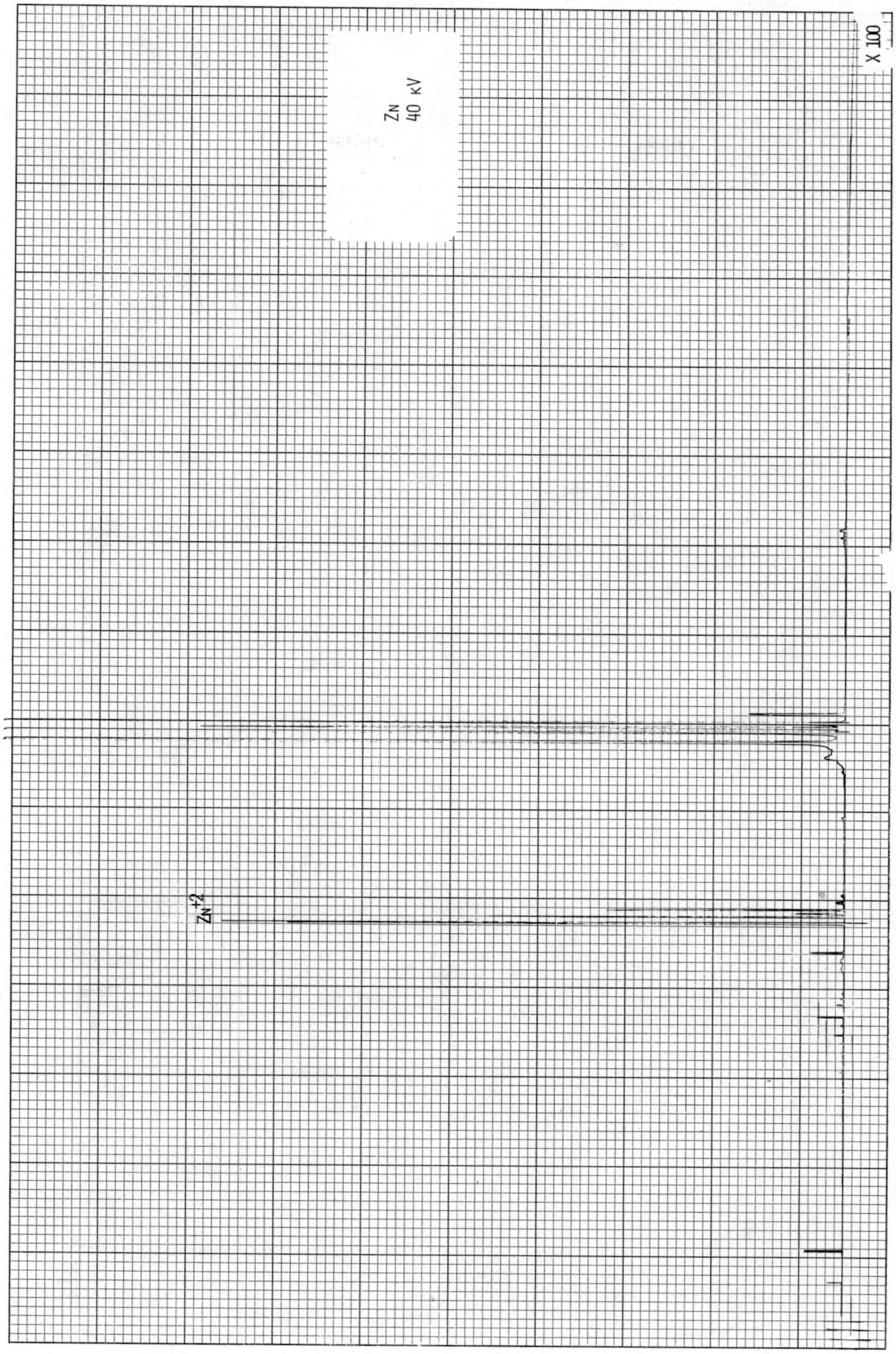
Zn
40 kV
X 100
Zn+2

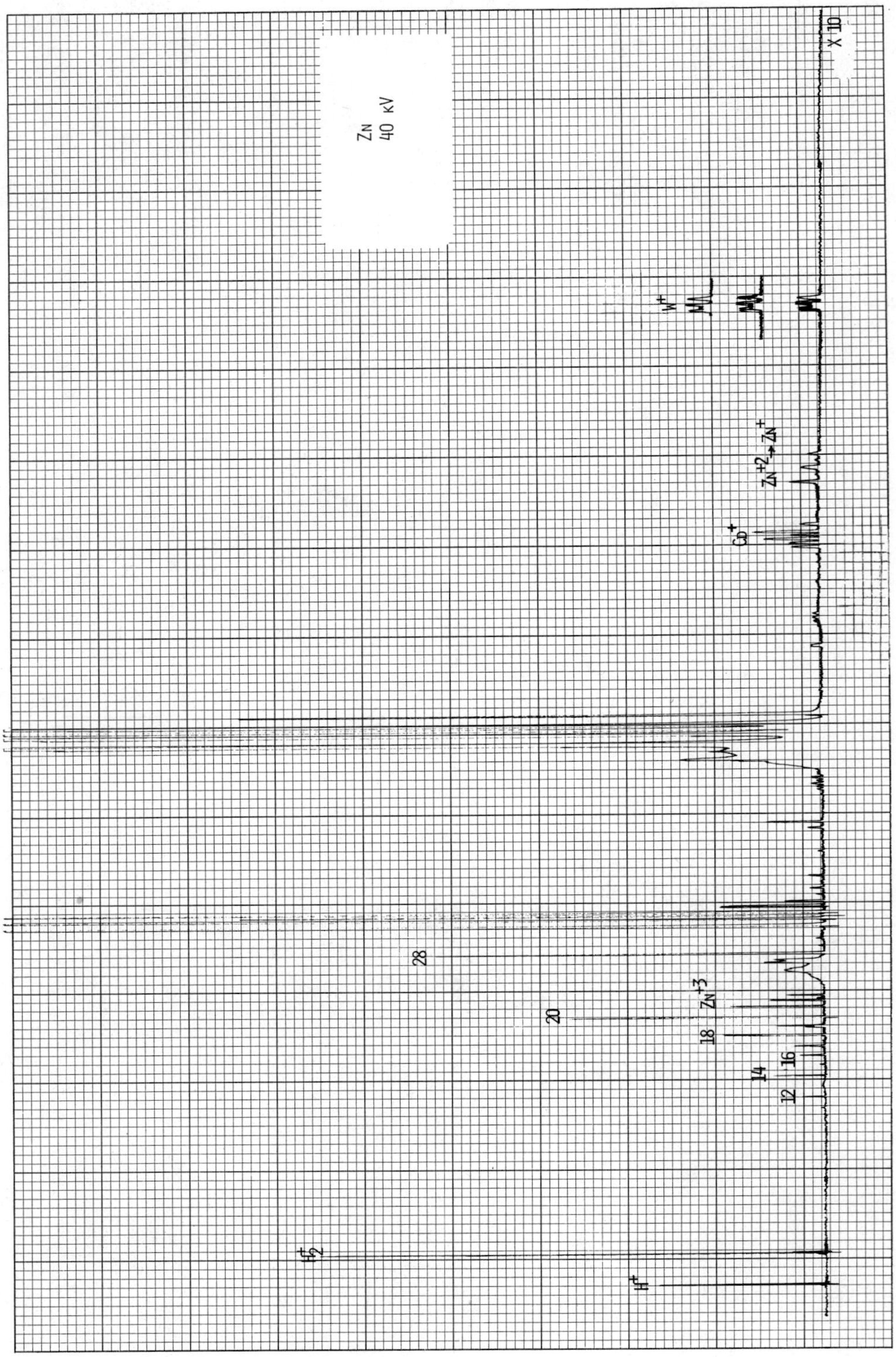

ZN
40 kV
X 10
W+
ZN+2→ZN+
CO+
ZN+3
28
20
18
14
16
12
H2+
H+

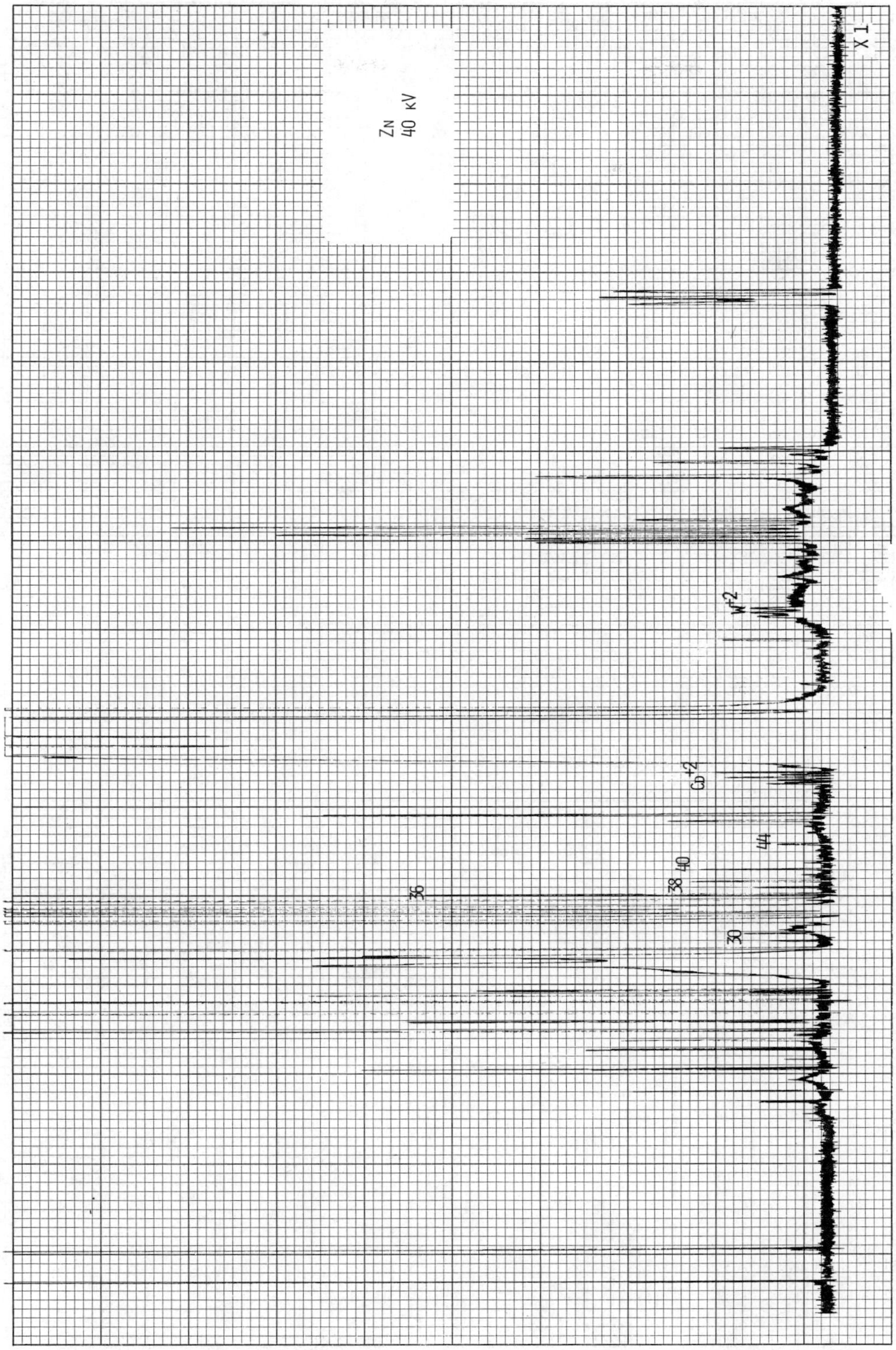
Zn
40 kV
X 1
36
38
40
44
30

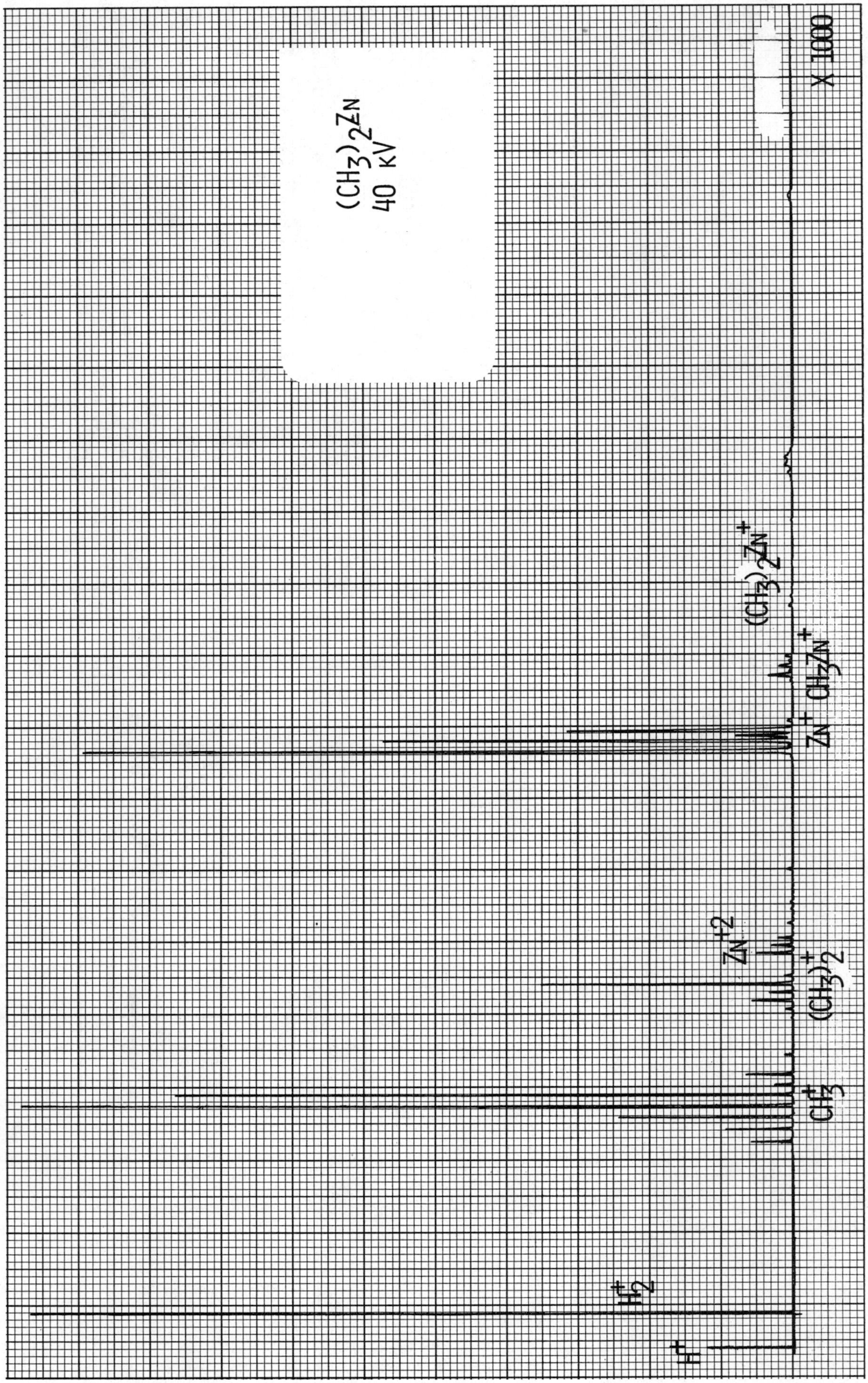
$(CH_3)_2Zn$
40 kV
X 1000
$(CH_3)_2Zn^+$
Zn^+
CH_3Zn^+
Zn^{+2}
$(CH_3)_2^+$
CH_3^+
H_2^+
H^+

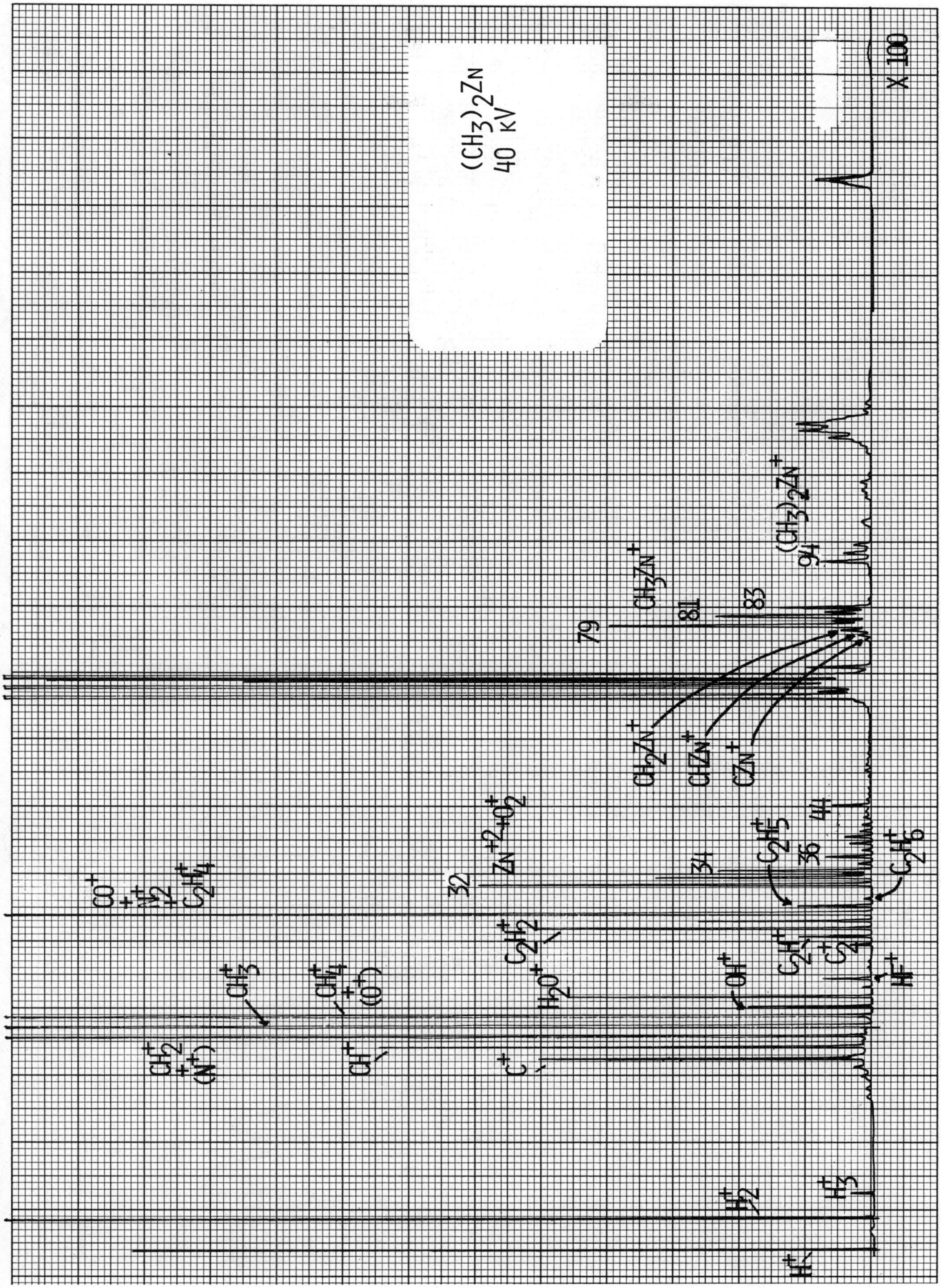

(CH3)2Zn
40 KV
X 100
79
81
83
94
CH3Zn+
(CH3)2Zn+
CH2Zn+
CHZn+
CZn+
32
34
36
44
Zn+2 + O2+
C2H5+
C2H6+
CO+ + N2+ + C2H4+
C2H2+
C2H+
C2+
H2O+
OH+
HF+
CH3+
CH4+ + (O+)
CH2+ + (N+)
CH+
C+
H2+
H3+
H+

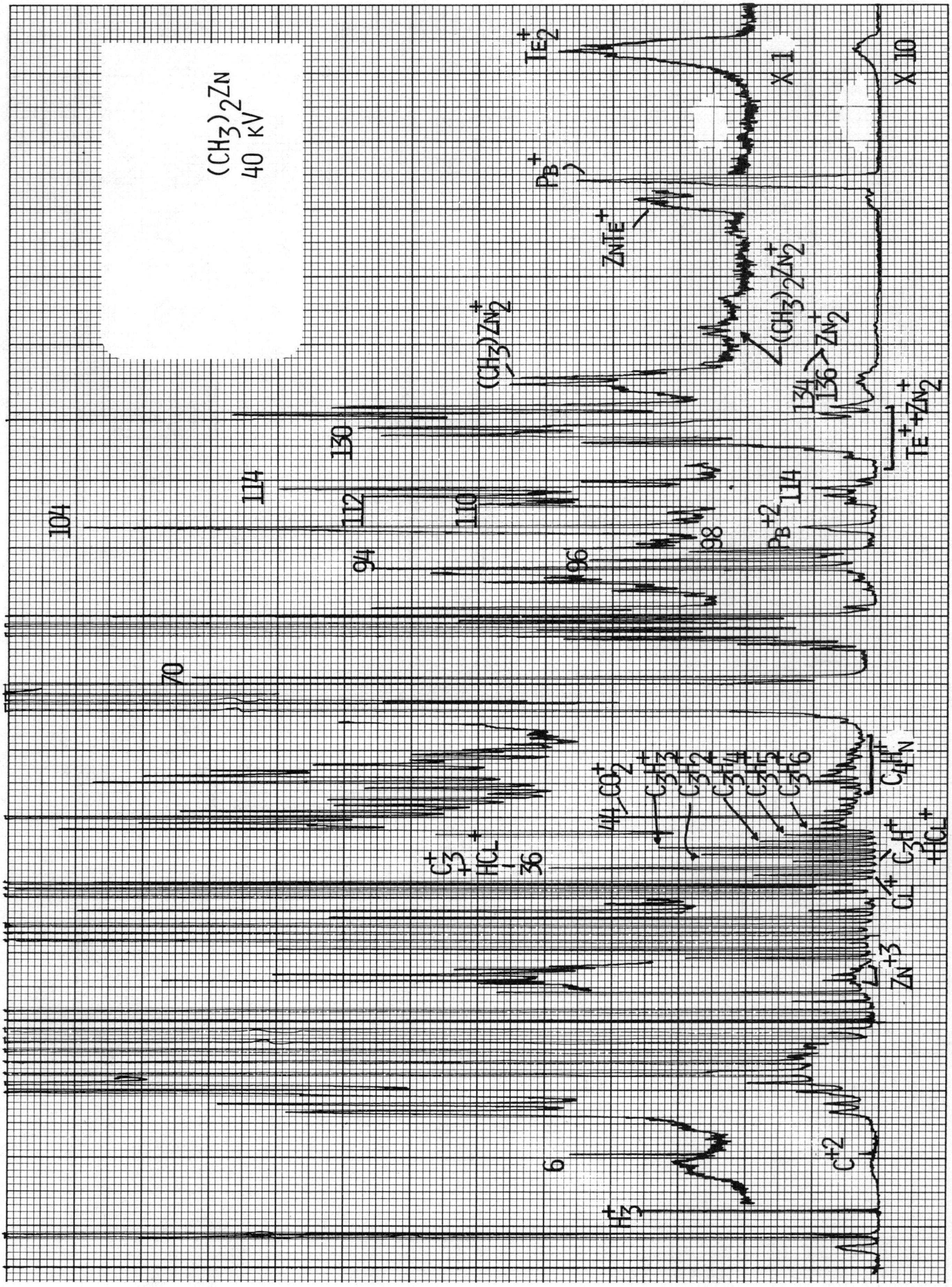
(CH3)2ZN
40 KV
TE2+
X 1
X 10
PB+
ZNTE+
(CH3)2ZN2+
(CH3)ZN2+
134
136
TE+ +ZN2+
130
114
104
112
110
94
96
98
PB+2
114
70
44 CO2+
C3H3+
C3H2+
C3H4+
C3H5+
C3H6+
C4HN+
C3+
+HCL+
36
CL+
C3H+
+HCL+
ZN+3
6
C+2
H3+

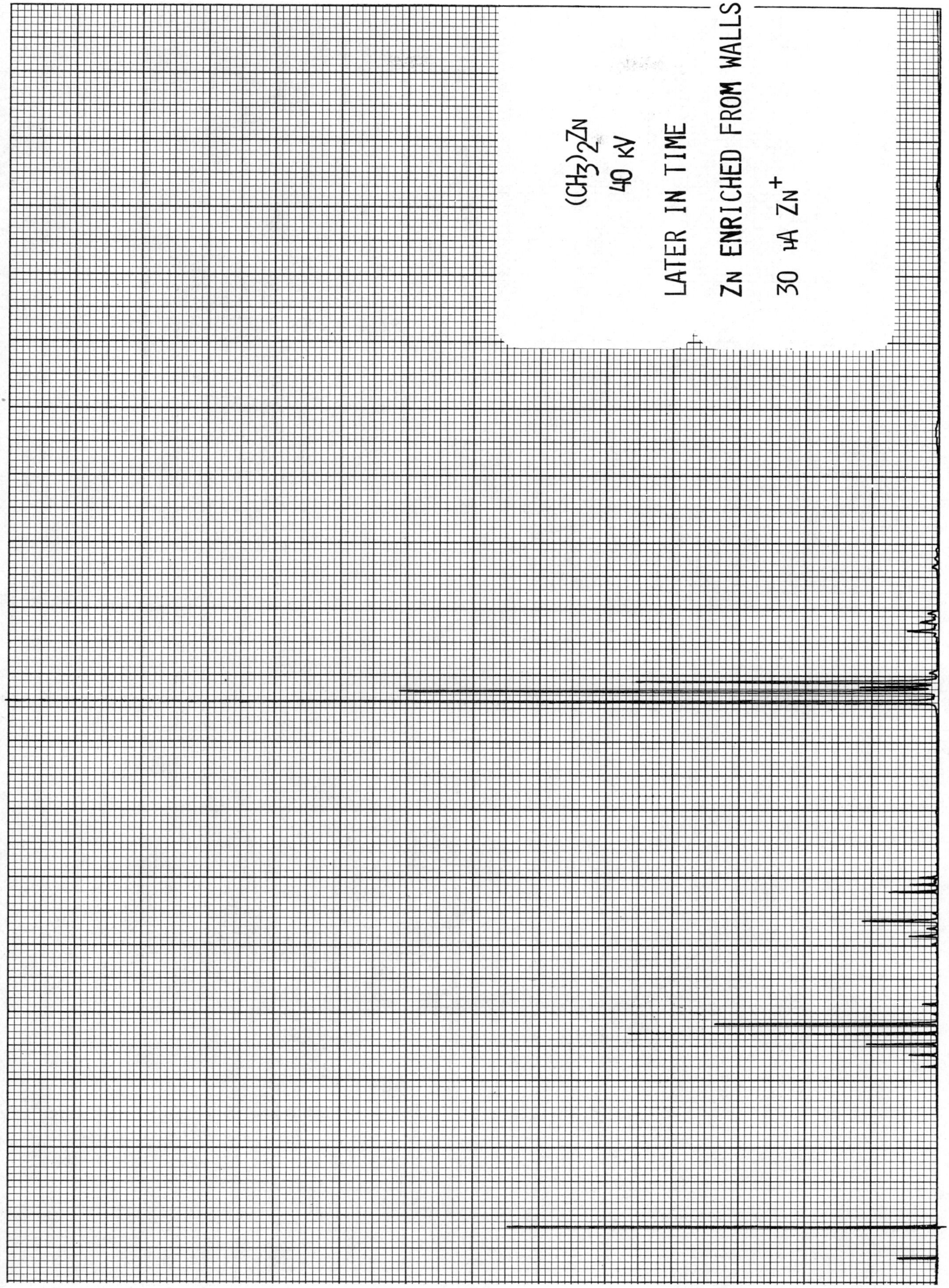

$(CH_3)_2Zn$
40 kV
LATER IN TIME
Zn ENRICHED FROM WALLS
30 µA Zn^+

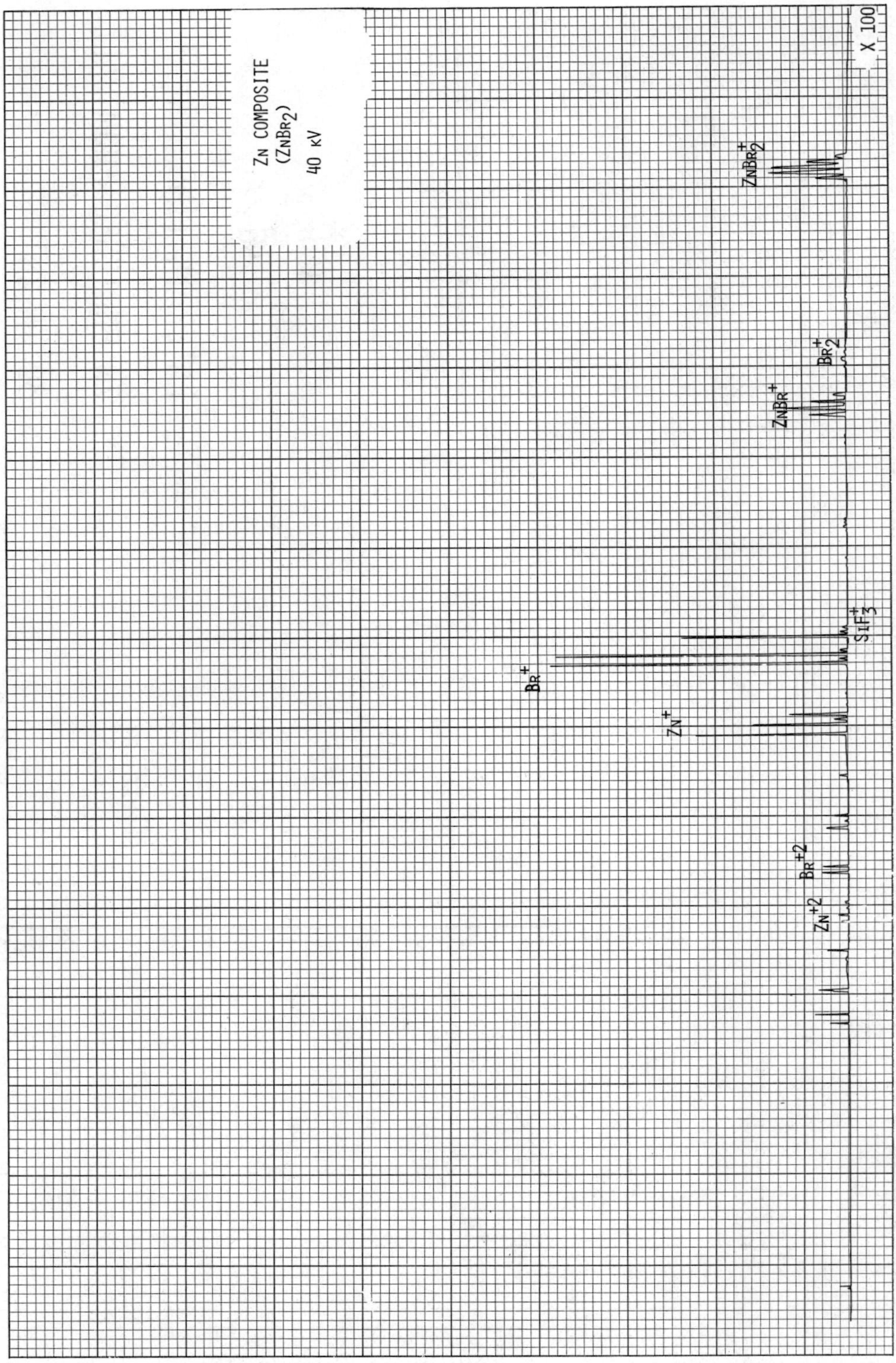

Zn COMPOSITE
($ZnBr_2$)
40 kV
X 100
$ZnBr_2^+$
$ZnBr^+$
Br_2^+
SiF_3^+
Br^+
Zn^+
Br^{+2}
Zn^{+2}

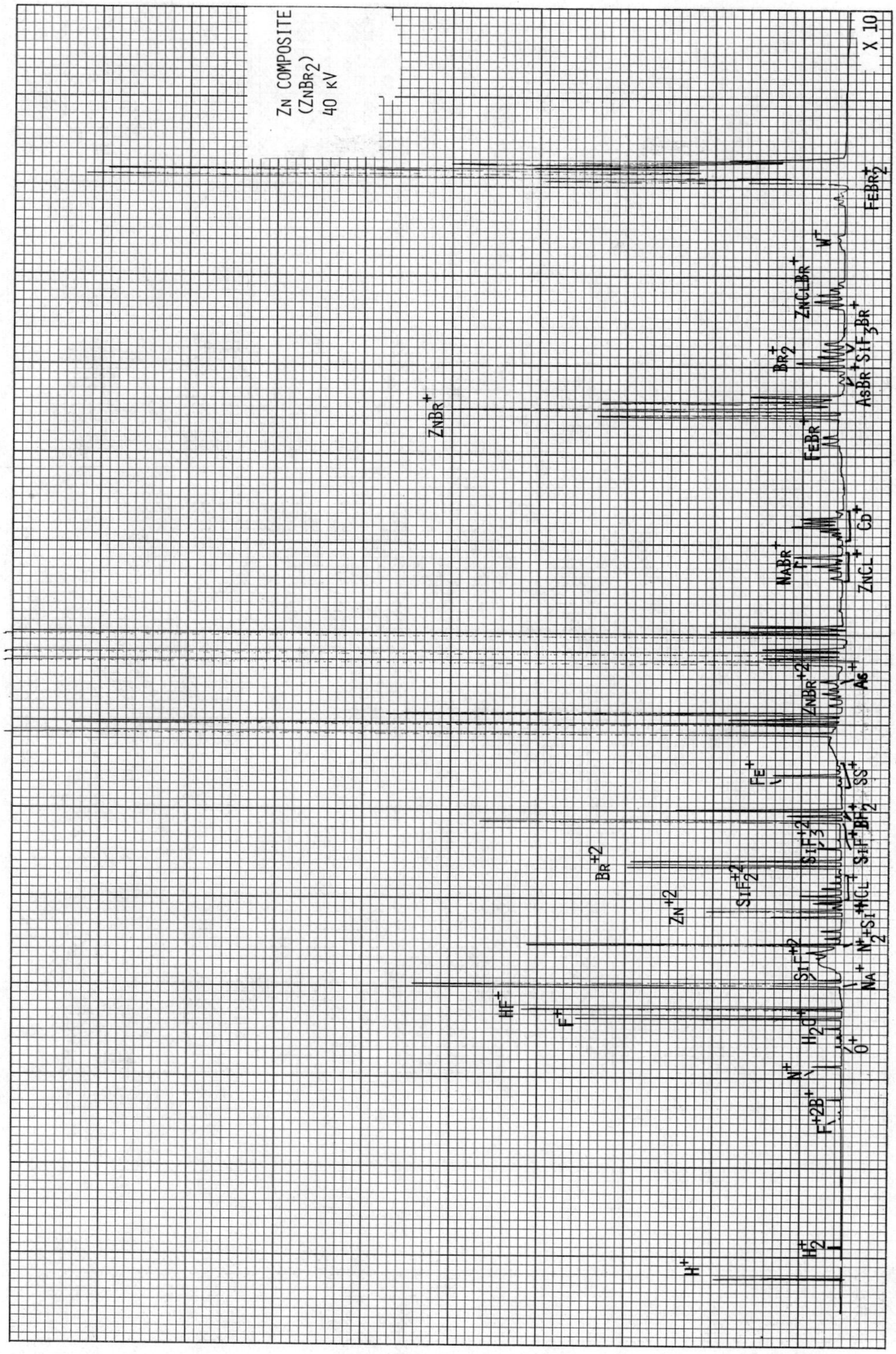
ZN COMPOSITE
(ZNBR2)
40 KV
X 10
H+
H2+
F+2B+
N+
H2O+
O+
HF+
F+
SIF+2
NA+
N2+SI+
ZN+2
SIF2+2
CL
BR+2
SIF3+2
SIF+
BF2+
FE+
SS+
ZNBR+2
AS+
NABR+
ZNCL+
CD+
ZNBR+
FEBR+
BR2+
ASBR+
SIF3BR+
ZNCLBR+
W+
FEBR2+

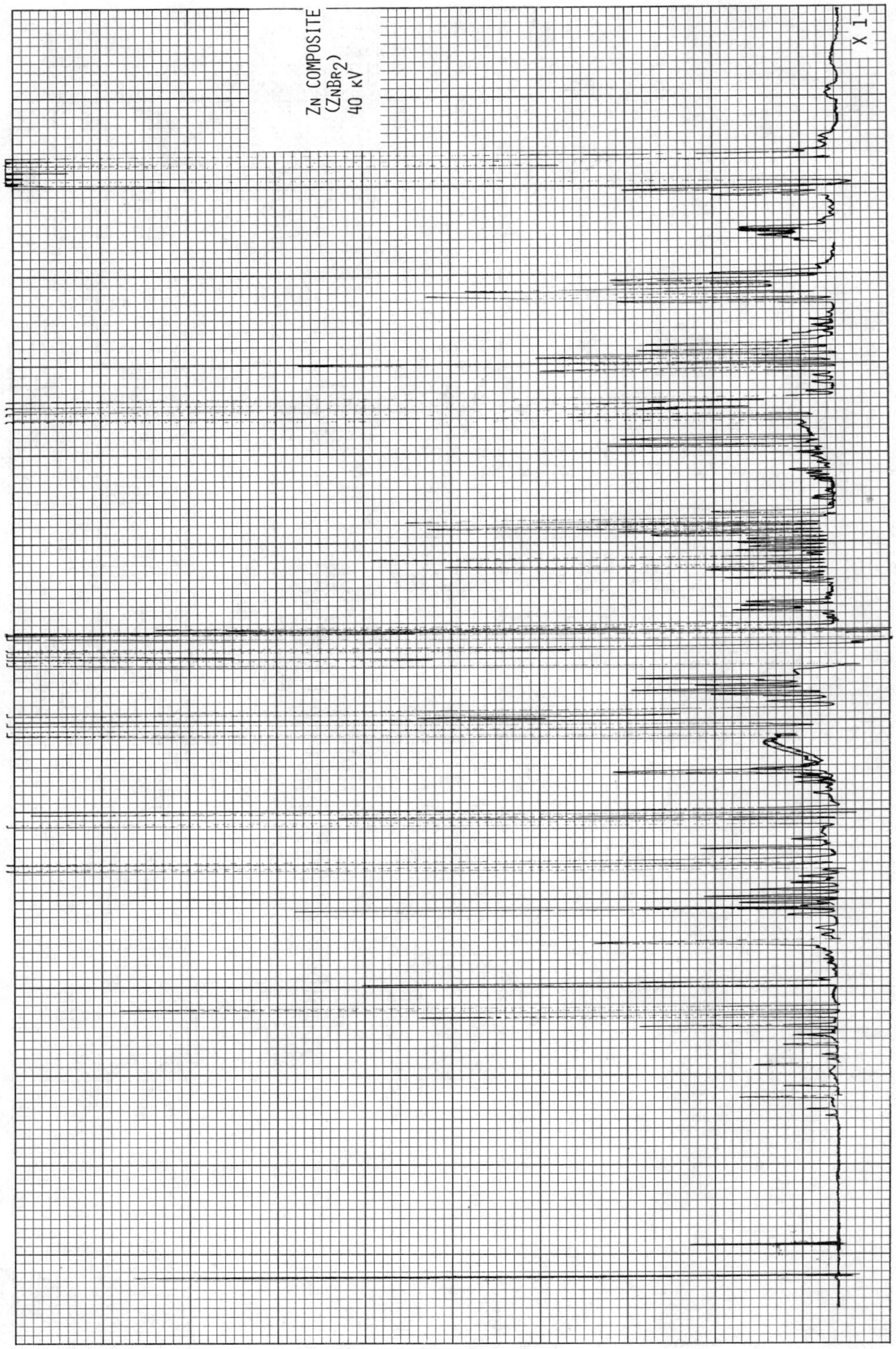
Zn COMPOSITE
($ZnBr_2$)
40 kV
X 1

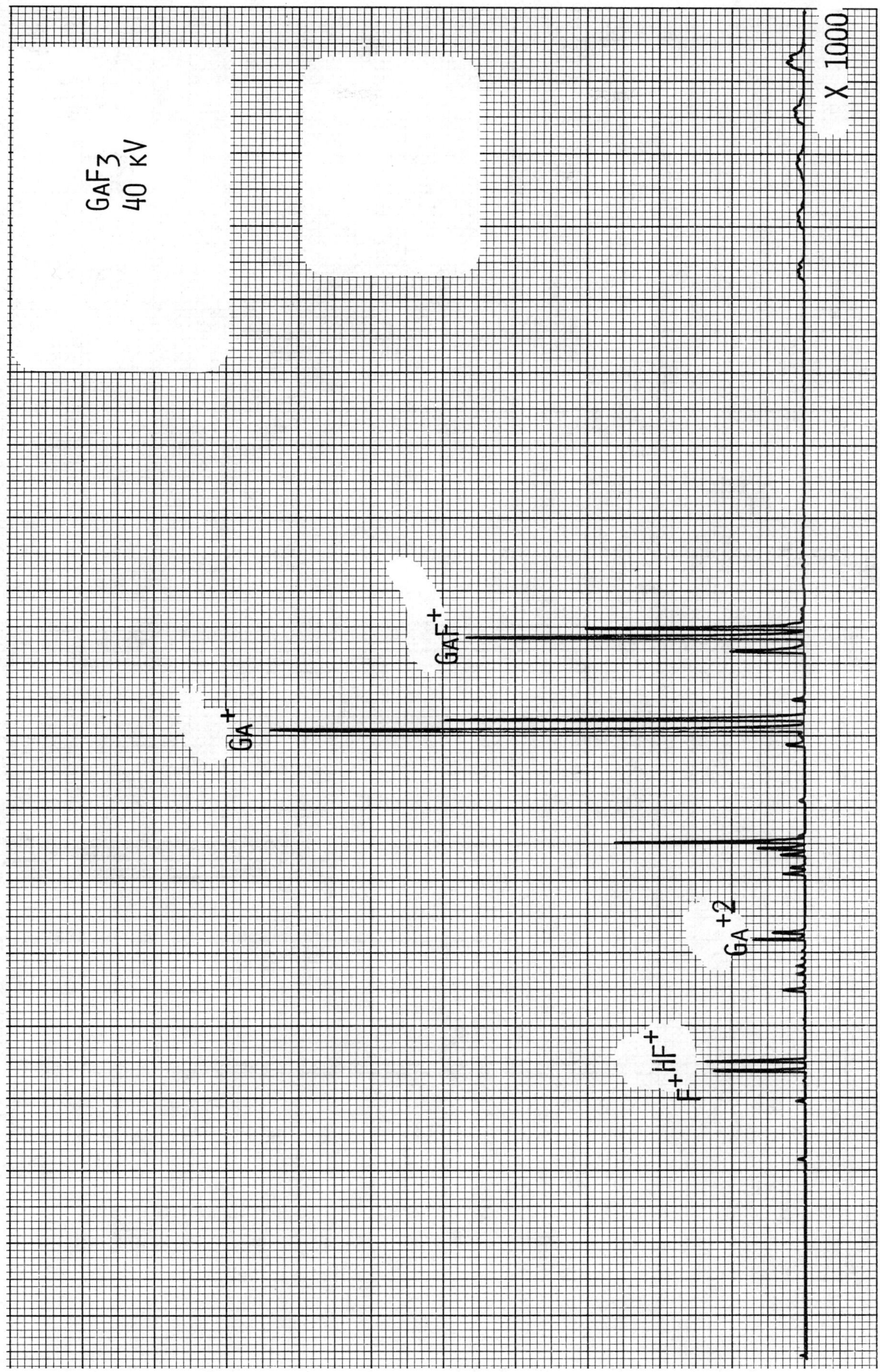
GAF3
40 KV
X 1000
GAF+
GA+
GA+2
F+HF+

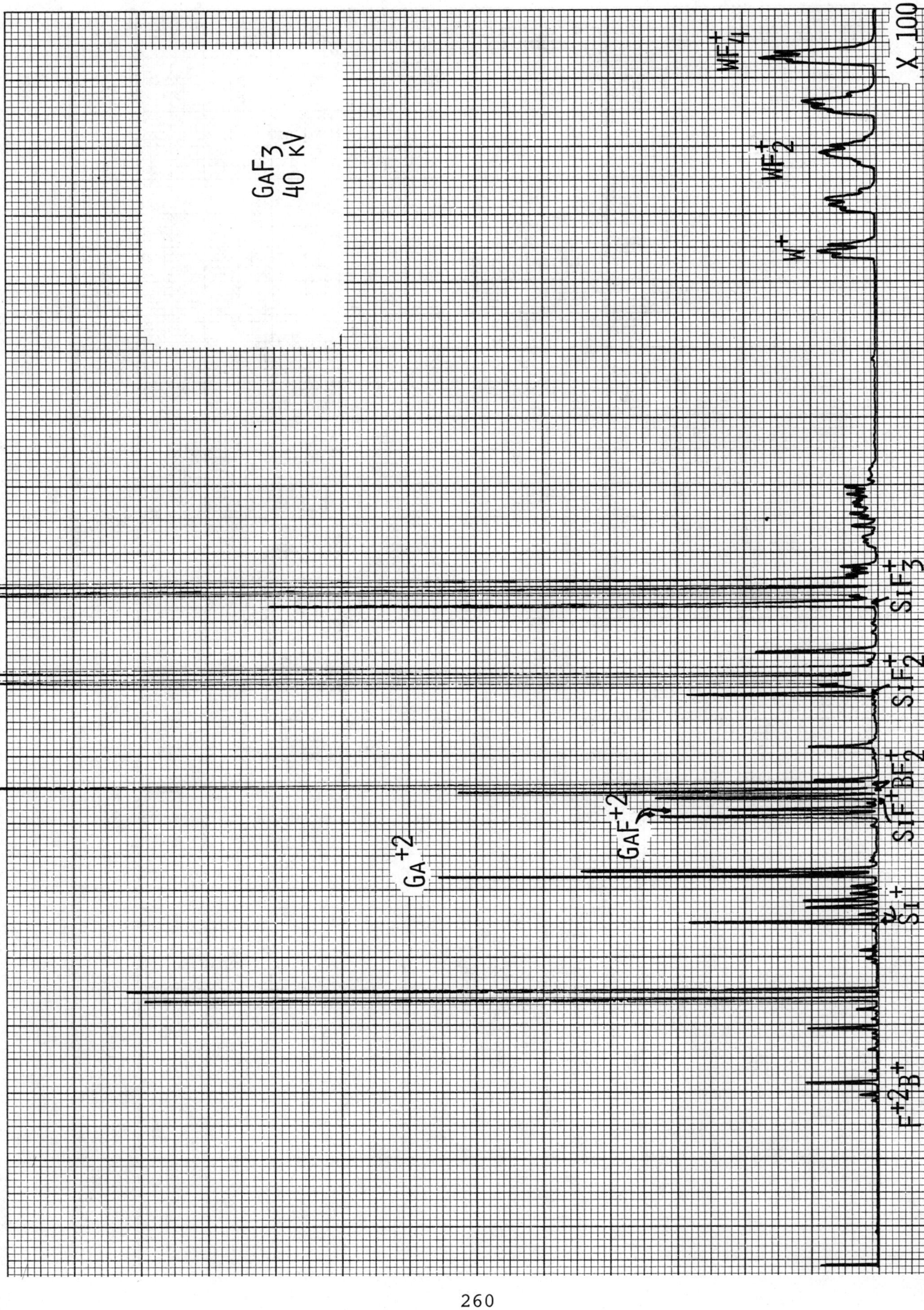
GaF$_3$
40 KV
X 100
WF$_4^+$
WF$_2^+$
W$^+$
SiF$_3^+$
SiF$_2^+$
BF$_2^+$
SiF$^+$
GaF^{+2}
Ga^{+2}
Si$^+$
F^{+2}
B$^+$

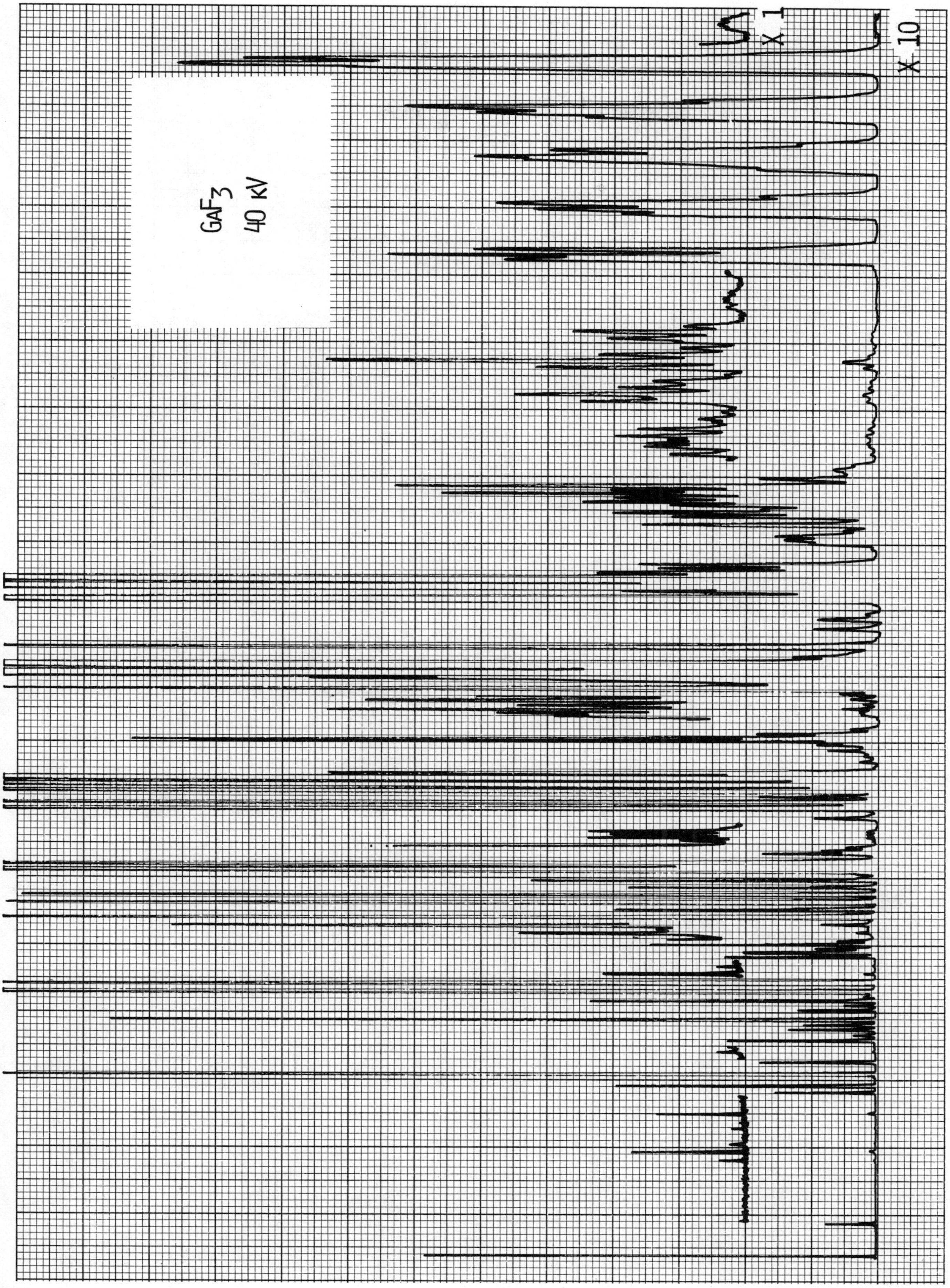

GaF3
40 kV
X 1
X 10

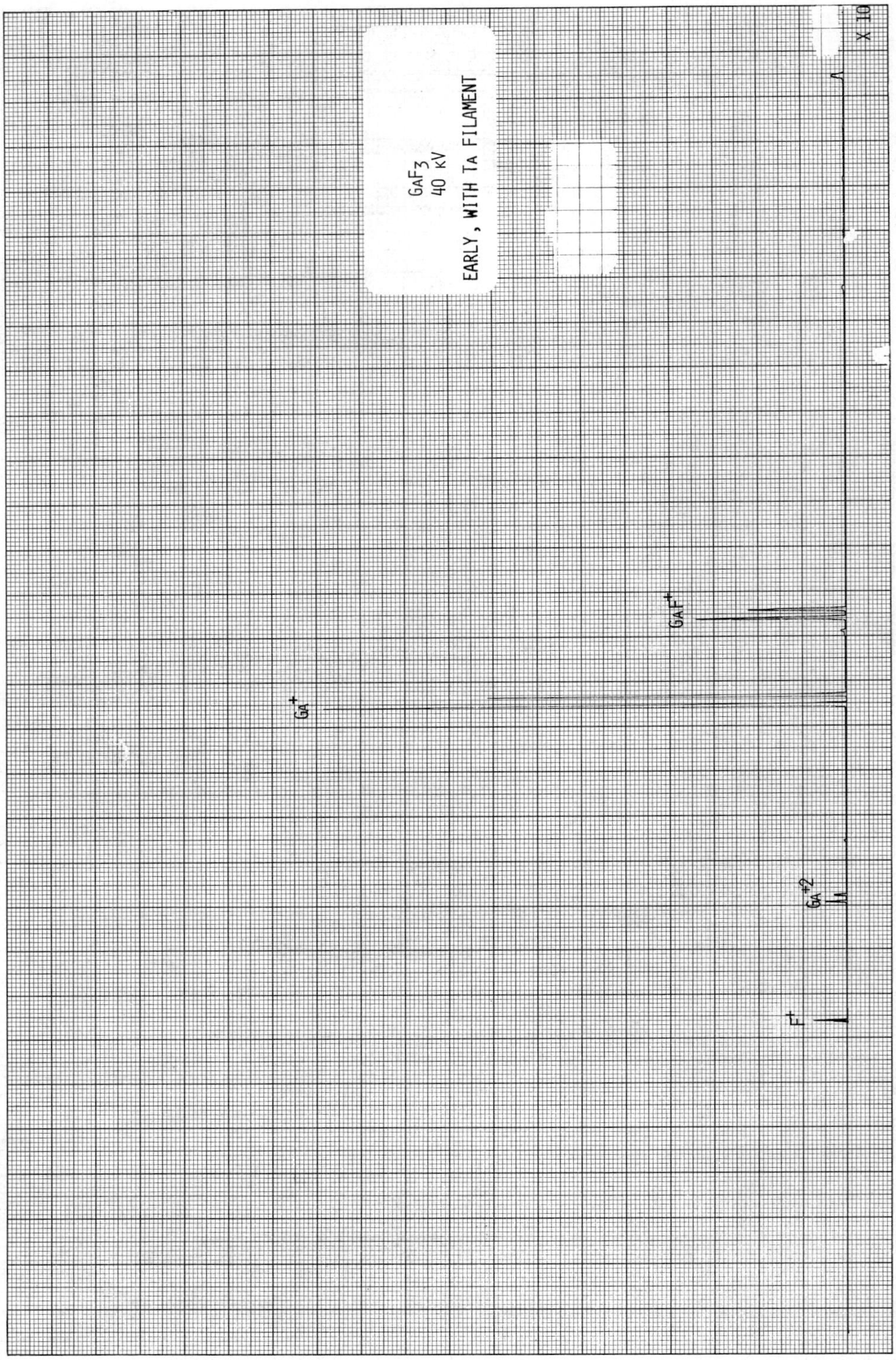
GaF3
40 kV
EARLY, WITH Ta FILAMENT
X 10
GaF+
Ga+
Ga+2
F+

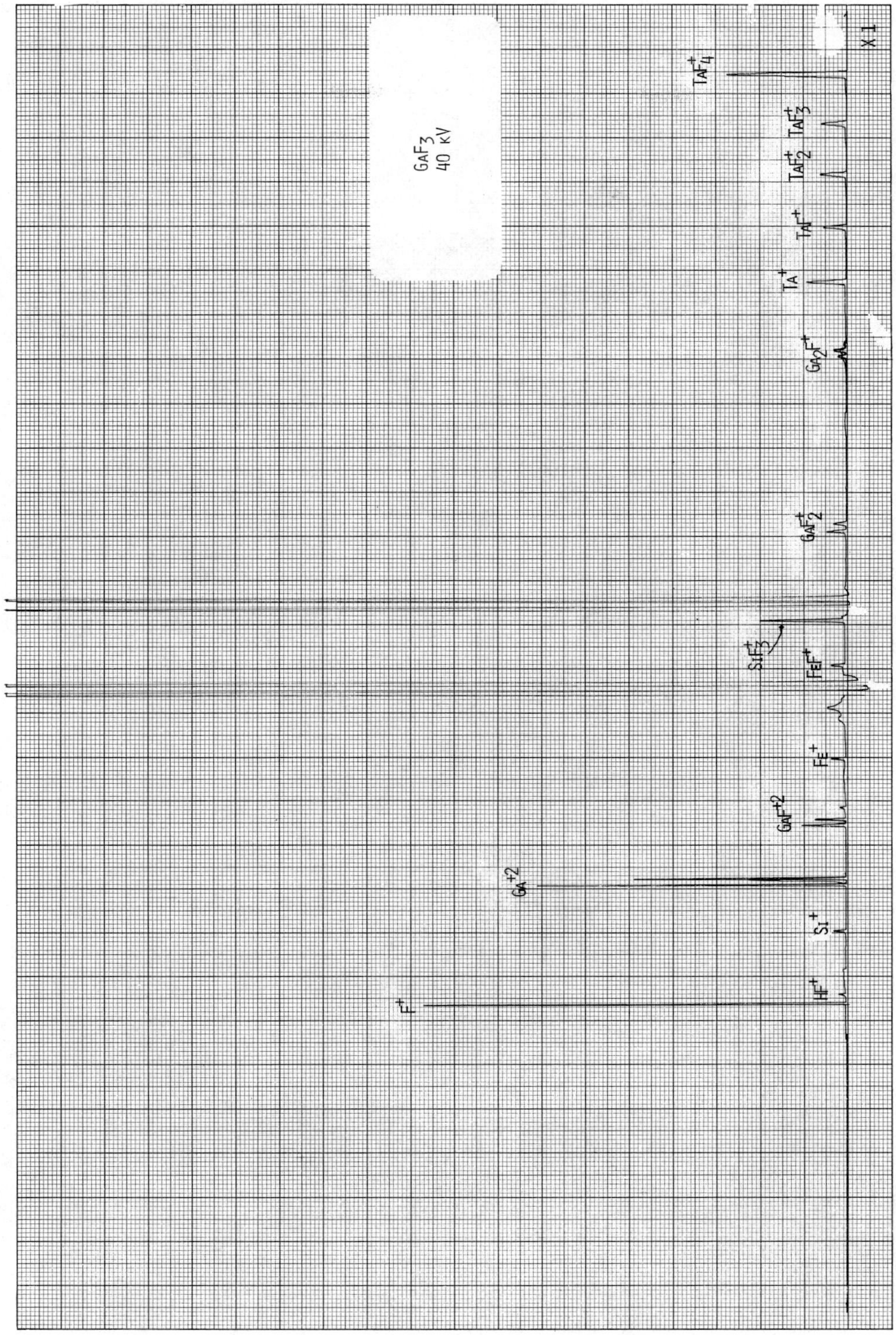
GaF3
40 kV
X 1
TaF4+
TaF3+
TaF2+
TaF+
Ta+
Ga2F+
GaF2+
SiF3+
FeF+
Fe+
GaF+2
Ga+2
Si+
HF+
F+

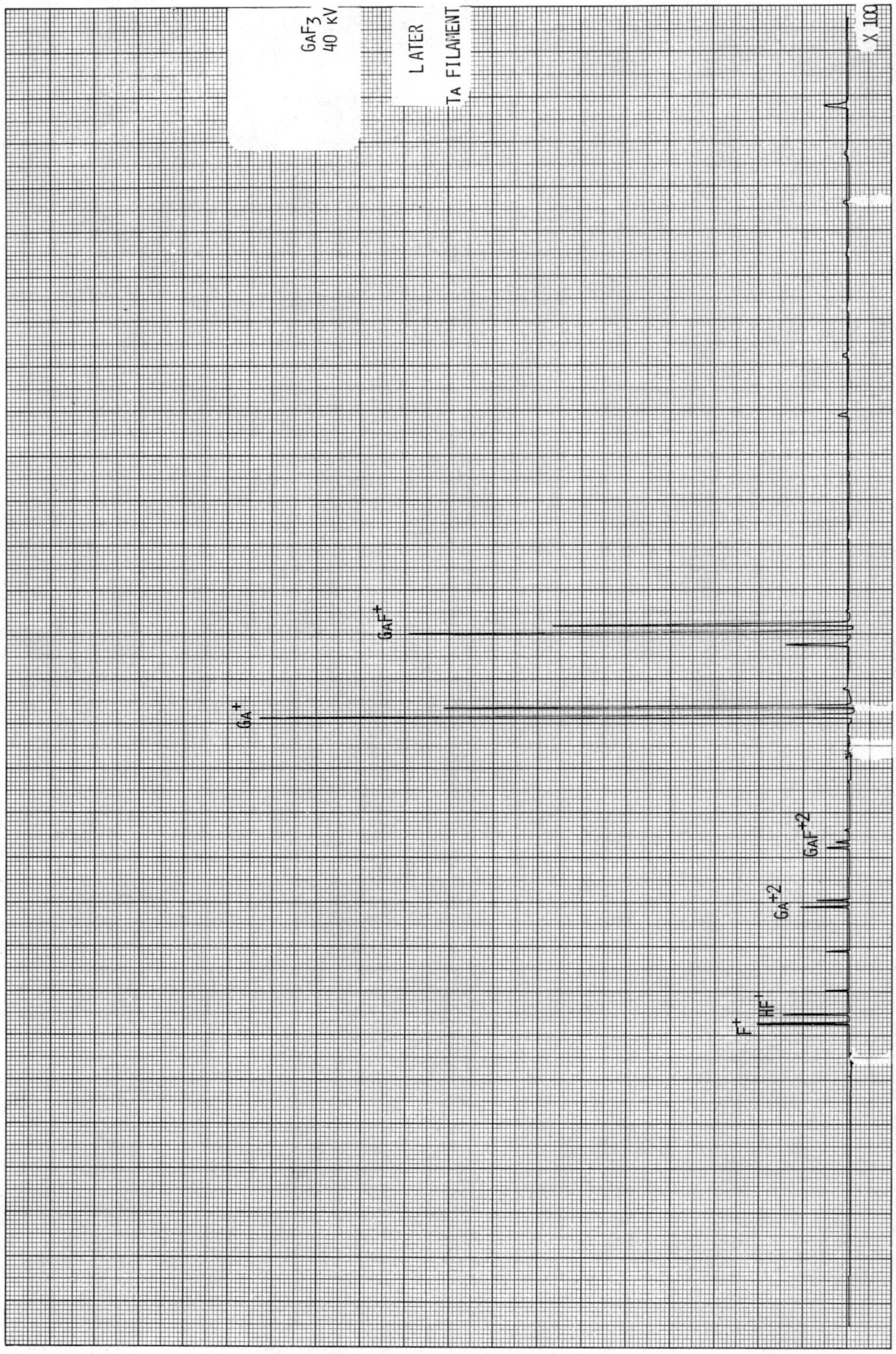
GaF3
40 kV
LATER
Ta FILAMENT
X 100
GaF+
Ga+
GaF+2
Ga+2
HF+
F+

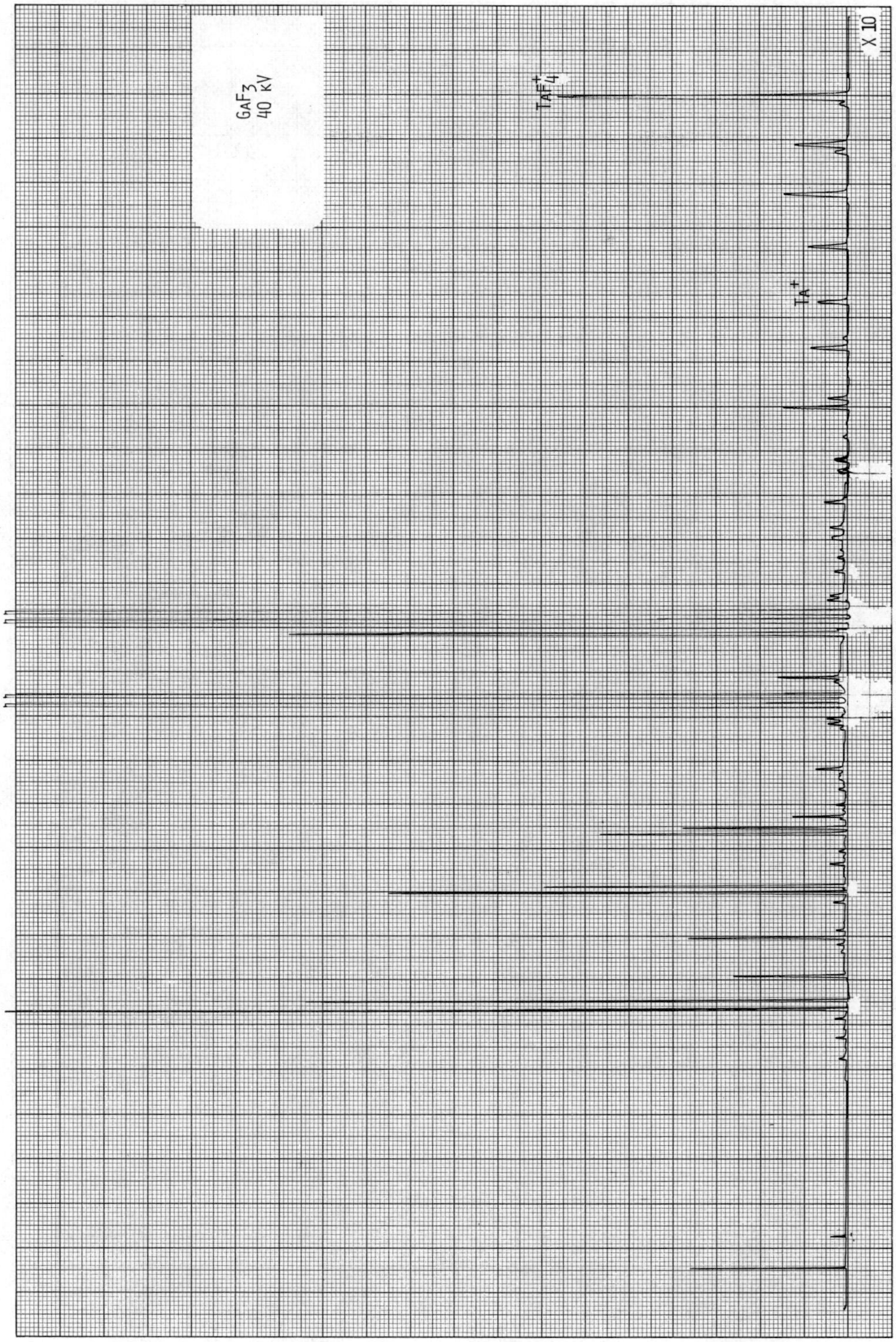
GAF3
40 KV
TAF4+
TA+
X 10

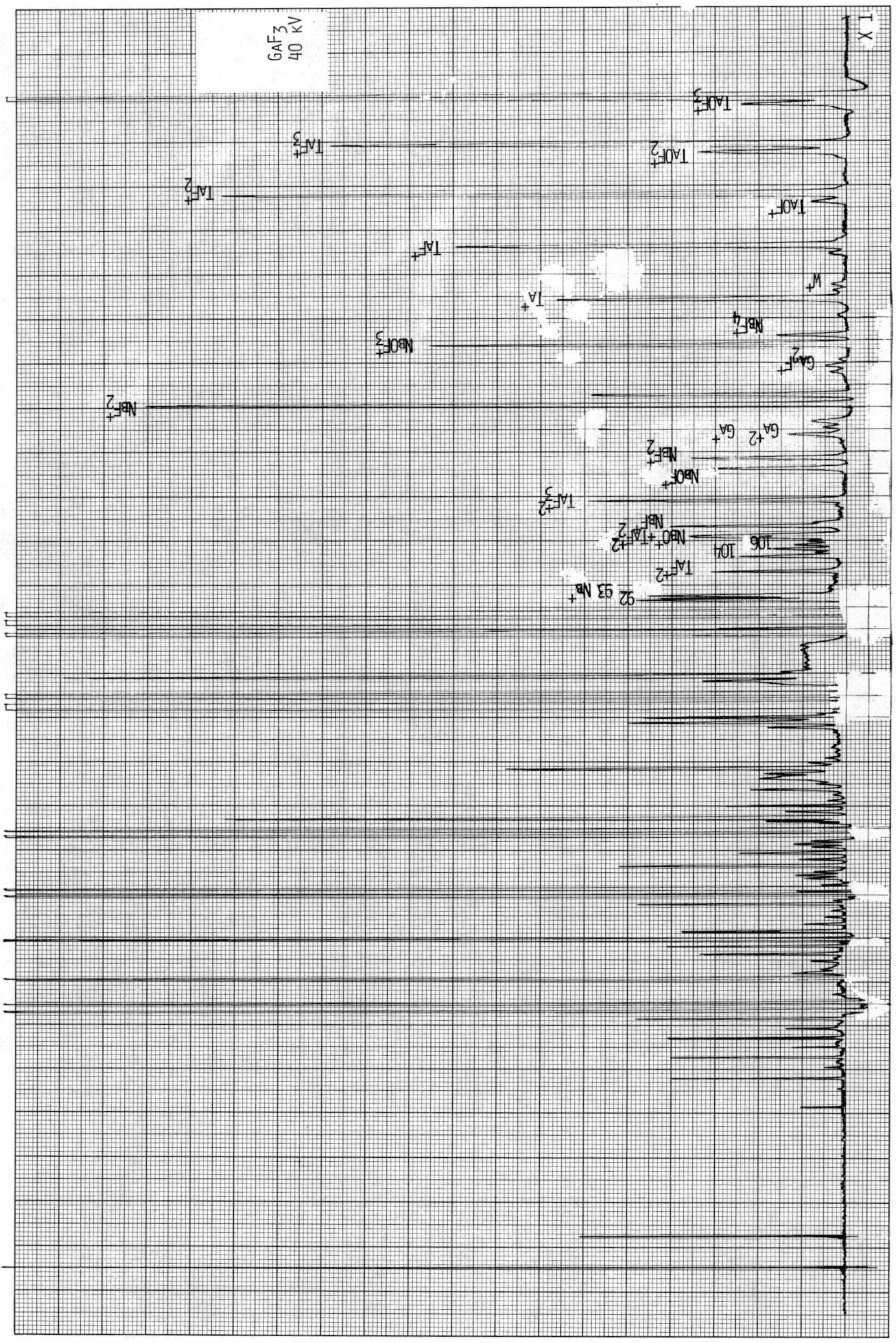
GaF3
40 kV

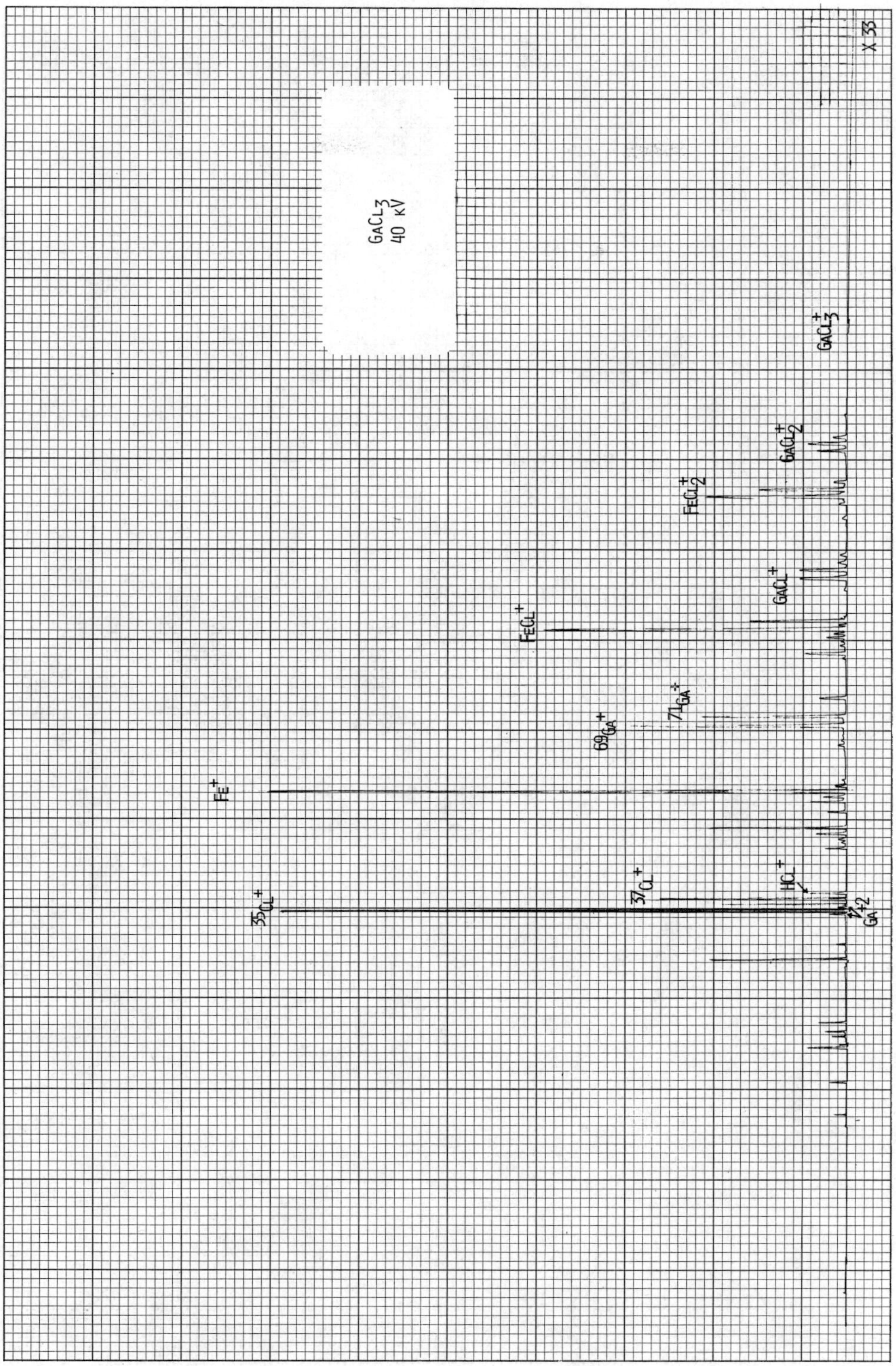

GACL3
40 KV
X 33
GACL3+
GACL2+
FECL2+
GACL+
FECL+
71GA+
69GA+
FE+
HC+
37CL+
35CL+

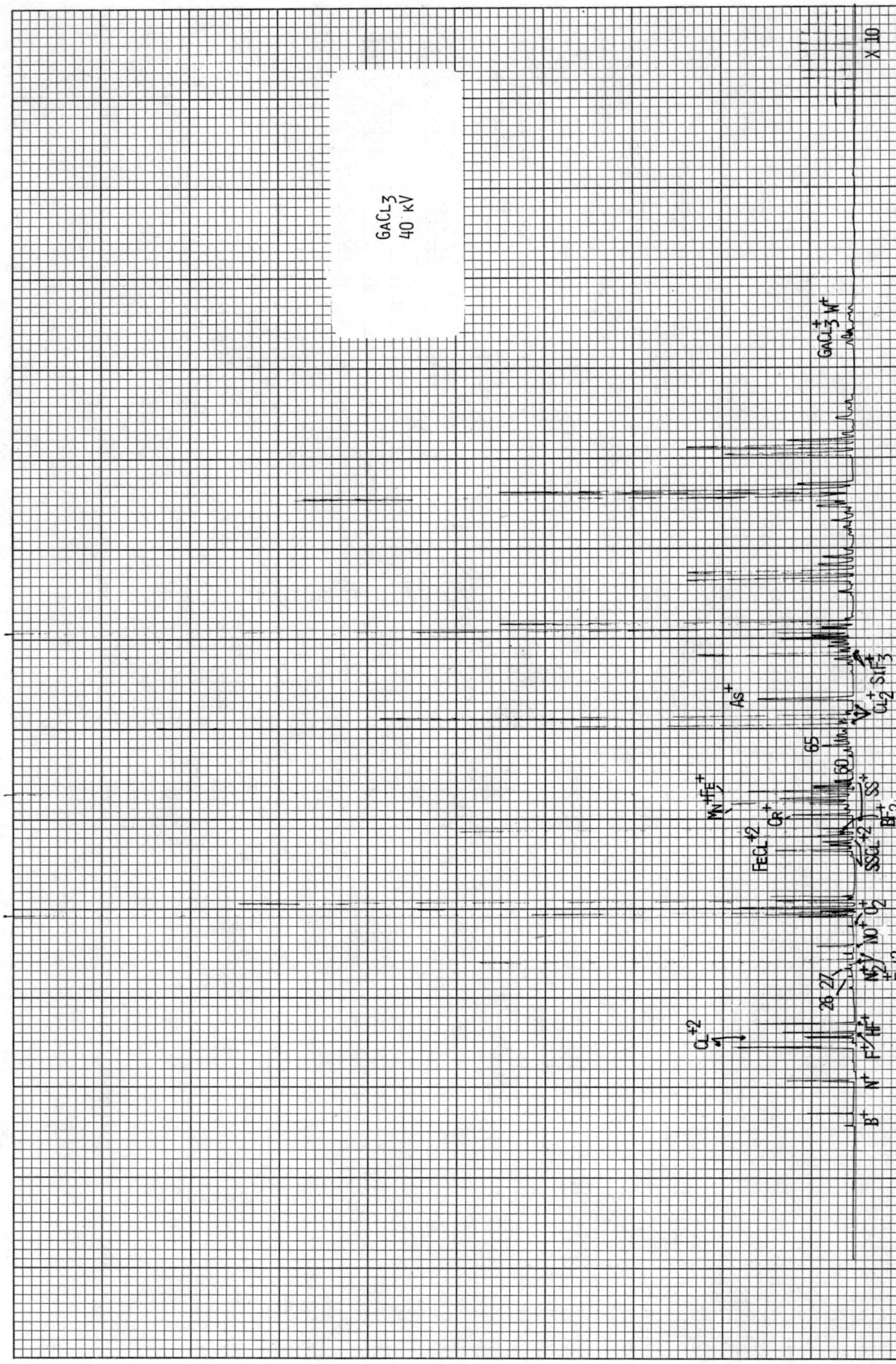
GACL3
40 kV
X 10
As+
65
60
26 27

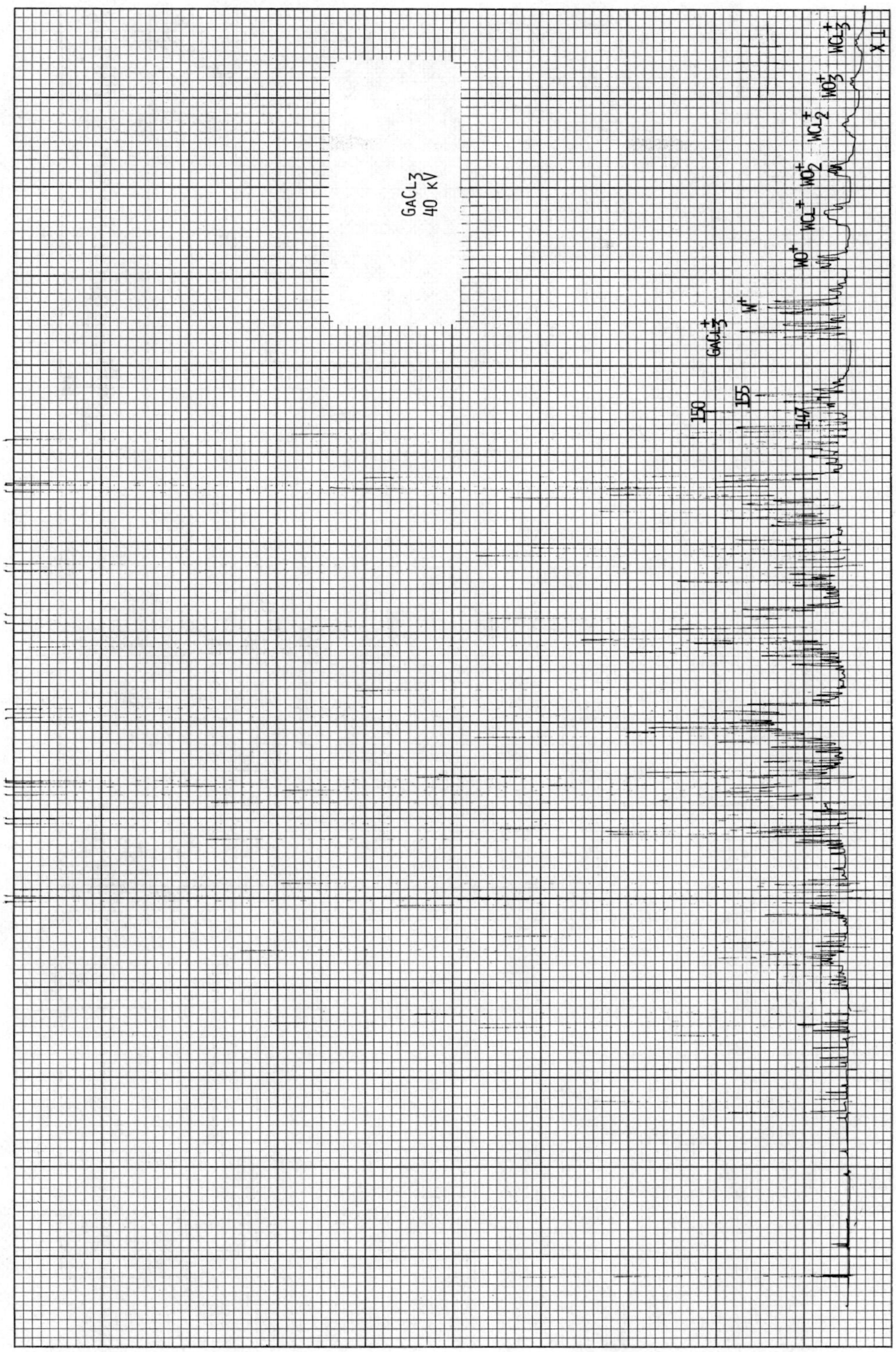
GaCl3
40 kV
150
155
GaCl3+
W+
147
WO+
WCl+
WO2+
WCl2+
WO3+
WCl3+
X 1

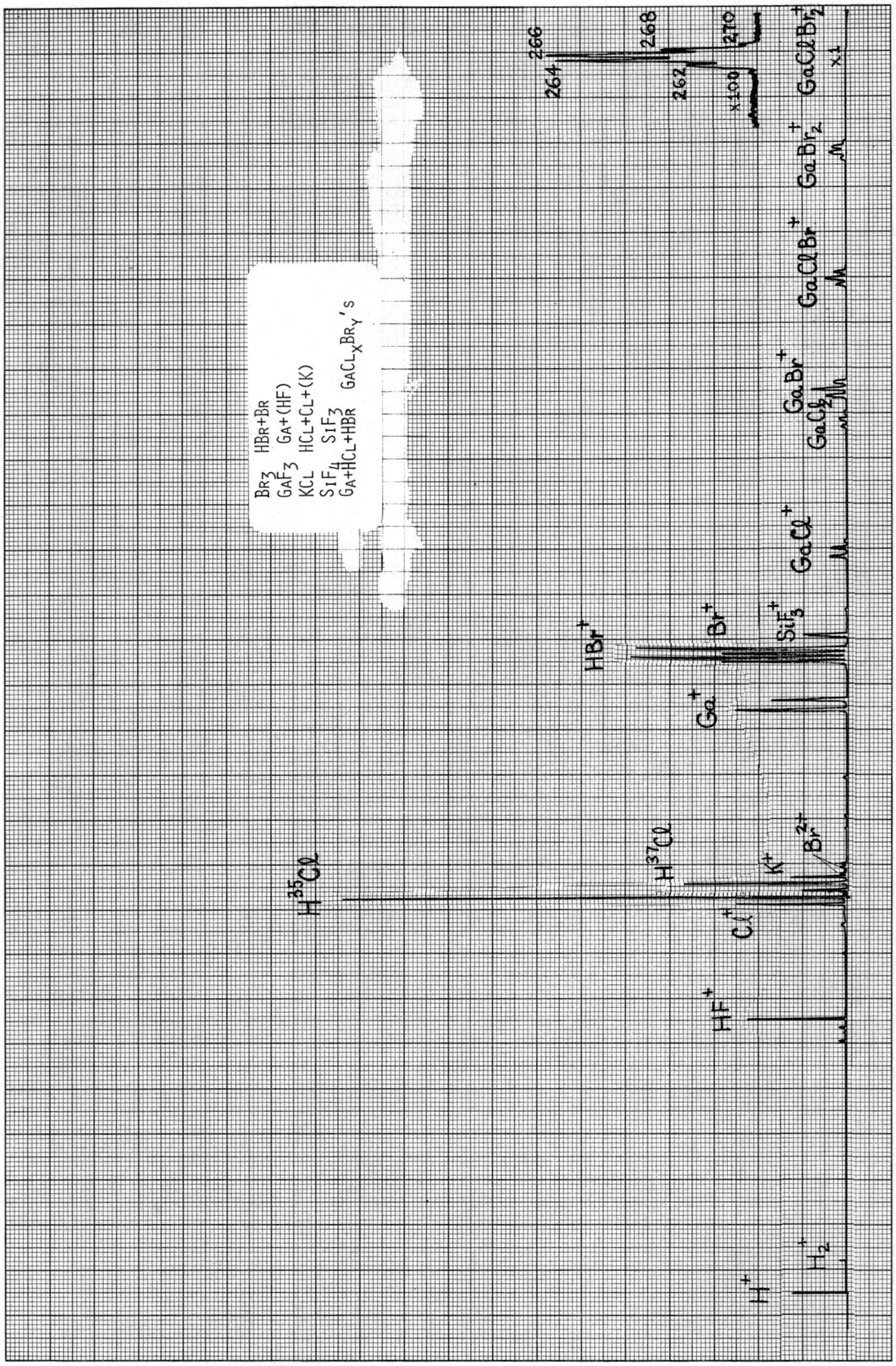

Br2 HBr+Br
GaF3 Ga+(HF)
KCl HCl+Cl+(K)
SiF4 SiF3
Ga+HCl+HBr GaClxBry's
H+
H2+
HF+
Cl+
H35Cl
H37Cl
K+
Br2+
Ga+
HBr+
Br+
SiF3+
GaCl+
GaCl2
GaBr+
GaClBr+
GaBr2+
GaClBr2+
x1
x100
262
264
266
268
270

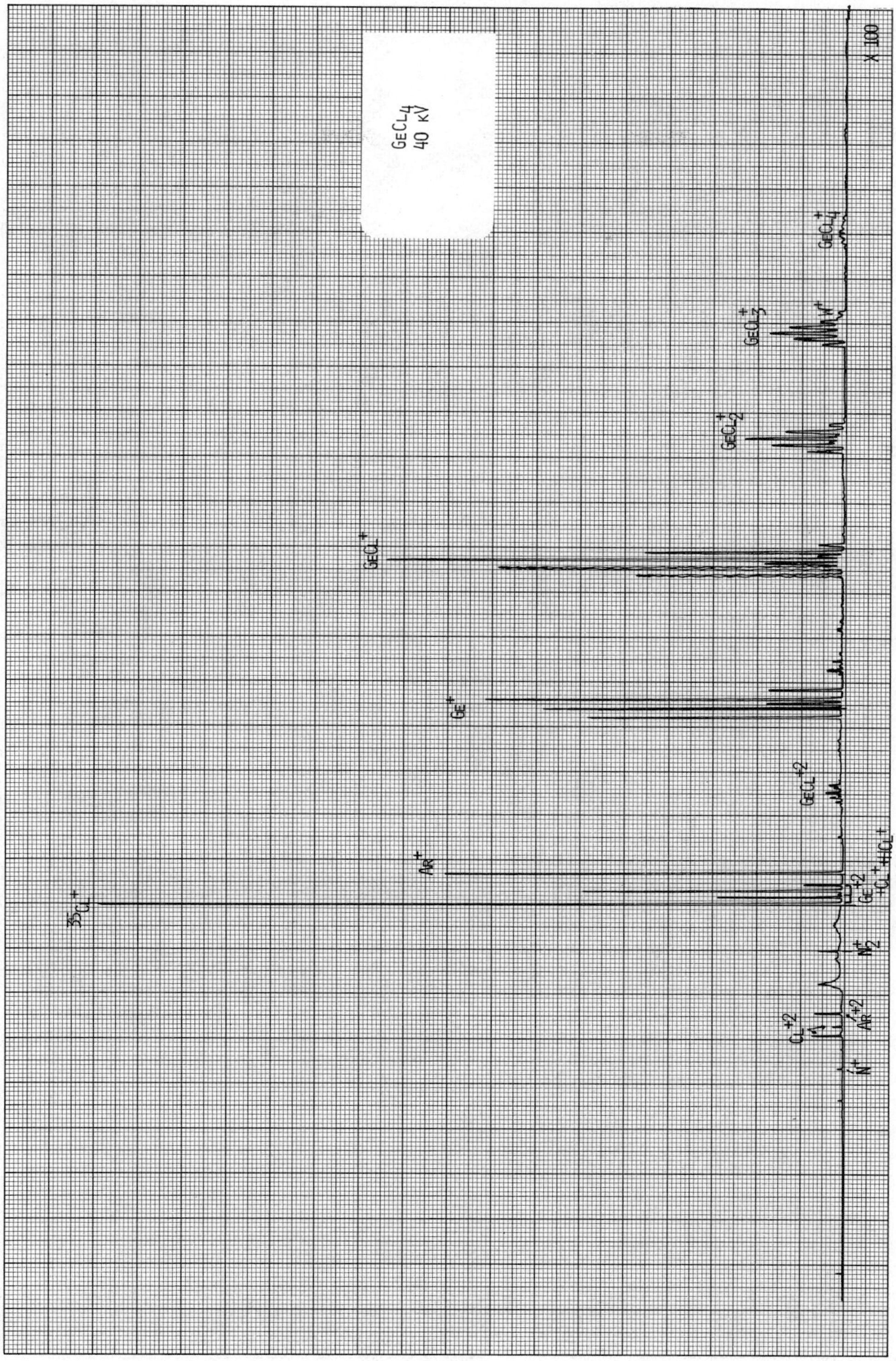

GeCl4
40 kV
X 100
GeCl4+
GeCl3+
GeCl2+
GeCl+
Ge+
GeCl+2
Ar+
35Cl+
Ge+2
Cl+ +HCl+
N2+
Cl+2
Ar+2
N+

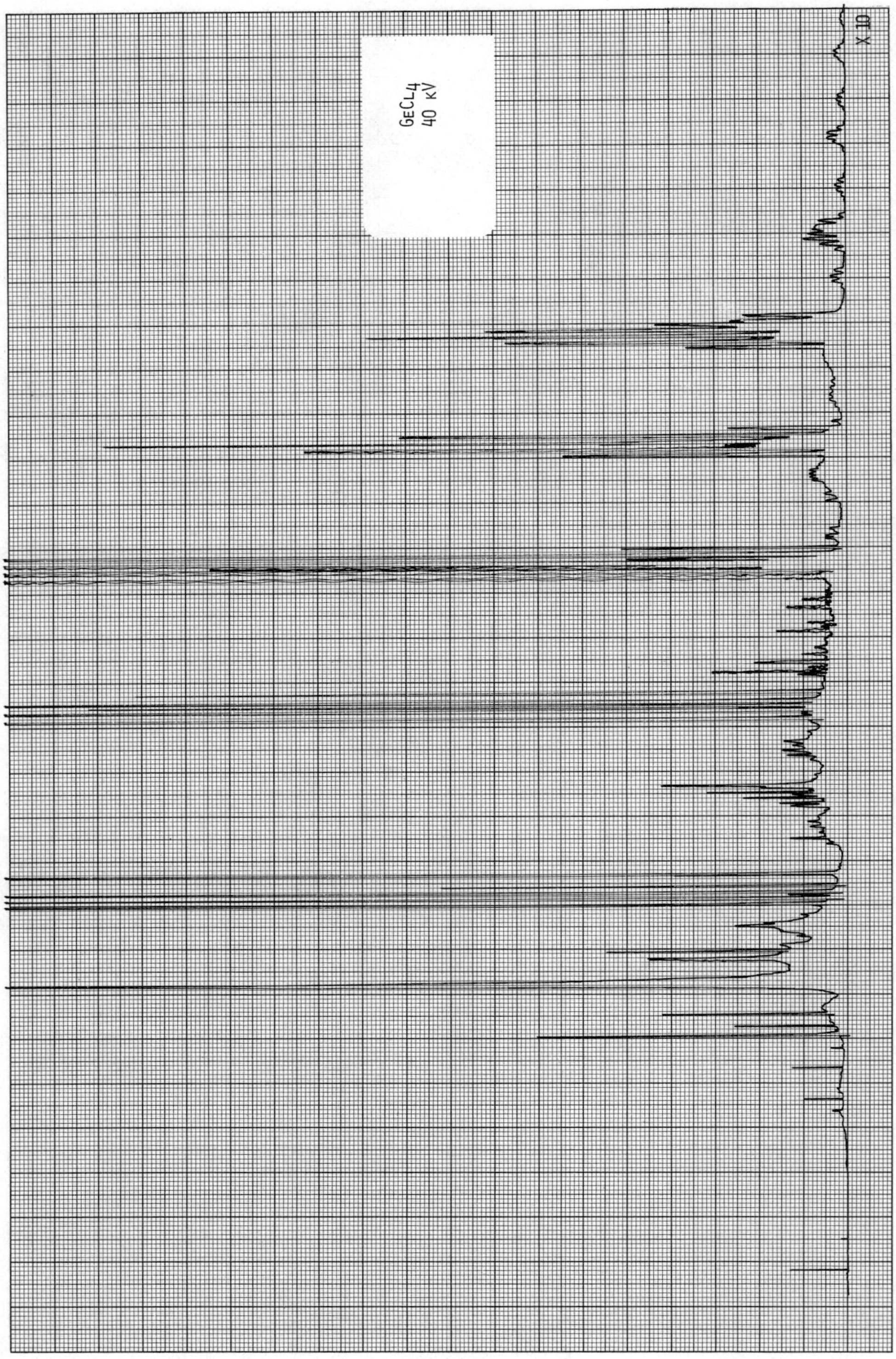
GeCl4
40 kV
x 10

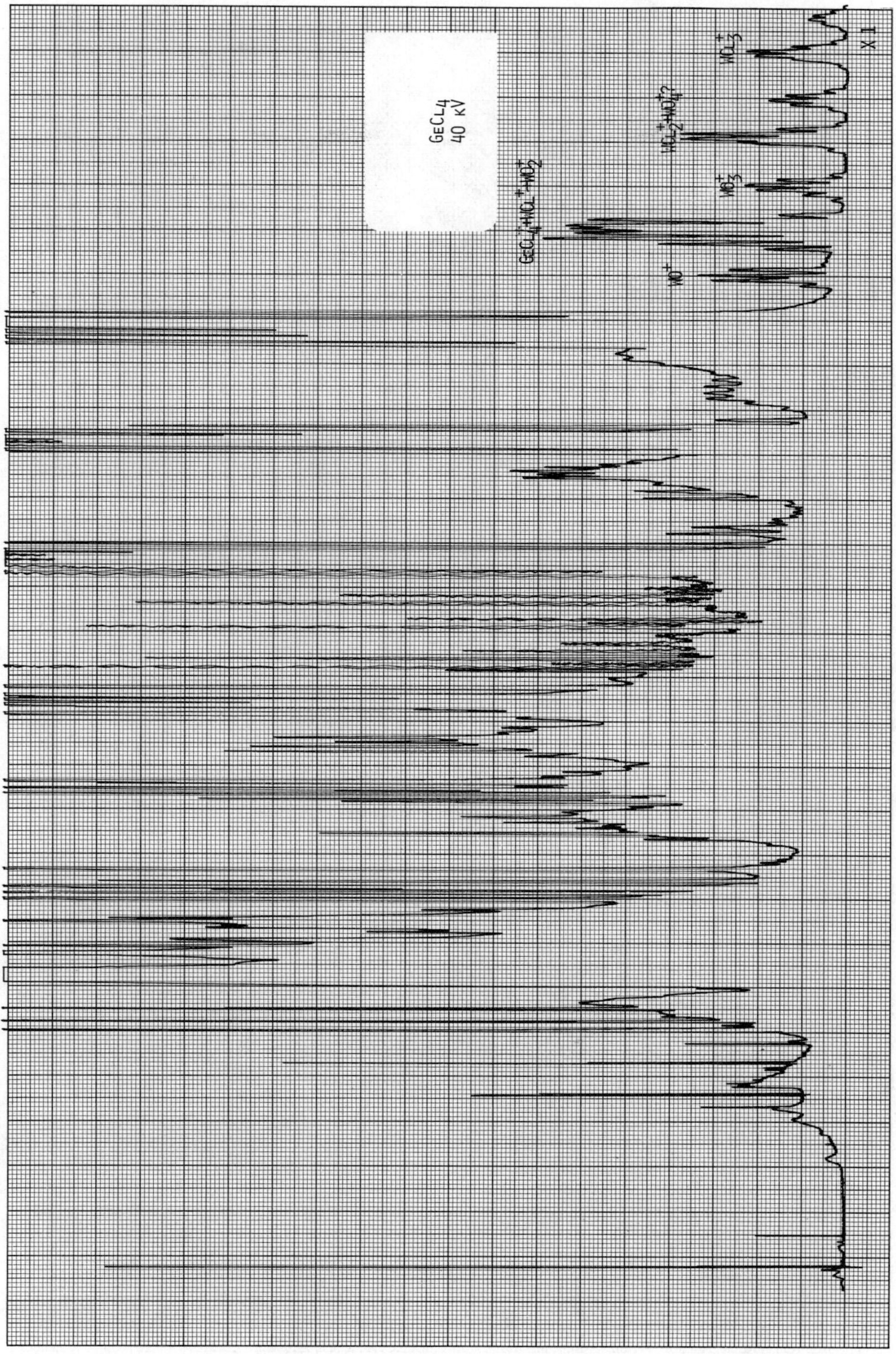
GeCl4
40 kV
X 1

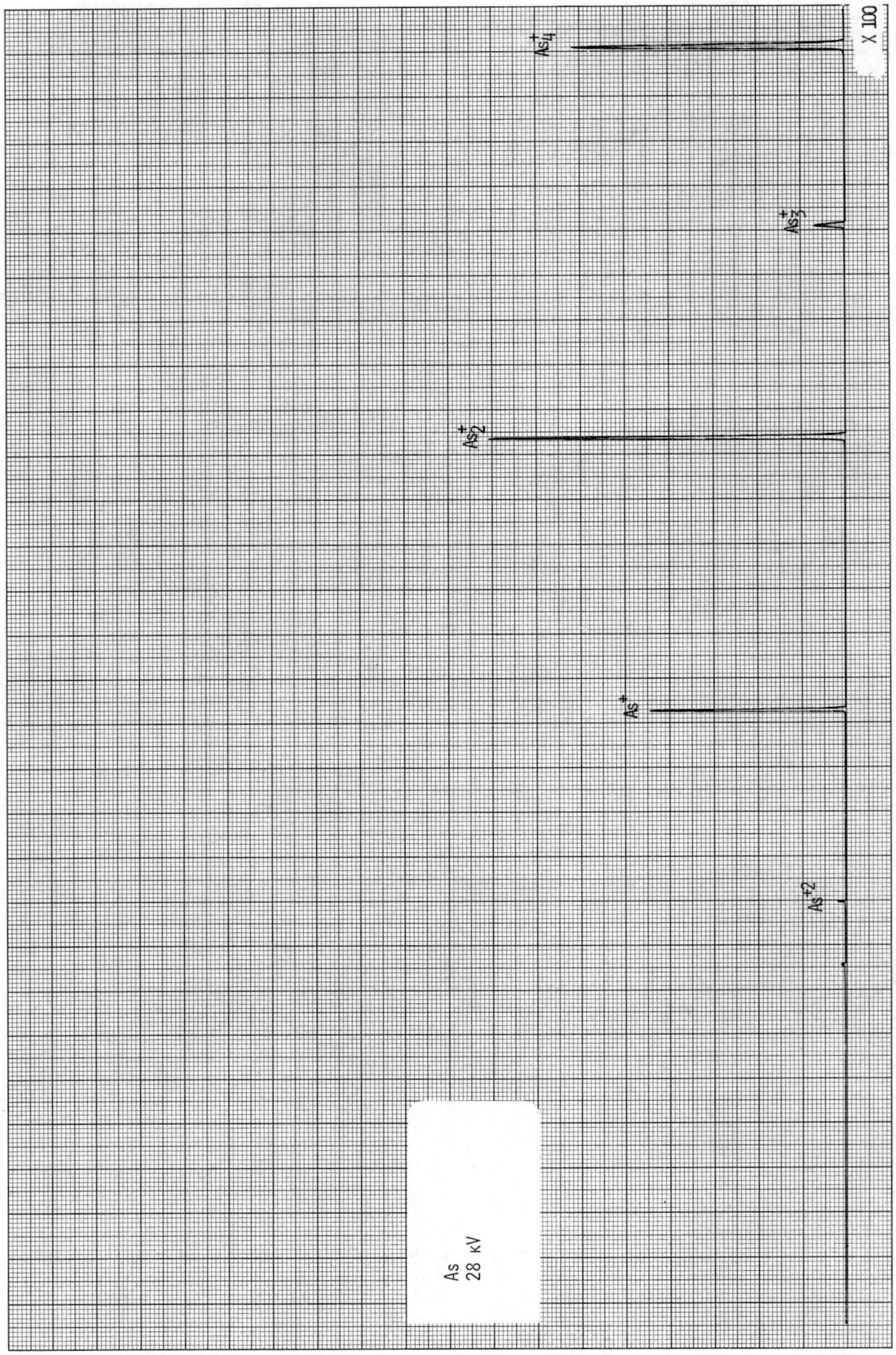
X 100
As4+
As3+
As2+
As+
As+2
As
28 kV

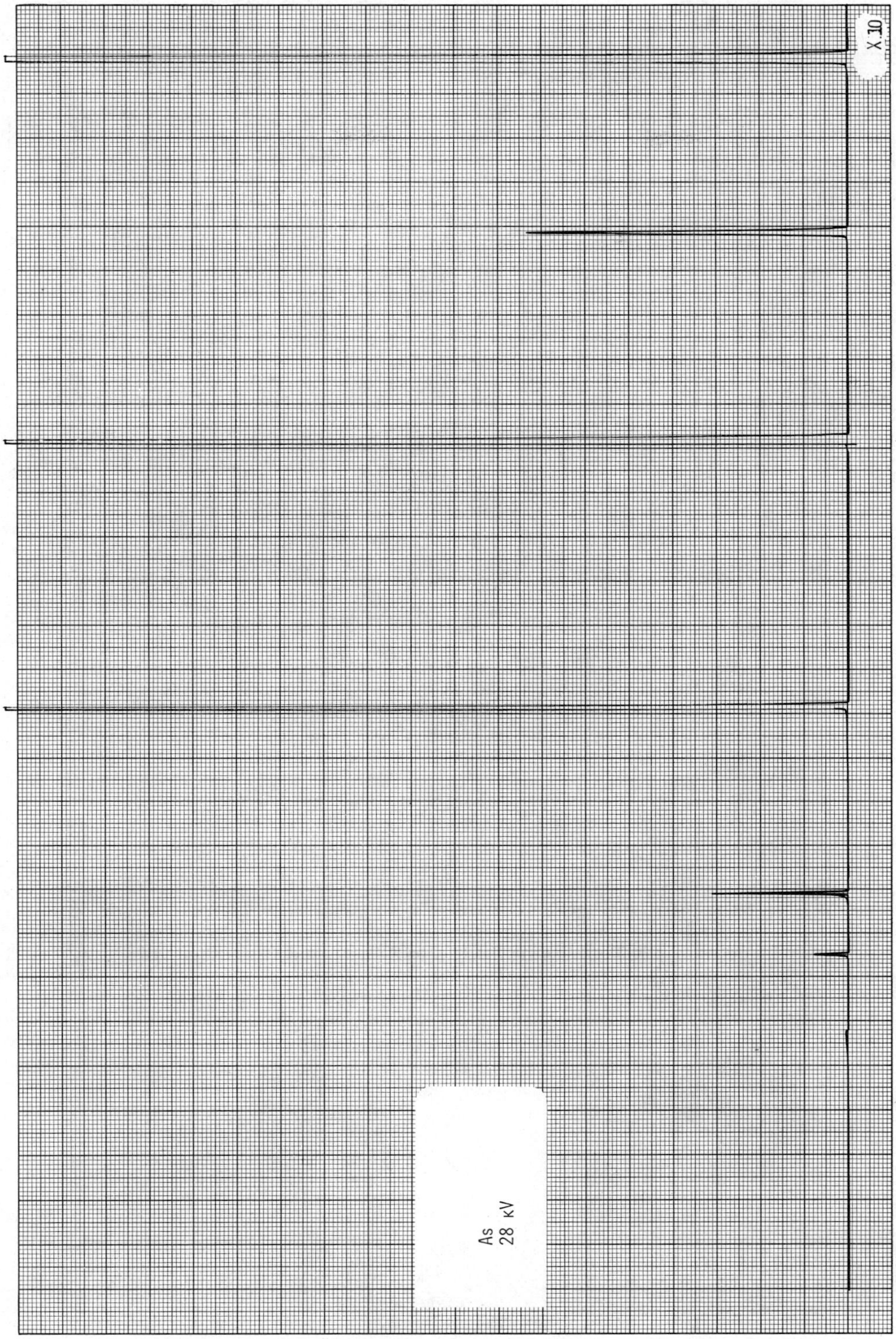
X 10
As
28 kV

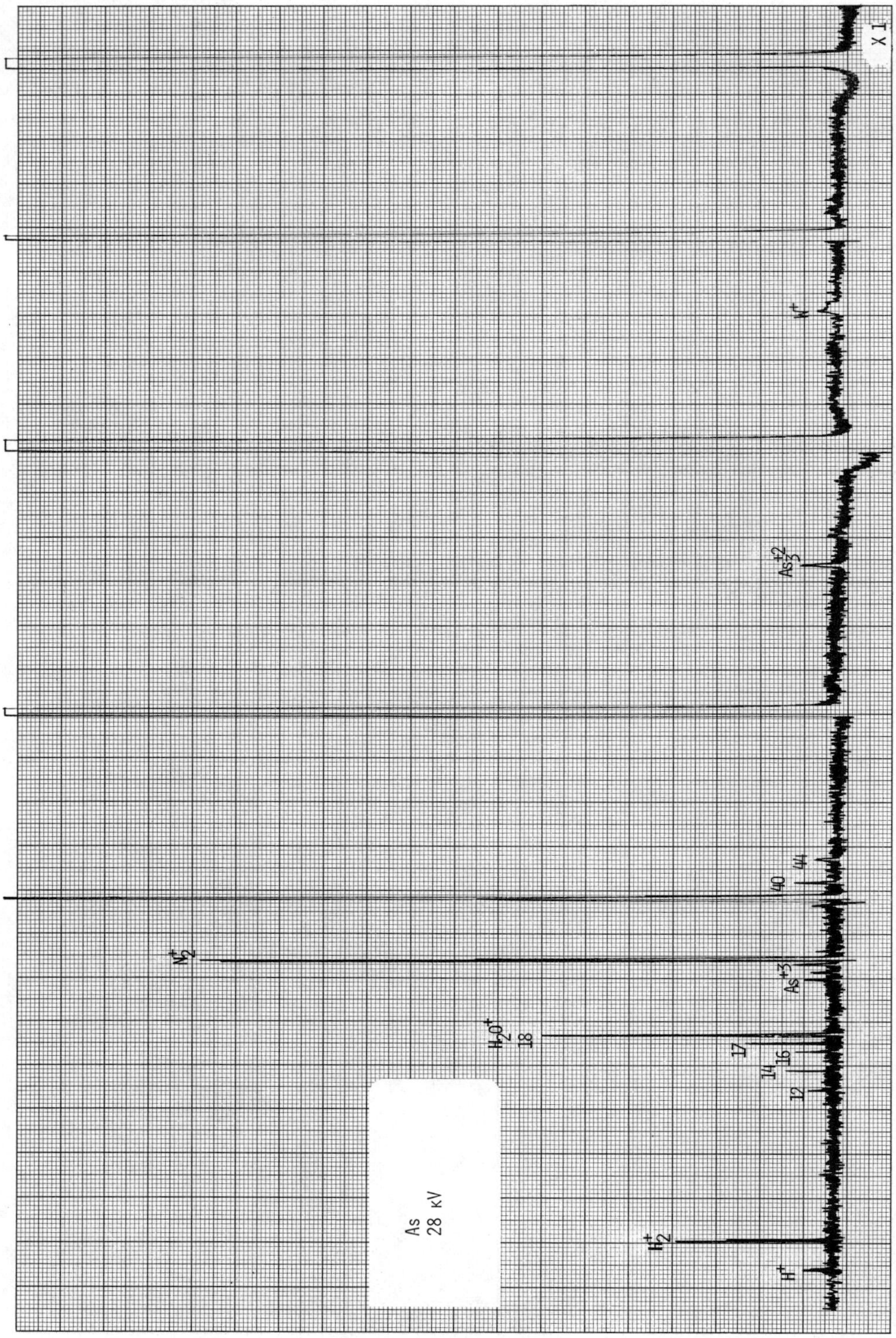
X 1
N+
As3+2
44
40
N2+
As+3
H2O+
18
17
14
16
12
As
28 kV
H2+
H+

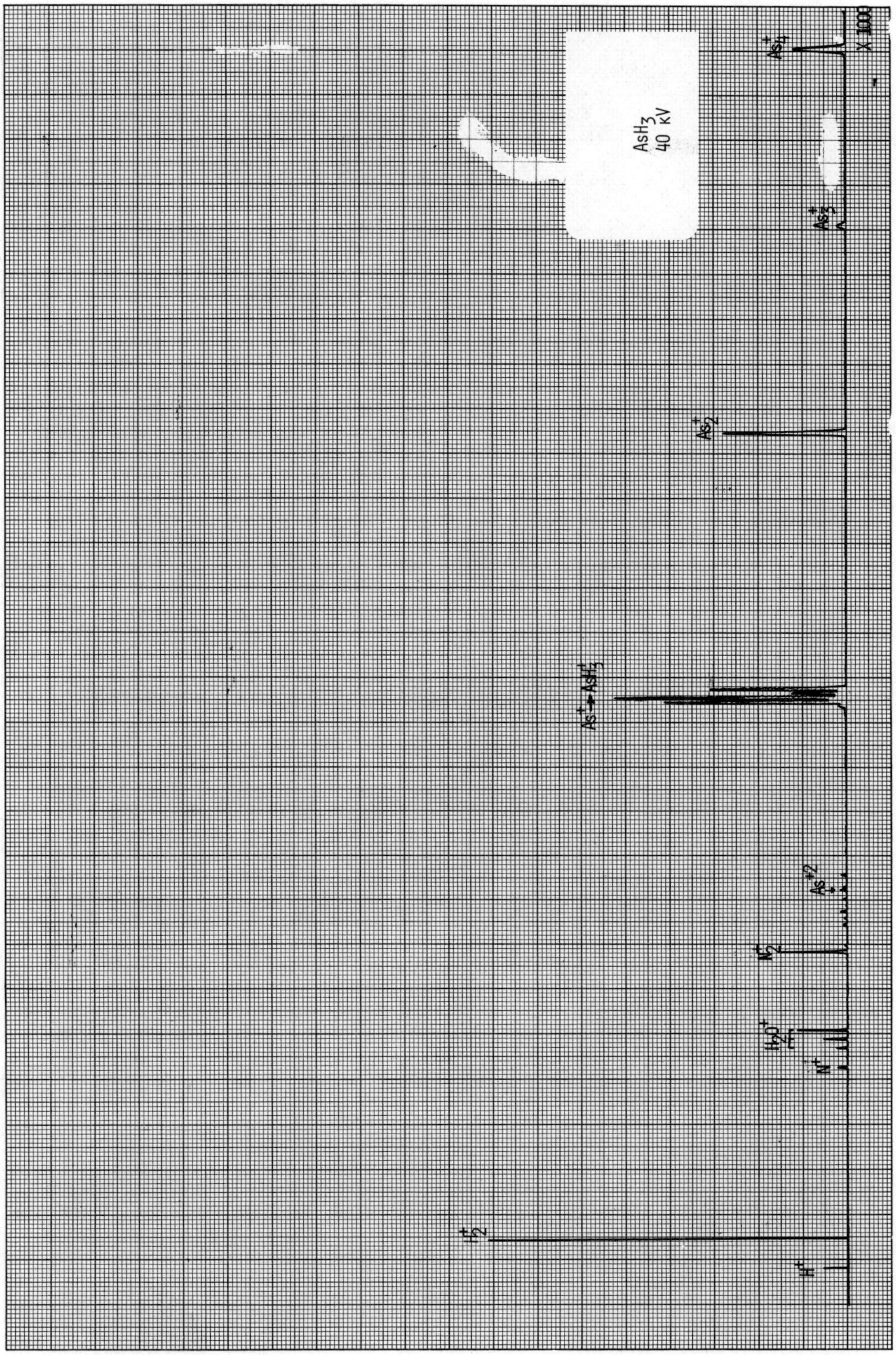

AsH3
40 kV
As4+
X 100
As3+
As2+
As+→AsH3+
As+2
N2+
H2O+
N+
H2+
H+

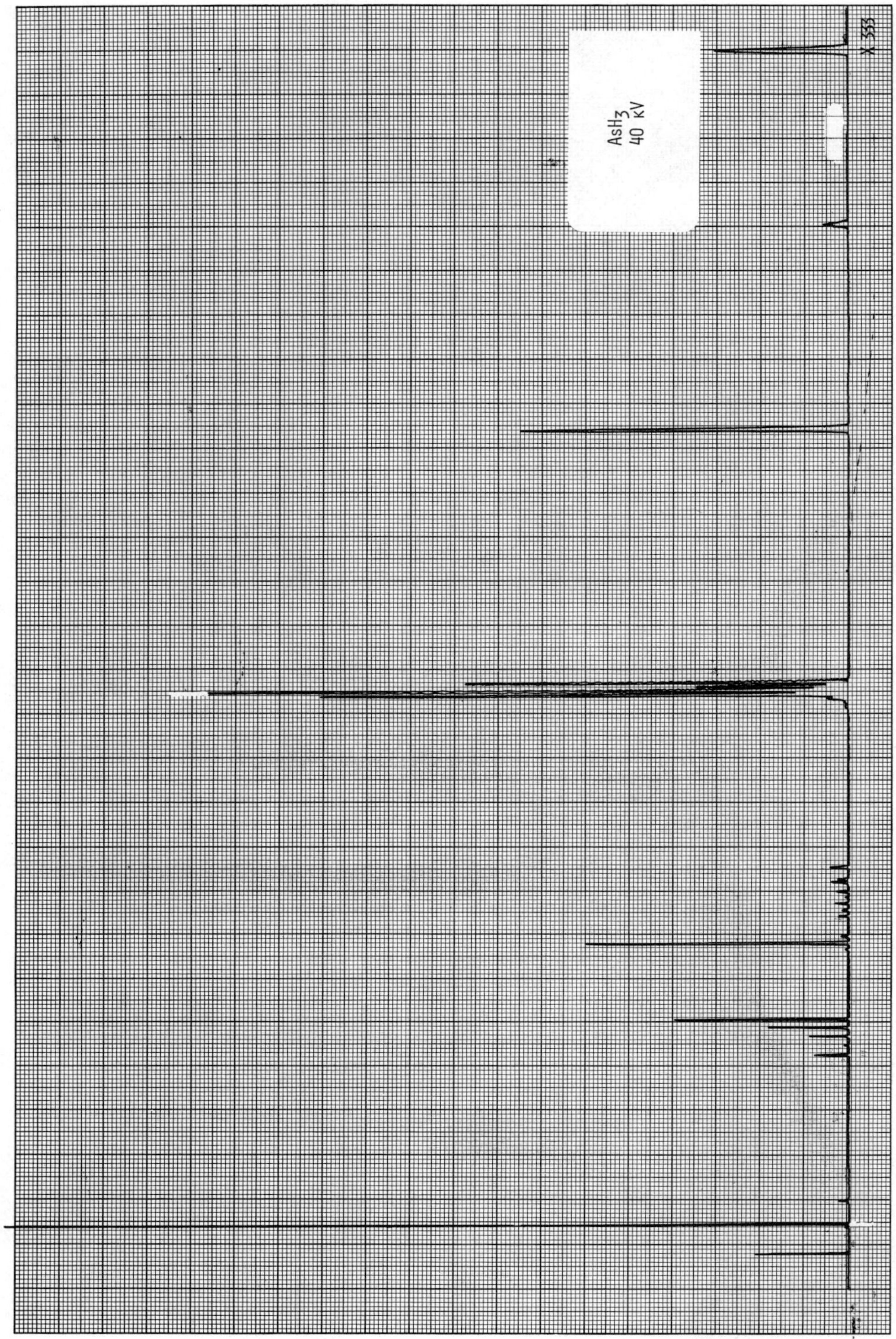
AsH3
40 kV
X 333

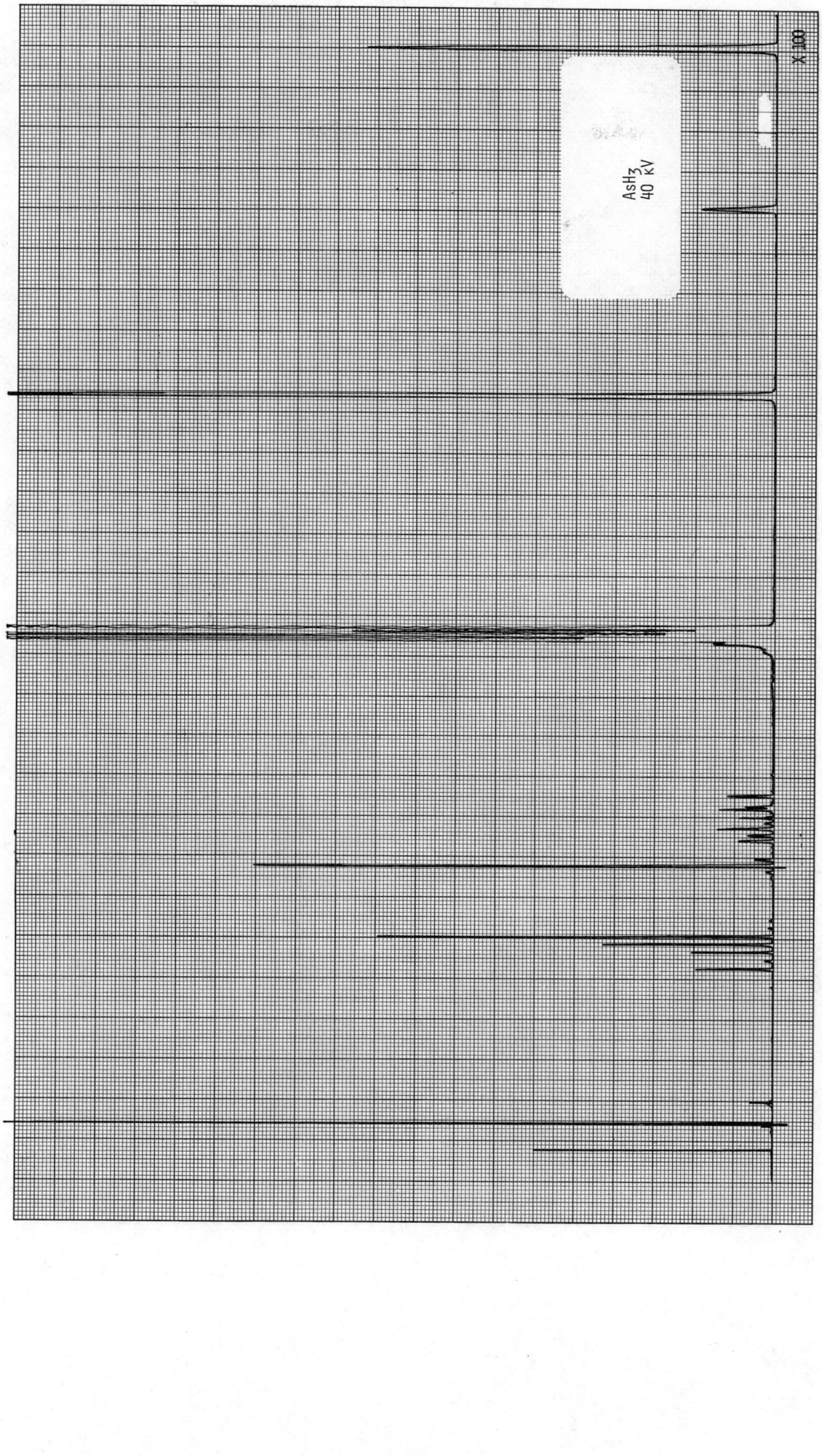

AsH$_3$
40 kV
X 100

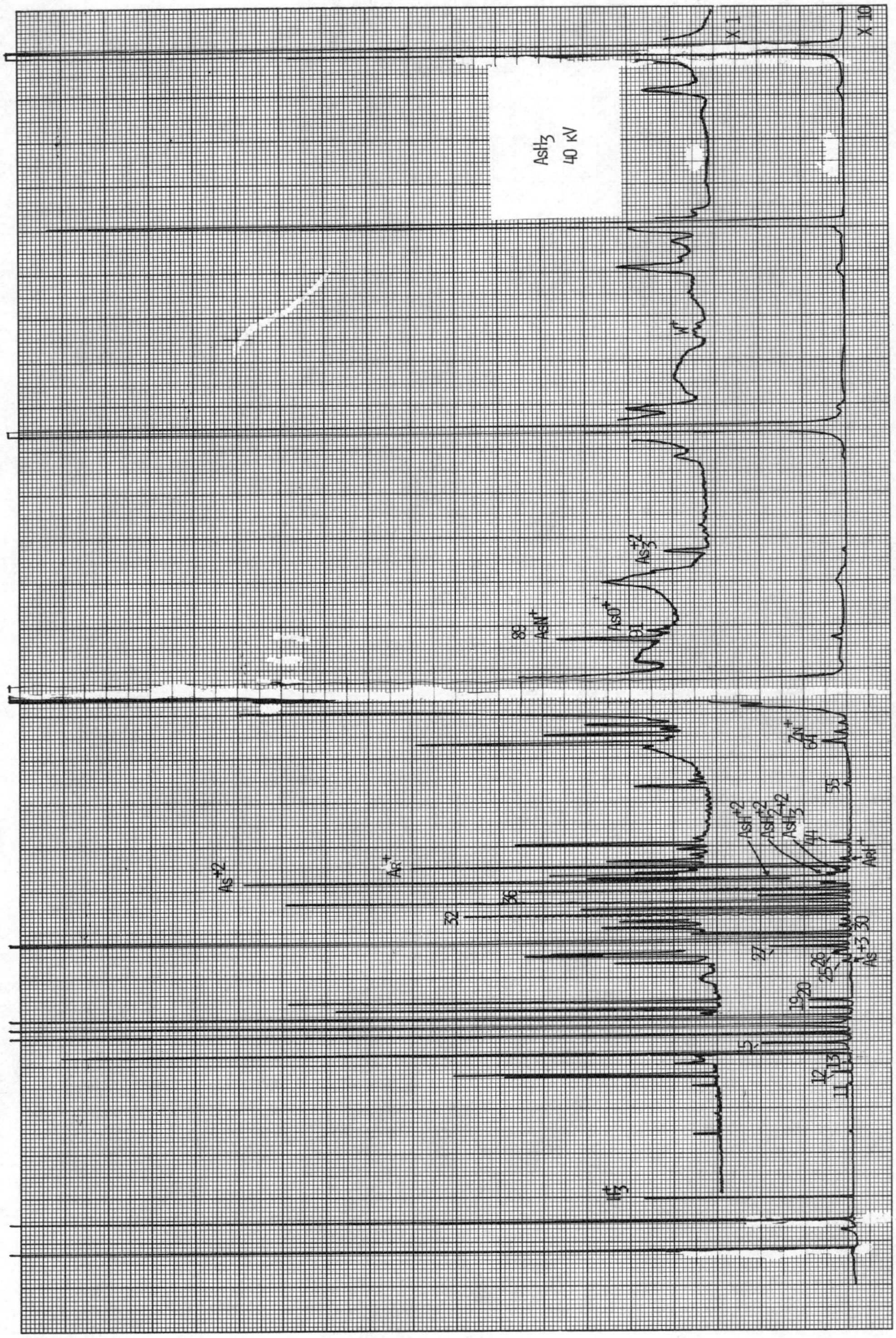
AsH$_3$
40 kV
X 1
X 10

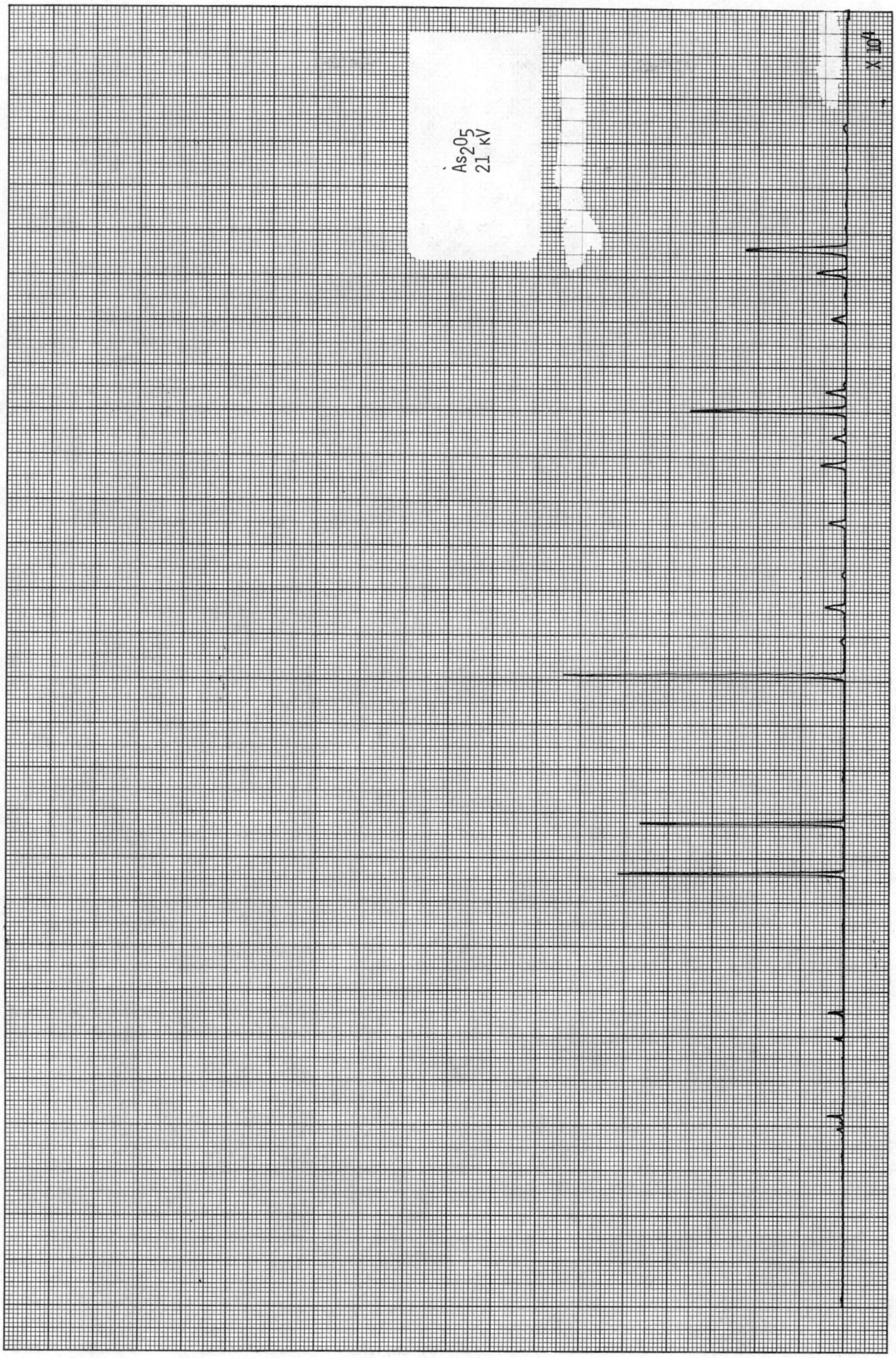

As_2O_5
21 kV
X 10^4

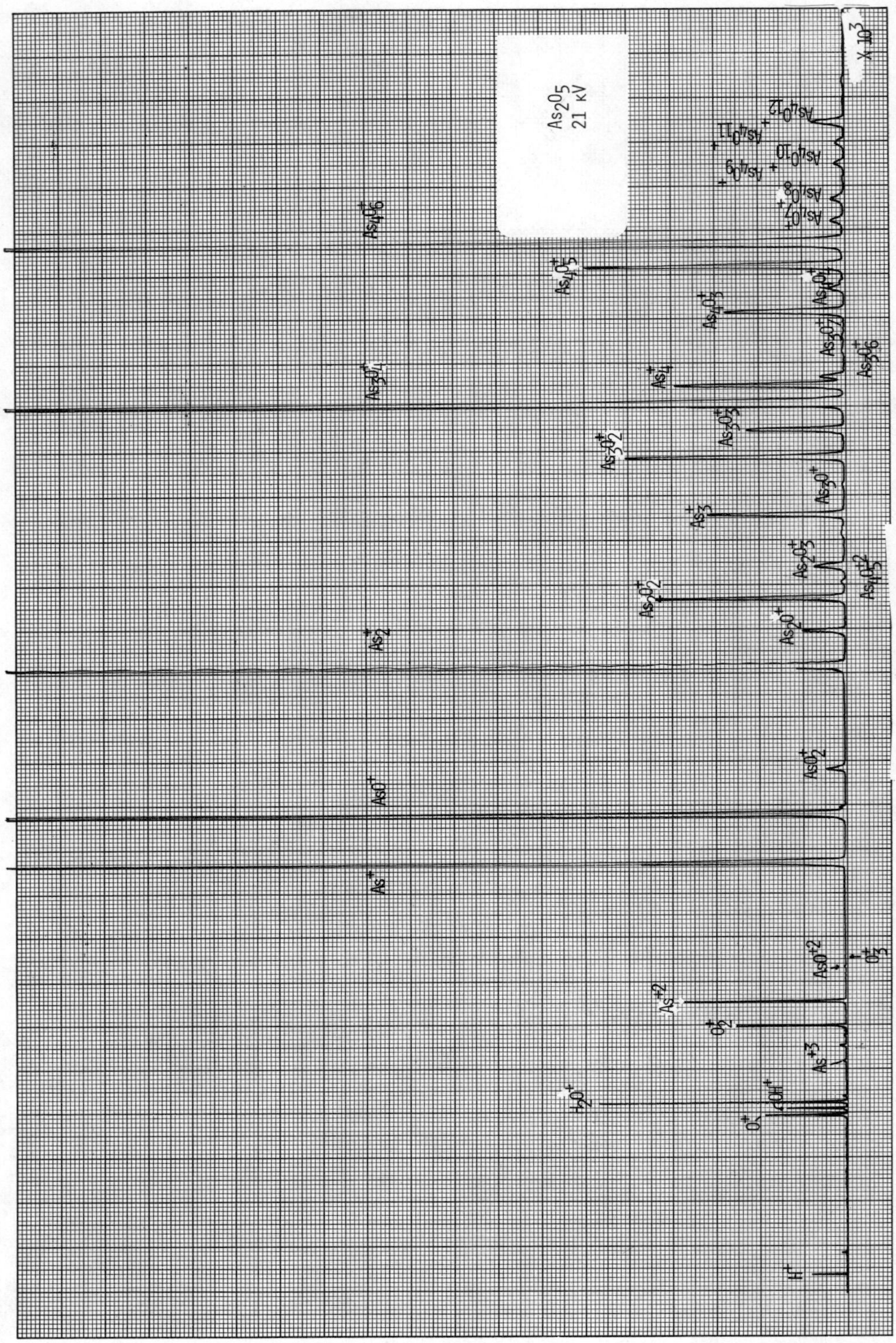
As2O5
21 kV
X 10^3
H+
O+
OH+
H2O+
As+3
O2+
As+2
AsO+2
O3+
As+
AsO+
AsO2+
As2+
As2O+
As2O2+
As2O3+
As3+
As3O+
As3O2+
As3O3+
As3O4+
As4+
As4O3+
As4O5+
As4O6+

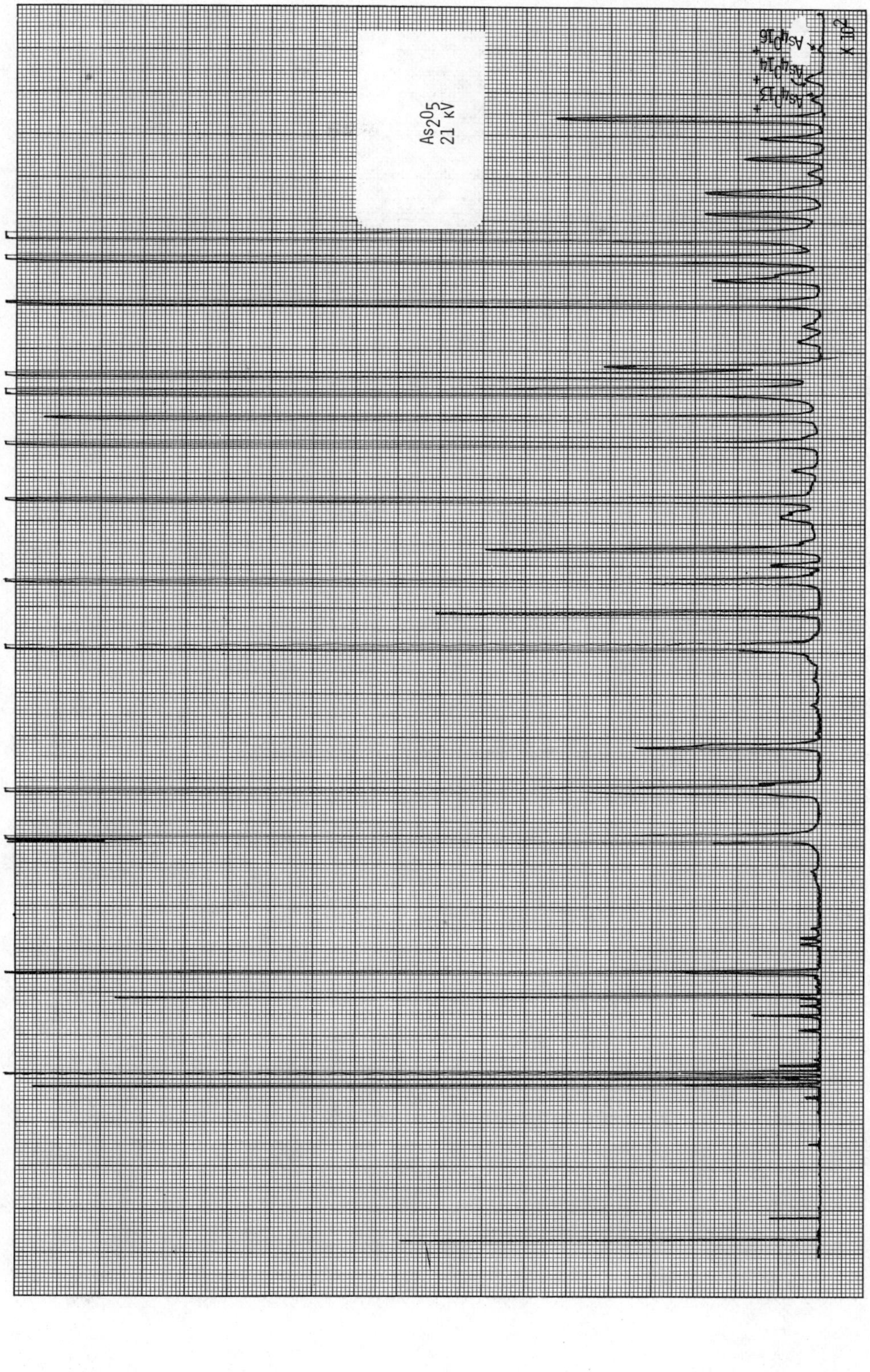
As2O5
21 kV
As4O13+
As4O14+
As4O16+
x 10^2

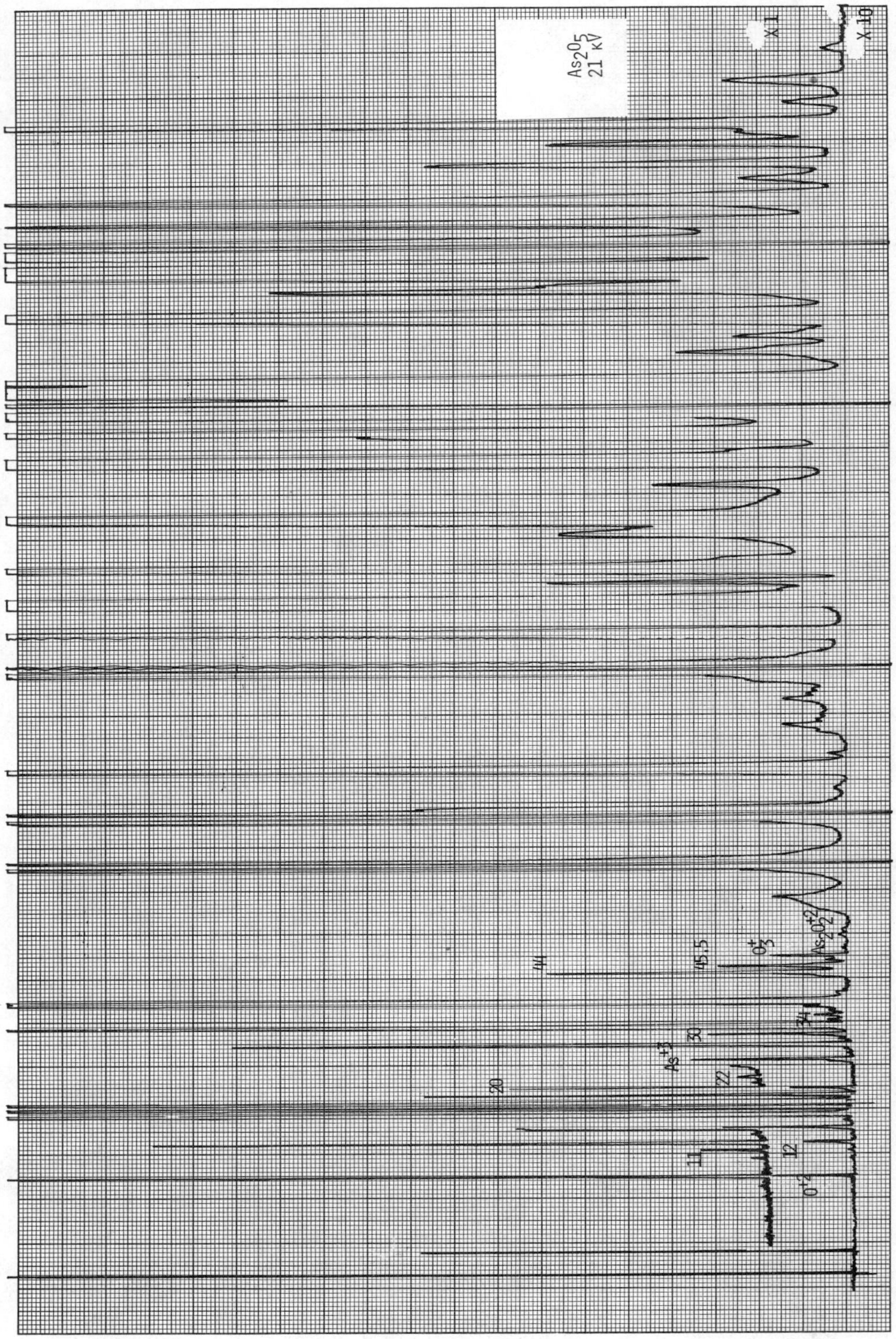

As2O5
21 kV
X 1
X 10
44
45.5
20
22
30
11
12

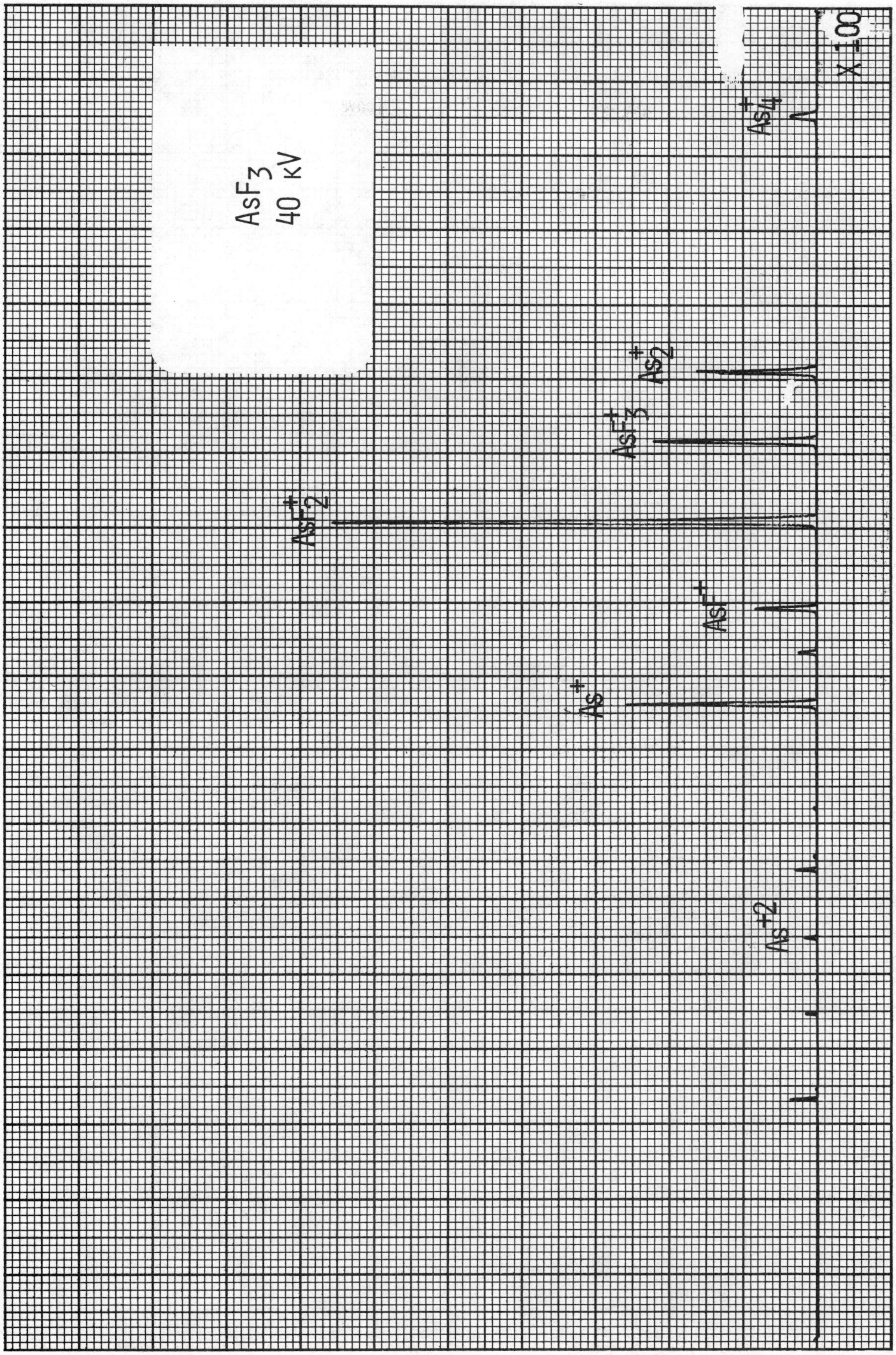

AsF_3
40 kV
X 100
As_4^+
As_2^+
AsF_3^+
AsF_2^+
AsF^+
As^+
As^{+2}

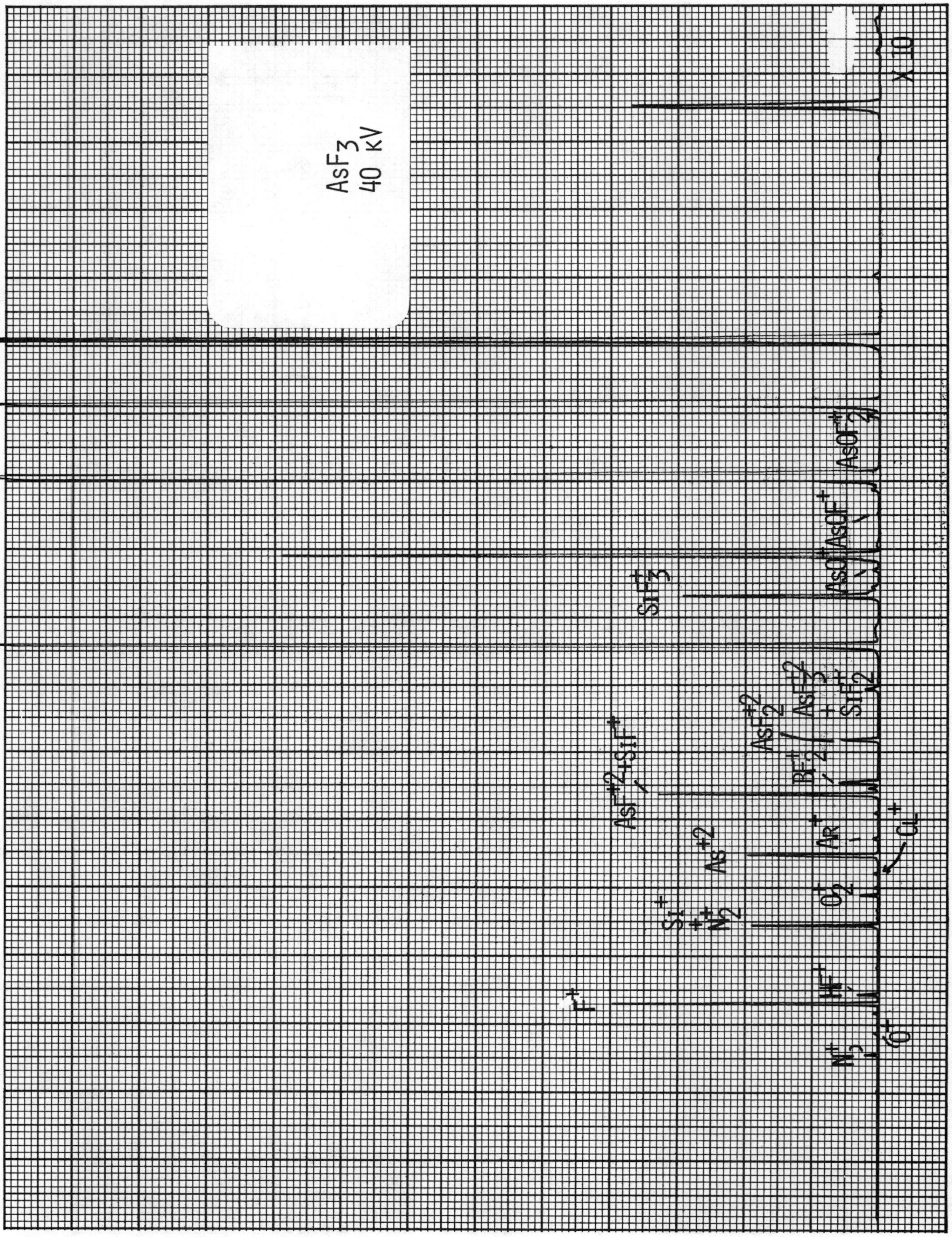
AsF$_3$
40 KV
X 10
AsOF$_2^+$
AsO$^+$
AsOF$^+$
SiF$_3^+$
SiF$_2^+$
AsF$_3^{+2}$
AsF$_2^{+2}$
BF$_2^+$
AsF^{+2}+SiF$^+$
Ar$^+$
Cl$^+$
As^{+2}
O$_2^+$
Si$^+$
N$_2^+$
F$^+$
HF$^+$
N$^+$
O$^+$

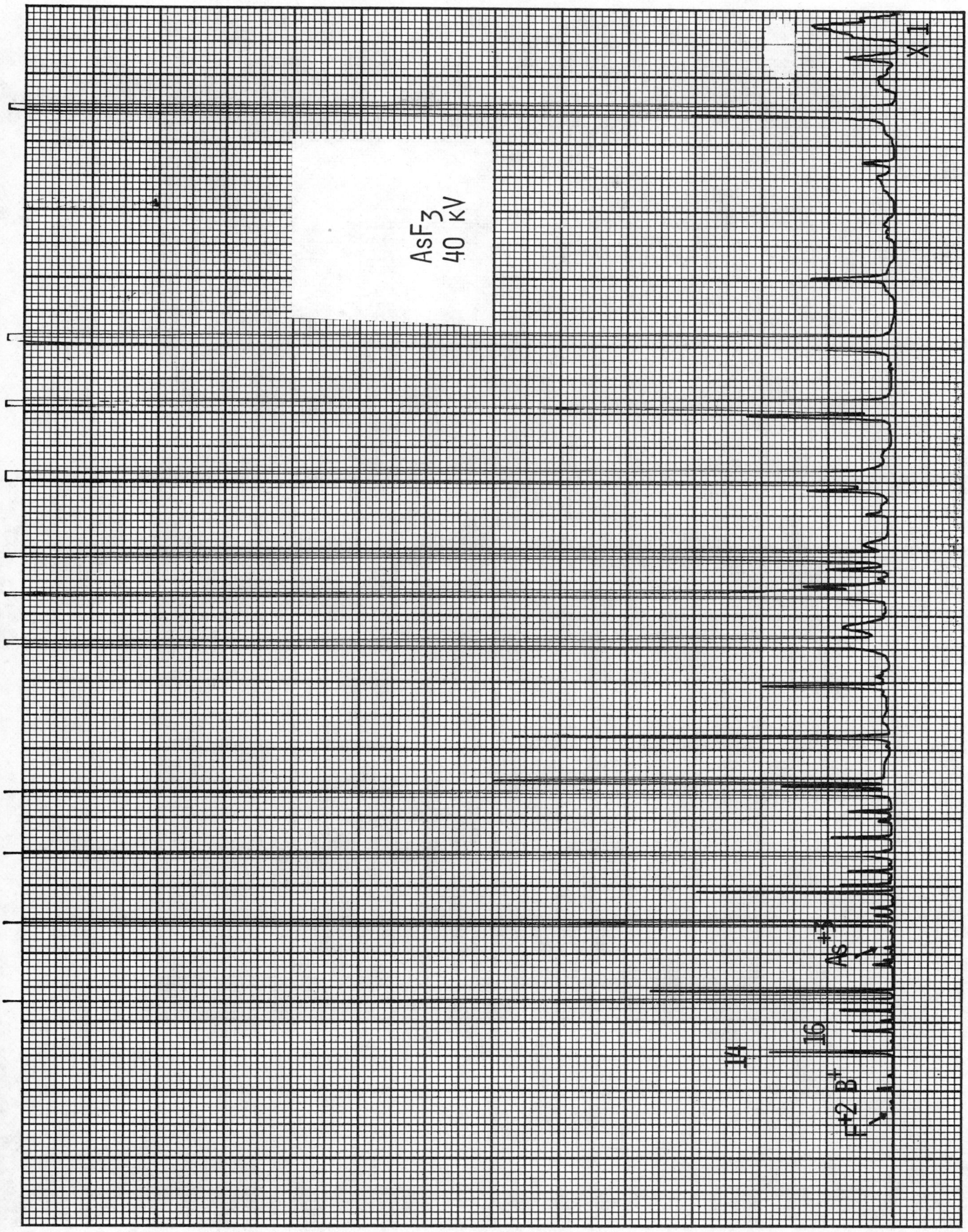
AsF3
40 kV
X 1
14
16
As+3
F+2
B+

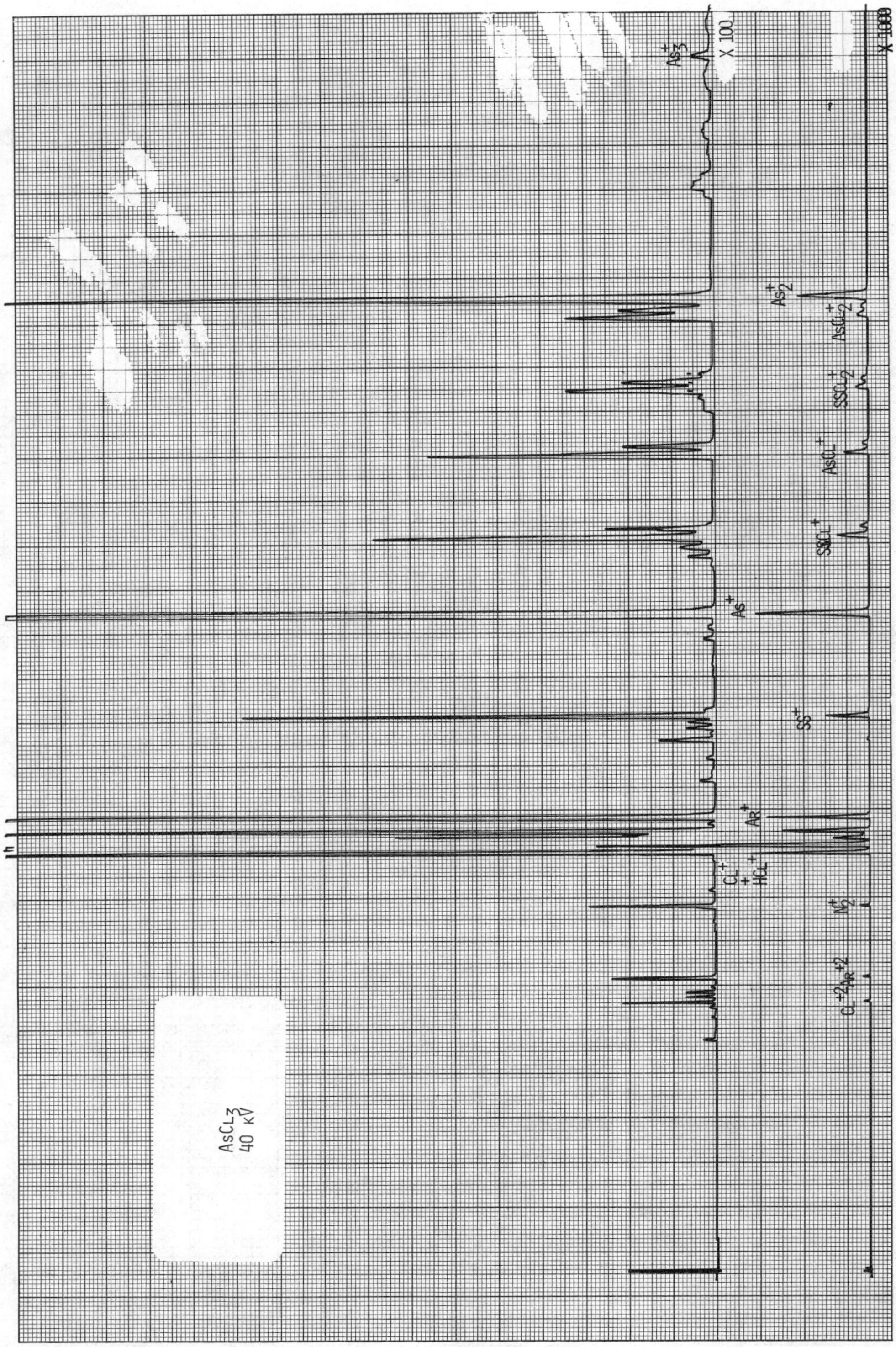
AsCl$_3$
40 kV
X 100
X 1000
As$_3^+$
As$_2^+$
AsCl$_2^+$
SSCl$_2^+$
AsCl$^+$
SSCl$^+$
As$^+$
SS$^+$
Ar$^+$
Cl$^+$ + HCl$^+$
N$_2^+$
Cl^{+2} Ar^{+2}

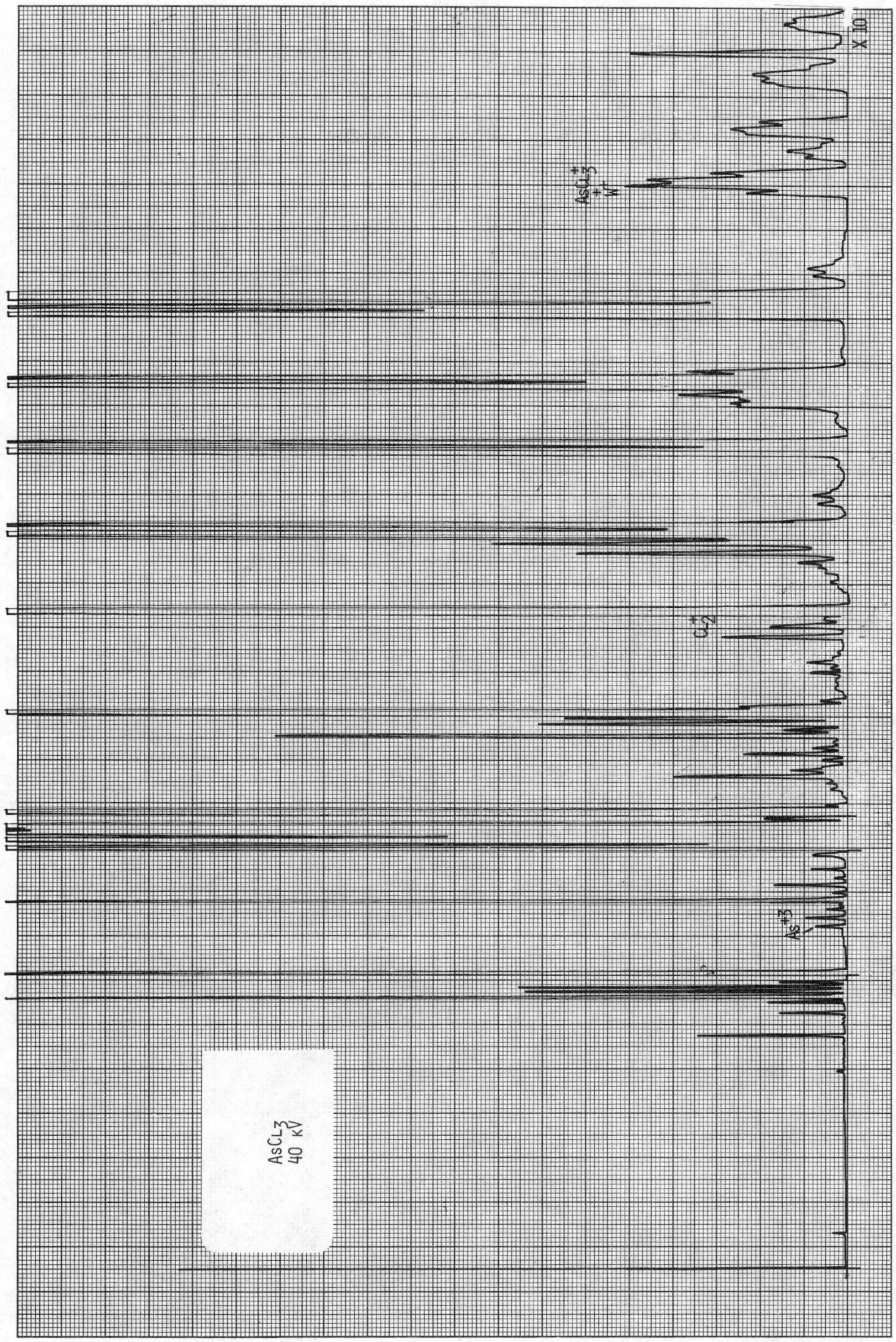
AsCl3
40 kV
X 10

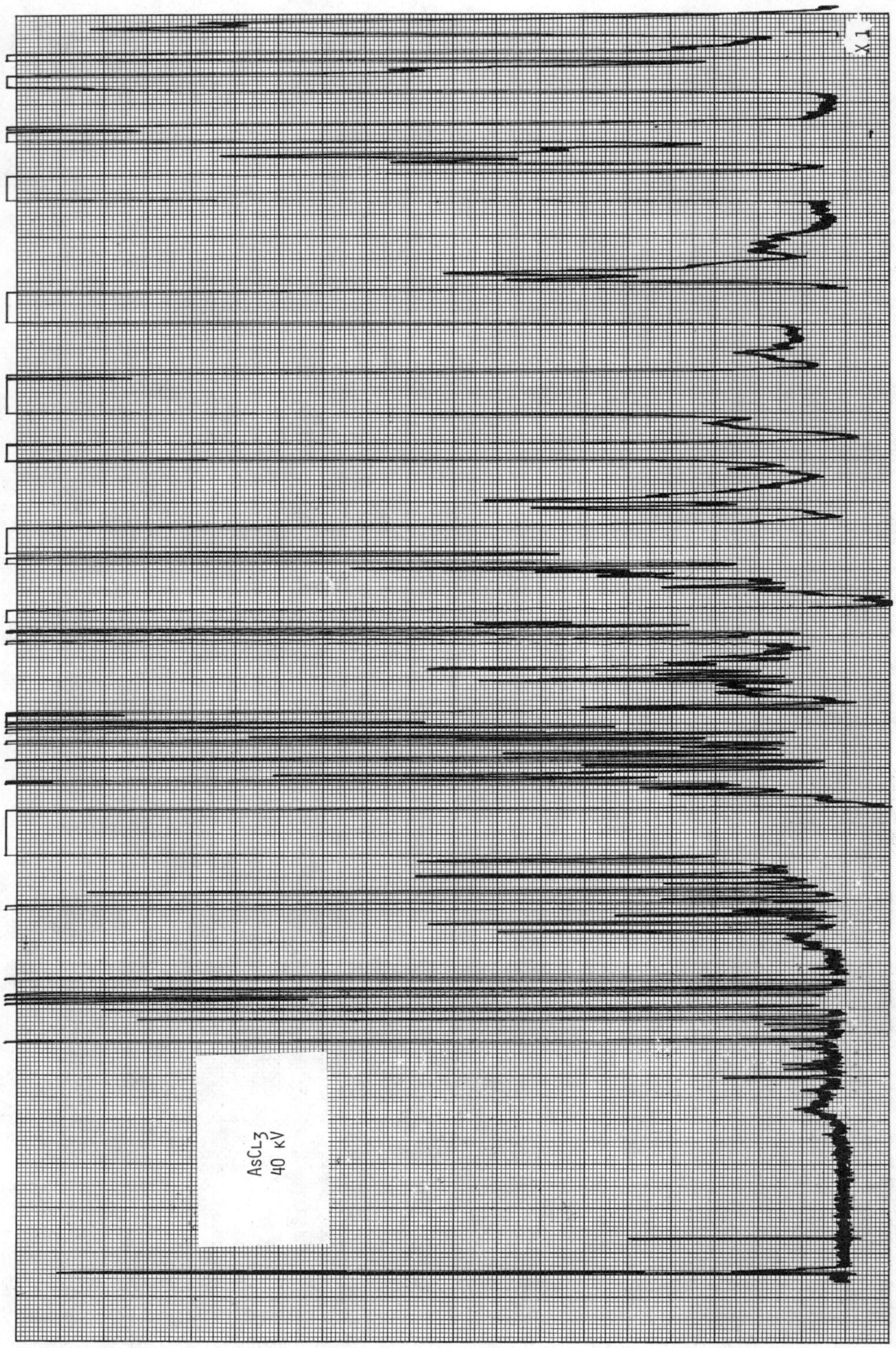
AsCl$_3$
40 kV
x1

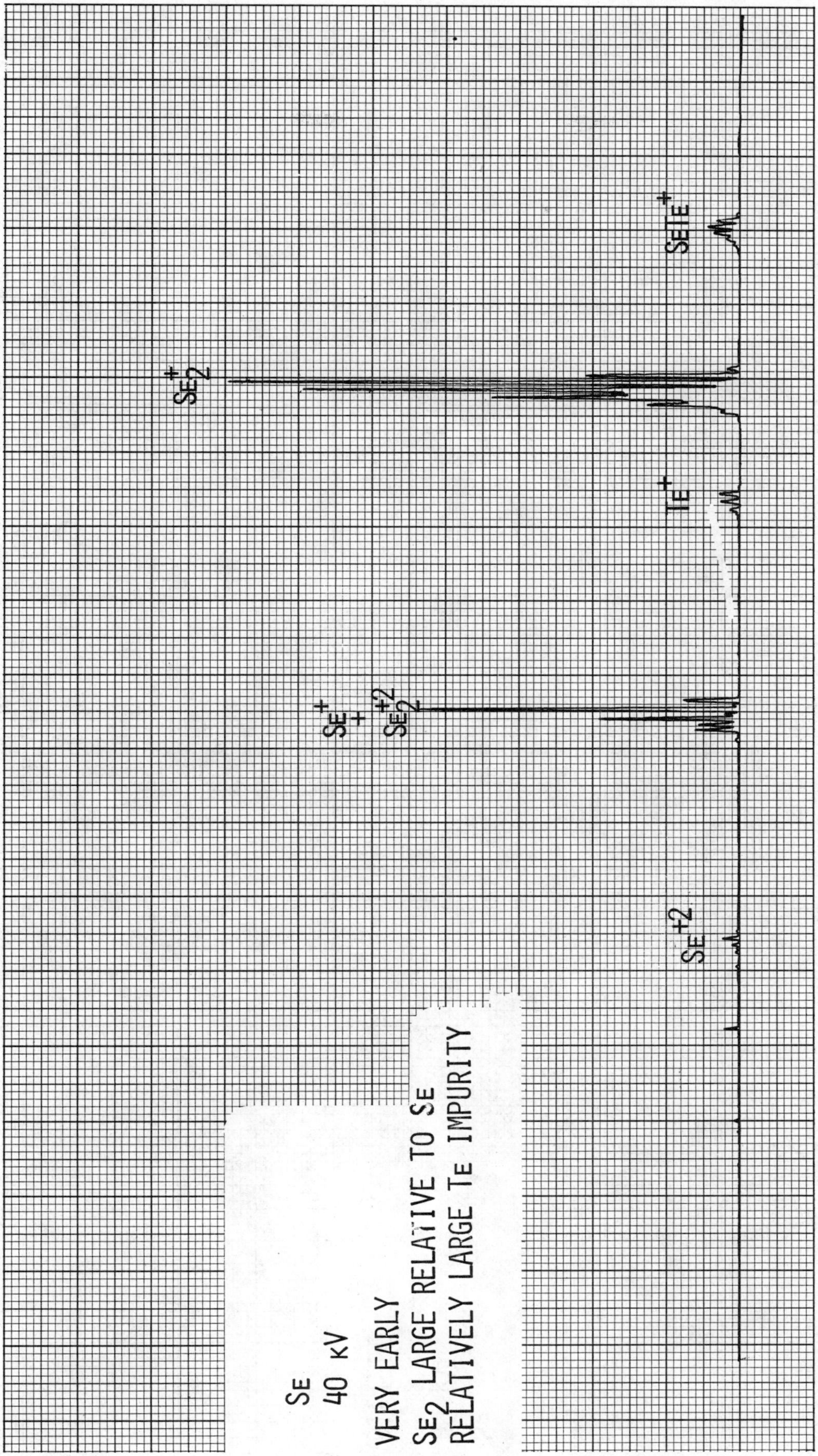
SE
40 KV
VERY EARLY
SE2 LARGE RELATIVE TO SE
RELATIVELY LARGE TE IMPURITY
SE+2
SE+ + SE2+2
TE+
SE2+
SETE+

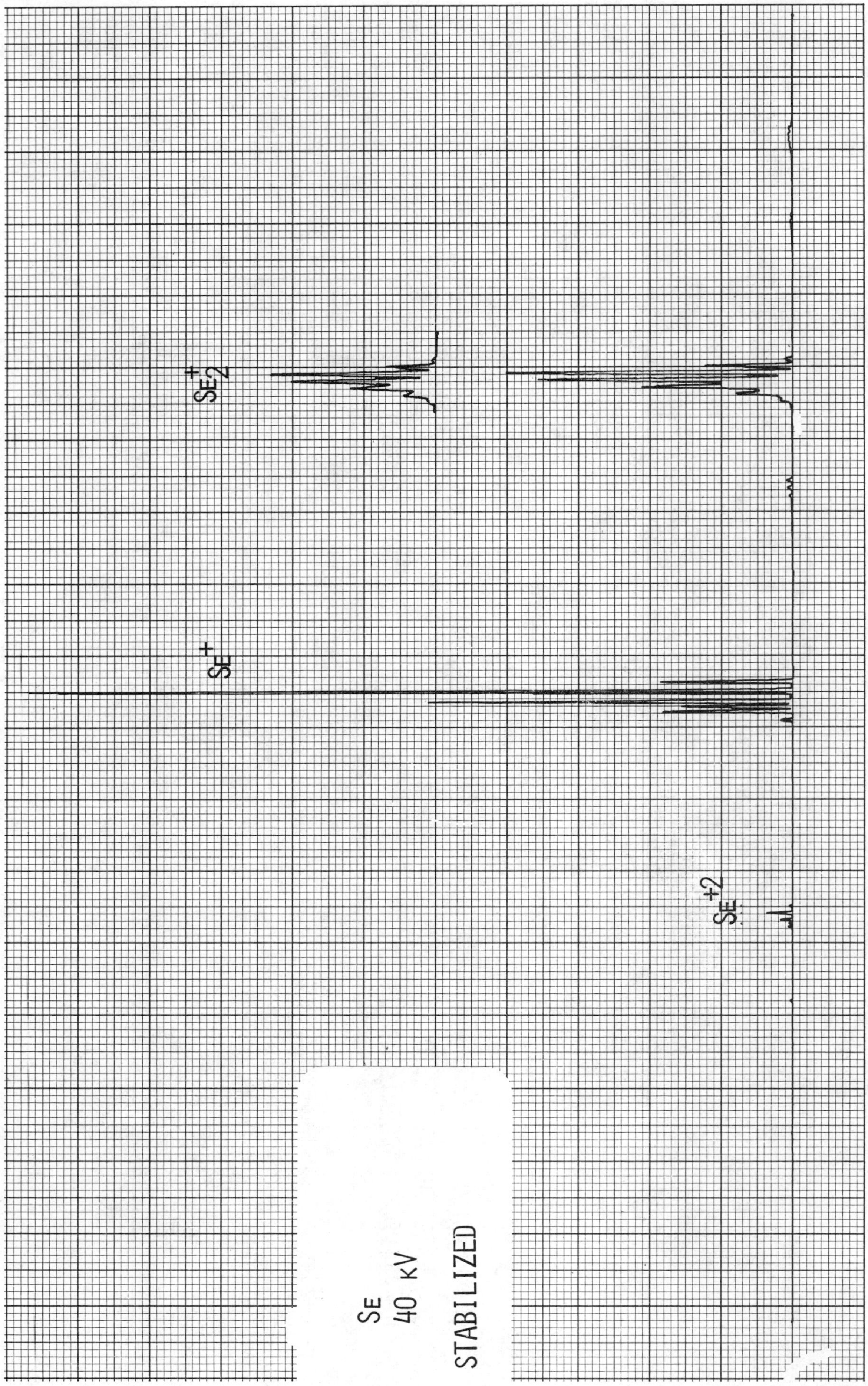

Se_2^+
Se^+
Se^{+2}
SE
40 kV
STABILIZED

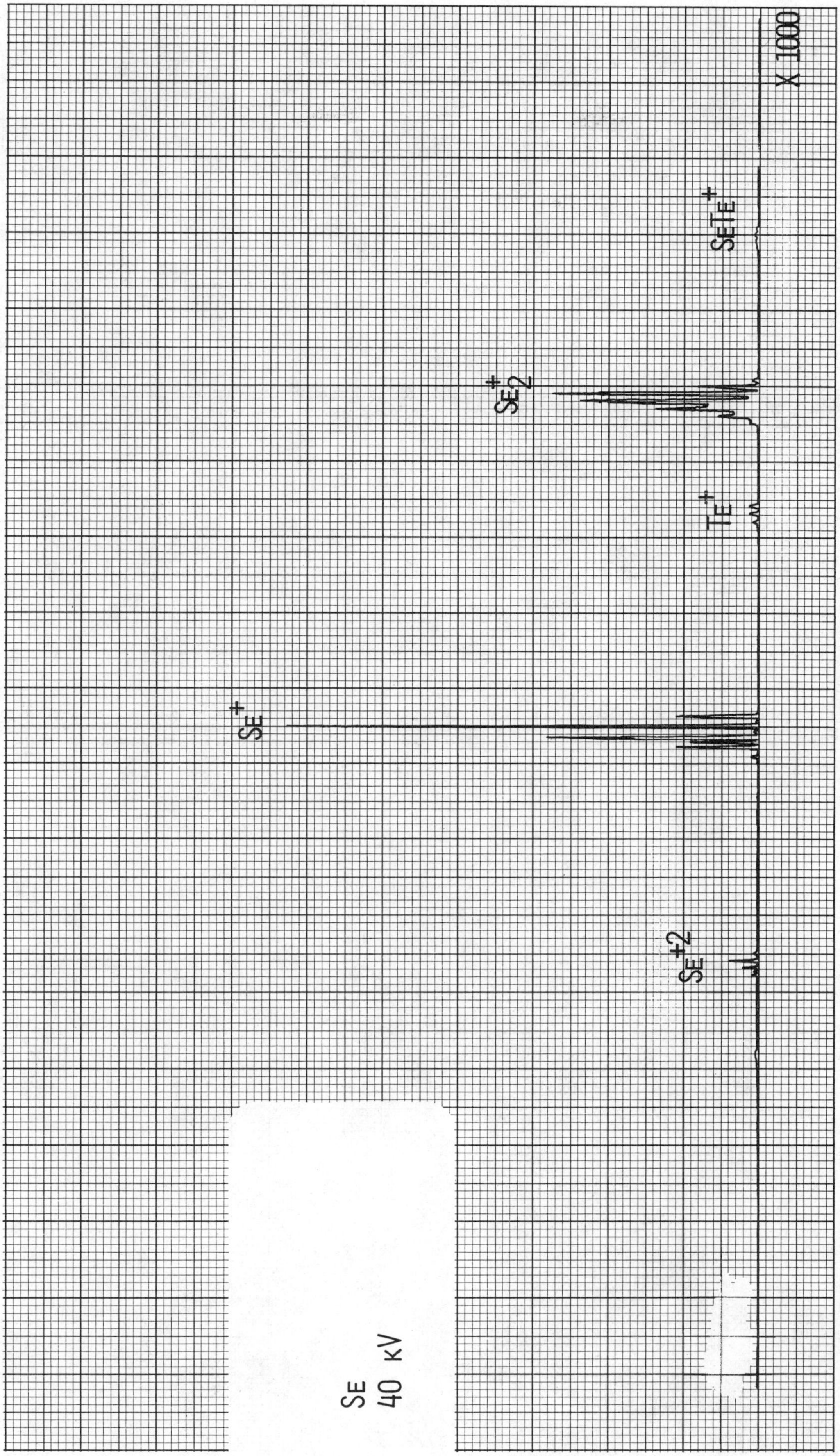
X 1000
SeTe$^+$
Se$_2^+$
Te$^+$
Se$^+$
Se^{+2}
Se
40 kV

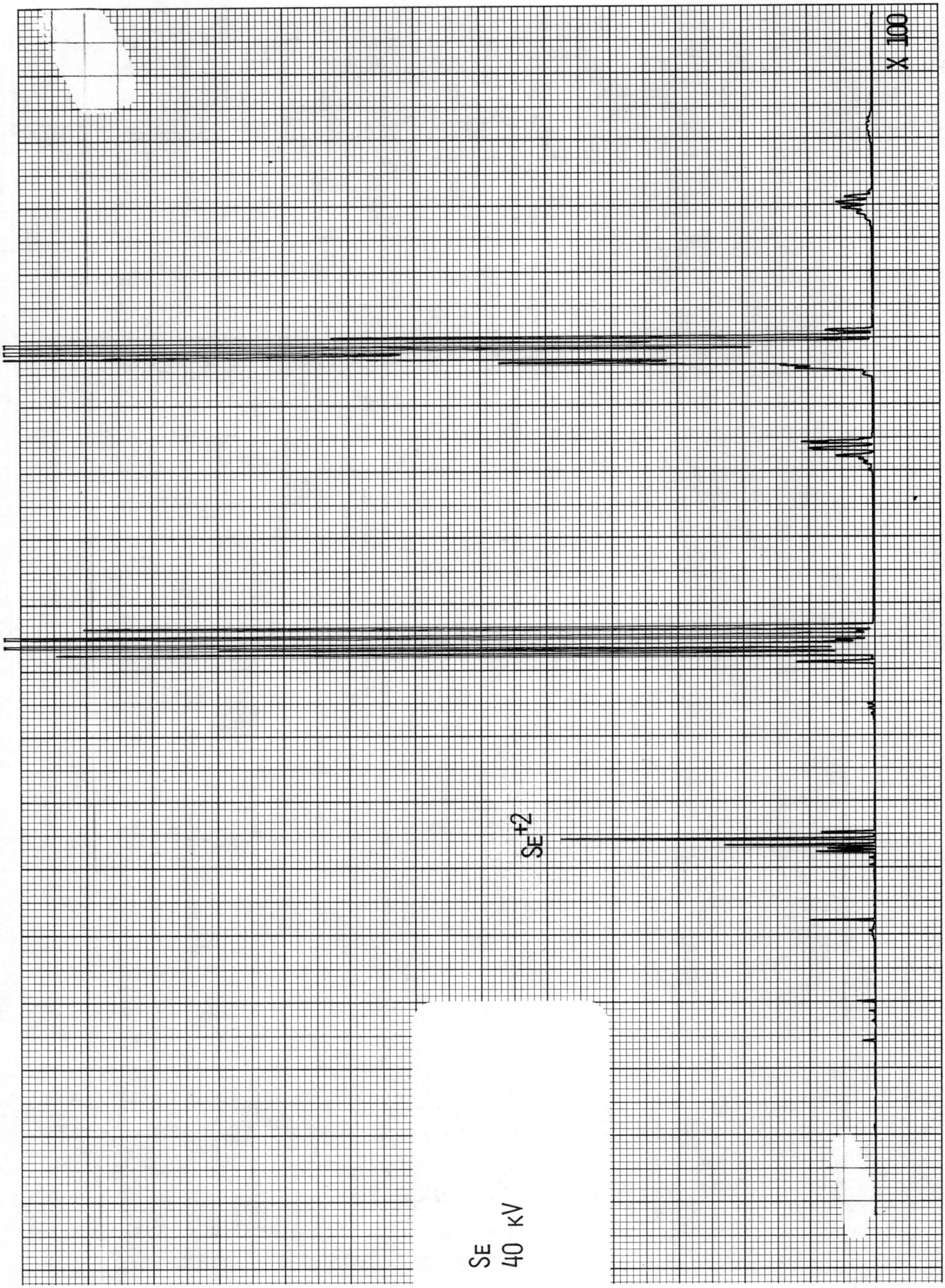
X 100
Se+2
Se
40 kV

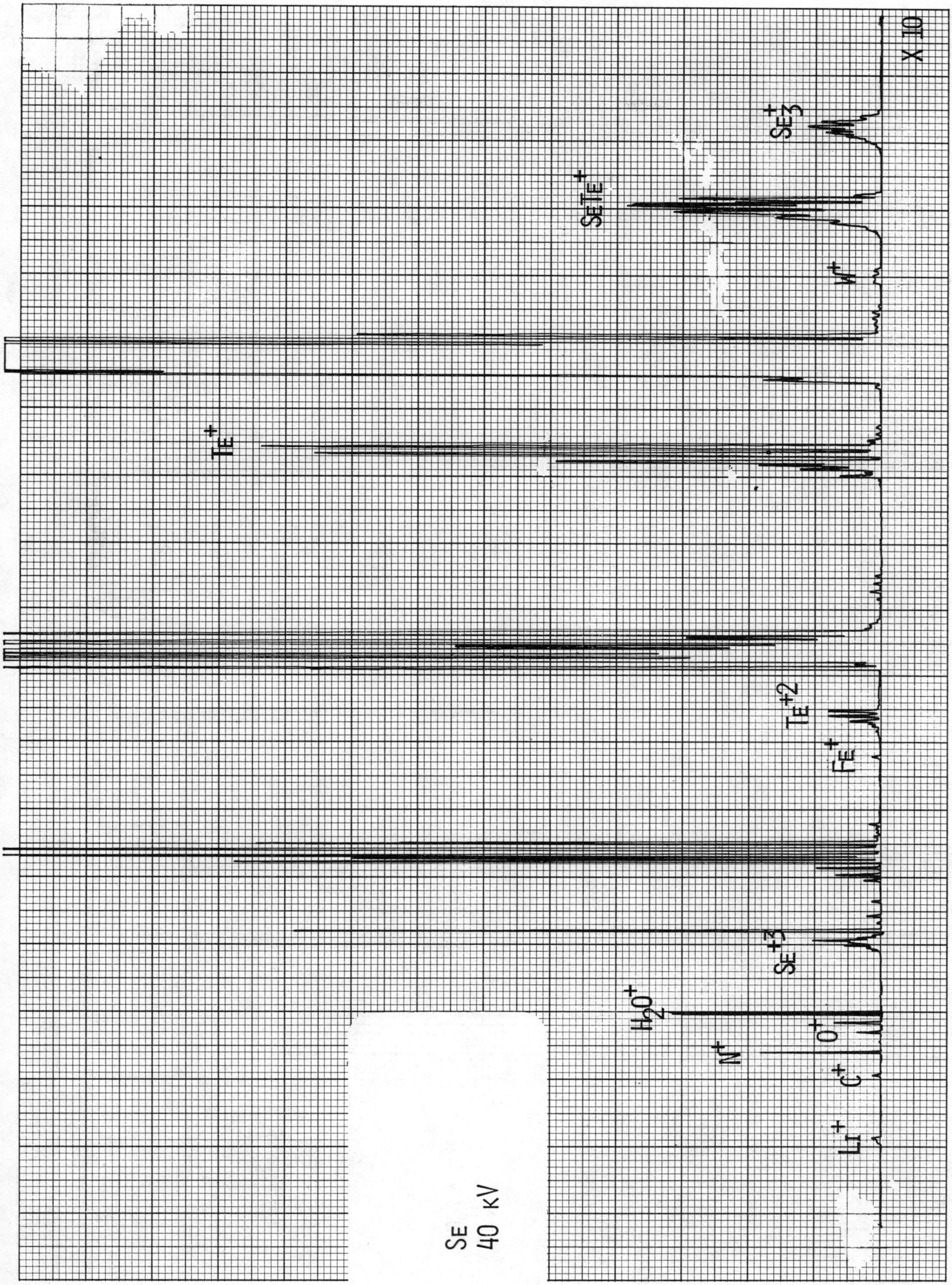
Se
40 kV
Li+
C+
N+
O+
H2O+
Se+3
Fe+
Te+2
Te+
W+
SeTe+
Se3+
X 10

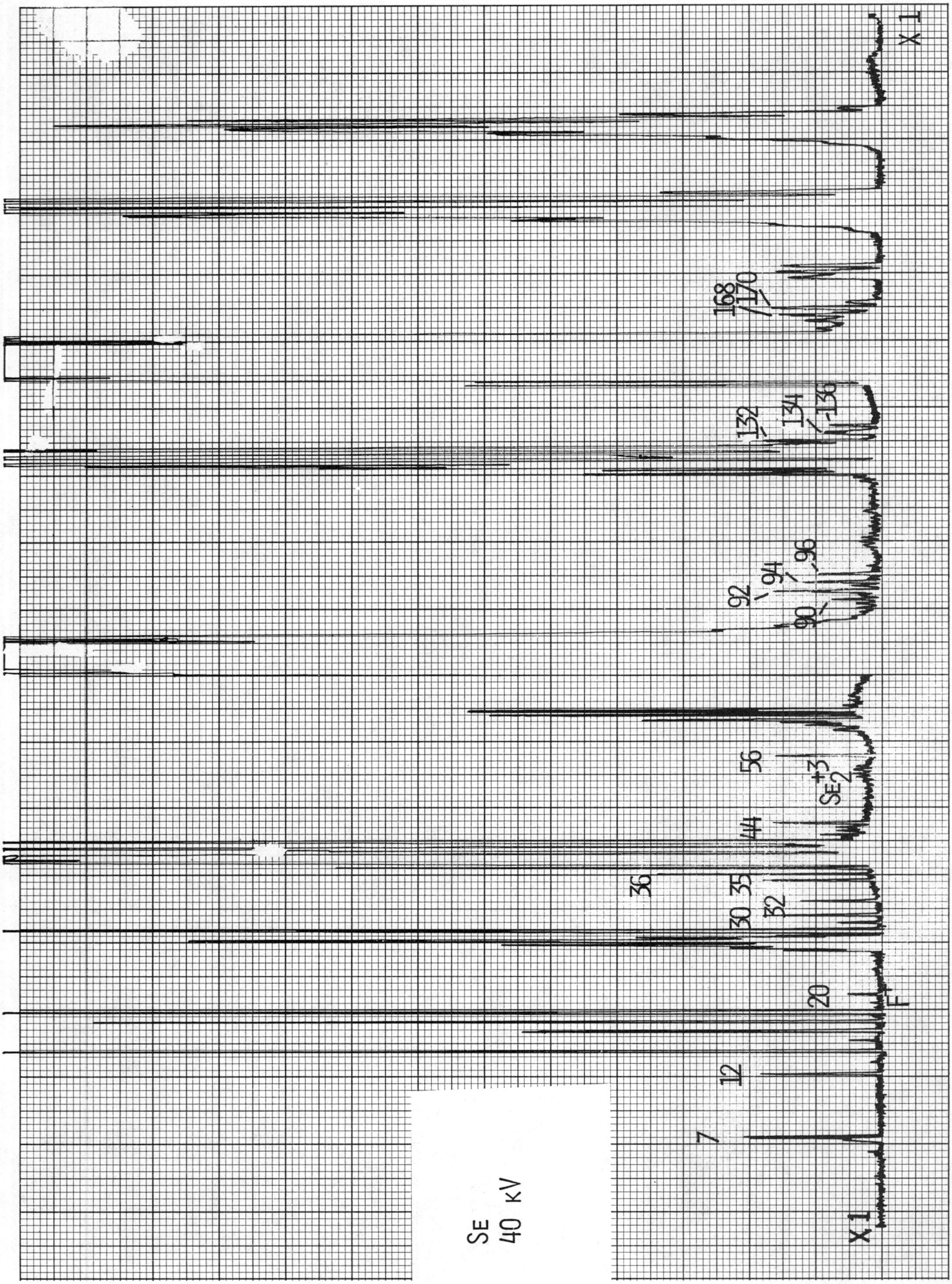
Se
40 kV
X 1
X 1
7
12
20
F+
30
32
35
36
44
56
Se_2^{+3}
90
92
94
96
132
134
136
168
170

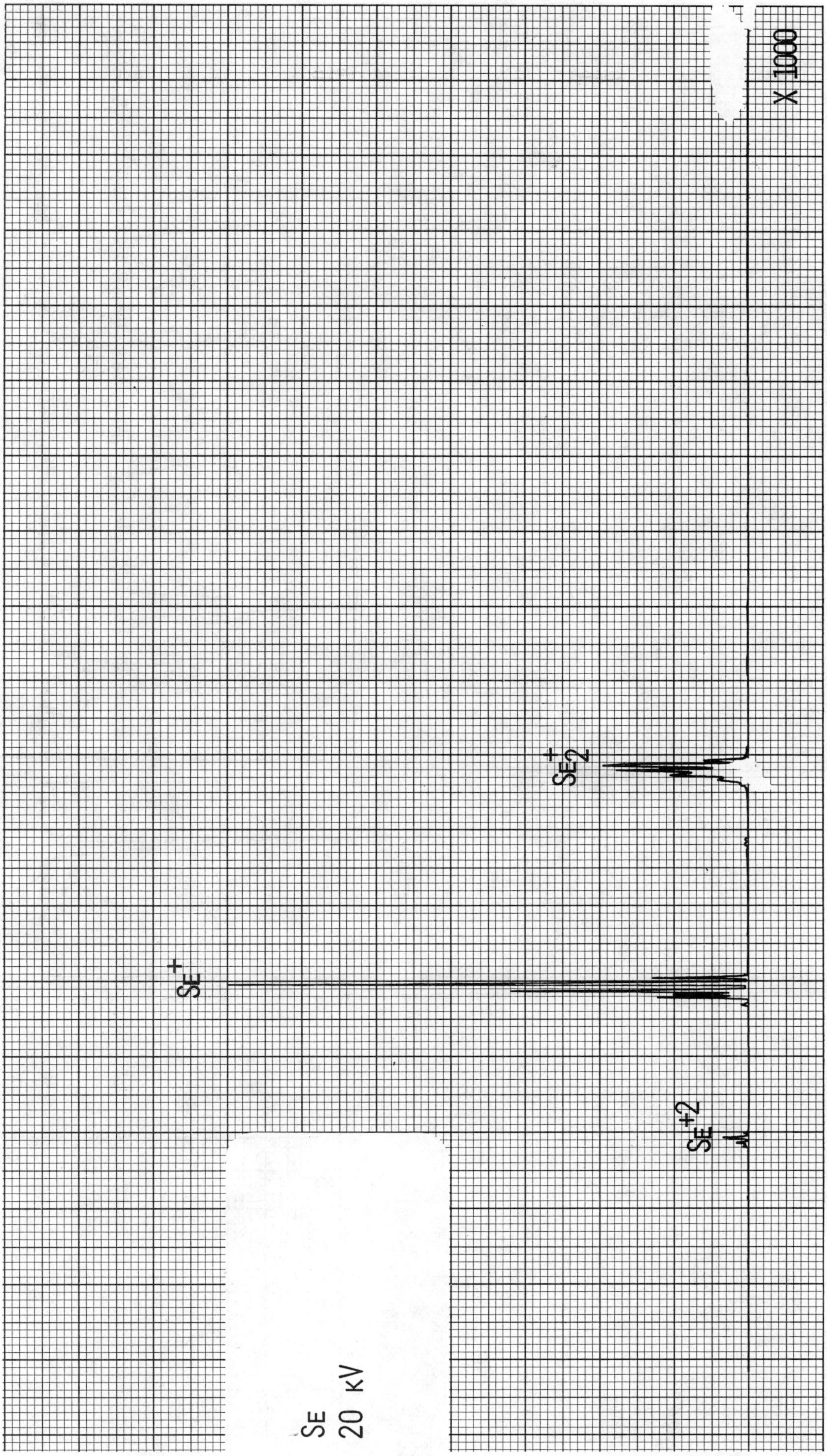
X 1000
SE2+
SE+
SE+2
SE
20 kV

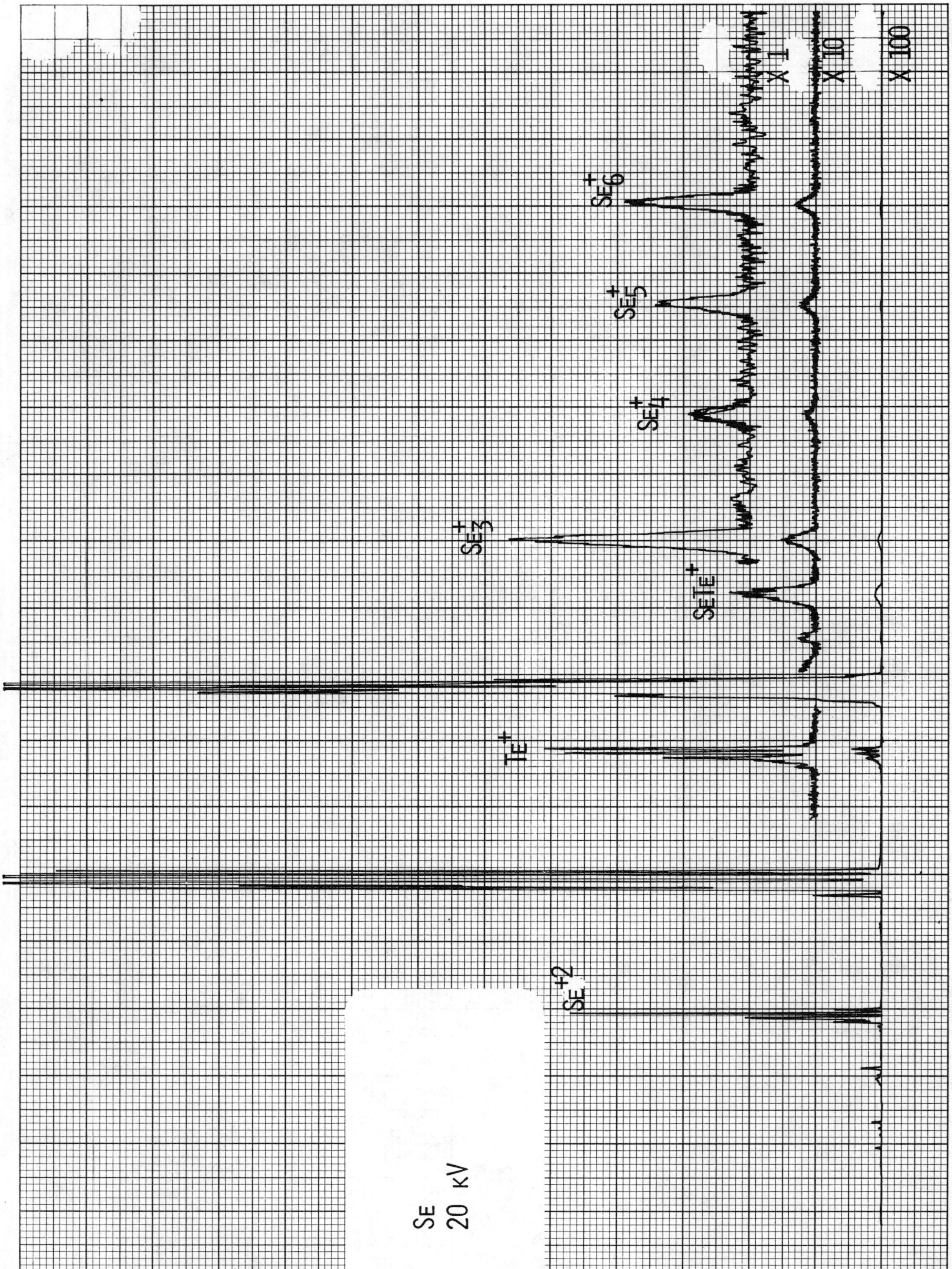
Se
20 kV
Se^{+2}
Te^+
$SeTe^+$
Se_3^+
Se_4^+
Se_5^+
Se_6^+
X 1
X 10
X 100

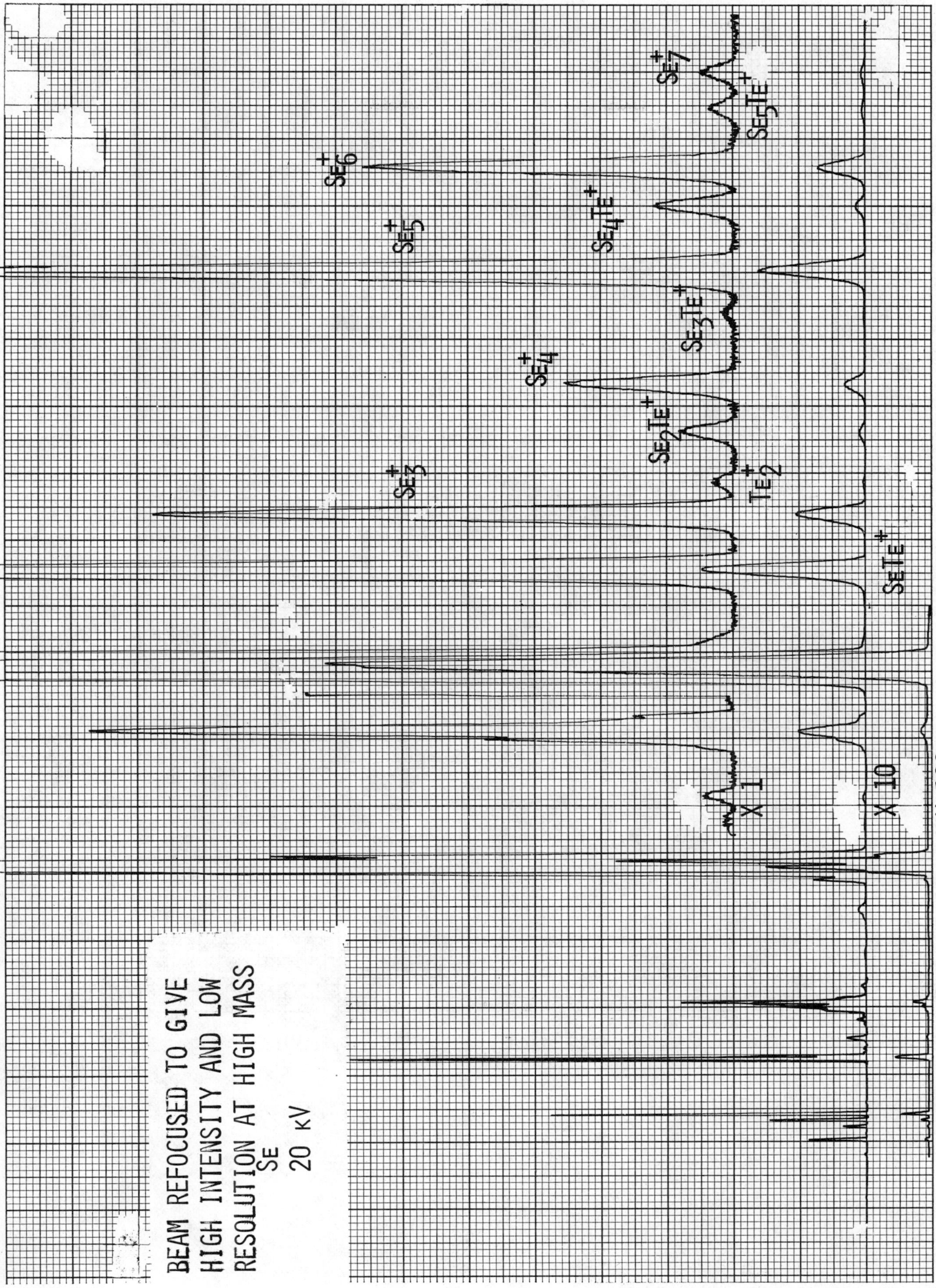
BEAM REFOCUSED TO GIVE
HIGH INTENSITY AND LOW
RESOLUTION AT HIGH MASS
SE
20 kV
X 1
X 10
X 100
Se_3^+
Se_4^+
Se_5^+
Se_6^+
Se_7^+
Te_2^+
$SeTe^+$
Se_2Te^+
Se_3Te^+
Se_4Te^+
Se_5Te^+

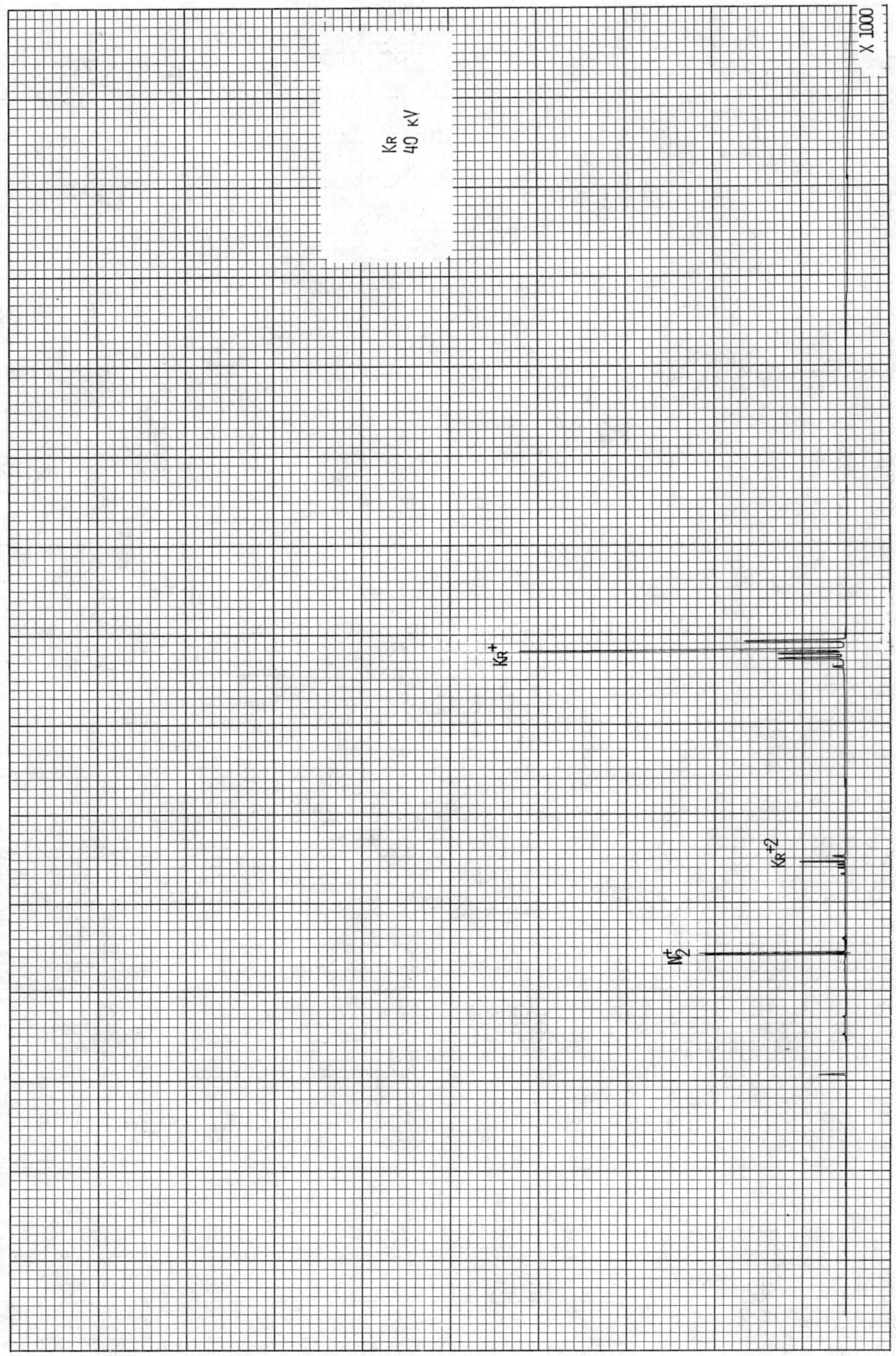

Kr
40 kV
X 1000
Kr+
Kr+2
N2+

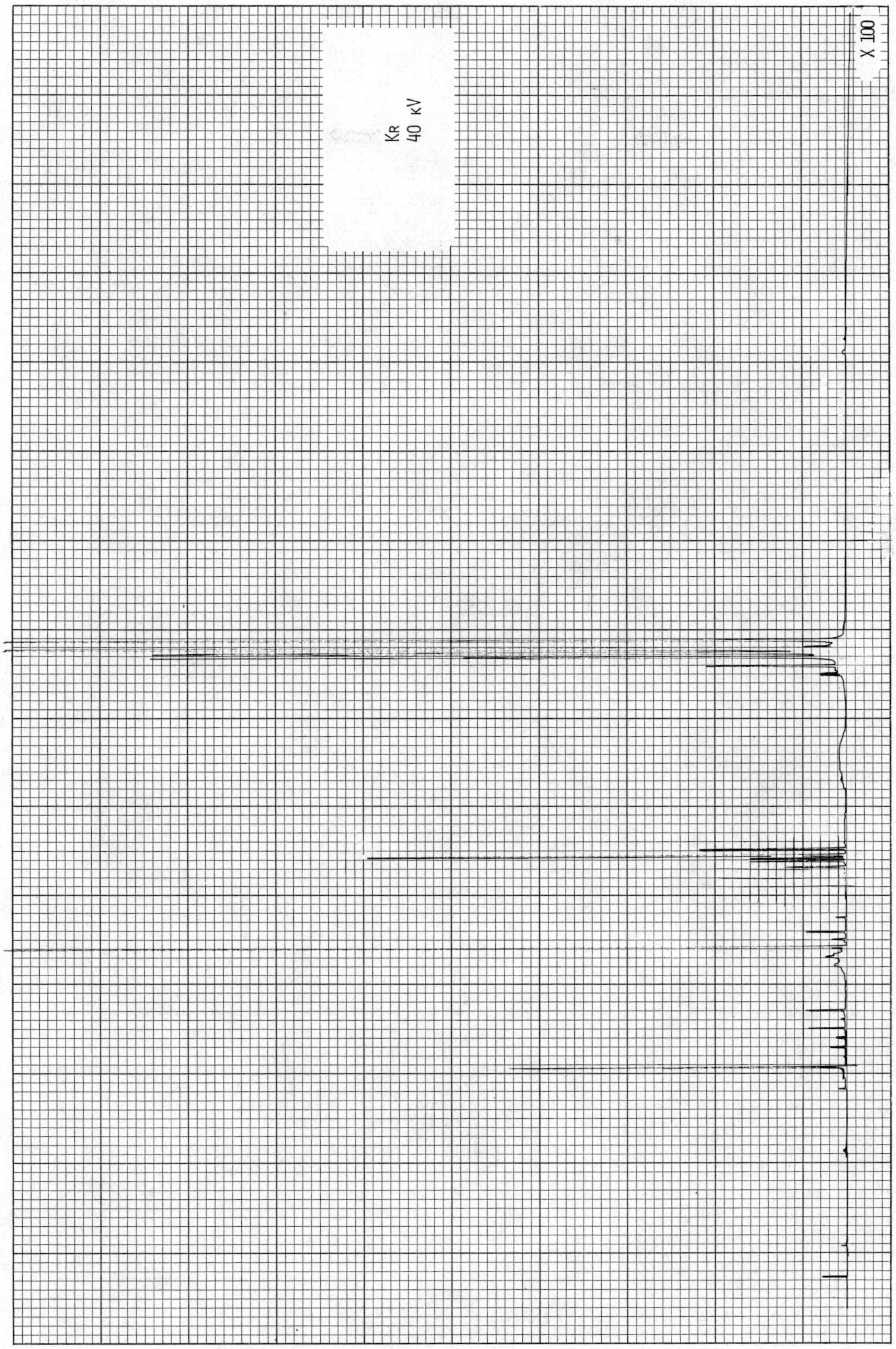
Kr
40 kV
X 100

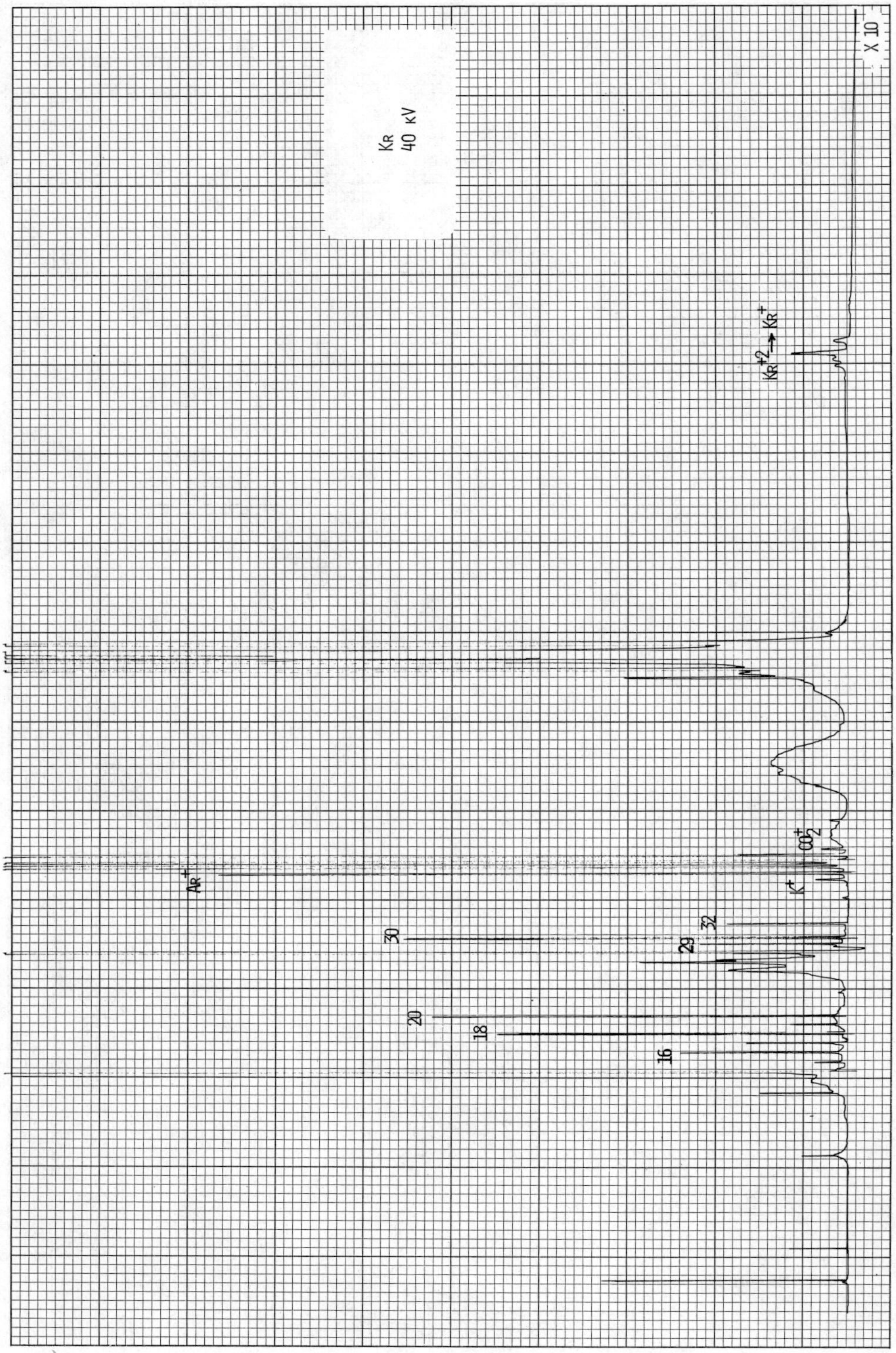
Kr
40 kV
X 10
Kr+2 → Kr+
Ar+
CO2+
K+
32
29
30
20
18
16

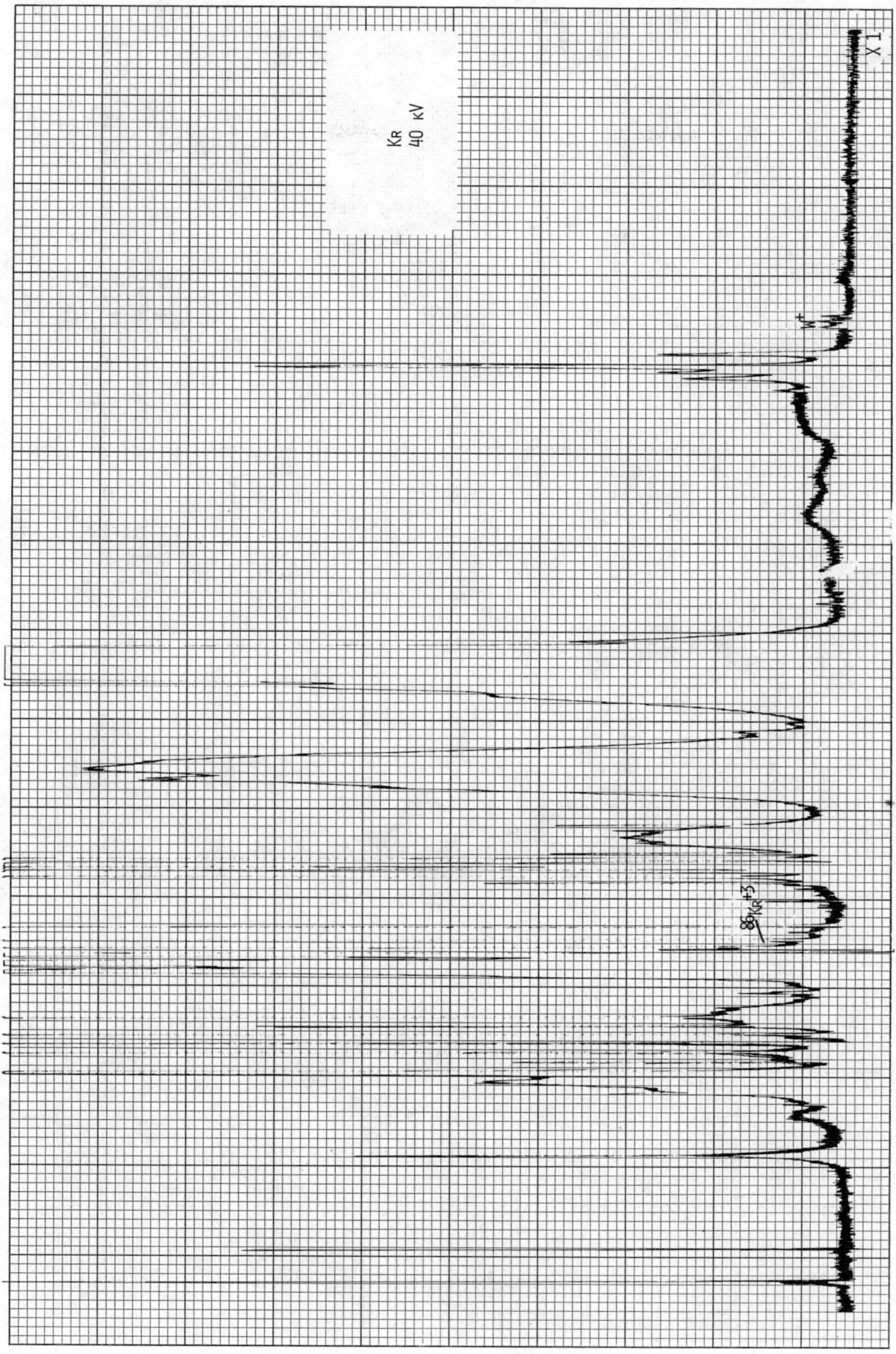

Kr
40 kV
X 1
86Kr+3

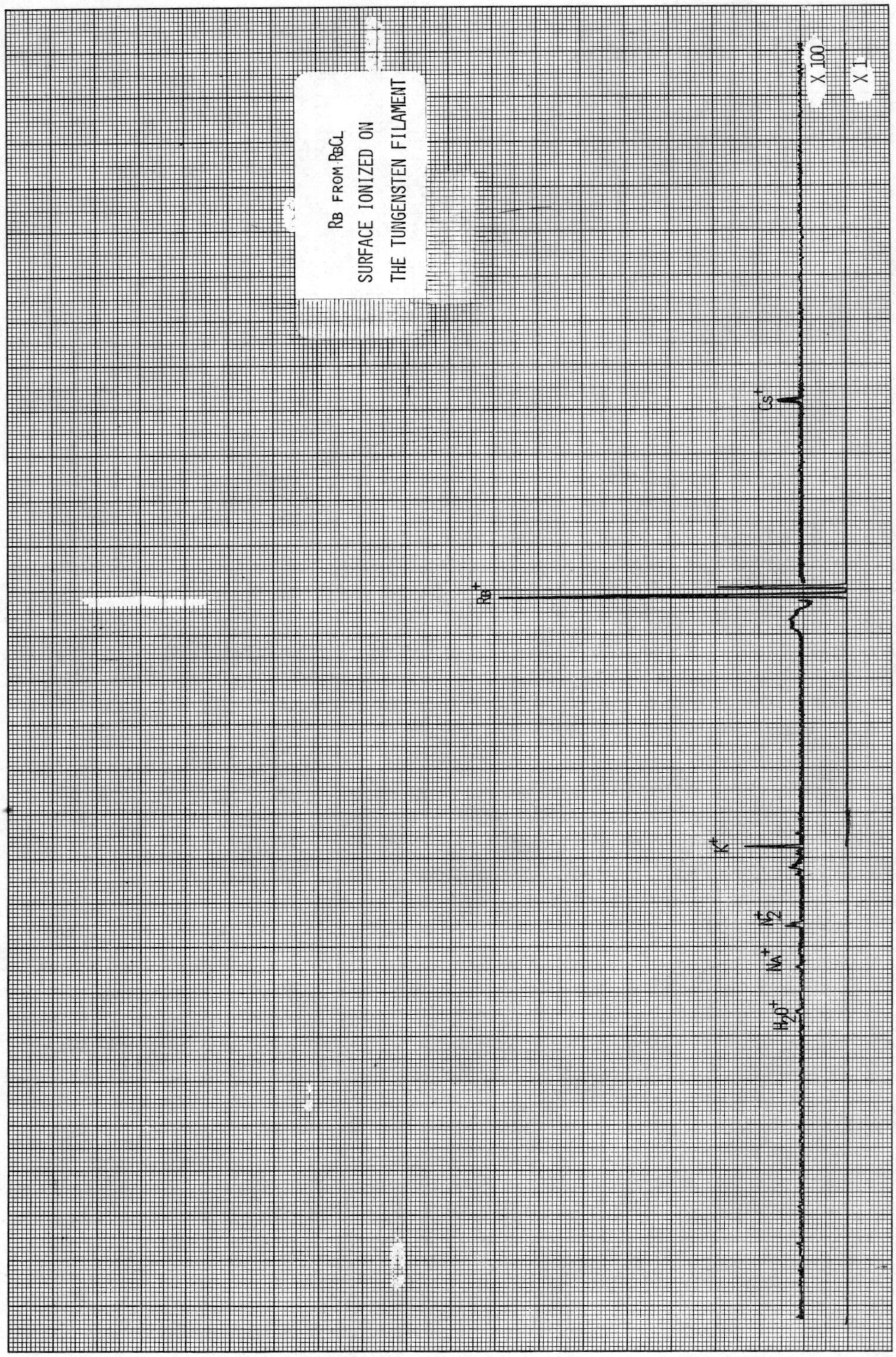
Rb FROM RbCl
SURFACE IONIZED ON
THE TUNGSTEN FILAMENT
X 100
X 1
Cs^+
Rb^+
K^+
N_2^+
Na^+
H_2O^+

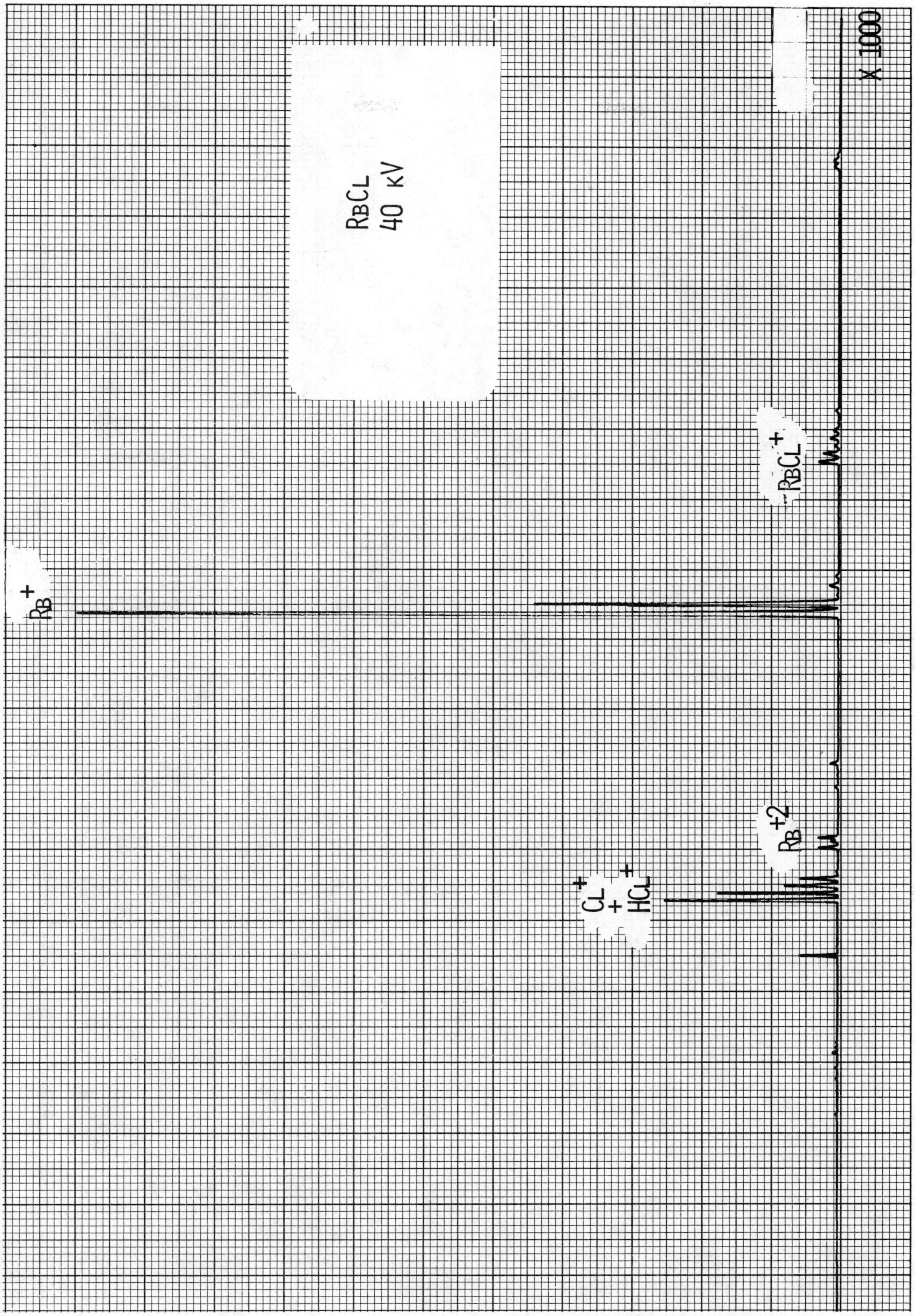
RbCl
40 kV
X 1000
Rb+
RbCl+
Rb+2
Cl+
HCl+

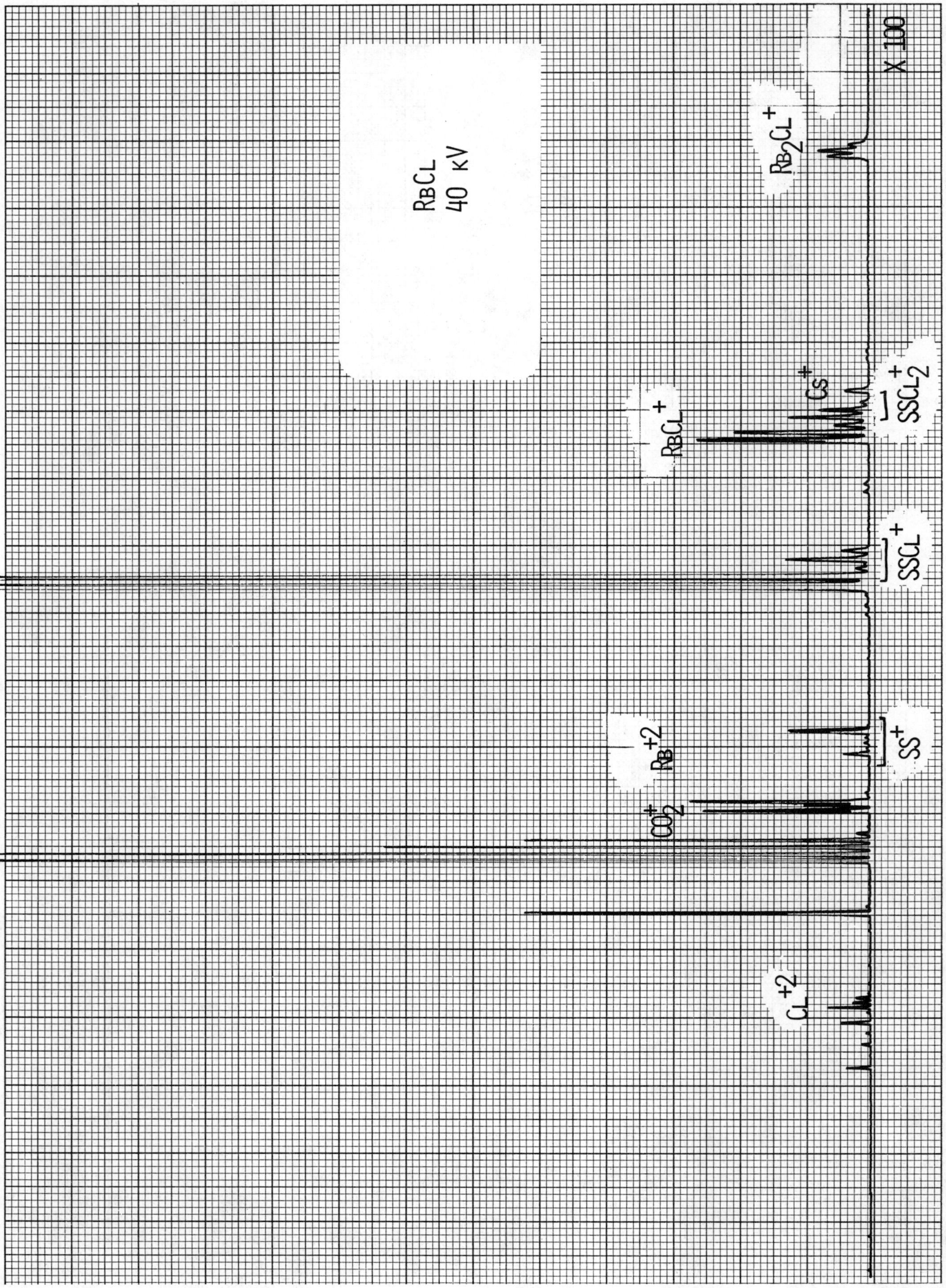
RbCl
40 kV
X 100
Rb2Cl+
Cs+
RbCl+
SSCl2+
SSCl+
Rb+2
SS+
CO2+
Cl+2

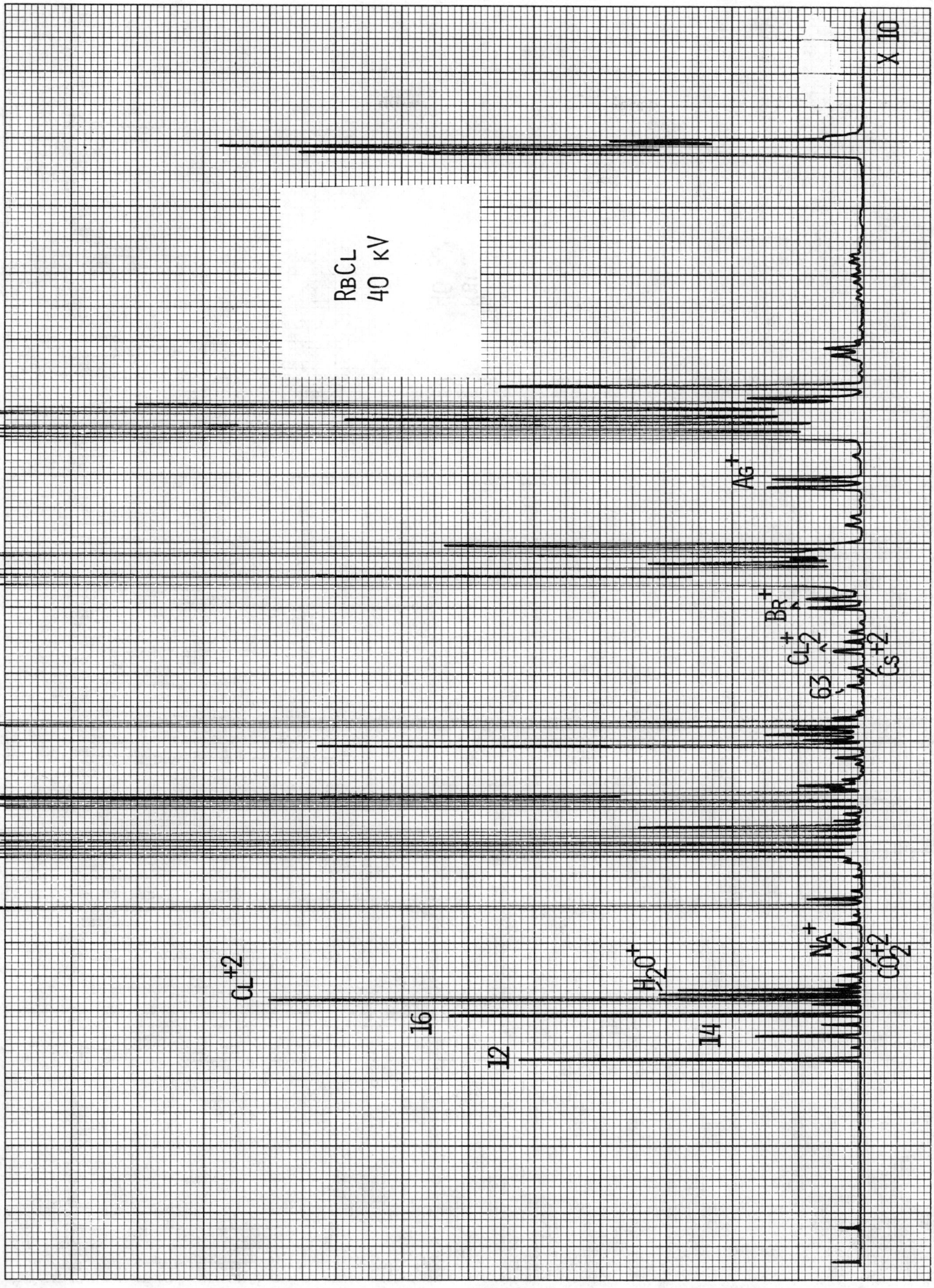

RbCl
40 kV
X 10
Ag+
Br+
Cl2+
63
Cs+2
Cl+2
16
12
14
H2O+
Na+
CO2+2

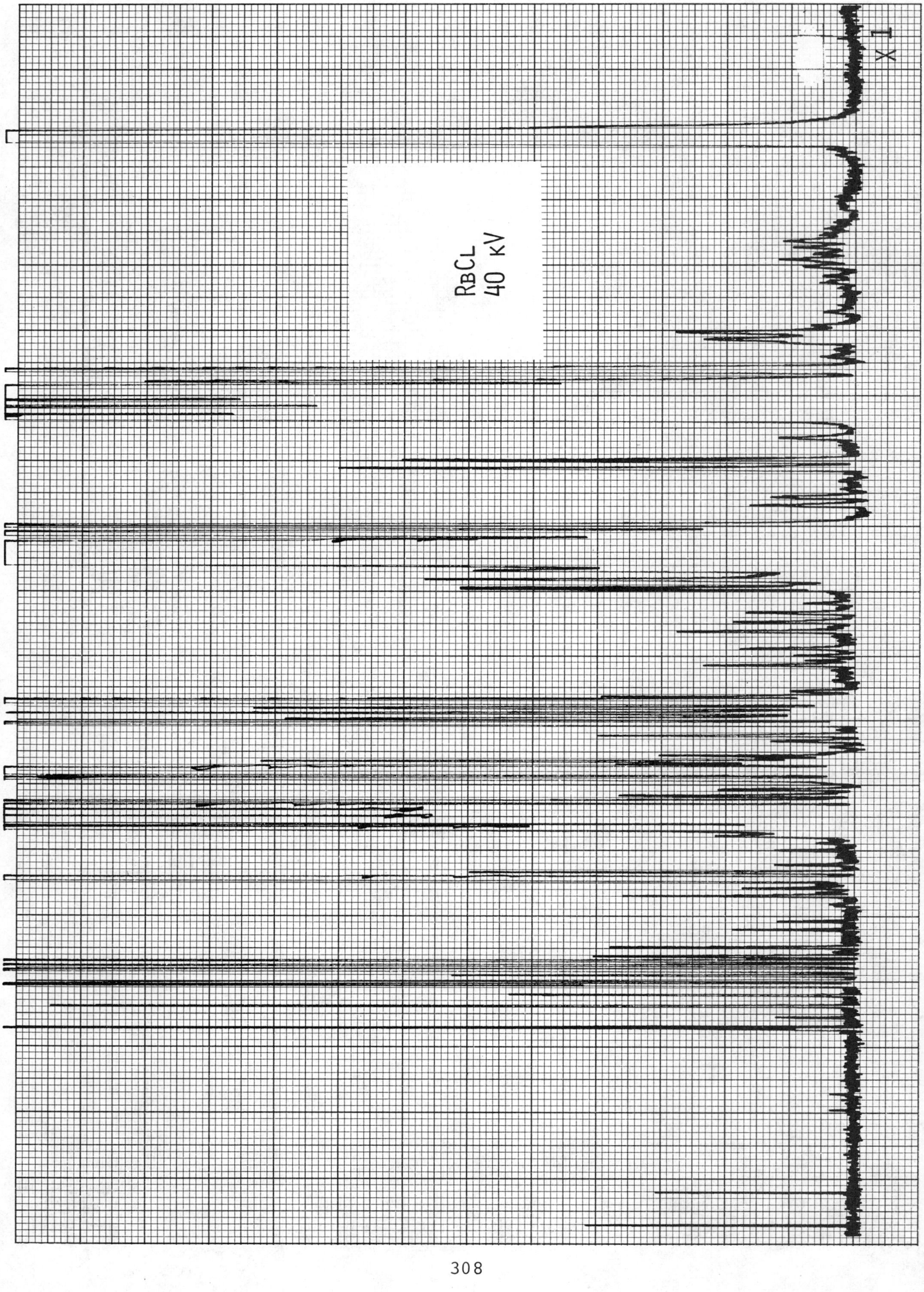
RbCl
40 kV
X 1

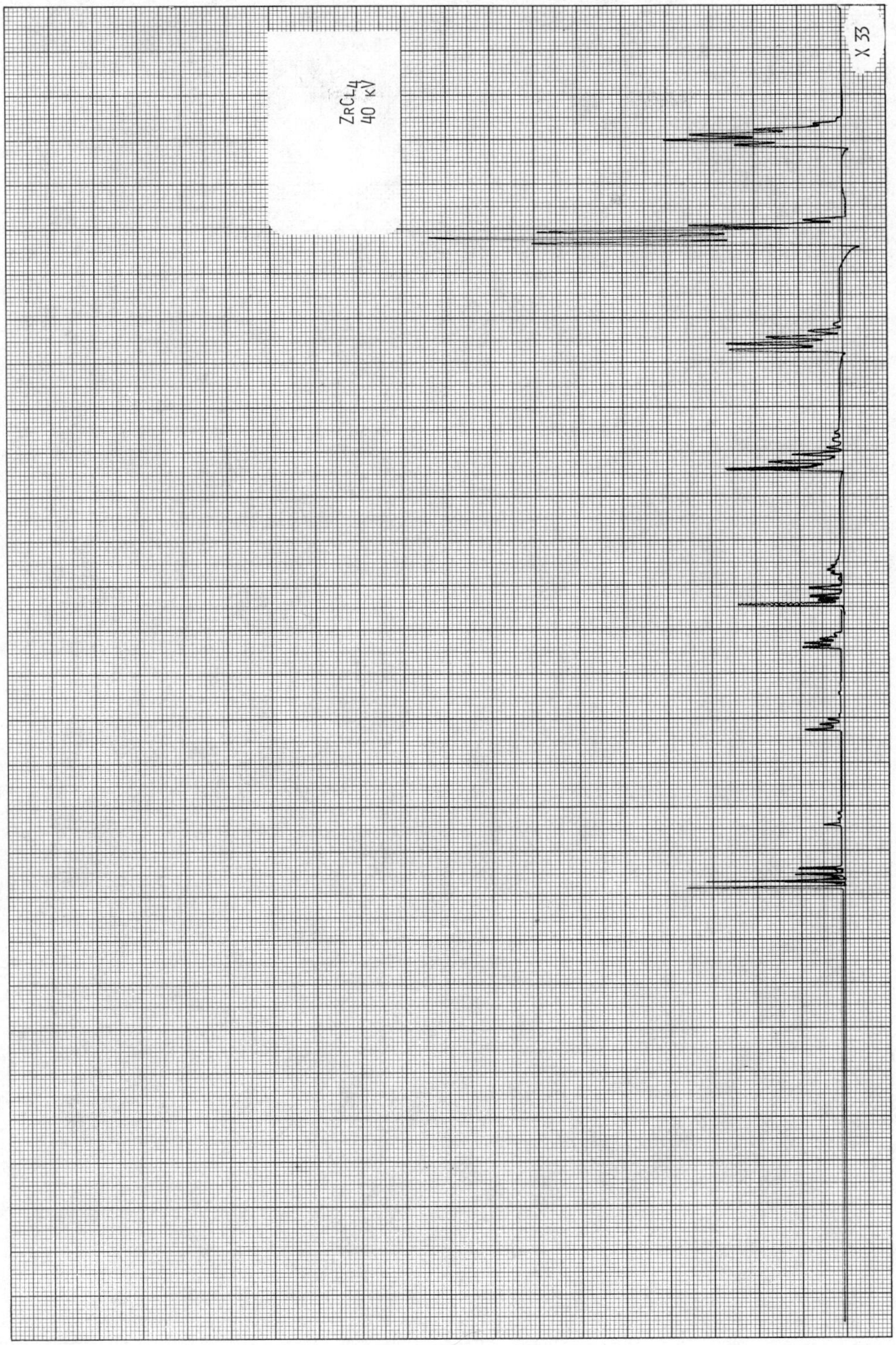
ZrCl4
40 kV
X 33

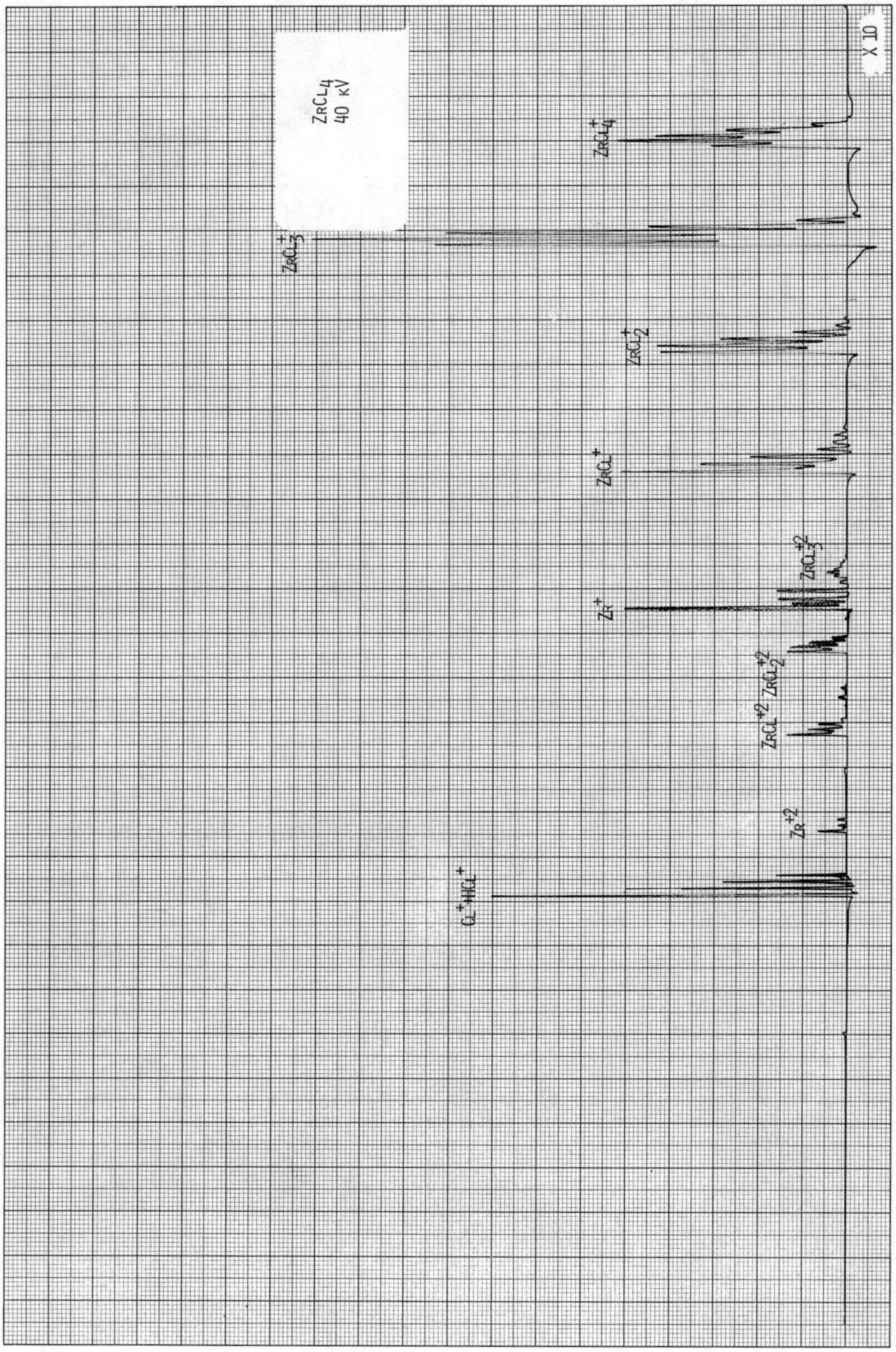

ZrCl4
40 kV
X 10
ZrCl4+
ZrCl3+
ZrCl2+
ZrCl+
ZrCl3+2
Zr+
ZrCl+2
ZrCl2+2
Zr+2
Cl+ HCl+

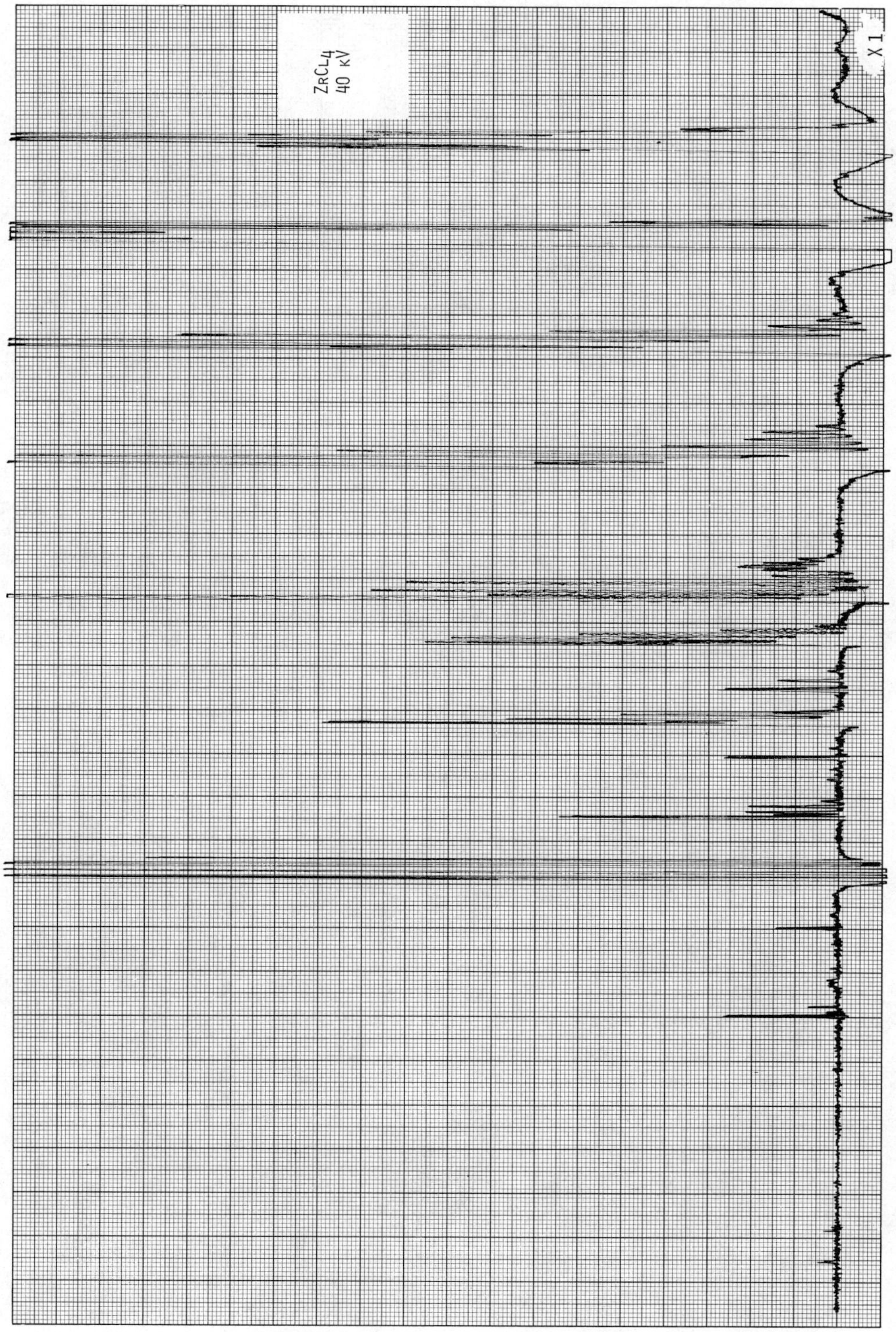
ZrCl4
40 kV
X 1

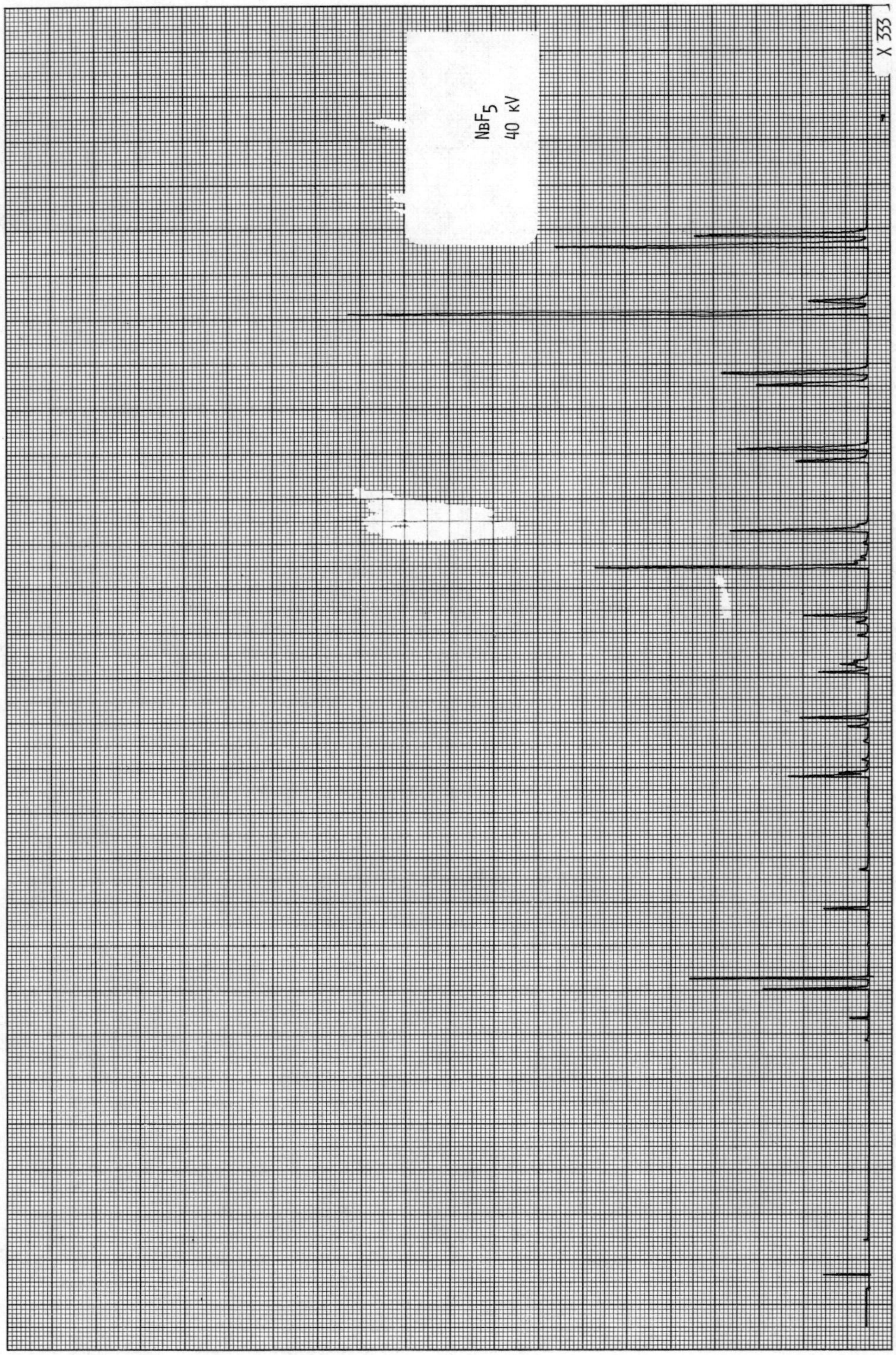

NbF_5
40 kV
X 333

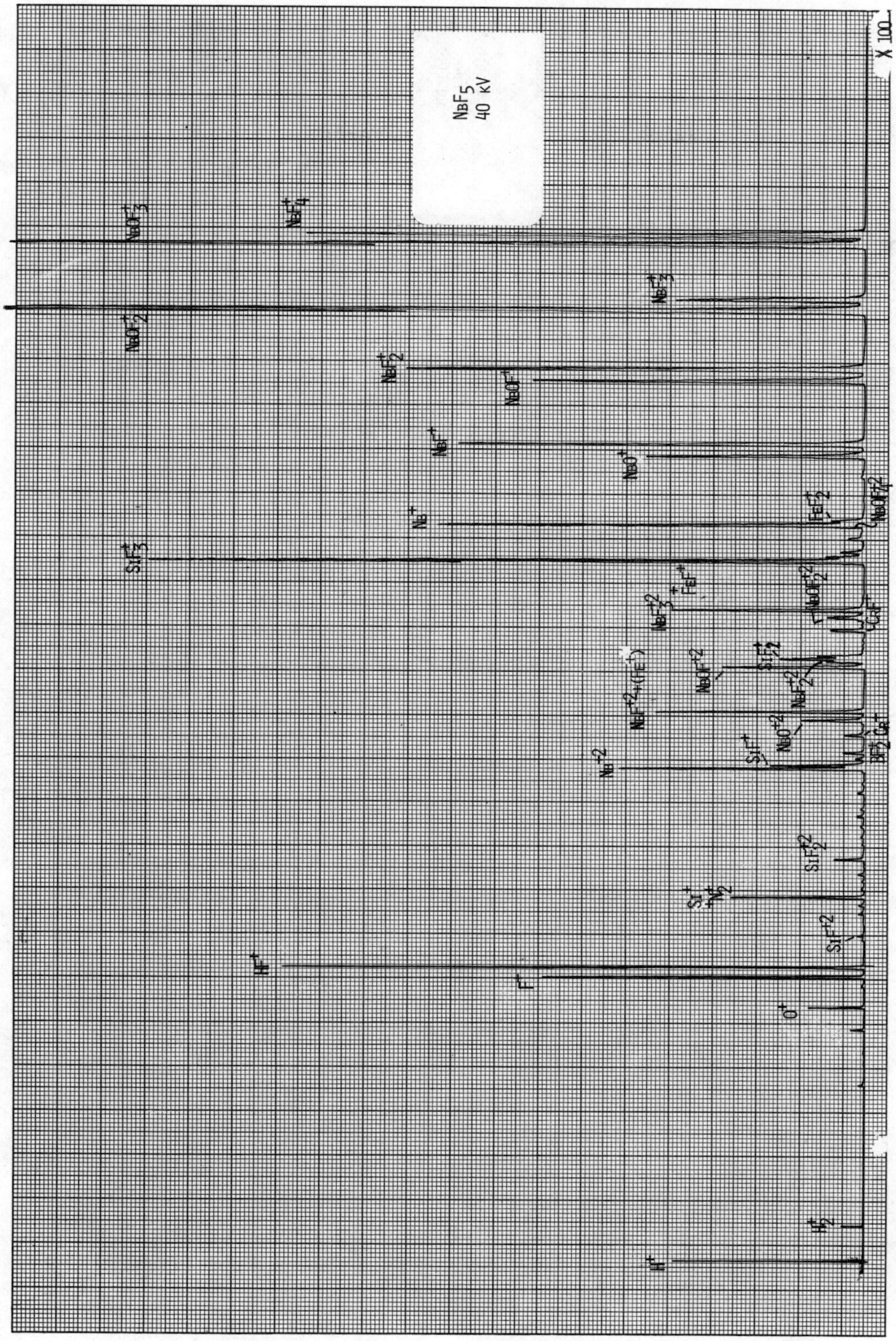
NbF5
40 kV
X 100
NbOF3+
NbF4+
NbF3+
NbOF2+
NbF2+
NbOF+
NbF+
NbO+
Nb+
SiF3+
Nb+2
SiF+2
HF+
F+
O+
H+

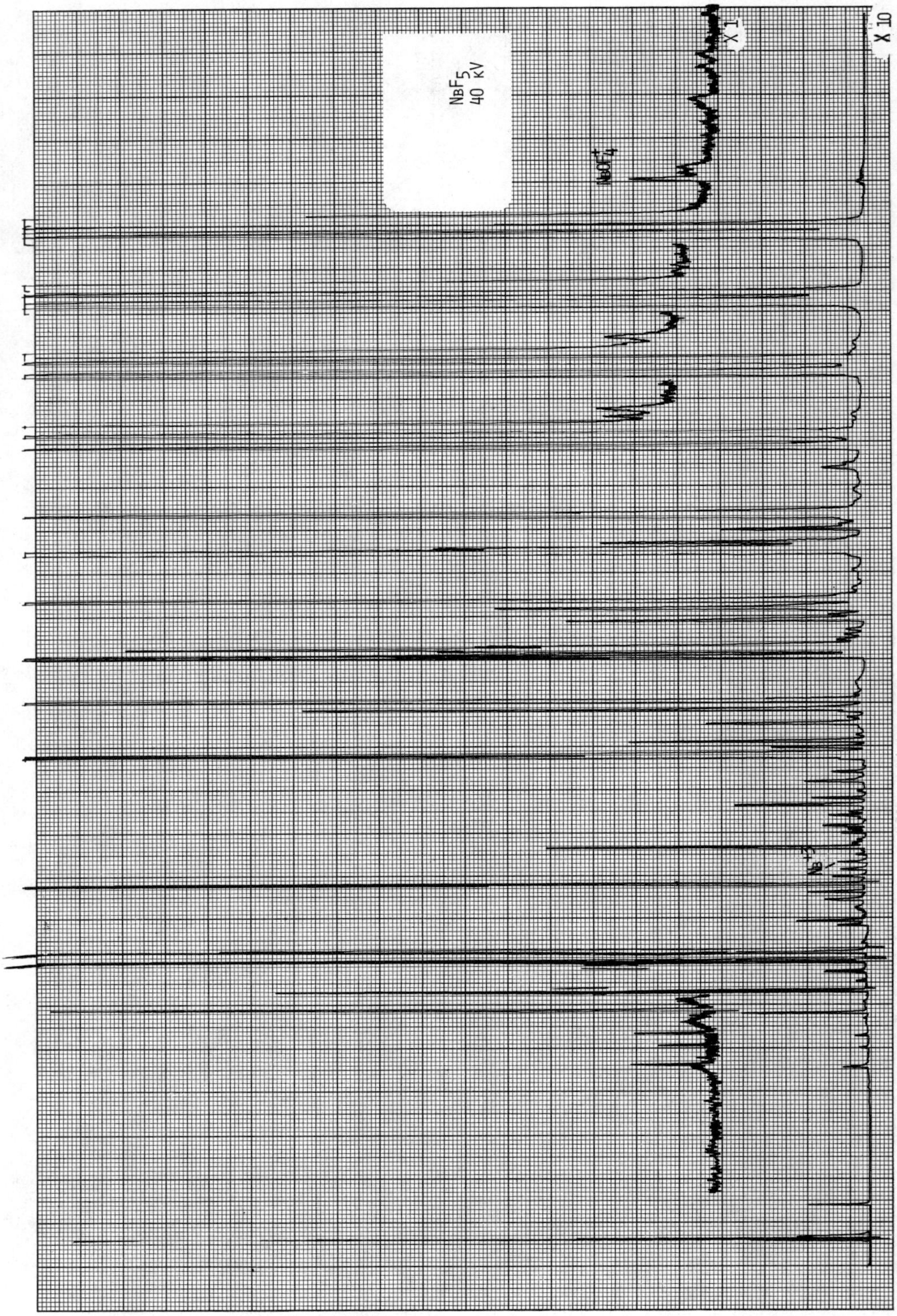

NbF_5
40 kV
X 1
X 10

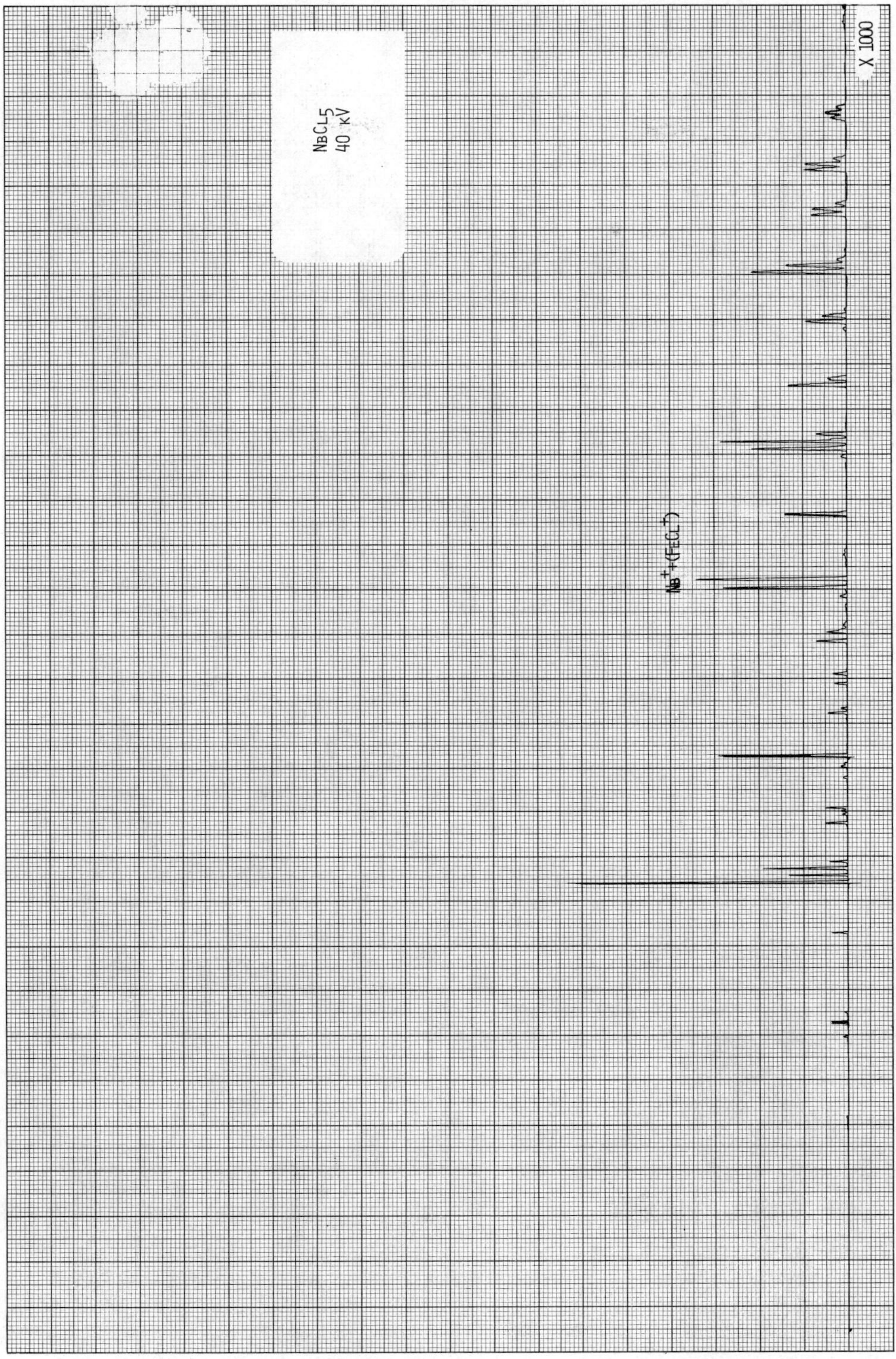
NbCl5
40 kV
X 1000
Nb+ + (FeCl+)

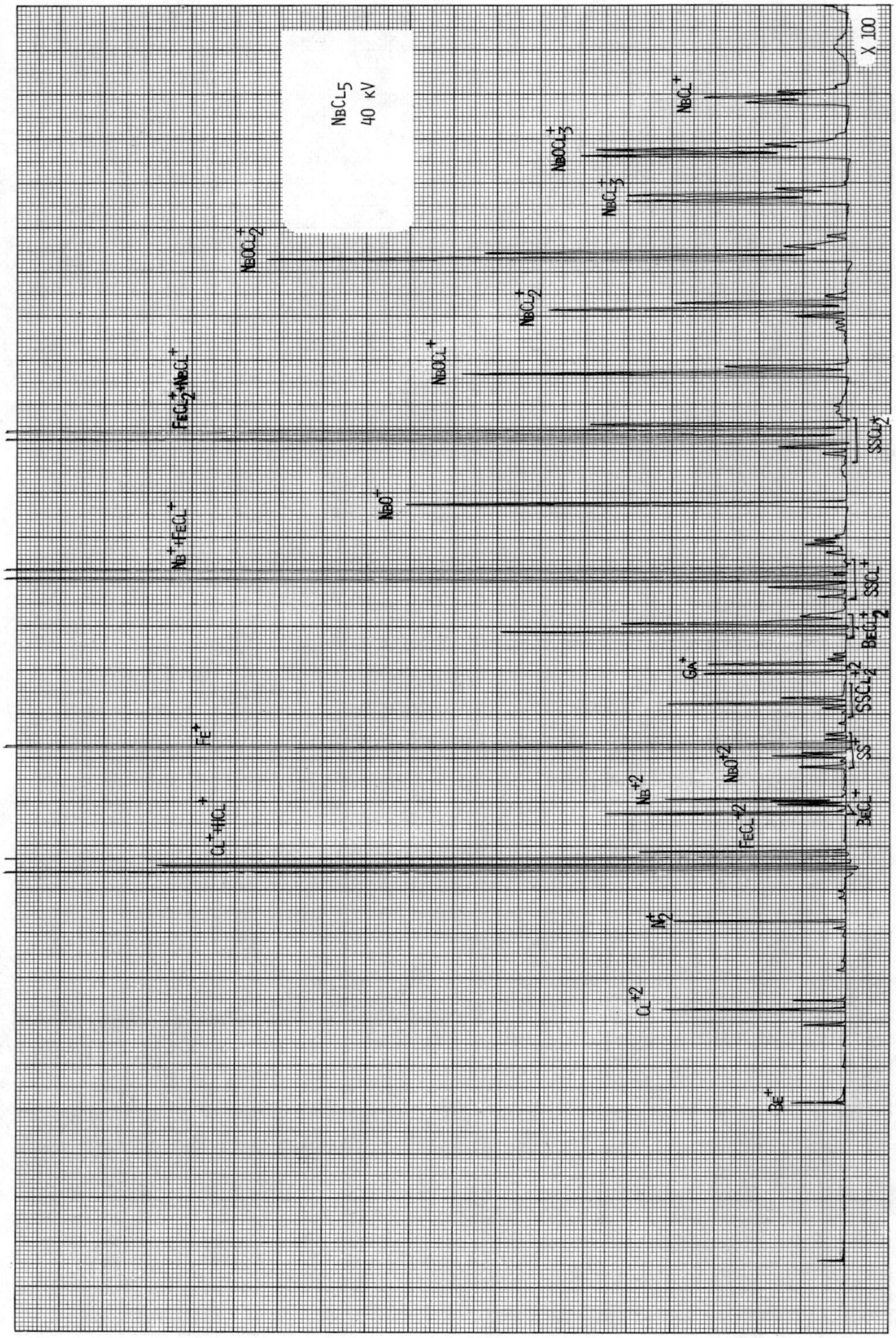

NbCl5
40 kV
X 100
NbCl+
NbOCl3+
NbCl3+
NbOCl2+
NbCl2+
NbOCl+
FeCl2+ + NbCl+
SSCl2+
NbO+
Nb+ + FeCl+
SSCl+
BeCl2+
Ga+
SSCl2+2
Fe+
SS+
NbO+2
Nb+2
BeCl+
Cl+ + HCl+
FeCl+2
N2+
Cl+2
Be+

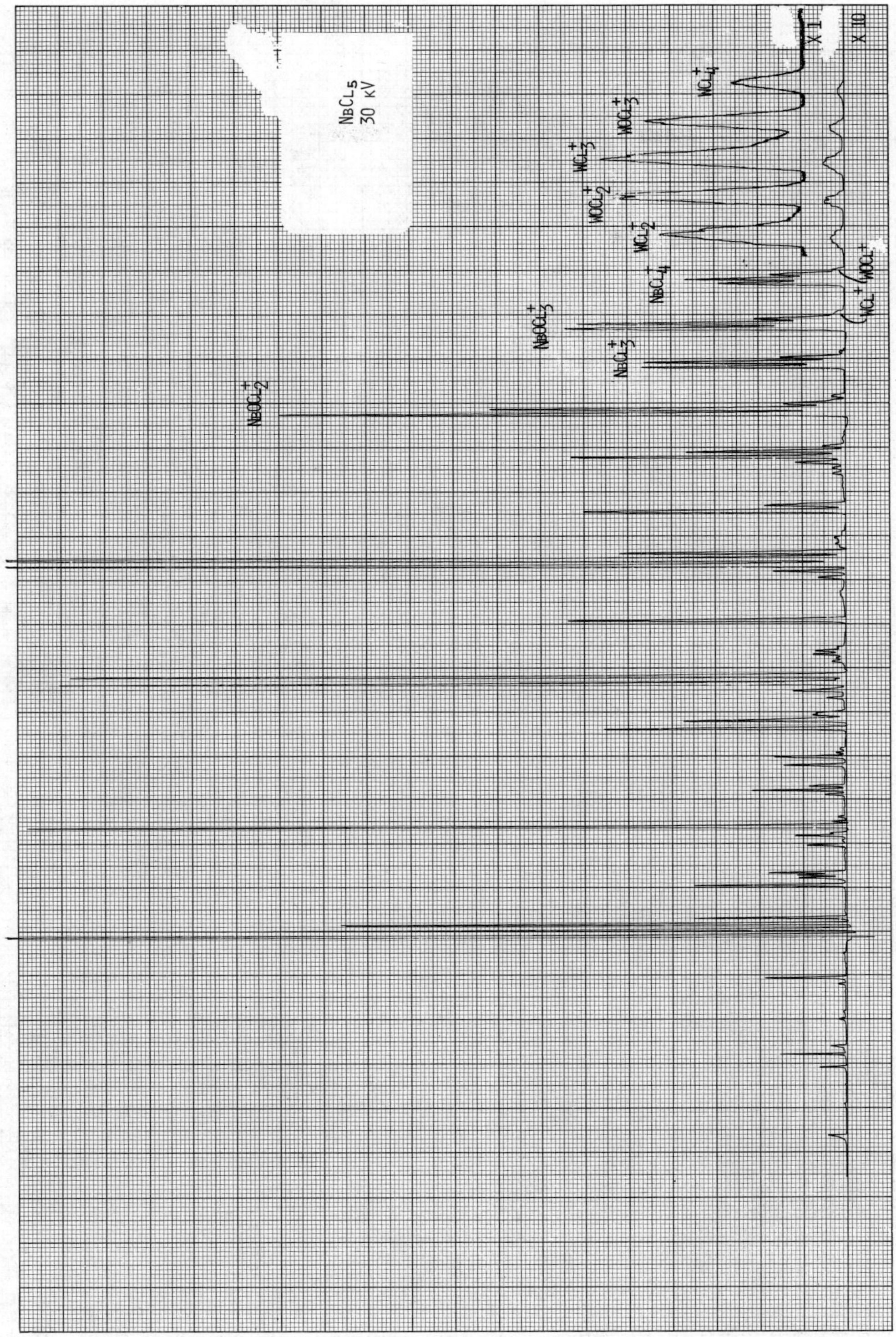

NbCl5
30 kV
NbOCl2+
NbOCl3+
NbCl3+
NbCl4+
WCl2+
WOCl2+
WCl3+
WOCl3+
WCl4+
WCl+/WOCl+
X 1
X 10

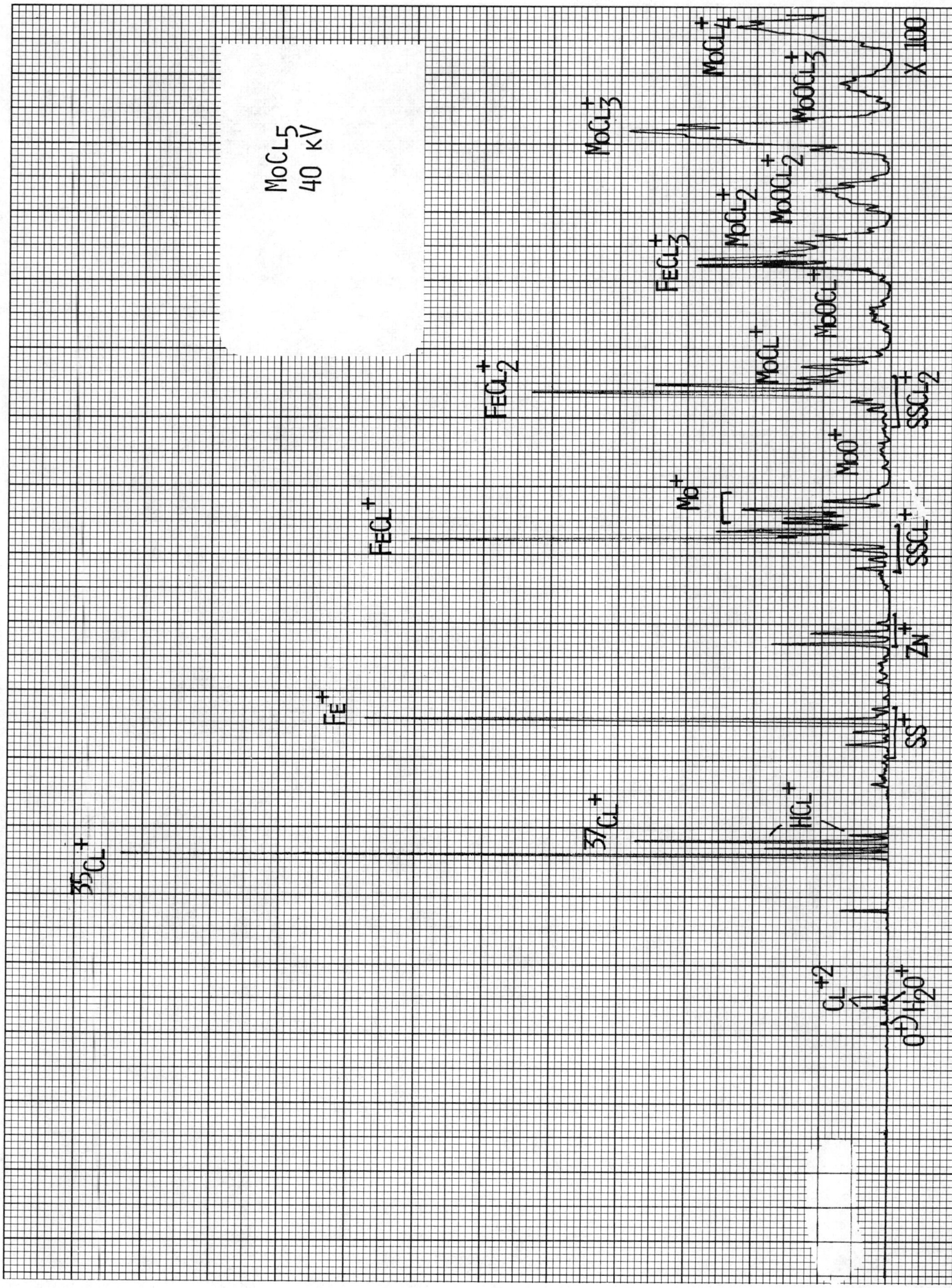
MoCl5
40 kV
X 100
MoCl4+
MoOCl3+
MoCl3+
MoOCl2+
MoCl2+
FeCl3+
MoOCl+
MoCl+
FeCl2+
SSCl2+
MoO+
Mo+
FeCl+
SSCl+
Zn+
Fe+
SS+
37Cl+
HCl+
35Cl+
Cl+2
O+
H2O+

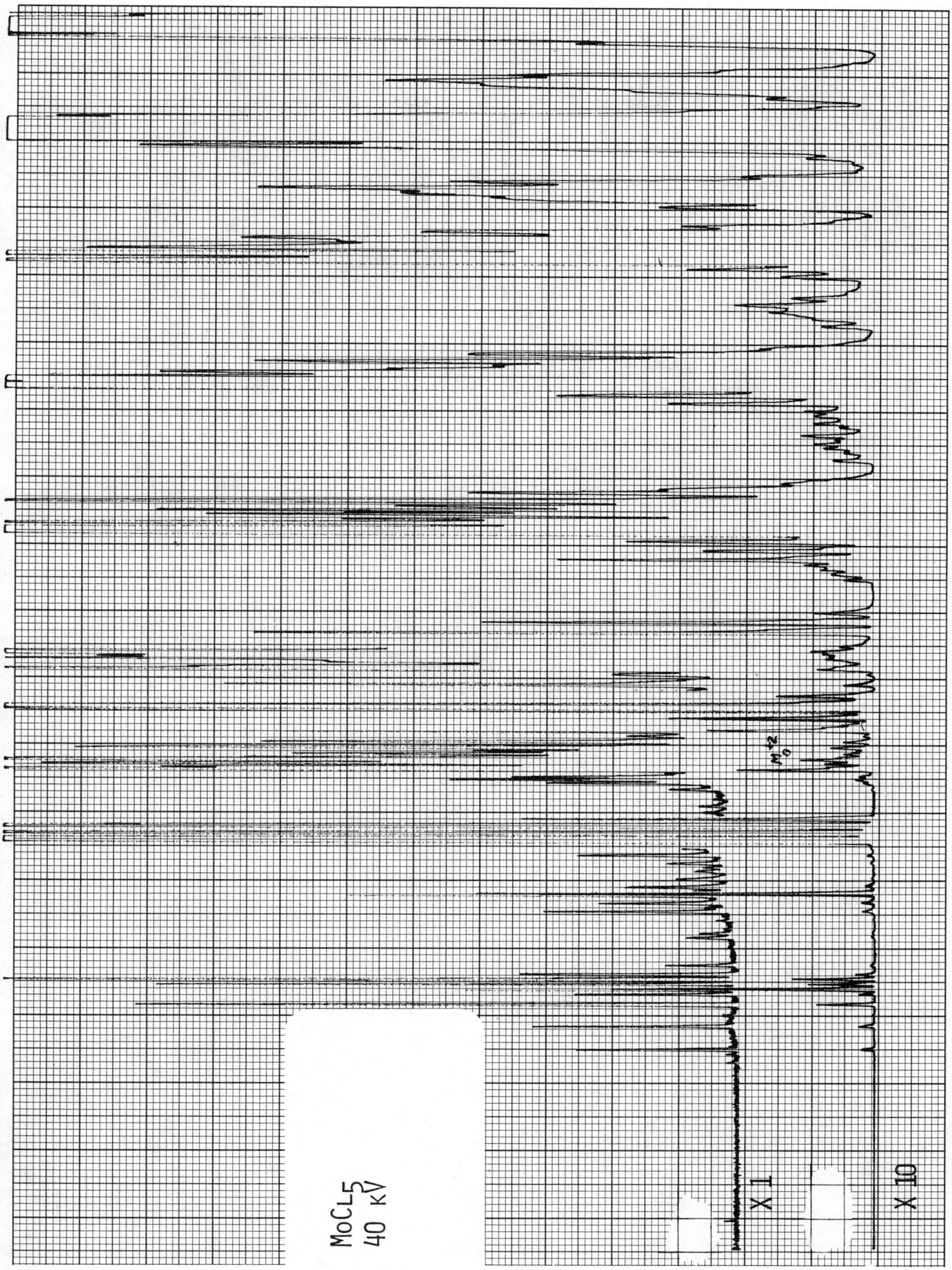
MoCl5
40 kV
X 1
X 10
Mo+2

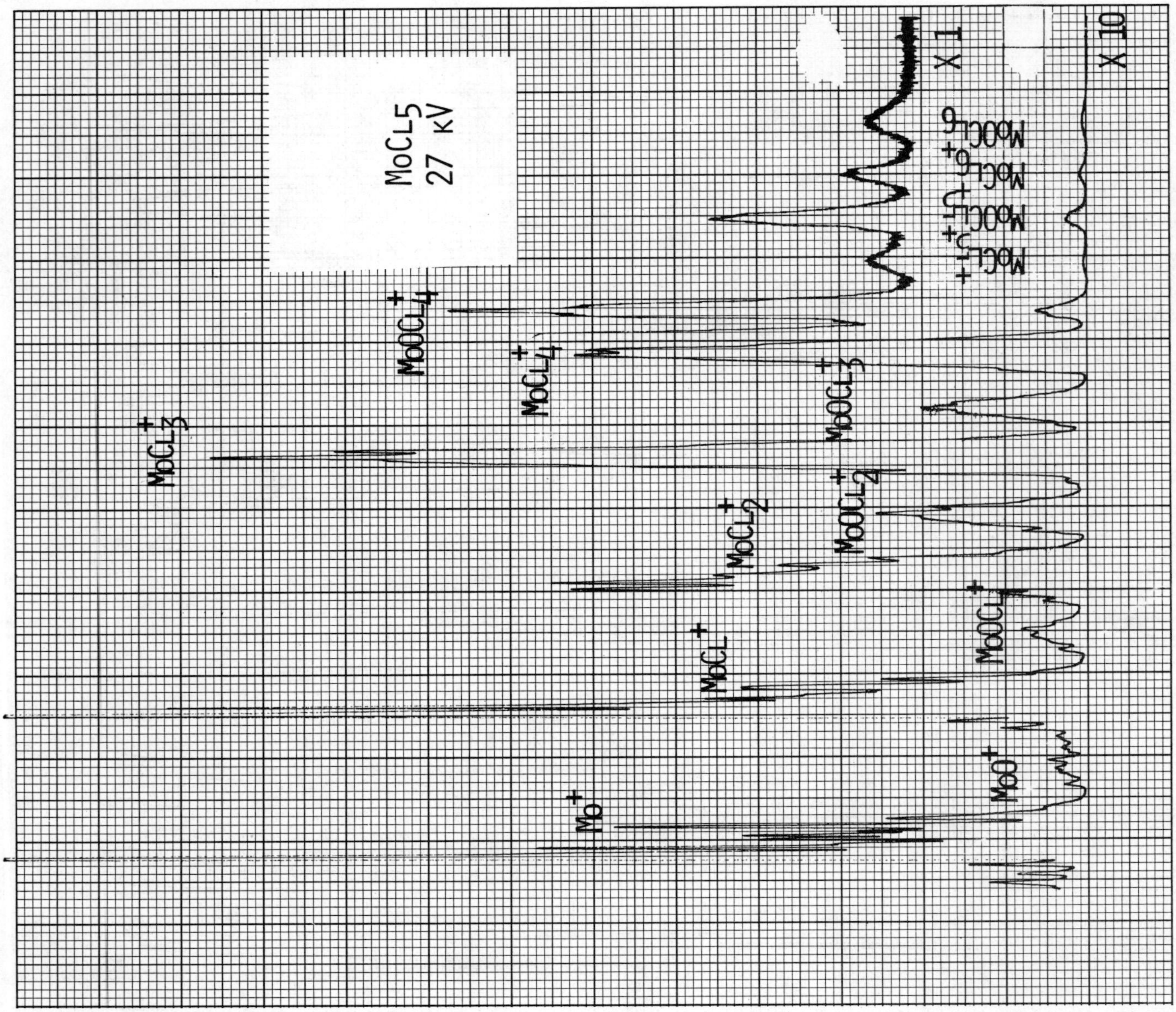
MoCL5
27 kV
MoCL3+
MoOCL4+
MoCL4+
MoOCL3+
MoCL2+
MoOCL2+
MoCL+
MoOCL+
Mo+
MoO+
X 1
X 10

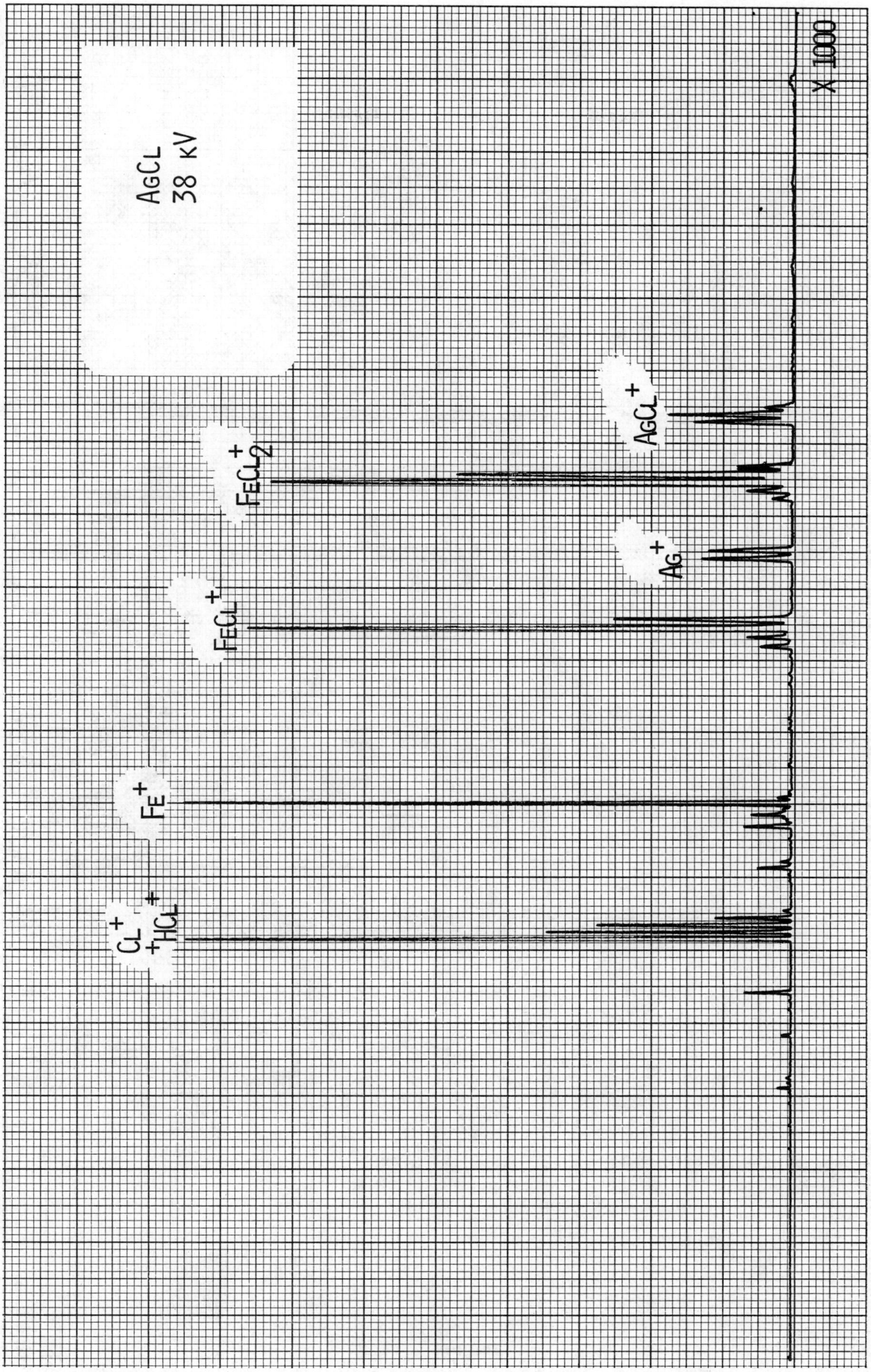

AgCl
38 kV
X 1000
Cl+
HCl+
Fe+
FeCl+
FeCl2+
Ag+
AgCl+

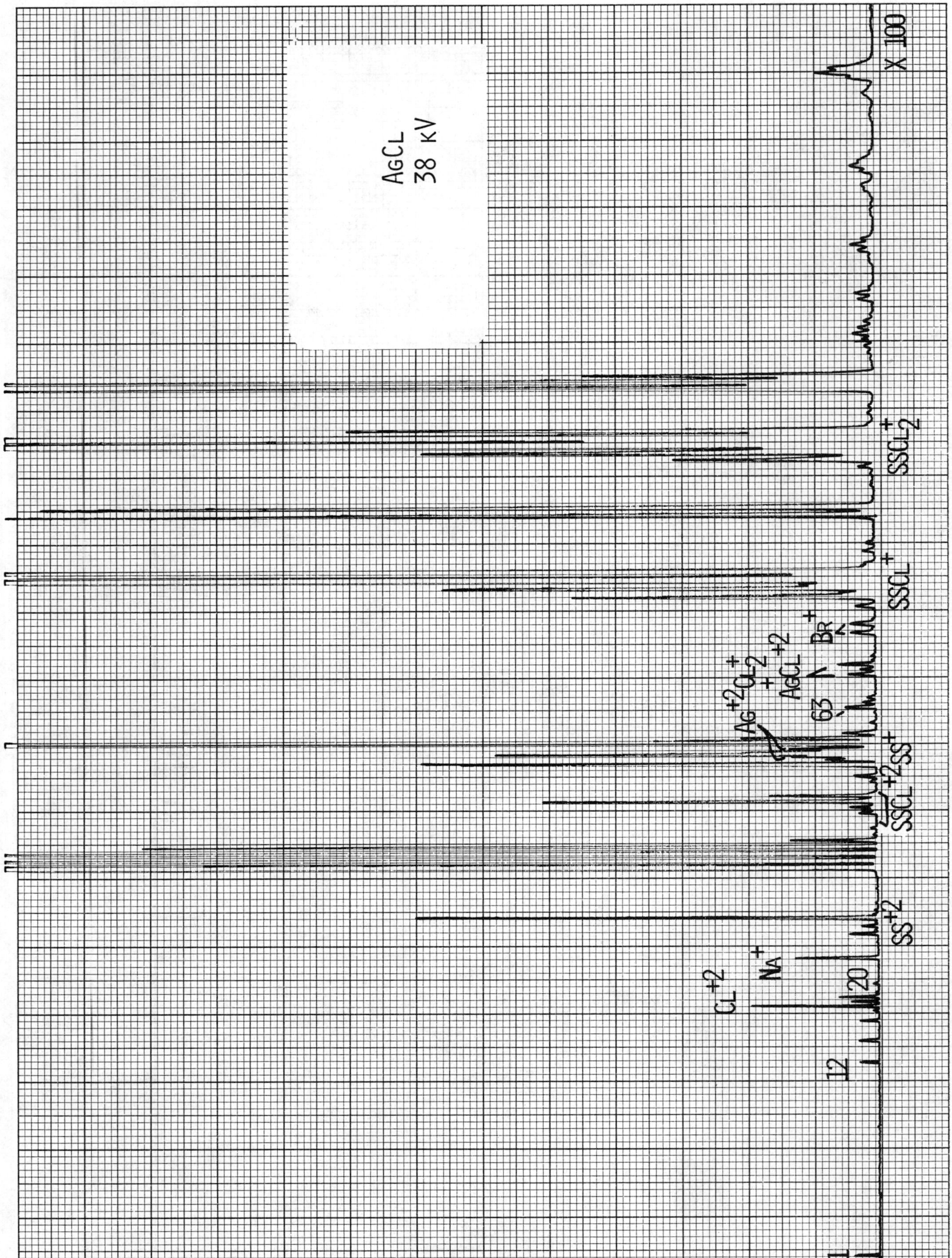
AgCl
38 kV
X 100
SSCl2+
SSCl+
Br+
AgCl+2
Cl2+
Ag+2
63
SS+
SSCl+2
SS+2
Na+
Cl+2
20
12
1

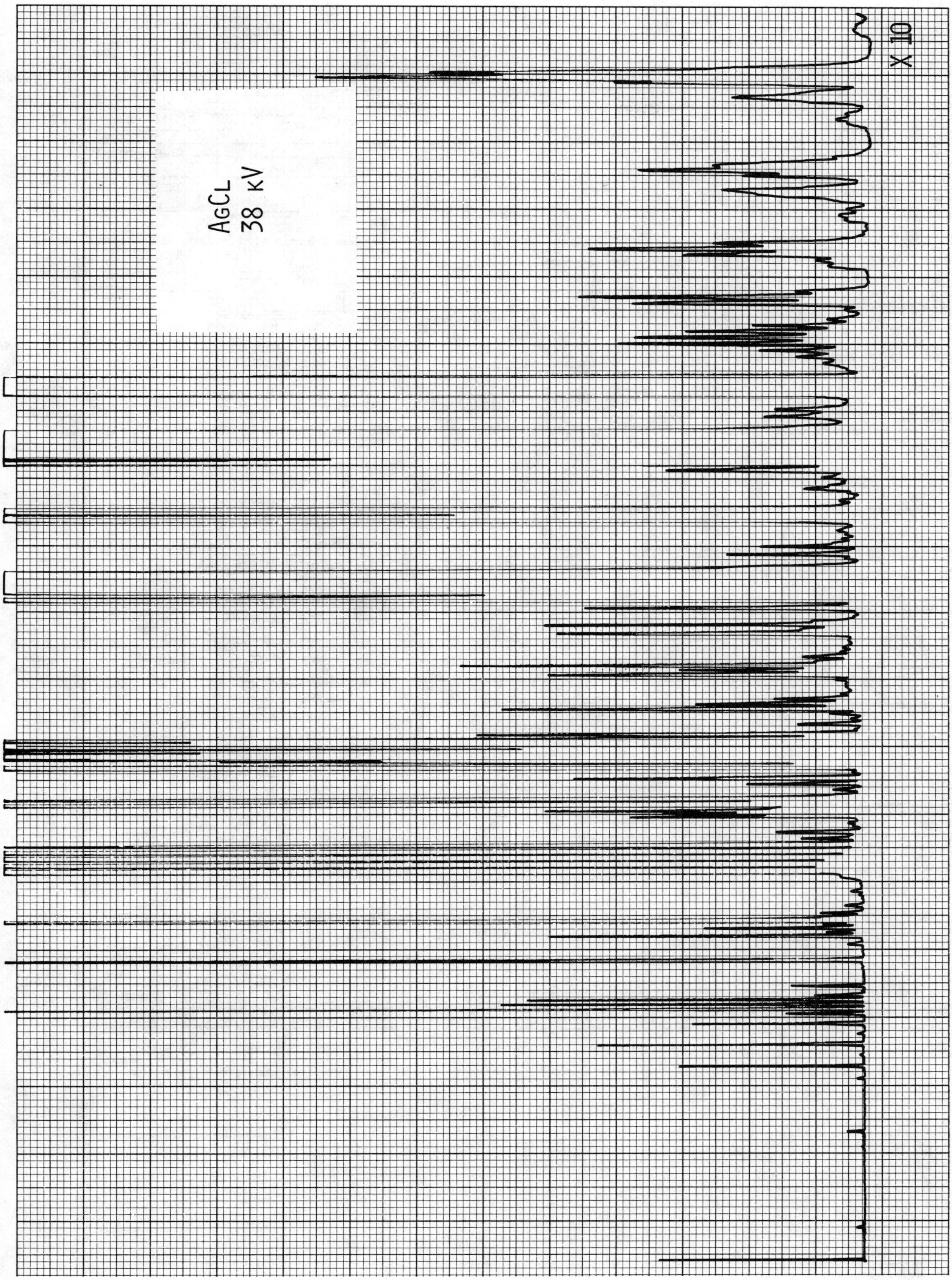
AgCl
38 kV
X 10

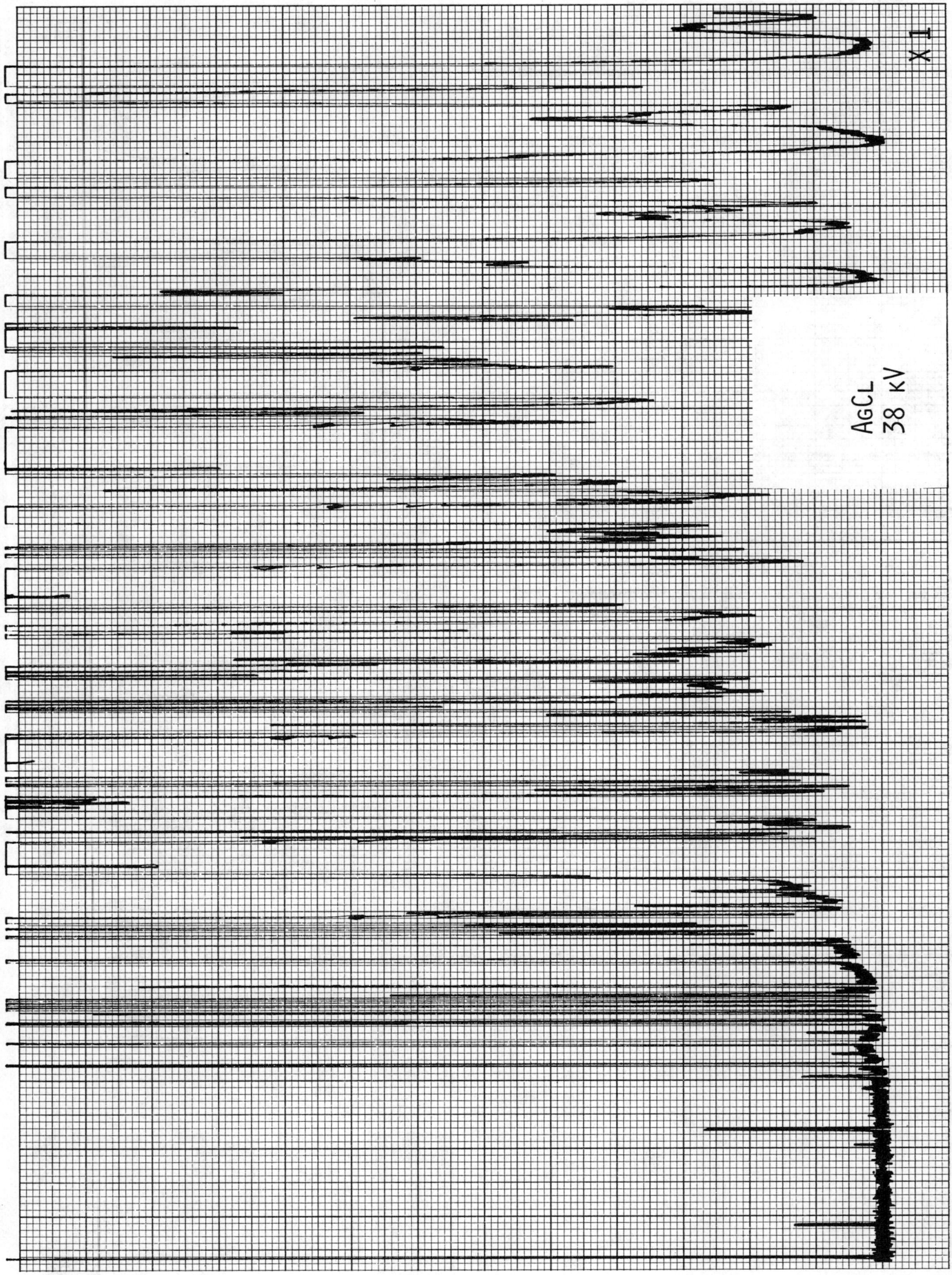
X 1
AgCl
38 kV

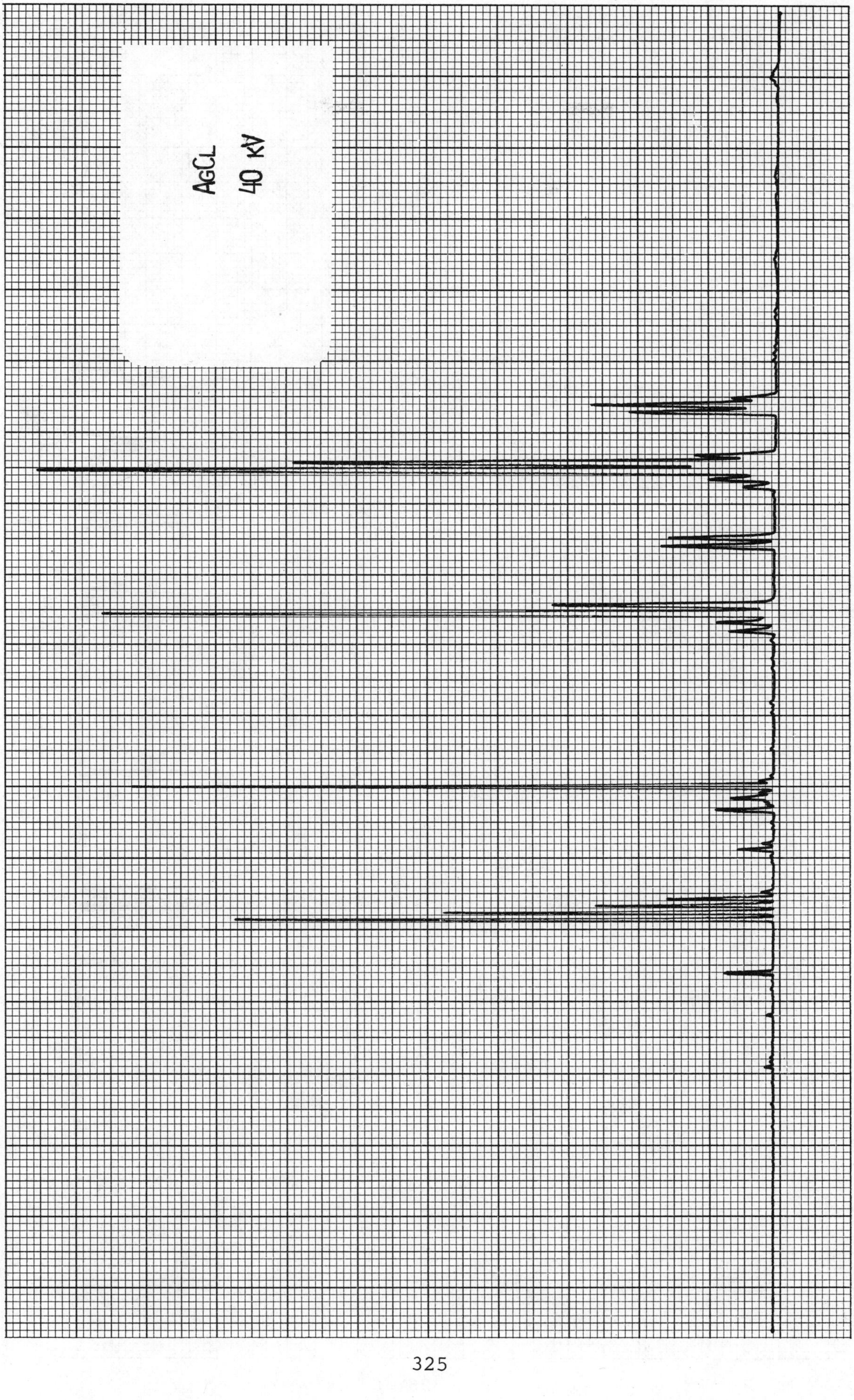
AgCl
40 kV

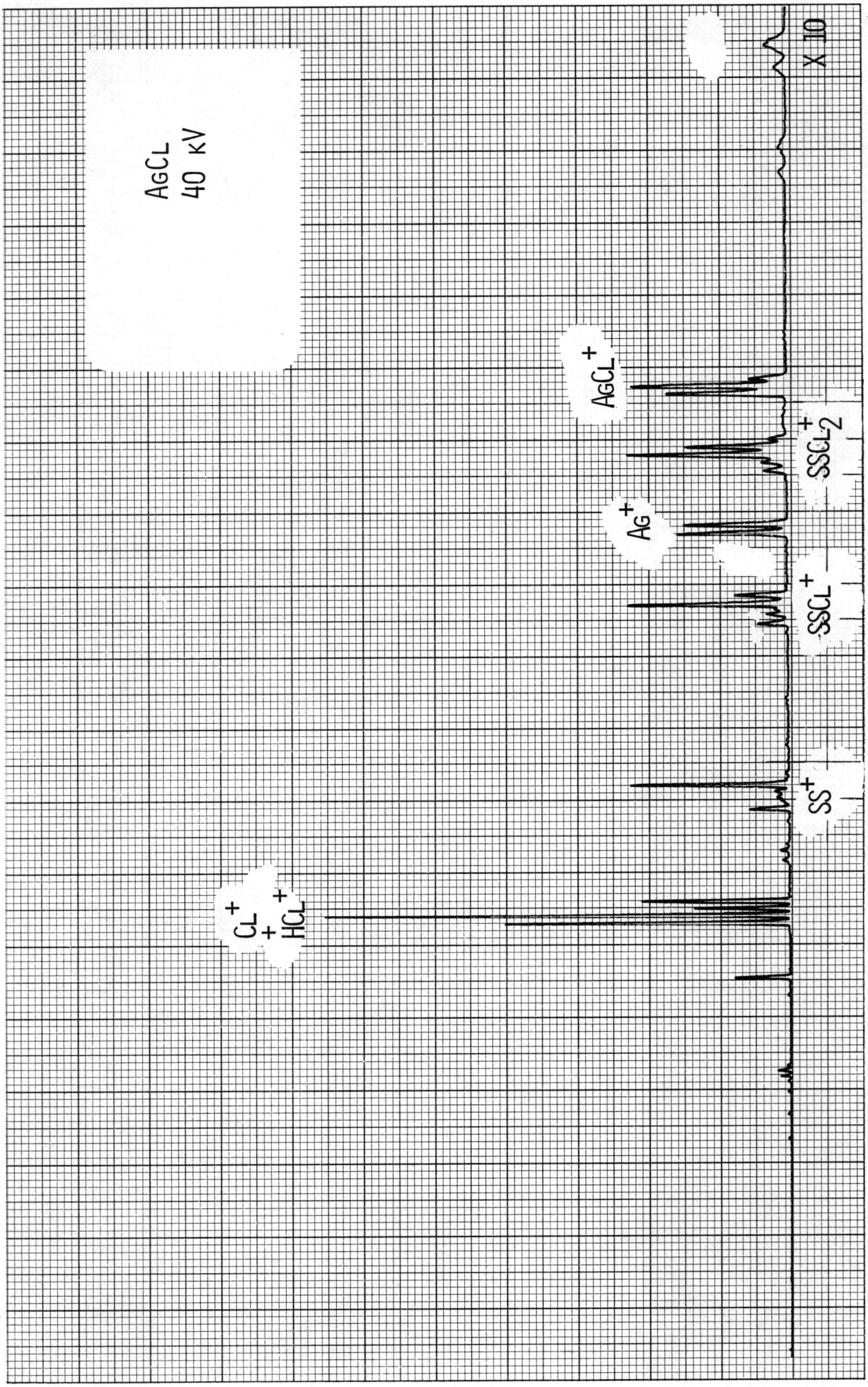
AgCl
40 kV
X 10
AgCl+
SSCl2+
Ag+
SSCl+
SS+
Cl+
HCl+

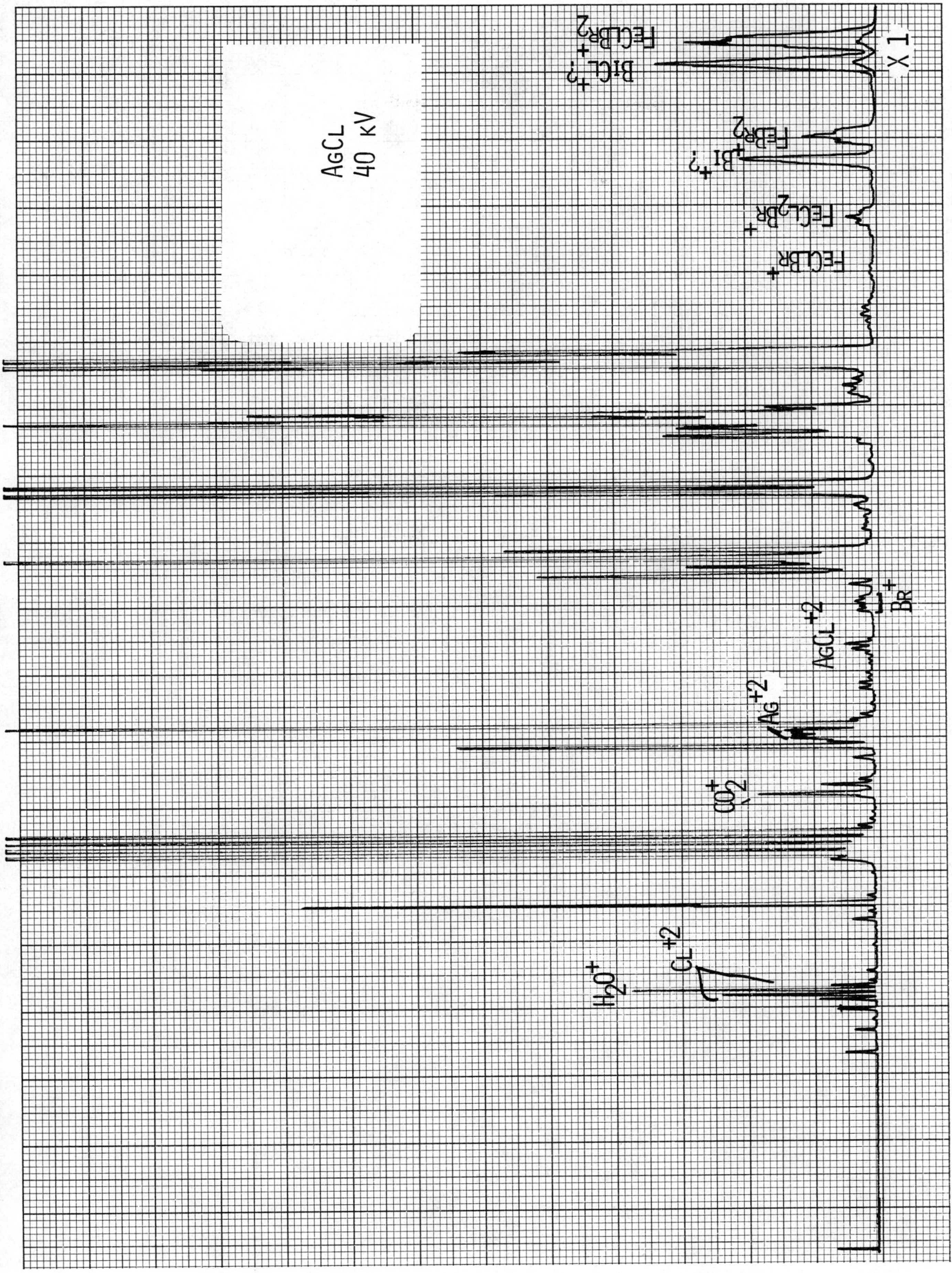
AgCl
40 kV
FeClBr2+
BiCl+?
FeBr2+
Bi+?
FeCl2Br+
FeClBr+
AgCl+2
Br+
Ag+2
CO2+
H2O+
Cl+2
X 1

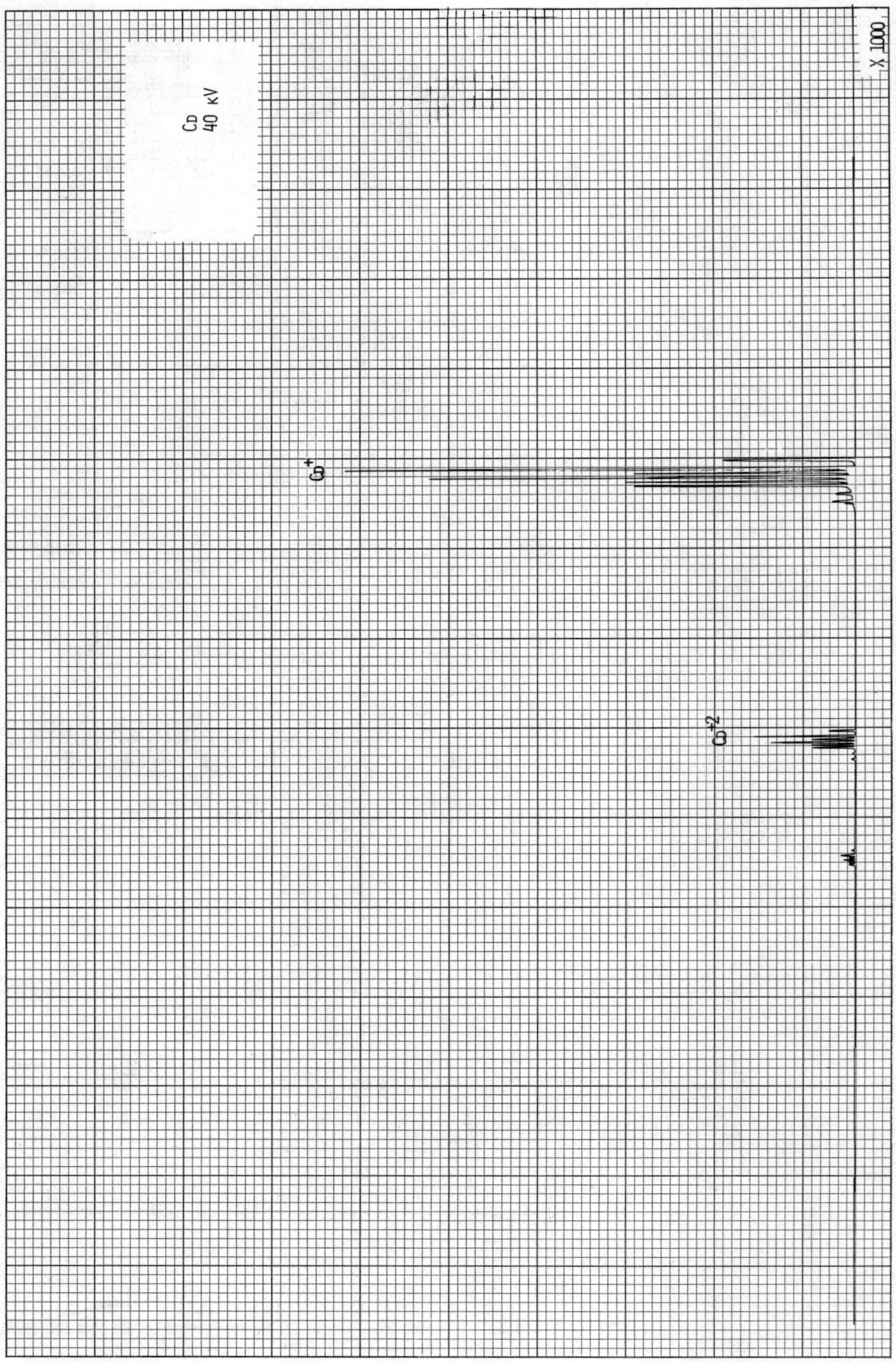
Cd
40 kV
X 1000
Cd+
Cd+2

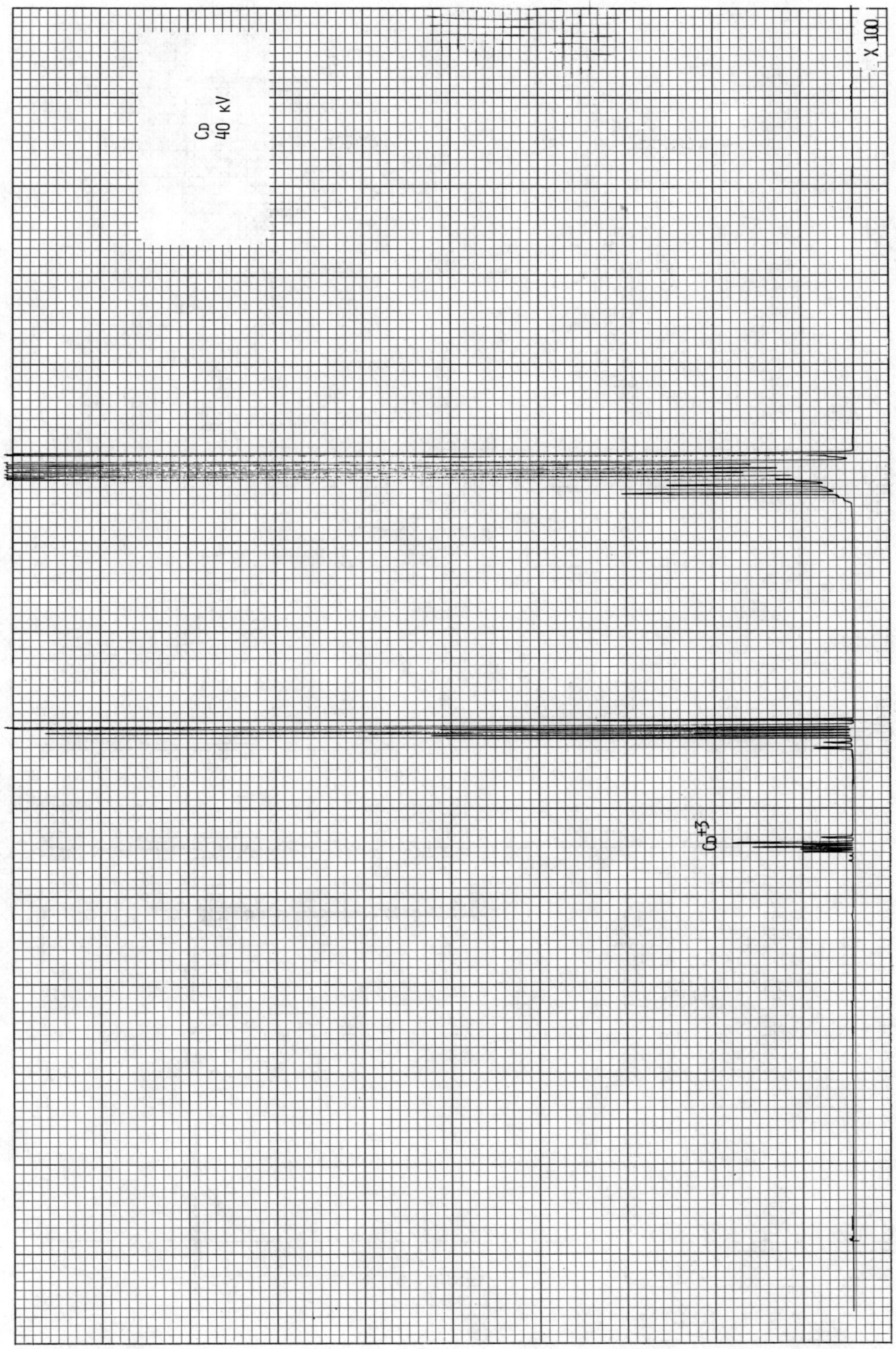
CD
40 kV
X 100
Cd+3

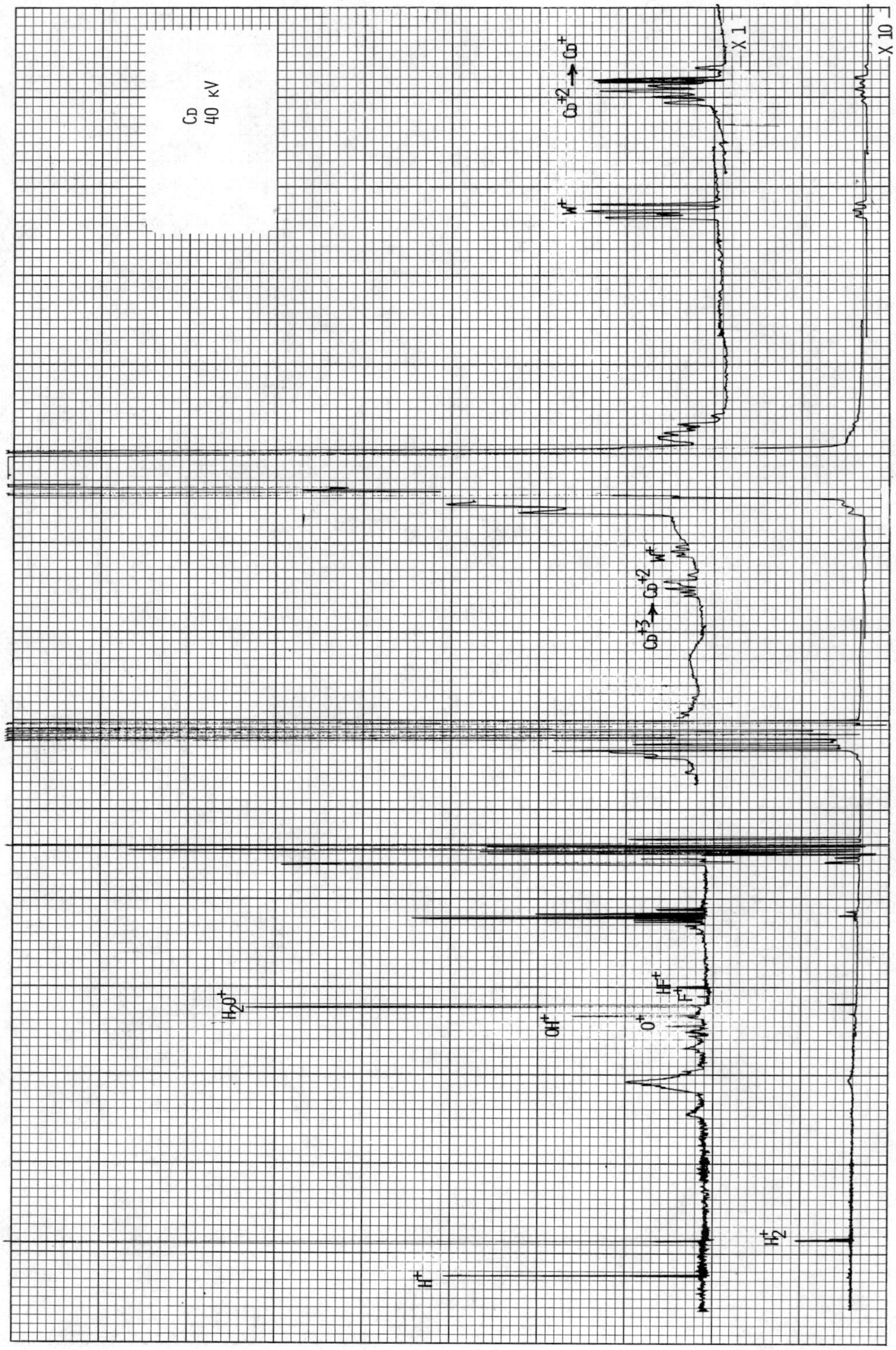
Cd
40 kV
W^+
$Cd^{+2} \rightarrow Cd^+$
X 1
X 10
$Cd^{+3} \rightarrow Cd^{+2}$
W^+
H_2O^+
OH^+
O^+
HF^+
F^+
H^+
H_2^+

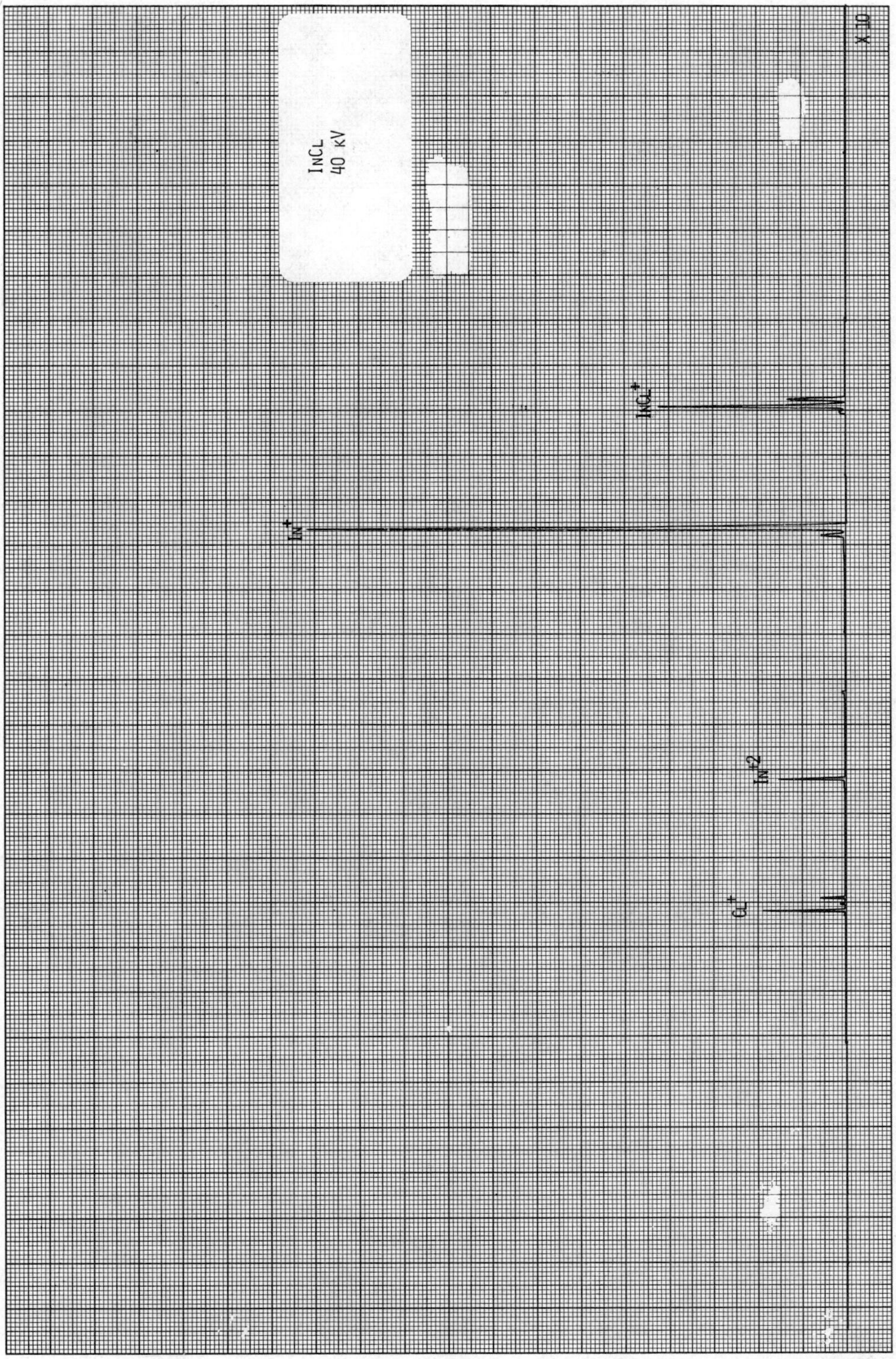
InCl
40 kV
In+
InCl+
In+2
Cl+
x 10

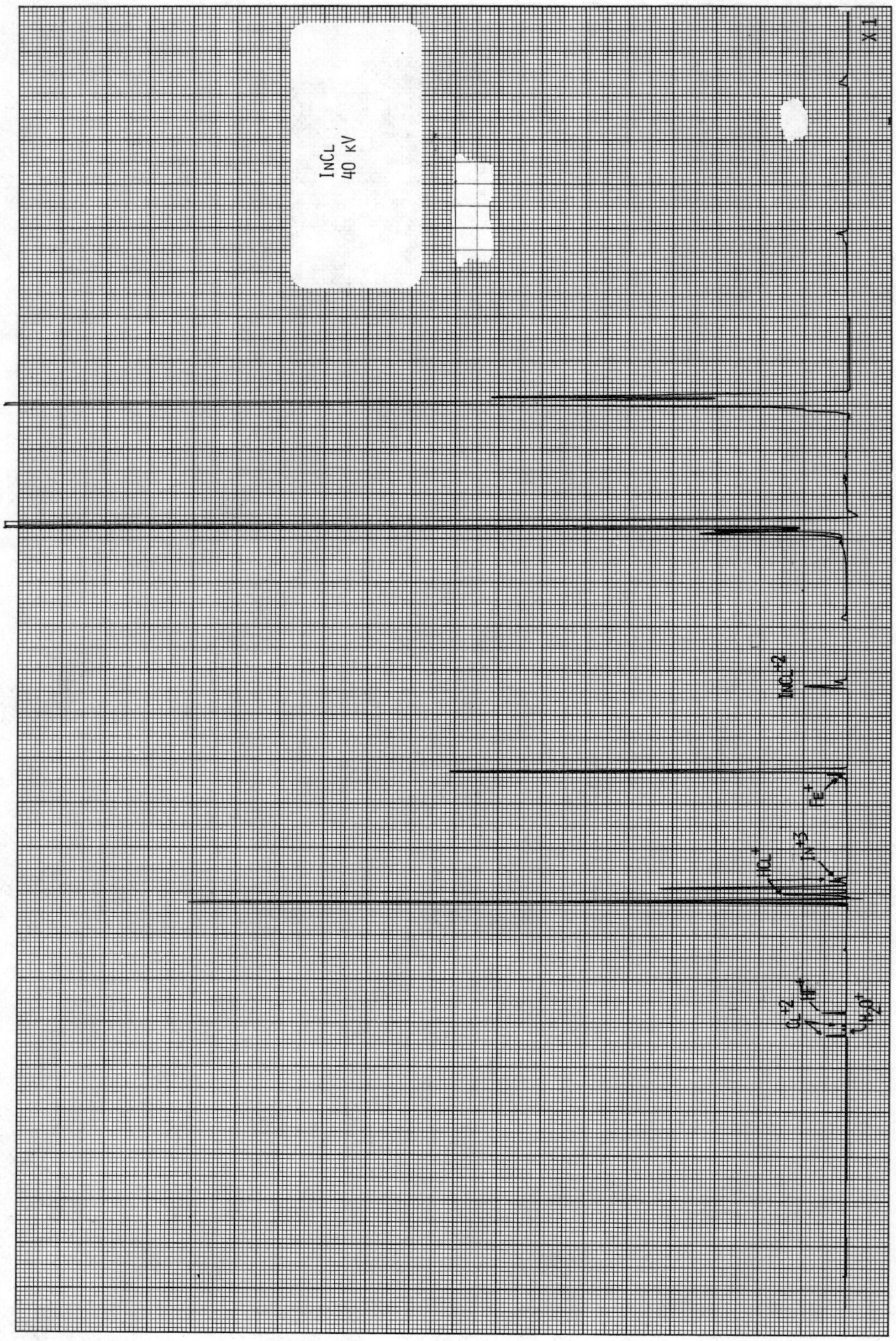

InCl
40 kV
X 1

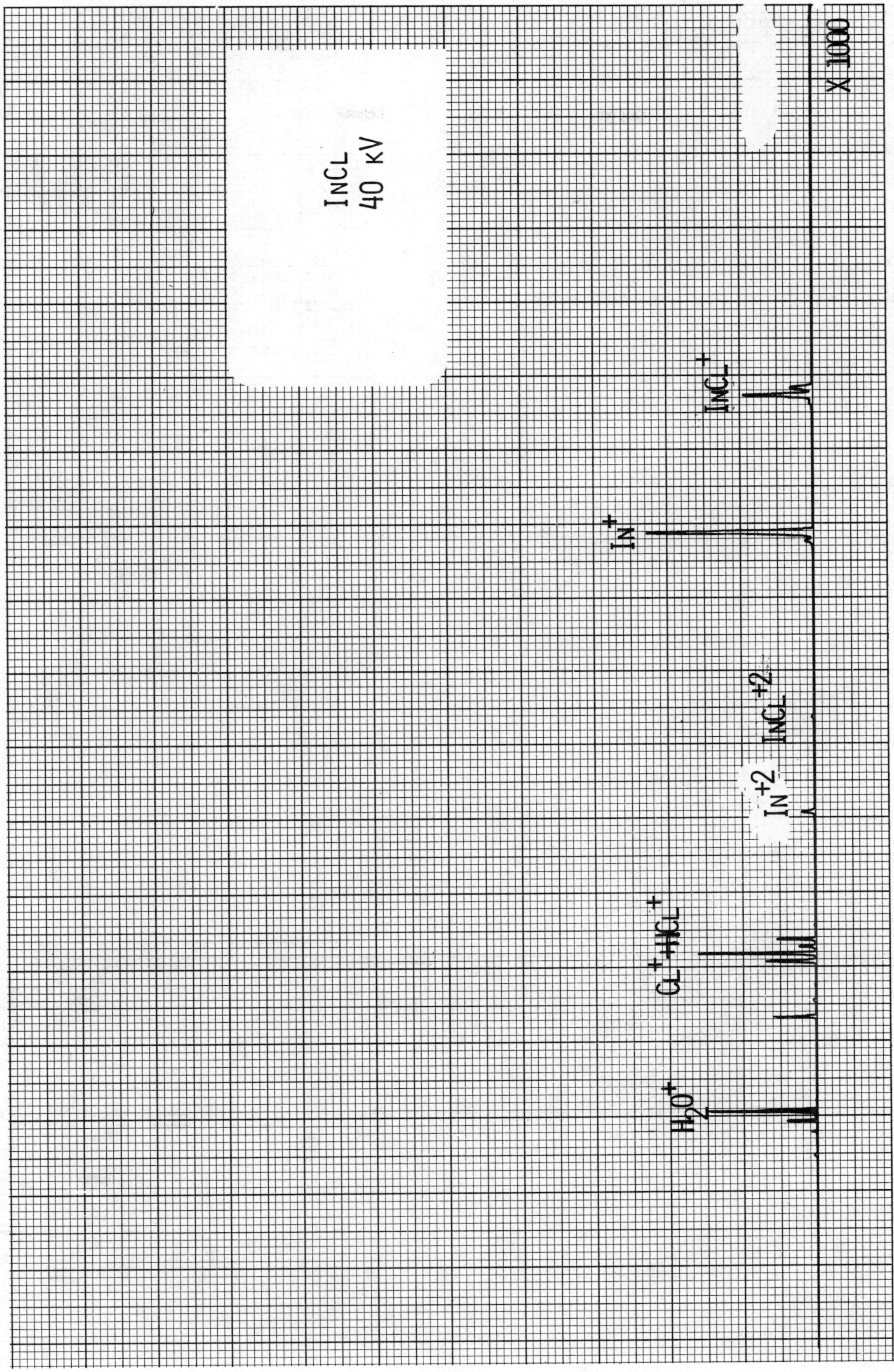
INCL
40 KV
X 1000
INCL+
IN+
INCL+2
IN+2
CL+ +HCL+
H2O+

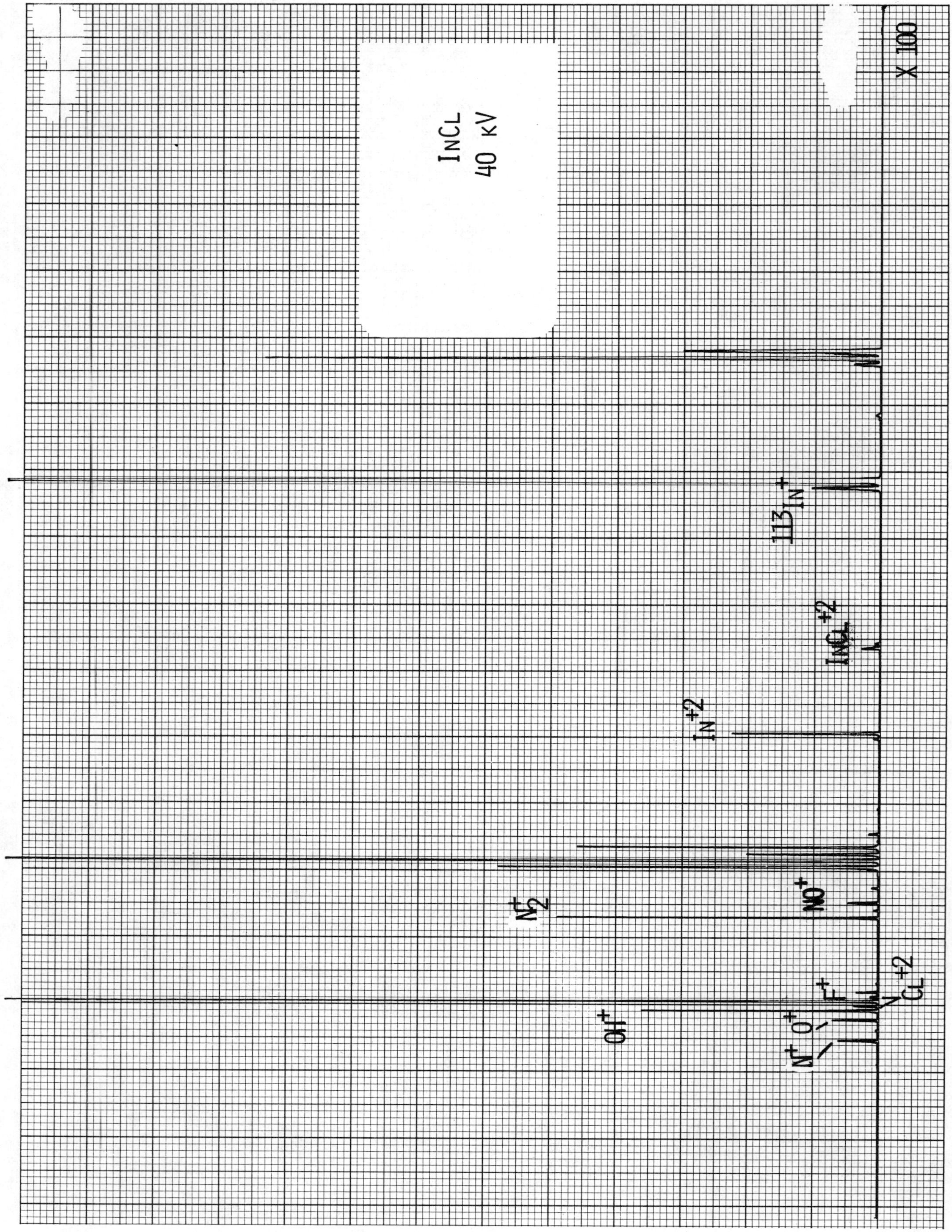
InCl
40 kV
X 100
113In+
InCl+2
In+2
N2+
NO+
OH+
N+
O+
F+
Cl+2

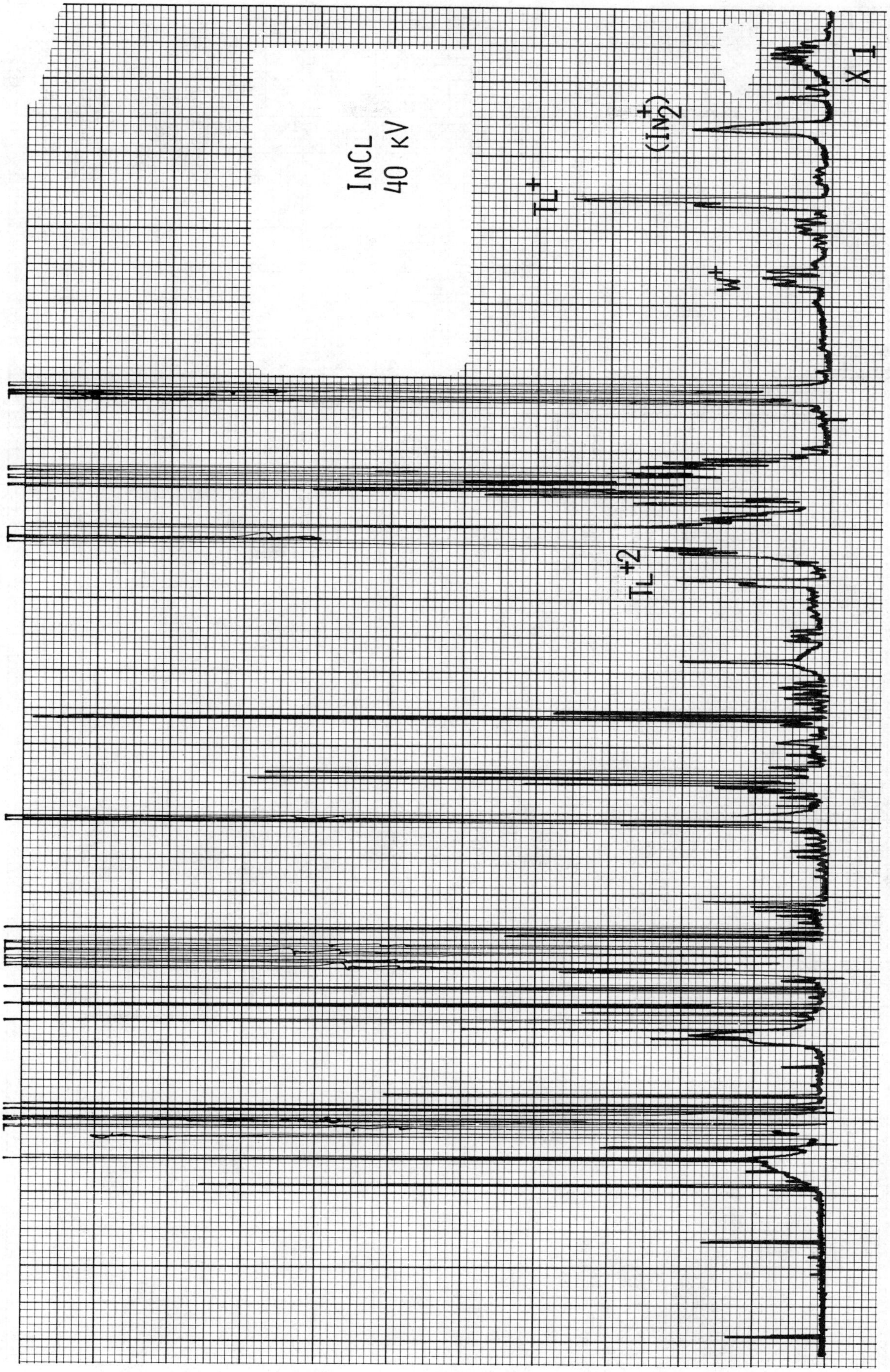
InCl
40 kV
Tl+
(In2+)
W+
Tl+2
X 1

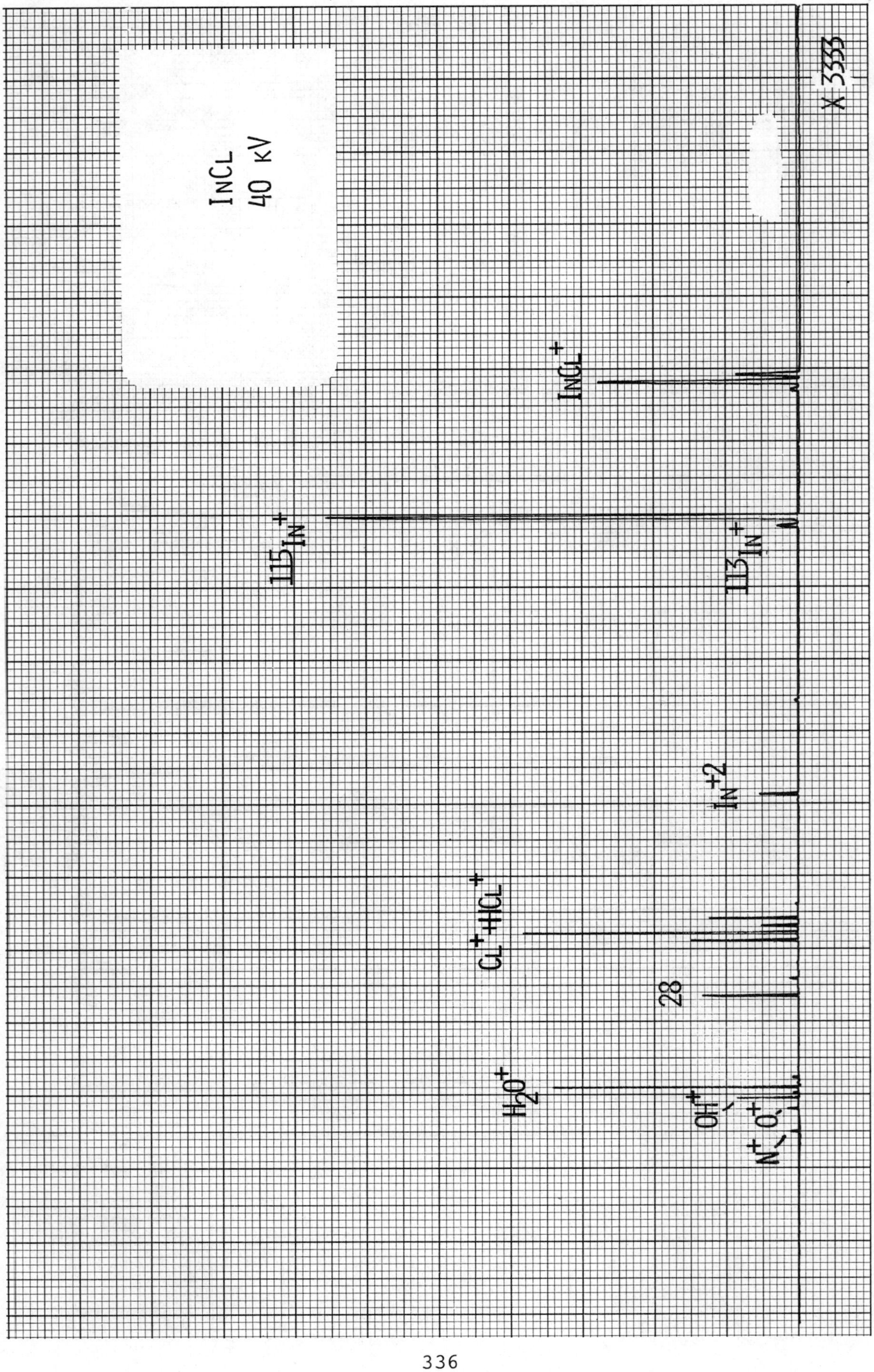
InCl
40 kV
X 3333
InCl⁺
115In⁺
113In⁺
In⁺2
Cl⁺+HCl⁺
28
H2O⁺
OH⁺
N⁺
O⁺

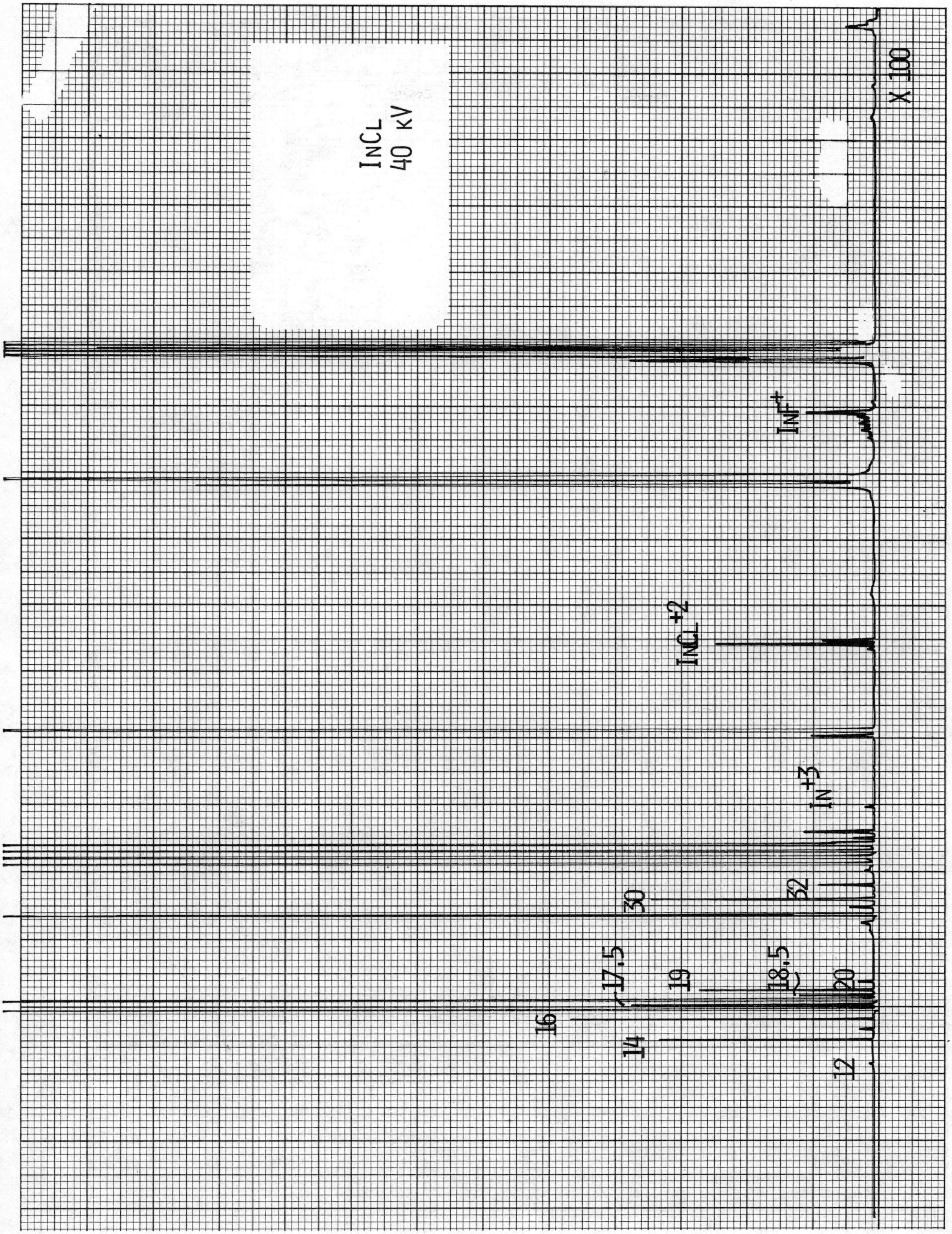

InCl
40 kV
X 100
InF+
InCl+2
In+3
30
32
17.5
19
18.5
20
16
14
12

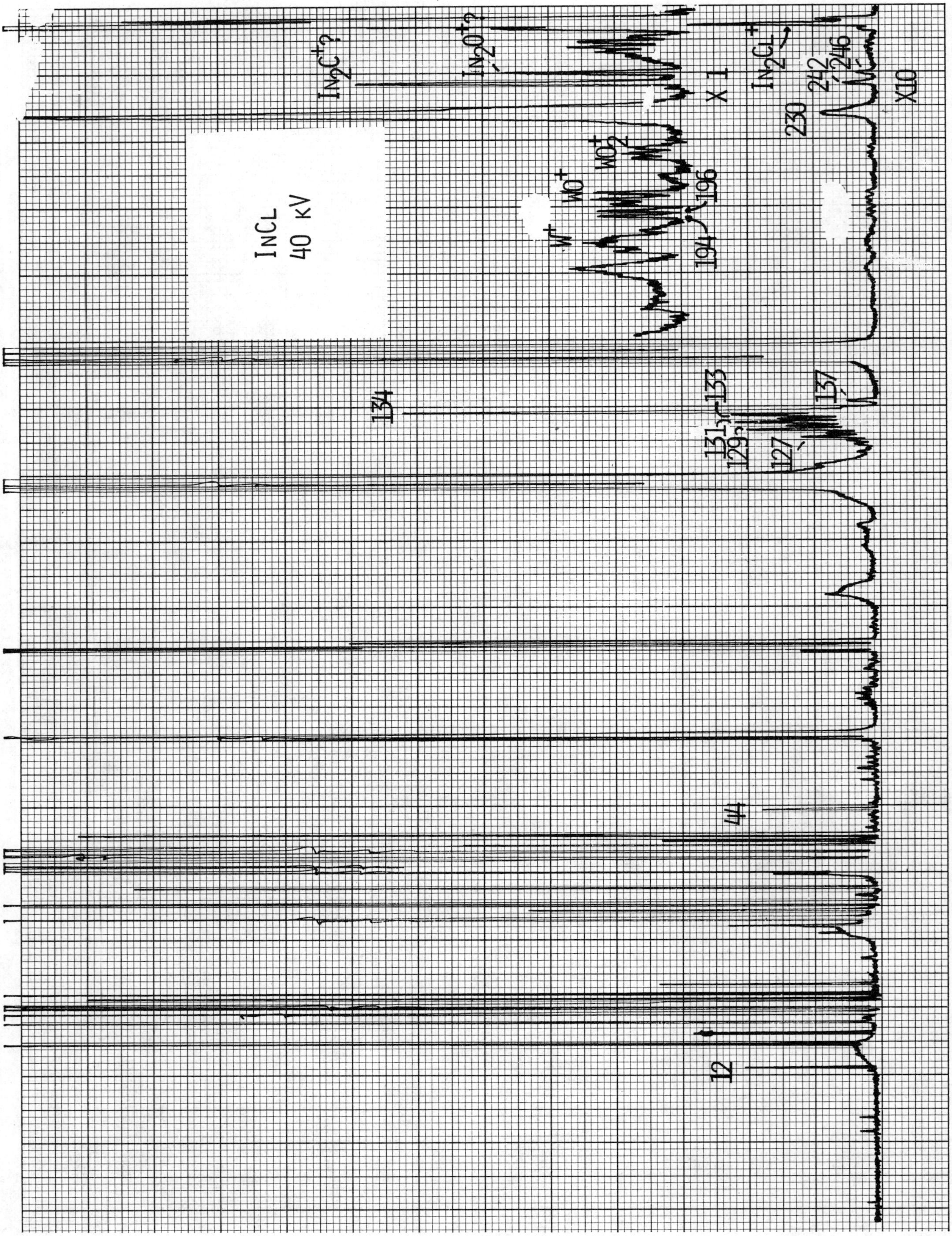
InCl
40 kV
In2C+?
In2O+?
X 1
In2Cl+
242
246
230
X10
W+
WO+
WO2+
194
196
134
131
133
129
127
137
44
12

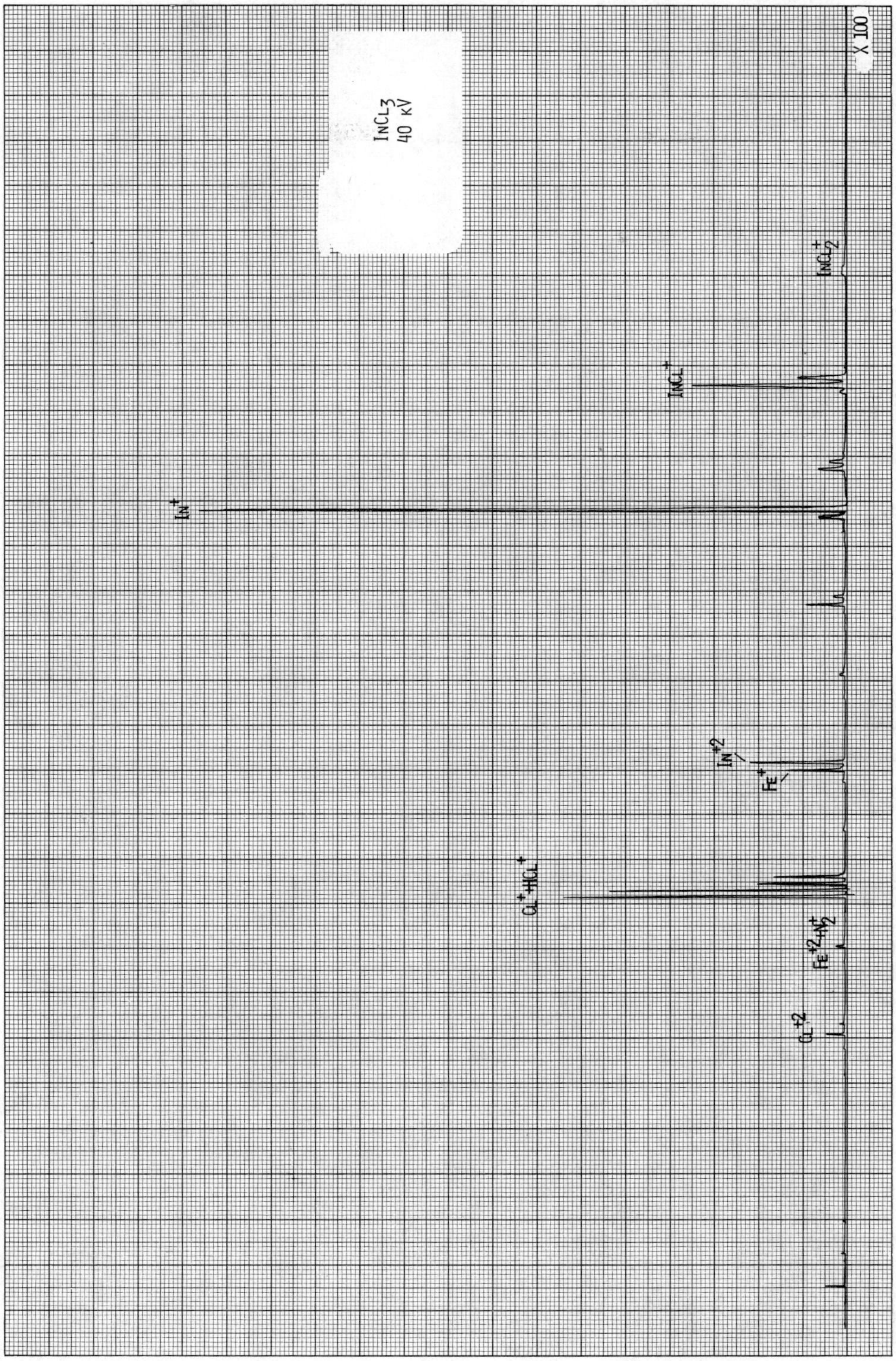

InCl3
40 kV
X 100
InCl2+
InCl+
In+
In+2
Fe+
Cl+ +HCl+
Fe+2 + N2+
Cl+2

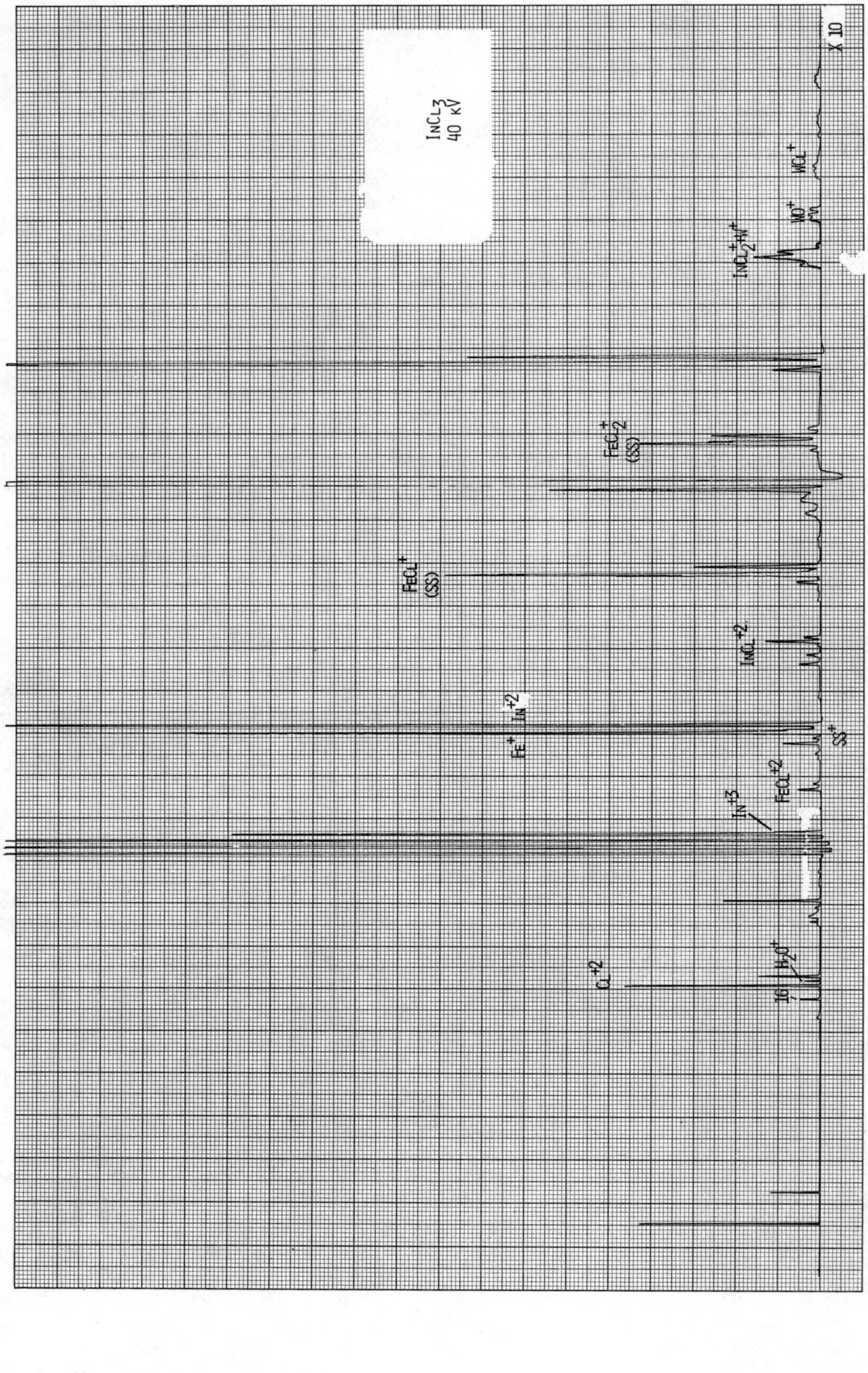
InCl3
40 kV
X 10
WCl+
WO+
InO2+ W+
FeCl2+
(SS)
FeCl+
(SS)
InCl+2
Fe+ In+2
SS+
In+3
FeCl+2
Cl+2
H2O+
16

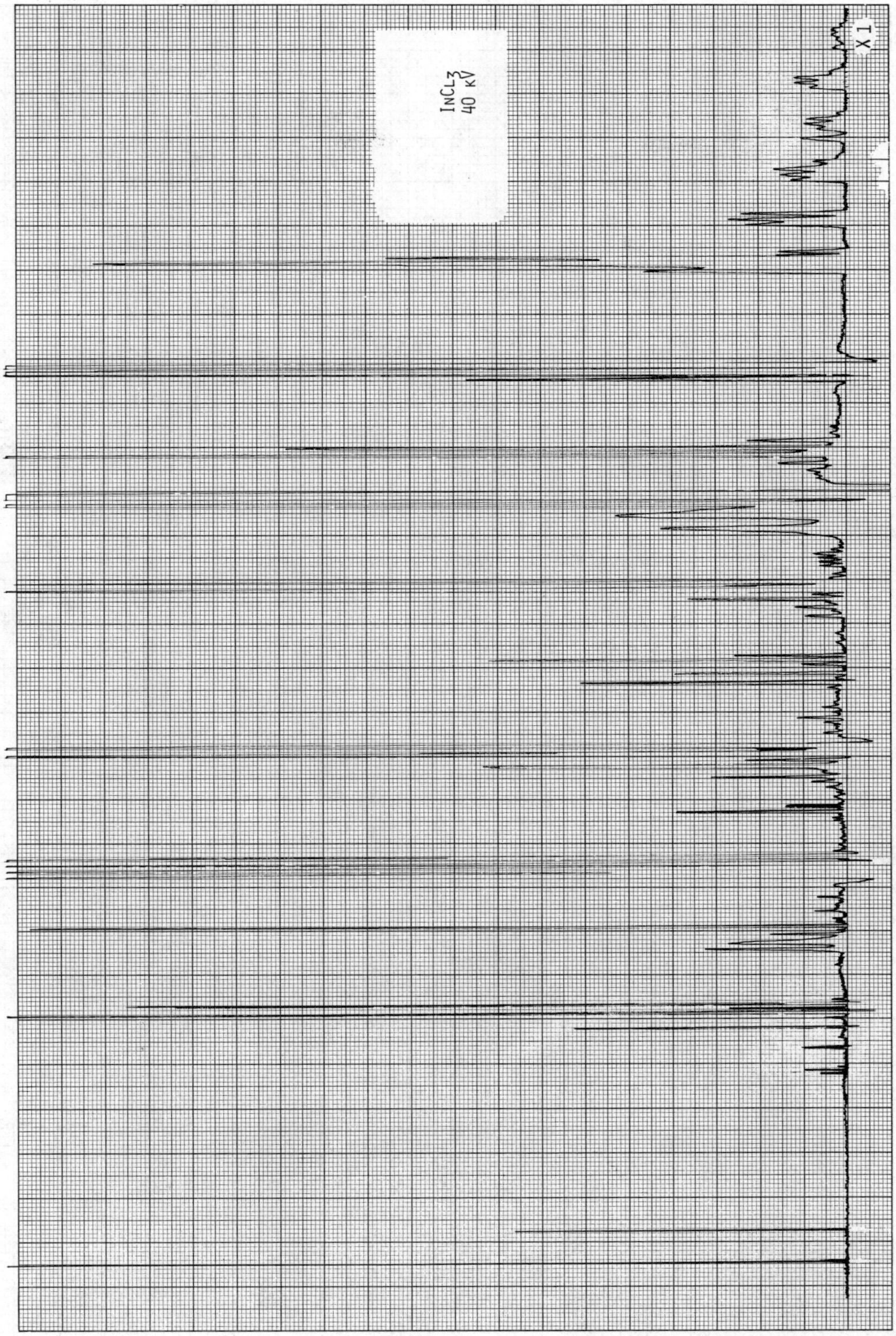
InCl3
40 kV
X 1

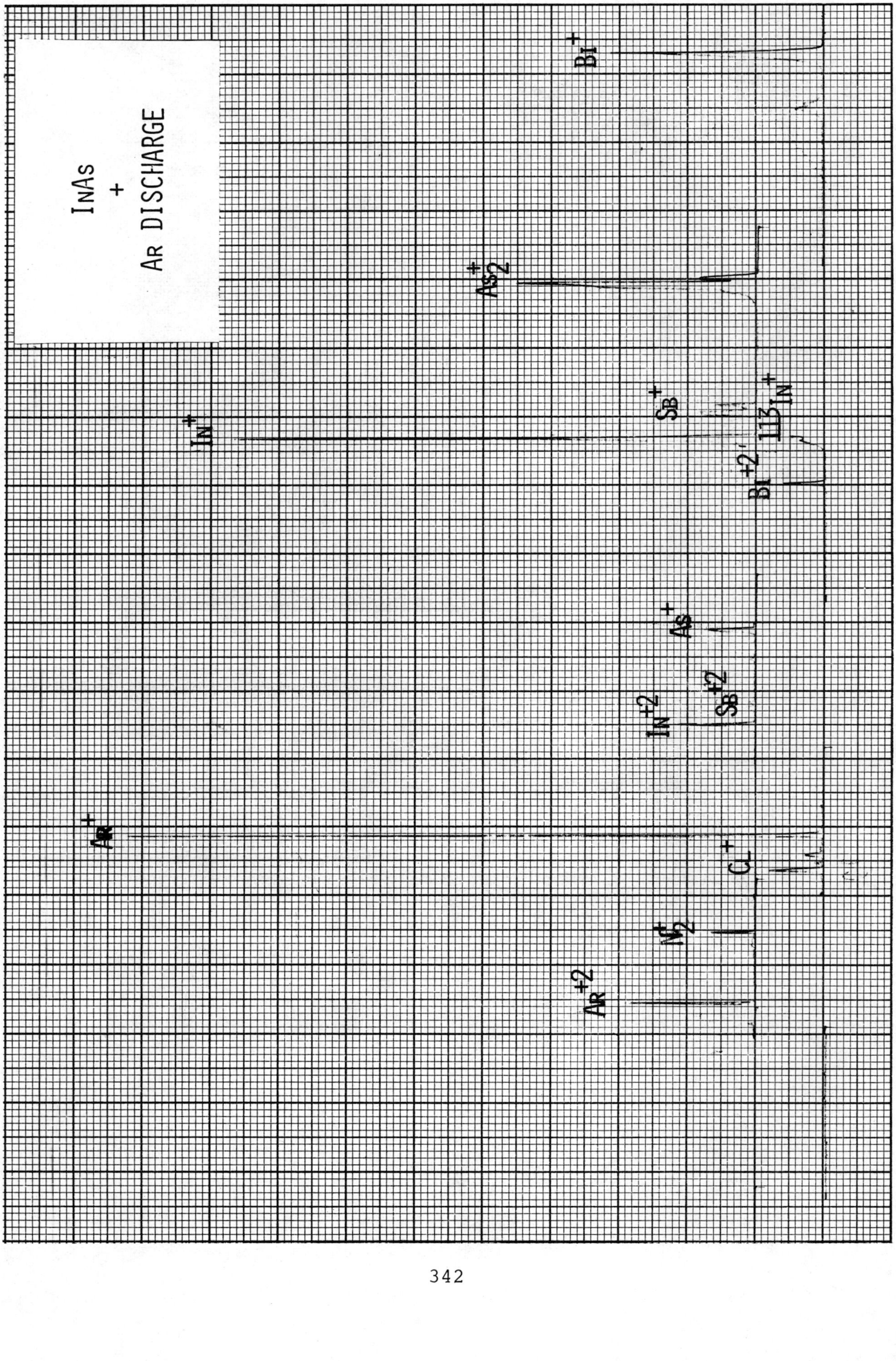

InAs
+
Ar DISCHARGE
Bi+
As2+
In+
Sb+
113In+
Bi+2
As+
In+2
Sb+2
Ar+
Cl+
N2+
Ar+2

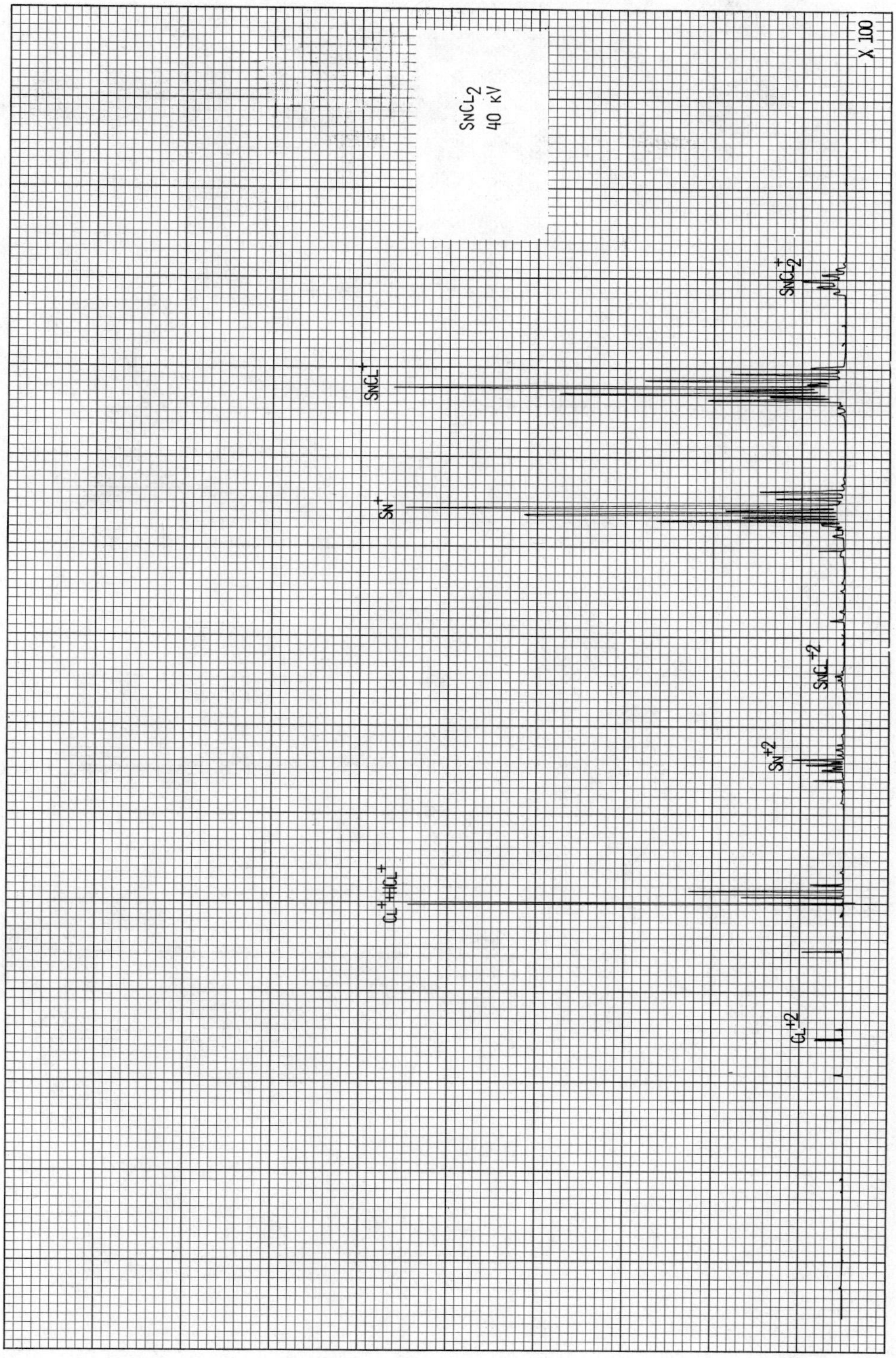
$SnCl_2$
40 kV
X 100
$SnCl_2^+$
$SnCl^+$
Sn^+
$SnCl^{+2}$
Sn^{+2}
$Cl^+ + HCl^+$
Cl^{+2}

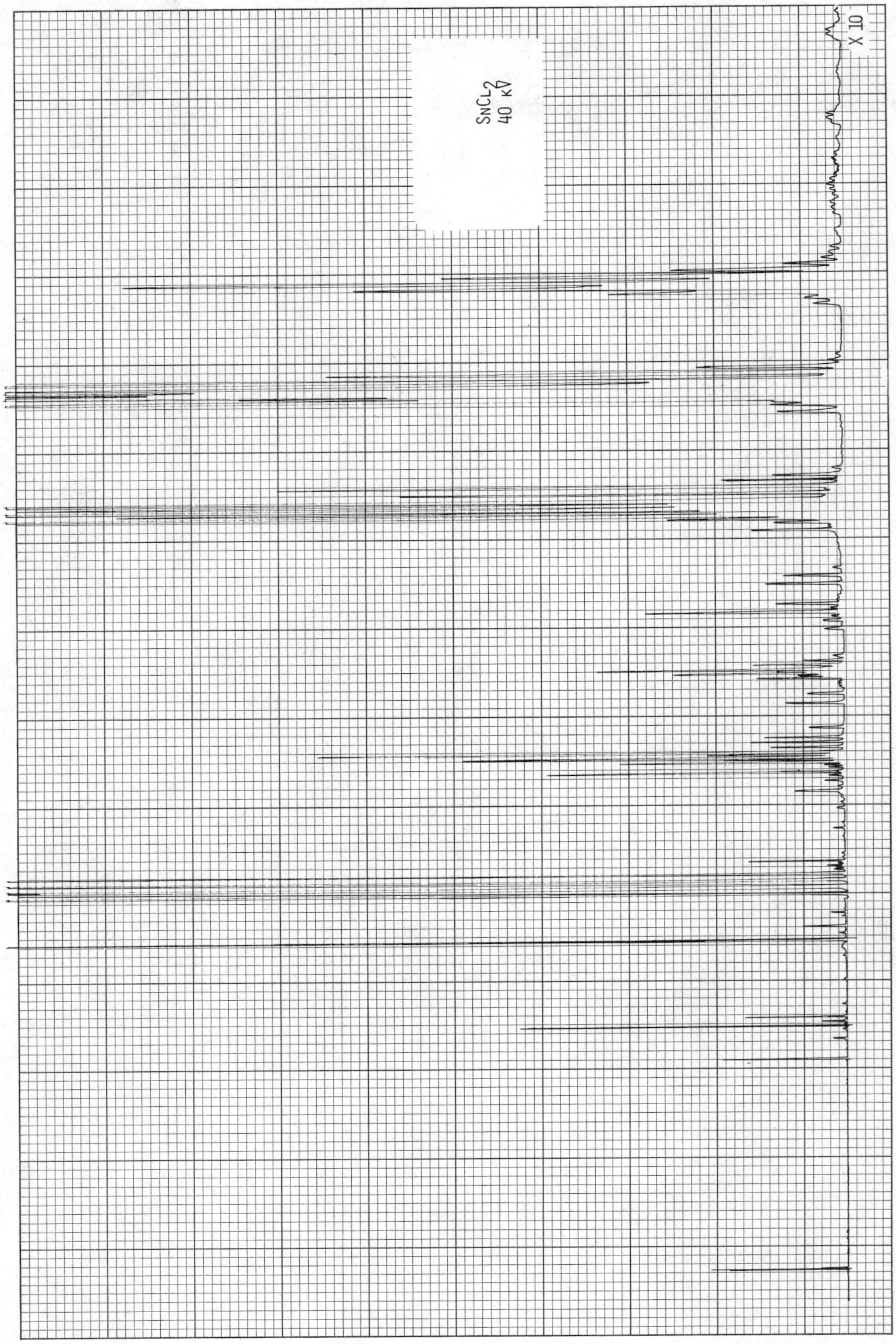
SnCl2
40 kV
X 10

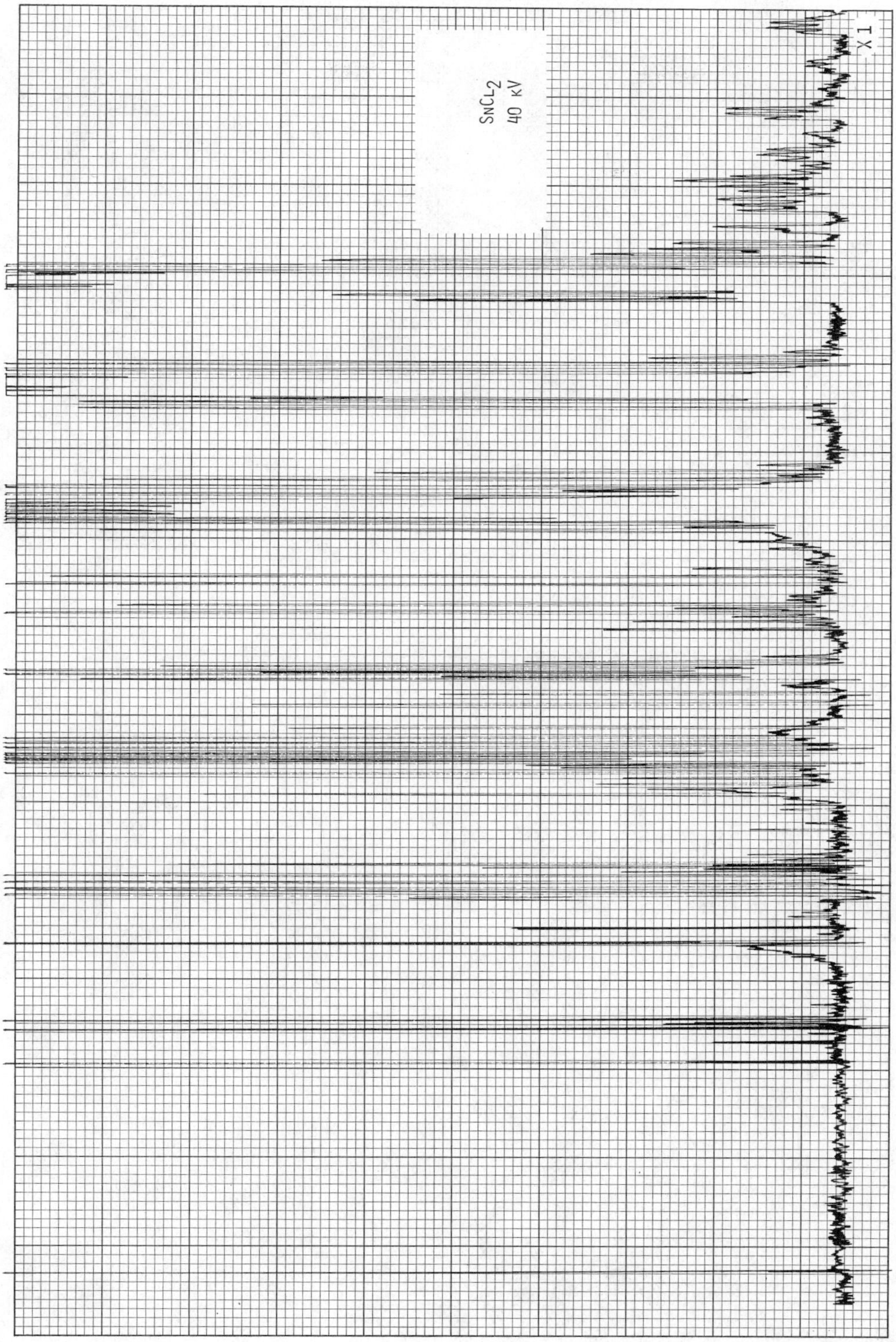
SnCl$_2$
40 kV
X 1

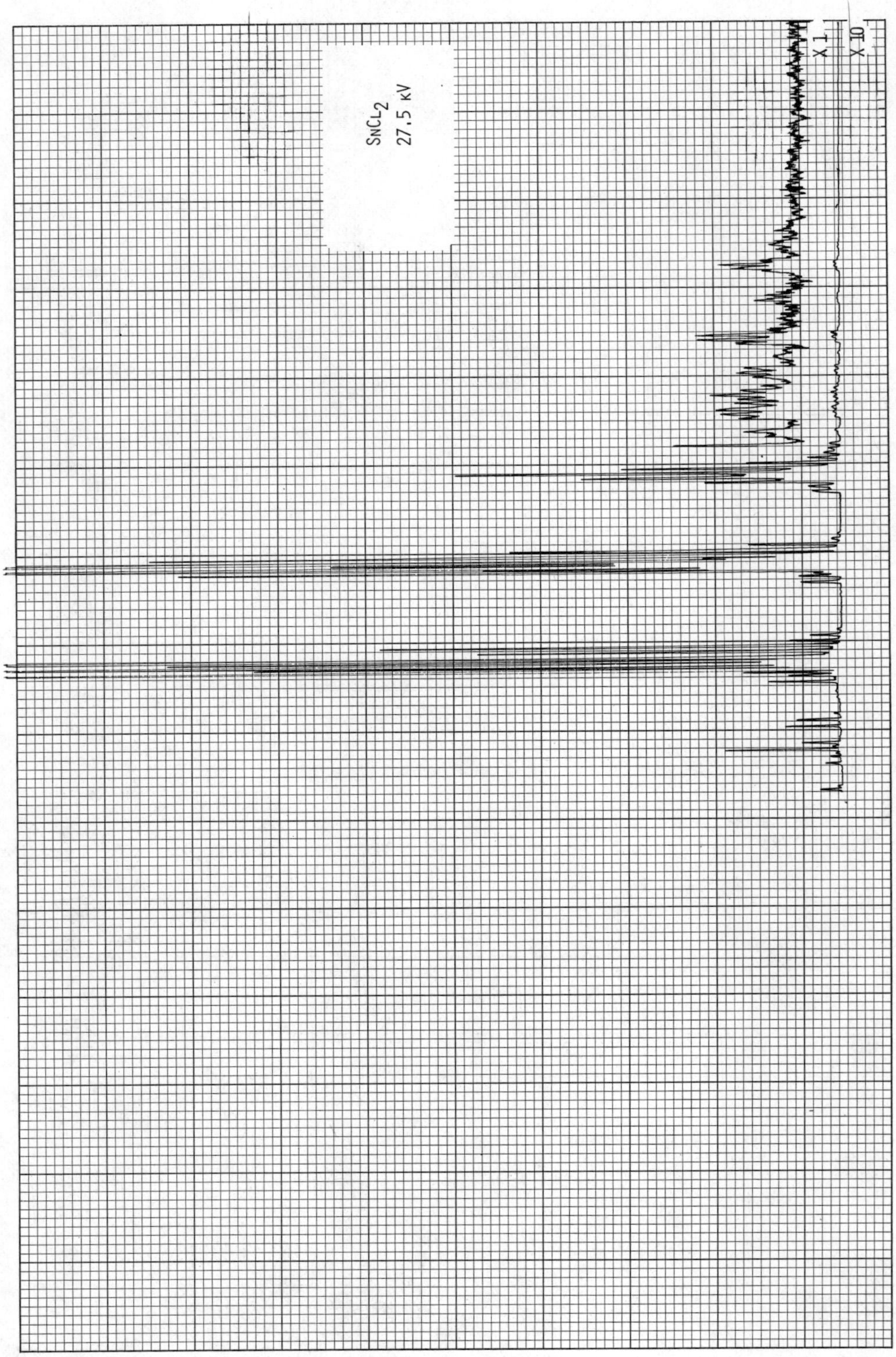
SnCl$_2$
27.5 kV
X 1
X 10

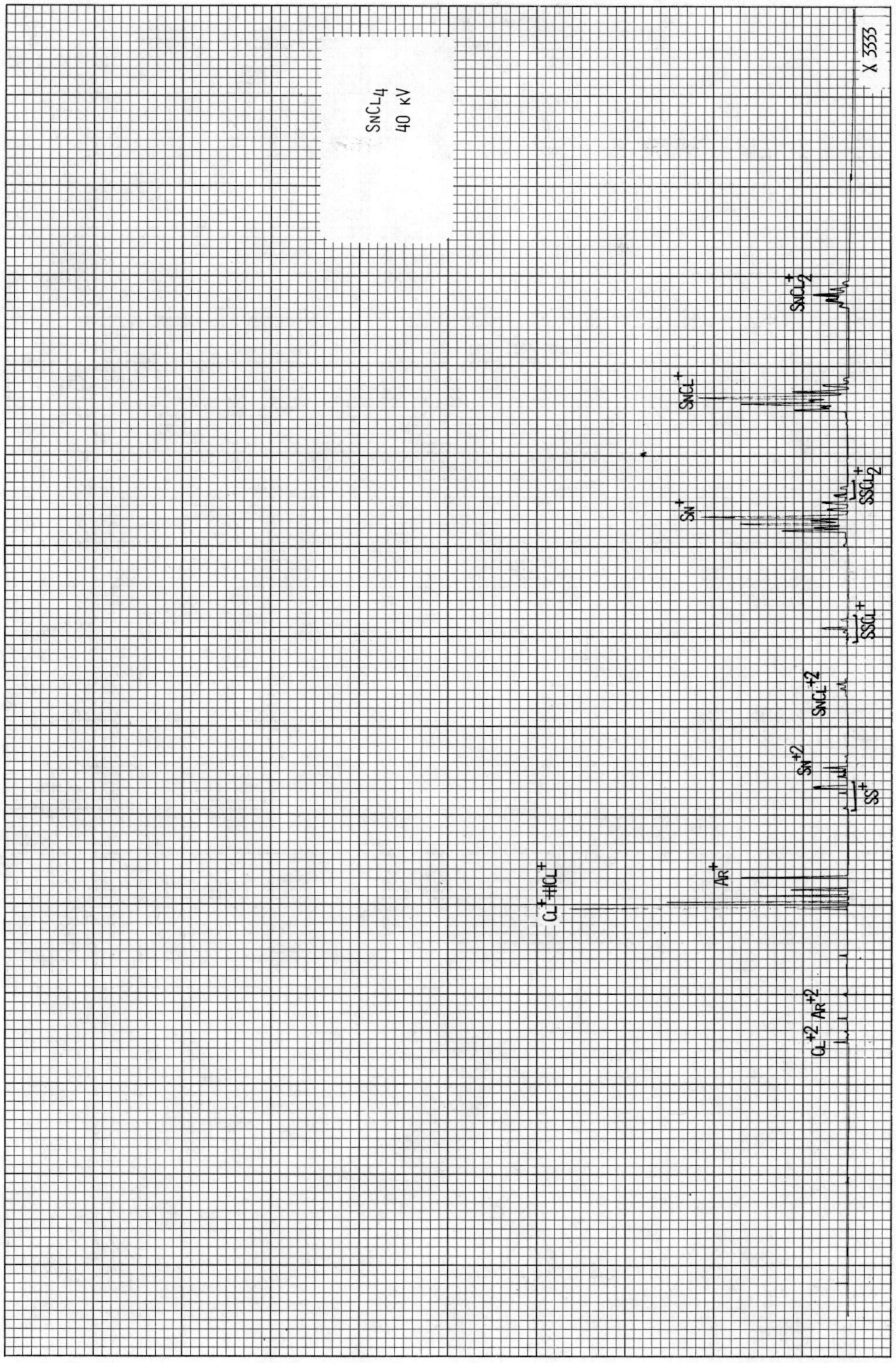
SnCl4
40 kV
X 3333
Cl+ +HCl+
Ar+
Sn+2
SS+
SnCl+2
SSCl+
Sn+
SSCl2+
SnCl+
SnCl2+
Cl+2
Ar+2

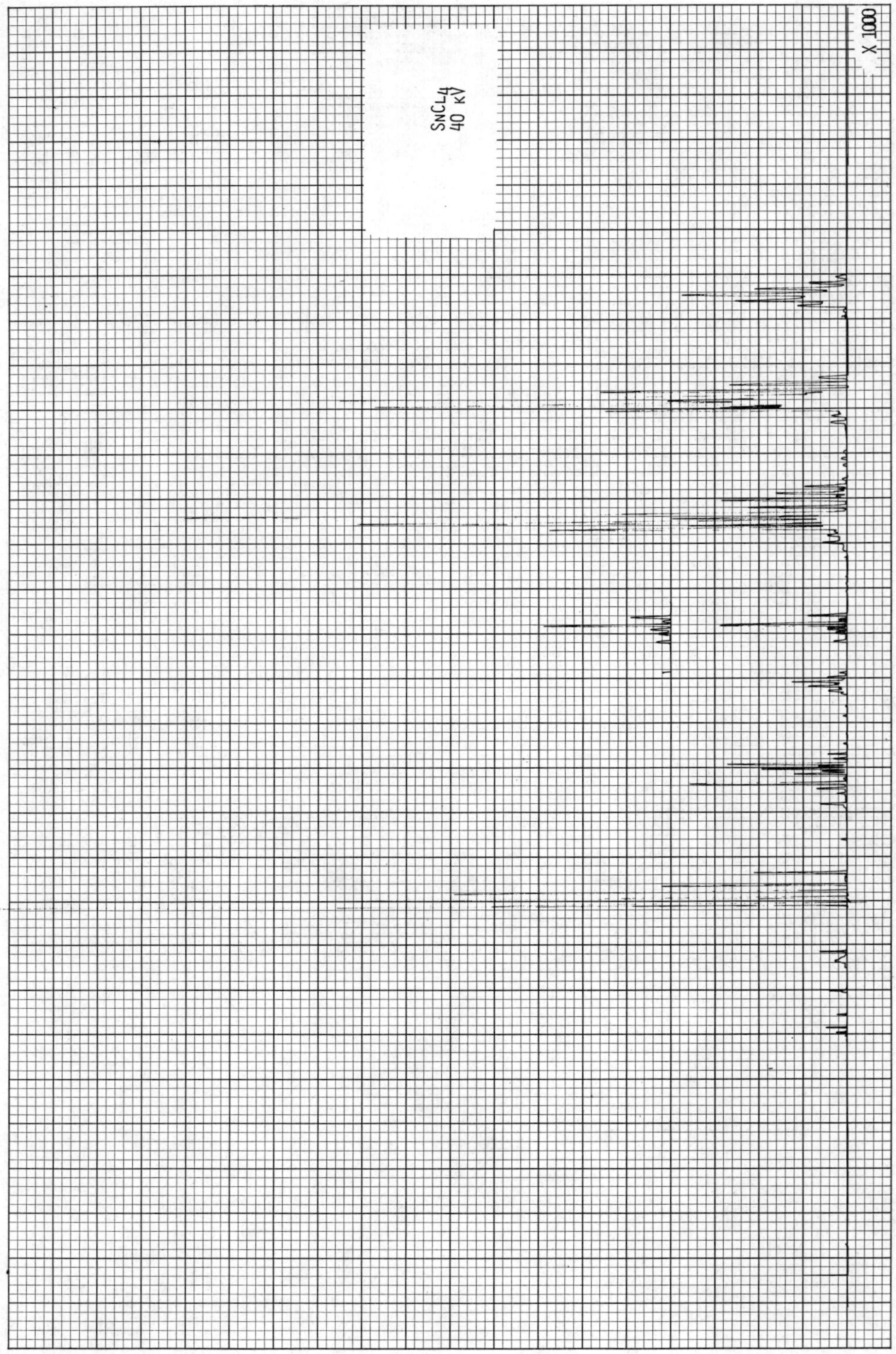
SnCl4
40 kV
X 1000

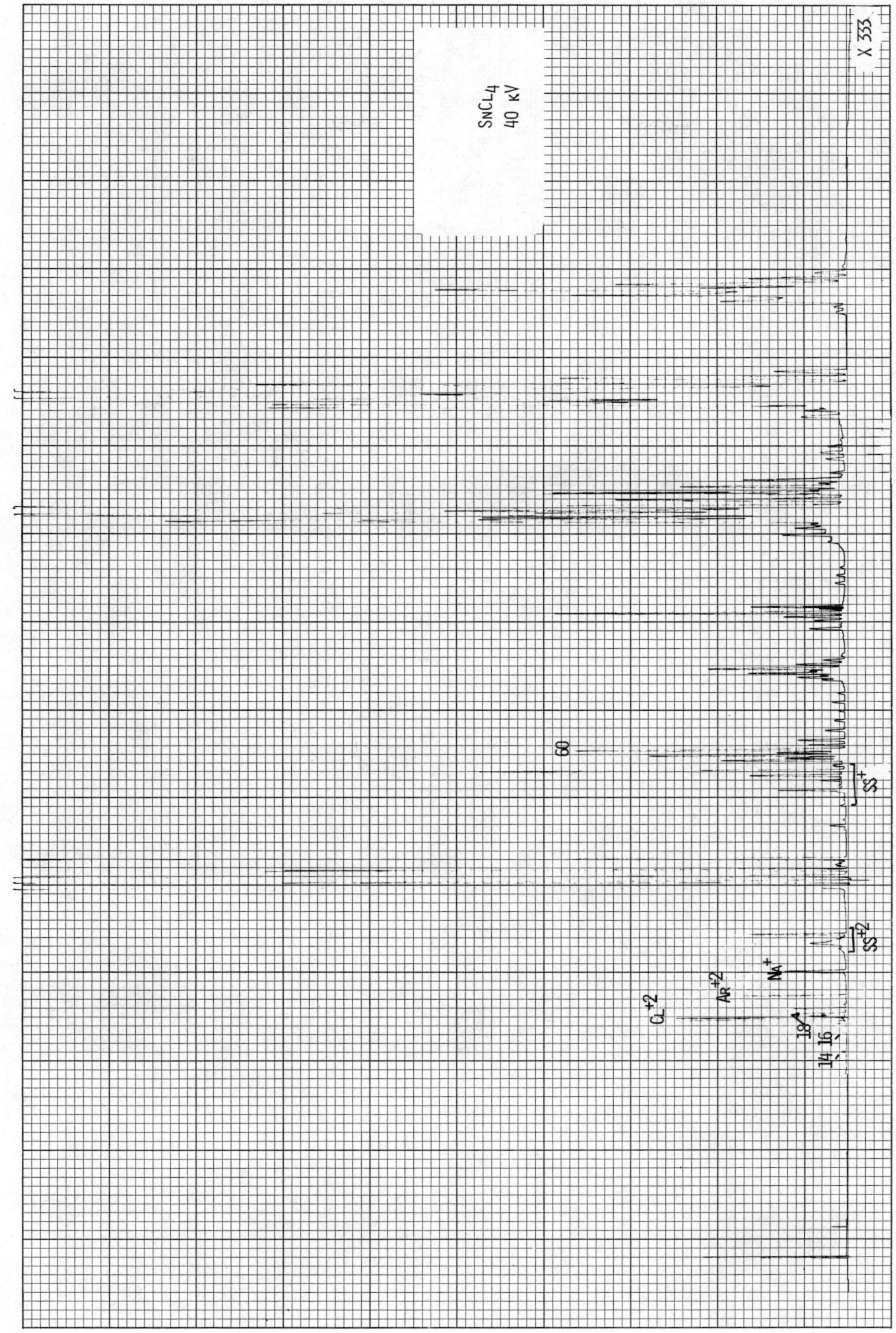
SnCl4
40 kV
X 333
60
SS+
SS+2
Na+
Ar+2
Cl+2
18
16
14

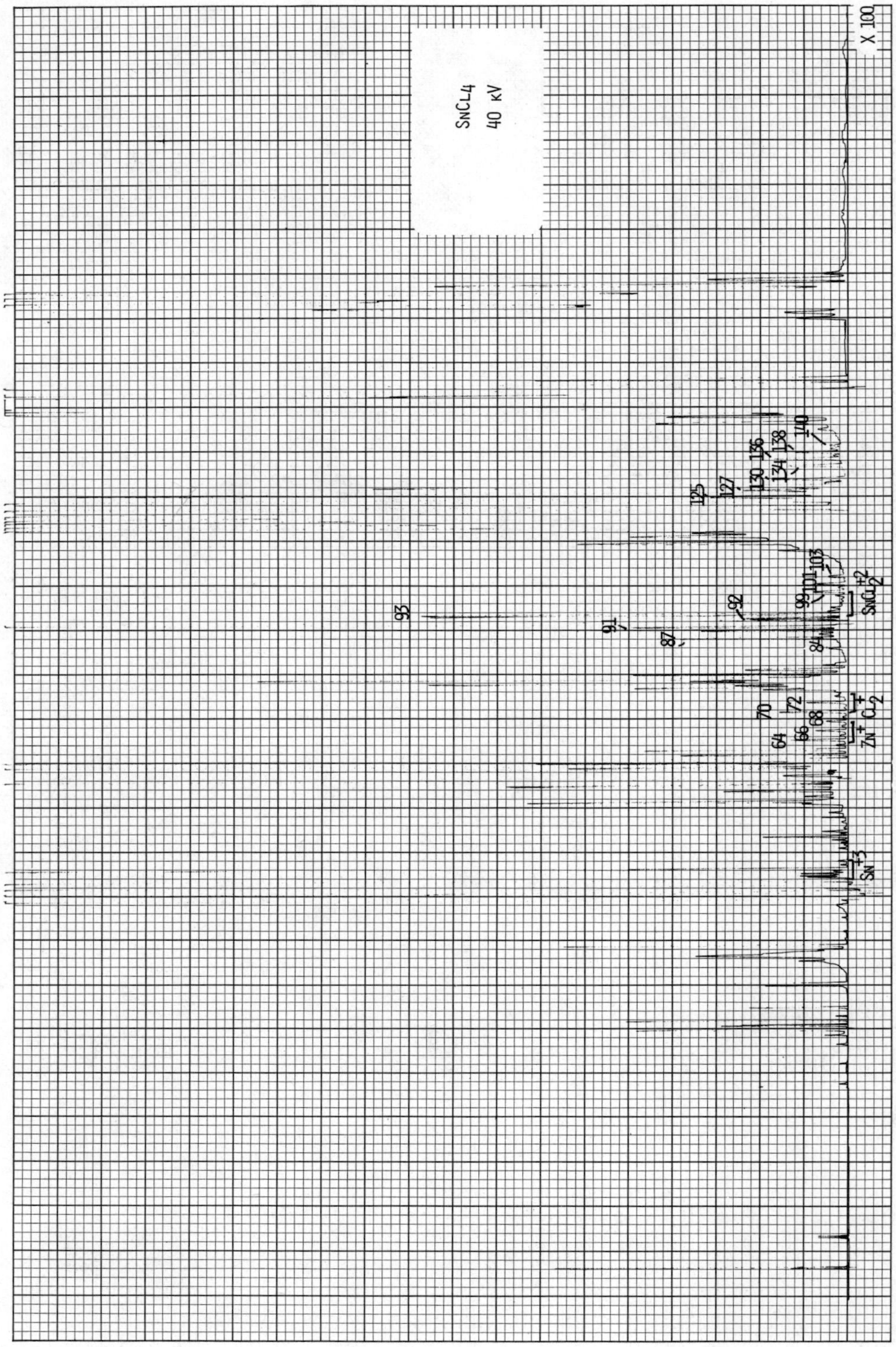
SnCl4
40 kV
X 100

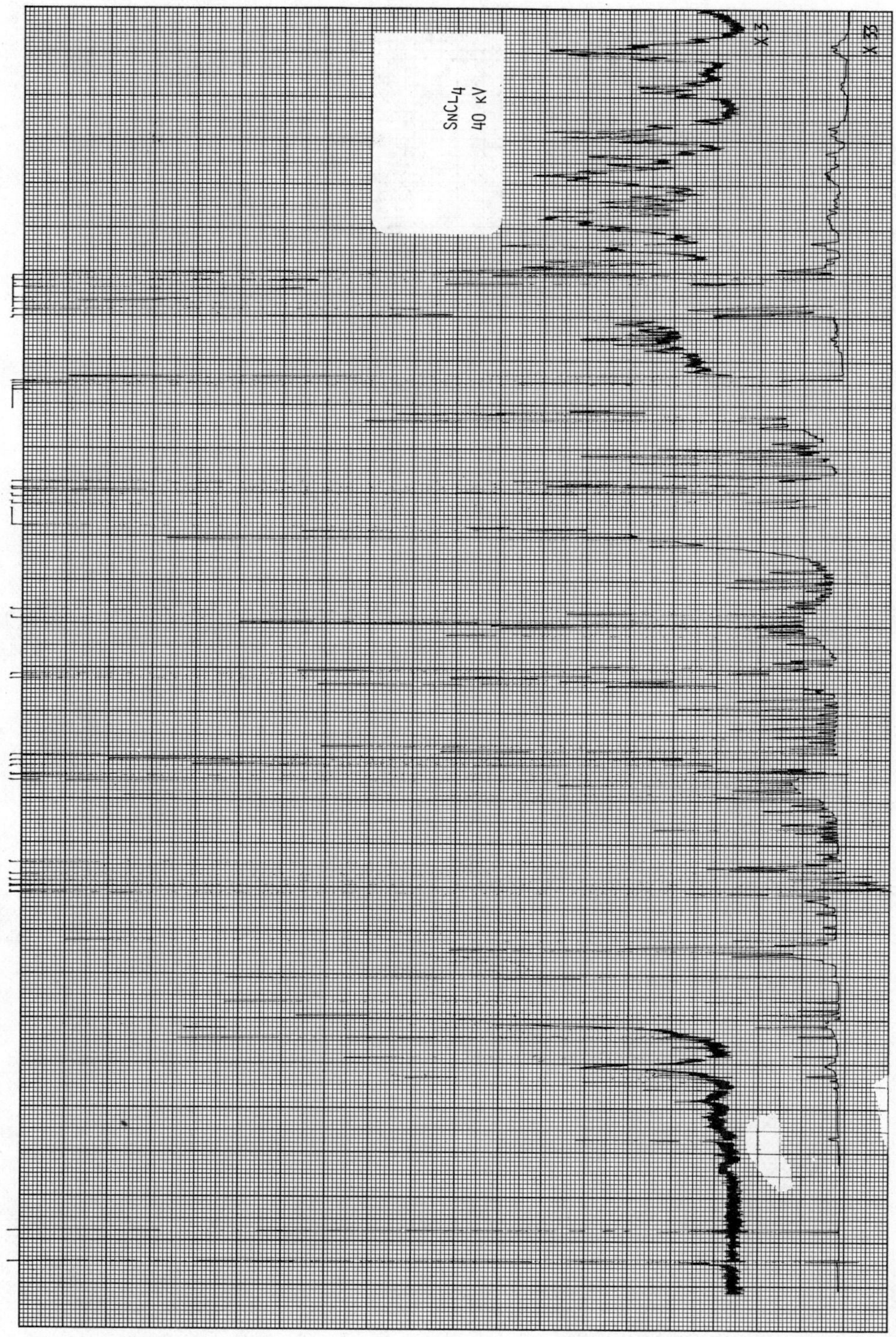
SnCl4
40 kV
X 3
X 33

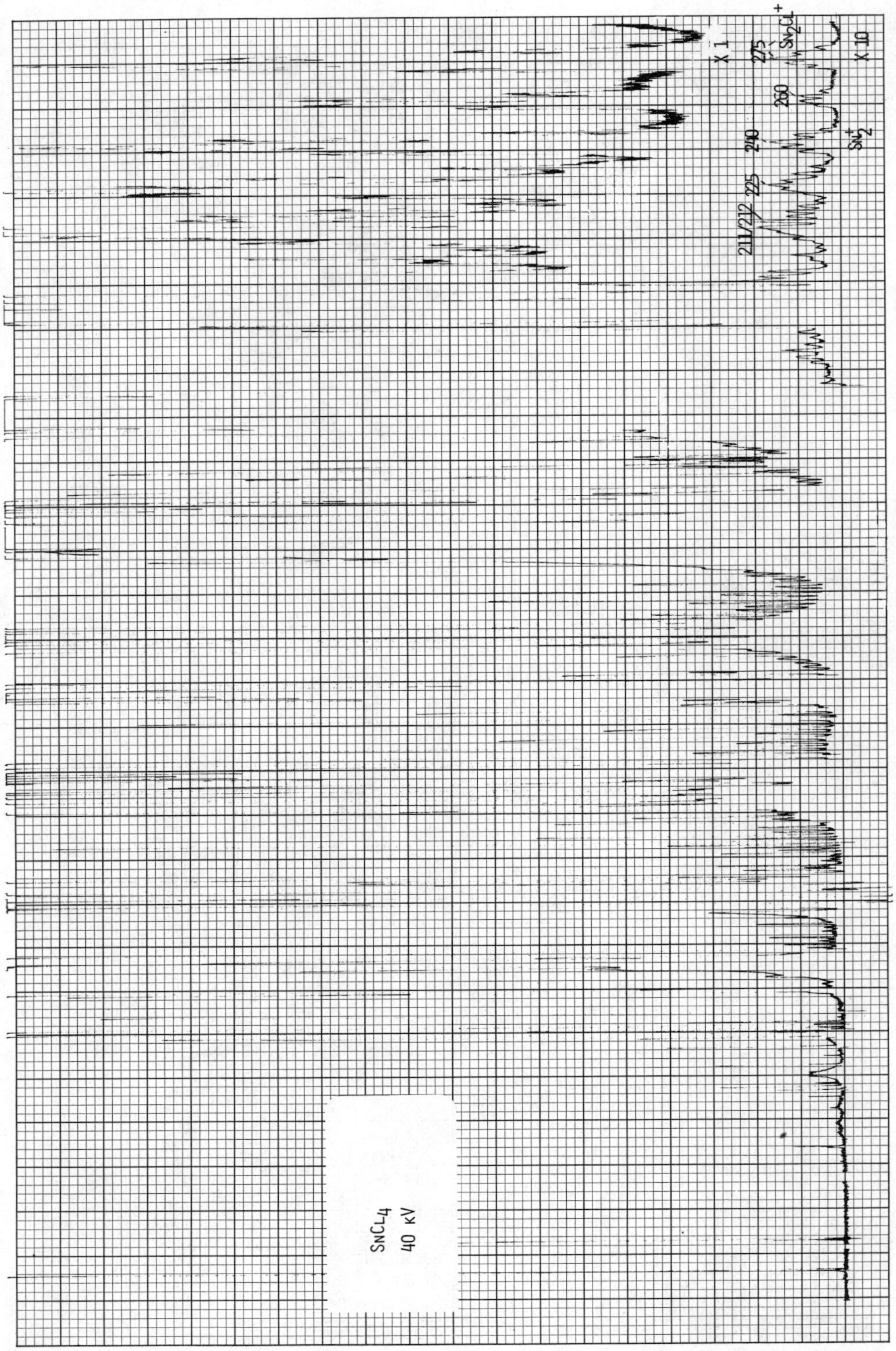
SnCl4
40 kV
X 1
X 10
211/212
225
240
260
275
Sn2Cl+
Sn2+

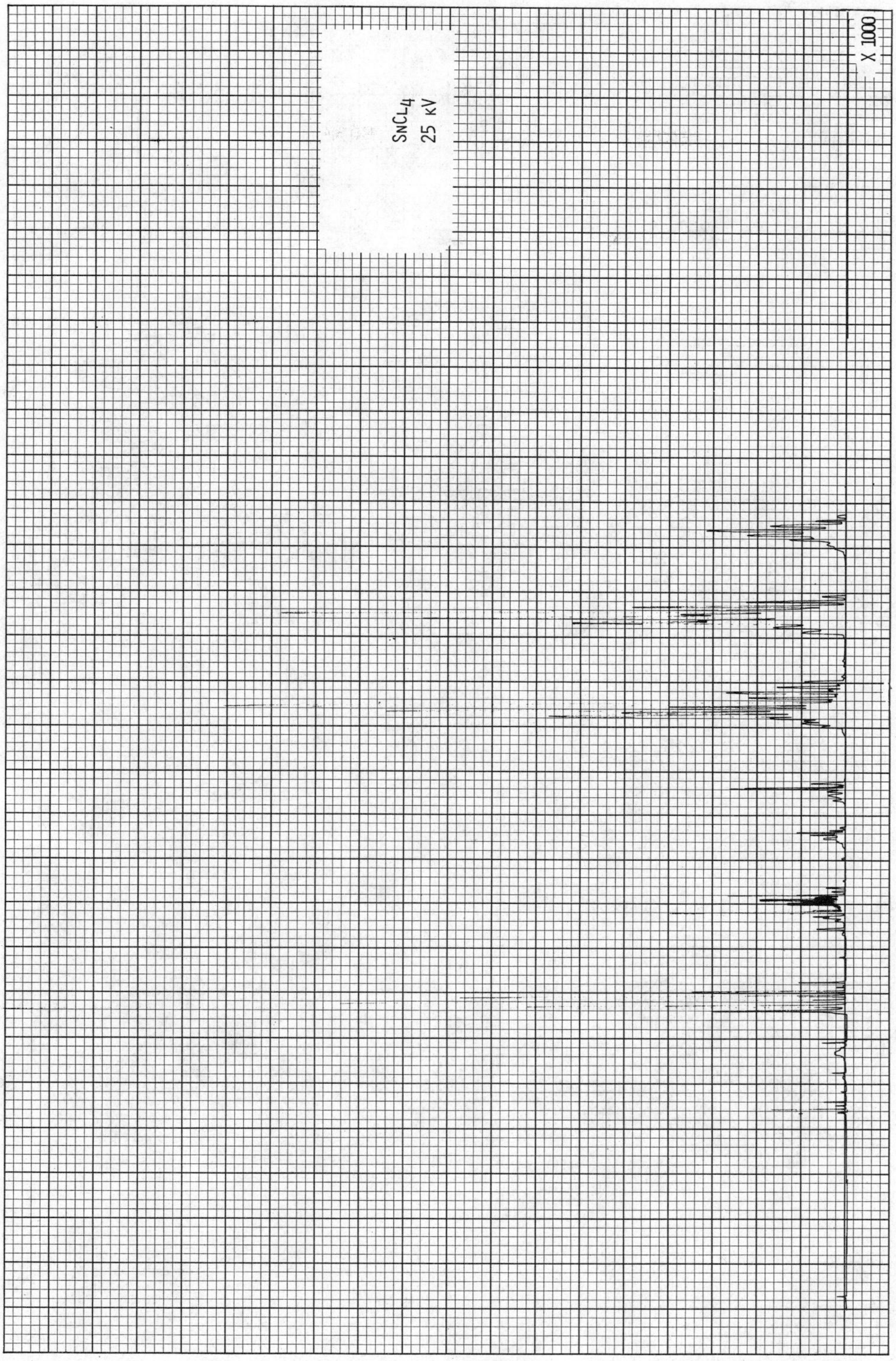
$SnCl_4$
25 kV
X 1000

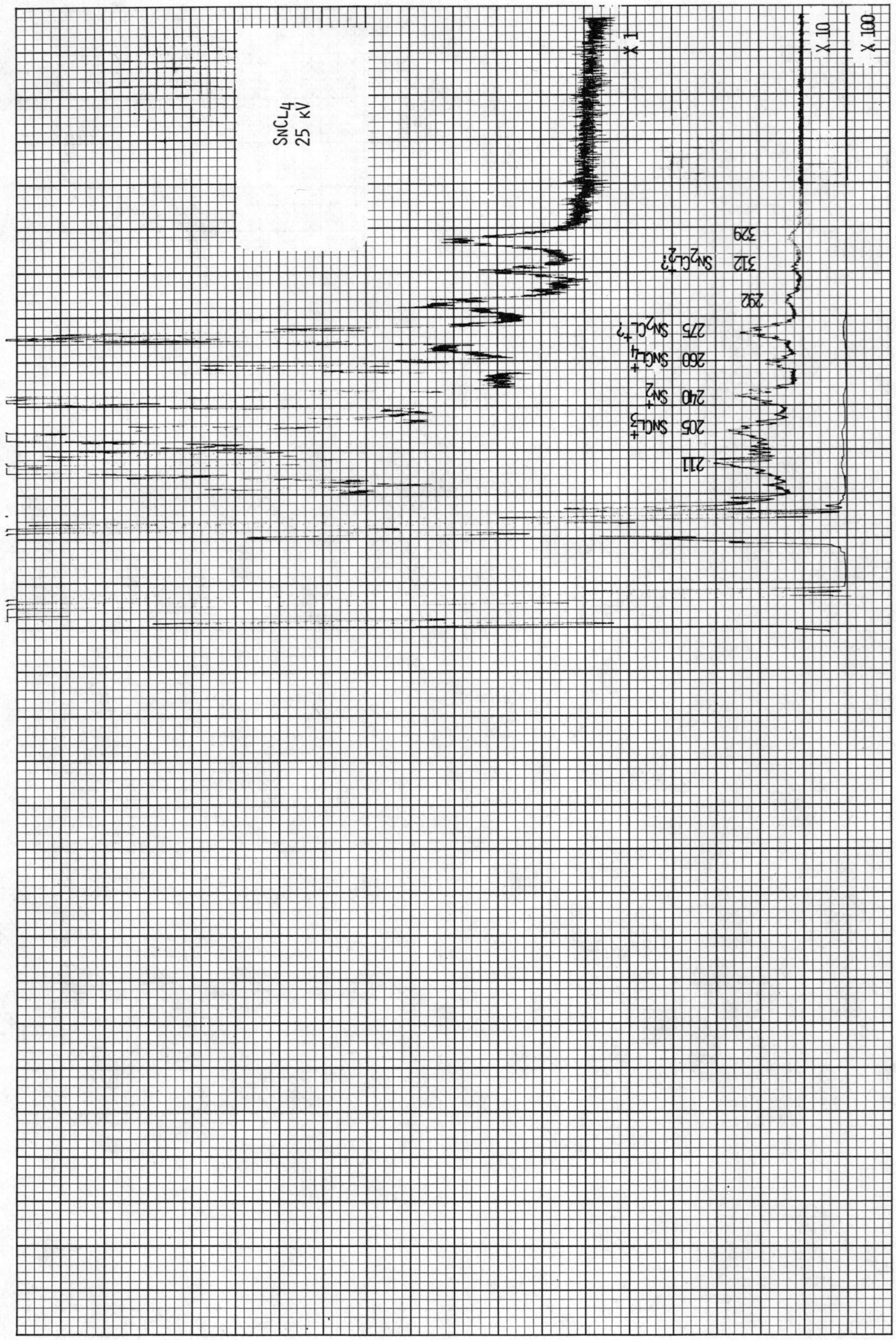
SnCl4
25 kV
X 1
X 10
X 100
329
312
292
275
260
240
205
211

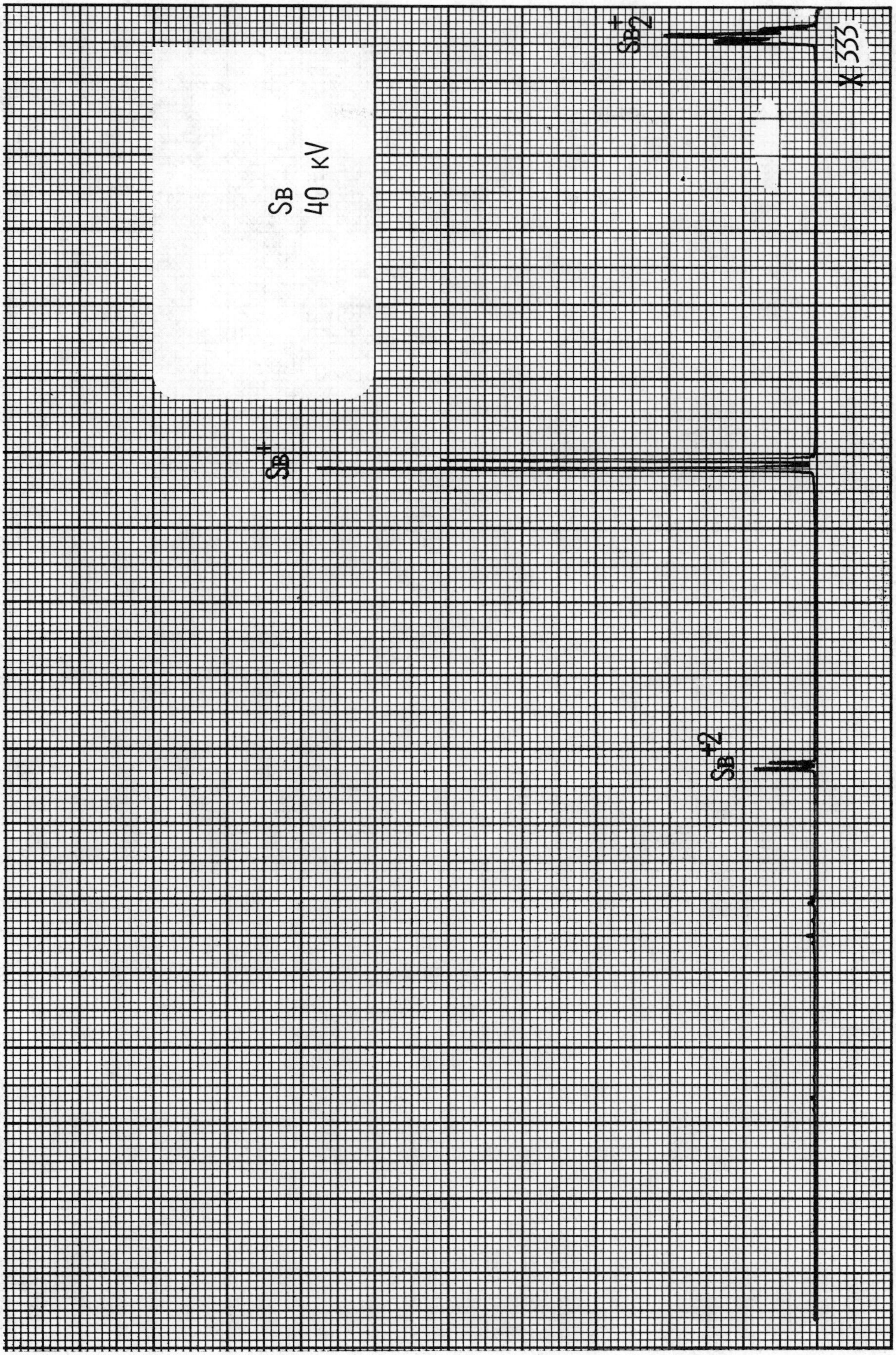
Sb
40 kV
Sb+
Sb2+
Sb+2
X 333

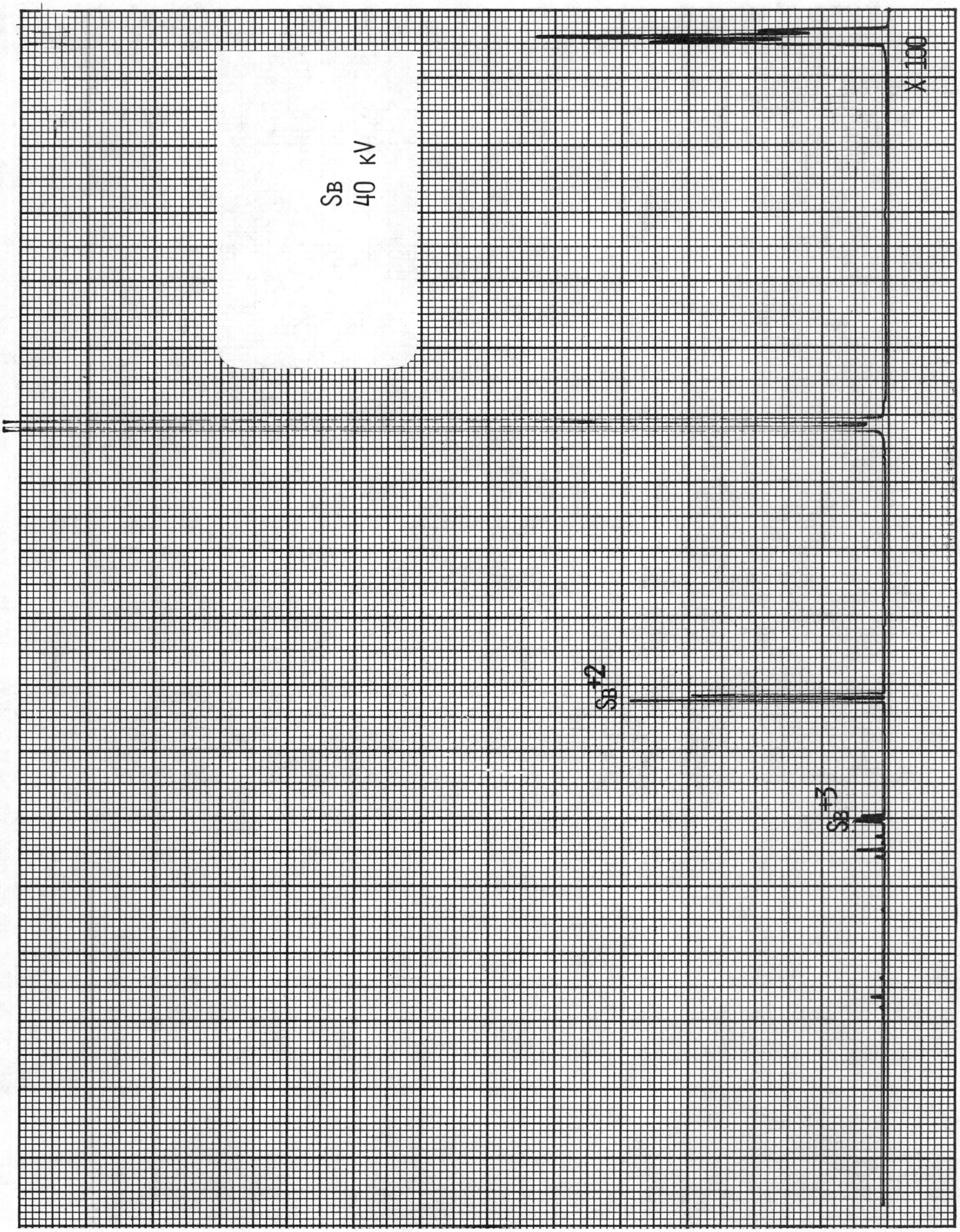
Sb
40 kV
X 100
Sb+2
Sb+3

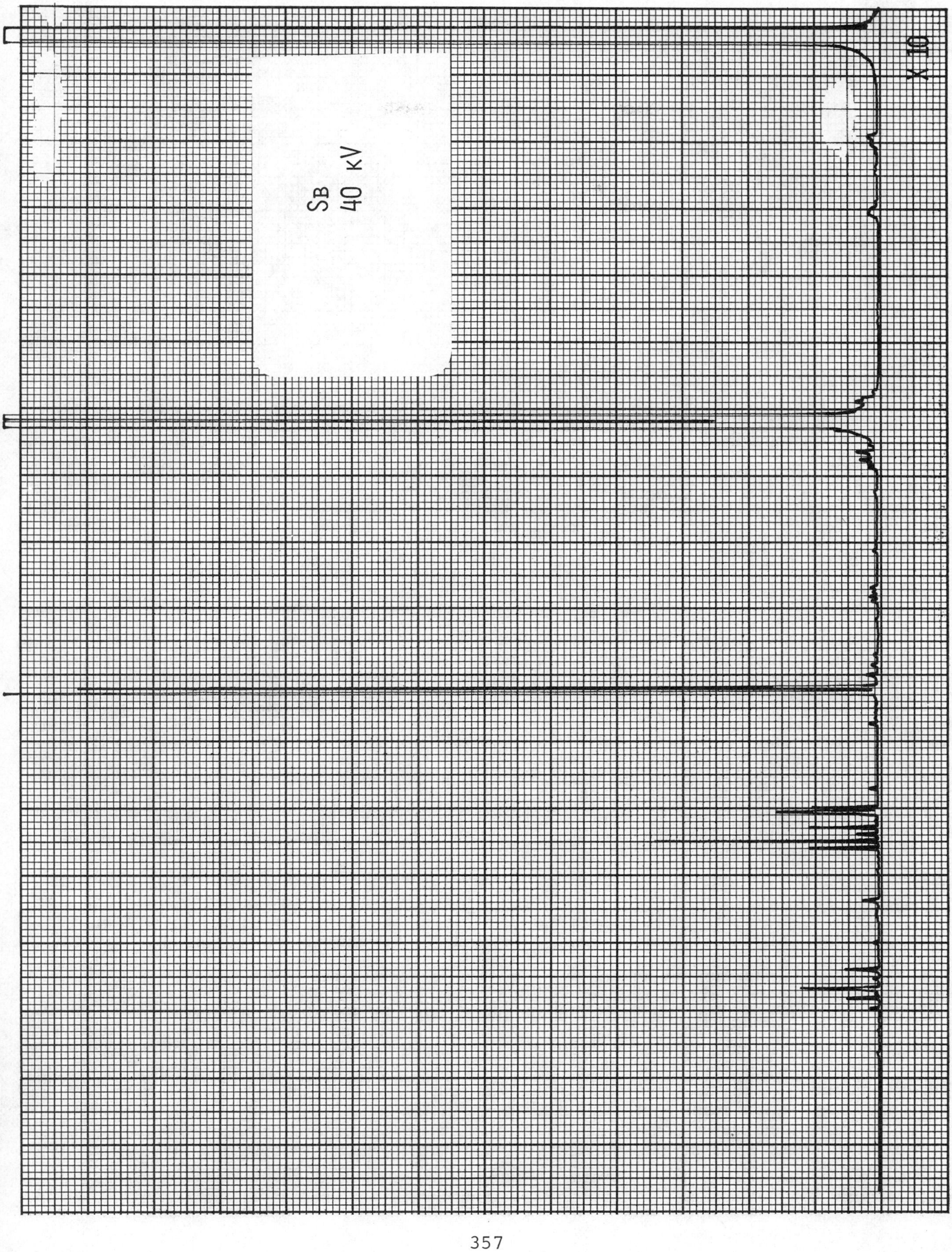
Sb
40 kV
X 10

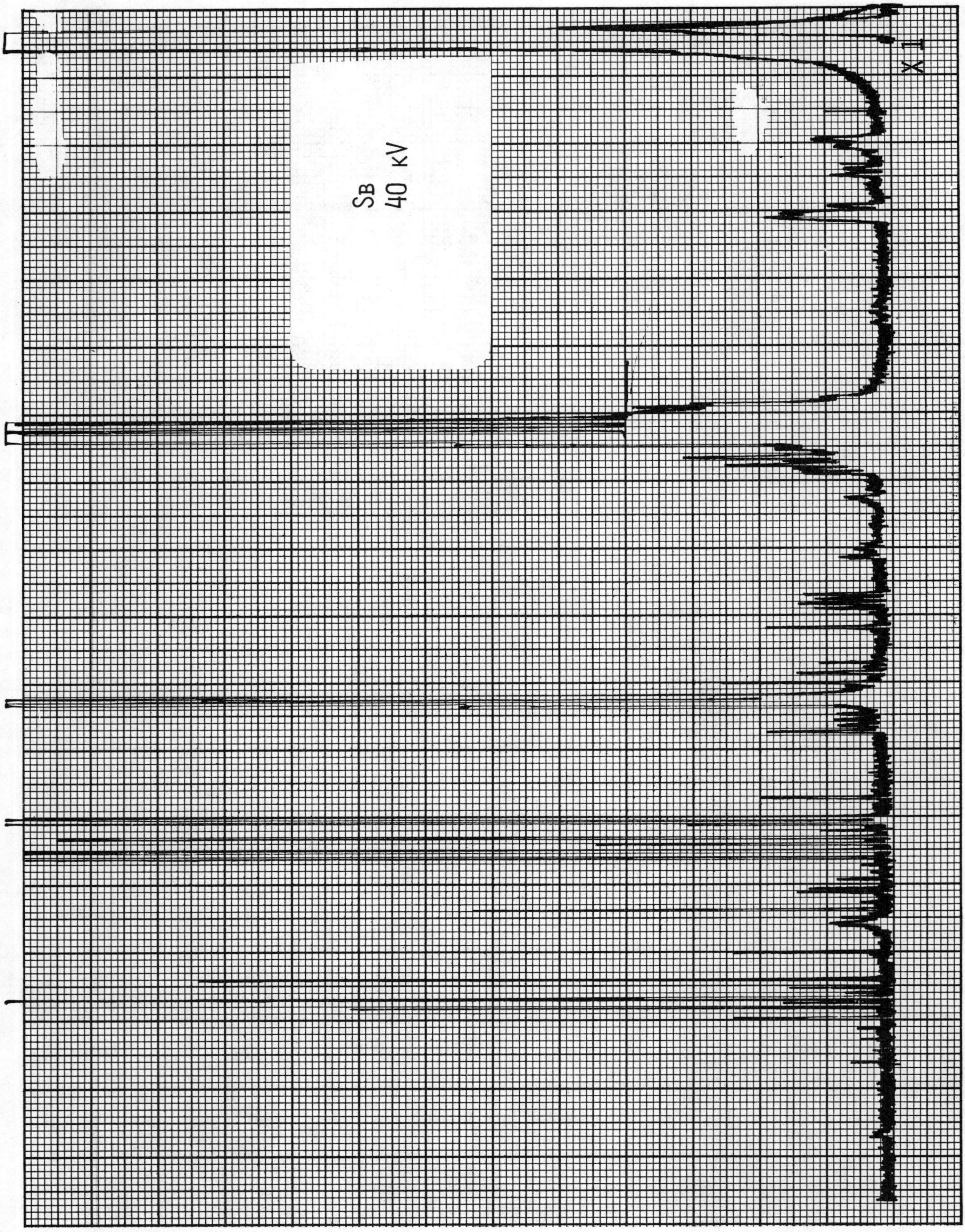
Sb
40 kV
X 1

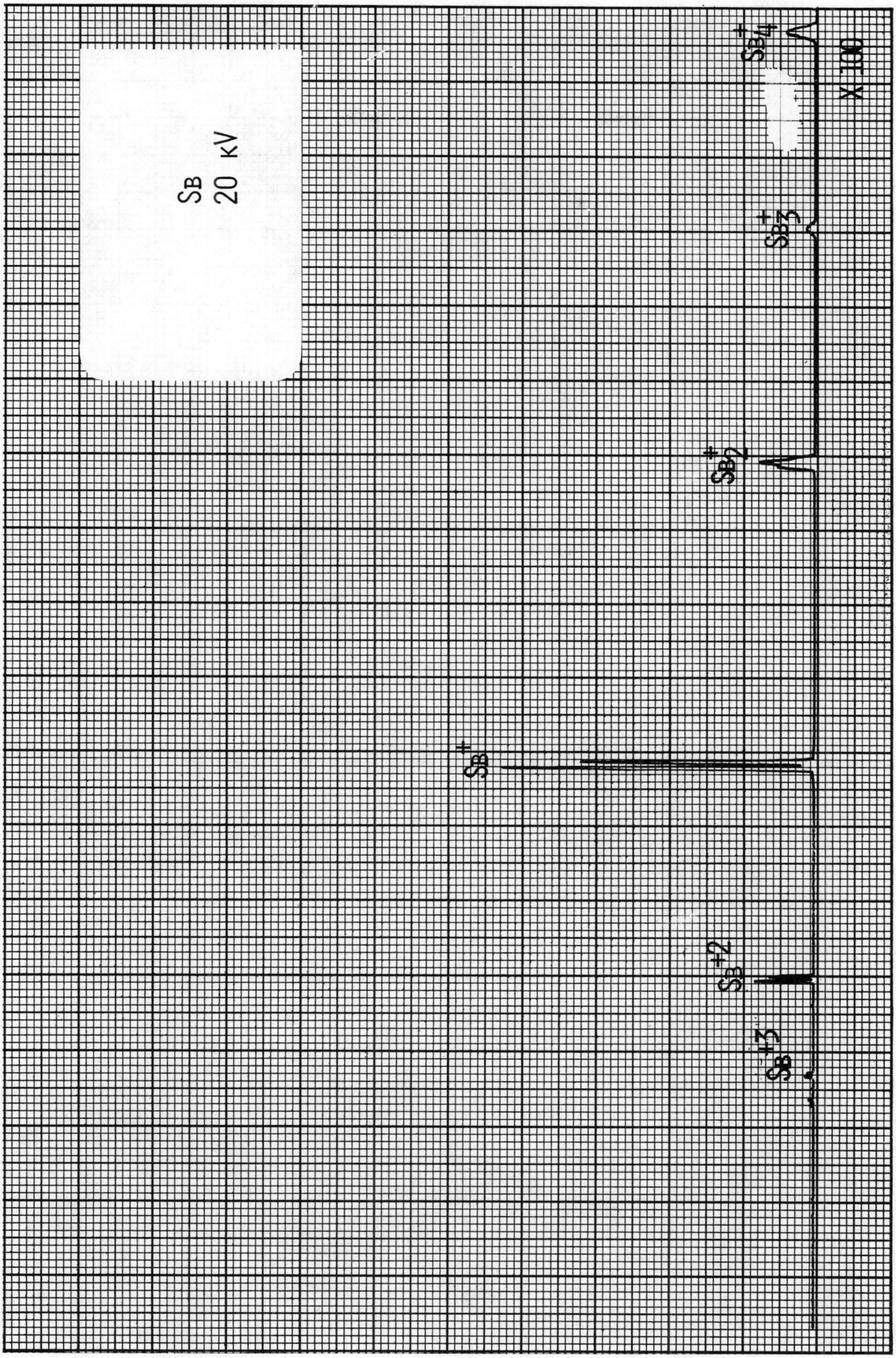
Sb
20 kV
X 100
Sb4+
Sb3+
Sb2+
Sb+
Sb+2
Sb+3

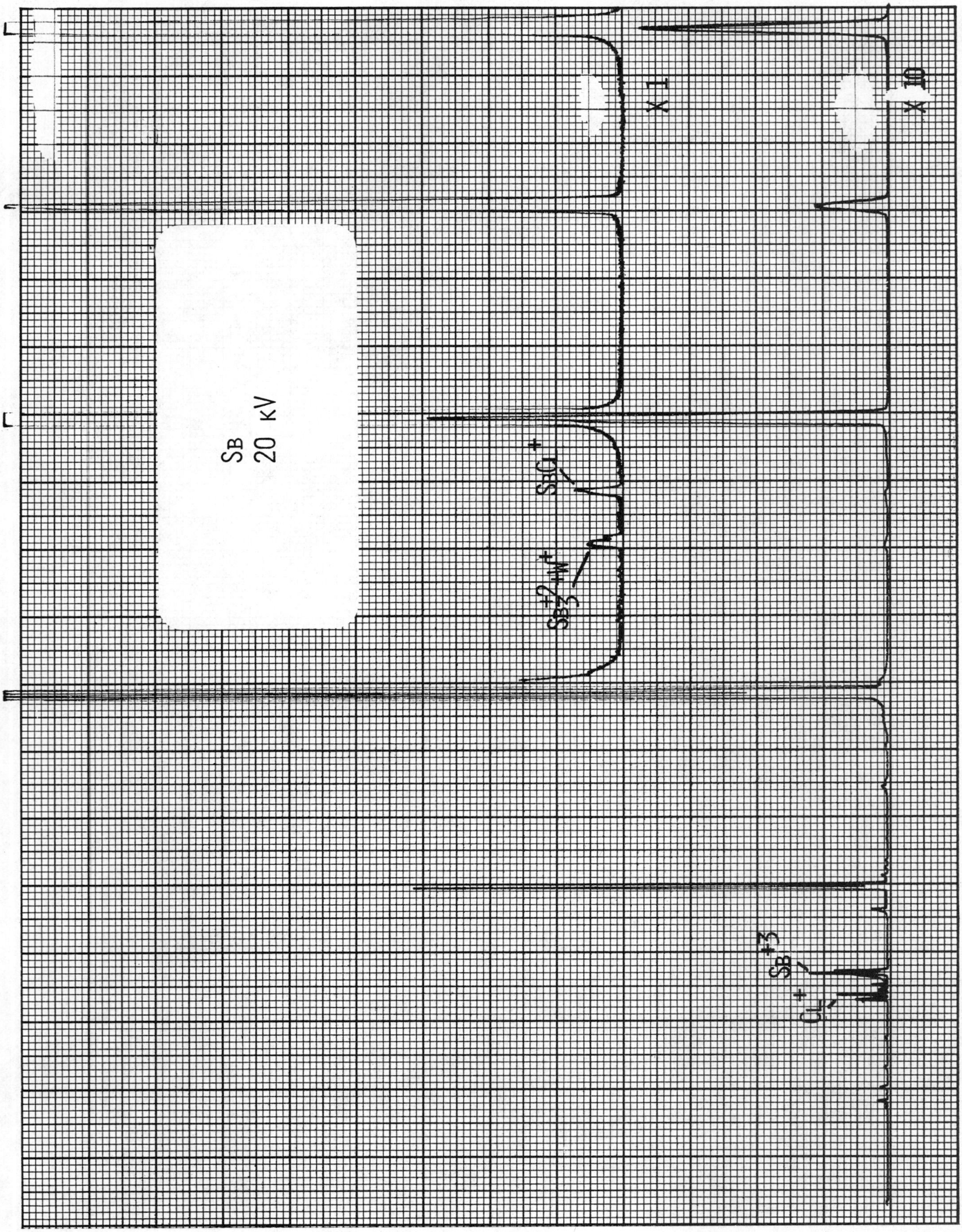
Sb
20 kV
x 1
x 10
SbCl+
Cl+
Sb+3

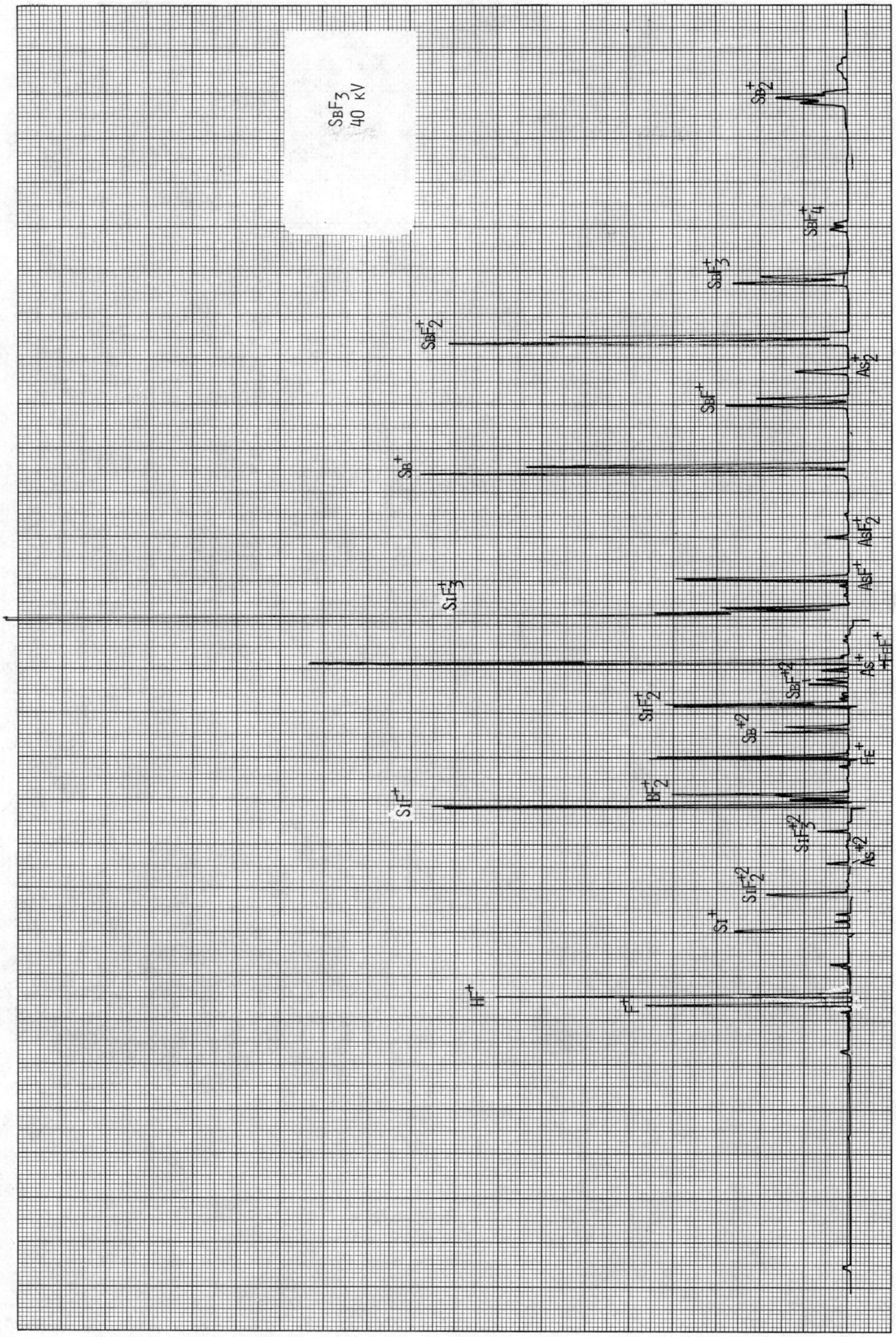

SbF_3
40 kV
Sb_2^+
SbF_4^+
SbF_3^+
SbF_2^+
As_2^+
SbF^+
Sb^+
AsF_2^+
AsF^+
SiF_3^+
As^+
FeF^+
SbF^{+2}
SiF_2^+
Sb^{+2}
Fe^+
BF_2^+
SiF^+
SiF_3^{+2}
As^{+2}
SiF_2^{+2}
Si^+
HF^+
F^+

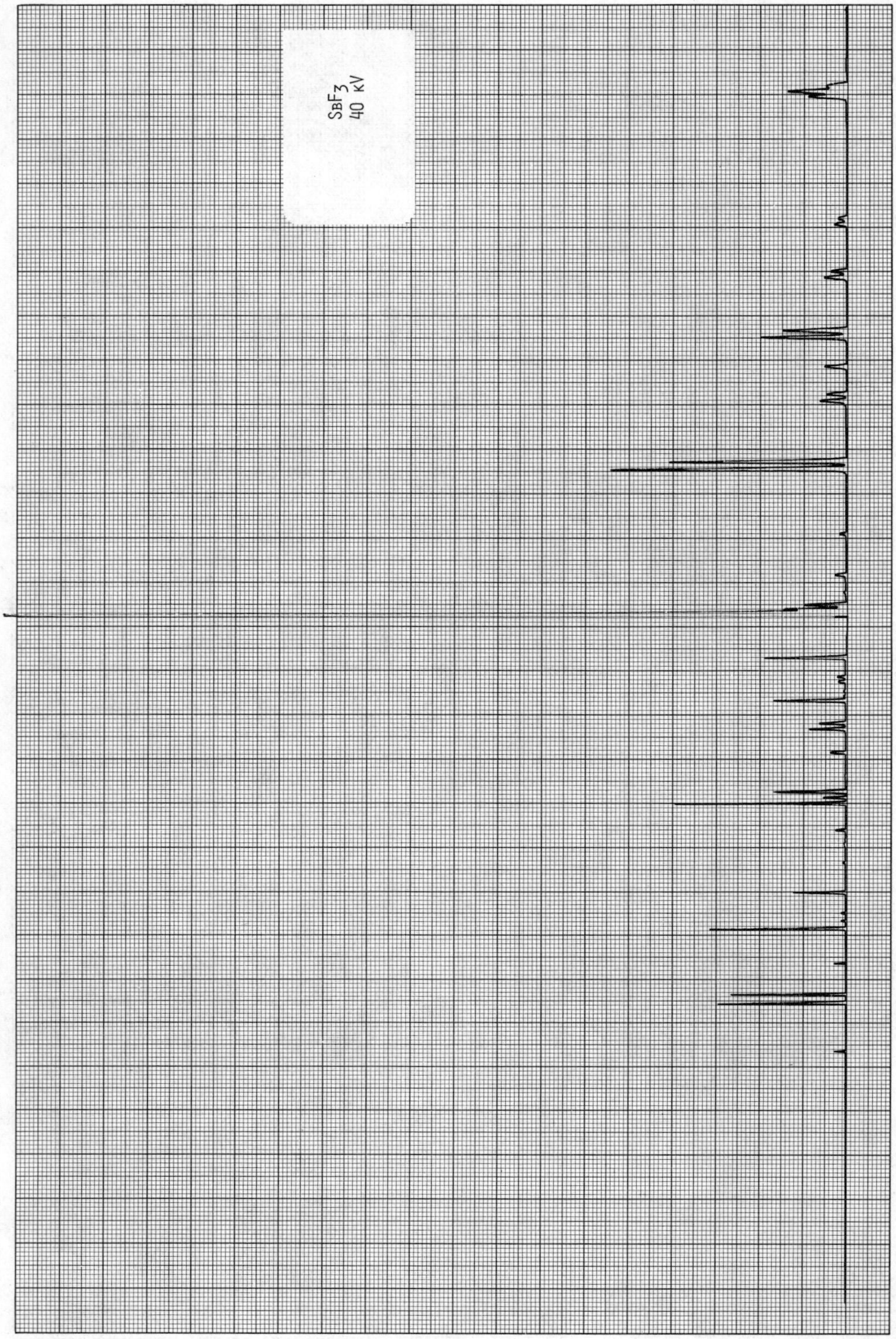
SbF_3
40 kV

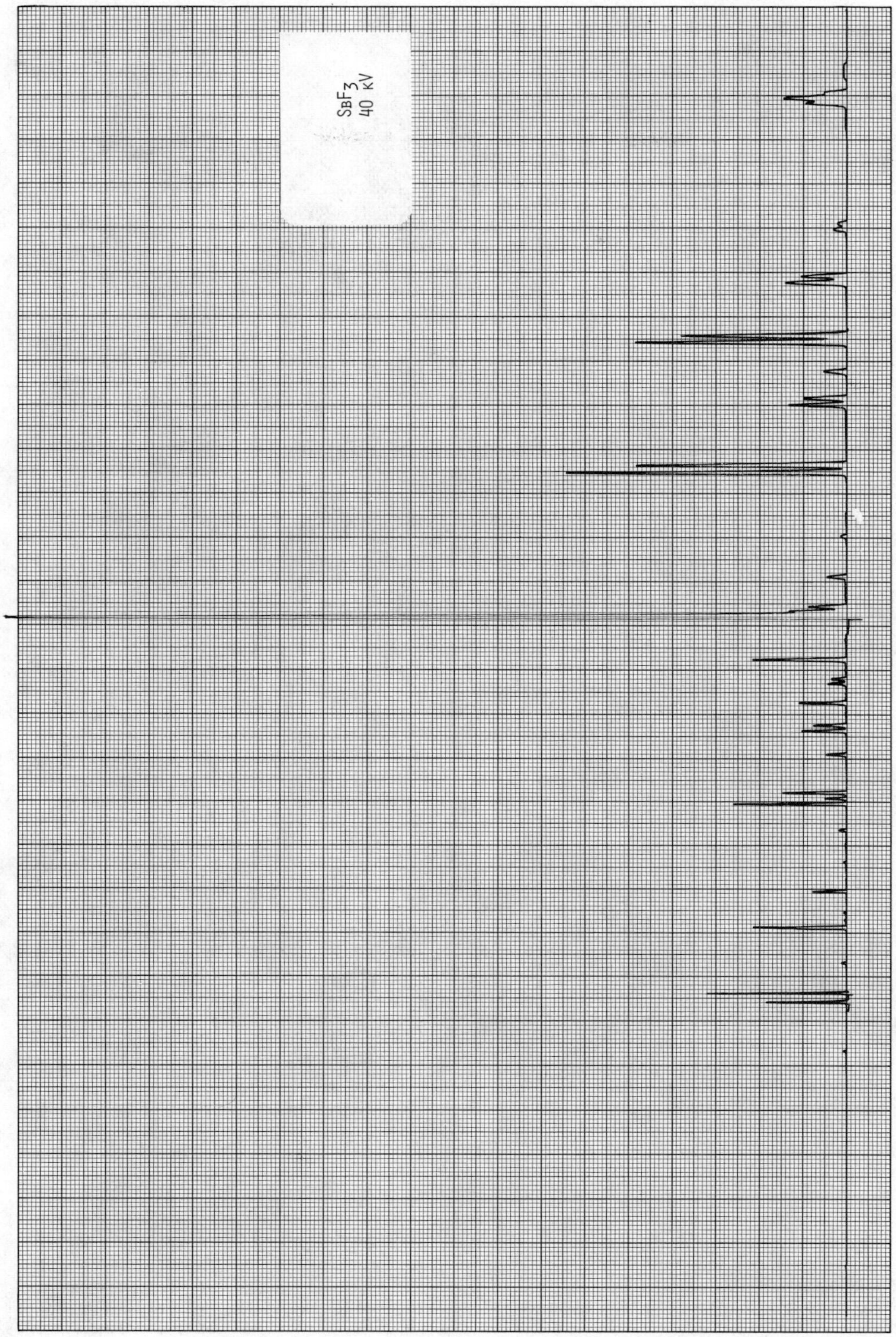
SbF_3
40 kV

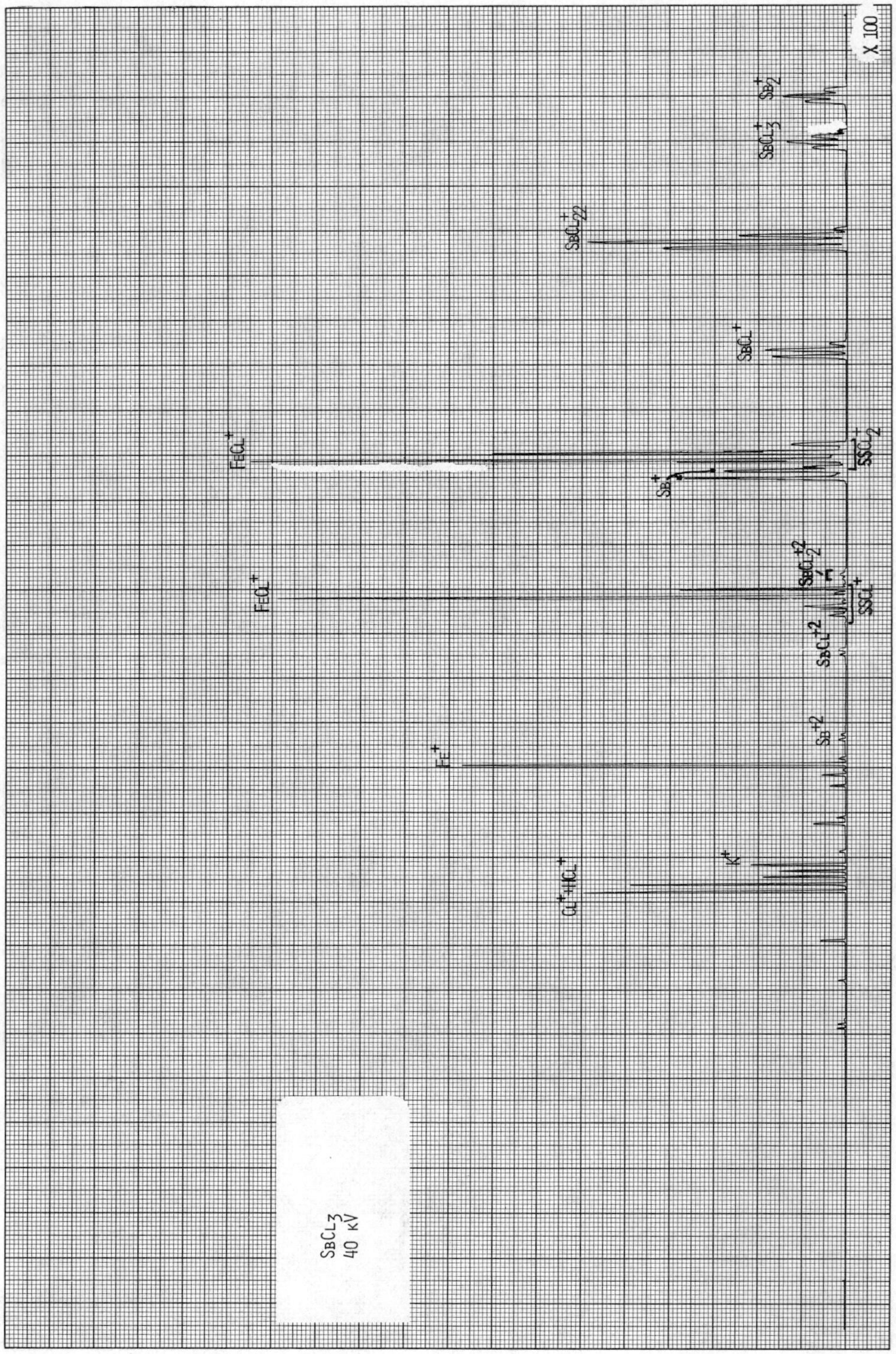

SbCl3
40 kV
X 100
FeCl+
Fe+
K+
Cl+HCl+

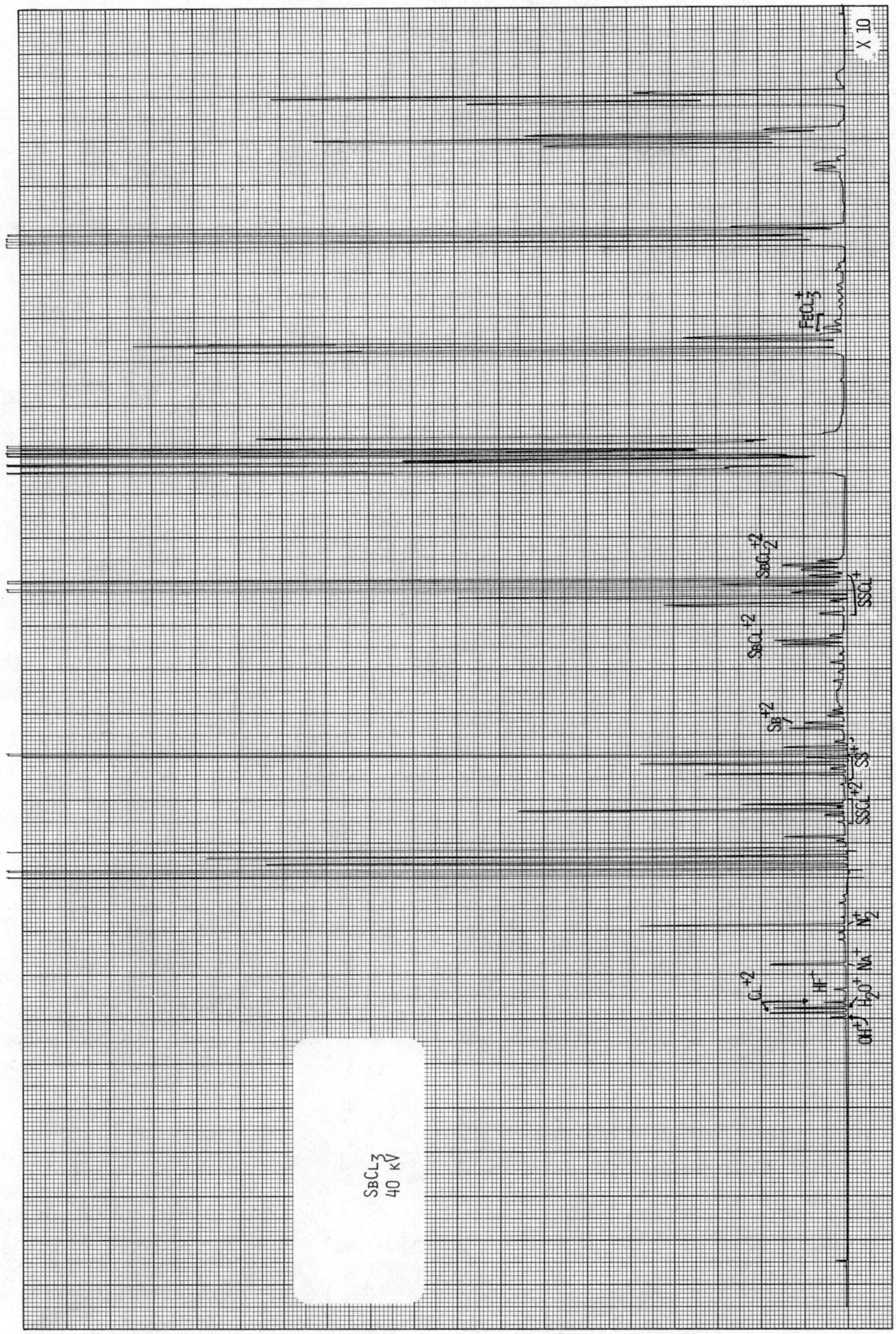
SbCl3
40 kV
X 10
FeCl3+
SbCl2+2
SbCl+2
SSCl+
Sb+2
SS+
SSCl+2
N2+
Na+
H2O+
HF+
OH+
C+2

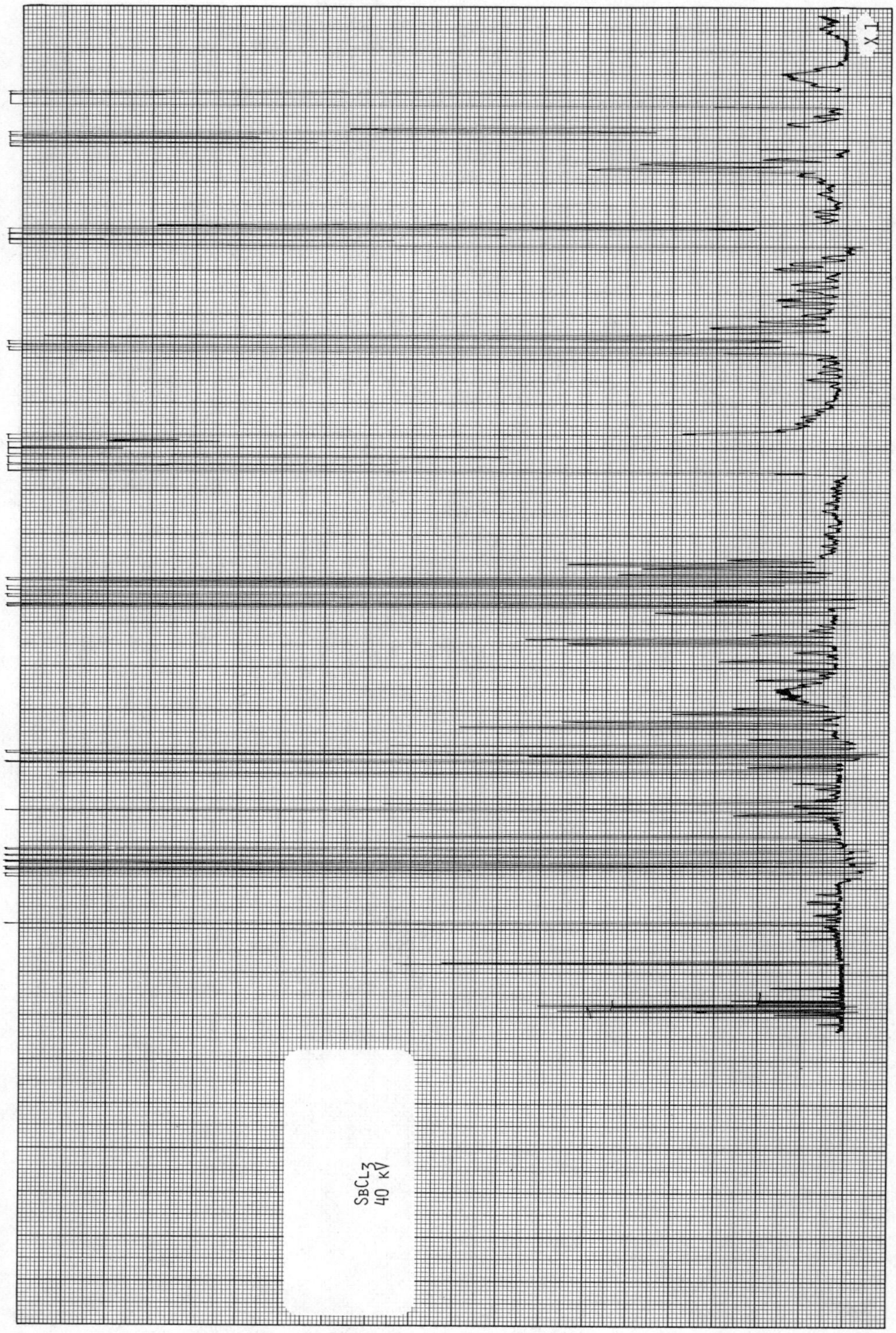
X1
SbCl3
40 kV

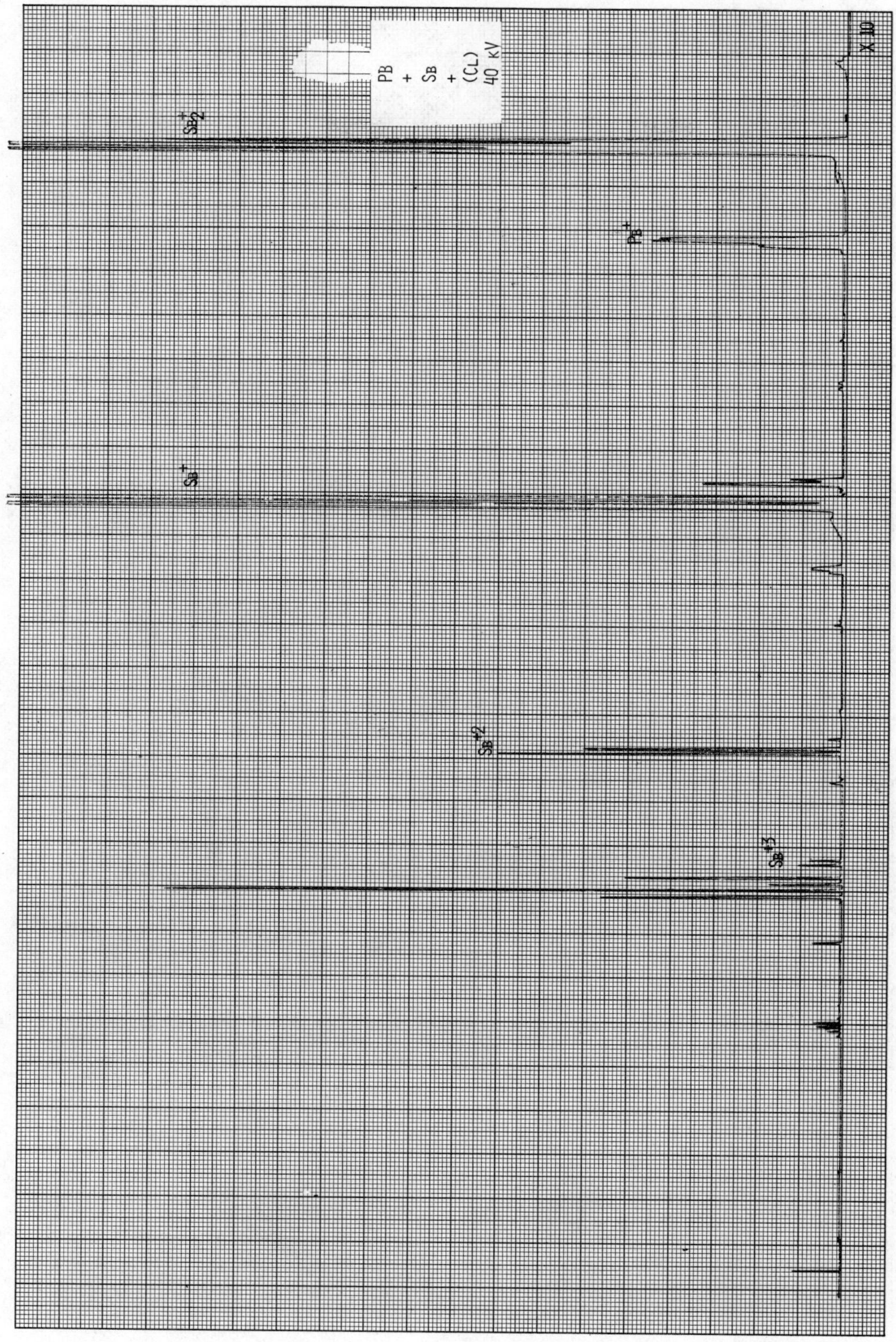
PB
+
SB
+
(CL)
40 kV
X 10
Sb_2^+
Pb^+
Sb^+
Sb^{+2}
Sb^{+3}

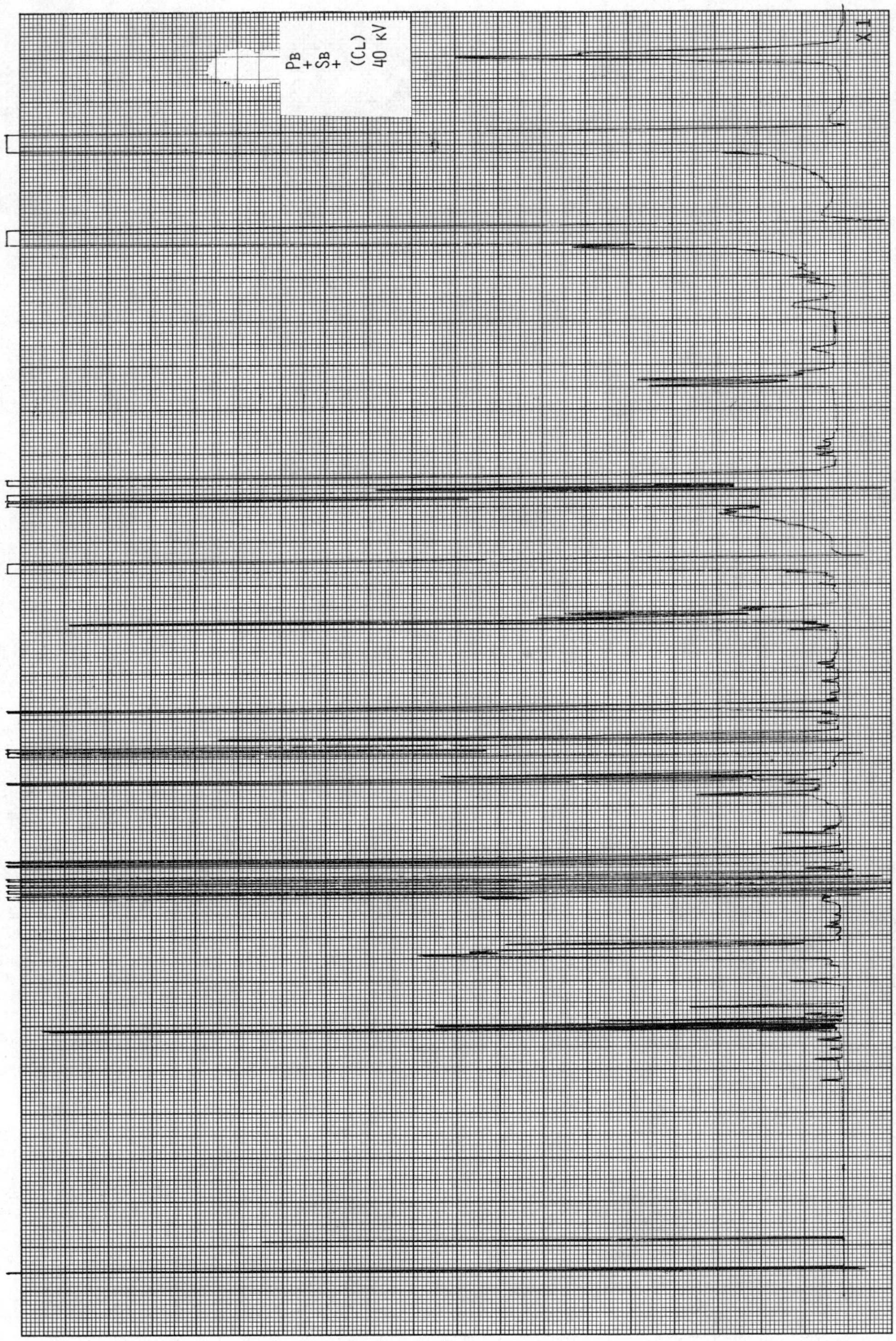
PB + SB + (CL) 40 kV
X1

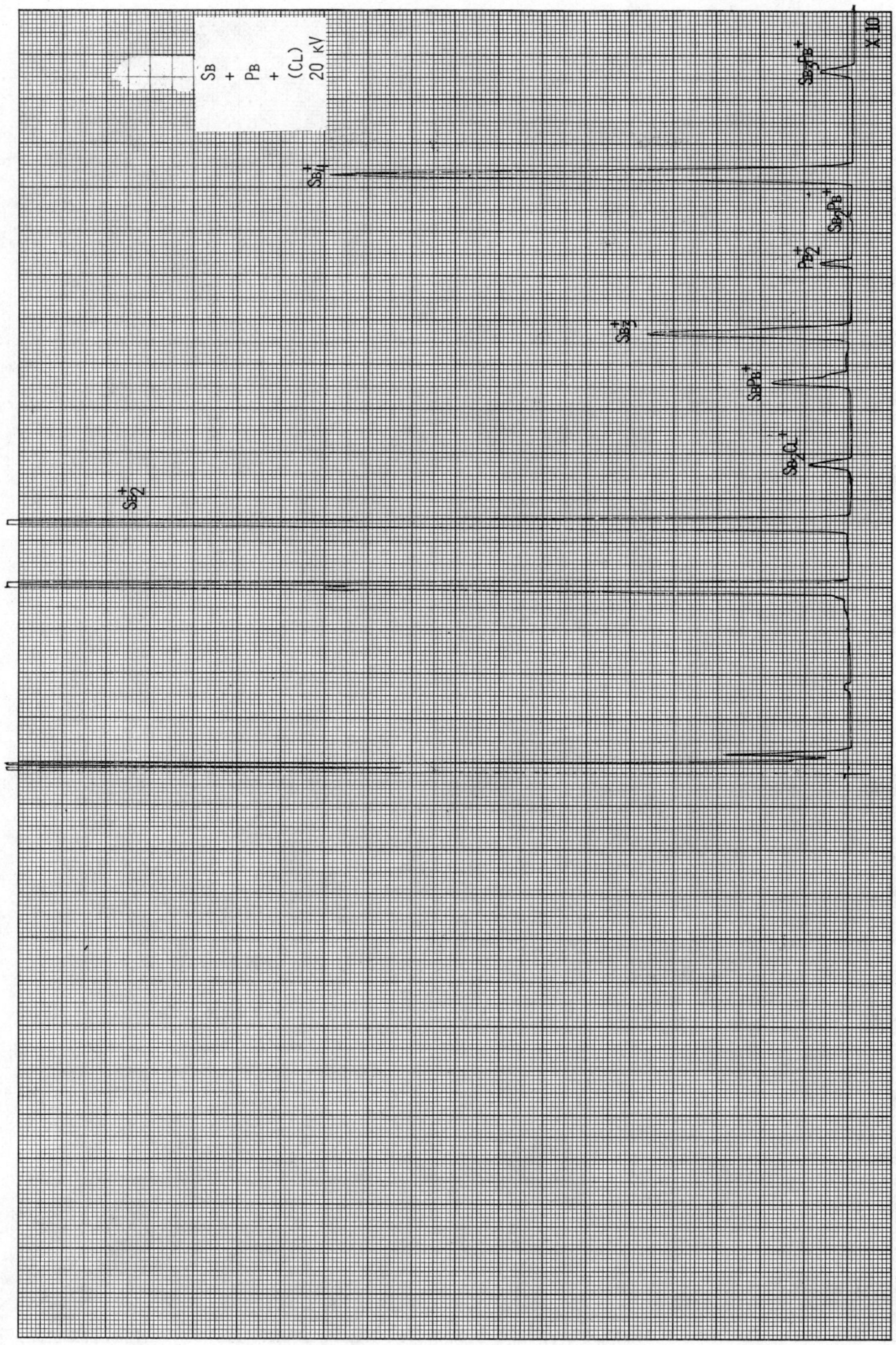
Sb
+
Pb
+
(CL)
20 kV
X 10

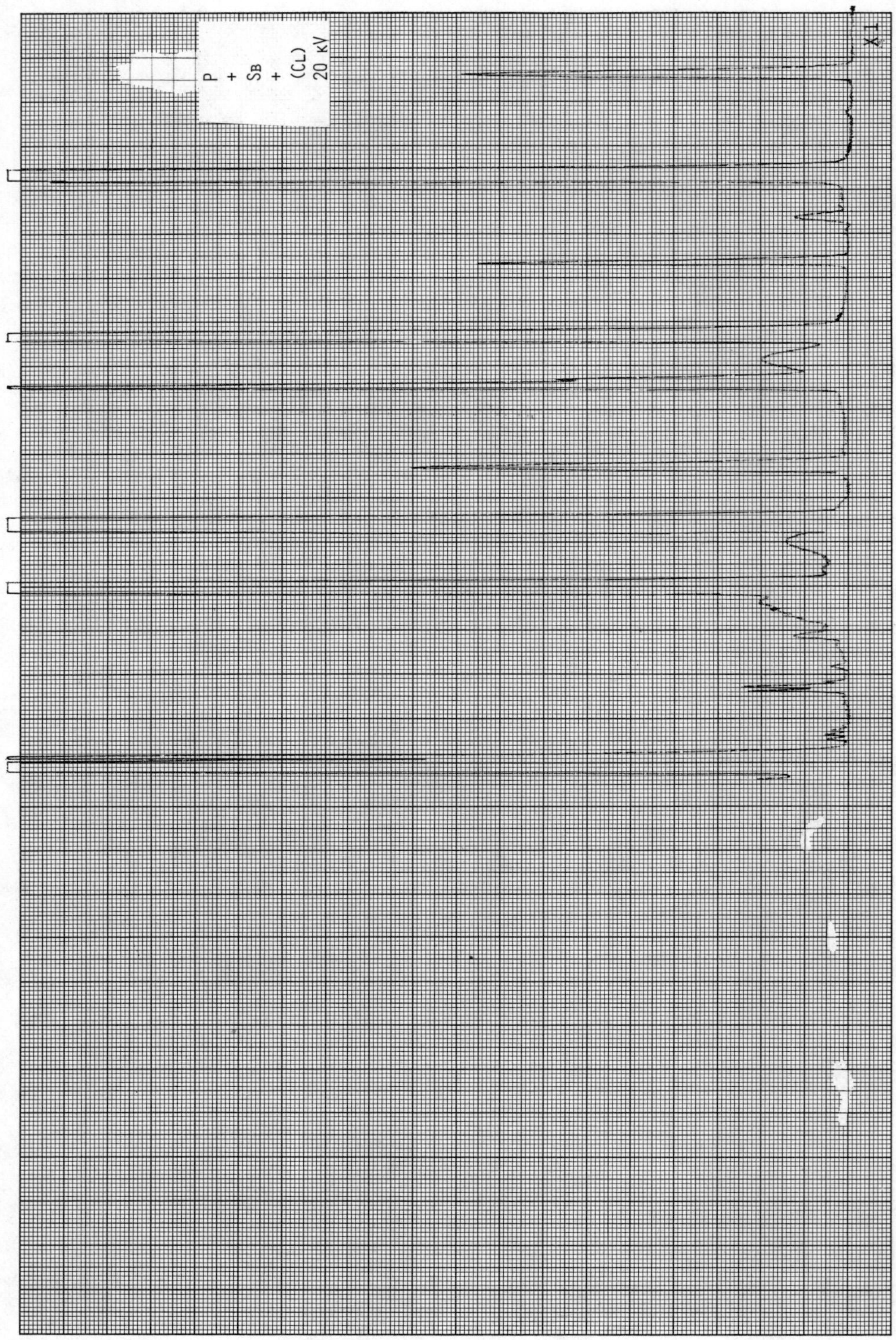
P
+
SB
+
(CL)
20 kV
X1

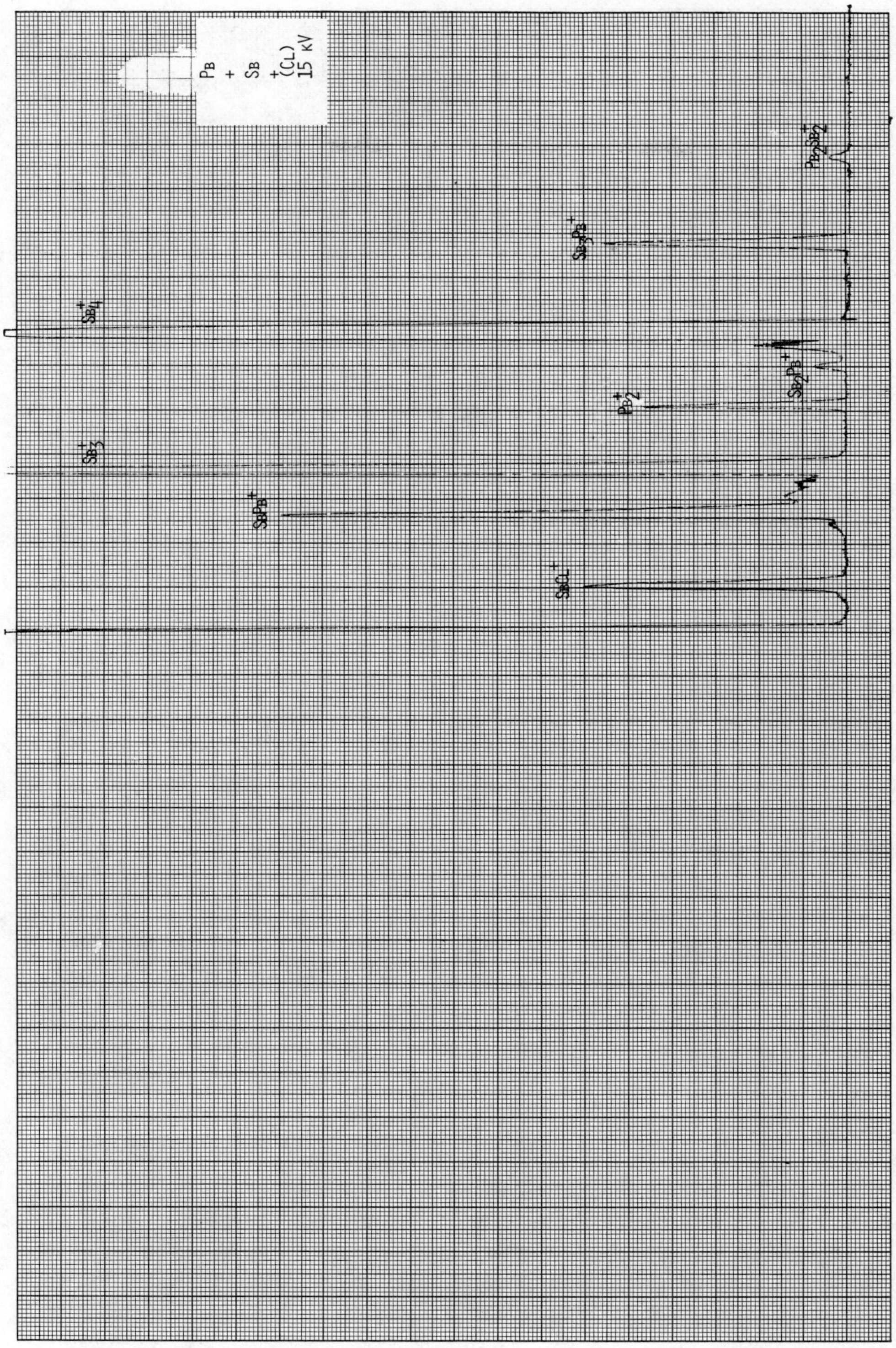
PB
+
SB
(CL)
15 kV
SB4+
SB3+
PB2+

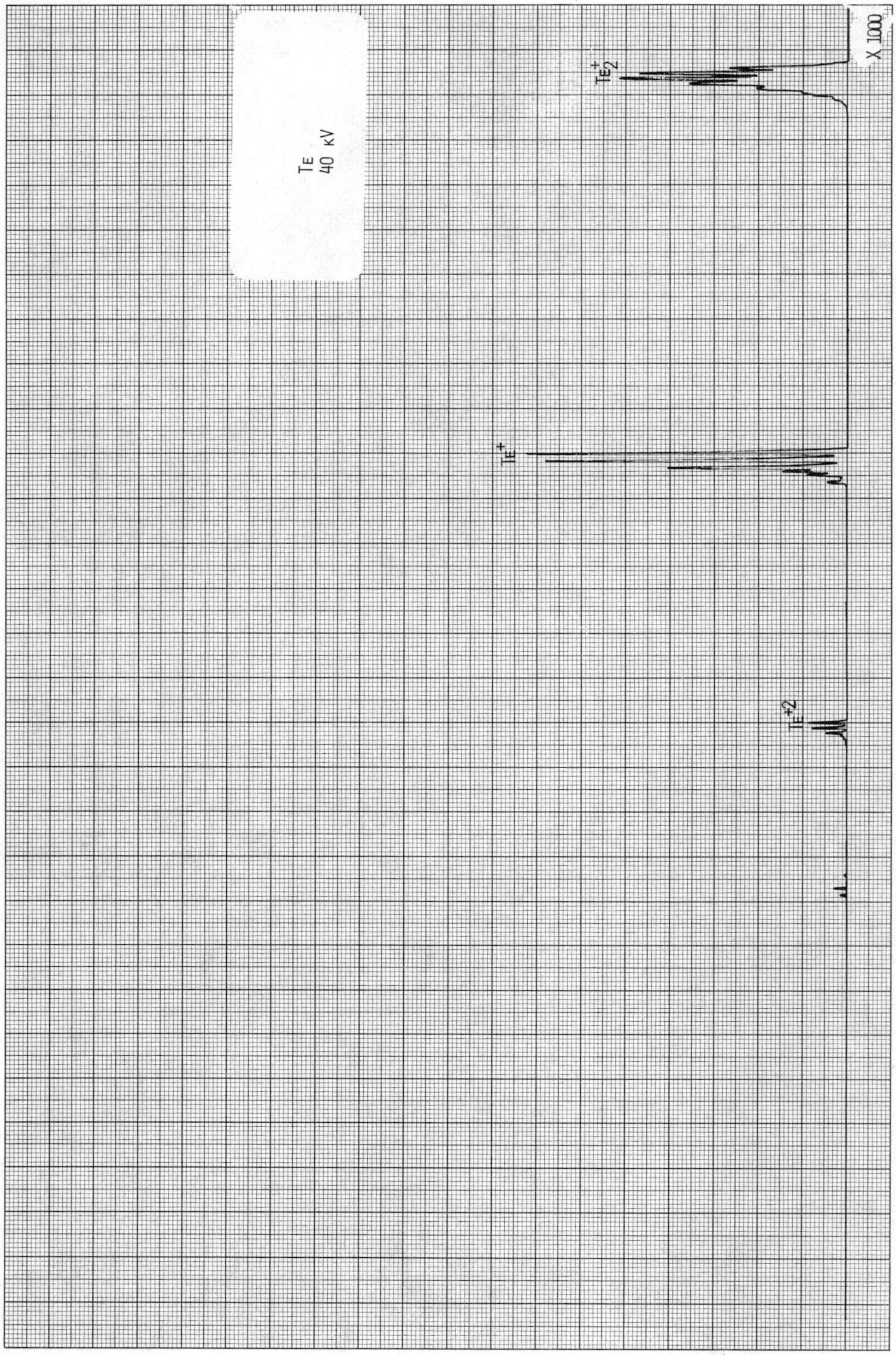
Te
40 kV
X 1000
Te_2^+
Te^+
Te^{+2}

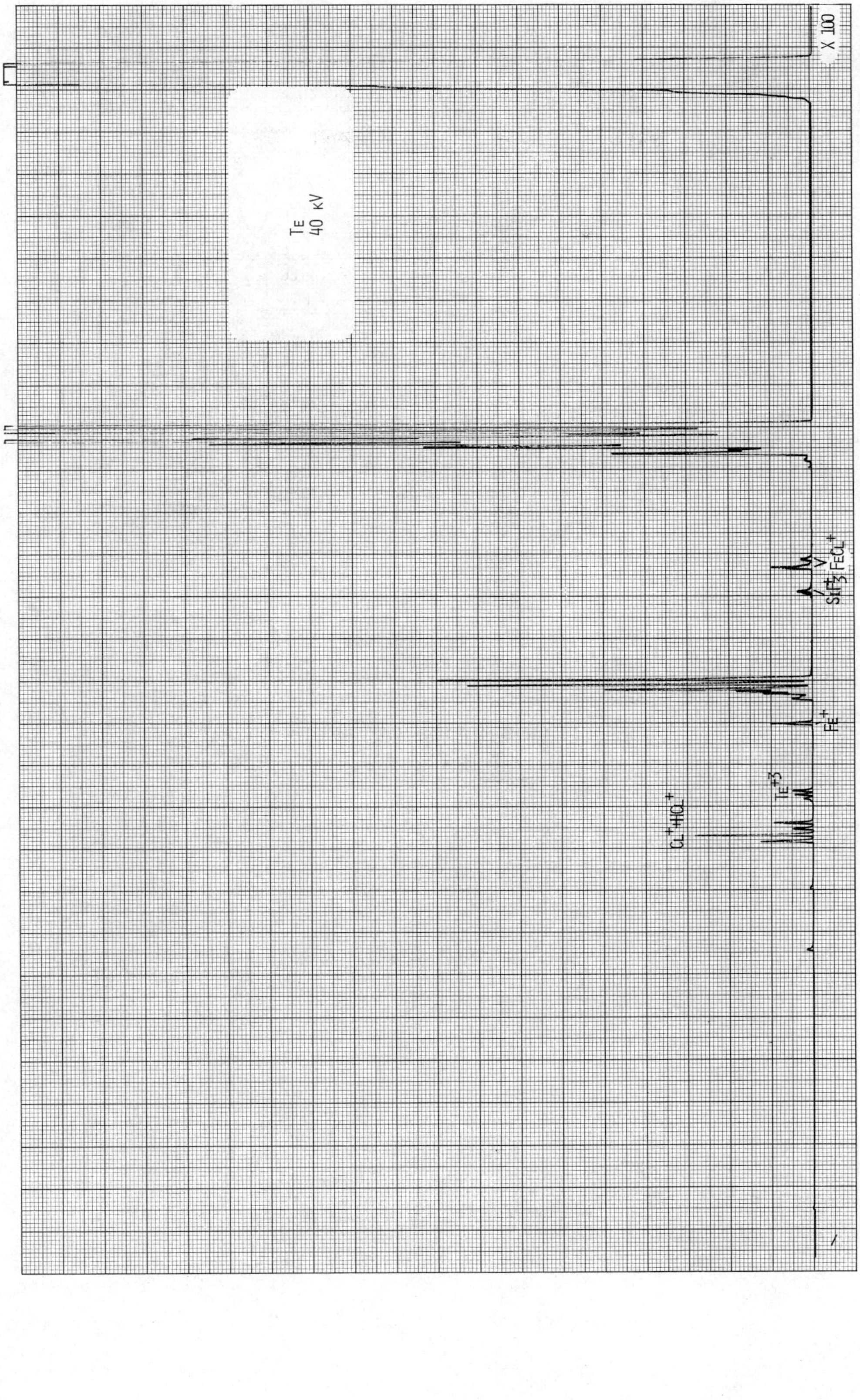
X 100
Te
40 kV
$FeCl_4^+$
Fe^+
Te^{+3}
$Cl^+ + HCl^+$

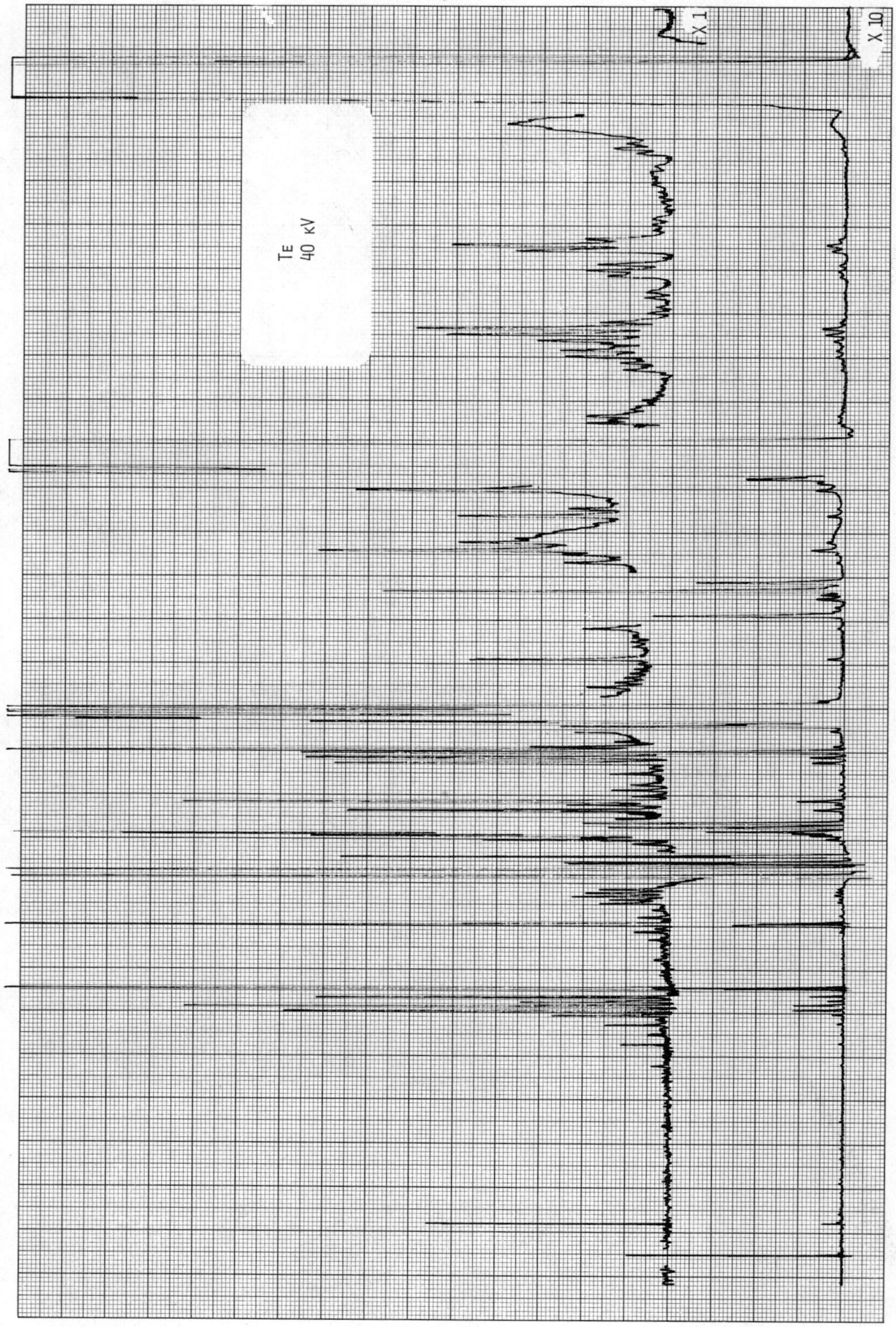
Te
40 kV
X 1
X 10

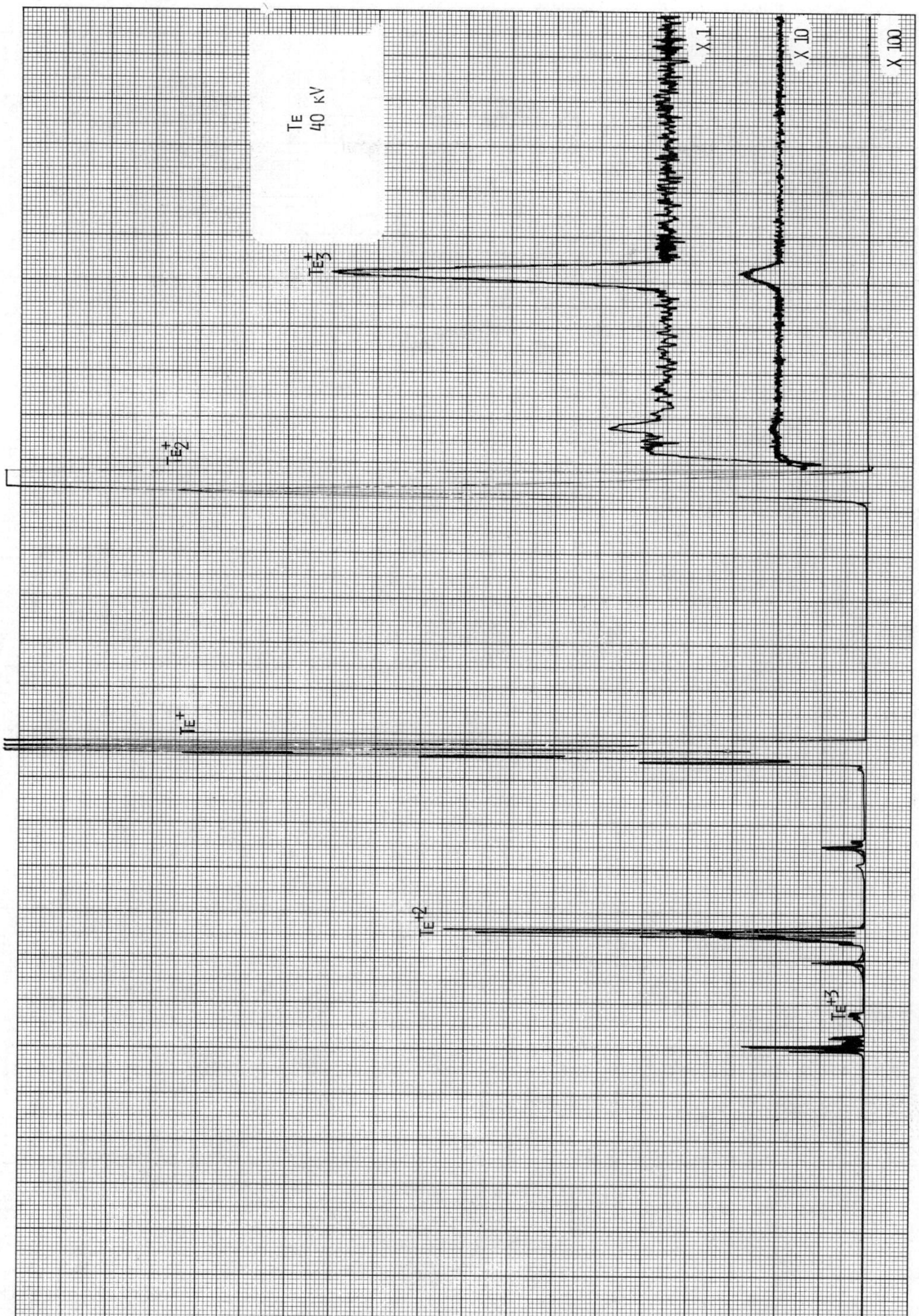
TE
40 kV
TE3+
TE2+
TE+
TE+2
TE+3
X 1
X 10
X 100

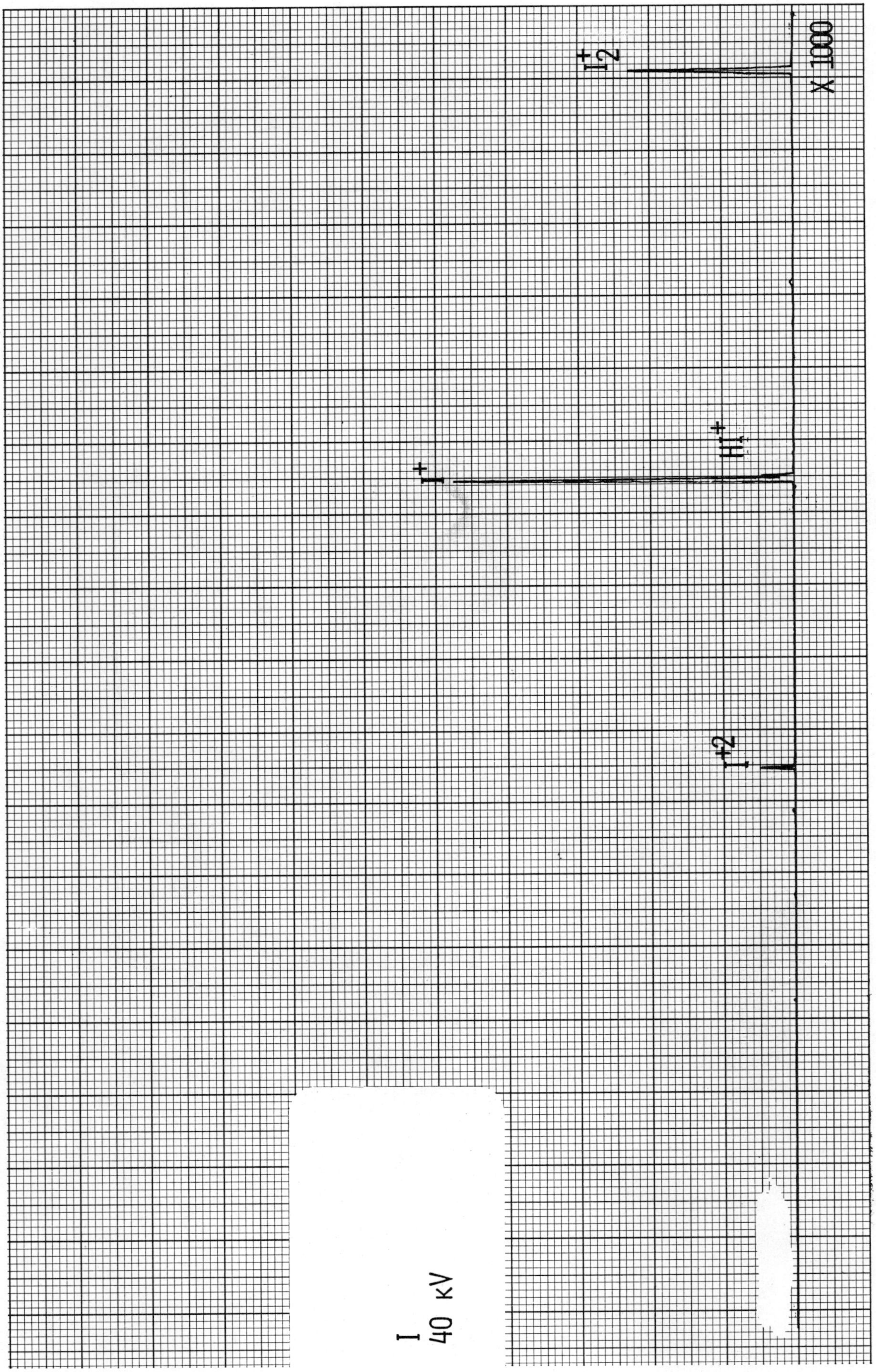

I_2^+
X 1000
HI^+
I^+
I^{+2}
I
40 kV

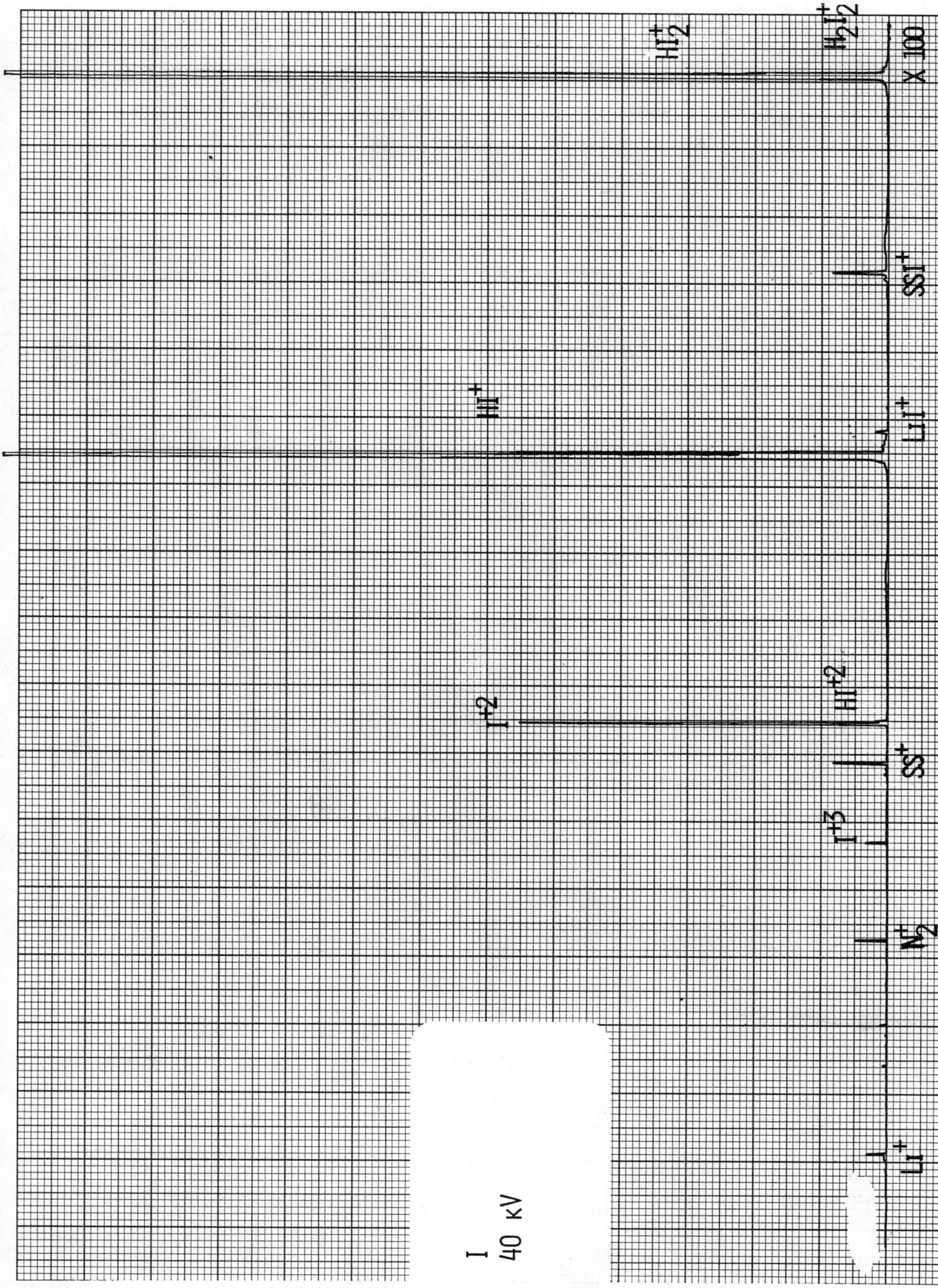

HI_2^+
$H_2I_2^+$
X 100
SSI^+
HI^+
LiI^+
I^{+2}
HI^{+2}
SS^+
I^{+3}
N_2^+
Li^+
I
40 kV

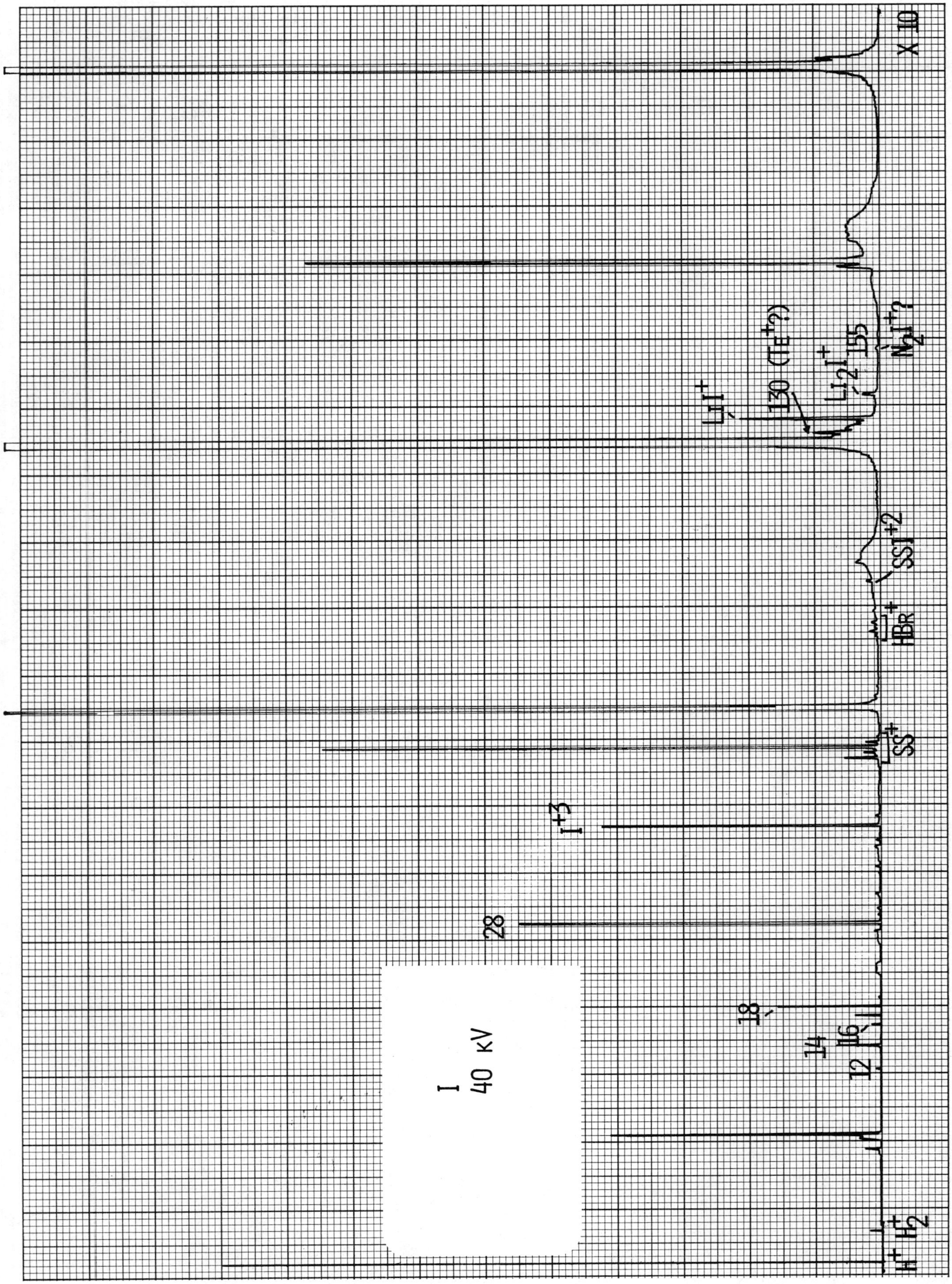
X 10
LiI+
130 (Te+?)
Li2I+
155
Na2I+?
SSI+2
HBr+
SS+
I+3
28
18
14
16
12
I
40 kV
H+ H2+

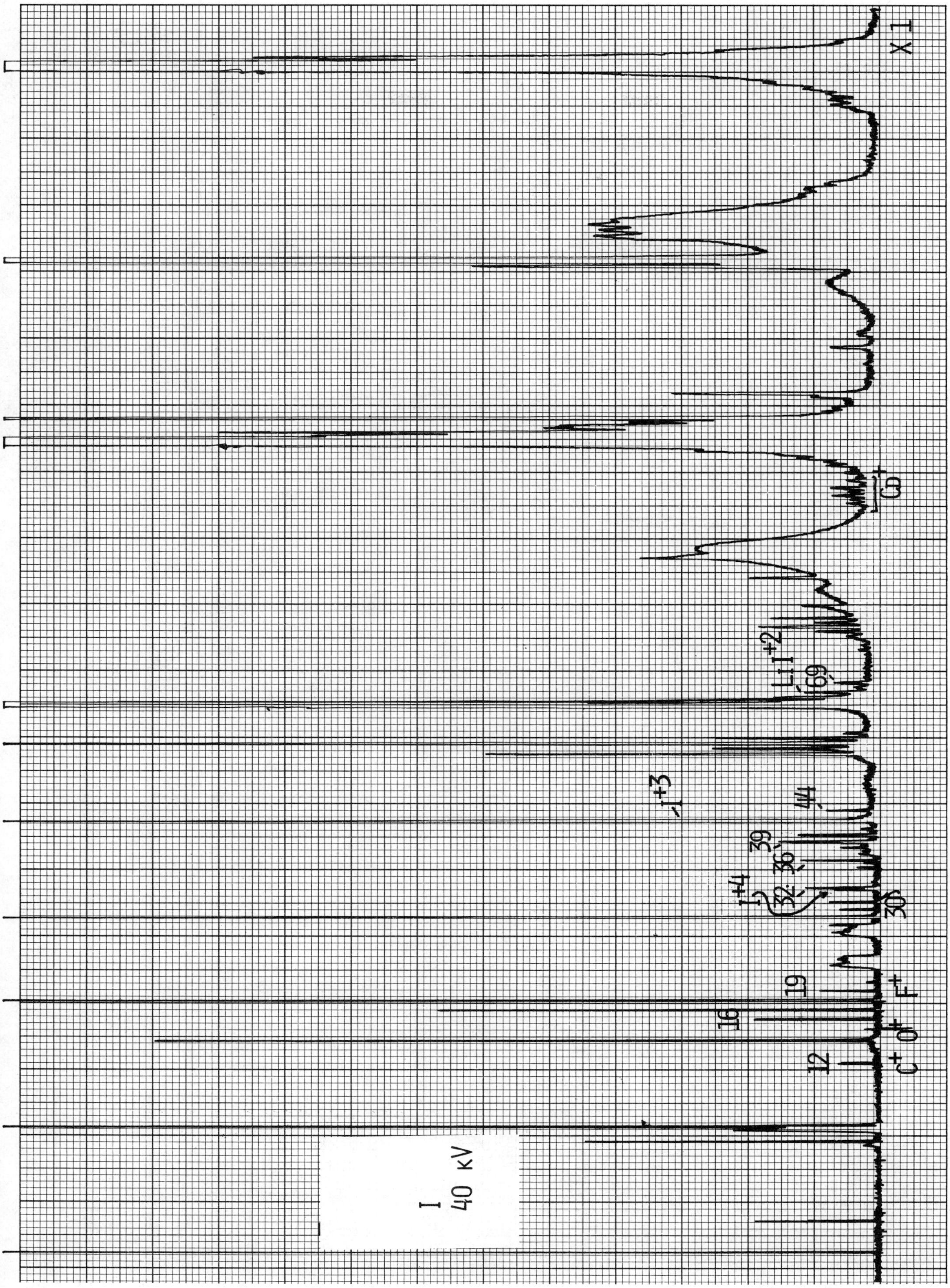

I
40 kV
X 1
12
16
19
30
32
36
39
44
69

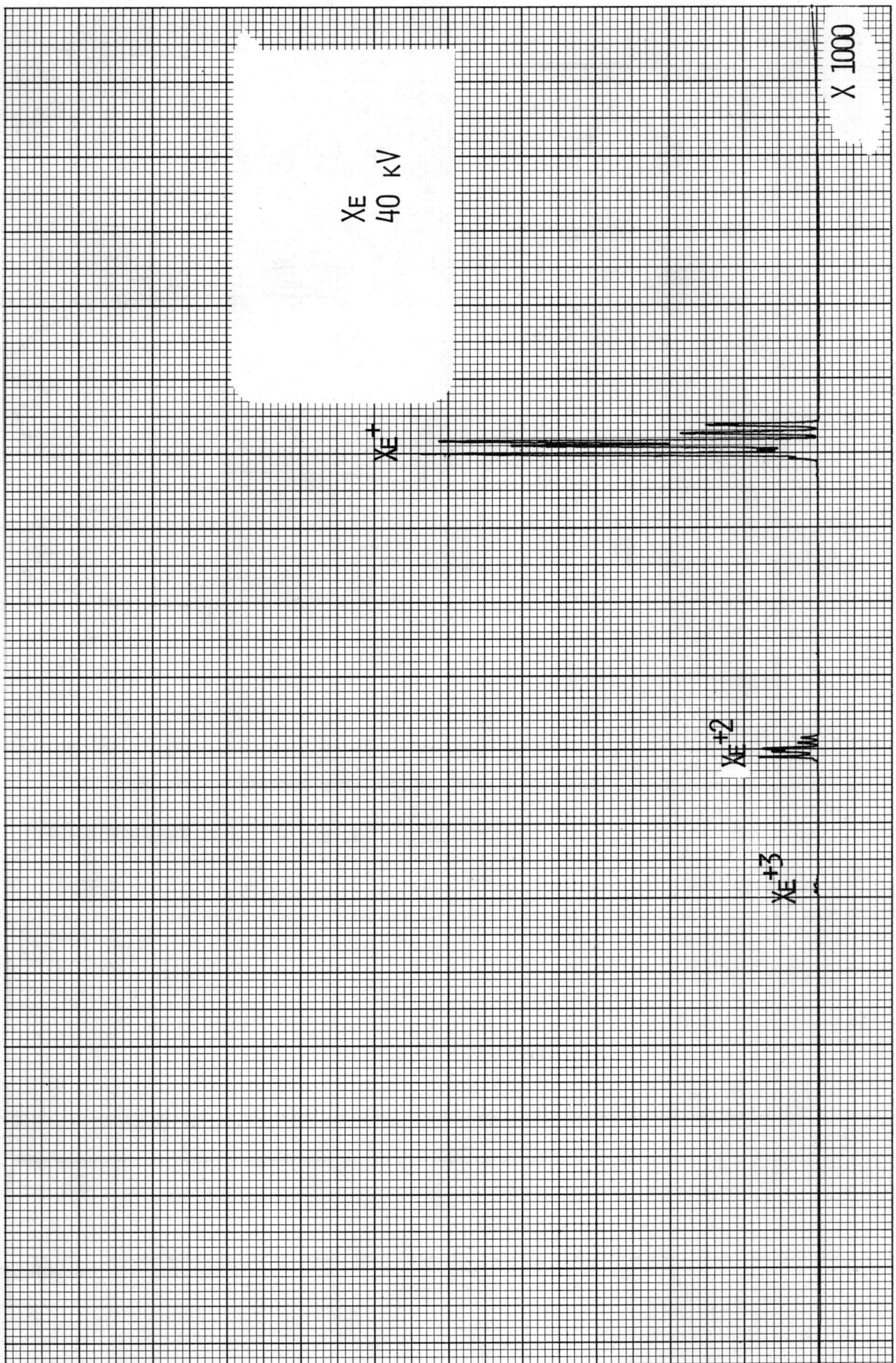
XE
40 kV
X 1000
XE+
XE+2
XE+3

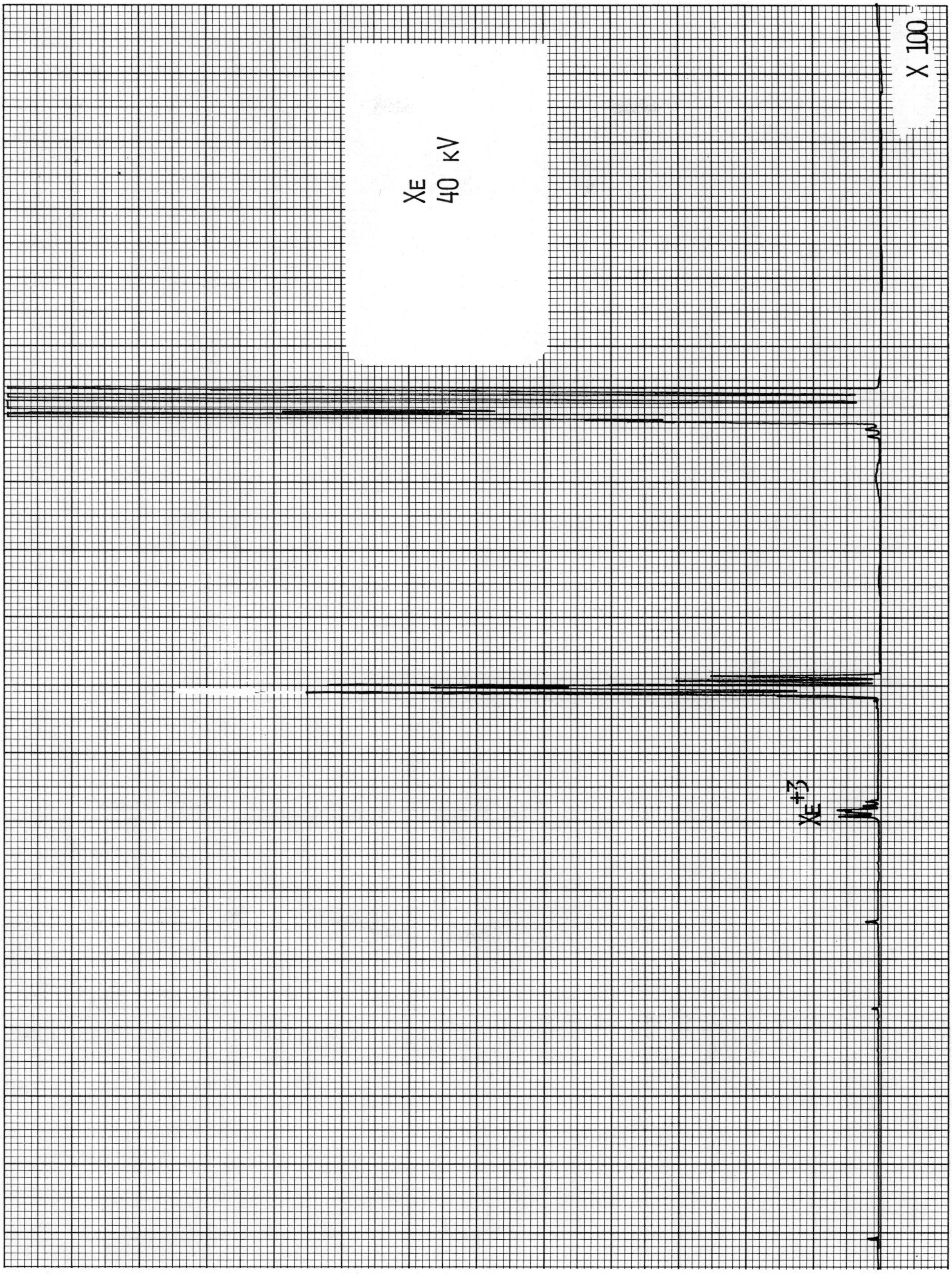
X 100
Xe
40 kV
Xe+3

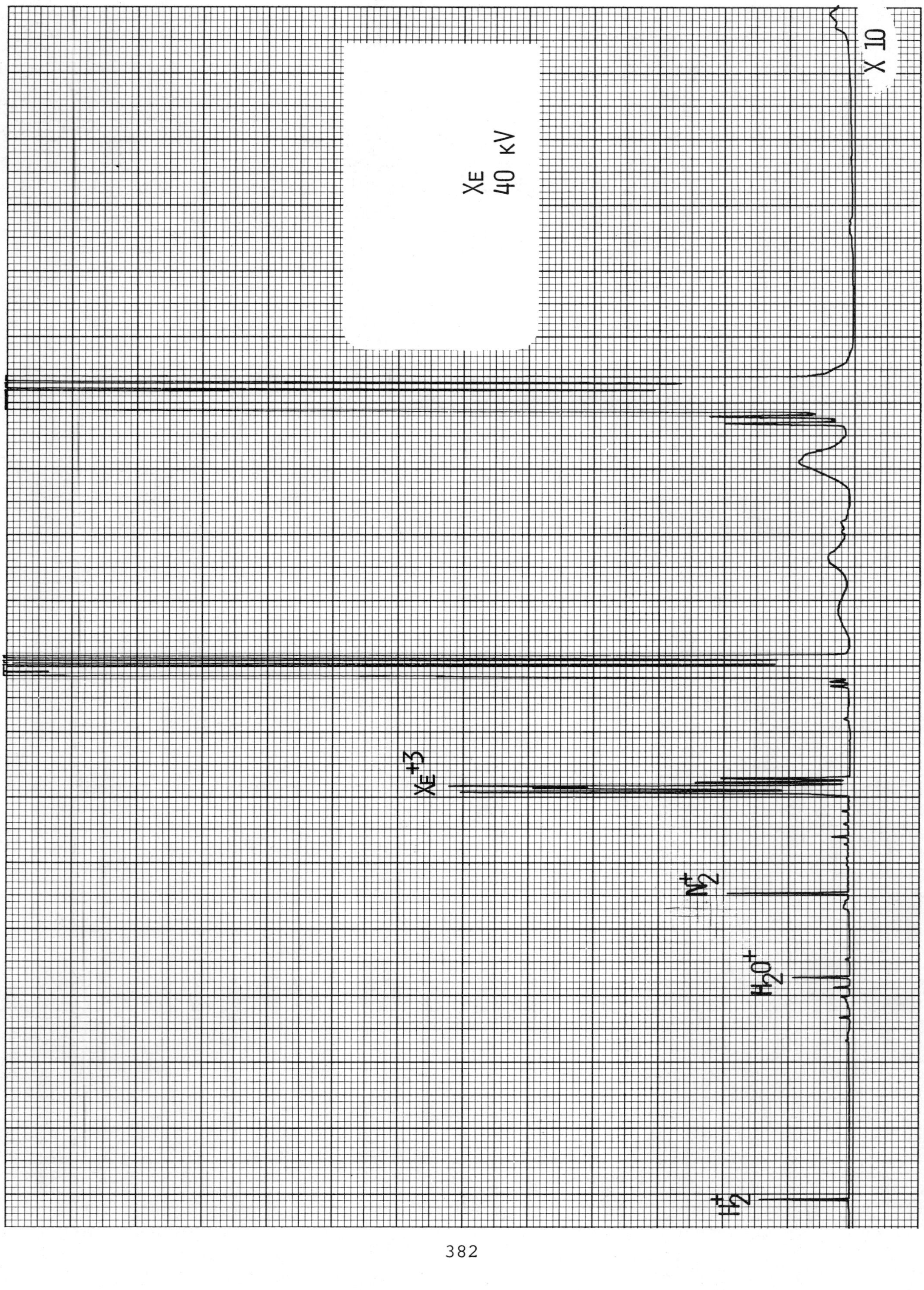
X 10
XE
40 kV
XE+3
N2+
H2O+
H2+

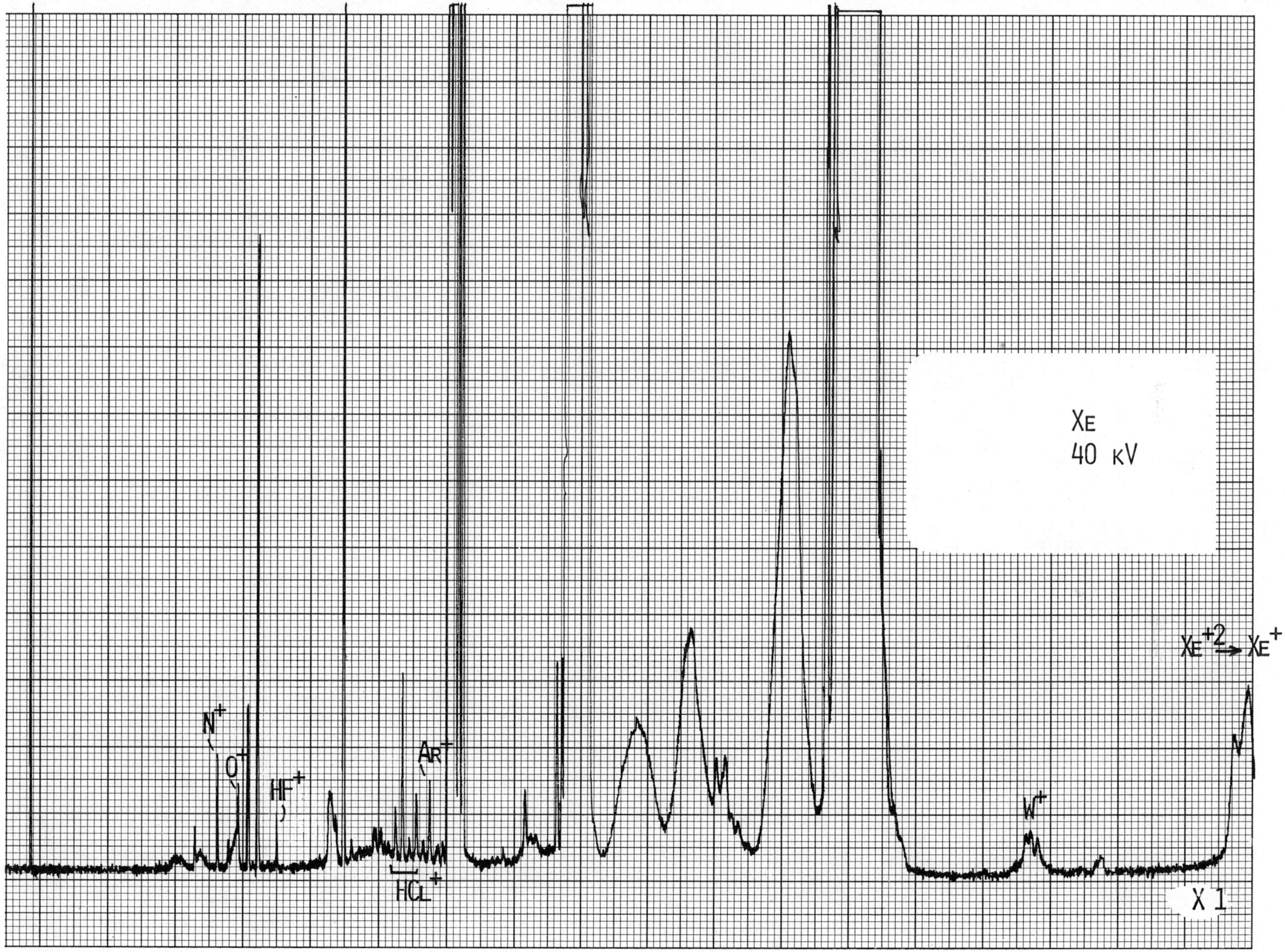

XE
40 KV
XE+2 → XE+
X 1
W+
AR+
HCL+
HF+
O+
N+

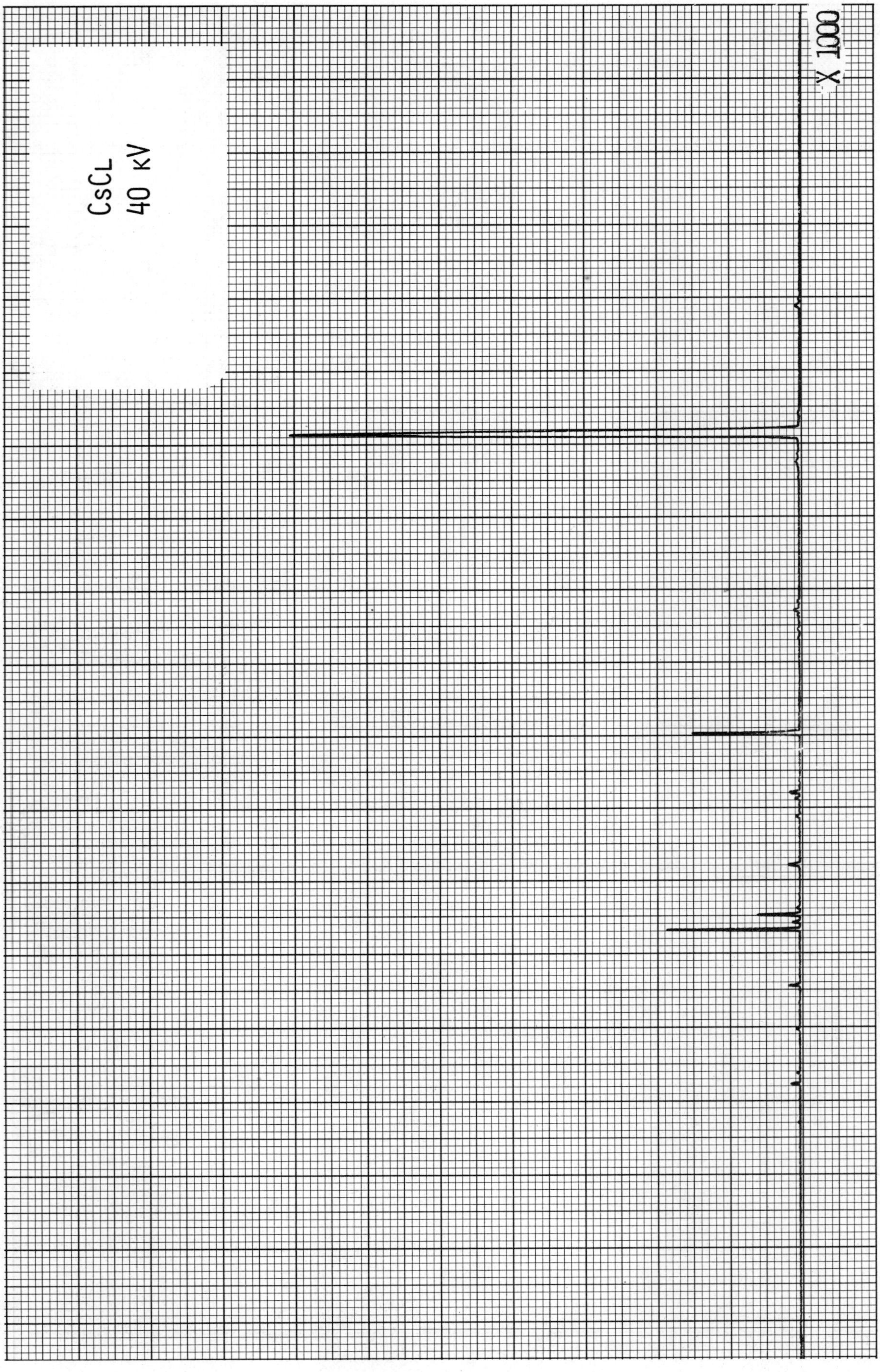
CsCL
40 kV
X 1000

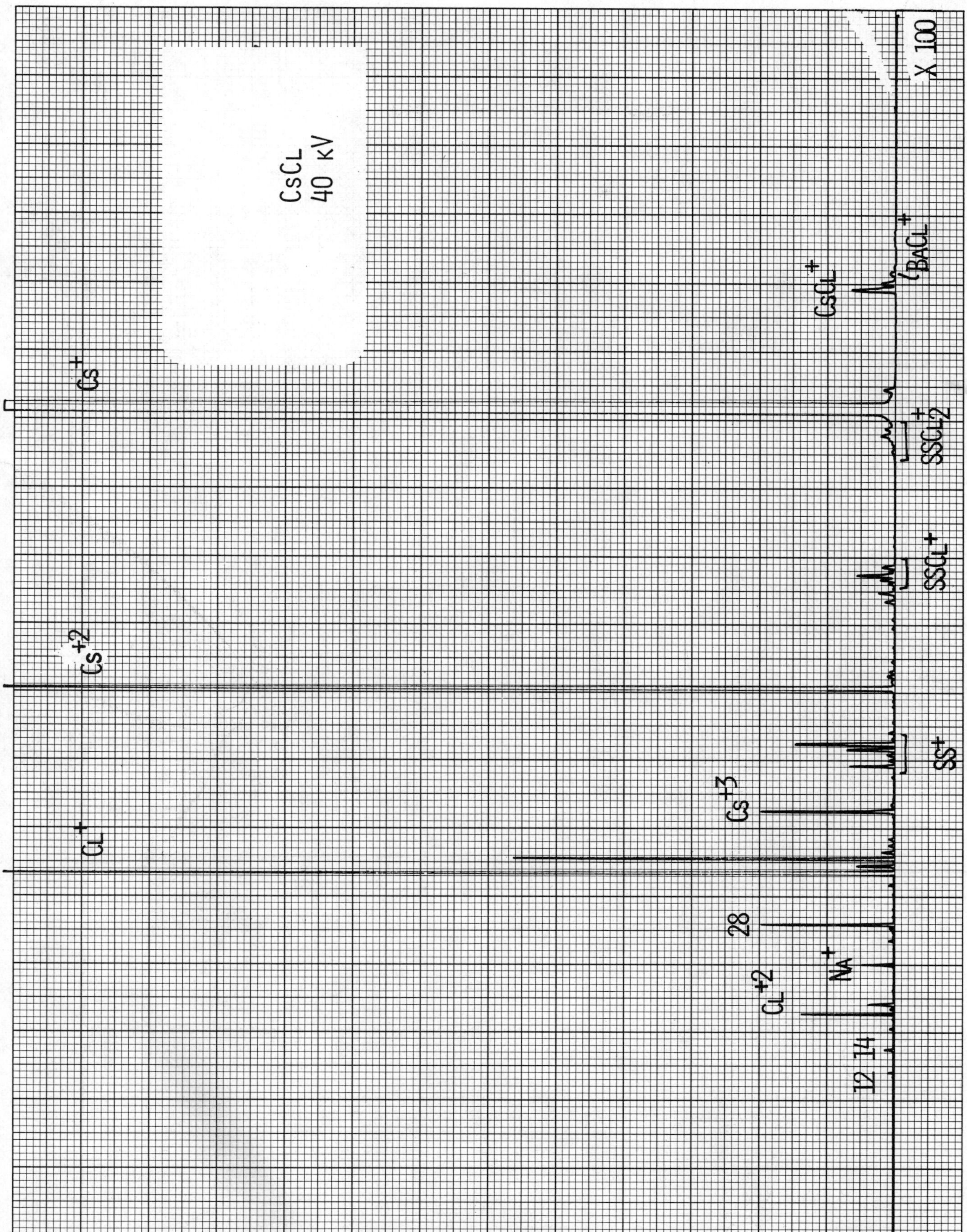
CsCl
40 kV
X 100
Cs+
Cs+2
Cl+
Cs+3
28
Na+
Cl+2
12 14
SS+
SSCl+
SSCl2+
CsCl+
BaCl+

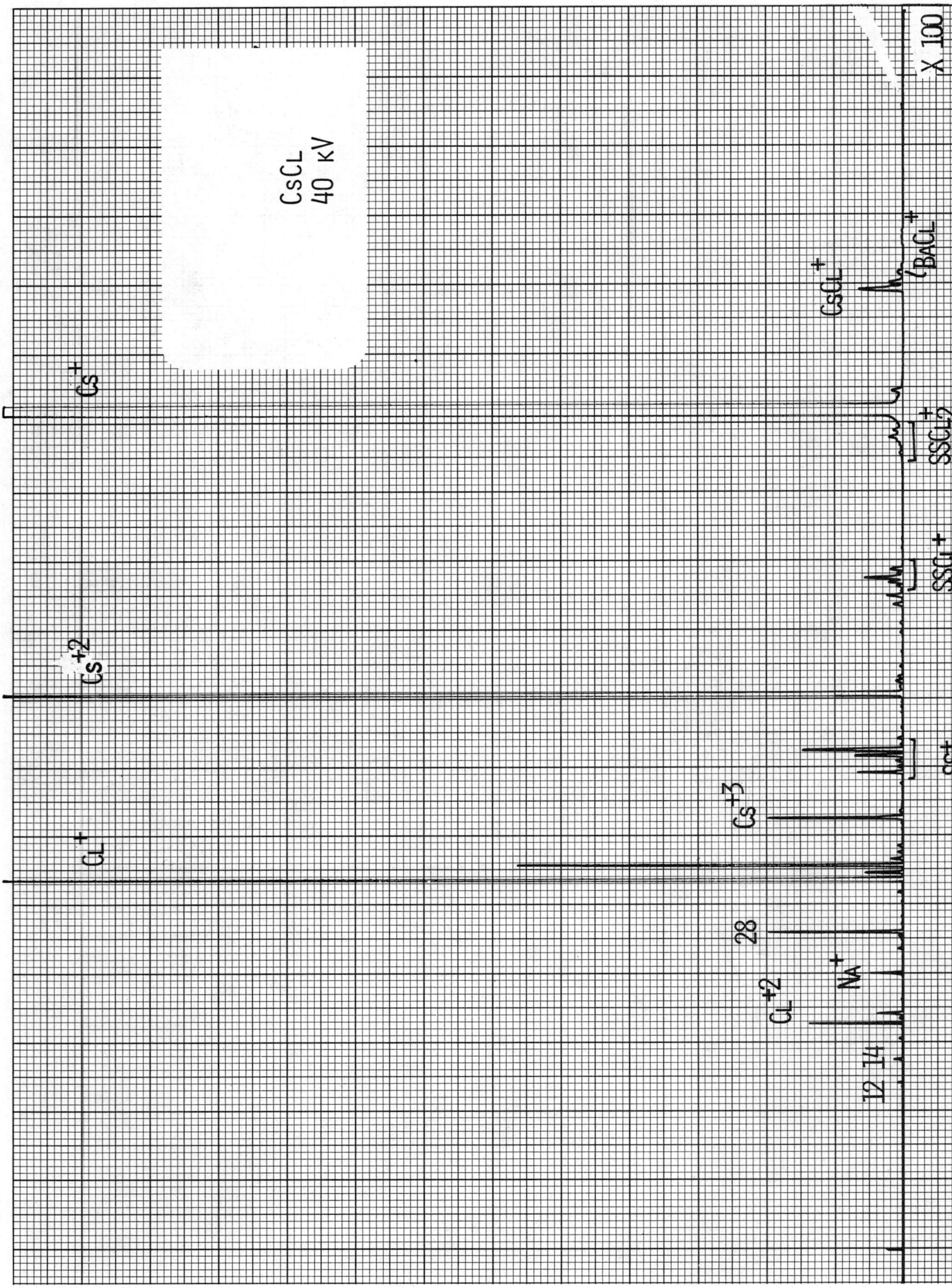
CsCL
40 kV
X 100
Cs+
Cs+2
Cs+3
CL+
CL+2
NA+
28
12 14
CsCL+
BACL+
SSCL2+
SSCL+
SS+

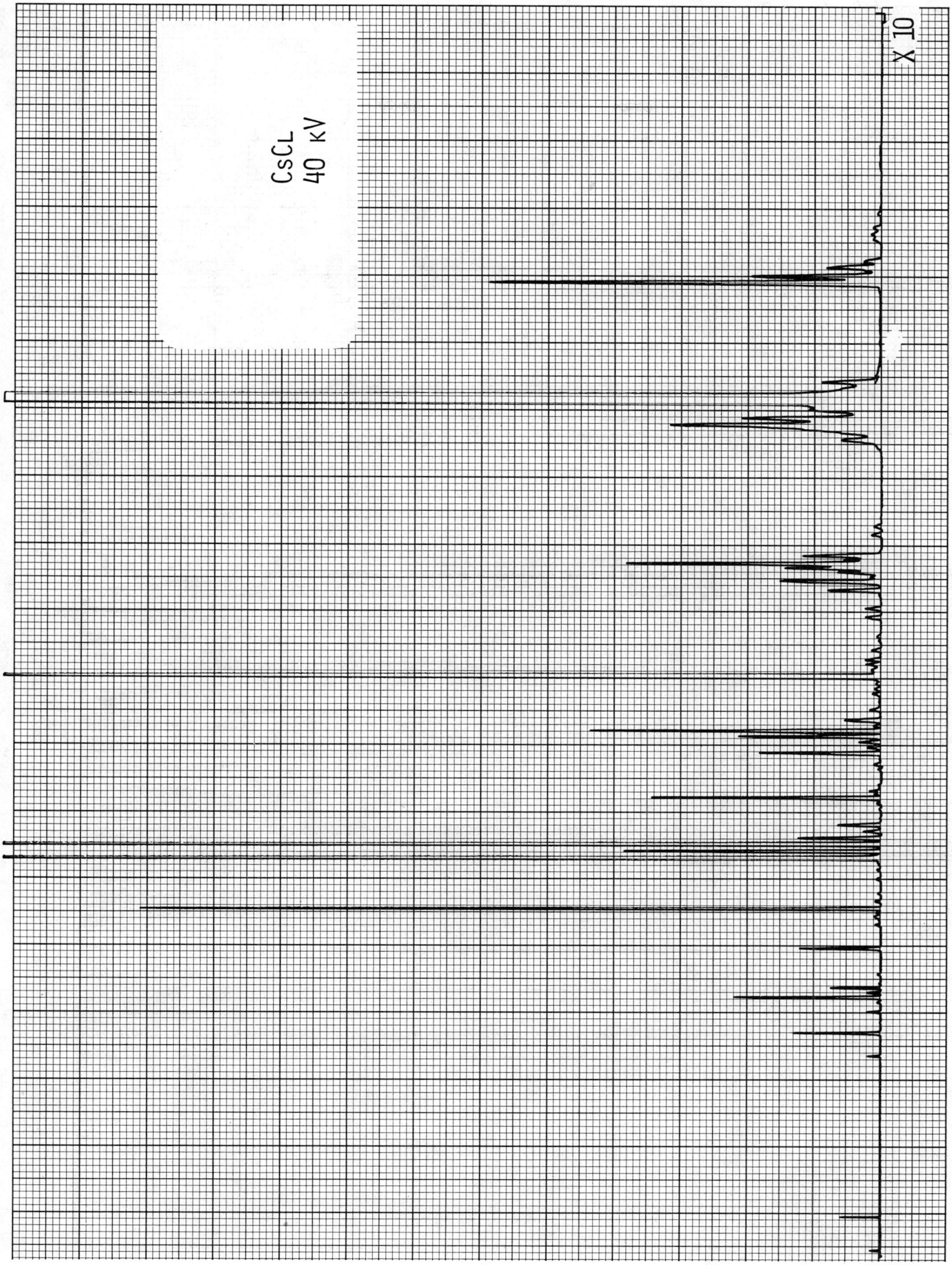
CsCL
40 kV
X 10

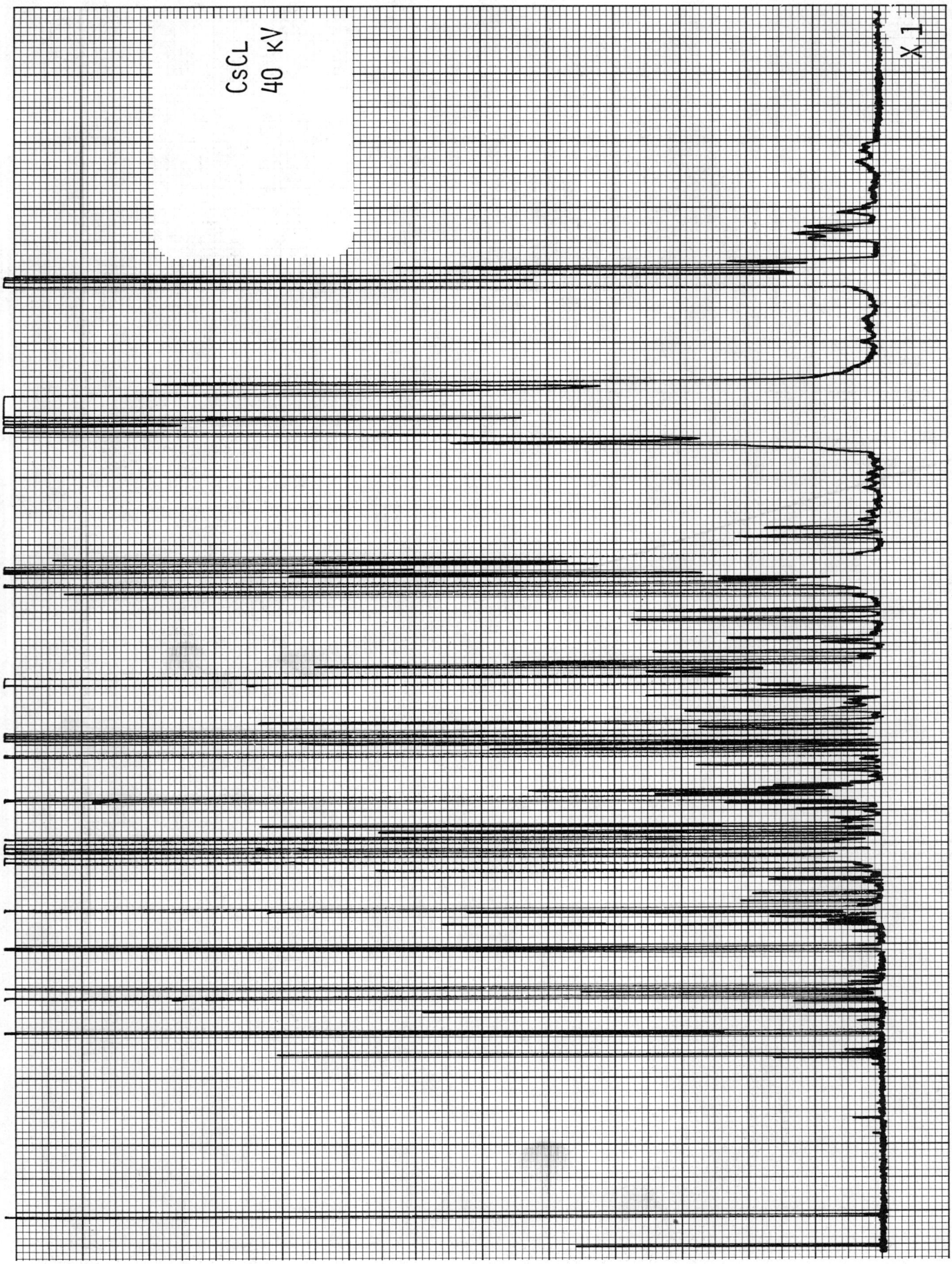
CsCL
40 kV
X 1

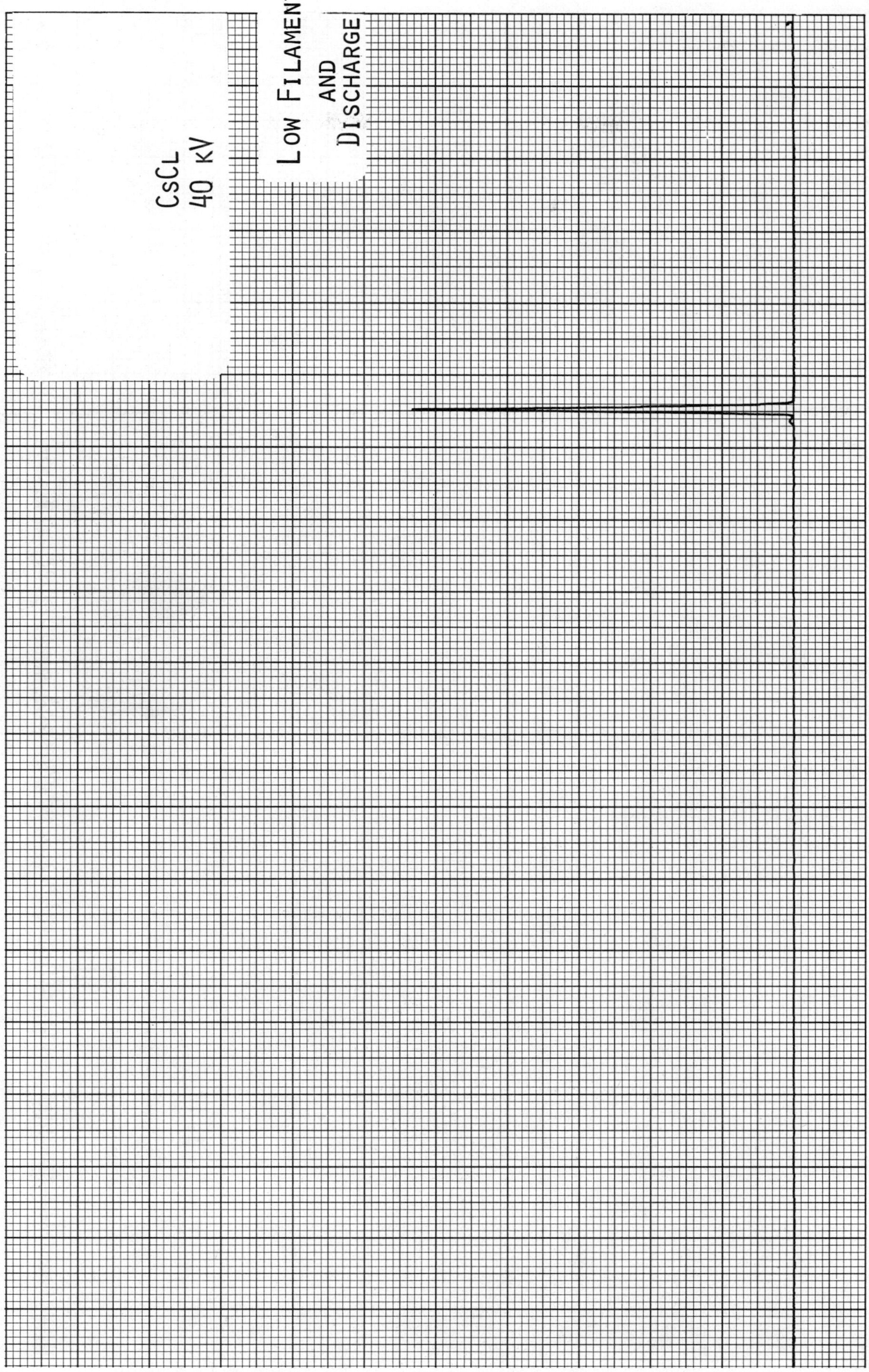
CsCl
40 kV
Low Filament
and
Discharge

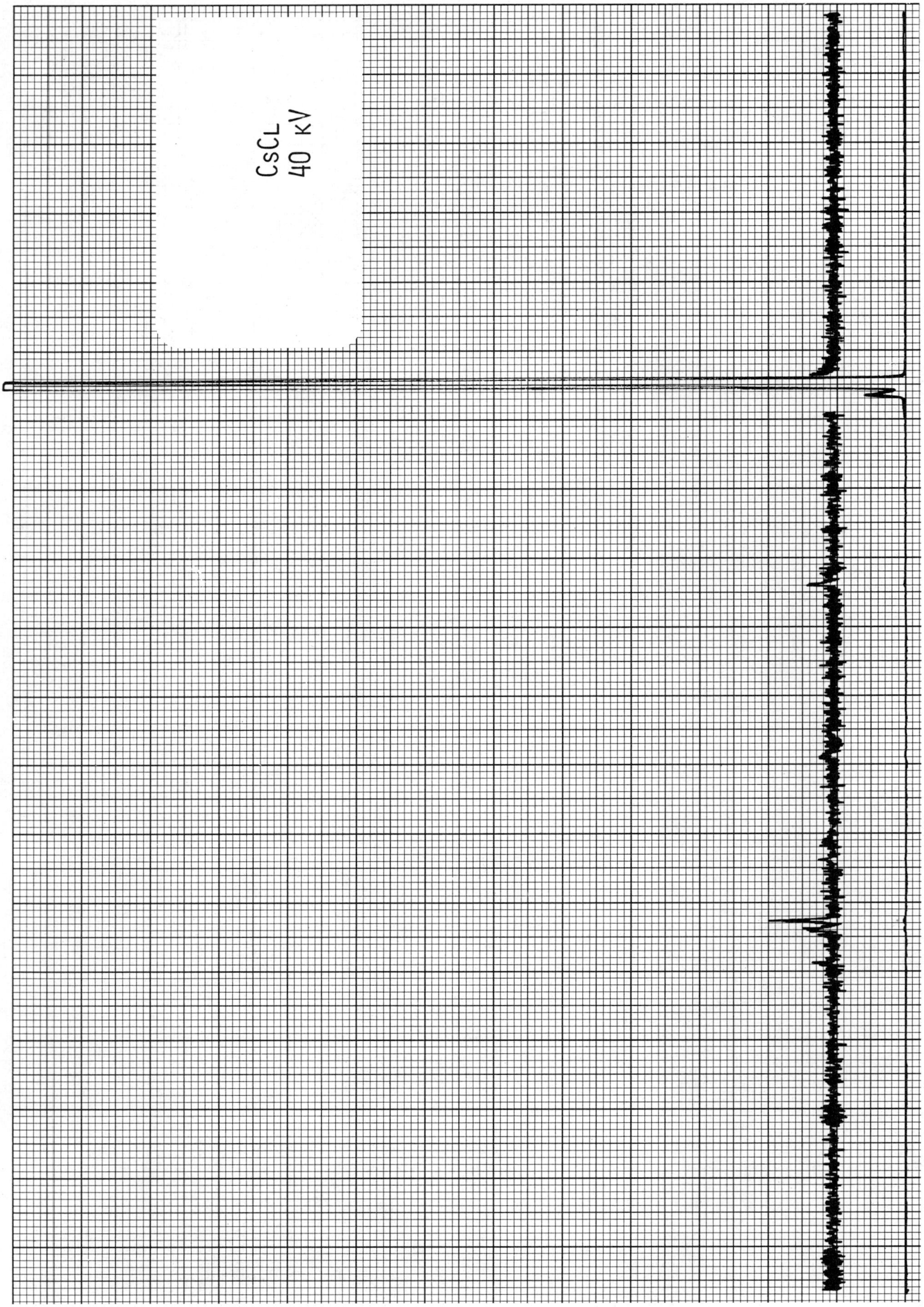
CsCL
40 kV

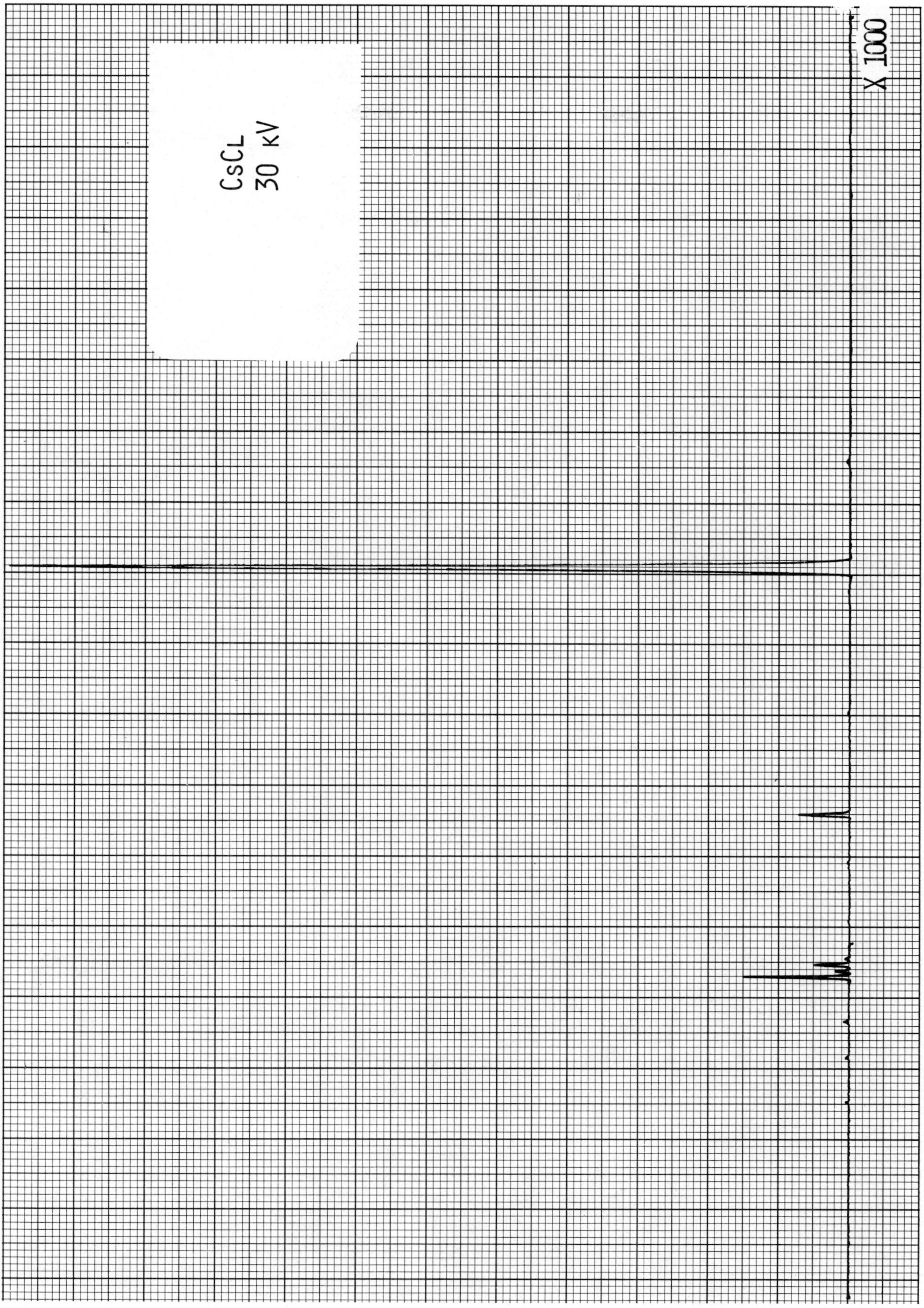
CsCL
30 kV
X 1000

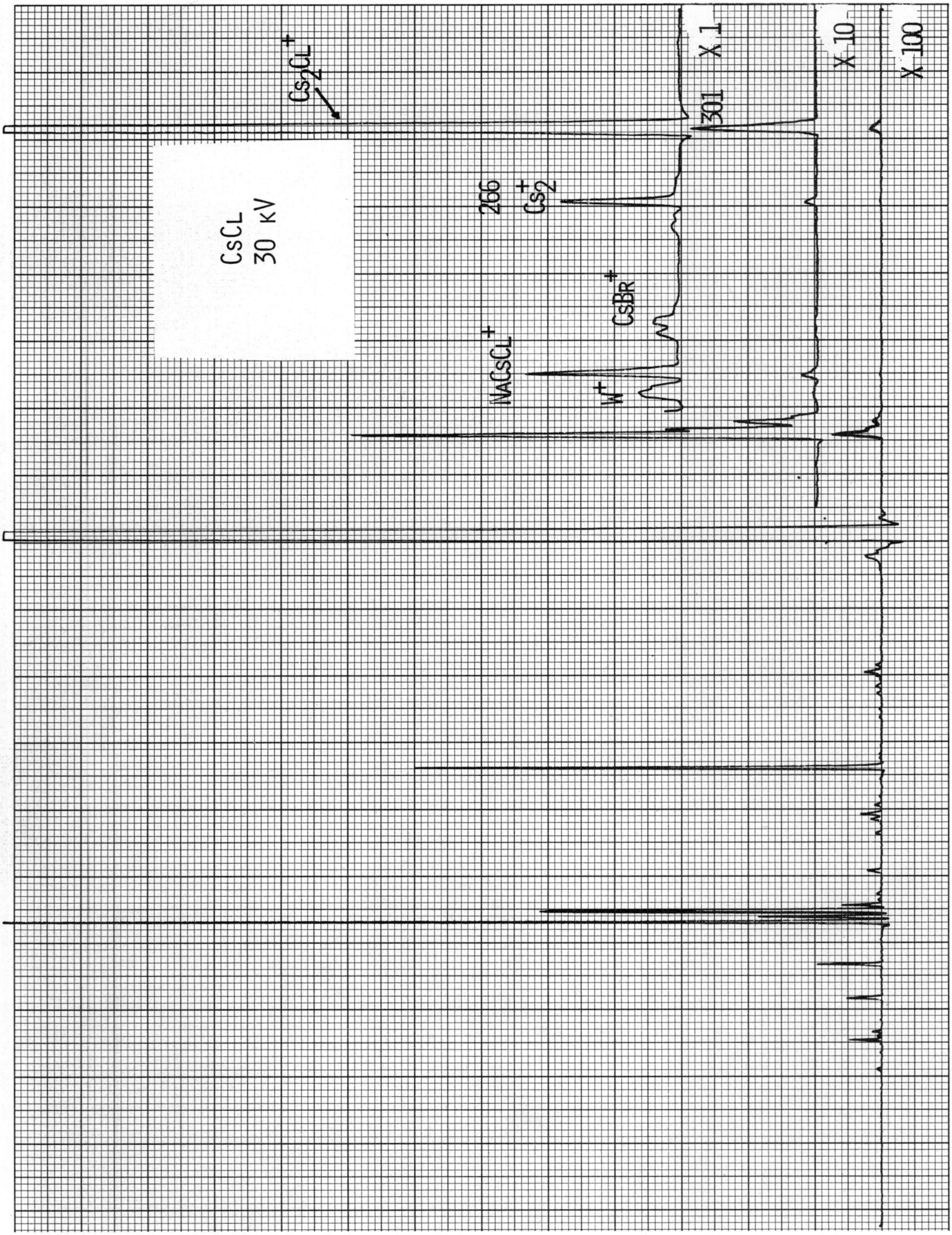
CsCL
30 KV
Cs_2CL^+
301
266
Cs_2^+
$CsBR^+$
$NACsCL^+$
W^+
X 1
X 10
X 100

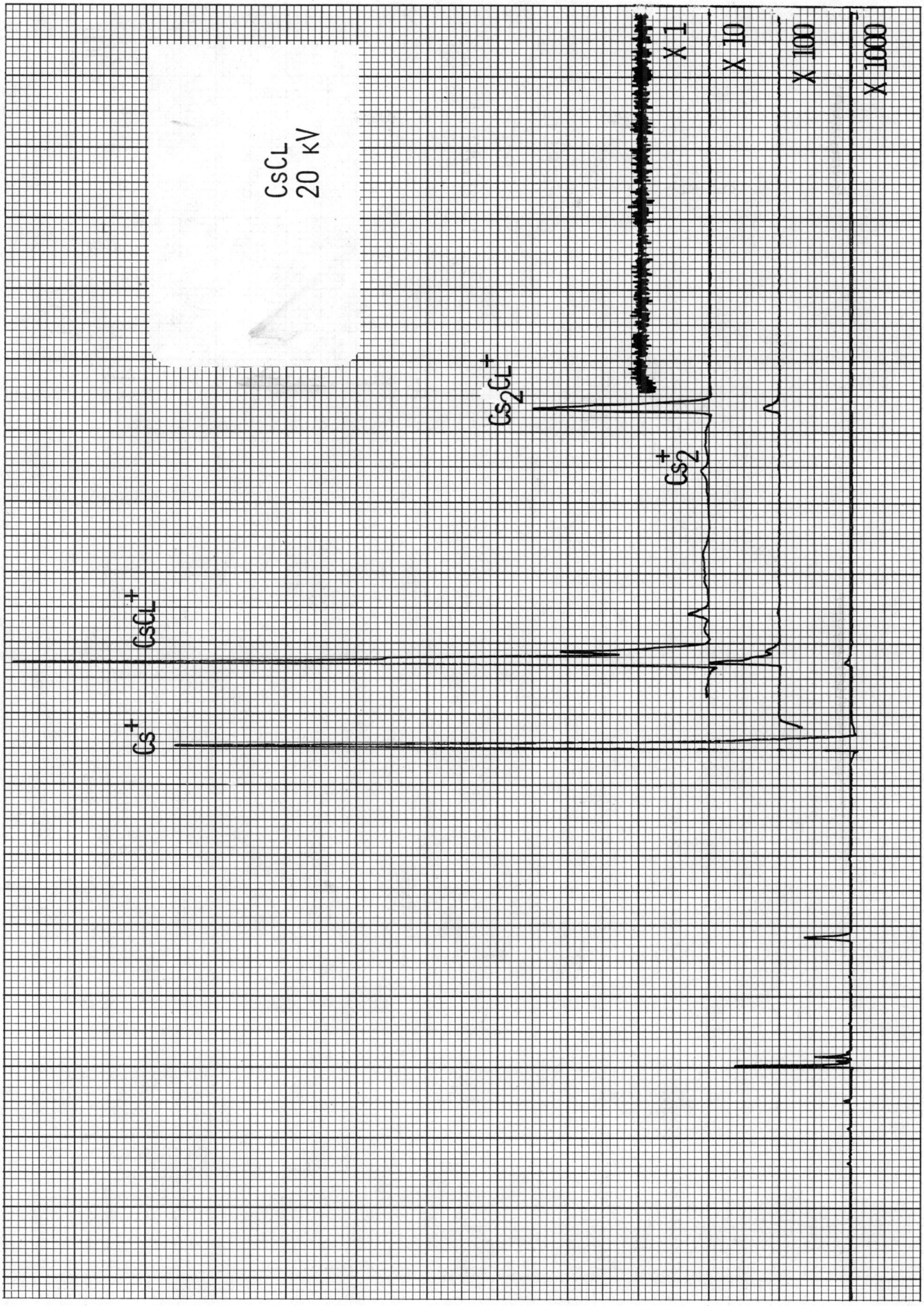

CsCl
20 kV
X 1
X 10
X 100
X 1000
Cs_2Cl^+
Cs_2^+
$CsCl^+$
Cs^+

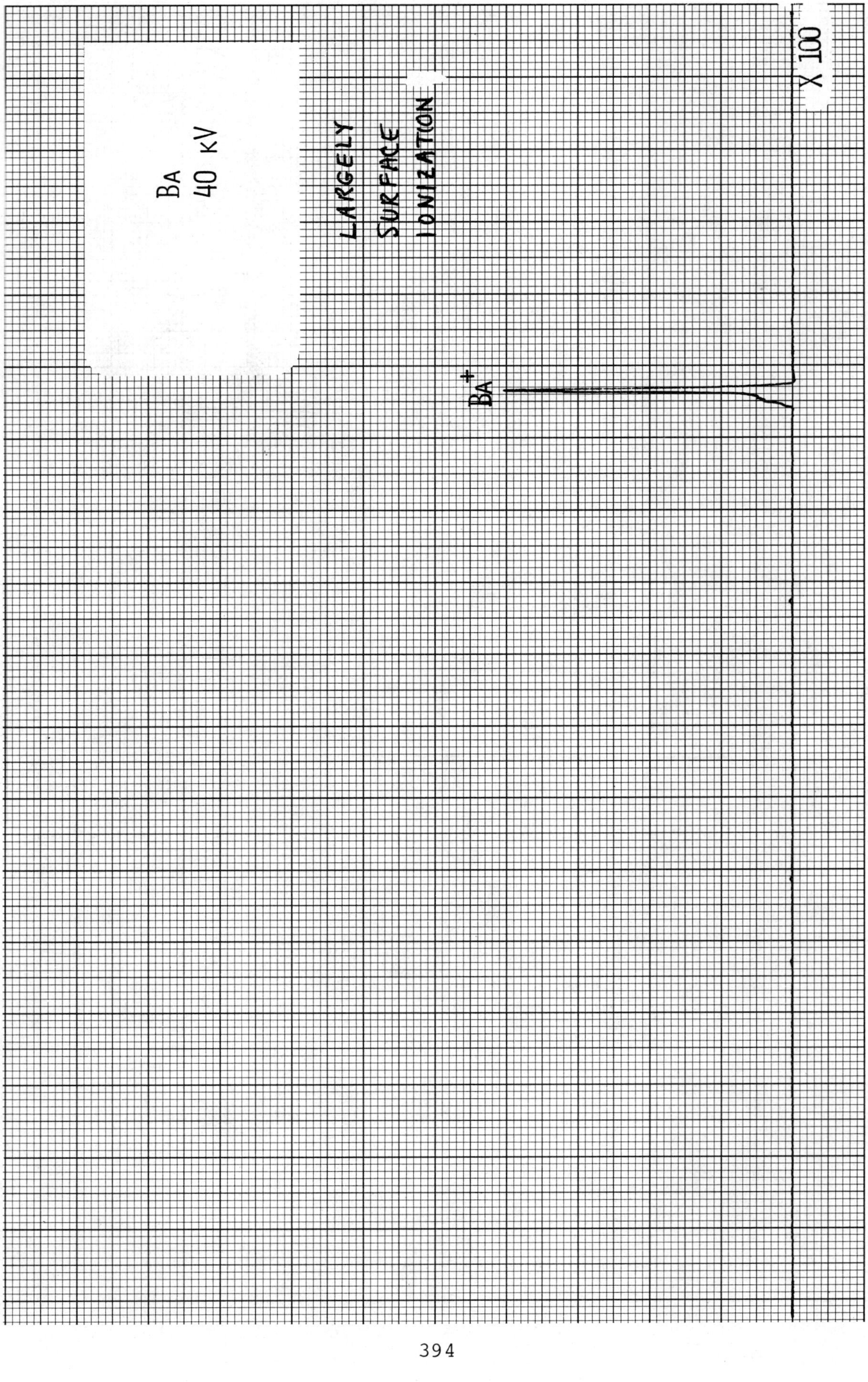
BA
40 kV
LARGELY
SURFACE
IONIZATION
BA+
X 100

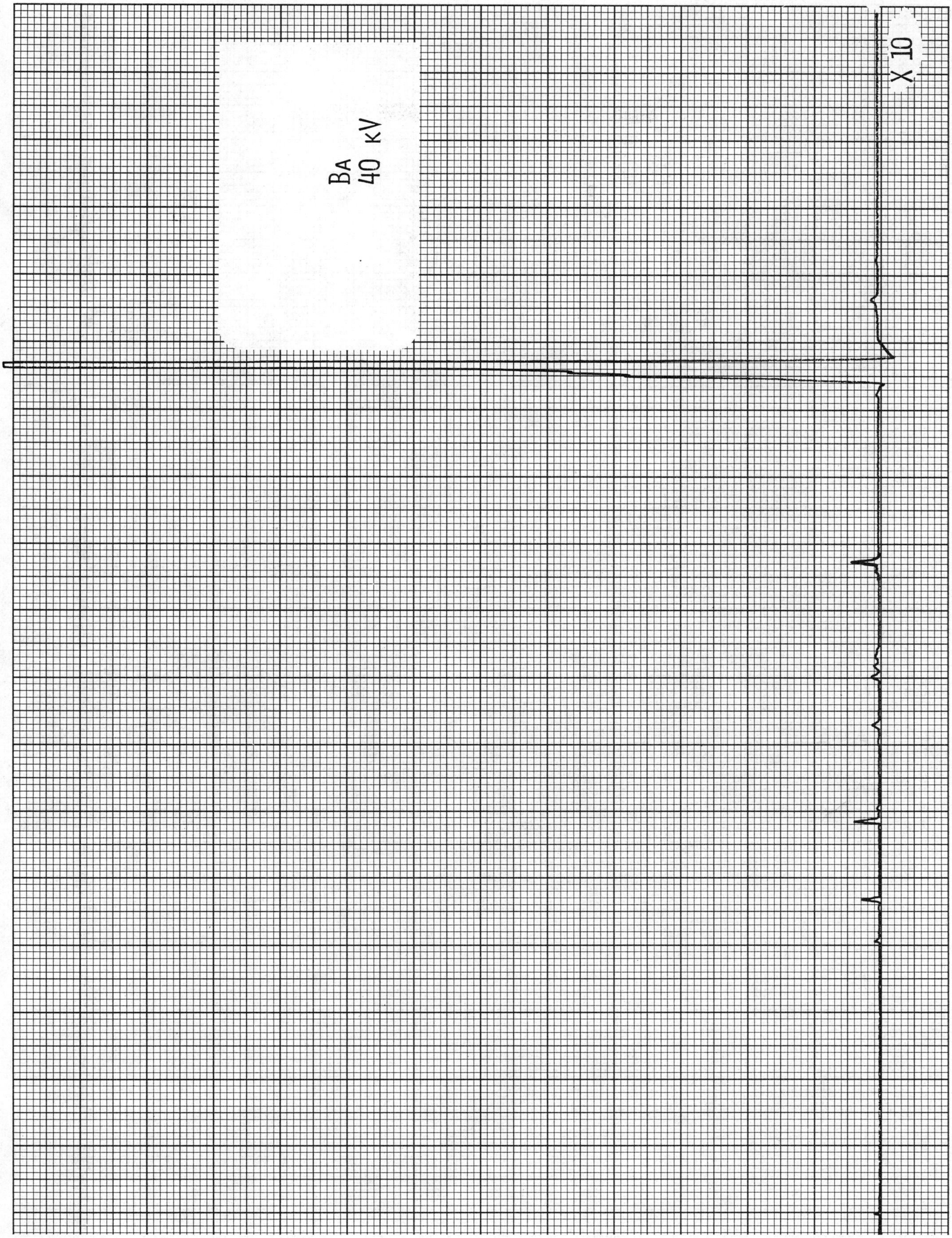
BA
40 kV
X 10

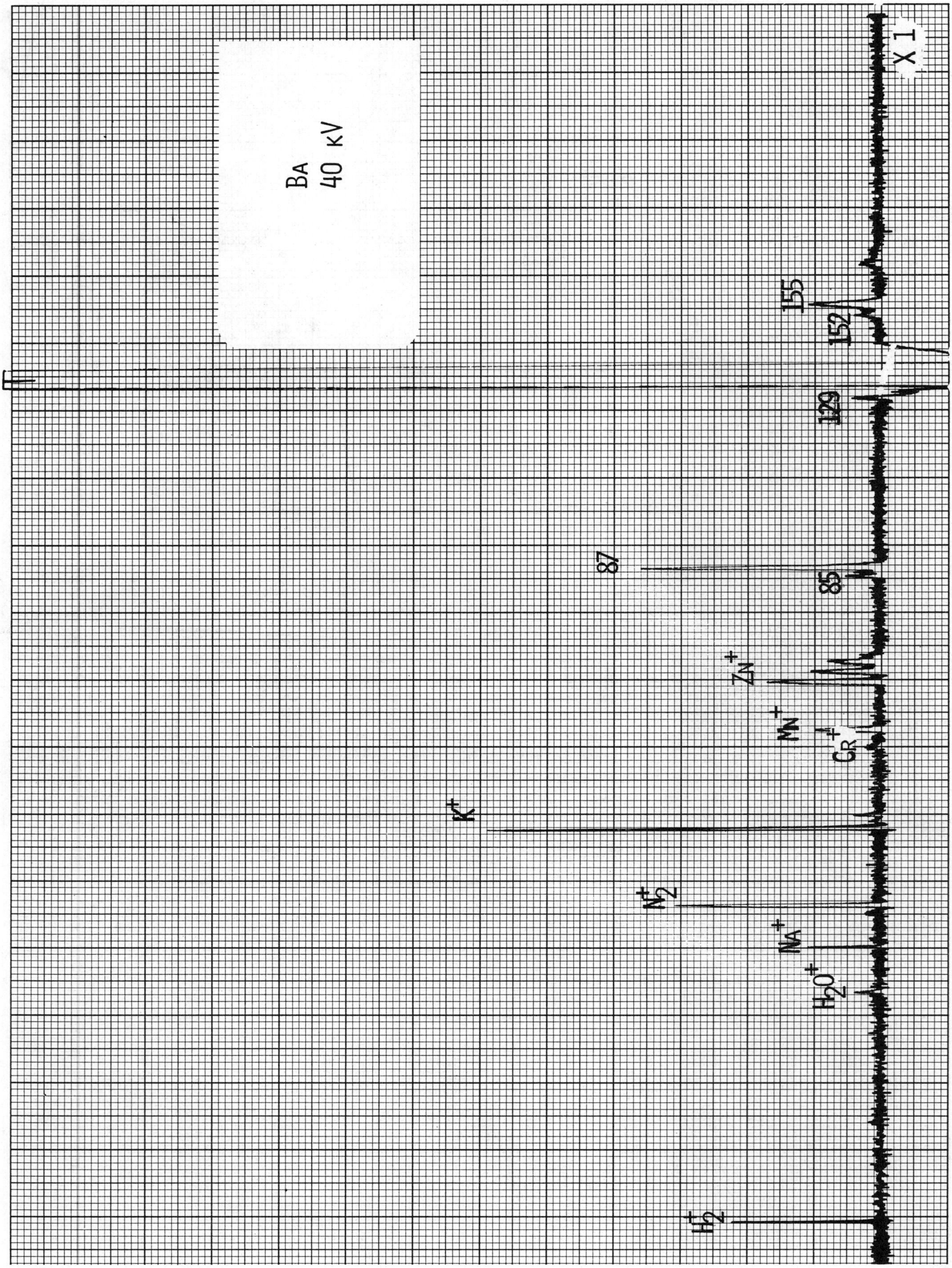
Ba
40 kV
X 1
155
152
129
87
85
Zn^+
Mn^+
Cr^+
K^+
N_2^+
Na^+
H_2O^+
H_2^+

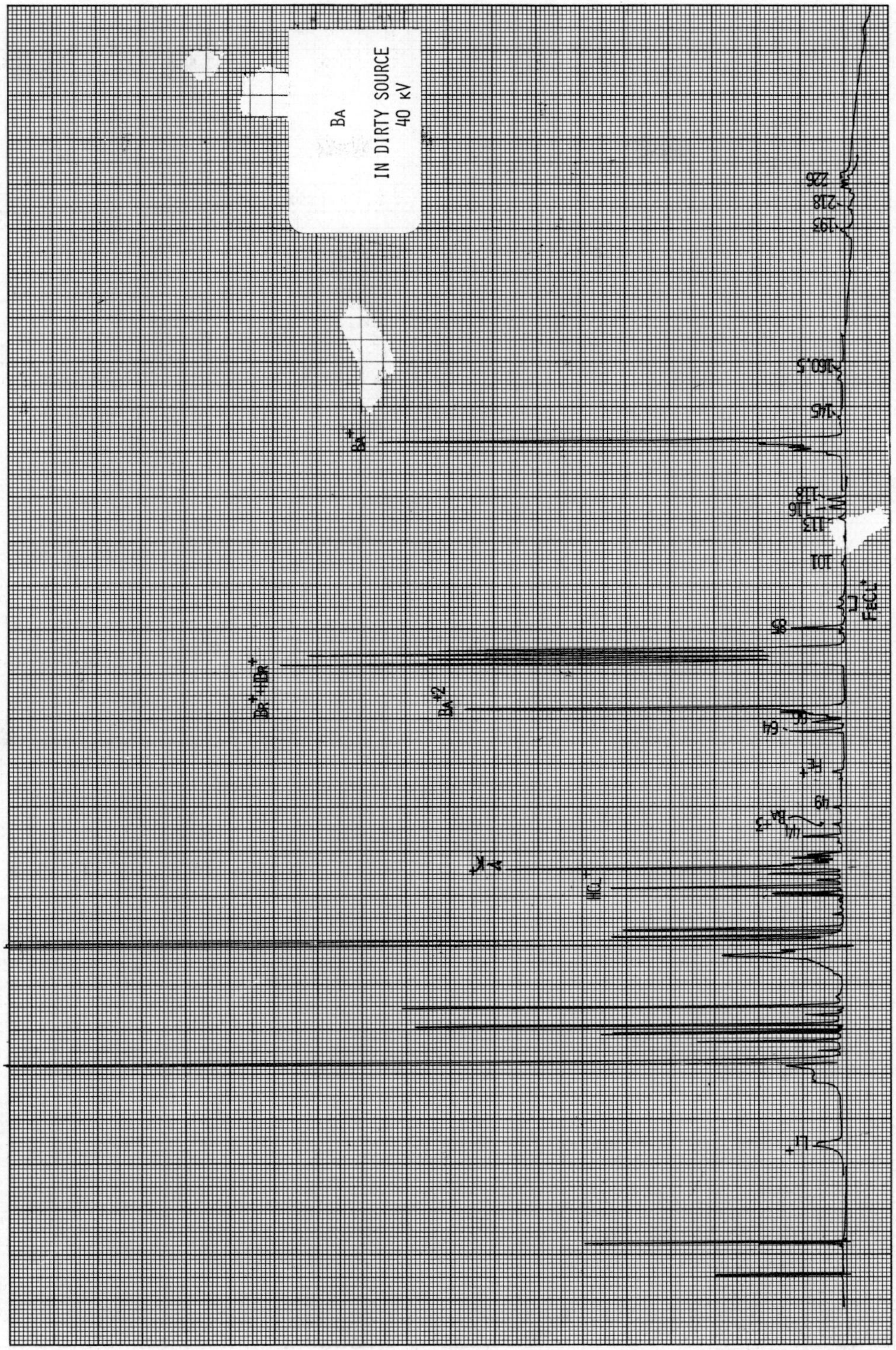
BA
IN DIRTY SOURCE
40 kV
226
218
195
160.5
146
BA+
118
116
113
101
FeCl+
96
Br+ +HBr+
BA+2
64
49
K+
HCl+
Li+

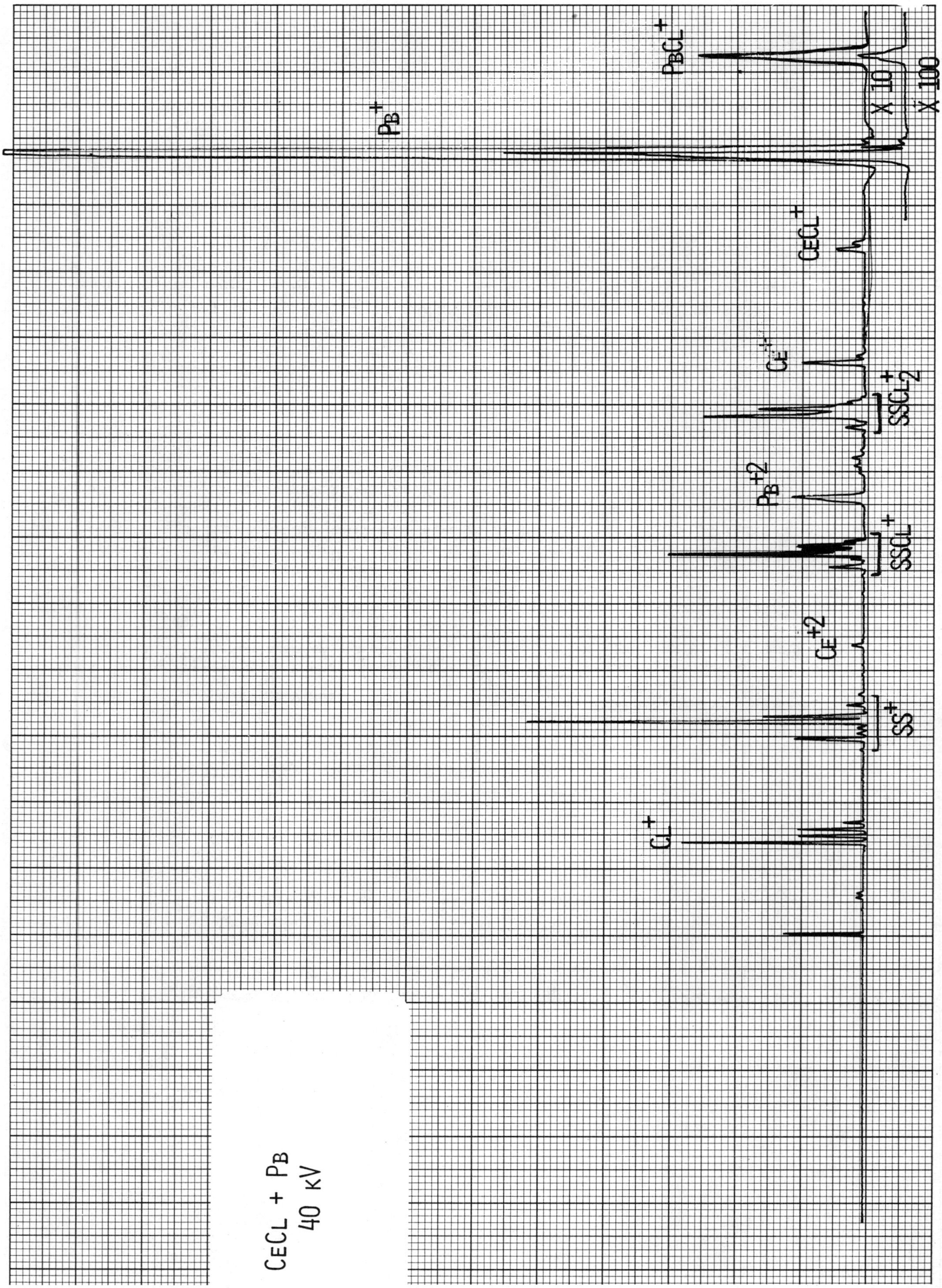

Pb^+
$PbCl^+$
X 10
X 100
$CeCl^+$
Ce^+
$SSCl_2^+$
Pb^{+2}
$SSCl^+$
Ce^{+2}
SS^+
Cl^+
CeCl + Pb
40 kV

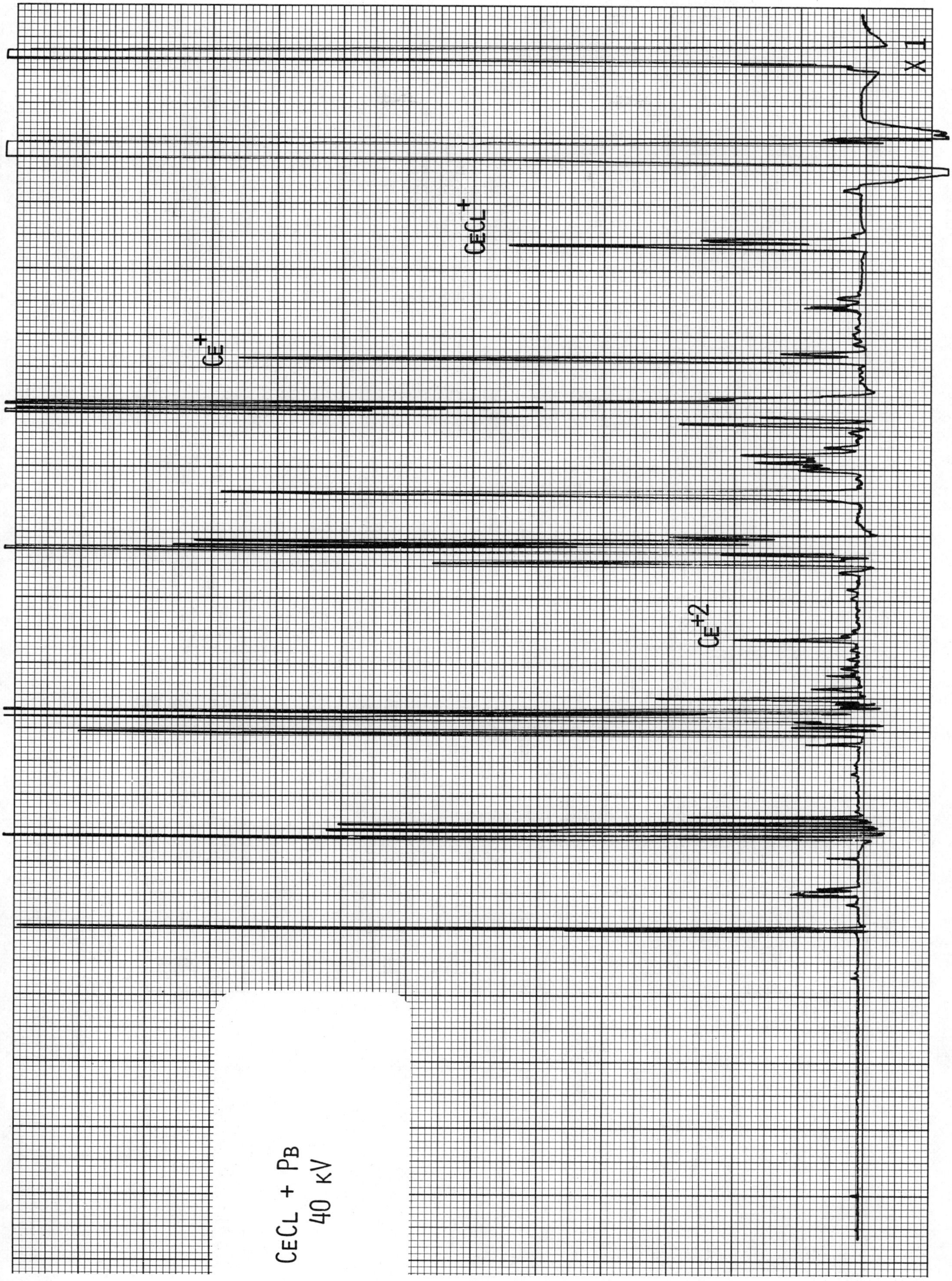
X 1
CeCl+
Ce+
Ce+2
CeCl + Pb
40 kV

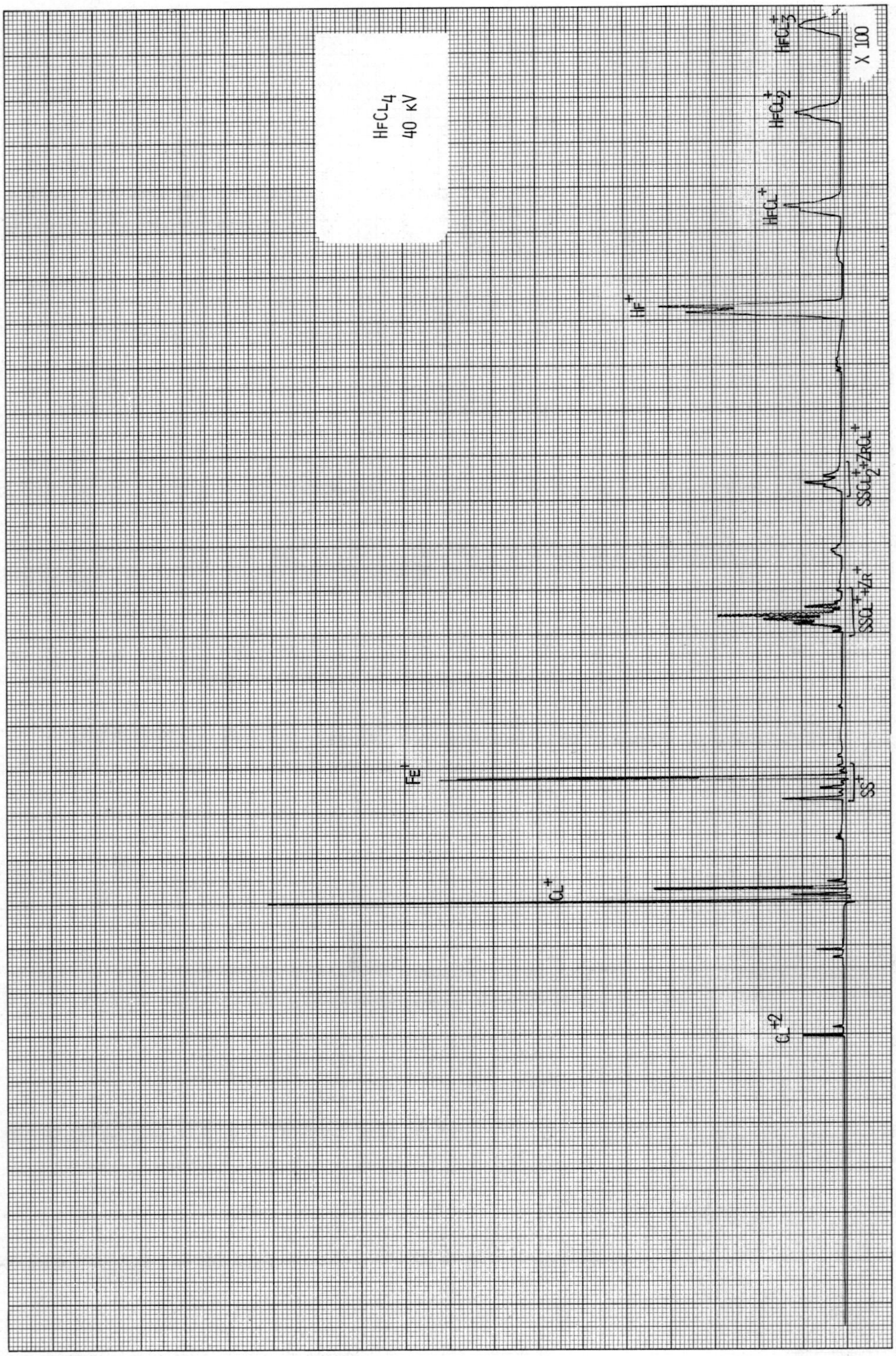
HfCl4
40 kV
HfCl3+
X 100
HfCl2+
HfCl+
Hf+
SSCl2+ + ZrCl+
SSCl+ + Zr+
Fe+
SS+
Cl+
Cl+2

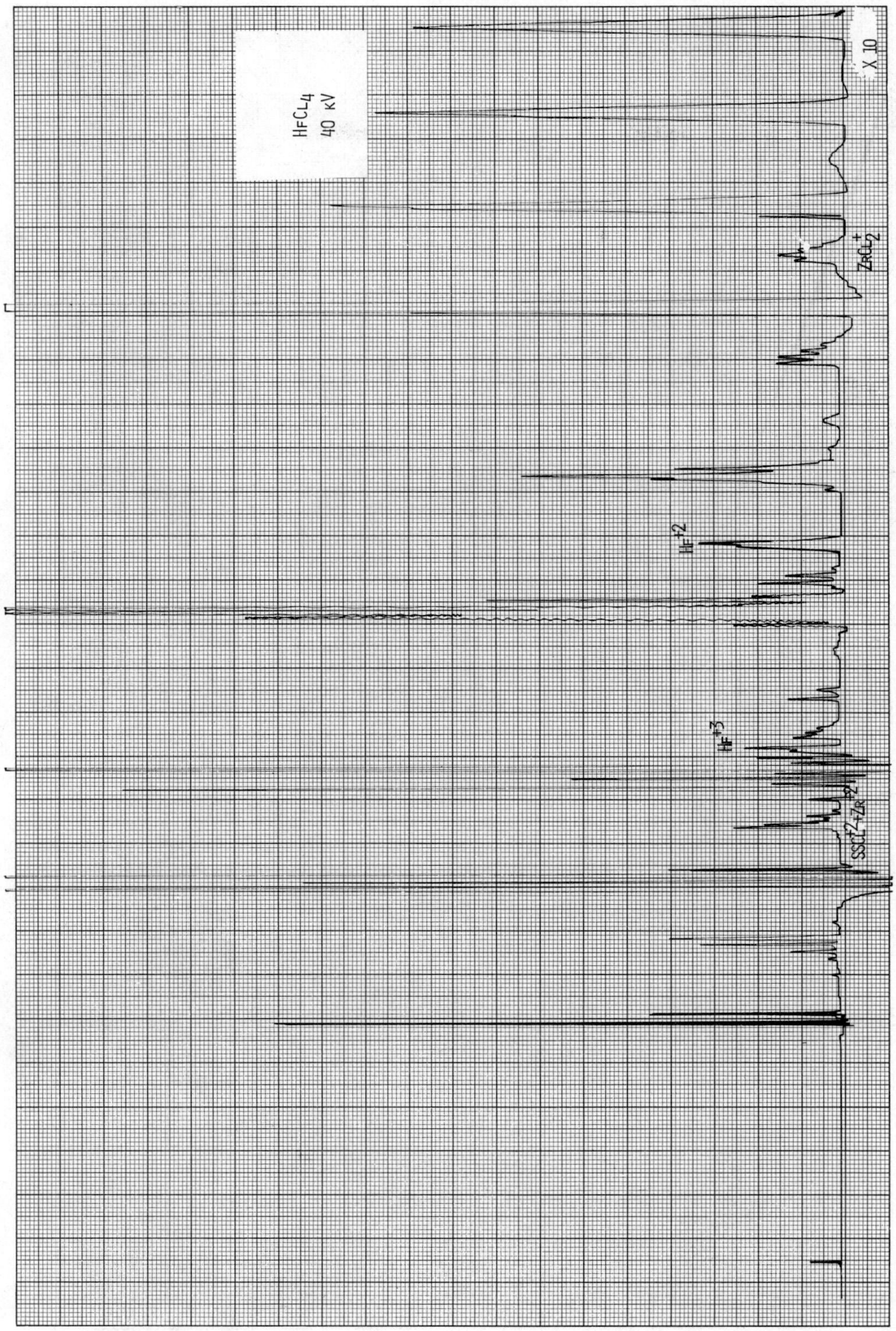
HfCl4
40 kV
X 10
ZrCl2+
Hf+2
Hf+3

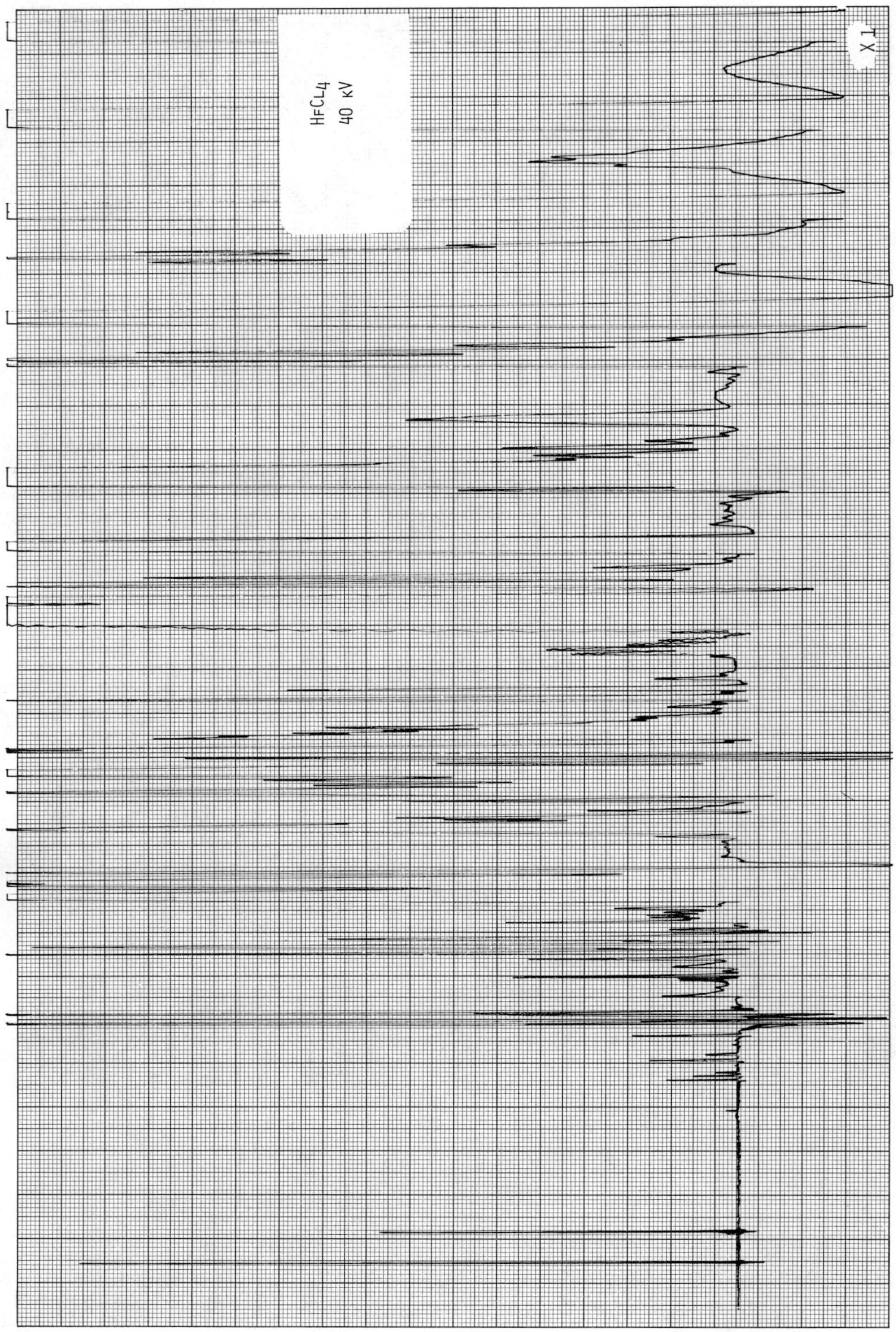
HfCl4
40 kV
X1

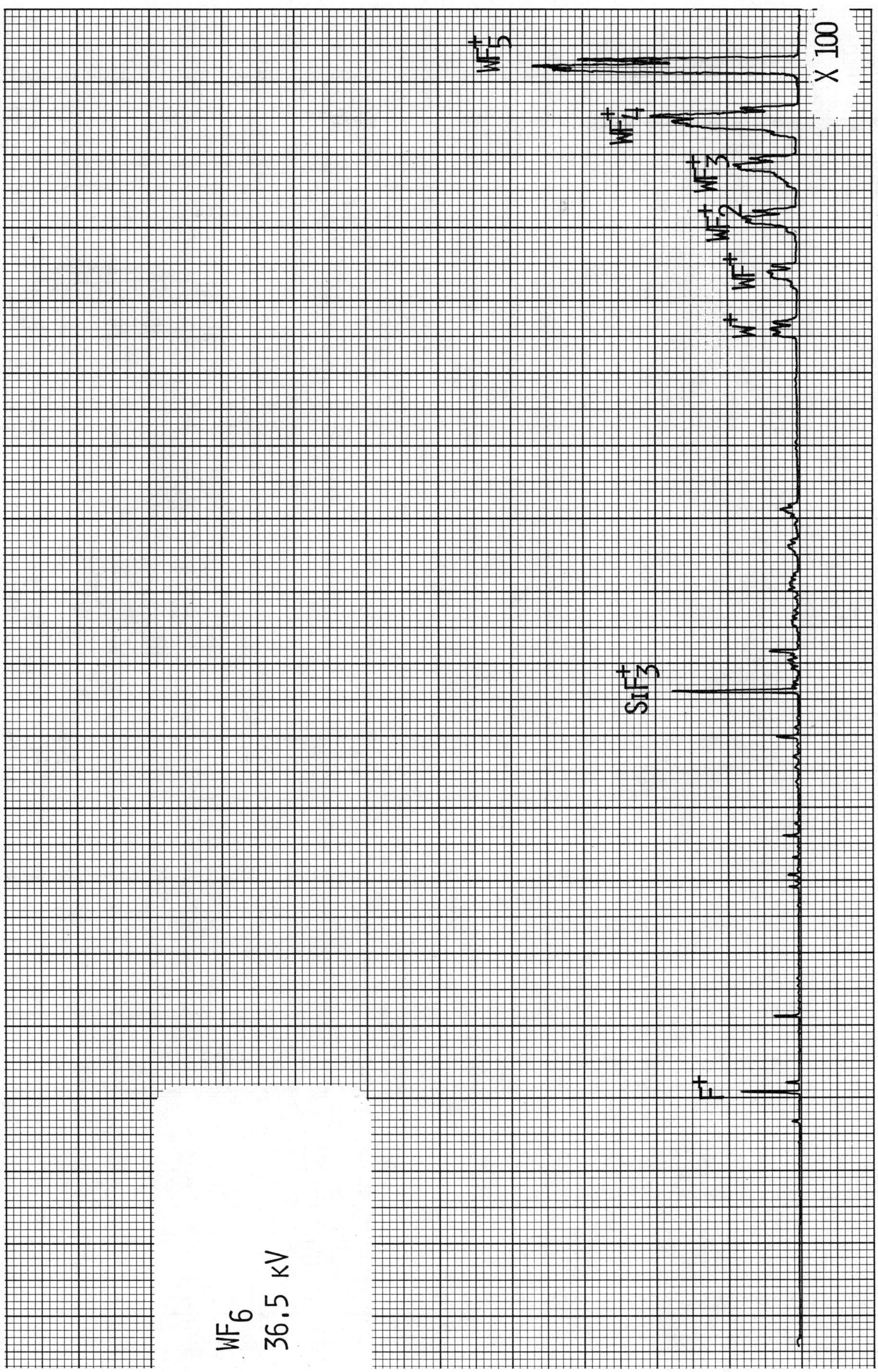
WF_5^+
WF_4^+
WF_3^+
WF_2^+
WF^+
W^+
SiF_3^+
F^+
X 100
WF_6
36.5 kV

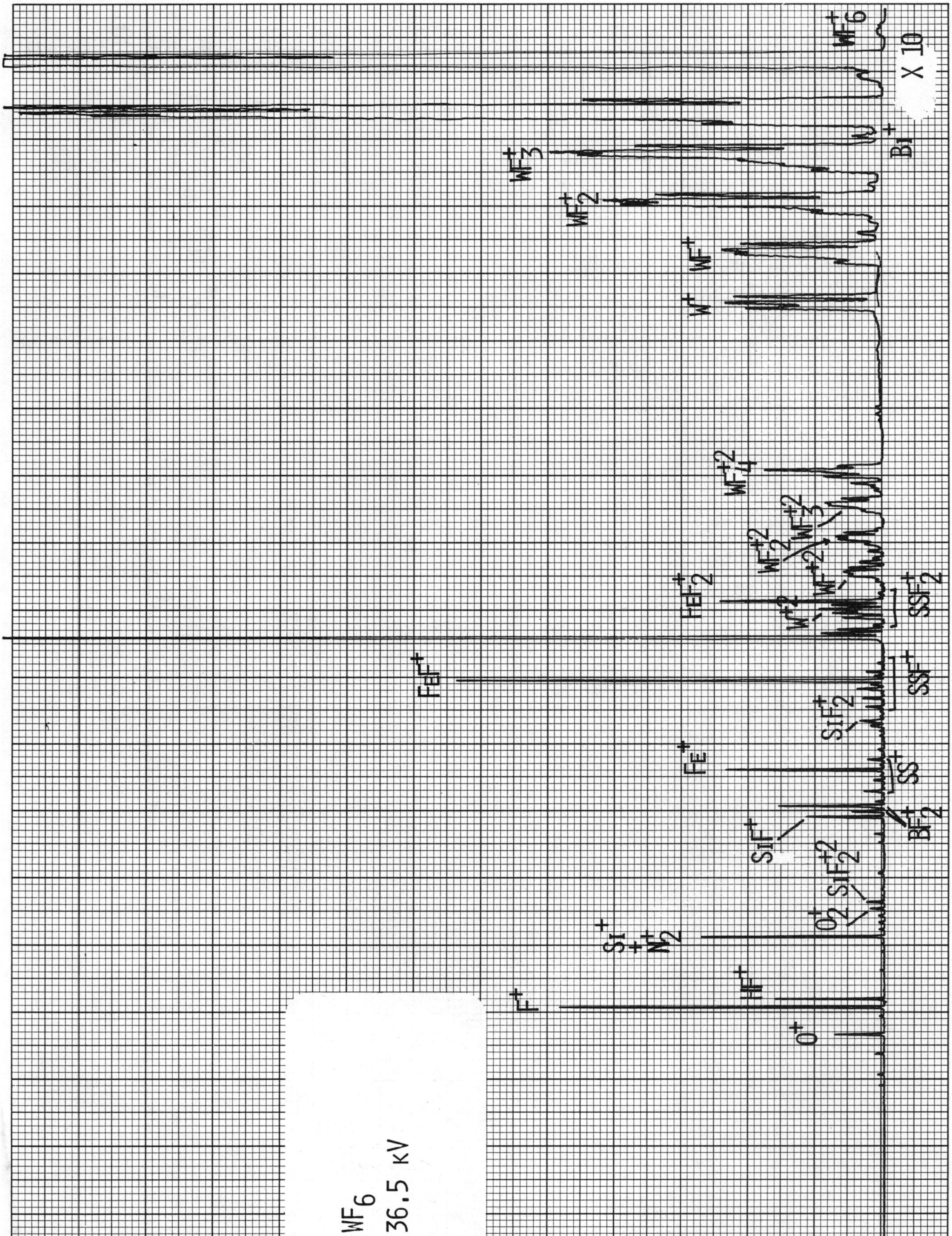

WF6
36.5 kV
X 10
WF6+
BI+
WF3+
WF2+
WF+
W+
WF4+2
WF3+2
WF2+2
WF+2
W+2
SSF2+
FEF2+
FEF+
SSF+
SIF2+
FE+
SS+
BF2+
SIF+
SIF2+2
O2+
SI+ + N2+
HF+
F+
O+

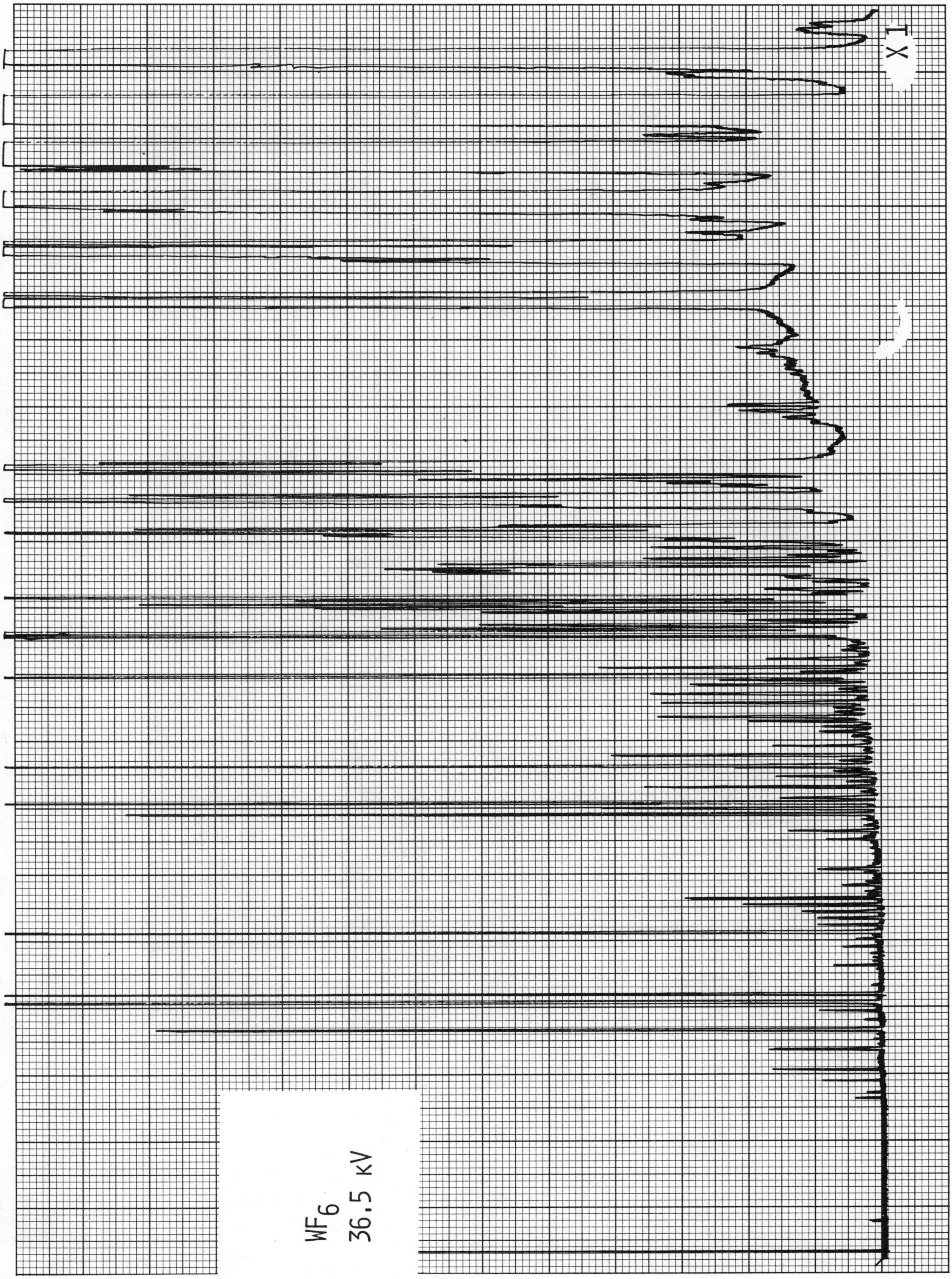
X 1
WF_6
36.5 kV

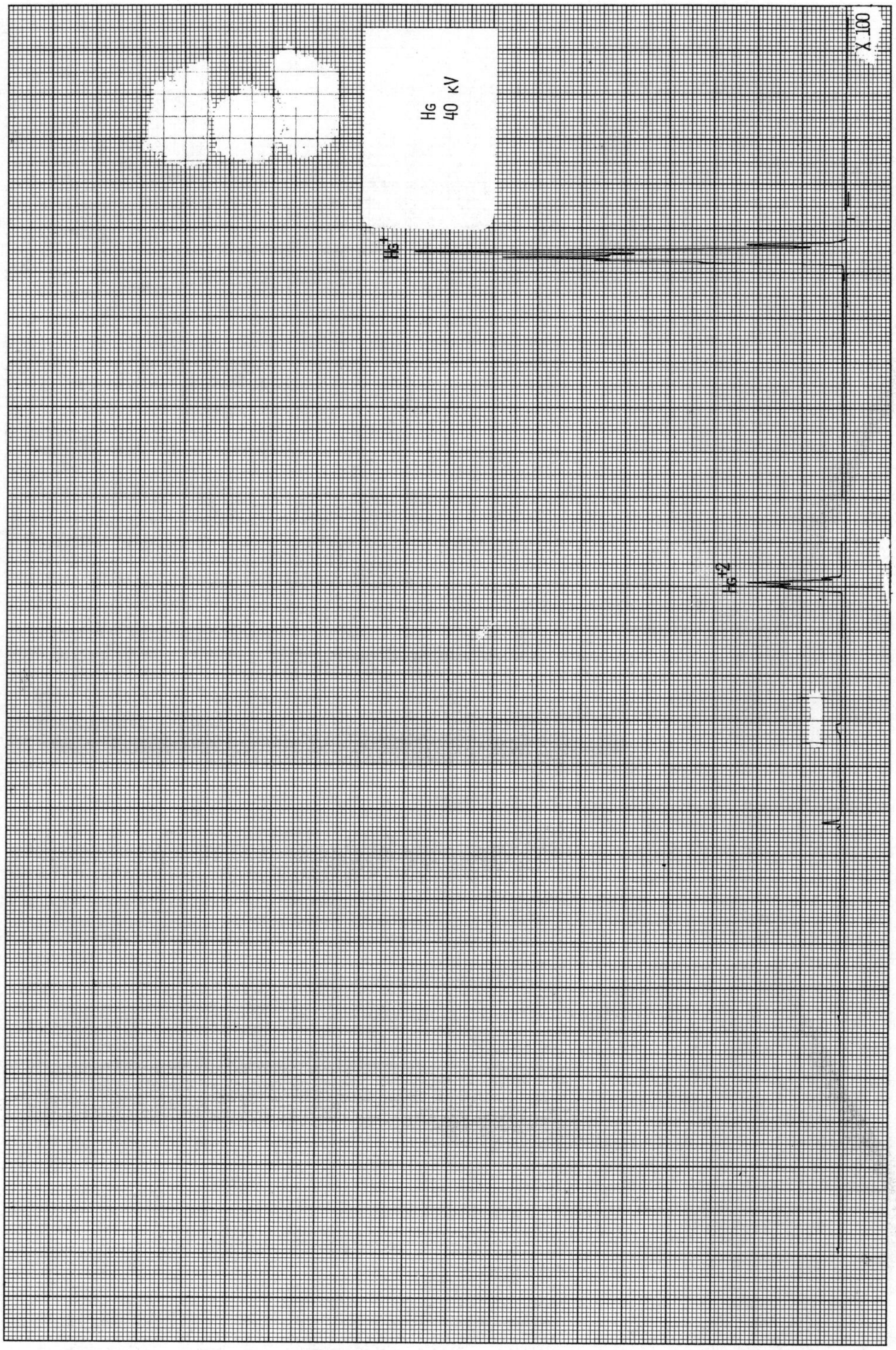

Hg
40 kV
X 100
Hg+
Hg+2

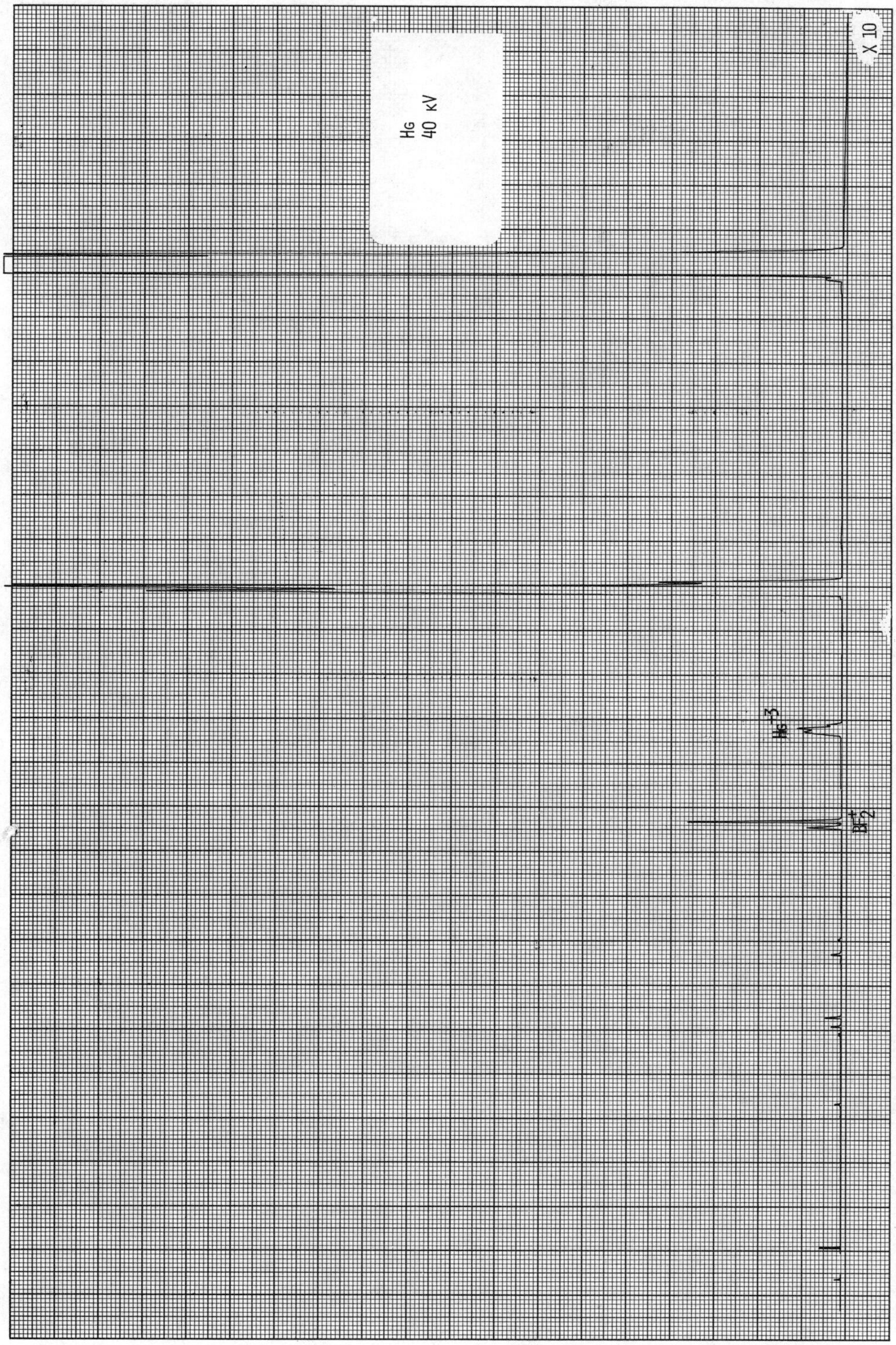
Hg
40 kV
X 10

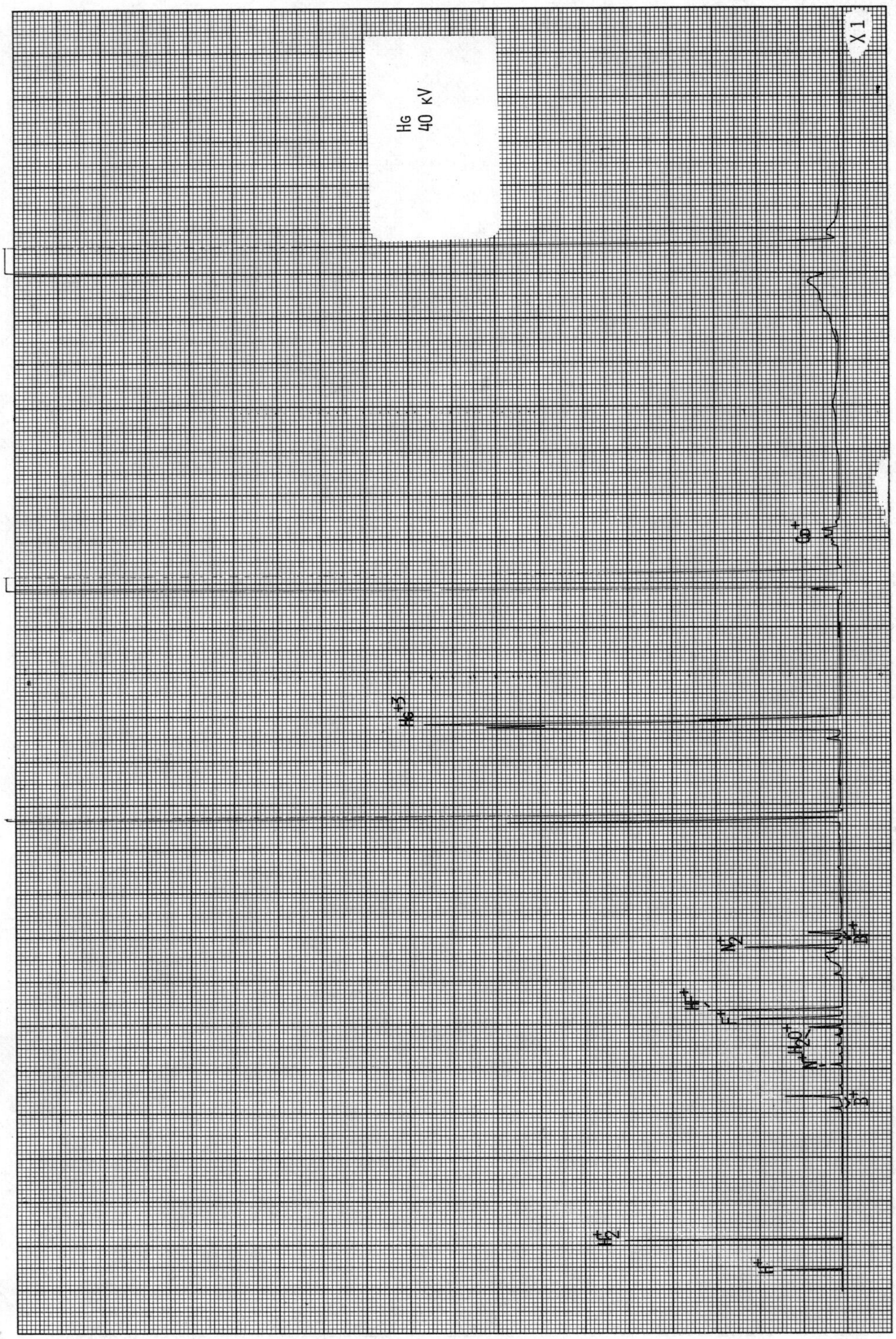
Hg
40 kV
X 1

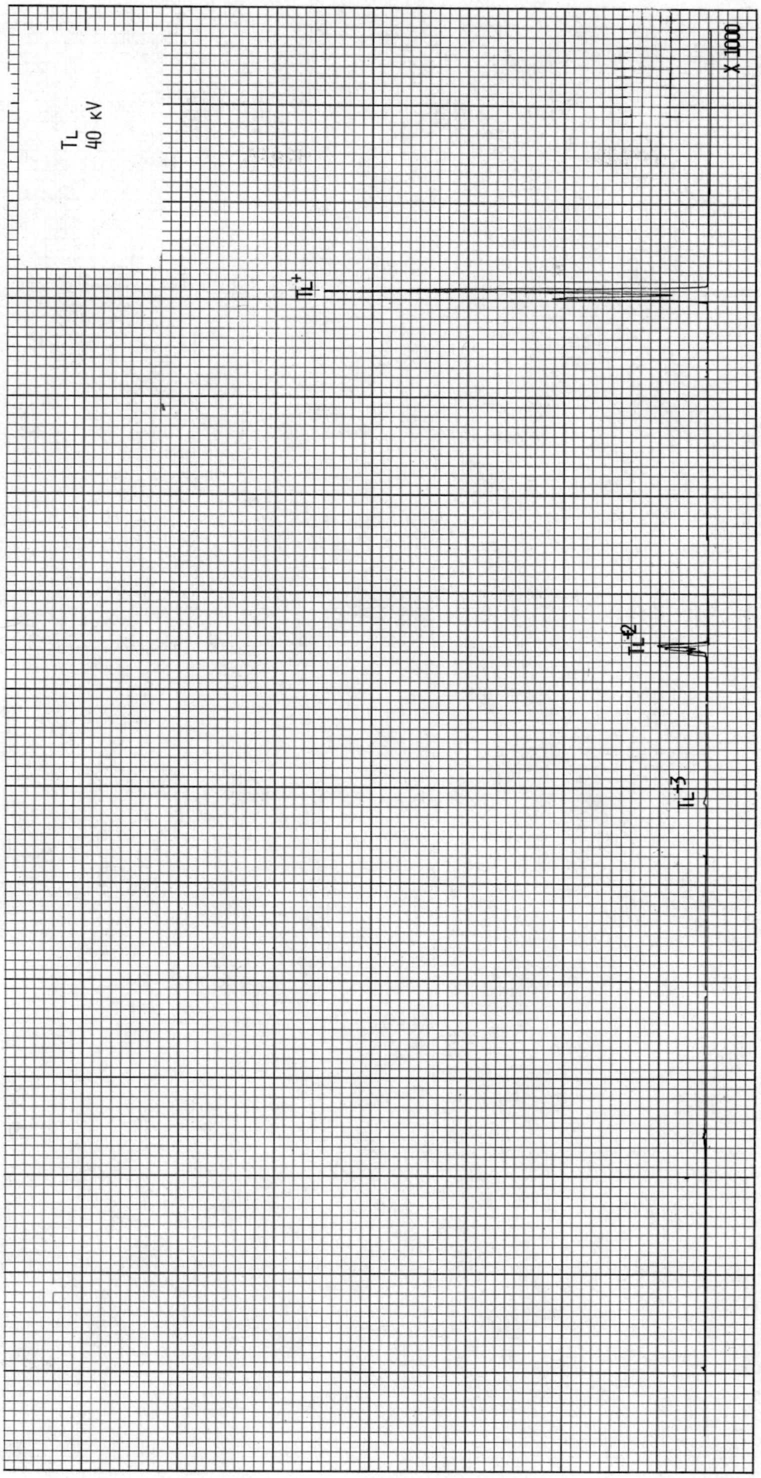
TL
40 kV
X 1000
TL+
TL+2
TL+3

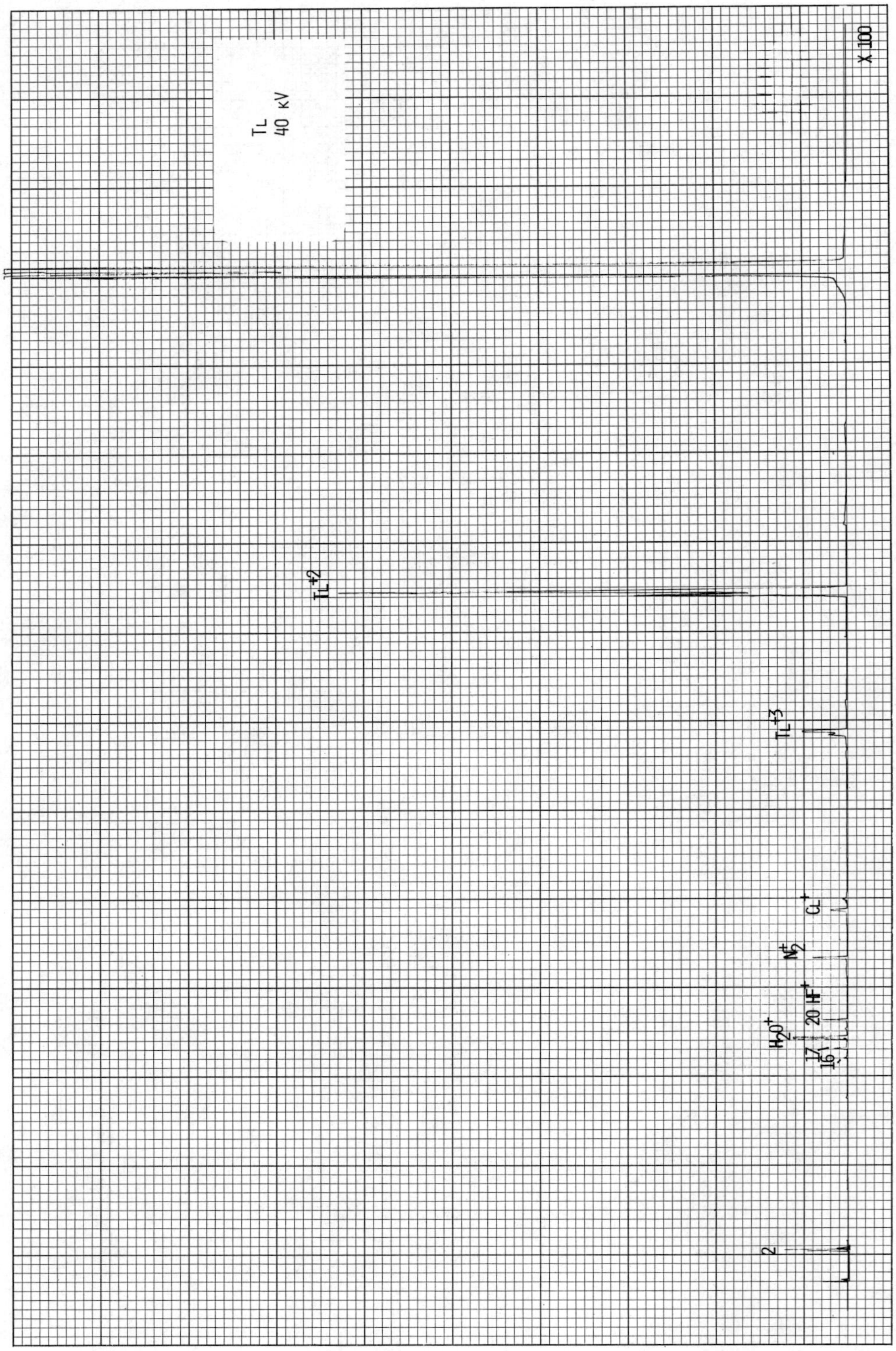
TL
40 kV
X 100
TL+2
TL+3
Cl+
N2+
20 HF+
H2O+
17
16
2

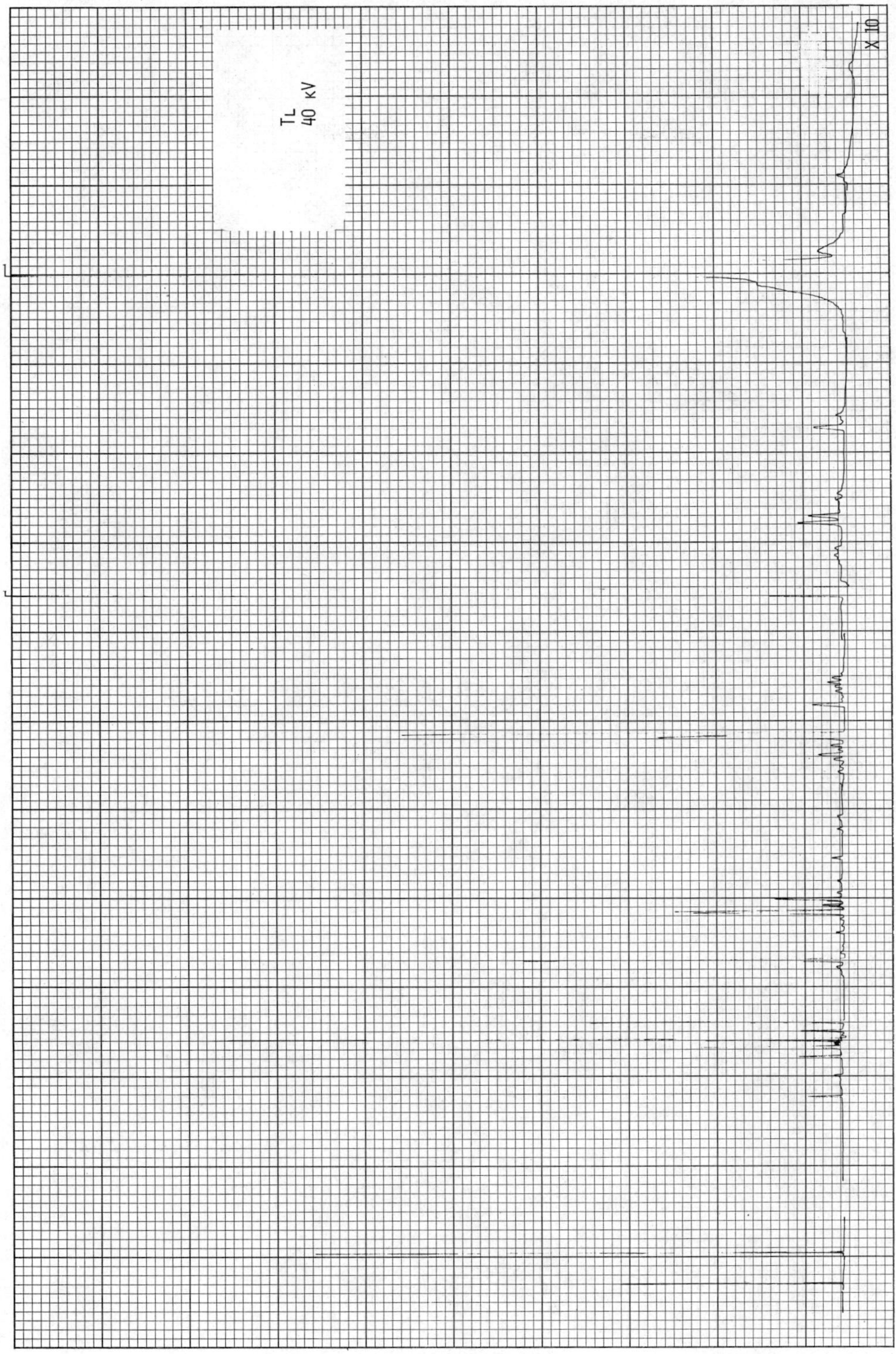
T_L
40 kV
X 10

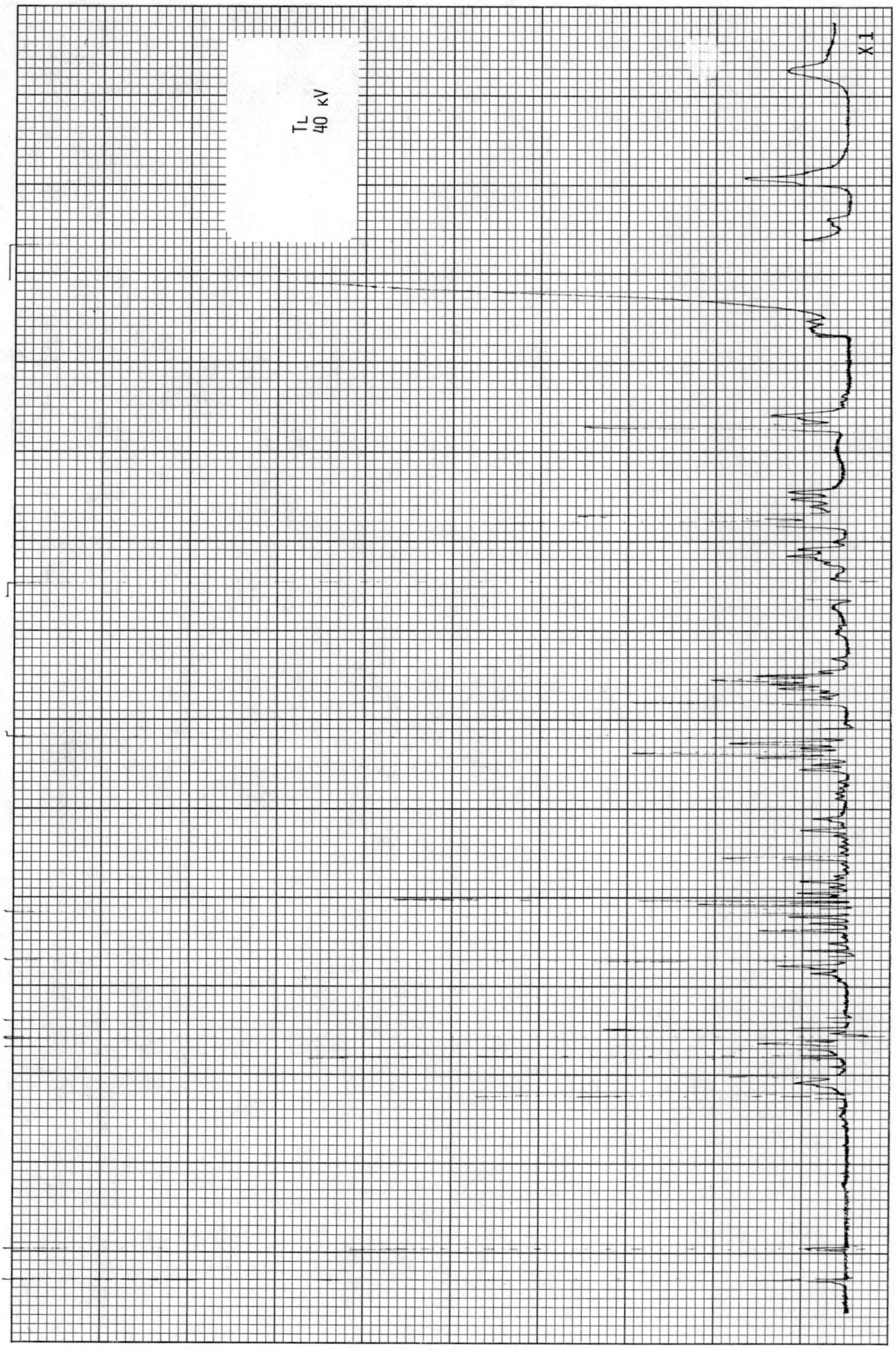
TL
40 kV
X 1

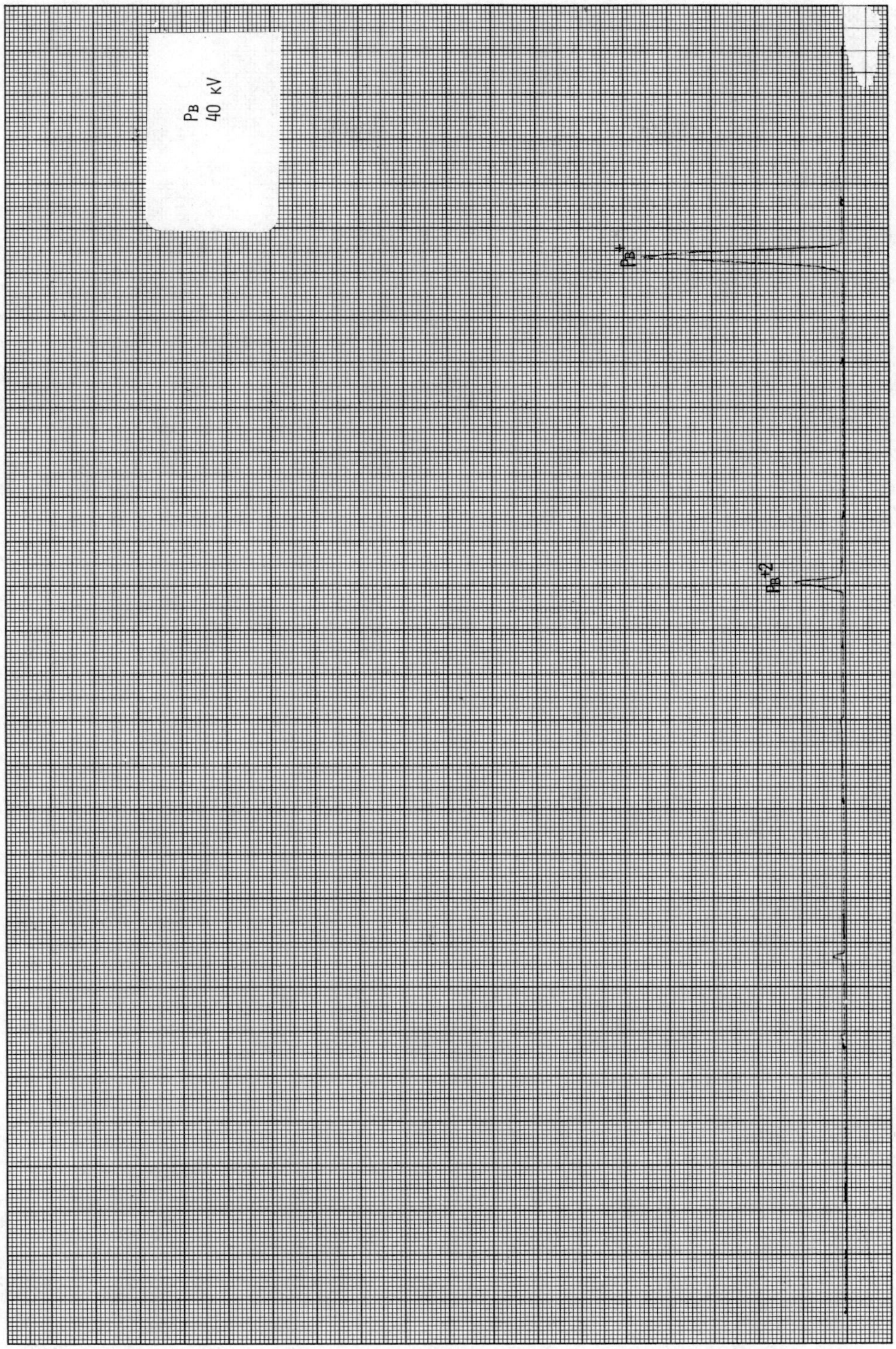
PB
40 kV
Pb+
Pb+2

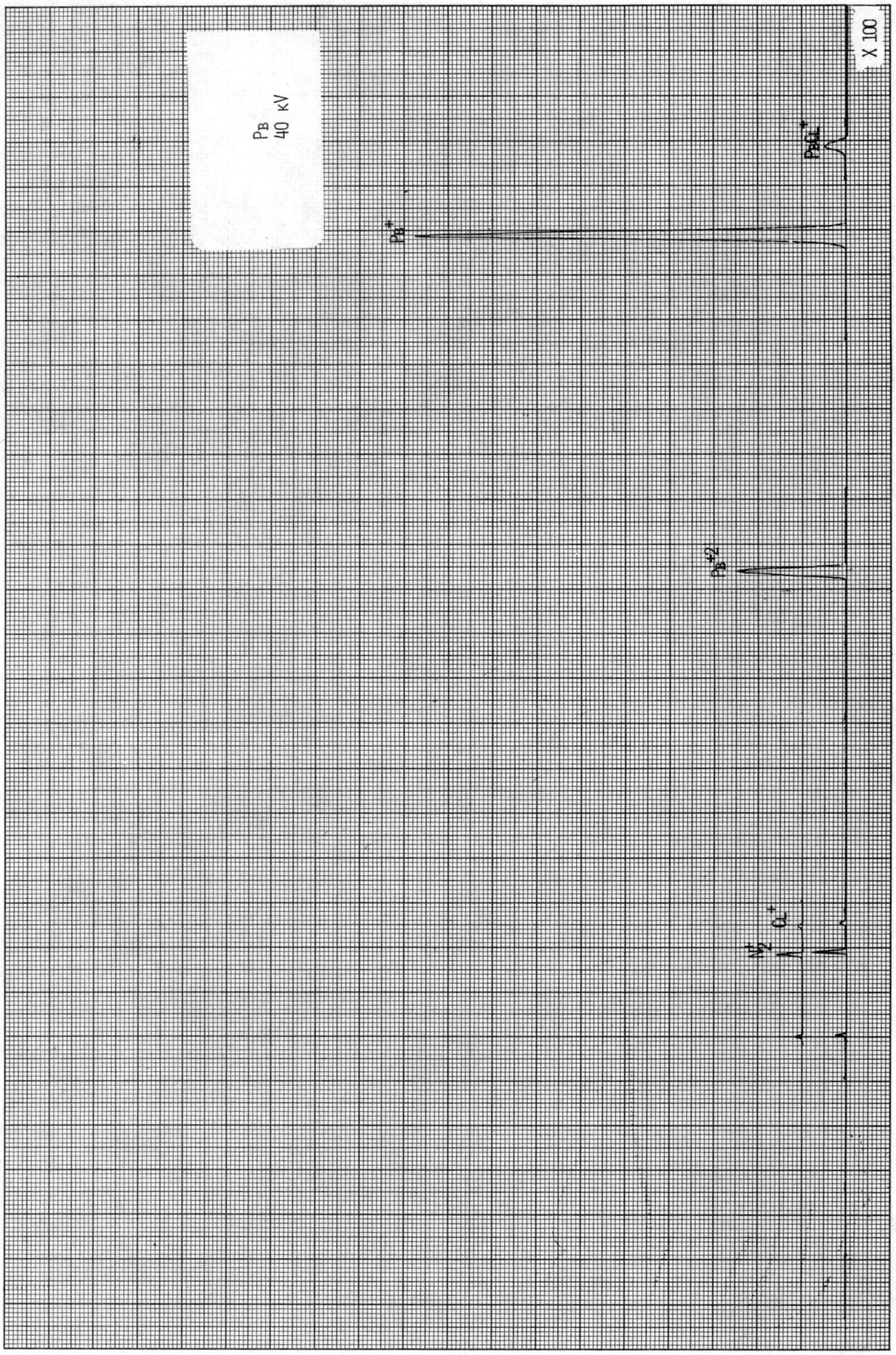

PB
40 kV
X 100

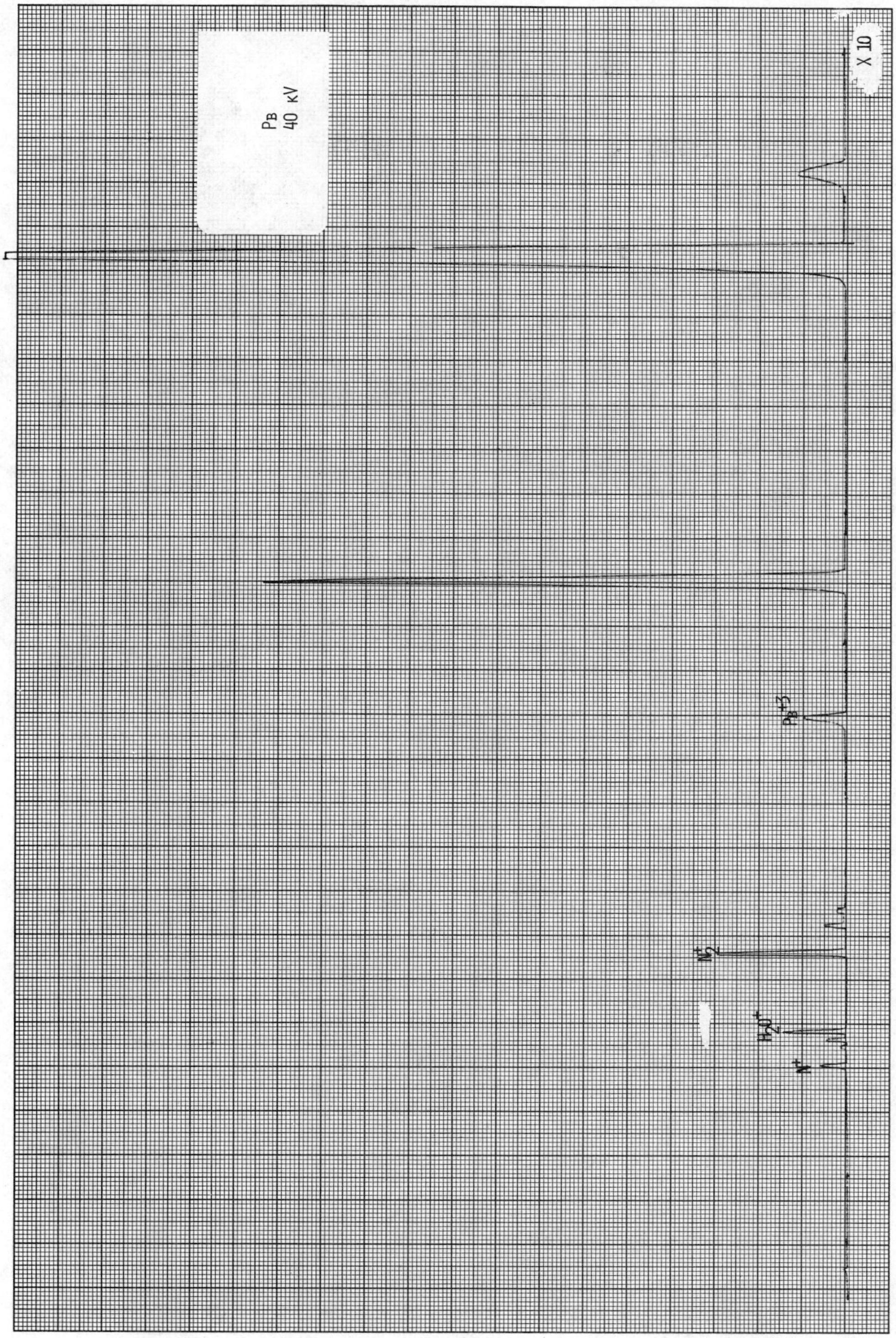

PB
40 kV
X 10
PB+3
N2+
H2O+
N+

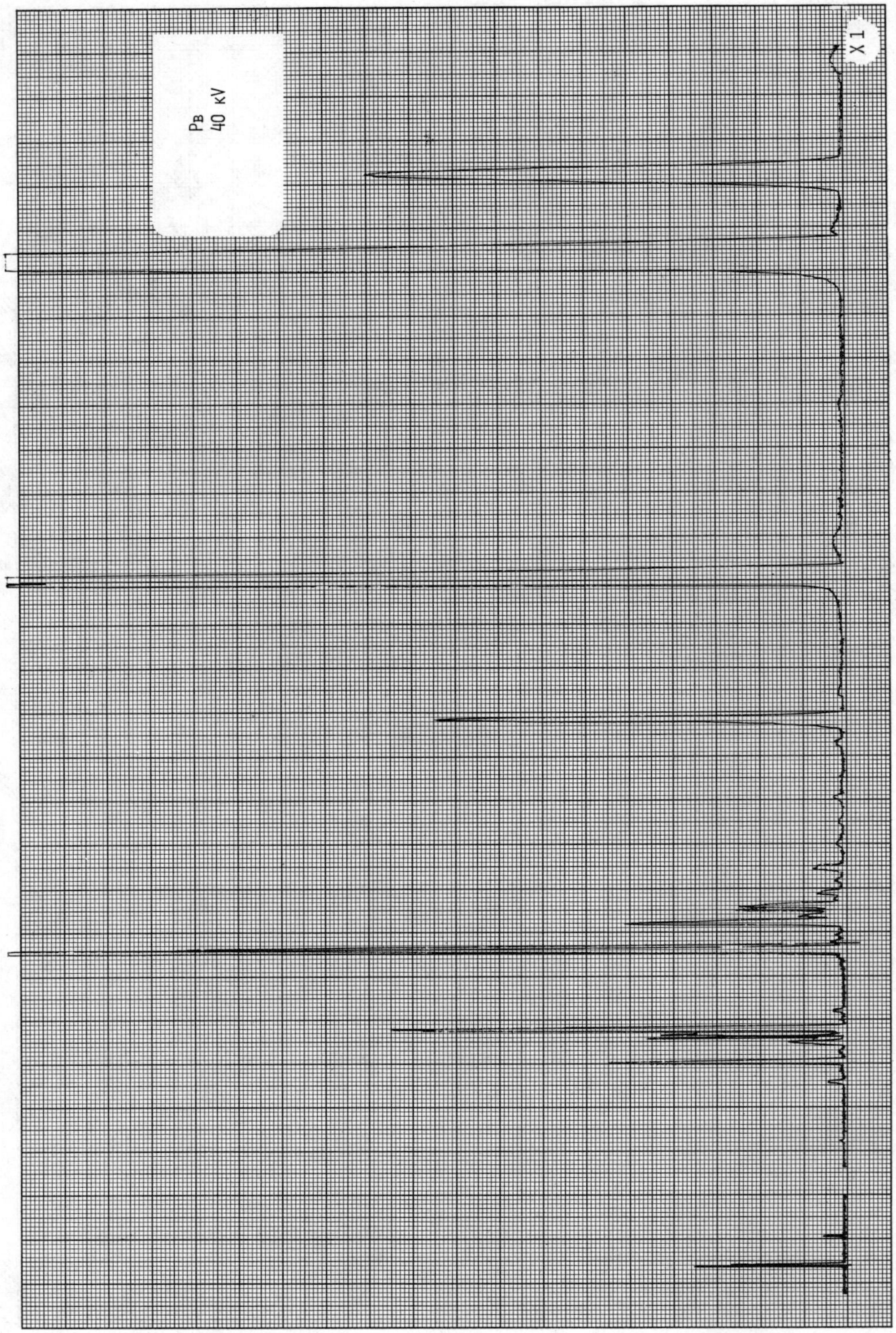
Pb
40 kV
X1

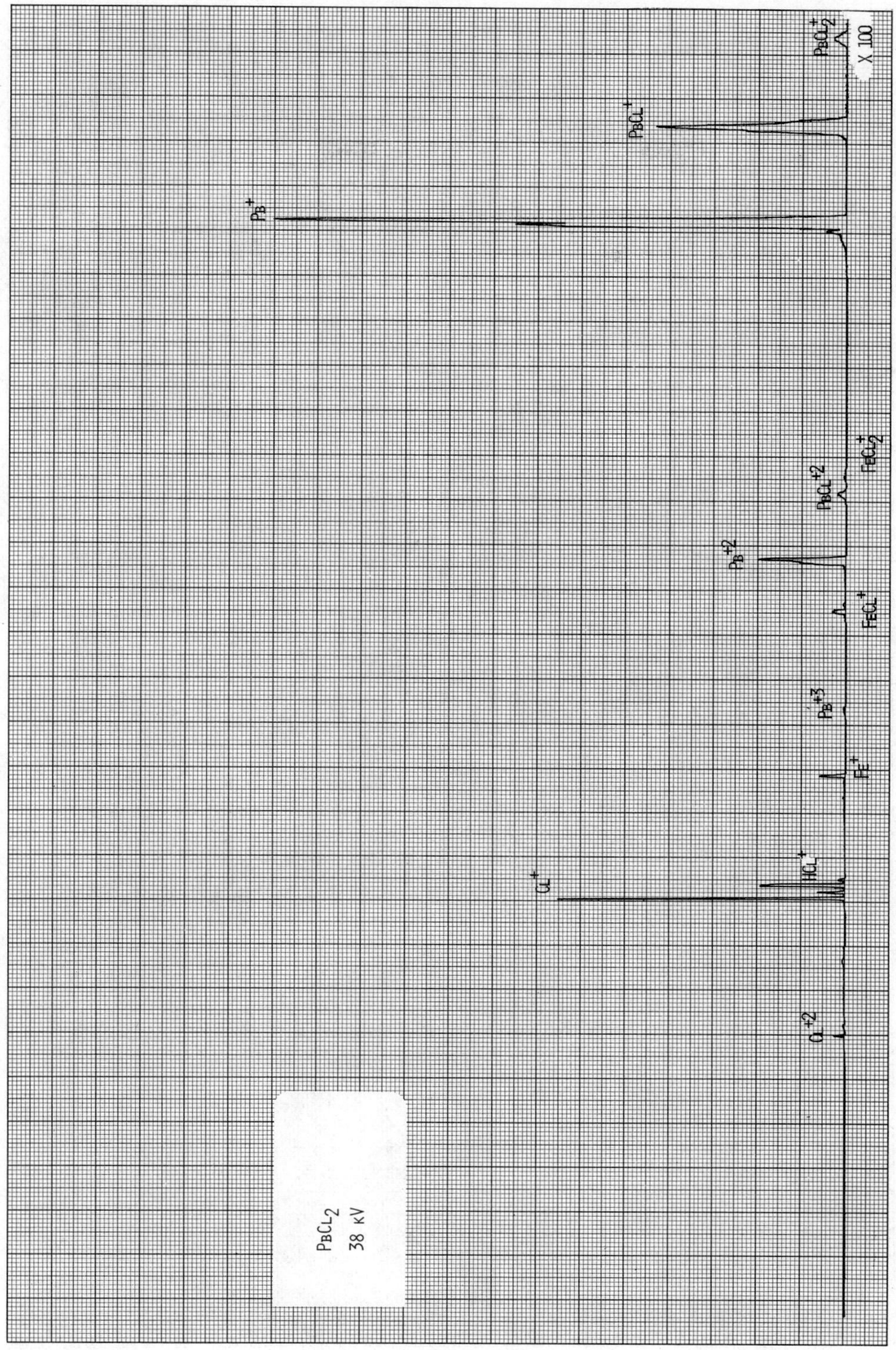
PbCl2
38 kV
Cl+2
Cl+
HCl+
Fe+
Pb+3
FeCl+
Pb+2
PbCl+2
FeCl2+
Pb+
PbCl+
PbCl2+
X 100

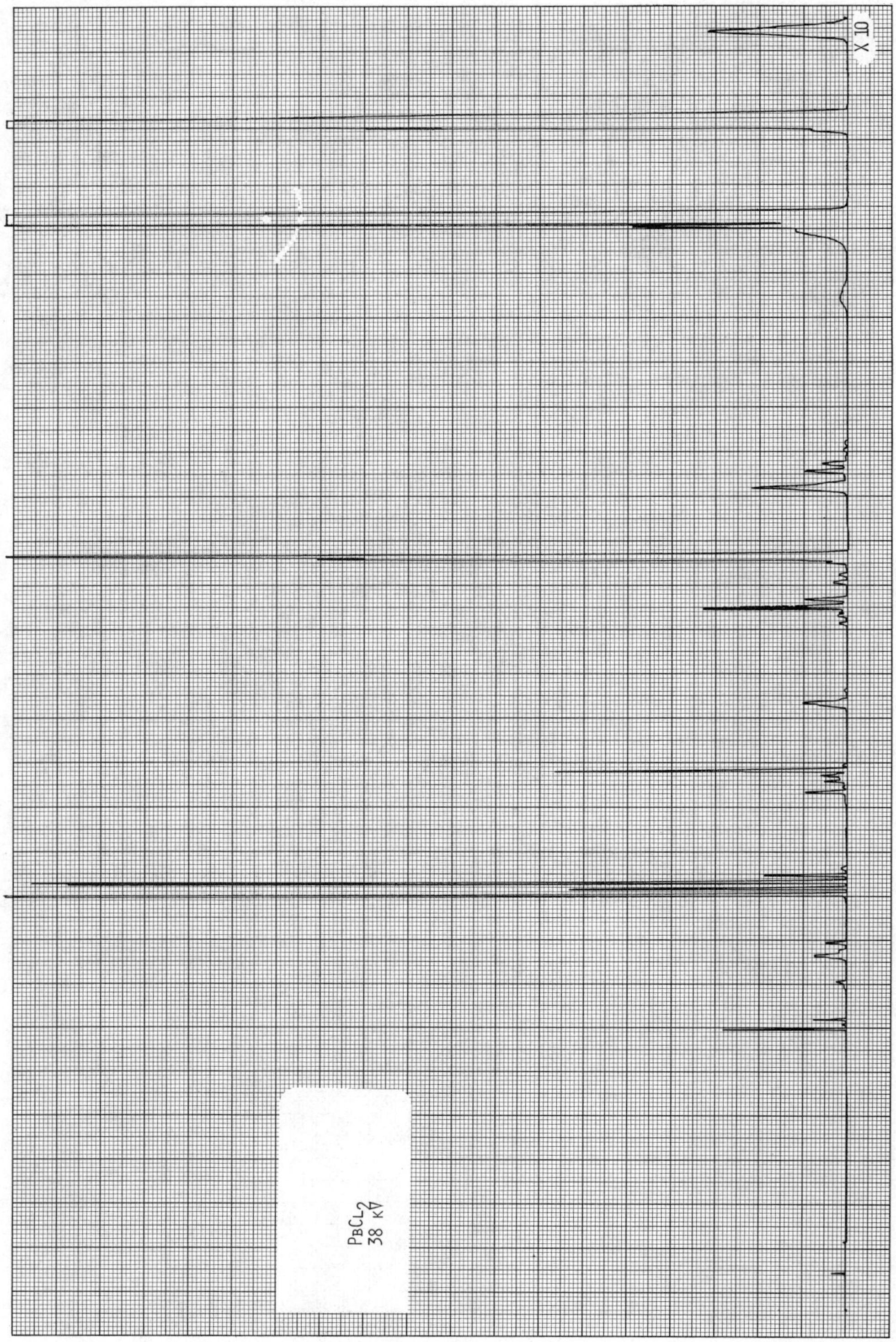
X 10
PbCl2
38 kV

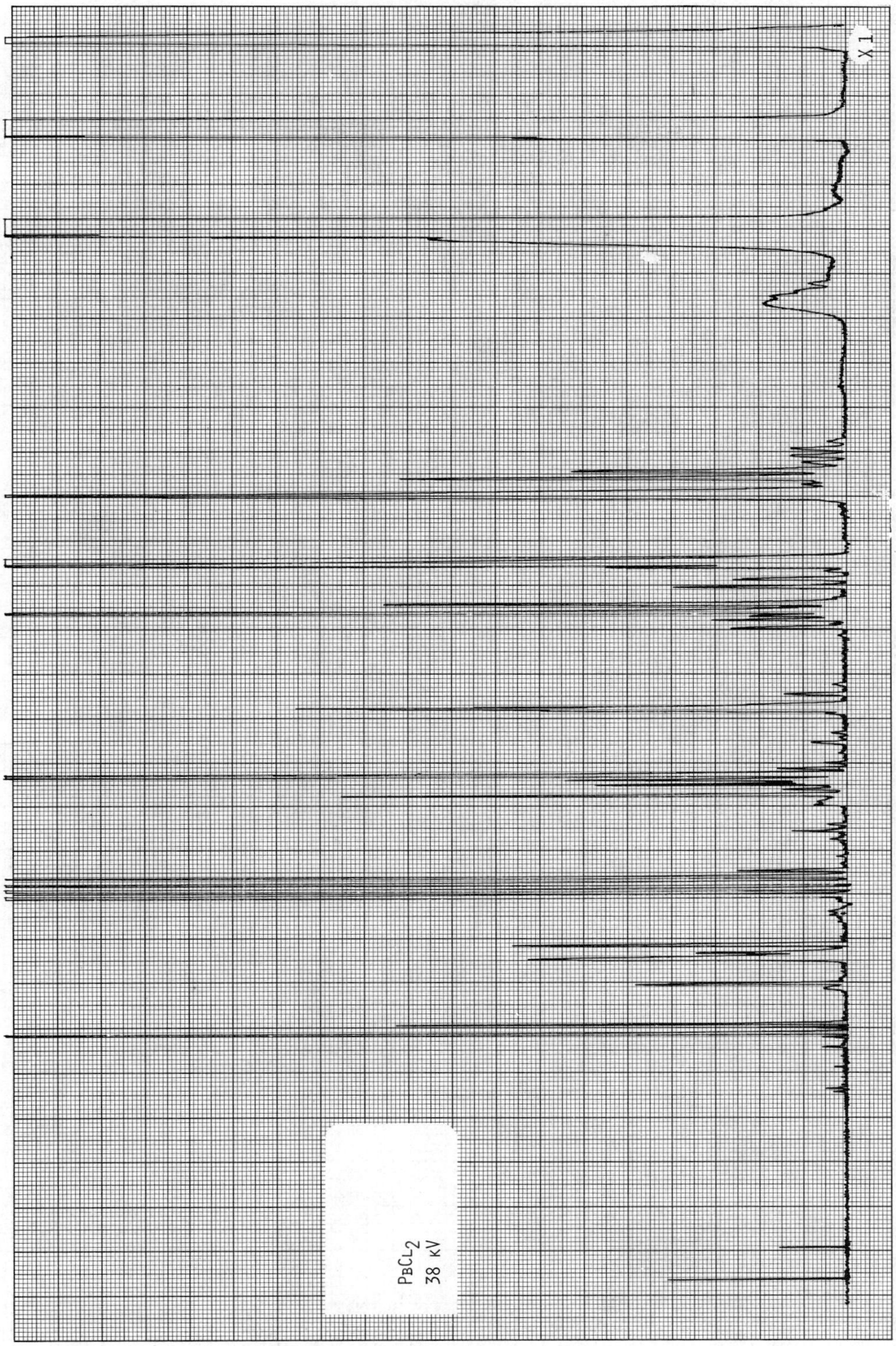

X 1
$PbCl_2$
38 kV

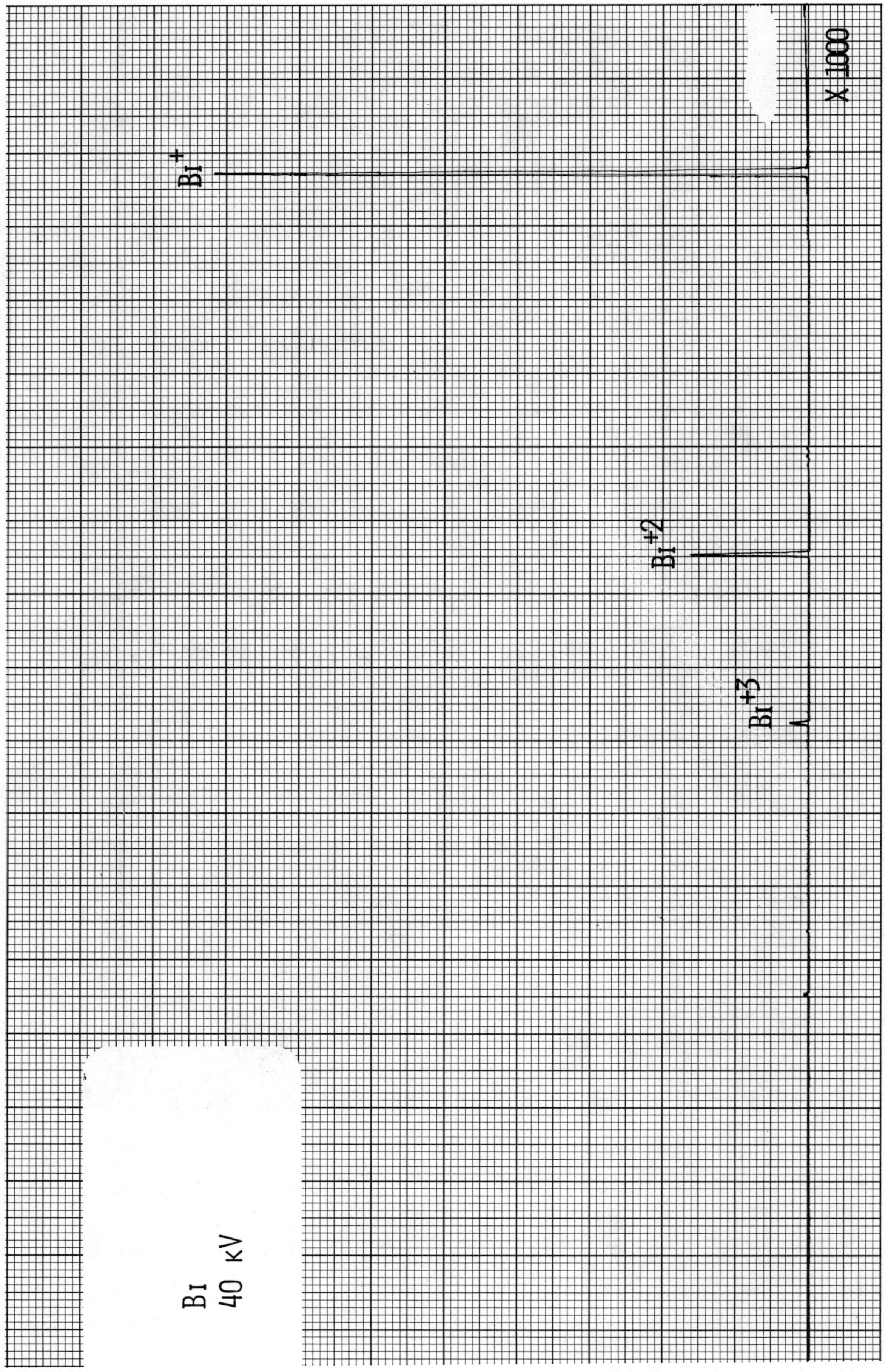
X 1000
Bi+
Bi+2
Bi+3
Bi
40 kV

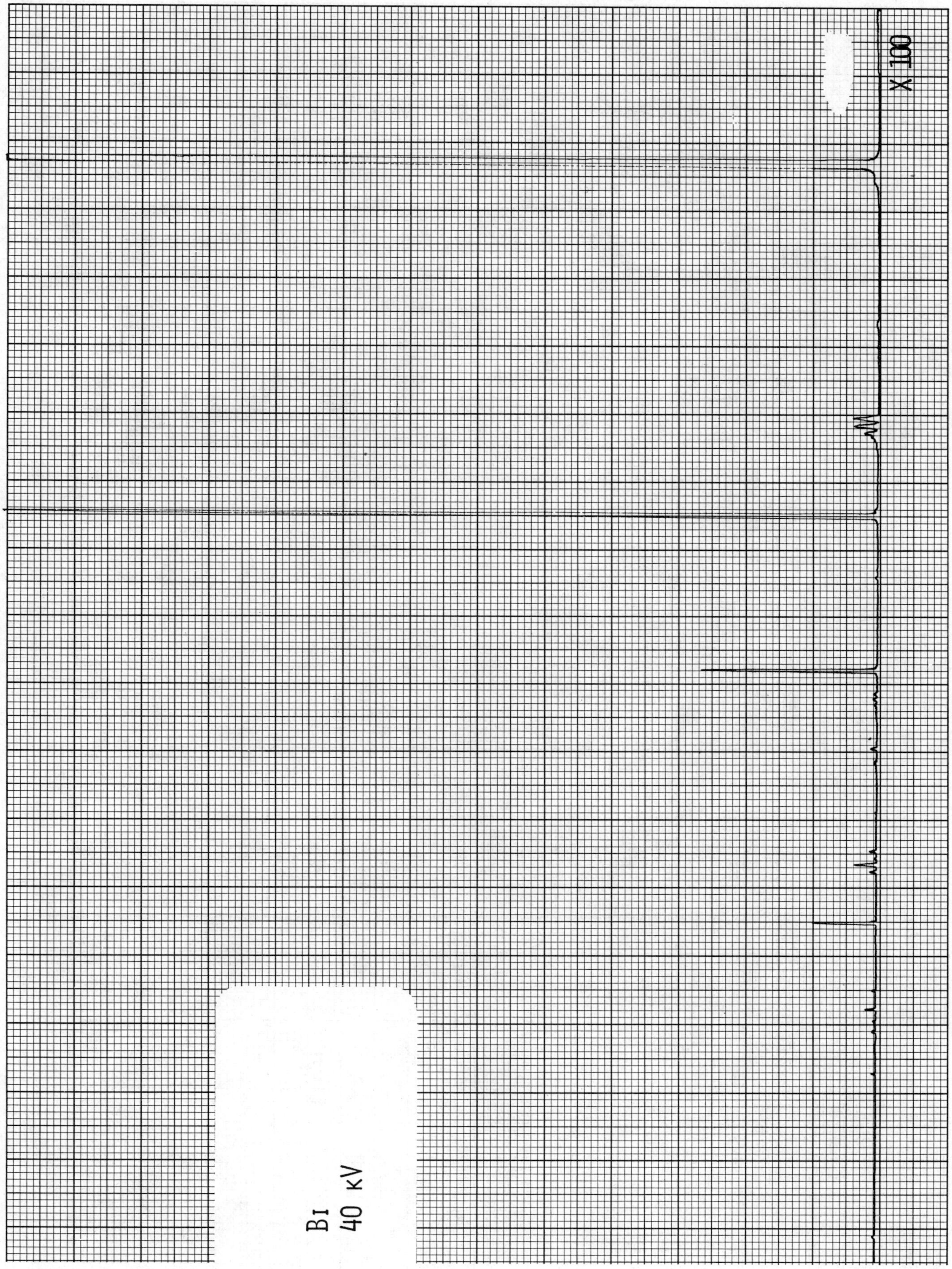
X 100
Bi
40 kV

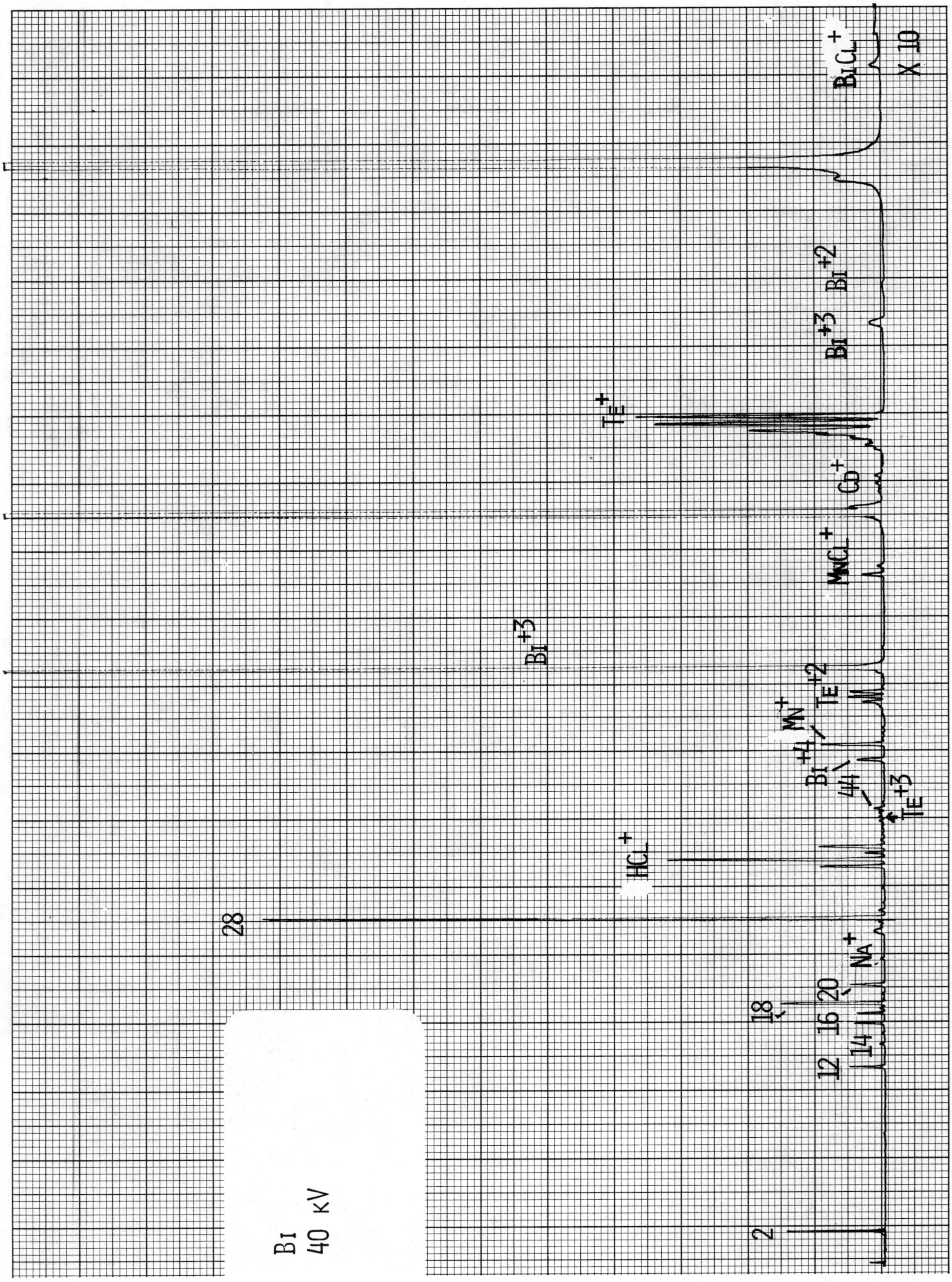
Bi
40 kV
2
12
14
16
18
20
Na+
28
HCl+
Te+3
4/1
Bi+4
Mn+
Te+2
Bi+3
MnCl+
Cd+
Te+
Bi+3
Bi+2
BiCl+
X 10

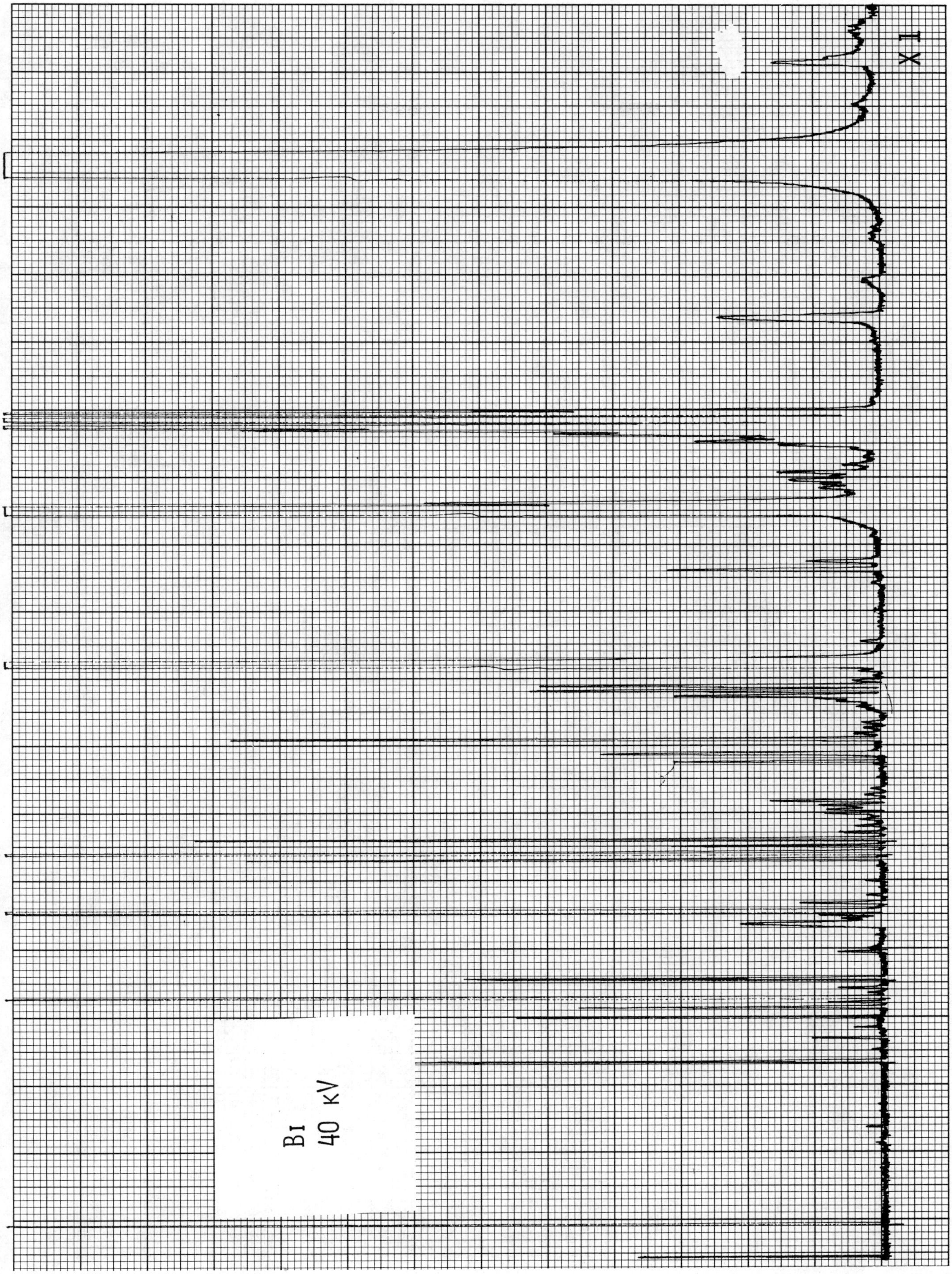
X1
Bi
40 kV

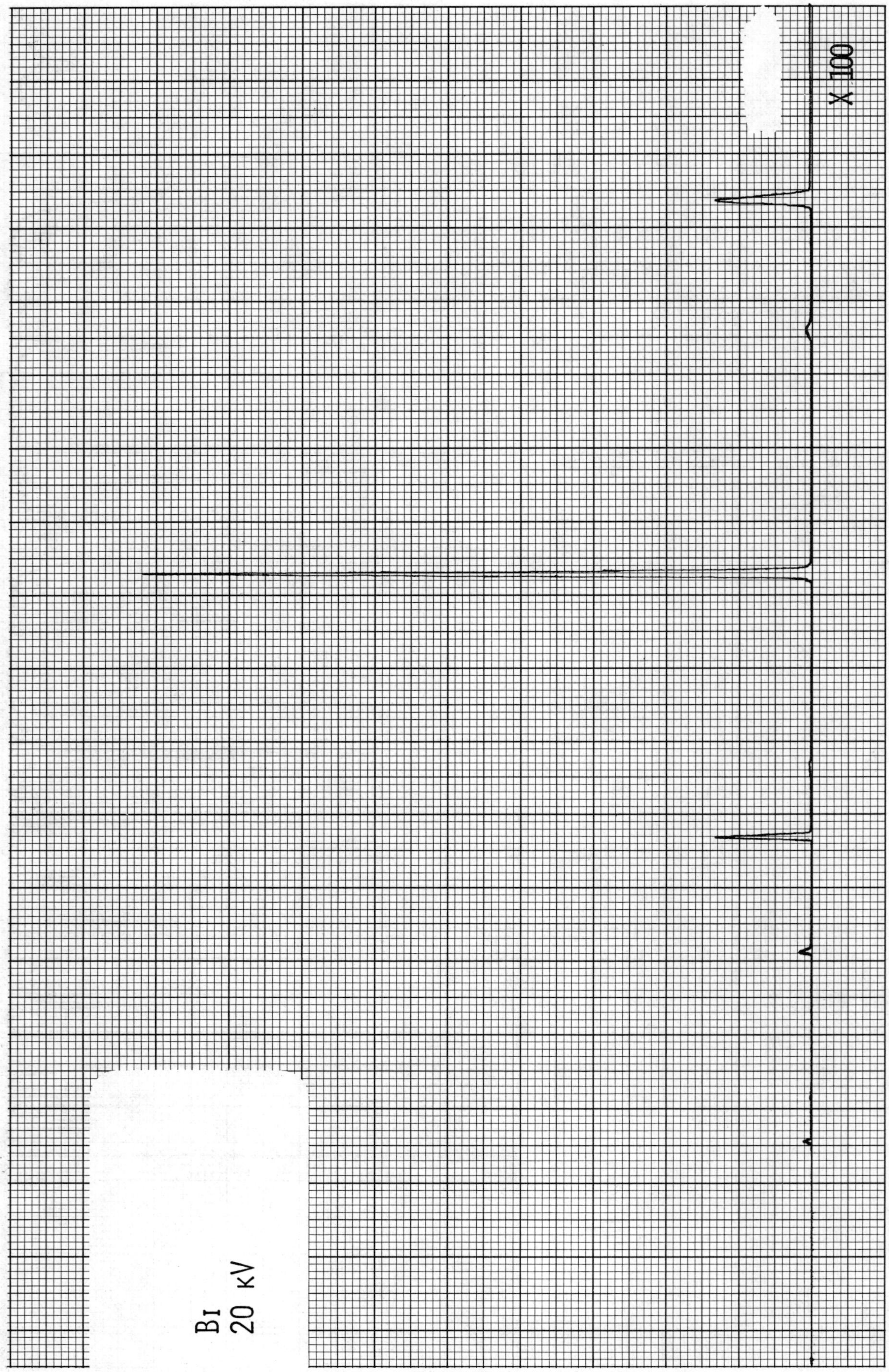
X 100
Bi
20 kV

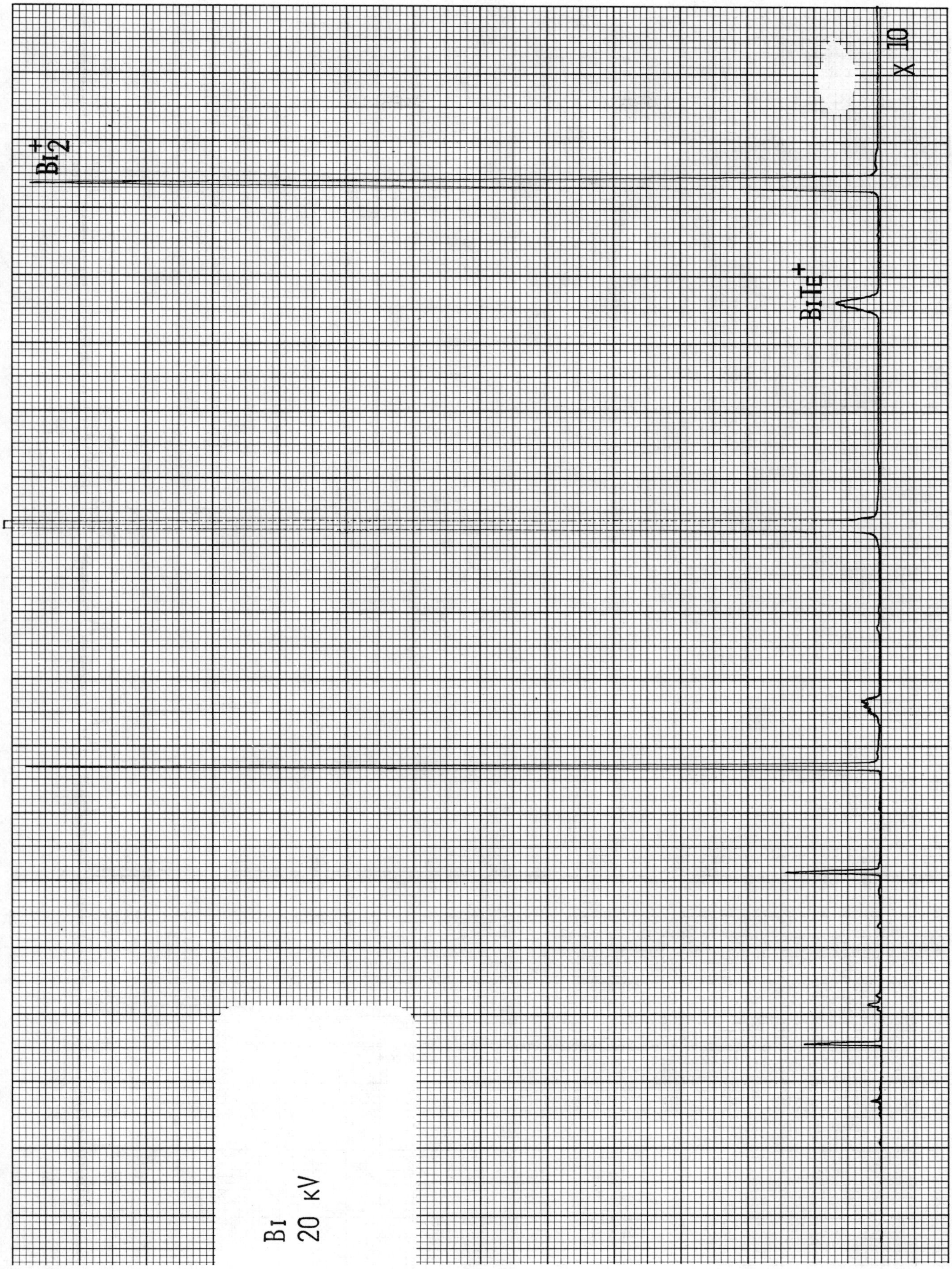
Bi_2^+
$BiTe^+$
X 10
Bi
20 kV

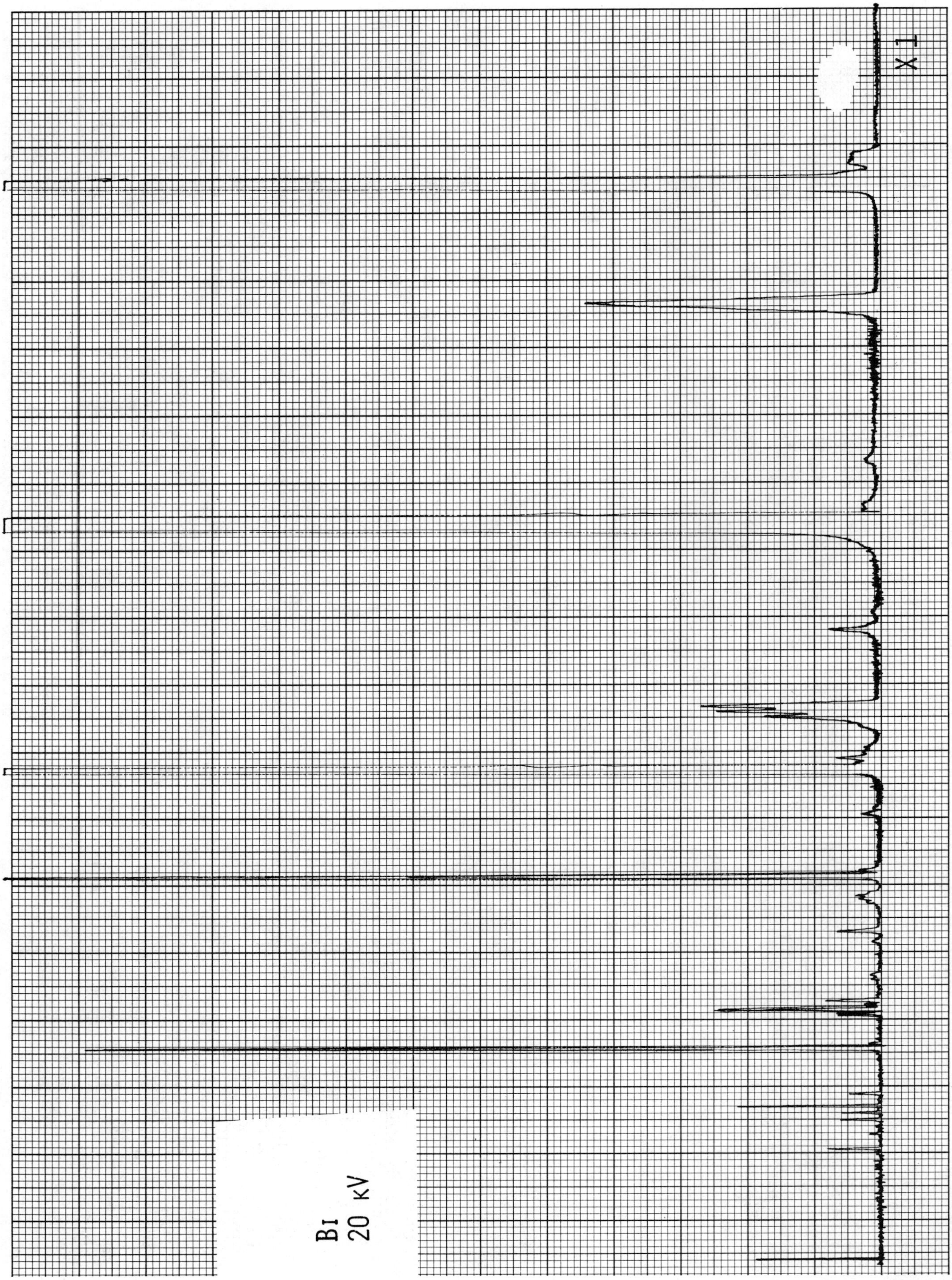
Bi
20 kV
X 1

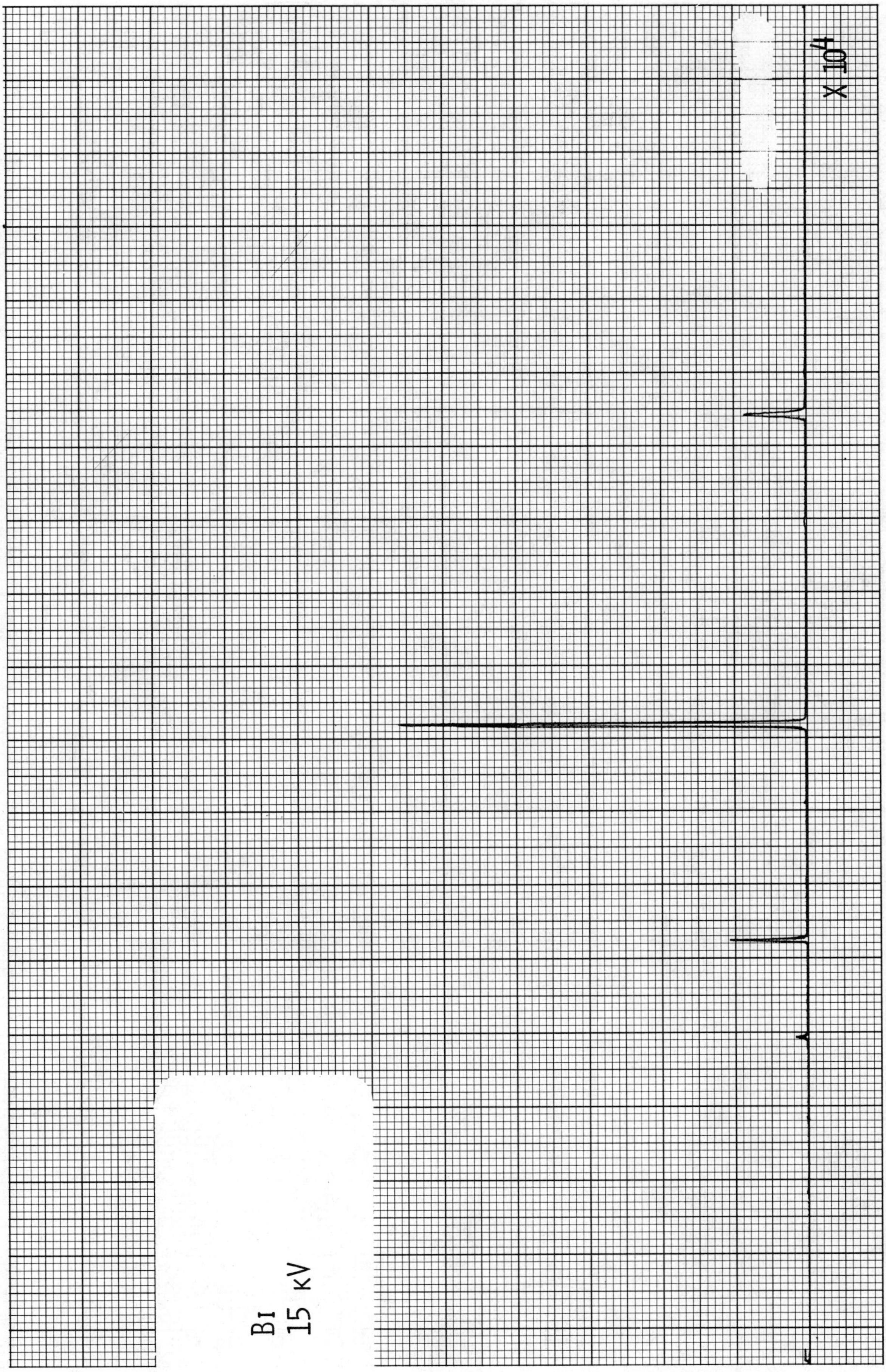
X 10^4
Bi
15 kV

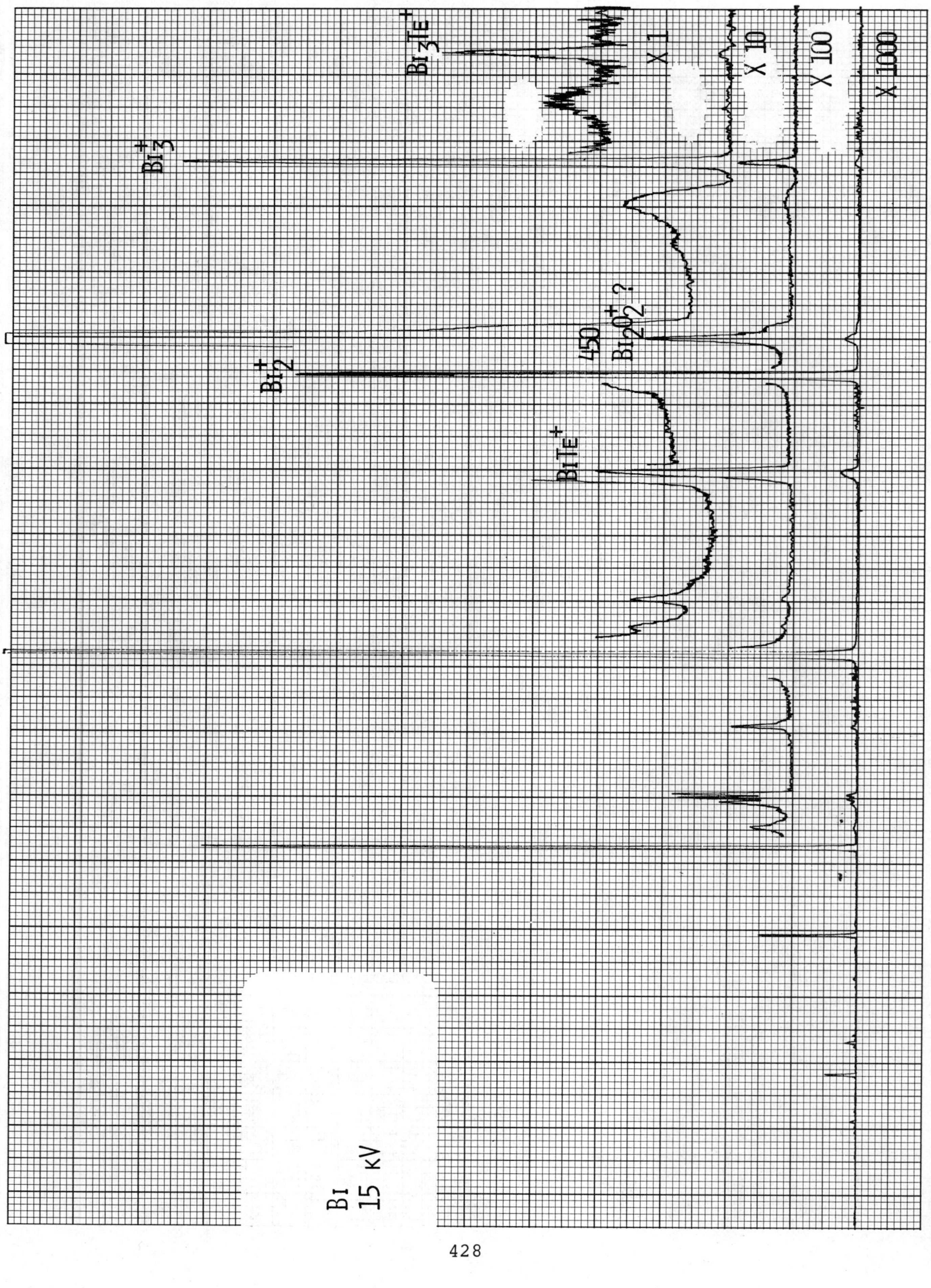
Bi
15 kV
Bi3+
Bi2+
BiTe+
Bi3Te+
450
Bi2O2+ ?
X 1
X 10
X 100
X 1000

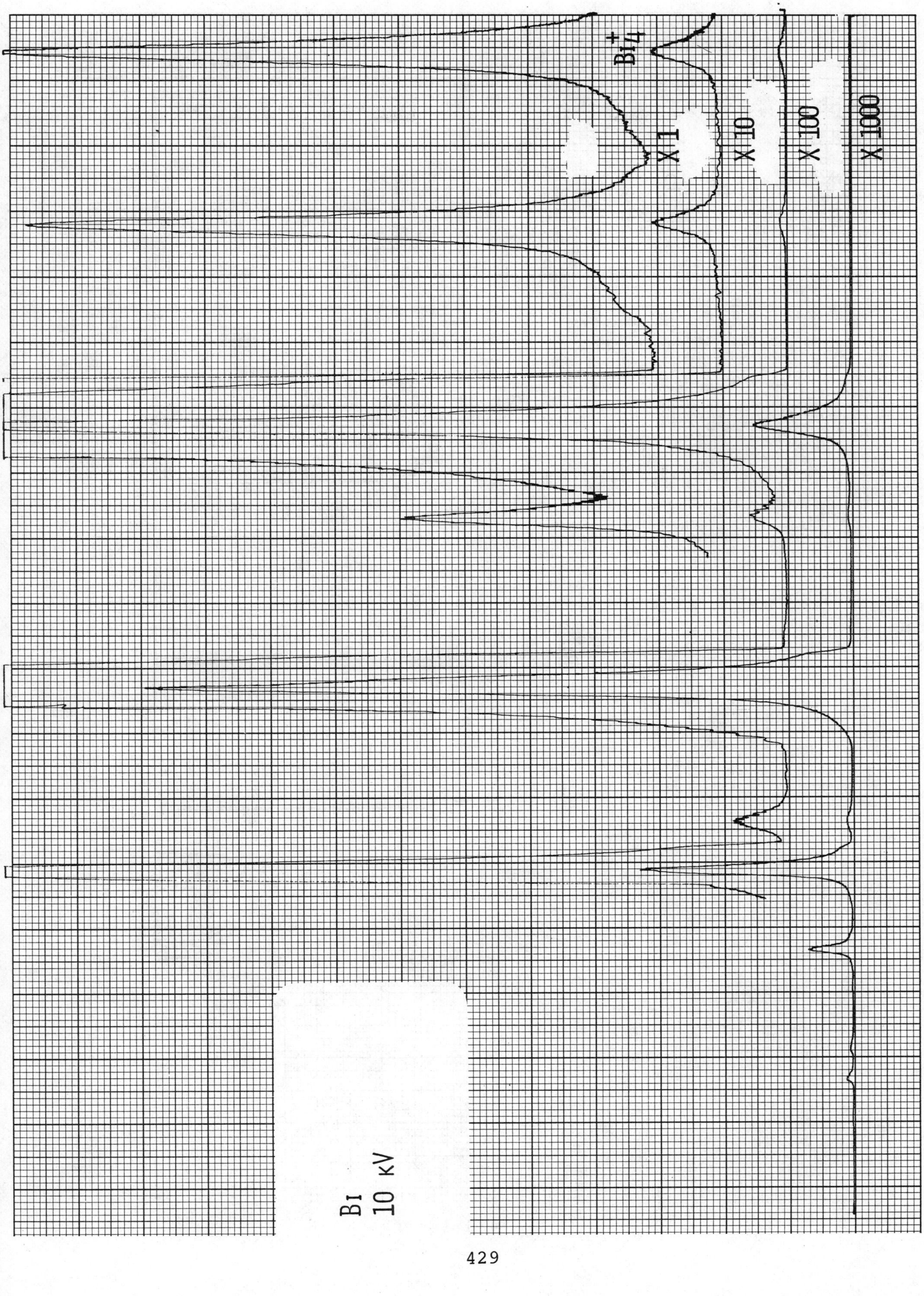

Bi_4^+
X 1
X 10
X 100
X 1000
Bi
10 kV

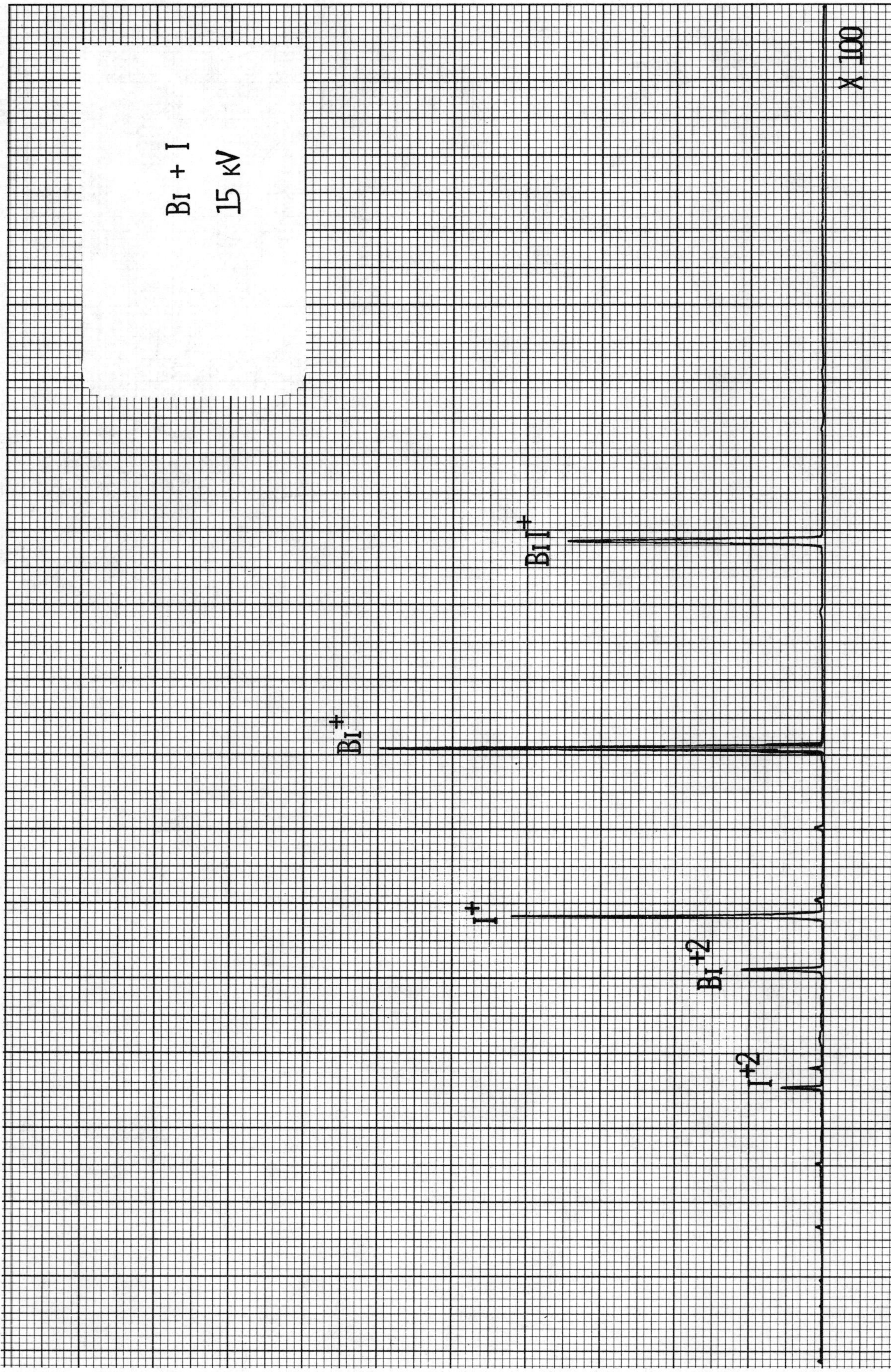
Bi + I
15 kV
X 100
BiI+
Bi+
I+
Bi+2
I+2

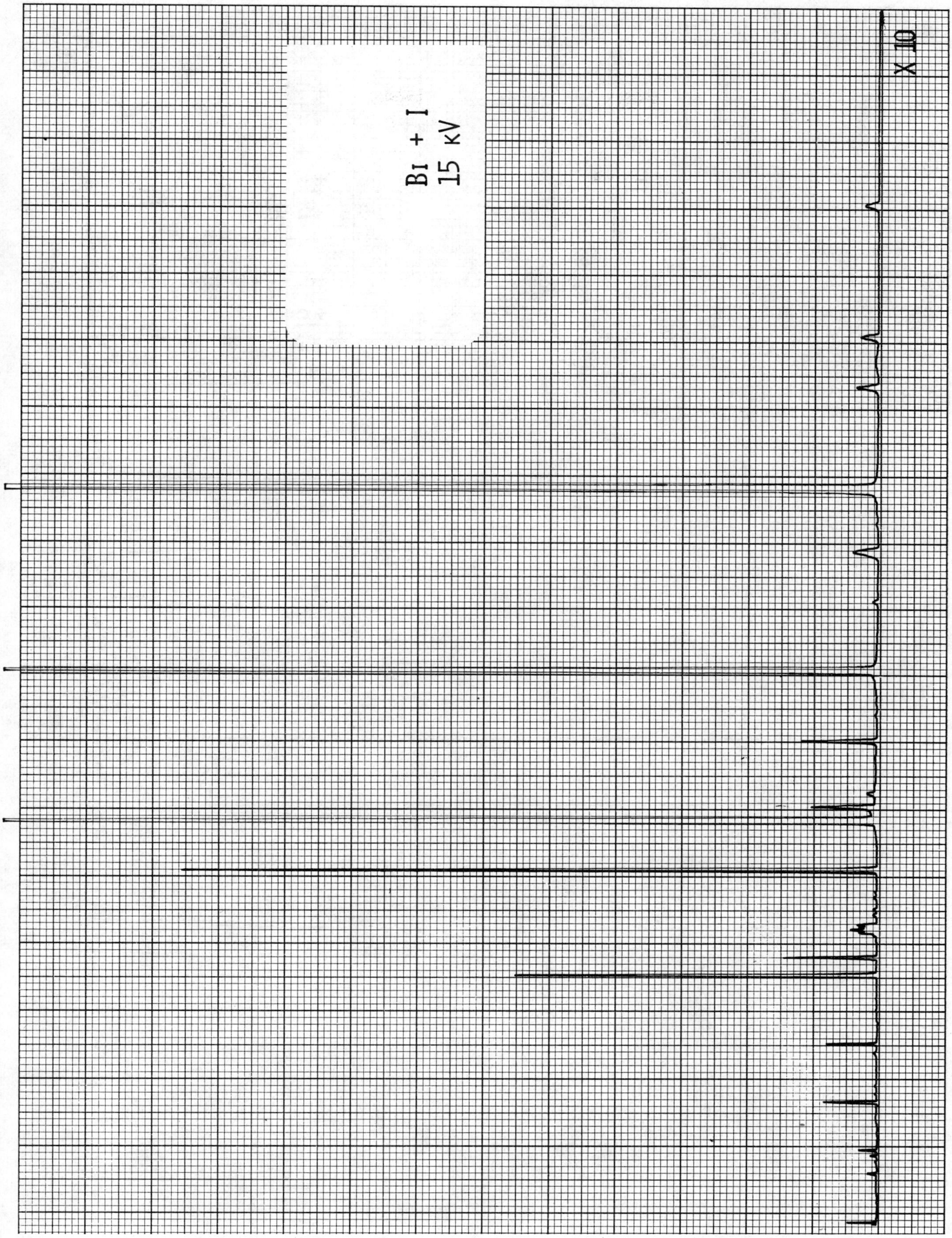
BI + I
15 kV
X 10

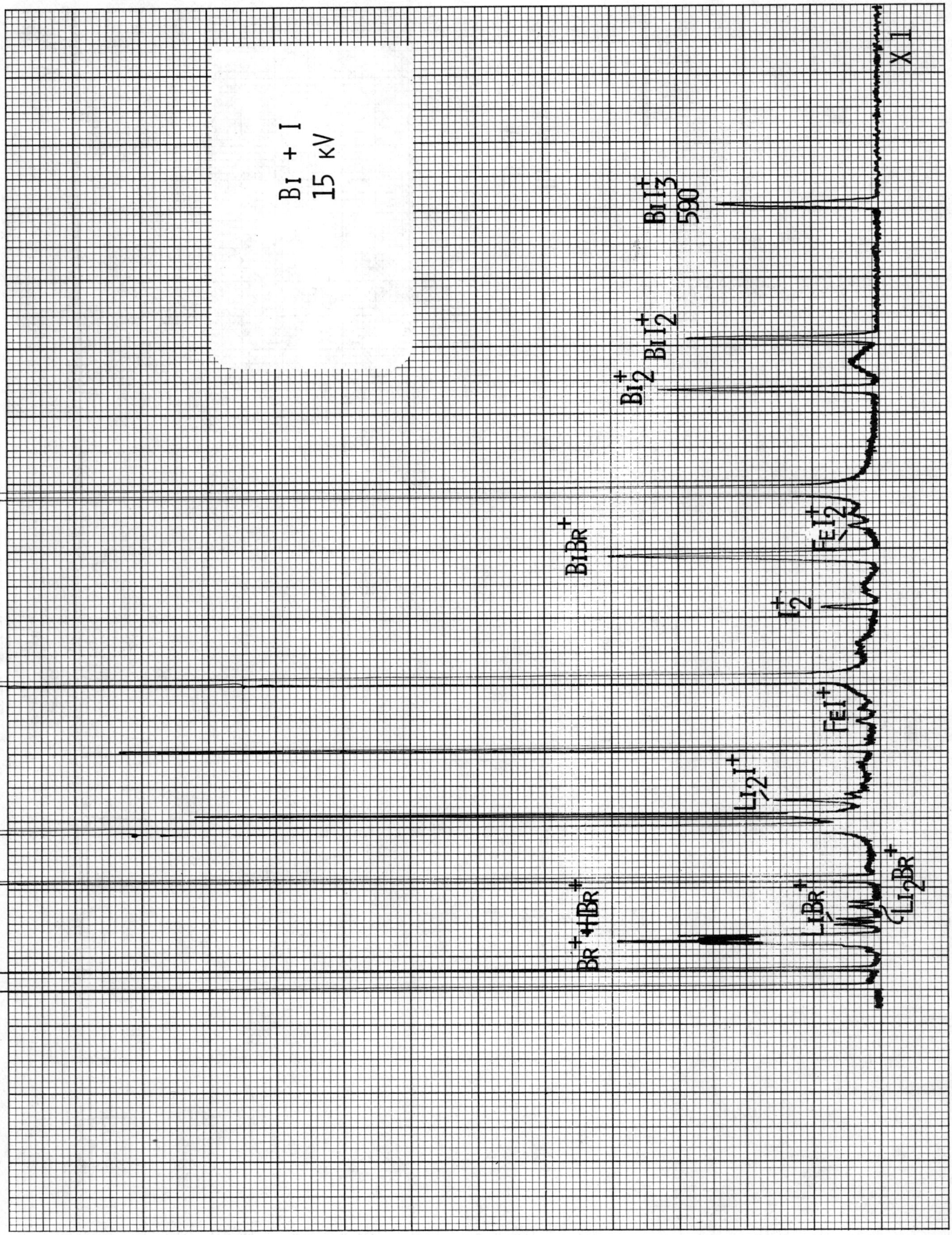
Bi + I
15 kV
X 1
BiI_3^+
590
BiI_2^+
Bi_2^+
$BiBr^+$
FeI_2^+
I_2^+
FeI^+
Li_2I^+
Br^+ +HBr^+
$LiBr^+$
Li_2Br^+